Plant Variation and Evolution

FOURTH EDITION

We are in the midst of a biological revolution. Molecular tools are now providing new means of critically testing hypotheses and models of microevolution in populations of wild, cultivated, weedy and feral plants. They are also offering the opportunity for significant progress in the investigation of long-term evolution of flowering plants, as part of molecular phylogenetic studies of the Tree of Life.

This long-awaited fourth edition, fully revised by David Briggs, reflects new insights provided by molecular investigations and advances in computer science. Briggs considers the implications of these for our understanding of the evolution of flowering plants, as well as the potential for future advances. Numerous new sections on important topics such as the evolutionary impact of human activities, taxonomic challenges, gene flow and distribution, hybridisation, speciation and extinction, conservation and the molecular genetic basis of breeding systems will ensure that this remains a classic text for both undergraduate and graduate students in the field.

David Briggs is Emeritus Fellow of Wolfson College at the University of Cambridge. He has a lifelong interest in evolution, genetics, conservation and taxonomy. He is also the author of *Plant Microevolution and Conservation in Human-Influenced Ecosystems* (Cambridge, 2009), which won the British Ecological Society's 2011 Marsh Book of the Year award.

S. Max Walters (1920–2005) was a leading British field botanist, and the author and editor of major works on the classification and identification of both wild and garden plants. He served as Director of Cambridge University Botanic Garden from 1973 until his retirement in 1983.

Plant Variation and Evolution

FOURTH EDITION

DAVID BRIGGS
Wolfson College, University of Cambridge

and

S. MAX WALTERS
Former Director of Cambridge University Botanic Garden

This edition revised by David Briggs

CAMBRIDGE
UNIVERSITY PRESS

Shaftesbury Road, Cambridge CB2 8EA, United Kingdom

One Liberty Plaza, 20th Floor, New York, NY 10006, USA

477 Williamstown Road, Port Melbourne, VIC 3207, Australia

314–321, 3rd Floor, Plot 3, Splendor Forum, Jasola District Centre, New Delhi – 110025, India

103 Penang Road, #05–06/07, Visioncrest Commercial, Singapore 238467

Cambridge University Press is part of Cambridge University Press & Assessment, a department of the University of Cambridge.

We share the University's mission to contribute to society through the pursuit of education, learning and research at the highest international levels of excellence.

www.cambridge.org
Information on this title: www.cambridge.org/9781107602229

First published by Weidenfeld and Nicolson 1969
Second and third editions published by Cambridge University Press 1984, 1997
Fourth edition 2016

A catalogue record for this publication is available from the British Library

Library of Congress Cataloging-in-Publication data
Briggs, D. (David), 1936– author. | Walters, S. M. (Stuart Max), author.
Plant variation and evolution / David Briggs, Wolfson College,
University of Cambridge, and S. Max Walters, formerly of Cambridge
University Botanic Garden.
Fourth edition. | New York : Cambridge University Press, 2016.
LCCN 2015038104 | ISBN 9781107602229
LCSH: Plants – Variation. | Plants – Evolution.
LCC QK983 .B73 2016 | DDC 580–dc23
LC record available at http://lccn.loc.gov/2015038104

ISBN 978-1-107-60222-9 Paperback

CONTENTS

PREFACE

In writing the earlier editions of *Plant Variation and Evolution*, with my great friend Max Walters (1920–2005), our approach to this complex subject was clearly set out in the preface to the Third Edition. 'When it was first proposed to establish laboratories at Cambridge, Todhunter, the mathematician, objected that it was unnecessary for students to see experiments performed, since the results could be vouched for by their teachers, all of them of the highest character, and many of them clergymen of the Church of England' (Bertrand Russell, 1931). While Russell's mischievously anti-clerical comments do not entirely reflect the views of Todhunter (Todhunter, 1873; Macfarlance, 1916), they do provoke us to take a critical look at the way scientific advances are made, in every historical period, through questioning the opinions of various 'authorities'. In many texts on evolution a judicious mixture of concepts, mathematical ideas and the results of laboratory and field experiments are combined in an elaborate pastiche to provide a more or less complete edifice. Perhaps one or two areas of uncertainty may be indicated, but the general impression is of a house well built, but awaiting the placing of the last few roof-tiles. Conversations with research biologists, however, quickly reveal a different picture. While the broad outlines of evolution are supported by an increasing body of evidence, almost nothing is completely settled: current views represent a provisional framework, and even some parts of the subject, long held to be clarified, are suddenly overturned by new discoveries. Teaching experience reinforces our view that students of science should be shown the way in which, slowly and painstakingly, our present partial pictures have been built up, how and to what extent they are testable by experiment and observation, and in what way they remain vague or defective. A healthy scepticism in the face of the complexities of organic evolution is the best guarantee of real progress in understanding its patterns and processes.

The aim of this new edition is to provide, as before, an authoritative introductory university text, while at the same time satisfying the general reader with a real interest in the subject, showing how the study of variation and evolution of flowering plants has developed over the last 400 years. This development has been increasingly scientific, leading to the realisation of the crucial importance of hypothesis and experiment.

Throughout the book, I have tried to provide a critical but concise overview of current excitements and advances, while at the same time paying attention to difficulties and uncertainties. Furthermore, I have intentionally introduced and shown the connection between many complex subjects, and have therefore provided references to important research papers and books, in order that the reader may build on the framework provided.

As in previous editions, I emphasise the logical and historical framework of early observation and experiment, which is almost wholly neglected in some university courses. Sapp (2003) stresses this important point when he writes, 'many teachers of science have noted, [that] scientific problems are usually much better understood from studying their history rather than their logic alone'. Accordingly this book shows how, building on historical foundations, modern investigative methods are providing new insights into past and present patterns of variation in nature and the processes that give rise to them.

ACKNOWLEDGEMENTS

First, I pay tribute to my co-author and mentor Max Walters, formerly Director of Cambridge University Botanic Garden. My family and I thank Max and his wife Lorna for lifelong friendship and many kindnesses.

I wish to thank those teachers, colleagues and friends who gave me encouragement and provided many life-changing opportunities: Ada Radford (my first biology teacher), Donald Pigott, David Valentine, Harry Godwin, Harold Whitehouse, Percy Brian, Richard West, John Burnett, Jack Harley, David Lewis and Peter Ayres.

Over the years, I have discussed many issues about evolution and conservation with a large number of friends and colleagues. To all of them I offer my thanks: John Akeroyd, Janis Antonovics, Elizabeth Arnold, John Barrett, David Bellamy, Alex Berrie, John Birks, May Block, Margaret Bradshaw, Tony Bradshaw, Arthur Cain, Arthur Chater, Judy Cheney, David Coombe, Gigi Crompton, Quentin Cronk, Jim and Camilla Dickson, Jeff Duckett, Trevor Elkington, Harriet Gillett, Peter Grubb, Mark Gurney, John Harper, Joe Harvey, John Harvey, Peter Jack, David Kohn, Andrew Lack, Vince Lea, Elin Lemche, Roselyne Lumaret, Terry Mansfield, Hugh McAllister, Pierre Morisset, Gina Murrell, Peter Orris, Philip Oswald, John Parker, Joseph Pollard, Duncan Porter, Chris Preston, Oliver Rackham, John Raven, Tom ap Rees, Peter Sell, Alison Smith, Betty Smocovitis, Edmund Tanner, Andrew and Jane Theaker, John Thompson, Lorna Walters, Alex Watt, David Webb, John West and Peter Yeo.

In particular I would like to thank those I taught in the Universities of Cambridge and Glasgow: their challenging questions stimulated me to write this book.

I am especially grateful to Joachim Kadereit, Andrew Lack, Chris Preston, Duncan Porter and Suzanne Warwick who offered comments on the first drafts of this book. I very greatly value their friendship, expert advice and encouragement.

I pay special tribute to my family for encouraging me to write this book, my parents Mabel and Tom Briggs, Nancy Briggs, Jonathan Briggs, Nicholas Oates, Alastair Briggs, Catherine, Miranda, Ella, Judith and Adrian Howe, Norman Singer and Geoffrey Charlesworth.

Without my wife Daphne's support, tolerance and unfailing commitment, this book would never have been written. With good humour and constant encouragement, she has helped me bring this project to fruition. I thank her for all her help, especially for hours of painstaking checking and proofreading.

I am very grateful for the friendly help, advice and encouragement I have received from the staff of Cambridge University Library and the Central Science Library. It is a pleasure to thank the staff of Cambridge University Press for their encouragement and assistance through all the stages of writing this new edition: Dominic Lewis, Megan Waddington, Hamish Adamson, Sarah Starkey, Charlotte Thomas and Ilaria Tassistro. Ken Moxham, my copy-editor, played an invaluable role in the final stages of the preparation of the book.

NOTE ON NAMES OF PLANTS

Scientific names are generally in accordance with the third edition of Stace, C. (2010) *New Flora of the British Isles* for British plants; and, for European plants not in the British flora, Tutin *et al.* (1964–93) *Flora Europaea*. In the cases not covered by either work we have used the name we believe to be correct. Where an author has used a non-current name, this is noted in brackets.

While some botanists continue to use long-standing names for important families, increasingly others use names derived from the type genus (e.g. Compositae/Asteraceae; Cruciferae/Brassicaceae, etc.). We give alternative names in the text.

ABBREVIATIONS

AFLP	amplified fragment length polymorphism
APG	Angiosperm Phylogeny Group
BGCI	Botanic Gardens Conservation International
BP	before the present
BSC	Biological Species Concept
CMS	cytoplasmic male sterility
CNI	cytonuclear incompatibilities
cpDNA	chloroplast DNA
ENM	ecological niche modelling
ESM	Earth System Model
FAO	Food and Agriculture Organization of the United Nations
FISH	fluorescence *in situ* hybridisation
GISH	genomic *in situ* hybridisation
GM	genetically modified
HGT	horizontal transfer of genetic information
IPCC	Intergovernmental Panel on Climate Change
ISH	*in situ* hybridisation
ISSRs	inter simple sequence repeats
IUCN	International Union for Conservation of Nature
K–T	Cretaceous–Tertiary (K–T) boundary
LGM	Last Glacial Maximum
mtDNA	mitochondrial DNA
MVP	Minimum Viable Population
Mya	million years ago
OTU	operational taxonomic unit
PCA	principal component analysis
PCO	principal coordinates analysis
PCR	polymerase chain reaction
ppb	parts per billion
ppm	parts per million
PVA	population viability analysis
QTLs	quantitative trait loci
RAPD	random amplified polymorphic DNA
RFLP	restriction fragment length polymorphism
SNP	single nucleotide polymorphism
SSRs	simple sequence repeats
SSSI	Site of Special Scientific Interest in UK
STR	short tandem repeat
UNEP	United Nations Environmental Programme
WGD	whole genome duplication
WCMC	World Conservation Monitoring Centre

1 Investigating plant variation and evolution

The endless variety of different organisms, in their beauty, complexity and diversity, gives to the biological world a special fascination. Some recognition of these different 'kinds' of organisms is a feature of all primitive societies, for the very good reason that humans had to know, and to distinguish, the edible from the poisonous plants, or the harmless from the dangerous animal. There is also a further dimension to the variation pattern. From folk taxonomies, the early product of human societies' need to understand, describe and use the plants and animals around them, the modern scientific biological classifications we use today have been developed. Different 'kinds' of organism are arranged in a hierarchy, each higher group containing one or more members of a lower group. We now distinguish more than a quarter of a million flowering plants, and these are arranged in a hierarchical-nested classification of species in genera, genera in families, families in orders etc.

One of the main issues to be examined in this book concerns the nature of species in this hierarchical system. Anyone familiar with the vegetation of an area, or the plants in a Botanic Garden, has to face a number of questions that have puzzled biologists. Is there any objective way of delimiting species? How can one account for the different degrees of intraspecific variation found in species? Why are certain species clearly distinct, while in other cases we find a galaxy of closely similar species, often difficult to distinguish from each other? Why is it that hybridisation occurs frequently in certain groups of plants and not in others? What is the significance of the hierarchies in classification and what is the origin of these groupings?

Historically, an examination of these questions, as we shall see in Chapter 2, produced a static picture of variation, based on a belief in the fixity of species. The hierarchy of classification reflected the plan of Special Creation. However, since 1859, with the publication of *On the Origin of Species*, all such studies have been made in the light of Darwin's profound generalisation of evolution by natural selection: populations and species vary in time and space and are part of a continuing process of evolution.
The hierarchical classifications of the pre-Darwin era are now reinterpreted in Darwin's' famous metaphor of the Tree of Life, with all organisms having a common origin in the very distant past.

Even though Darwin's theory has not always been accepted, it has had a tremendous impact on all fields of biology. Nowadays, the fact of evolution is sometimes taken for granted by many, in part because of the wealth of evidence assembled by Darwin and other scientists. While there is often an uncritical acceptance of the theory, it is also important to acknowledge that there has also been widespread rejection of the theory of evolution (Chapter 2). Implicit in Darwin's ideas is the assumption that evolution is still taking place. Therefore, in this book we look not only at issues raised by intraspecific variation and patterns of variation, but also the evidence for evolution, particularly experimental evidence for natural selection and other processes in populations, often called 'microevolution'.

Historical evidence reveals how biologists have long been attempting to understand the patterns of variation in organisms (Chapters 2–5). As we shall see, an ever-widening range of techniques has been devised, including morphological studies, hybridisation experiments, cultivation trials and biometrical and cytogenetical investigations. By this means our understanding of evolution at or below the species level has been greatly increased. While these investigations have yielded many insights and provided a framework of concepts and hypotheses, 'historic' techniques had significant limitations. But in the last few years unparalleled advances in our understanding of *all* aspects of plant evolution have been achieved by the use of molecular approaches. As these tools have been developed and refined, we have been provided with the means of critically testing a very wide range of hypotheses. Many recent advances have come through laboratory-based studies. While readily admitting the importance of such experiments and observations, many key insights have come from studying evolution out of doors, whether in the wild, cultivated or weedy habitats. It is increasingly clear that the basic raw material for studies of variation and microevolution exists in every country of the world, not only in 'natural', unspoiled vegetation but also from the study of communities radically altered by human activities. In fact some of the most important insights into microevolution have come from studies of introduced plants, agricultural crops, weeds, and the vegetation of areas subject to pollution. Such studies have revealed how certain species have adapted successfully in agricultural and urban industrial landscapes.

The past few decades have been a period of great excitement and achievement, as molecular, and associated computational, methods have been successfully applied in many areas of plant evolutionary biology. Thus, major insights have been obtained into the underlying basis of variation (Chapter 6); plant breeding systems (Chapter 7); population variability, the effects of chance and gene flow (Chapters 8–11); hybridisation and speciation (Chapters 12–14), threats of species extinction (Chapter 18), and conservation biology (Chapter 20). Often, this fascinating new information could not have been revealed by any other means, and there is huge potential for further insights, as the molecular revolution in biology gathers pace.

Our account will also consider how molecular investigations are providing major advances in our understanding of plant evolution in geological timescales, so-called macroevolution (Chapter 16). Present evidence suggests that the universe is c.10–16.5 billion years old, with Earth's geological history beginning of the order of 4.6 billion years ago. Life began to evolve *c.*4 billion years ago (Graur & Li, 2000). A basic question to be considered is whether macroevolution is just microevolution writ large: or whether other processes are involved (Bateman, 1999).

Since the publication of Darwin's *Origin*, evolution of biodiversity has been widely represented as the Tree of Life (Chapter 16). Very considerable progress has recently been made in devising 'phylogenetic trees' and other diagrams to explore evolution on longer timeframes. As we shall see, the study of plant biogeography has also been revolutionised by molecular investigations and advances in computer modelling, and these approaches are providing major new insights into both recent and past evolutionary history (Chapter 17).

This book also examines an issue of great international concern. For decades, there has been growing evidence that more than 10% of the world's flora is endangered by human activities (Briggs, 2009). Guided by our knowledge of pattern and process in microevolutionary change, we consider the effectiveness or otherwise of the historic conservation measures taken to prevent this loss of biodiversity (Chapter 20).

In considering the prospects for successful conservation of biodiversity, the potentially catastrophic scenarios revealed by the projections of future climate change must be seriously considered (IPCC, 2013). Climate change is not the only looming threat to biodiversity. Ecosystems are also endangered by the interactions between climate change and other significant factors: habitat loss, the non-sustainable use of resources, pollution, population growth, invasive species etc. (Chapter 18).

Taxonomists were the first biologists to study plant variation. An important theme of this book, therefore, is the impact that evolutionary investigations have had on the theory and practice of taxonomy, especially studies of phylogenetic relationships (Chapter 16). For instance, there is disagreement about whether/how current plant classifications, based on Linnaean principles, should/

could be modified in the light of the patterns apparent in phylogenetic trees (Ereshefsky, 2001). Also, attempts to find DNA sequences that will allow a unique barcode to be defined for each species are provoking intense debate (Wheeler, 2008).

A major concern is the effect that the present dominance of molecular approaches has had on taxonomy and other traditional fields of research, studies that have made such major contributions to our understanding of plant evolution. Whilst freely acknowledging the progress that has been made through the application of molecular tools and computer modelling, in Chapter 19, I discuss the reports of a number of authorities that studies in taxonomy etc. have recently been under pressure or indeed in decline (see Wheeler, 2008; Stuessy, 2009). Amongst the possible reasons for this decline are issues of funding and the way the subjects are taught. As a consequence of the support given to molecular investigations, are some areas being starved of attention and/or resources? Is a form of Darwinian selection taking place amongst subject areas allied to plant evolution? Some take this view. For instance, Sapp (2003) considers that 'the history of biology is regarded as a contest over what questions are important, what answers are acceptable, what phenomena are interesting, and what techniques are most useful'.

Faced with the evidence of these changes in status and support, how might traditional approaches be revitalised? Will the 'road ahead' for studies of plant evolution remain a 'broad church' with integrated contributions from a wide range of disciplines (Chapter 20), or will the study of evolution increasingly develop into a branch of molecular biology?

2 From Ray to Darwin

In 1660 Robert Sharrock, Fellow of New College, Oxford, wrote a book titled *History of the Propagation and Improvement of Vegetables by the Concurrence of Art and Nature*. He was concerned in its early pages to debate a live issue of the day (Bateson, 1913), namely:

> It is indeed growen to be a great question, whether the transmutation of a species be possible either in the vegetable, Animal or Minerall Kingdome. For the possibility of it in the vegetable; I have heard Mr Bobart and his Son often report it, and proffer to make oath that the Crocus and Gladiolus, as likewise the Leucoium, and Hyacinths by a long standing without replanting have in his garden changed from one kind to the other.

The Bobarts were both professional botanists. Sharrock investigated their claim, and found 'diverse bulbs growing as it were on the same stoole, close together, but no bulb half of the one kind, and the other half of the other'. In this age we find it hard to understand a belief in the possibility of transformation of Crocus into Gladiolus. Our reason for disbelief is partly concerned with the nature of evidence; we are not satisfied with the test for the alleged transmutation and would not have been content merely to examine the crowded underground parts. Another reason, however, relates to current ideas of the nature of species. We have a different notion of species from that of the seventeenth century.

Ray and the definition of species

It was the English naturalist John Ray (1628–1705) who was probably the first man to seek a scientific definition of species (Raven, 1950; Oswald & Preston, 2011). In his definition is an implied rejection of the sort of transmutation of species claimed by the Bobarts of Oxford, although in other passages in Ray's work he does not wholly dismiss the possibility of transmutation. For instance, he cites as reliable the case of cauliflower seed supplied by a London dealer, which on germination produced cabbage. Richard Baal, who sold the seed, was tried for fraud and ordered by the court at Westminster to refund the purchase money and pay compensation (De Beer, 1964).

Ray's views on species were published in 1686 in *Historia plantarum*. He wrote (trans. Silk in Beddall, 1957):

> In order that an inventory of plants may be begun and a classification of them correctly established, we must try to discover criteria of some sort for distinguishing what are called 'species'. After a long and considerable investigation, no surer criterion for determining species has occurred to me than distinguishing features that perpetuate themselves in propagation from seed.

He is concerned to define species as groups of plants that breed true within their limits of variation. This definition of species, based as it is partly upon details of the breeding of the plant, was a great advance upon older ideas, which relied entirely upon consideration of the external form.

Ray was also very interested in intraspecific variation. In his letters to various friends (collected by Lankester, 1848), he noted several striking variants of common plants discovered on his journeys around Britain. For example, at Malham in Yorkshire he noticed white-flowered as well as the normal blue-flowered

Jacob's Ladder (*Polemonium caeruleum*), and from other localities he reported white-flowered Foxglove (*Digitalis purpurea*), double-flowered specimens of Water Avens (*Geum rivale*) and white-flowered Red-rattle (*Pedicularis palustris*). Ray also made observations on a prostrate variant of Bloody Cranesbill (*Geranium sanguineum* var. *prostratum*). He wrote to a friend: 'Thousands hereof I found in the Isle [Walney] and have sent roots to Edinburgh, York, London, Oxford, where they keep their distinction.' This report on the constancy of this distinct variant of *Geranium sanguineum* in cultivation is of particular interest, and is referred to again in Chapter 4.

We may learn more of Ray's ideas on the nature of species and intraspecific variation by examining a discourse given to the Royal Society on 17 December 1674 (Gunther, 1928). In this, he expresses his concern that great care should be taken in deciding what constitutes a species and what variation is insufficient for specific distinction. He shows, for instance, that within a species there might occur individuals different from the normal in one or more of the following characters: height, scent, flower colour, multiplicity of leaves, variegation, doubleness of flower, etc. Plants differing by such 'accidents', as Ray calls them, should not be given specific status (Cain, 1996). He records the origin of one notable variant in his own garden: 'I found in my own garden, in yellow-flowered Moth-Mullein (*Verbascum*), the seed whereof sowing itself, gave me some plants with a white flower'. Concerning other variants Ray suggests that they are caused by growing plants under unnatural conditions, for example, a rich or a poor soil, extreme heat, and so on.

He concludes his analysis of specific differences and the problem of intraspecific variation as follows:

> By this way of sowing [rich soil, etc.] may new varieties of flowers and fruits be still produced in infinitum, which affords me another argument to prove them not specifically distinct; the number of species being in nature certain and determinate, as is generally acknowledged by philosophers, and might be proved also by divine authority, God having finished his work of creation, that is, consummated the number of species, in six days.

Ray's views on the origin of specific and intraspecific variation are here laid bare. Given sufficient regard for the variation patterns of a particular group of plants, a botanist should be able to avoid elevating 'accidental' variants to the level of the species (Cain, 1999). Species themselves were, for Ray, all created at the same time, and all therefore of the same age. That new species can come into existence Ray denies, as this is inconsistent with the account of the Creation given in Genesis. This idea is again expressed in a passage written towards the end of his life: 'Plants which differ as species preserve their species for all time, the members of each species having all descended from seed of the same original plant' (Stearn, 1957).

Thus, Ray, an ordained minister himself, firmly upholds the doctrine of special creation. Such views were almost universally accepted in the seventeenth century, Protestants being particularly influenced by the works of Milton. However, Ray's views also took account of contemporary philosophical ideas (C. D. Preston, personal communication).

The Great Chain of Being

A very powerful idea underlay the attitudes of philosophers and theologians throughout Classical and Medieval Europe when they attempted to understand the world.

> The ancient *Scala Naturae*, or 'Great Chain of Being', that originated with Aristotle attempted to classify animals in relations to a hierarchical ladder or stair-way of nature, in which matter and living things were arranged according to their structural complexity and function. The ladder of nature was most often depicted graphically as a list of entities arranged vertically, with the most basic elements like air, water,

> earth and fire at the bottom and man, angels, and God at the top. (Pietsch, 2012)

Significantly, in some representations, humans were not all on the same rung of the ladder: males were on the rung above females (Dawkins, 2009). Lower organisms including plants were at the base of the chain. The relationship between God and Nature was one of the great philosophical questions that underlay the rise of natural science in the sixteenth and seventeenth centuries (see Lovejoy, 1966; Cain, 1996, 1999, 2008). The acceptance by great pioneer biologists, such as Ray, of the natural world as rational and available to observation and experiment marked the beginning of modern science. Pietsch (2012) reviews the wide range of pre-Darwinian representations of the relationships of different organisms, including tree-like structures. The rigid timeless plan of the *Scala Naturae* finally gave way to the evolutionary Tree of Life (Darwin, 1859).

Linnaeus

In our examination of historical aspects of the subject, we must next study Linnaeus (Carl von Linné, 1707–78), the great Swedish systematist, who made extremely important contributions (Goerke, 1989; Koerner, 1999; Blunt, 2004; Harnesk, 2007). He too, in *Critica botanica* (1737), championed the idea of the fixity of species:

> All species reckon the origin of their stock in the first instance from the veritable hand of the Almighty Creator: for the Author of Nature, when He created species, imposed on his Creations an eternal law of reproduction and multiplication within the limits of their proper kinds. He did indeed in many instances allow them the power of sporting in their outward appearance, but never that of passing from one species to another. Hence today there are two kinds of difference between plants: one a true difference, the diversity produced by the all-wise hand of the Almighty, but the other, variation in the outside shell, the work of Nature in a sportive mood. Let a garden be sown with a thousand different seeds, let to these be given the incessant care of the Gardener in producing abnormal forms, and in a few years it will contain six thousand varieties, which the common herd of Botanists calls species. And so I distinguish the species of the Almighty Creator which are true from the abnormal varieties of the Gardener: the former I reckon of the highest importance because of their Author, the latter I reject because of their authors. The former persist and have persisted from the beginning of the world, the latter, being monstrosities, can boast of but a brief life. (trans. Hort, 1938)

The approaches of both Ray and Linnaeus were typological; they upheld the Greek philosophical view that beneath natural intraspecific variation there existed a fixed, unchangeable type of each species. It was the job of botanists to see these 'elemental species': 'natural variation' was in a sense an illusion.

We see also in the passage quoted above that Linnaeus had a very similar attitude to intraspecific variation to that of Ray. Stearn (1957), in an interesting analysis of the origin of Linnaeus' views, draws attention to his love for gardening and his experience as personal physician and superintendent of gardens to George Clifford, a banker and director of the Dutch East India Company. During this period of his life, Linnaeus, working on his great illustrated book on the plants in Clifford's gardens – the *Hortus Cliffortianus* – lived at Hartekamp, near Haarlem, in the centre of the Dutch bulb-growing area. Here thousands of varieties of Tulips and Hyacinths were grown. At this time, Linnaeus wrote the *Critica botanica*, and no doubt his personal observations at the time prompted the following outburst: 'Such monstrosities, variegated, multiplied, double, proliferous, gigantic, wax fat and charm the eye of the beholder with protean variety so long as gardeners perform daily sacrifice to their idol: if they are neglected these elusive ghosts glide away and are gone.'

Other observations of Linnaeus in the *Critica botanica* show his familiarity with variation in wild plants and his experimental approach to problems. For instance, he studied flower colour, noting that purple flowers tend to fade after a few days, turning to a bluish colour; but '... sprinkle these fading flowers with any acid, and you will recover the pristine red hue'. Concerning aquatic plants he notes: 'Many plants which are purely aquatic put forth under water only multifid leaves with capillary segments, but above the surface of the water later produce broad and relatively entire leaves. Further, if these are planted carefully in a shady garden, they lose almost all these capillary leaves, and are furnished only with the upper ones, which are more entire.' As an example Linnaeus gives *Ranunculus aquaticus folio rotundo et capillaceo*, the aquatic species of *Ranunculus* to which we refer later.

Linnaeus was particularly interested in cultivation and its effect upon plants:

> *Martagon sylvaticum* is hairy all over, but loses its hairiness under cultivation. Hence plants kept a long while in dry positions become narrow-leaved as *Sphondylium, Persicaria* ... Hence broad-leaved plants, when grown for a long while in spongy, fertile, rich soil have been known to produce curly leaves, and have been distinguished as varieties ... the following have been distinguished as '*crispum*': *Lactuca, Sphondylium, Matricaria*, etc.

The early botanical work of Linnaeus is extremely important in the history of ideas about species and variation. He championed firmly the reality, constancy and sharp delimitation of species. He was also concerned to refute the Ancient Greek idea of transmutation of species, which was still widely believed in his day. In *Critica botanica* he wrote:

> No sensible person nowadays believes in the opinion of the Ancients, who were convinced that plants 'degenerate' in barren soil, for instance, that in barren soil Wheat is transformed into Barley, Barley into Oats, etc. He who considers the marvellous structure of plants, who has seen flowers and fruits produced with such skill and in such diversity, and who has given more credence to experiments of his own, verified by his own eyes, than to credulous authority, will think otherwise.

Linnaeus is immortalised for botanists by his great work *Species plantarum* (1753), in which are described in a concise and methodical fashion all the approximately 5,900 species of plants then known. In classifying these species, Linnaeus grouped species into genera and genera into classes on the basis of the number and arrangement of their stamens. These groups were then subdivided into orders on the number of pistils. This classification, which Linnaeus acknowledged as artificial, was to be a preliminary to a more natural classification – of which he only produced a fragment – based on overall resemblance. Linnaeus' classification of the plant kingdom did not yield a 'Chain of Being' but a hierarchical classification that he likened to the pattern of countries on a map (Jonsell, 1978; Bowler, 1989a, b).

However, Linnaeus' concept of the species seems to have been subject to change as his experience grew. In early works, and most explicitly in the theoretical *Philosophia botanica* (1751, now available in a new translation by Freer, 2003), he stresses the clear distinction between *species*, which were constituted as such by the Creator from the beginning, and mere *varieties*, which may be induced by changed environmental conditions, or raised by the art of gardeners. Nevertheless, not infrequently in *Species plantarum*, there are comments which show that Linnaeus did not always find specific distinctions clear: for example, under *Rosa indica* we find that 'the species of *Rosa* are with difficulty to be distinguished, with even greater difficulty to be defined; nature seems to me to have blended several or by way of sport to have formed several from one' (Stearn, 1957). It is even true that Linnaeus speculates, in a few cases, on the possible evolutionary derivation of one species from another in the pages of *Species plantarum*. Thus, under *Beta vulgaris*, we find, after a list of seven agricultural crop varieties, the fascinating statement: 'Probably born of *B. maritima* in a foreign country'. *B. maritima*, the Wild Beet (now called *B. vulgaris* ssp. *maritima*), is given separate treatment as a distinct species! This and several other cases are interestingly discussed by Greene (1909), who points out that there is good evidence to support the view that the dogmatic 'special creation' statements of *Philosophia botanica* and similar writings of Linnaeus did not, even in his earlier days, represent Linnaeus' real views, but were diplomatic writings to satisfy the 'orthodox ecclesiastics who, in his day, ruled the destinies of all seats of learning in Sweden'.

If he was orthodox on these matters in the main works that established his academic and scientific reputations, Linnaeus allowed himself much more freedom in several of the 186 dissertations which his research students, following the medieval rules of disputation, had to defend in Latin. It is clear from these writings that Linnaeus came to believe less rigidly in the fixity of species. For instance, in 1742 a student brought to him, from near Uppsala, an unusual specimen of Toadflax (*Linaria vulgaris*). The flower was not of the usual structure but had five uniform petals and five spurs. Experiments showed that the plant bred true and Linnaeus called it *Peloria* (Fig. 2.1). After close study Linnaeus decided that *Peloria* was a new species, which had arisen from *L. vulgaris* (Linnaeus, 1744).

He also considered that certain other species might have arisen as a result of hybridisation (see Linnaeus (1749–90; Eriksson, 1983). In *Plantae hybridae* (1751), records are given of 100 plants that might be regarded as hybrids. In *Somnus plantarum* (1755) we read: 'The flowers of some species are impregnated by the farina (pollen) of different genera, and species, inasmuch that hybridous or mongrel plants are frequently produced, which if not admitted as new species, are at least permanent varieties.' Later, in the summer of 1757, Linnaeus made what might be considered to be the first scientifically produced interspecific hybrid, between the Goatsbeards

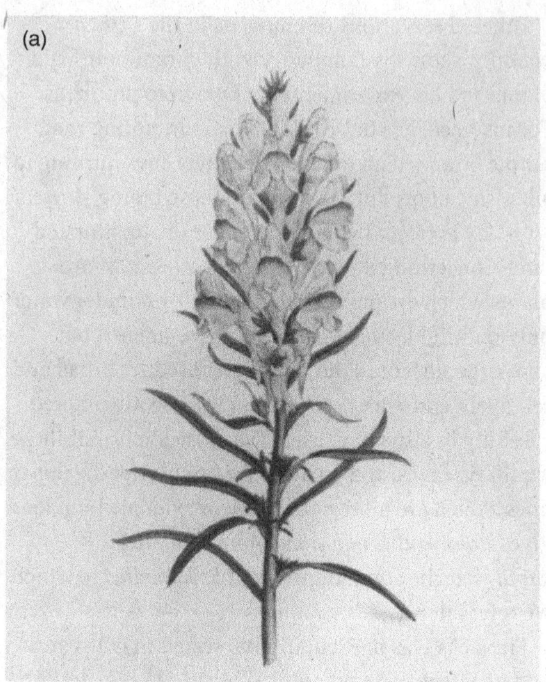

Fig. 2.1. *Linaria vulgaris* (a) and its *Peloria* variant (b). ((Illustrations by Sowerby in Boswell Syme, 1866.)

Tragopogon pratensis (yellow flowers) and *T. porrifolius* (violet flowers). Ownbey (1950), who studied *Tragopogon* species introduced in America, gives the following details of Linnaeus' experiment. After removing the pollen from the flower-heads of *T. pratensis* early in the morning, Linnaeus sprinkled the stigmas with pollen of *T. porrifolius* at about 8 a.m. The flower-heads were marked, the seed eventually harvested and subsequently planted. The first-generation hybrid plants flowered in 1759, producing purple flowers yellow at the base. Seed of the cross, together with an account of the experiment and its bearings upon the problems of the sexuality of plants, formed the basis for a contribution to a competition arranged by the Imperial Academy of Sciences at St Petersburg. Linnaeus was awarded the prize in September 1760. It is of great historical interest that the seed sent by Linnaeus was planted in the Botanic Garden in St Petersburg, where the progeny flowered in 1761. Here it was examined by the great hybridist Kölreuter, who concluded that 'the hybrid Goatsbeard . . . is not a hybrid plant in the real sense, but at most only a half hybrid, *and indeed in different degrees*'. It is also interesting that the second generation progeny produced by the inter-crossings of Linnaeus' hybrid plants clearly showed segregation of different types, a very early record of genetic segregation which we discuss in Chapter 4.

We see how Linnaeus came to believe that, as in the case of *Peloria*, certain species had arisen from others in the course of time, and also that new species could arise by hybridisation. There is, however, contemporary evidence against Linnaeus' views (Glass, 1959). Adanson, an eighteenth-century French botanist whose originality has only recently been appreciated, tested *Peloria* more fully than Linnaeus. He found that *Peloria* specimens supplied by Linnaeus to the Paris Jardin des Plantes were not stable, producing flowering stems with both 'peloric' and normal flowers. Germination of seed of these plants often gave normal progeny as well as 'peloric'. Adanson concluded that the plant was a monstrosity, not a new species. He came to similar conclusions in two other cases, after experiments with an entire-leaved strawberry (*Fragaria*) discovered by the horticulturalists Duchesnes and son at Versailles in 1766, and the famous laciniate plant of *Mercurialis annua* discovered by Marchant in 1715. There was also evidence against the origin of new species by hybridisation. Kölreuter made a large number of crosses in Tobacco (*Nicotiana*) and other genera. True-breeding new species were not produced by hybridisation; indeed the hybrids were often almost completely sterile, and even when they were fertile there was great variation in the progeny.

Returning to the writings of Linnaeus, we find that in later life he also gave further thought to the origins of the patterns of variation in plant groups. He speculated on the Creation as follows (*Fundamenta fructificationis*, 1762–3, trans., quoted from Ramsbottom, 1938):

> We imagine that the Creator at the actual time of creation made only one single species for each natural order of plants, this species being different in habit and fructification from all the rest. That he made these mutually fertile, whence out of their progeny, fructification having been somewhat changed, Genera of natural classes have arisen as many in number as the different parents, and since this is not carried further, we regard this also as having been done by His Omnipotent hand directly in the beginning; thus all Genera were primeval and constituted a single Species. That as many Genera having arisen as there were individuals in the beginning, these plants in course of time became fertilised by others of different sort and thus arose Species until so many were produced as now exist . . . these Species were sometimes fertilised out of congeners, that is other Species of the same Genus, whence have arisen Varieties.

Linnaeus ascribes here almost an evolutionary origin to present-day species, genera having been formed at the Creation, species-formation being a more recent process (see Erikkson, 1983; Linroth, 1983). This most important change in Linnaeus' views relates to his hybridisation studies. He appears to have been convinced in later life that species can arise by hybridisation, and moved away from the idea of a fixed number of species all created at the same moment in time. Linnaeus' early views on the fixity of species received wide circulation in Europe in his main works, *Critica botanica, Systema naturae* and *Species plantarum*, while his more mature views, presented in the dissertations, did not have such a wide readership. So it is not surprising that even today he is often credited with rigid views on the question.

Buffon and Lamarck

In the mid-eighteenth century zoologists, too, were considering special creation. Linnaeus' contemporary, the French zoologist Buffon (1707–88), had also started his career with orthodox beliefs: 'We see him, the Creator, dictating his simple but beautiful laws and impressing upon each species its immutable characters.' Later, in 1761, however, he speculated on the mutability of species: 'How many species, being perfected or degenerated by the great changes in land and sea, . . . by the prolonged influences of climate, contrary or favourable, *are no longer what they formerly were*?' (Osborn, 1894).

The speculative ideas of Buffon and others remained untested by experiment; the majority of botanists and zoologists, engaged as they were in the late eighteenth century on the naming and classification of the world's flora and fauna, believed in the fixity of species. This belief was indeed so firmly held by naturalists that Cuvier (1769–1832), who had studied many fossil animals, accounted for extinct species by postulating a series of great natural catastrophes, which wiped out certain intermediate species. Cuvier believed that there had been only one Creation, and that after each disaster the Earth was repopulated by the offspring of the survivors. The last catastrophe was the Great Flood recorded in Genesis.

The doctrine of fixity of species was not without its critics in the nineteenth century (see Corsi, 1988; Ruse, 2013). Lamarck (1744–1829), in his *Philosophie zoologique* (1809), attacked the belief that all species were of the same age, created at the beginning of time in a special act of Creation. He believed, much as Ray and Linnaeus did, that species could be changed by growth in different environments, but he also believed that modifications in plant structure brought about by environmental change were inherited (Elliot, 1914):

> In plants, . . . great changes of environment . . . lead to great differences in the development of their parts . . . and these acquired modifications are preserved by reproduction among the individuals in question, and finally give rise to a race quite distinct from that in which the individuals have been continuously in an environment favourable to their development . . . Suppose, for instance, that a seed of one of the meadow grasses . . . is transported to an elevated place on a dry, barren and stony plot much exposed to the winds, and is there left to germinate; if the plant can live in such a place, it will always be badly nourished, and if the individuals reproduced from it continue to exist in this bad environment, there will result a race fundamentally different from that which lives in the meadows and from which it originated.

Thus Lamarck believed that a normally tall plant, dwarfed by growth at high altitude, would produce dwarf offspring. His belief in such an inheritance of acquired characters, which closely parallels the writings of Erasmus Darwin (1731–1802), formed the basis of his evolutionary speculation: one species evolved into another as hereditary changes arose in a plant under the impact of environmental variation. Lamarck, who suffered ill-health at the end of his life and was totally blind for the last 10 years, did not make any experimental investigations in search of evidence for his hypothesis (Jordanova, 1984). He did, however, cite a number of possible cases of apparent change of species brought about by environmental agency. For example:

> So long as *Ranunculus aquatilis* is submerged in the water, all its leaves are finely divided into minute segments; but when the stem of this plant reaches the surface of the water, the leaves which develop in the air are large, round and simply lobed. If several feet of the same plant succeed in growing in a soil that is merely damp without any immersion, their stems are then short, and none of their leaves are broken up into minute divisions, so that we get *Ranunculus hederaceus*, which botanists regard as a separate species.

In this interesting quotation we see that Lamarck puts quite a different interpretation upon variation exhibited by aquatic *Ranunculus* species, from that of Linnaeus, who considered such changes in leaf characters part of intraspecific variation. We consider modern interpretations of this variation in Chapter 6, and also reconsider the present standing of Lamarckism at various points in the book (Jablonka & Lamb, 1999).

Darwin

Our ideas on evolution are based on the work of Charles Darwin (1809–82). In 1825, following in the footsteps of his father, Charles began medical studies. But he was unable to complete his qualifications at the University of Edinburgh, as he was appalled by surgery before the use of chloroform. In 1828, he entered Christ's College, Cambridge, to study to become a minister of the Church of England. Here, he greatly enjoyed the botanical lectures and excursions of Professor John Henslow; Darwin became 'the man who walks with Henslow' (Walters & Stowe, 2001). Through Henslow, he was introduced to Professor Adam Sedgwick who stimulated Darwin's developing interest in geology. Janet Browne (1995) records the part that Henslow also played in Darwin's decision to become a companion to Captain Fitzroy, on the round-the-world voyage of the survey ship *Beagle* (December 1831–October 1836). Throughout the voyage, Darwin studied many aspects of the geology, natural history and social conditions not only in South America, but also in New Zealand, Australia and South Africa. He collected many specimens.

In an attempt to understand his observations made on the voyage of the *Beagle*, Darwin began a series of notebooks on transmutation, the first dated July 1837 (De Beer, 1960; Hodge, 2009, 2013), and he also wrote a sketch of his views in 1842 and a longer 'essay' in 1844 (Francis Darwin, 1909a, b). A new edition of the notebooks has been published by Barrett *et al.* (1987). In attempting to understand the development of Darwin's ideas about natural selection, scholars have examined these sources in minute detail (see Becquemont, 2009; Hodge, 2009). Some of the evidence comes also from a study of Darwin's correspondence, a first complete edition of which is in preparation (Burkhardt *et al.*, 1985–).

Many historians have investigated the origin and development of Darwin's ideas, as revealed by his writings and annotations in his books (Smith, 1960; Schweber, 1977; Kohn, 1985a; Bowlby, 1990b; Desmond & Moore, 1991; Browne, 1995, 2002), researches which suggest the crucial role of such writers as Lyell, Comte, Adam Smith, Quetelet and Malthus. Darwin delayed publication of his work, partly out of respect for his wife's strong religious beliefs, and also possibly because he wished to collect further information relevant to a theory of evolution (De Beer, 1963). It has also been suggested that he was concerned about the likely social consequences of his theory (Desmond & Moore, 1991). In 1844 a famous book *Vestiges of the Natural History of Creation*, advocating an evolutionary interpretation of nature, was published anonymously. It was later revealed that the author was Robert Chambers, a publisher based in Edinburgh. The book, of which there were 10 editions, proposed that 'new species and the ascent of life were planned linear developments, controlled by natural law as preordained by God' (Sapp, 2003). The volume received strong condemnation from many religious conservatives and this may have contributed to the delay in the publication of Darwin's ideas (Schweber,

1977). Considering this issue, Browne (1995) suggests that Darwin

> got into the habit of letting it be known that he was working on some large project on species without really coming to grips with the inevitability of putting it before the public. It was far easier to carry on collecting facts, to keep busy, to say the work was unfinished, than it was to stop. If he stopped he would feel obliged to present it. Underneath, there ran the incessant fear of being judged, the fear of other people's outrage.

Browne (2002) also observes that 'Darwin seemingly closed his mind to the possibility that other thinkers might be moving along the same road as he and that any one of them might come up with the same answer.'

Much to his dismay, in 1858, Darwin received an essay from the naturalist Wallace that set out a hypothesis almost identical to his own (for details of Wallace's books, articles etc. see Wallace-online.org/; www.nhm.ac.uk provides an archive of his letters). It has been suggested that Darwin plagiarised Wallace's ideas (Brackman, 1980; Brooks, 1983; Davies, 2008). For some the possibility of plagiarisation is still a live issue, but these accusations have been firmly rebutted by Kohn (1981, 1985b) and Beddall (1988). The whole question is discussed at length in the Introduction to Volume 7 of the Correspondence (Burkhardt & Smith, 1991).

At a meeting of the Linnean Society on 1 July 1858 the work of Darwin and Wallace was first presented in public. The papers were presented in alphabetical order by author as was customary at the Linnean Society for double contributions (Browne, 2002). This arrangement helped Darwin's friends in their efforts to resolve the delicate question of priority (Burkhardt & Smith, 1991).

First, unpublished extracts from Darwin's essay of 1844 were read. Then, to establish priority, an excerpt from a letter from Darwin to Professor Asa Gray, written in September 1857, was presented. Finally, Wallace's essay was read out. The contributions of Darwin and of Wallace, together with a letter from Lyell and Hooker explaining the historical background, have been reprinted in Jameson (1977). Darwin and Wallace did not have identical views (Bowler, 1990). Harper (1977, 1983) considers that while Wallace emphasised the role of inanimate forces, such as climate and soils, Darwin stressed the competitive interactions of organisms. Neither Darwin nor Wallace was present at the meeting. Furthermore, Wallace, being in the Dutch East Indies, was not consulted about the arrangement.

Bowler (1989b) discusses the thorny question of whether Wallace was fairly treated or whether, as many authors have hinted, 'history has been less than just to him'. Reflecting on this point, Browne (2002) observes that Wallace 'accepted the lesser role of co-discoverer that was thrust upon him. Perhaps he realised there was little else he could do . . . Even so, he probably felt a stab of regret to find he was not alone in his ideas. To his lasting credit, he never afterwards displayed the smallest flicker of resentment.'

Is Wallace now a forgotten figure? Many of his contemporaries reacted to his 'eccentric ideas and beliefs including spiritualism, opposition to vaccination, and a then deeply unfashionable support for socialism' (Bowler, 2008). However, recent studies of his writings confirm the considerable intellectual legacy of Wallace, who made major contributions to evolutionary studies, the geographical distribution of species and the evolution of animal coloration (see Smith & Beccaloni, 2008). Bowler (2008) emphasises that 'providing alternatives to the triumphalist story of Darwinism is laudable enough in this age of obsessive celebrity-worship, but Wallace deserves better than the routine use of his name by iconoclasts seeking to undermine Darwin's position in the pantheon of science'.

The main strands of the hypothesis of Darwin and Wallace may be summarised as follows:

1. Plants and animals vary. Darwin recognised two sorts of intraspecific variation: discontinuous variants (sports, monstrosities, jumps, saltations) and continuous variations (small, slight or individual differences, deviations or modifications). In his letter to Gray, Darwin wrote '*Natura non*

facit saltum', making clear his view that it was individual differences and not saltations that were important in evolution.

2. Because of the fecundity of organisms there would be a geometrical increase in numbers unless checked. Such natural checks occur. Darwin and Wallace both acknowledged a debt to Malthus in their understanding of natural checks to population increase.
3. As a consequence of these checks, only those individuals survive which have an inherent advantage over others in the population.
4. These better-fitted organisms, surviving this 'natural selection', pass on their 'advantage' to a proportion of their offspring.
5. Selection continues over thousands of generations, and in a rapidly changing environment new variants take the place of the original organisms.

The principal ideas of Darwin's hypothesis are set out in the following quotations from the extract read at the Linnean Society meeting.

> De Candolle, in an eloquent passage, has declared that all nature is at war, one organism with another, or with external nature . . . It is the doctrine of Malthus applied in most cases with tenfold force . . . Reflect on the enormous multiplying power *inherent and annually in action* in all animals; reflect on the countless seeds scattered by a hundred ingenious contrivances, year after year, over the whole face of the land; and yet we have every reason to suppose that the average percentage of each of the inhabitants of a country usually remains constant. Finally, let it be borne in mind that this average number of individuals (the external conditions remaining the same) in each country is kept up by recurrent struggles against other species or against external nature (as on the borders of the Arctic regions, where the cold checks life), and that ordinarily each individual of every species holds its place, either by its own struggle and capacity of acquiring nourishment in some period of its life, from the egg upwards; or by the struggle of its parents . . . with other individuals of the *same* or *different* species. But let the external conditions of a country alter . . . Now, can it be doubted, from the struggle each individual has to obtain subsistence, that any minute variation in structure, habits or instincts, adapting that individual better to the new conditions, would tell upon its vigour and health? In the struggle it would have a better *chance* of surviving; and those of its offspring which inherited the variation, be it ever so slight, would also have a better *chance*. Yearly more are bred than can survive; the smallest grain in the balance, in the long run, must tell on which death shall fall, and which shall survive. Let this work of selection . . . go on for a thousand generations, who will pretend to affirm that it would produce no effect . . .

Darwin then goes on to give an example:

> If the number of individuals of a species with plumed seeds could be increased by greater powers of dissemination within its own area (that is, if the check to increase fell chiefly on the seeds), those seeds which were provided with ever so little more down, would in the long run be most disseminated; hence a greater number of seeds thus formed would germinate, and would tend to produce plants inheriting the slightly better-adapted down. (Darwin & Wallace, 1858)

After the meeting of the Linnean Society, Darwin spent the next few months writing the text that was eventually published in 1859 under the title *On the Origin of Species by Means of Natural Selection*. Darwin saw the *Origin* as an introduction to a series of works. Sapp (2003) observes that the prior publication of Chambers's *Vestiges* proved to 'be of great value to Darwin . . . the book's critics supplied him with a list of objections, which he took care to address in the *Origin*'. The book itself, which provided insights into so many facets of biology, provoked very considerable controversy (Vorzimmer, 1972; Jones, 1999), and, reworking the material to accommodate the views of critics and in the development of his own ideas, Darwin produced six editions in all. Sometimes substantial changes were made, as can be seen in the *variorum* text of the *Origin* produced by Peckham

(1959). How the *Origin* evolved through its different editions (1859–72) has been intensively studied. Hoquet (2013) provides a very useful review. Endersby (2009) points out that for Darwin the *Origin* was 'a work in progress'. In 1872, a review of the book, in the newly founded scientific journal *Nature*, applauded 'the way in which Darwin drew attention to his own errors and his willingness to publicly change his mind'.

For details of Darwin's life as well as biographical details of those involved in the development and publication of his ideas, for instance, Asa Gray, Hooker, Lyell, Wallace, Desmond & Moore (1991) and Browne (1995, 2002), should be consulted. For a wider perspective on the impact of Darwinism, see for example, Ruse (2008), Dawkins (2009) and Jones (2009). These volumes were produced in or around 2009, to celebrate the 200th anniversary of Darwin's birth and the 150th anniversary of the publication of *On the Origin of Species*.

We rightly give credit to Darwin for establishing, against considerable opposition, a plausible mechanism in natural selection to explain organic evolution. As is usually the case in the history of ideas, however, a careful reading of the literature reveals a number of statements of 'selectionist' ideas long before Darwin and Wallace. Indeed, Zirkle (1941), in a remarkably interesting and little-quoted paper, provides abundant evidence of such ideas, tracing them back to the writings of Empedocles (495–435 BC) and Lucretius (99–55 BC). In the sixth edition of the *Origin* Darwin himself provides a short historical review in which he makes it clear that the idea of natural selection had occurred to others and indeed had been published, for example by Dr W. C. Wells (1818) in 'An account of a white female, part of whose skin resembles that of a Negro' and by Patrick Matthew (1831) in his book *On Naval Timber and Arboriculture*. Stott (2012) has recently provided a splendid wide-ranging study of first evolutionists.

Many of the criticisms raised by contemporary critics, almost all of which are of interest to modern biologists, are very involved. Mivart (1871) produced a fascinating book that reviews all the contemporary difficulties. How these difficulties are viewed by present-day experts is considered in great detail by an army of Darwin scholars: see, for example, Vorzimmer (1972), Jones (1999) and the books by Dawkins (2003, 2005, 2009).

Here is a brief treatment of some of the most important issues raised in Darwin's lifetime.

The role of saltations in evolution

As Darwin pointed out, discontinuous variants – sports, monstrosities, etc. – are often sterile and therefore of little or no consequence in evolution. However, various biologists including Harvey (who studied a mutant *Begonia*, 1860), Huxley (in his review of the first edition of the *Origin*) and Asa Gray, all suggested that saltations may be important in evolution.

> When you suppose one species to pass, by insensible degrees into another, so many facts of variation support your view that it does not seem very improbable; but where a generic limit has to be passed, bearing in mind how *persistent* generic differences are, I think we require a *saltus* (it may be a small one) or a real break in the chain, namely, a sudden divarication. (Unpublished letter from Harvey to Darwin, quoted by Vorzimmer, 1972)

The mechanisms of heredity

In many ways this was the most important problem raised by Darwin's critics. For evolution to take place there must be selection of favoured varieties, such variants on crossing leaving better-adapted offspring. Darwin was unable to understand the mechanism of heredity, believing in a type of blending inheritance (see Chapter 4). Fleeming Jenkin, Professor of Engineering at Edinburgh University, writing anonymously in *The North British Review*, June 1867, showed a very serious weakness in Darwin's argument. He pointed out that if a rare variant favoured by natural selection appeared in a population, it would cross with the more abundant less-favoured plants in the population. Its hereditary advantage would then be lost in blending inheritance. How did favoured genetic variants ever become abundant? Darwin was never able satisfactorily to answer this criticism. Apparently Jenkin was not alone in appreciating this particular difficulty, for Vorzimmer (1972) argues that before Jenkin's review several biologists, including Darwin himself, were aware of the implication of blending inheritance.

The effect of chance

Jenkin also drew attention to another important problem – the effect of chance on the survival of favoured variants. He discusses his ideas in relation to hares.

> [L]et us here consider whether a few hares in a century saving themselves by this process [burrowing] could, in some indefinite time, make a burrowing species of hare. It is very difficult to see how this can be accomplished, even when the sport is very eminently favourable indeed; and still more difficult when the advantage gained is very slight, as must generally be the case. The advantage, whatever it may be, is utterly out-balanced by numerical inferiority. A million creatures are born; ten thousand survive to produce offspring. One of the million has twice as good a chance as any other of surviving; but the chances are fifty to one against the gifted individuals being one of the hundred survivors. No doubt, the chances are twice as great against any one other individual, but this does not prevent their being enormously in favour of *some* average individual. However slight the advantage may be, if it is shared by half the individuals produced, it will probably be present in at least fifty-one of the survivors, and in a larger proportion of their offspring; but the chances are against the preservation of any one 'sport' in a numerous tribe.

The limits of variation

While not the first to raise this problem, Jenkin, in the following quotations, points to what many biologists saw as an important difficulty in Darwin's theory.

> If we could admit the principle of a gradual accumulation of improvements, natural selection would gradually improve the breed of everything, making the hare of the present generation run faster, hear better, digest better, than his ancestors ... Opinions may differ as to the evidence of this gradual perfectibility of all things, but it is beside the question to argue this point, as the origin of species requires not the gradual improvement of animals retaining the same habits and structure, but such modification of those habits and structure as will actually lead to the appearance of new organs. We freely admit, that if an accumulation of slight improvements be possible, natural selection might improve hares as hares, and weasels as weasels, that is to say, it might produce animals having every useful faculty and every useful organ of their ancestors developed to a higher degree; more than this, it may obliterate some once useful organs when circumstances have so changed that they are no longer useful, for since that organ will weigh for nothing in the struggle of life, the average animal must be calculated as though it did not exist.
>
> We will even go further: if, owing to a change of circumstances, some organ becomes pre-eminently useful, natural selection will undoubtedly produce a gradual improvement in that organ, precisely as man's selection can improve a special organ ... Thus, it must apparently be conceded that natural selection is a true cause or agency whereby in some cases variations of special organs may be perpetuated and accumulated, but the importance of this admission is much limited by a consideration of the cases to which it applies: ... Such a process of improvement as is described could certainly never give organs of sight, smell, or hearing to organisms which had never possessed them. It could not add a few legs to a hare, or produce a new organ, or

even cultivate any rudimentary organ which was not immediately useful to an enormous majority of hares ... Admitting, therefore, that natural selection may improve organs already useful to great numbers of a species, does not imply an admission that it can create or develop new organs, and so originate species.

The origin of complex organs and structures

Many biologists, assuming a useless incipient stage in the development of complex organs (for example, the eye), could not see how such structures could have evolved as a consequence of natural selection either of small individual differences or of saltations. Similar difficulties are encountered, for example, in the evolution of complex floral structures. For instance, the extraordinary adaptations shown by many Orchids by which cross-pollination is brought about by particular insect visitors to the flower, were used by Darwin as evidence, for the variety and complexity of adaptation, to be explained by natural selection (Darwin, 1862, 1877b). Writers such as Mivart (1871), however, who were opposed to any completely selectionist interpretation of evolution, were inclined to turn Darwin's 'evidence' against himself. Thus, Darwin had mentioned the remarkable Orchid *Coryanthes speciosa* in which the labellum, modified to a bucket with a lip, is filled with water secreted by special glands. Pollination involves the visiting bees in an involuntary bath from which they can only rescue themselves by crawling through the narrow passage at the lip and thus effecting pollination. As Mivart observes: 'Mr Darwin gives a series of the most wonderful and minute contrivances ... structures so wonderful that nothing could well be more so, except the attribution of their origin to minute, fortuitous and indefinite variations.'

This book is about evolution of plants, but a number of important questions about complex structures and their origin are briefly considered. Some people believe in the creation stories in religious books, and reject Darwinism in all its forms. Others, while acknowledging modern views of the age of the Earth etc., believe in 'Intelligent Design', a 'hypothesis that in order to explain life it is necessary to suppose the action of an unevolved intelligence' (Dembski & Ruse, 2007b). Supporters of Intelligent Design consider that the eye, the ear, flight in birds, flagella movement in bacteria, biochemical pathways in plants and animals etc. are systems of *such interlocking irreducible complexity* that they could not have evolved through natural selection alone. They take the view that these examples, and others, point to the 'purposeful arrangements of parts' indicating the existence of design and that 'life was designed by an intelligence' (see Behe, 1996).

Proponents of Intelligent Design, approaching the subject from a fundamentalist religious perspective, insist that these ideas should be taught in science classes alongside Darwinian evolutionary biology (Phy-Olsen, 2010). In response, 'the American Association for the Advancement of Science (the organisation that publishes *Science*) has declared officially that in its opinion Intelligent Design is not so much bad science as no science at all and accordingly has no legitimate place in the science classrooms of the United States' (Dembski & Ruse, 2007b).

While freely acknowledging that our current understanding is often partial and imperfect, most biologists reject the idea of Intelligent Design. Cases of supposed 'irreducible complexity' are being researched using molecular tools to study the biochemistry and genetics of biological development. The notion that complex systems have evolved by Darwinian evolution by modification of existing structures, parallel processing and changes in function has increasing support (see Weber & Depew, 2007).

Most scientists are sceptical about Intelligent Design. For example in considering the origin of complexity, Sapp (2003) notes that:

according to Darwinian evolutionary theory, there is no design in the natural world, no preconceived plan. Organisms evolve in a makeshift or contingent manner in relation to changing ecological conditions. New organs do not suddenly appear that seem to have been specially created for some purpose.

The appearance of species results from numerous forces that combine at a certain epoch in a certain place. Had the conditions been different, the natural world would be different today: nothing is necessary, nothing is purposeful, and nothing therefore is beyond investigation.

However, Sapp continues: 'evolutionary theory and rationalist explanation do not necessarily preclude the concept of God: some evolutionists may be agnostic or they may invoke God to explain the origins of the natural laws through which life evolves'.

For valuable overviews of the various facets of the extremely controversial and often acrimonious debate on the origins and significance of biological complexity, see Johnson (1993), Pigliucci (2002), Ruse (2003), Young & Edis (2004) and Dembski & Ruse (2007a).

Darwin and Lamarck

Returning to the contemporary responses to Darwin's work, Endersby (2009) notes that

> it is often claimed that Darwin responded to some of the criticisms levelled at his theory by placing less emphasis on natural selection and more on the inheritance of acquired characteristics. Because this latter mechanism is primarily (but not entirely accurately) associated with Lamarck, it is therefore claimed that the *Origin of Species* becomes more 'Lamarckian' in successive editions.

Darwin does indeed introduce Lamarckian elements into his discussion of use and disuse of characters and the origin of domesticated plants and animals, making reference to the 'direct action of physical conditions of life'. However, Endersby's analysis 'suggests that the *Origin* does not become steadily or progressively more Lamarckian over succeeding editions; it is really only the sixth edition where there is a modest shift'.

The role of isolation

Whilst it may be argued from Darwin's writings (Mayr, 1963) that he saw isolation as providing a 'favourable circumstance' for speciation, it was the German naturalist Wagner (1868), pointing to the probable loss of favourable variants in 'blending' inheritance, who argued that spatial isolation was a necessary condition of speciation. This important difference of opinion is discussed in some detail later.

The age of the Earth

The modern biologist may find it difficult to realise the extent to which controversies about the age of the Earth and its rocks preceded and eventually made possible the Darwinian revolution. The study of fossils, or 'figured stones' as they were first called, undoubtedly produced questioning about the age of the Earth as early as the middle of the sixteenth century. By the end of the eighteenth century, when Hutton published his *Theory of the Earth*, geology was recognised as a separate and quite well-based science, and the Genesis account of the origin of the Earth was already under great strain (see Lieberman & Kaesler, 2010).

After overcoming naive Church opposition, which offered a date of 4004 BC for the Biblical Creation, Darwin's ideas were, however, faced with much more sophisticated criticism from the eminent physicist William Thompson (later Lord Kelvin), who estimated the probable age of the Earth from a range of evidence including calculations of average rate of heat loss based upon measurable temperature increases down boreholes and mines. In his famous paper (Kelvin, 1871) 'On geological time', read at a meeting of the Geological Society of Glasgow in 1868, Kelvin calculated that the consolidation of the crust of the

Earth took place at a maximum of 400 Mya. Later, he and other scientists also made further estimates. Burchfield (1990) summarises Kelvin's developing views as follows:

> By 1868, he was convinced that sufficient evidence existed to justify limiting the assumed duration of life on earth, if not the earth's total age, to no more than 100 million years. In 1876 he was willing to accept an upper limit of only 50 million years for the earth's age, and by 1881 a limit of 20 to 50 million. Finally, in 1897 he declared that the earth's age was nearer 20 million than 40 million, and embraced Clarence King's estimate of 24 million as the best available.

Kelvin's findings proved very difficult for Darwin. In a letter to Wallace in 1869 he wrote: 'Thompson's views on the recent age of the world have been for some time one of my sorest troubles.' Had there been sufficient geological time for the slow evolutionary processes postulated by Darwin (see Burchfield, 1990)?

We now know that Kelvin's estimate was inaccurate as he was unaware of radioactivity (Sapp, 2003). Heat is generated from the decay of radioactive elements such as radium, and modern estimates of the age of the Earth, taking into account all sources of radioactivity, reveal that the Earth is of the order of 4.6 billion years old. This estimate is compatible with the Darwinian concept of the slow evolution of life.

Human origins

In the last chapter of the *Origin*, 'Recapitulation and conclusion', Darwin wrote: 'Light will be thrown on the origin of man and his history.' This single sentence is the only explicit reference to human origins in the book. Endersby (2009) points out that 'Darwin may have hoped to avoid the question, but it inevitably became a central focus of debate as soon as the *Origin* appeared. Darwin returned to the question in greater detail in *The Descent of Man* (1871b), by which time the fact of evolution had become much more widely accepted.' In the *Descent*, Darwin concluded that

> man is descended from some less organised form. The grounds upon which this conclusion rests will never be shaken, for the close similarity between man and the lower animals in embryonic development, as well as innumerable points of structure and constitution ... Now when viewed by the light of our knowledge of the whole organic world, their meaning is unmistakable. The great principle of evolution stands up clear and firm ... He who is not content to look, like a savage, at the phenomena of nature as disconnected, cannot any longer believe that man is the work of a separate act of creation ... the close resemblance of the embryo of man to that, for instance, of a dog – the construction of his skull, limbs and whole frame with that of other mammals ... all point in the plainest manner to the conclusion that man is the co-descendant with other mammals of a common progenitor... man is descended from a hairy, tailed, quadruped, probably arboreal in its habits, and an inhabitant of the Old World.

Darwin's book on human origins was published twelve years after the *Origin*, when he was 62. It might be concluded that his ideas on the subject came later in his life, but in the view of Desmond & Moore (2009) this would be completely incorrect. They point out that Darwin and his relatives were part of the anti-slavery movement, that, on the *Beagle* voyage, Darwin lived for a time in slave countries and that in several countries he witnessed the plight of indigenous peoples under colonial rule. In a close examination of Darwin's published writings, papers and letters, Desmond & Moore conclude that 'human evolution wasn't his last piece in the evolution jigsaw; it was the *first*. From the outset Darwin concerned himself with the unity of humankind. The notion of "brotherhood" grounded his evolutionary enterprise. It was there in his first musings on evolution in 1837.' They continue:

> Darwin's starting point was the abolitionist belief in blood kinship, a 'common descent' ... It implied a single origin for black and white, a shared ancestry.

And this was *the* unique feature of Darwin's particular brand of evolution. Life itself was made up of countless trillions of sibling 'common descents', not only black and white, but among all races, all species, through all time, all joined up in bloodlines back to a common ancestor.

The nature of specific difference

Turning to another important issue, it is clear that Darwin's view of the species, based as it was on a thorough study of living organisms as well as on the pertinent literature, is very different indeed from that of Ray and Linnaeus.

In chapter two of the sixth edition of the *Origin* we read:

> Hence, in determining whether a form should be ranked as a species or a variety, the opinion of naturalists having sound judgement and wide experience seems the only guide to follow. We must, however, in many cases, decide by a majority of naturalists, for few well-marked and well-known varieties can be named which have not been ranked as species by at least some competent judges.

He was impressed in his study of the variability of plants and animals by how difficult it was, in many groups, to delimit species, and gives many examples. In polymorphic groups he notes: 'With respect to many of these forms, hardly two naturalists agree whether to rank them as species or as varieties. We may instance *Rubus, Rosa* and *Hieracium* amongst plants.' He also considered the opinion of such great taxonomists as de Candolle, who, completing his monograph of the Oaks of the world, wrote:

> They are mistaken, who repeat that the greater part of our species are clearly limited, and that the doubtful species are in a feeble minority. This seemed to be true, so long as a genus was imperfectly known, and its species were founded upon a few specimens, that is to say, were provisional. Just as we come to know them better, intermediate forms flow in, and doubts as to specific limits augment.

Darwin concludes:

> Certainly no clear line of demarcation has as yet been drawn between species and sub-species – that is, the forms which in the opinion of some naturalists come very near to, but do not quite arrive at, the rank of species: or, again, between sub-species and well-marked varieties, or between lesser varieties and individual differences. These differences blend into each other by an insensible series . . .

In chapter nine, on hybridism, he notes, after examining the extensive writings of Gärtner, Kölreuter, etc., that even though exceptions are known: 'First crosses between forms, sufficiently distinct to be ranked as species, and their hybrids, are very generally, but not universally, sterile.' In the section on intraspecific crosses he writes: 'It may be urged, as an overwhelming argument, that there must be some essential distinction between species and varieties, inasmuch as the latter, however much they may differ from each other in external appearance, cross with perfect facility, and yield perfectly fertile offspring'. Notwithstanding these views on the crossing of different groups, one is impressed on reading the *Origin* by the absence of any definition of species incorporating both morphological and crossing information. Why Darwin provides no definition of species is very clear from the discussion of species in the concluding chapter:

> When the views advanced by me in this volume, and by Mr Wallace, or when analogous views on the origin of species are generally admitted, we can dimly foresee that there will be a considerable revolution in natural history. Systematists will be able to pursue their labours as at present; but they will not be incessantly haunted by the shadowy doubt whether this or that form be a true species. This, I feel sure and I speak after experience, will be no slight relief. The endless disputes whether or not some fifty species of British brambles are good species will cease.

> Systematists will have only to decide (not that this will be easy) whether any form be sufficiently constant and distinct from other forms, to be capable of definition; and if definable, whether the differences be sufficiently important to deserve a specific name. This latter point will become a far more essential consideration than it is at present: for differences, however slight, between any two forms, if not blended by intermediate gradations, are looked at by most naturalists as sufficient to raise both forms to the rank of species. Hereafter we shall be compelled to acknowledge that the only distinction between species and well-marked varieties is, that the latter are known, or believed, to be connected at the present day by intermediate gradations whereas species were formerly thus connected. Hence, without rejecting the consideration of the present existence of intermediate gradations between any two forms, we shall be led to weigh more carefully and to value higher the actual amount of difference between them. It is quite possible that forms now generally acknowledged to be merely varieties may hereafter be thought worthy of specific names; and in this case scientific and common language will come into accordance. In short, we shall have to treat species in the same manner as those naturalists treat genera, who admit that genera are merely artificial combinations made for convenience. This may not be a cheering prospect; but we shall at least be freed from the vain search for the undiscovered and undiscoverable essence of the term species.

From this passage it is abundantly clear that Darwin considered the taxonomist's task in recognising and naming species as a severely practical one, to be decided on criteria of degree of discontinuity and clarity of descriptive diagnosis. Undoubtedly he was influenced by his close study of domesticated plants and animals, and his views were not very acceptable to most practising taxonomists working with wild plants and animals, who continued to be impressed by the high proportion of apparently clear-cut entities to be described and named in the natural world. To them, species were 'real' in a way that genera were not.

Classification

Darwin's ideas about evolution provoked him to a new and radical reappraisal of classification, and the results of this are set out in chapter 14 of the *Origin*. First, Darwin notes that 'Organic beings, like all other objects, can be classed in many ways, either artificially by single characters, or more naturally by a number of characters'. Clearly, specimens in a diverse collection may be grouped on flower colour, habit or any other single characters. As we have seen Linnaeus' 'Sexual System' was clearly artificial. While his system was highly successful in practical terms, Linnaeus himself recognised that his work would be superseded by natural classifications, in which plants were classified on the basis of many characters. In the period between the *Species plantarum* and the *Origin*, much progress was made in constructing 'Natural Systems', major contributions being made by Adanson, B. and A. L. de Jussieu, Robert Brown, de Candolle and others (Davis & Heywood, 1963; Cronquist, 1988).

As a prelude to stating his own views, Darwin notes that 'Some authors look at it [the Natural System] merely as a scheme for arranging together those living objects which are most alike, and for separating those which are most unlike ... But many naturalists think that something more is meant by the Natural System; they believe that it reveals the plan of the Creator'. Darwin then presents his own ideas at length, of which the following is a brief quotation.

> 'The Natural System is founded on descent with modification ... community of descent is the hidden bond which naturalists have been unconsciously seeking, and not some unknown plan of creation, or the enunciation of general propositions, or the mere putting together and separating objects more or less alike... Thus, the natural system is genealogical in its arrangement, like a pedigree: but the amount of modification which the different groups have undergone has to be expressed by ranking them under different so-called genera, sub-families, sections, orders and classes.

The implications of these views about branching phylogenetic trees will be discussed in later chapters.

Tests of specific difference

We have examined so far in this chapter a number of ideas about what constitutes a species. Linnaeus stressed the morphological difference between species whilst Darwin, considering both external morphology and the results of hybridisation experiments, found the species difficult to define. Many other botanists were interested in the species problem in the mid-nineteenth century, and tests of specific rank were devised.

At first, it seemed that hybridisation experiments might provide an objective guide as to whether a plant was a species or a variety. A number of scientists supported this view, in particular Professor Godron of the University of Nancy. In 1863 he published the following opinion (see Roberts, 1929). If two given plants could be crossed without difficulty giving fertile offspring, they were to be called varieties of one species. If, on the other hand, two plants crossed with difficulty, if sterility barriers existed between different plants, then such plants were to be considered different species. Further, crossing between plants of different genera was impossible. The categories of variety, species and genus were therefore to be determined by crossing experiments. Godron's rigid ideas, which were based upon his own work as much as on the extensive publications of earlier hybridists, contrast sharply with the cautious views of Darwin. Of other botanists interested in the problems of defining species experimentally one must mention Professor von Nägeli of Vienna, with whom Mendel fruitlessly corresponded (see Chapter 4). He published a massive review of hybridisation in 1865, noting in particular the difficulties in the sort of ideas published by Godron.

A second test of species is associated with the name of Alexis Jordan of Lyons in France. He considered that cultivation experiments, with progeny testing, provided an objective means of distinguishing species. In 1864 he published a great many of his results. He is perhaps best remembered for his work on *Erophila verna*, in which he described 53 'elementary species', each retaining its distinctive characters in cultivation and coming true from seed. An even greater number could easily have been described, for he indicates that he had more than 200 distinct lines of *E. verna* in cultivation (Fig. 2.2). His experiments were not confined to this taxon, and it is of considerable interest to note the large number of 'elementary species' he described in several common genera, as shown in Table 2.1.

Table 2.1. **Numbers of 'elementary species' in various taxa published by Jordan in 1864**

Arabis	23	*Iberis*	23
Biscutella	21	*Ranunculus*	25
Erophila	53	*Thalictrum*	47
Erysimum	26	*Thlaspi*	21

Jordan was followed by others in his practice of describing 'elementary species' within Linnaean species: for instance, Wittrock in Sweden working with *Viola tricolor*, and later de Vries (1905) experimenting with *Oenothera* species. This practice was condemned by many botanists, as it led to an inordinate number of new plant names.

The idea of using only a single line of experimental evidence as a test of specific rank did not meet with universal approval. For example, Hoffmann, Professor of Botany in the University of Giessen, carried out a large number of experiments with many taxa. He observed plants closely in the wild and also carried out cultivation and crossing experiments. In a review of his researches (1881) he considered the many different lines of evidence that could be used in judging specific rank. Not only did he study the performance of plants in cultivation and the results of crossing plants, but he also took into account geographical distribution and the extent of hybridisation *in the wild*.

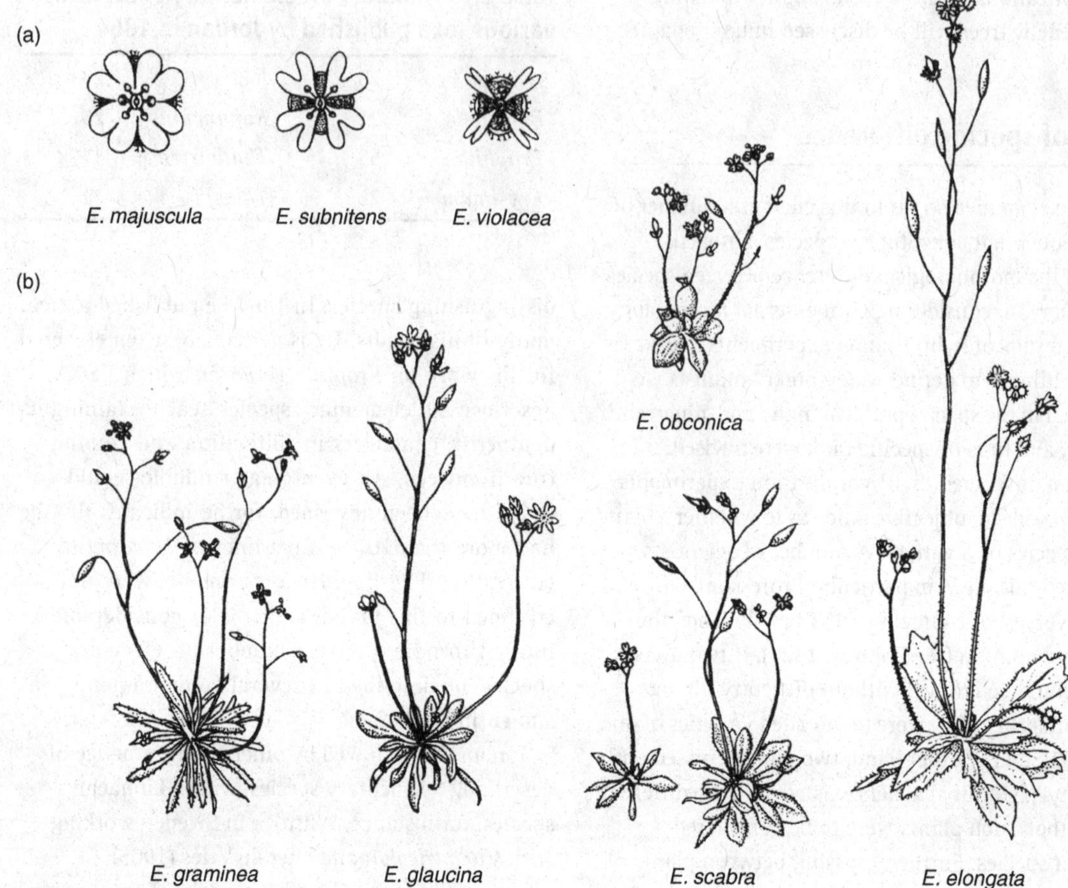

Fig. 2.2. 'Elementary species' in *Erophila verna.* (a) Enlargements of flowers showing petal variation (×1.6). (b) Habit variation (×0.75). (From Rosen, 1889.)

This historical review of species and their variation has brought us almost to the end of the nineteenth century, and it is in this period that there emerged two new aspects of the study of variation. First, the statistical examination of biological variation: some of the results of this work are the subject of the next chapter. Secondly, following the epoch-making rediscovery in 1900 of Mendel's work on heredity, the science of genetics made its appearance.

3 Early work on biometry

In the second half of the nineteenth century, as Darwinism was making its impact upon biology, an interesting new approach to biological variation, especially intraspecific variation, was being examined. Instead of trying to describe variation patterns in words, the investigators, examining large samples of organisms, collected numerical data and subjected them to statistical analysis. In the following account we discuss selected themes: for comprehensive studies of the subject see Pearson & Kendall (1970), Kendall & Plackett (1977), Porter (1986) and Stigler (1986).

The first worker to study natural variation statistically, a science that became known as biometry, was probably the Belgian Quetelet (1796–1874). He wrote a famous series of letters on the subject to his pupil, the Grand Duke of Saxe-Coburg and Gotha. Later in the century, Darwin's cousin, Francis Galton (1822–1911), made notable contributions to the statistical investigation of variation and inheritance, ideas that were developed by many other investigators, especially in relation to social issues (see Bowler, 1989; Keynes, 1993; Gillham, 2001; Bulmer, 2003).

Biometrics and eugenics

While Darwin and others considered humankind to be a single species, others, for example the American biologist Louis Agassiz (1807–73), took the view that there were several species of humans. In the developing fields of anthropology and sociology, biometrical methods were increasingly employed to examine racial variation.

Darwinian concepts provoked many questions with social implications. In 1871, in his book *The Descent of Man*, Darwin wrote:

> With savages, the weak in body or mind are soon eliminated; and those that survive commonly exhibit a vigorous state of health. We civilized men, on the other hand, do our utmost to check the process of elimination; we build asylums for the imbecile, the maimed, and the sick; we institute poor-laws; and our medical men exert their utmost skill to save the life of every one to the last moment . . . Thus the weak members of civilised societies propagate their kind. No one who has attended to the breeding of domestic animals will doubt that this must be highly injurious to the race of man. (Darwin, 1871)

Darwin continues:

> The surgeon may harden himself whilst performing an operation, for he knows that he is acting for the good of his patient; but if we were intentionally to neglect the weak and helpless, it could only be for a contingent benefit, with a great and certain present evil. Hence we must bear without complaining the undoubted bad effects of the weak surviving and propagating their kind.

Galton took a different view, in promoting the concept of eugenics.

> Its first object is to check the birth-rate of the Unfit, instead of allowing them to come into being, though doomed in large numbers to perish prematurely. The second object is the improvement of the race by furthering the productivity of the Fit by early marriages and healthful rearing of their children. Natural selection rests upon excessive production and wholesale destruction; Eugenics on bringing no more individuals into the world than can be properly cared for, and those only of the best stock. (Galton, 1908)

Darwin's libertarian viewpoint contrasts sharply with that of Galton, who wrote: 'What Nature does blindly, slowly, and ruthlessly, man may do providently, quickly, and kindly... The improvement of our stock seems to me one of the highest objects that we can reasonably attempt' (Galton, 1904).

Cowan (1972) is of the opinion that Galton was so convinced in his eugenic ideas that his biometrical and statistical techniques were developed in the hope of persuading the scientific community of the importance of eugenics. In the following decades both Karl Pearson and R. A. Fisher (see Bennett, 1983; Norton, 1983) were also actively concerned with eugenics and this may have provided strong motivation for their own biometrical and statistical investigations.

Although Galton writes of a 'kindly' approach, this was not the outcome of later developments in eugenics, for Bulmer (2003) concludes that 'enthusiasm for eugenics is strongly correlated with disregard for individual rights'. Thus, in many countries in the early twentieth century sterilisation programmes were instituted for the feeble minded, insane, alcoholic, criminal and unemployable (see Bashford & Levine, 2012; Kühl, 2013). These were forerunners of the notorious eugenic actions of the Nazis, in which millions were killed – Jews, gypsies, homosexuals, the mentally ill and others (Weiss, 2013).

Returning to the early years of biometrical investigations, it is now clear from these comments why much of the early work on biometry was concerned with racial differences, measuring criminals, students etc. It is not the intention in this book, which deals with plant variation and evolution, to explore these important issues. However, we should acknowledge the powerful ideas underlying the measurement of variation, and note that the extremely interesting investigations carried out on plants were a sideline to the early interest in eugenics.

Biometrical studies of plants

In presenting the botanical findings, it is important to examine first the general characteristics of this new approach. Instead of contenting themselves with the study of a few herbarium specimens or cultivated plants, the early biometricians took large samples, often using living material of common species. These samples were then carefully scrutinised and measured, as we read in Davenport (1904): 'Having settled upon the general conditions of race, sex, locality, age, which the individuals to be measured must fulfill, take the individuals methodically at random and without possible selection of individuals on the basis of the magnitude of the character to be measured.' Finally, having collected the samples and obtained the numerical data, the worker performed the analysis, observing Quetelet's precept that statistics must be collected without any preconceived ideas and without neglecting any numbers (Quetelet, 1846).

These early studies of biological material established the important point that there are two main kinds of intraspecific variation. Firstly, much of the variation is discontinuous. If, for instance, one is examining the number of chambers in a capsule, the number of seeds in a fruit, the number of leaves on a plant – in fact any variation in the number of parts – then the numbers found must be integers. One never discovers 14.5 undamaged peas in a pod. Often, in considering variation in the number of parts – so-called meristic variation – a more or less complete series of members is found. For instance, Pearson (1900) noticed, in a cornfield in the Chiltern Hills, England, that Poppies (*Papaver rhoeas*) had different numbers of stigmatic bands on the capsule. He collected a very large sample of 2268 capsules: the frequency of different numbers of stigmatic bands is given in Table 3.1. In other instances of discontinuous variation, however, only two or a few strikingly different variants are found. For instance, the Opium Poppy (*Papaver somniferum*) may or may not have a dark spot at the base of the petal; Groundsel (*Senecio vulgaris*) has either radiate or non-radiate capitula; and Foxglove (*Digitalis purpurea*) may have white or red flowers, and hairy or glabrous stems. On the other hand, a second type of variation – continuous variation – is also common in plants. In considering variation in such characters as height, weight, leaf length and root spread, any value is possible within a given range. There are no breaks in the variation for particular characters.

Table 3.1. **Calculation of the mean, variance, standard deviation and coefficient of variation in number of stigmatic bands in capsules of the Poppy (*Papaver rhoeas*). (Data from Pearson, 1900)**

Number of bands x	Frequency, f	fx	Difference from mean $x-\bar{x}$	Square of difference $(x-\bar{x})^2$	$f\times$ square of difference $f(x-\bar{x})^2$
5	1	5	−4.8	23.04	23.04
6	12	72	−3.8	14.44	173.28
7	91	637	−2.8	7.84	713.44
8	295	2360	−1.8	3.24	955.80
9	550	4950	−0.8	0.64	352.00
10	619	6190	+0.2	0.04	24.76
11	418	4598	+1.2	1.44	601.92
12	195	2340	+2.2	4.84	943.80
13	54	702	+3.2	10.24	552.96
14	25	350	+4.2	17.64	441.00
15	5	75	+5.2	27.04	135.20
16	3	48	+6.2	38.44	115.32
	2268	22 327			5032.52

$$\text{Mean} = \frac{22\,327}{2268} = 9.8$$

$$s^2 = \frac{\sum f(x-\bar{x})^2}{n-1} = \frac{\sum d^2}{n-1} = \frac{5032.52}{2267} = 2.2 \qquad s = 1.49$$

$$\text{Coefficient of variation} = \frac{s}{\bar{x}} \times 100 = \frac{1.49}{9.8} \times 100 = 15.2\%$$

Strikingly discontinuous variation patterns, as in white or red-purple flower colour in *Digitalis*, presented little difficulty in examination or classification. The analysis of arrays of data, however, whether of discontinuous or continuous variates, posed somewhat more complex problems. With an array of data, how was it possible to show numerically where the bulk of the variation lay; how could a numerical estimate or spread of the data within the sample be obtained and, further, how could the variability of two samples be compared? Using Pearson's data for variation in stigmatic band number in *Papaver rhoeas* (Table 3.1), we may now briefly examine some of the statistics employed by the early biometricians.

Commonest occurring variation in an array

Sometimes a knowledge of the mode, or most frequent class, and the median, or middle value of an array, is a useful indication of where the bulk of the variation lies in a sample. These are, however, less useful than the arithmetic mean, $\bar{x}$. This is calculated quite simply by summing (Σ) the observed values (x) and dividing by the number of observations, n.

$$\text{Mean} = \bar{x} = \frac{\sum x}{n} \qquad (1)$$

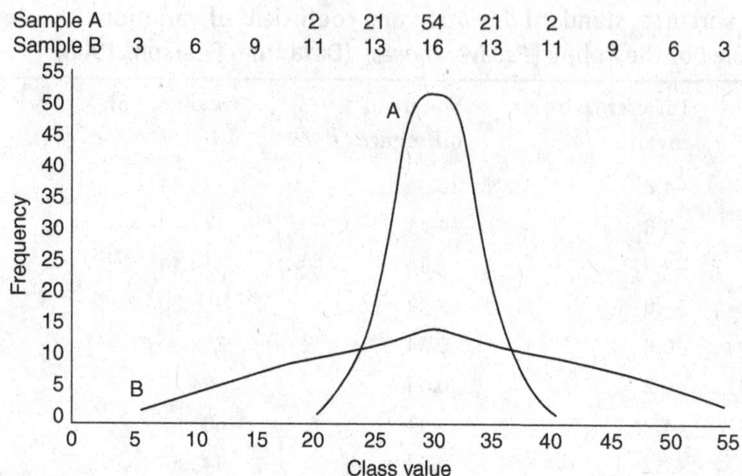

Fig. 3.1. Hypothetical frequency distribution for two population samples with the same mean. Sample B is much more variable than sample A. (From Srb & Owen, 1958.)

Estimates of dispersion of the data

Values of the mean give no indication of the variation within a sample. Identical means may be obtained if the data are all clustered very closely to the mean or if the data are markedly above and below the mean (Fig. 3.1). It is clearly very important to have an estimate of the degree of dispersion of the data within a particular sample.

There are several possible ways of examining dispersion. Early biometricians often noted the extreme values of the array, or alternatively they calculated how much each value of the array differed from the mean and, after summing the differences, calculated an *average deviation from the mean*. They also used a statistic, now seldom if ever calculated, called the *probable error*. (Details of this calculation may be found in statistics books.) In more recent times, however, dispersion has been estimated by calculating the variance, s^2, and standard deviation, s.

The variance, s^2, is calculated by summing the squares of the deviations of all the observations from their mean (d^2) and dividing by $n - 1$.

$$s^2 = \frac{\sum d^2}{n-1} \quad (2)$$

Except where samples are small, d^2 is more readily calculated by employing the equations

$$\sum d^2 = \sum (x - \bar{x})^2 \quad (3)$$

$$\sum (x - \bar{x})^2 = \sum x^2 - \frac{\left(\sum x\right)^2}{n} \quad (4)$$

In calculating the variance, s^2, it is important to note (and statistics books should be consulted for justification) that the divisor is $n - 1$. The standard deviation, s, is found by obtaining the square root of the right-hand side of equation 2. The calculation of variance and standard deviation for Pearson's Poppy data is given in Table 3.1. The variance is a valuable statistic, giving a measure of the dispersion of the data about the mean. It is used a good deal in more complex statistics, where different populations are being compared. The standard deviation too is a useful measure of dispersion, especially as the 'spread' of the data is here expressed in the same units as the mean. (The probable error – the statistic estimating dispersion that was often calculated by early biometricians – is 0.6745 times the standard deviation.) Now that the variance and standard deviation values have been calculated, how are they to be interpreted? Before we examine this point, let us look at early work on the visual representation of arrays of data.

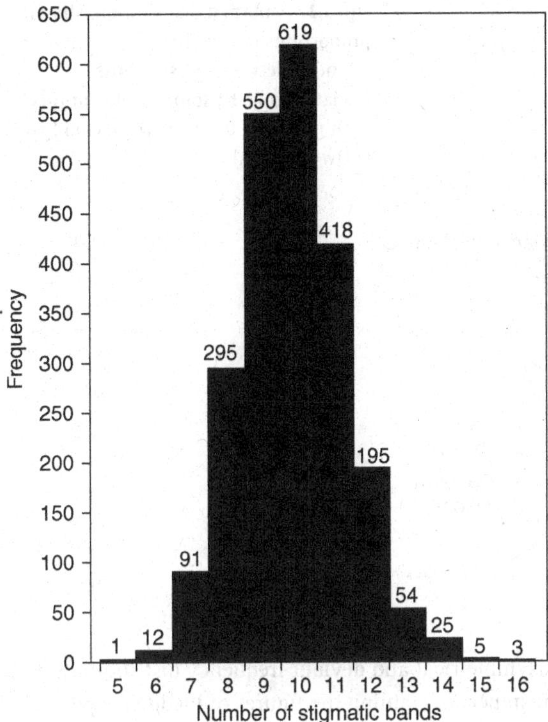

Fig. 3.2. Histogram of Pearson's data (Table 3.1) for variation in the number of stigmatic bands in a sample of capsules of *Papaver rhoeas.* Such histograms were often used in early biometrical studies. Campbell (1967), discussing the use of histograms, suggests that they should be used only in cases of continuous variation. For examples of meristic or other discontinuous variation, the frequency of each class should be indicated by a vertical line on the graph.

Histograms, frequency diagrams and the normal distribution curve

Most people find it easier to comprehend the significance of data expressed visually rather than numerically. The variation in *Papaver* may be expressed as $\bar{x} = 9.8$, $s^2 = 2.22$, $s = 1.49$, or it may be represented in the form of a diagram. Histograms and plotted curves were frequently employed in early biometrical studies. The distribution of the values for stigmatic band number in *Papaver* has been plotted as a histogram in Fig. 3.2; the distribution is roughly bell-shaped, being almost symmetrical about the mean value. Small irregularities in the distribution are the result of small sample size; a closer fit to a bell-shaped curve would result from an even larger set of data for stigmatic band number.

The results for *Papaver* are an example of a very common frequency distribution in biological material: the 'normal' or Gaussian distribution, the latter after Gauss (1777–1855), one of the investigators of this type of distribution.

In the last decades of the nineteenth century, approximately normally distributed variation was demonstrated in a great range of biological materials. Davenport (1904), in the second edition of his book *Statistical Methods with Special Reference to Biological Variation*, first published in 1889, provides a very valuable survey of early biometrical results, giving references and details of scores of botanical and zoological examples.

As approximately normal distributions are frequently encountered in biological material, it is important to look at some of their properties. First, Fig. 3.3 shows that in a normal curve the median, mode and mean of the array fall at the same point.

Second, and of great importance, is the relation of standard deviation to the curve. We have outlined above how to calculate variance and standard deviation. Now, how precisely does knowledge of the standard deviation help us to understand the dispersion of the data within the sample? Examining Fig. 3.3 we see that about two-thirds (68.26% to be exact) of the total variation under a normal curve falls within the range 'mean ± one standard deviation'. Twice the standard deviation on each side of the mean excludes about 5% of the variation, 2.5% in each tail of the normal curve. For different sets of data that are normally distributed, different values of the standard deviation will be found. Thus, if we have a large amount of variability in a sample, a wide curve corresponding to the large standard deviation will be obtained. Whatever the width of the curve, however, the 'mean ± one standard deviation' always contains 68.26% of the variation. We can see now how the standard deviation is so useful in indicating dispersion. One last point remains to be considered. How is the appropriate normal curve to

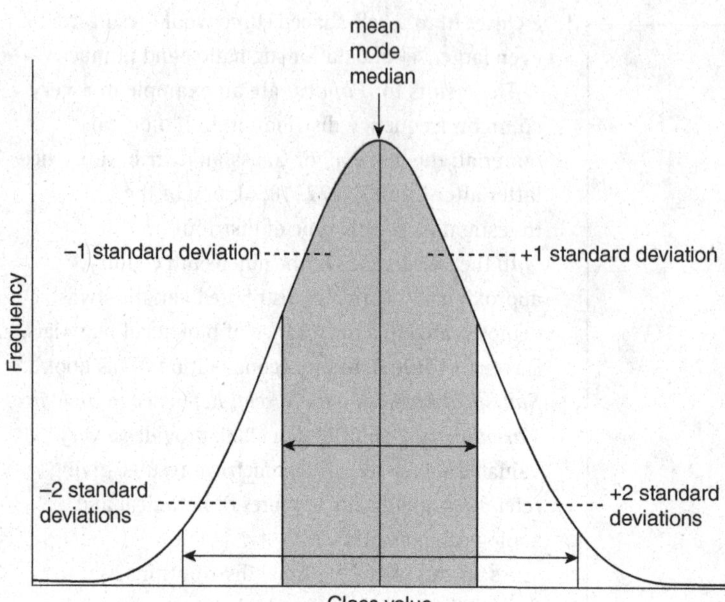

Fig. 3.3. A normal distribution, showing proportions of the distribution that are included between ±1 standard deviation and ±2 standard deviations with reference to the mean. (From Srb & Owen, 1958.)

be fitted to a histogram? It cannot, of course, be drawn 'by eye'. Recourse to a statistics book will indicate full details, and it is sufficient for our purposes to note that it involves the substitution of values of the mean and standard deviation into the equation for the normal curve.

Other types of distribution

Not all the biometrical studies of plant materials gave normally distributed variation, however. De Vries (1894) was one of the first to point to deviant distributions, calling attention to what he called 'Half-Galton' curves. Table 3.2 gives sets of data for the number of compartments in the fruit of Sycamore (*Acer pseudoplatanus*), and petal number in Marsh Marigold (*Caltha palustris*) and Silverweed (*Potentilla anserina*), which in each case approximates to half a normal curve. In *Caltha* and *Acer*, the 'right-hand half' of the curve is represented, whilst in *Potentilla* only the 'left-hand half' is found.

Other researches of this period, especially those dealing with numbers of plant parts, revealed further asymmetrical and deviant frequency distributions. For instance, examining the figures of Pledge for petal frequency in the Buttercup *Ranunculus repens* (Table 3.3), we see that the frequency distribution when plotted would have a long tail to the right: such a curve is described as positively skewed. (A curve, with a long tail to the left, is said to be negatively skewed.) The data for sepal numbers, collected in the same study, also depart from a normal distribution, in this case by being too tightly bunched together. Such a distribution is said to be leptokurtic (high-peaked). Sometimes flat-topped curves (platykurtic) have been discovered (Bulmer, 1967; Rayner, 1969; David, 1971).

Comparison of different arrays of data

By visual inspection, it is often possible to see that a group of plants is more variable in, say, height than in flower size, and the problem of investigating this biometrically particularly fascinated Pearson. In the late 1890s he first devised a statistic known as the *coefficient of variation.* Easy to calculate, it is merely the ratio of the standard deviation to the mean.

Table 3.2. **Half-Galton curves (de Vries, 1894)**

Caltha palustris	Petal number	5	6	7	8
	Frequency	300	87	25	4
Acer pseudoplatanus	Number of fruit compartments	2	3	4	
	Frequency	50	17	3	
Potentilla anserina	Petal number	3	4	5	
	Frequency	6	537	1819	

Table 3.3. ***Ranunculus repens.* (Data of Pledge, 1898, in Vernon, 1903)**

	3	4	5	6	7	8	9	10	11	12	13
Sepal frequency	1	20	959	18	2						
Petal frequency		8	706	145	72	38	15	7	7	1	1

In order to have a scale of reasonable-sized numbers, the resulting coefficient is usually expressed as a percentage:

$$\text{C (coefficient of variation)} = \frac{s}{\bar{x}} \times 100\%$$

An important property of the coefficient is that, as it is calculated as a *ratio*, direct comparison of different coefficients is possible. This even applies when the original figures were calculated in different units, as in metres, inches, grams, etc. Table 3.4 shows some data for human height and weight.

Taking the figures for height first, a comparison of means is impossible in certain cases, as some of the measurements are in inches and others in centimetres. Direct comparison of coefficients of variation is, however, possible, and we can see that English criminals and US recruits show a similar degree of variation in height. There are small differences in height between male and female students and between males and females at birth. Considering the information for variation in weight, again we have differences between male and female at birth and at college. Finally, as high values of the coefficient of variation indicate greater variation for a particular character, we can see that there is a much greater variation in weight than in height in the samples examined.

Coefficients of variation continue to be very useful in the study of variation. Fig. 3.4 shows the coefficients calculated by Gregor (1938) for different parts of the Sea Plantain, *Plantago maritima*. These data illustrate convincingly a fact known before Linnaeus, namely that floral parts are generally less variable than vegetative parts.

Complex distributions

Other biometrical studies in the 1890s revealed more complex frequency distributions. Some of the results of Professor Ludwig (1895) of Greiz in Germany may be used as an illustration. He counted the numbers of ray-florets in 16,800 heads of the Ox-eye Daisy, *Leucanthemum vulgare* (*Chrysanthemum leucanthemum*), collected from Greiz, Plauen, Altenberg and Leipzig between the years 1890 and 1895.
The frequency distribution he obtained was not of the 'normal' type but had several peaks (Table 3.5).

He obtained similar multimodal distributions for many species: for example, in number of ray-florets in Daisy (*Bellis perennis*), number of disc-florets in Yarrow (*Achillea millefolium*) and number of flowers in the umbel of Cowslip (*Primula veris*).

In collecting his *Leucanthemum* (*Chrysanthemum*) data, Ludwig records some interesting differences in ray-floret numbers in different localities. Mountain plants showed 'peaks' at 8 and 13, while lowland

Table 3.4. **Variation in human height and weight: mean = $\bar{x}$, coefficients of variation = C. (Data of Pearson and others, Davenport, 1904)**

		n	$\bar{x}$	C
Height				
English upper middle class	male	683	69.215 in	3.66
English criminals		3000	166.46 cm	3.88
US recruits		25 878	170.94 cm	3.84
Cambridge University	male	1000	68.863 in	3.66
students	female	160	63.883 in	3.70
English newborn babies	male	1000	20.503 in	6.50
	female	1000	20.124 in	5.85
Weight				
Cambridge University	male	1000	152.783 lb	10.83
students	female	160	125.605 lb	11.17
English newborn babies	male	1000	7.301 lb	15.66
	female	1000	7.073 lb	14.23

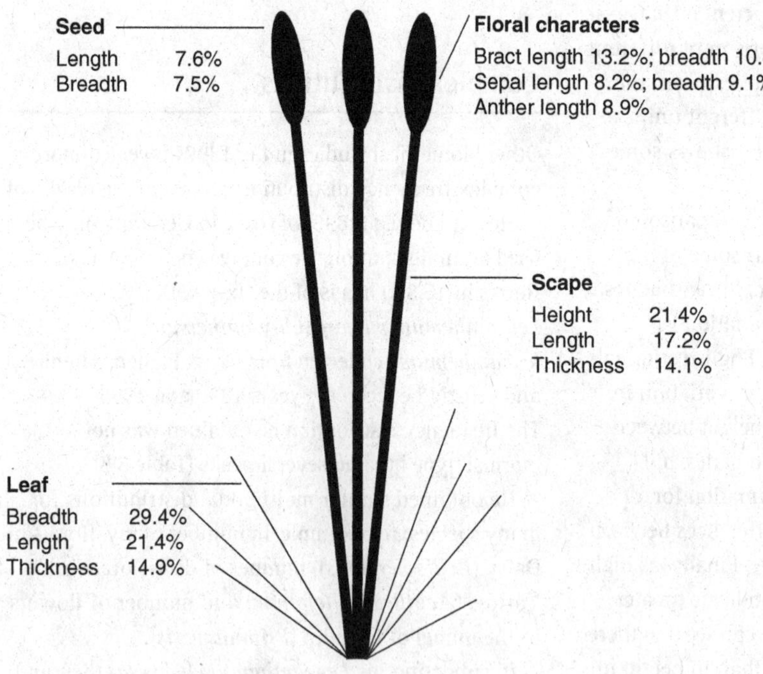

Fig. 3.4. Coefficients of variation in *Plantago maritima*: typical mean values from a number of populations. (From Gregor, 1938.)

Table 3.5. **Variation in the number of ray-florets in *Leucanthemum vulgare* (*Chrysanthemum leucanthemum*)**

Number	Ludwig (1895)[a]	5 July 1901 Tower (1902)[b]	30 July 1901 Tower (1902)[b]	Total Tower (1902)[b]
7	2			
8	9			
9	13			
10	36			
11	65			
12	148		1	1
13	427		8	8
14	383		3	3
15	455		6	6
16	479	1	8	9
17	525	0	9	9
18	625	0	8	8
19	856	2	12	14
20	1568	8	19	27
21	3650	17	26	43
22	1790	23	11	34
23	1147	22	10	32
24	812	21	10	31
25	602	22	8	30
26	614	19	5	24
27	375	16	4	20
28	377	14	6	20
29	294	12	4	16
30	196	10	2	12
31	183	16	4	20
32	187	18	2	20
33	307	29	1	30
34	346	20	1	21
35	186	6	0	6
36	64	6	0	6
37	28	0	0	0
38	16	0	0	0
39	16	2	0	2
40	14			
41	0			
42	3			
43	2			
Total	16 800	284	168	452

[a] Ludwig (1895): Plants from Greiz, Plauen, Altenberg and Leipzig, 1890–5.

[b] Tower (1902): Plants from Yellow Springs, Ohio, USA (two collections and total from same locality).

plants on the other hand had a 'peak' at 21 ray-florets. In fertile soil Ludwig often found a strong 'peak' at 34. He considered that these variations between plants from different areas were the result of nutritional factors. What interested Ludwig most of all about the *Chrysanthemum* was the presence (in the results) of clear peaks at 8, 13, 21 and 34 ray-florets. These numbers, he pointed out, belong to the famous Fibonacci sequence of numbers discovered by Leonardo Fibonacci of Pisa in the twelfth century. The sequence runs 0, 1, 1, 2, 3, 5, 8, 13, 21, 34, 55, 89, 144 . . ., each term being the sum of the two terms which precede it. It represents a set of whole numbers that satisfy almost exactly an exponential growth curve. Not all Ludwig's results gave such clear peaks at the Fibonacci numbers, and he was hard-pressed to explain peaks at 11 and 29, which he discovered in certain samples. Nevertheless, he believed that the Fibonacci sequence of numbers was important in understanding complex patterns of variation.

Certain other biometricians, notably Weldon (1902b), were sceptical about Ludwig's claim for the Fibonacci sequences. They pointed to the fact that plants from different areas had been amalgamated in collecting the data. Plants from a single locality often gave a different picture of the variation; for instance, counts for *Chrysanthemum* ray-florets from Keswick, England, in 1895 gave an approximately normal distribution. Weldon also pointed out that sampling at different times could have an important influence upon the results. This is well illustrated in the results of Tower (1902). He collected, at the beginning and end of July 1901, two sets of *Chrysanthemum* plants from a locality at Yellow Springs, Ohio, USA. His results show clearly that early flowers have more ray-florets than those produced later in the season. Tower went on to show that it was not a question of different plants in flower at the beginning and end of July; marked plants continued to flower throughout the summer, producing flowers with different numbers of parts at different times in the season. Different peaks for ray-floret numbers are found in early July (22, 33) compared with those found later in the month (13, 21). It is interesting to note that it is only in the amalgamated data that these peaks are found, and also that the highest peak is not at the Fibonacci number of 34, as found by Ludwig, but at 33. Such results as these cast some doubt upon the importance of the Fibonacci numbers, indicating that the location of peaks in a complex distribution, far from conforming to a mathematical sequence, was greatly influenced by the method of collecting the data. The precise results obtained would depend upon whether all the plants were collected at the same stage of maturity, a point particularly difficult to ascertain if data for plants from widely different localities and ecological conditions were amalgamated.

Local races

Close study of local variation in the species occupied the attention of many early biometricians. For instance, Ludwig (1901) made a special analysis of variation in the Lesser Celandine *Ficaria verna* (*Ranunculus ficaria*). He showed that plants from different localities had different numbers of carpels and stamens. Details of the two most dissimilar populations, from Gais and Trogen, are given in Table 3.6. Clearly Gais has plants with more carpels and stamens than Trogen.

Ludwig called these local populations, characterised by different mean numbers of floral parts, *petites espèces* or 'local races'. Until this time the term 'local race' had been used rather loosely for plants from particular areas used for biometrical study or experiments, but Ludwig sought to demonstrate the reality of 'local races', using biometrical evidence. In his view these races could be distinguished on the basis of the mean number of floral organs, amalgamation of data for a number of races giving a multimodal distribution curve.

Ludwig's views were again challenged by British and American biometricians, particularly by Lee (1902) who, using the data of MacLeod on *Ficaria verna* (*Ranunculus ficaria*), pointed to the great seasonal variation in floral parts (Table 3.7). Her criticism of Ludwig's 'local races' is particularly telling as the variation in early and late

Table 3.6. **Mean number of stamens and carpels in *Ranunculus ficaria* (Ludwig, 1901)**

	Mean number		Standard deviation
Gais	Stamens	23.8250	2.8872
(80 plants)	Carpels	18.1125	4.2885
Trogen	Stamens	20.3682	3.8234
(385 plants)	Carpels	13.2635	3.0606

Table 3.7. **MacLeod's data for seasonal variation in floral parts in a population of *Ranunculus ficaria* (Lee, 1902)**

	Mean number		Standard deviation
Early flowers	Stamens	26.7313	3.7609
(268 plants)	Carpels	17.4478	3.8942
Late flowers	Stamens	17.8633	3.2984
(373 plants)	Carpels	12.1475	3.3878

flowers from a single locality covers almost the entire range between the Gais and Trogen plants.

This criticism of Ludwig's results did not clinch the issue, however, as there was earlier work by Burkill (1895) on two dissimilar *Ficaria verna* (*Ranunculus ficaria*) populations in which large differences in mean numbers of floral parts were maintained (although not completely) on later sampling on the same site (Table 3.8). The reality of 'local races' was an important issue in the early volumes of the journal *Biometrika*, which was launched in 1901. In an editorial (1: 304–6, 1902) it was contended that the polymorphism found in most results was spurious. It was difficult to defend the notion that each peak of a complex distribution represented a 'local race', especially as peaks often disappeared as sample size was increased. Another important point concerned sampling techniques. It was stressed that random sampling was essential, a point perhaps neglected by early workers. Further, the problem of what constitutes a locality was raised, and the validity of putting together data for samples taken from different areas was questioned. Finally, the editorial stressed the difficulties of seasonal variation and environmental effects, and concluded that a species is not broken up into 'local races'.

Returning once again to variation in *Ranunculus ficaria*, we find the same conclusion is reached by Pearson *et al.* (1903) in a paper in *Biometrika*, which draws together published records, together with new results of variation in floral parts in different areas of Europe. The tables of data are too large for inclusion here, but the following conclusions were drawn from the extensive statistics. 'Local races' could not be distinguished by the number of floral parts, and the influence of the environment and seasonal variation would seem to be sufficient to mask any difference due to 'local races'.
The problem of how to eliminate seasonal and environmental variables from experimental studies was not seriously investigated until later, as we shall see in Chapter 4.

Correlated variation

Many early biometricians examined closely a further aspect of variation, namely the simultaneous

Table 3.8. **Variation in *Ranunculus ficaria* (Burkill, 1895)**

	Date of collection	Number of flowers	Mean number of stamens	Mean number of carpels
Cambridge	3 March	32	22.87	13.41
(under trees)	16 April	75	19.49	11.95
Cayton Bay				
(open field, top of cliffs)	31 March	100	38.24	32.32
	4 May	43	30.67	25.72

variation in pairs of characters. For instance, Pearson was interested in the relation of measurements of different parts of the human body. Suppose we consider body height and its relation to forearm length. It may be that there is some relation between the two variables or they may be independent. Three different situations are possible:

1. The taller the person the longer the forearm.
2. The taller the person the shorter the forearm.
3. A tall person is as likely to have a long or a short forearm as is a short person.

The first situation is one of positive correlation, the second of negative correlation, whilst if the last were discovered we should conclude that there was no correlation between the traits.

In investigating correlation, a statistic called the correlation coefficient (r) is often calculated. It is not necessary for our purposes to give the formulae and details of calculation, which may be found in any statistics book. What is important is the way in which r values indicate correlation or lack of it. $r = +1$ indicates complete positive correlation; $r = -1$ signifies complete negative correlation. If $r = 0$, then correlation is absent.

In biological material, perfect correlation – either positive or negative – is very rare; the various degrees of positive and negative correlation which are often found are indicated by figures which lie between $r = +1$ and $r = -1$.

Table 3.9. **Correlation coefficients in *Ranunculus ficaria* (Davenport, 1904)**

Numbers of	Values of r
Sepals to petals	+0.34 to −0.18
Sepals to stamens	+0.06 to +0.02
Sepals to carpels	+0.25 to +0.03
Petals to stamens	+0.38 to +0.22
Petals to carpels	+0.35 to +0.19
Stamens to carpels	+0.75 to +0.43

Examining the relation between stature and forearm length, Pearson demonstrated positive correlation: in one case $r = +0.37$. A number of botanical situations were also studied at this time. Among the problems investigated was the correlation in the size of leaves in the same rosette in *Bellis perennis* (Verschaffelt, 1899), and correlation between pairs of measurements of leaves and fruits of various species (Harshberger, 1901). The sort of figures obtained for correlation in the floral parts of plants may be illustrated with data, summarised from various authors, on *Ficaria verna* (*Ranunculus ficaria*) (Table 3.9). Clearly there is a stronger correlation between numbers of stamens and carpels than between other organs.

Correlation coefficients – and a further method of studying the association of pairs of measurements known as regression analysis – were used, particularly by Galton and Pearson, for studying heredity. It is a matter of common experience that tall fathers tend to have tall sons, and that short fathers usually have short sons. The association is by no means complete, however. Galton examined the situation biometrically, analysing data from a large number of human families (Galton, 1889): 'Mr Francis Galton offers 500L in prizes to those British Subjects resident in the United Kingdom who shall furnish him, before May 15 1884, with the best Extracts from their own Family Records'. Galton sifted through particulars of 205 couples of parents with their 930 adult children of both sexes. He examined his data carefully, looking for association between the characteristics of parent and offspring. In many cases r values proved to be positive: as high as $r = +0.5$ for height of parents and offspring. We shall examine Galton's interpretation of these results in Chapter 4.

Problems of biometry

In this short survey of early biometrical work a number of problems remain to be examined. In our opening remarks we indicated that there are two main types of variation found on sampling. Arrays of data may be obtained showing either discontinuous or continuous variation. Also there may be found markedly discontinuous patterns of variation with two or more very distinct non-overlapping categories. The reality of these distinct groups is important, as they figure widely in genetic work. As we shall show in Chapter 4, Mendel's work on genetics, published in 1866 and rediscovered in 1900, involved crossing peas with different-coloured cotyledons (green or yellow), or plants of different height (tall or dwarf). Early geneticists crossed glabrous and hairy plants of *Biscutella laevigata* (Saunders, 1897), and *Silene* spp., especially *S. dioica* and *Silene latifolia* (*S. alba*) (de Vries, 1897; Bateson & Saunders, 1902). Among the biometricians it was Weldon (1902a), an opponent of Mendelism, who pointed out a certain ambiguity in defining discontinuities. For instance, he showed that if a large range of cultivated pea stocks was examined, it was found that there was a continuous range of cotyledon colour from green to yellow. It was impossible to sort into green and yellow categories. Similarly, he also showed that there was an enormous range of hairiness in *Silene* species and that it is very hard to accept a classification into glabrous or hairy variants. The important point to bear in mind, however, is the scale of the operation; it may be that general discontinuities do not occur, but marked discontinuities in limited collections and in the progeny from carefully controlled crosses certainly exist. When we read of Mendel crossing tall and dwarf peas, yellow and green peas, it is as well to remember that he deliberately chose stocks with markedly contrasting characters and that, even though there would have been variation in, say, height in his tall and dwarf stocks – perhaps normally distributed variation – there was no overlap in the distribution curves of tall and short plants.

A further problem raised by early biometrical work is that of the significance of differences between sets of numerical data. For instance, the coefficient of variation for weight in Cambridge University students (Table 3.4) shows that females ($C = 11.17\%$) show greater weight variation than males ($C = 10.83\%$). The difference in values is, however, quite small. Now, is this result due to differences in sample size? There were few female students in Cambridge in the 1890s, and there was difficulty in getting even 160 measurements. Or is the variation due to chance? Would further samples taken in different years give the same basic pattern of greater weight variation in female students?

This type of problem is widespread in biometry. Is there any statistically significant difference in the frequency distributions of two sets of data? Do the peaks in a multimodal distribution reveal a true polymorphism or is it the result of sample size or chance? Questions of this type are now tackled by applying statistical significance tests. In Chapter 8 we shall go further into these problems; it is sufficient at this point to note that most of these tests, and indeed many advances in the investigation of complex multivariate data sets, came into being because biometricians wrestled with the problems of interpreting and analysing variation from biological material.

Finally, we must return to another issue: the vexed question of the underlying basis of variation, which fascinated and puzzled early workers. What part of the variation was due to environmental variation and what part was genetic? In the next chapter we examine this issue.

4 Early work on the basis of individual variation

In the last chapter we saw how the early biometricians found great difficulty in analysing some of their data because they were unable to decide which part of the variation had a genetic basis and which part was environmentally induced. For animal studies it was Galton (1876) who appreciated the unique value of twins in investigations of the relative roles of nature and nurture in the development of the individual. To study genetic and environmental effects in plants, specimens selected for comparison may be cultivated under a standard set of environmental conditions. Experiments, both historical and recent, have been performed on the assumption that residual differences between plants of the same species, collected from the same or different habitats and grown under such standard conditions, might be considered to have a genetic basis. What follows is a brief survey of early studies. In Chapter 8 we will discuss in some detail the design and interpretation of garden experiments.

It is very interesting to see how cultivation techniques have developed as methods of analysing variation in plants. Experimental cultivation of plants undoubtedly arose as an adjunct to gardening and horticulture, and in Chapter 2 we saw how Ray, collecting the striking prostrate variant of *Geranium sanguineum* from Walney Island, demonstrated its constancy by cultivating plants in different gardens. The most valuable of these experimental tests were undoubtedly those of a comparative nature. For instance, Mendel cultivated two variants of the Lesser Celandine *Ficara verna* (*Ranunculus ficaria*), which he called *Ficaria calthaefolia* and *F. ranunculoides*, and reported to Dr von Niessl that each remained distinct (Bateson, 1909).

In a paper of quite remarkable scope, Langlet (1971) has reviewed the extent to which foresters in the eighteenth and nineteenth centuries were using experimental cultivation to study adaptive variation in some of the widespread forest trees of Europe. He cites, for example, the neglected (and largely unpublished) work of Duhamel du Monceau, Inspector-General of the French Navy, who, around the time when Linnaeus published his *Species plantarum* (1753), brought together an impressive collection of samples of Scots Pine, *Pinus sylvestris*, from Russia, the Baltic countries, Scotland and Central Europe, and established the first experimental provenance tests for any wild plant. This early development of what we could now call 'genecology' is understandable because of the economic and military importance of the timber supply, but the neglect by most modern writers of the further expansion of such studies in the nineteenth century is less easy to explain and probably, as Langlet suggests, is in part due to the fact that much of this forestry research was published in German. Darwin himself, of course, was greatly interested in the variation of cultivated plants; but forestry differed from agriculture and horticulture, as Langlet shrewdly observes, because its source material was almost entirely the wild species not already subject to artificial selection.

These examples show the importance of simple cultivation of carefully examined material, comparing performance in the wild with that in culture, and comparing also the behaviour of samples of the same or closely related species in the same garden. The method of comparative cultivation, whether seeds or plants are collected from the wild, permits us to investigate the basis of variation patterns. It is easy with hindsight to get a false impression of the ideas of the past and here is a case in point. Even though ideas about the balance between genetic and environmental variation are implicit in some of the writings of the nineteenth century and even discernible in the work of Linnaeus, an explicit statement came only with the researches of the Danish botanist Johannsen (1909), carried out in the years 1900–7. He worked with dwarf beans of the species *Phaseolus vulgaris*, which is naturally self-fertilising.

Phenotype and genotype

Johannsen obtained commercial seeds of the variety 'Princess' and grew nineteen of them, each from a different source, in an experimental garden. The progeny from each of these beans had a different mean seed weight, and Johannsen inferred that these differences were genetic. From each of these 19 original beans, he established a separate line by self-fertilisation, growing up to six generations of daughter beans. For each line he raised a sub-line by selecting heavy seeds at each generation and a separate sub-line in which light seeds were selected. Very great care was taken to label the plants, and in each generation the mean seed weight for a line was calculated separately for progeny from heavy and light mother beans.

Table 4.1 gives the results for two lines. Johannsen found that for a particular line in any one year the mean seed weight for progenies from light and heavy beans did not differ significantly. From each of the 19

Table 4.1. **Two pure lines of *Phaseolus vulgaris* (Johannsen, 1909)**

Mean weight (grams) of selected small seeds	Mean weight of progeny		Mean weight (grams) of selected large seeds	Year	Mean weight (grams) of selected small seeds	Mean weight of progeny		Mean weight (grams) of selected large seeds
30			40		60			70
	36	35		1902		63	65	
25			42		55			80
	40	41		1903		75	71	
31			43		50			87
	31	33		1904		55	57	
27			39		43			73
	38	39		1905		64	64	
30			46		46			84
	38	40		1906		74	73	
24			47		56			81
	37	37		1907		69	68	
	Pure line'A'					Pure line'B'		

original beans a pure line was established, selection having no effect upon mean seed weight.
The implication of these results may be more readily understood later, when it will be shown that habitual self-fertilising leads to genetic invariability. Even though the pure lines from the 19 beans were each genetically uniform, Johannsen found great differences in individual bean weights, approximately normally distributed, giving slightly different mean values for a line in different years. He attributed these differences to the effects of the environment.

These experiments led him to define clearly the distinction between genetic and environmental effects upon an organism. Of first importance were the hereditary properties of an individual – the *genotype* – which were largely fixed at fertilisation.
The appearance, or *phenotype*, of particular individuals of the same genotype might, however, be different because of environmental factors, e.g. two seeds may have the same genotype but have very different weights because of the position in which they developed in the pod. Even though Johannsen's results were obtained with a habitually self-fertilising species, there is no reason to doubt that the concept of genotype and phenotype is of general validity.

Transplant experiments

Besides the rather simple cultivation experiments we have examined so far, nineteenth-century botanists also investigated, through transplant and transfer experiments, the degree of adaptation that a plant showed when placed in a habitat different from that in which it was collected in the wild. Not only were they interested in what we now call changes in phenotype of a plant but also in the persistence of any changes that occurred during the experiment.

As part of a general study of adaptation Bonnier studied many European plants. His experimental technique is of special interest as he used cloned material. Experimental plants were allowed to grow to a convenient size. They were then divided into pieces, and these pieces – 'ramets'– were transplanted into experimental beds at different altitudes in the Alps, the Pyrenees and in Paris. His alpine sites were not gardens: ramets were planted into natural vegetation, protected sometimes by fencing. No fertiliser was added and no watering of the plants took place. In the first reports of his experiments (begun in 1882) he showed how 'alpine' ramets grew into very dwarf compact plants with very vivid flowers, in comparison with 'lowland' ramets (Fig. 4.1). In the 1890s, in a series of largely neglected papers, he published a great deal about the physiological and anatomical adaptation of these plants.

In 1920 Bonnier presented a summary of his researches and claimed that in the course of his experiments certain lowland species became modified to such a degree that they were transformed into related alpine and subalpine species or subspecies. This claim, which, Bonnier notes, supports the ideas of Lamarck, is of very great interest, and if true would have a profound effect upon the interpretation of natural variation patterns. It is worthy of note that Bonnier did not publish his conclusions in his *earlier* papers. Writing in 1890, he does not mention any transmutation of *Lotus corniculatus* into *L. alpinus*, although he had grown plants for some years both in the Pyrenees and in the Alps. He merely reported the dwarfing of the alpine clones in comparison with lowland ones.

Bonnier's claims were supported by the researches of Clements working in Colorado and California, who made a large series of clone-transplant experiments. In these experiments, too, it was asserted that lowland species had been transformed into alpine ones by growth at high altitude. *Chamerion angustifolium* (*Epilobium angustifolium*) was considered to have been changed into *Chamerion latifolium* (*E. latifolium*), and Clements claimed that the grasses *Phleum alpinum* and *P. pratense* could be reciprocally converted (see Clausen, Keck & Hiesey, 1940).

Before examining the alleged transformations we should note that a number of central European botanists, notably Nägeli and Kerner, had been carrying out similar experiments and had come to

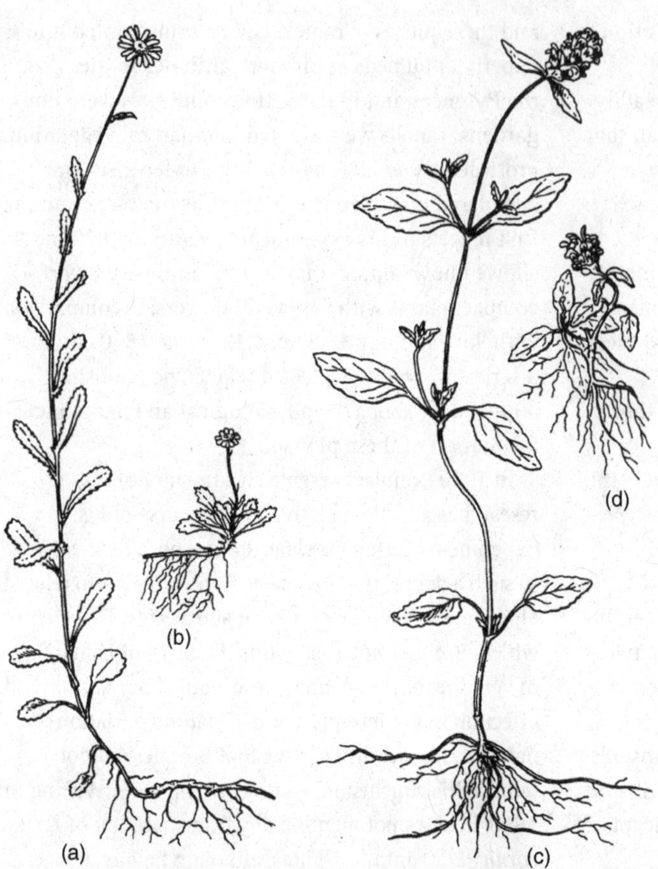

Fig. 4.1. Two examples of Bonnier's transplant experiments, showing the dwarfing effect of cultivation of ramets of the same clone at high altitudes. (a) & (b) Lowland and mountain *Leucanthemum vulgare* (*Chrysanthemum leucanthemum*). (c) & (d) Lowland and mountain *Prunella vulgaris.* (From Bonnier, 1895.)

different conclusions. Nägeli was one of the first to study alpine populations in experimental gardens. He brought a wide range of alpine plants into cultivation at the Botanic Garden in Munich, and many changed their appearance greatly. This was particularly true of species of the genus *Hieracium.* Small alpine plants grown at Munich on rich soil became very large, much-branched plants. Nägeli was most interested to discover, however, that the acquired characters disappeared when plants were transplanted to gravelly soil within the garden, and the specimens again assumed the appearance of alpine plants.

Kerner, Professor of Botany in Vienna, carried out many transplant and reciprocal sowing experiments using an alpine garden at Blaser at 2195 m in the Tyrol, and the Botanic Gardens at Vienna and Innsbruck. He discovered that, for many species, if seeds were grown in two contrasting environments, dwarf plants with more vivid flowers were produced in alpine conditions. He noted the parallel case of more vivid colours in snails and spiders transferred to alpine conditions from the lowlands. Writing of his experiments in his famous book *The Natural History of Plants, their Forms, Growth, Reproduction and Distribution* (1895), Kerner noted:

> *in no instance was any permanent or hereditary modification in form or colour observed* . . . They [the modifications] were also manifested by the descendants of these plants *but only as long as they grew in the same place as their parents.* As soon as the seeds formed in the Alpine region were again sown in the beds of the Innsbruck or Vienna Botanic Gardens the

plants raised from them immediately resumed the form and colour usual to that position. [author's italics]

Kerner, therefore, came to very different conclusions from Bonnier and Clements as to the nature of the changes that had taken place in the material planted at high altitude. Since Kerner's experiments thousands of experimental plantings have been carried out, deservedly the most famous being those of Clausen, Keck & Hiesey (1940) in California. No evidence of transformation of the kind claimed by Bonnier and Clements has been discovered. The most reasonable explanation for their anomalous results is that their experimental areas became invaded by the related alpine species, which were growing naturally at these high altitudes.

From these observations it can be seen that experimental cultivation can be of very great value in investigating variation in plants. Simple cultivation tests, in which a range of material is grown under standard conditions, in conjunction with crossing experiments, may reveal genetic differences between the stocks under investigation. Transfer and transplant experiments, properly carried out with special care in labelling and organisation, will give information upon the plasticity and adaptation of different plants. Especially useful are clone-transplants, as the performance of material of a single individual is investigated in different environments. In this respect Bonnier's experiments were to be preferred to those of Kerner who often used seeds. Seeds, except in special circumstances (see Chapter 7), may be genetically heterogeneous, and raise difficulties in interpretation not present in the clone-transplant method.

Mendel's experiments

Let us now suppose that cultivation and transplant experiments in a particular instance have established a *prima facie* case of genetic difference between two plants. What is the nature of this difference? Our present knowledge of heredity stems from the various experiments of Mendel, which he carried out over many years. Mendel, an Augustinian friar of the monastery of St Thomas at Brünn (now Brno in the Czech Republic), reported his work on crossing Garden Peas in two papers to the Natural History Society of Brünn on 8 February and 8 March 1865, and the proceedings of these meetings were subsequently published in the *Transactions* of the Society in 1866.

Even though Mendel may be credited with the discoveries leading to the establishment of genetics, in many elementary textbooks the accounts of his work lack historical perspective. There is a wealth of pre-Mendelian experiments in hybridisation (Roberts, 1929), although it is true that early workers often had different objectives from those of Mendel. Kölreuter and Linnaeus investigated the phenomenon of sex in plants. Others, such as Laxton and the de Vilmorins, tried to improve varieties of plants of horticultural and agricultural importance. Another group of hybridists, as we saw in Chapter 2, were trying to find criteria for the experimental definition of species, using the data from experimental and natural hybridisation. Darwin was extremely interested in all aspects of hybridisation and published a book on the effects of self- and cross-fertilisation in plants.

Many of the findings of Mendel were, in fact, anticipated by earlier hybridists, although they were not connected into a coherent theory (see Zirkle, 1966). Kölreuter, for example, discovered that *Nicotiana paniculata* × *N. rustica* and the reciprocal cross gave identical hybrids. He also had crosses that showed dominance: *Dianthus chinensis* (normal flowers) × *D. hortensis* (double flowers) resulted in dominance of double flowers. The phenomenon of segregation was also known long before Mendel's day.

Turning now to discuss the main points of Mendel's contribution, we might ask what are the ways in which his approach to the problem of heredity differed from those of his predecessors? The contemporary preoccupation with species led to many interspecific crossing experiments. For the purpose of elucidating the mechanisms of heredity, species-crosses are not very helpful because species differ in innumerable characters, and in a number of generations a bewildering array of hybrid variants may appear. Before Mendel, hybridists did not in general concern themselves with the numbers of progeny of different sorts, and sometimes they did not even keep separate the progeny from different plants or different generations. In Mendel's paper we see that he is aware of previous work in the field – the experiments of Kölreuter, Gärtner, Herbert, Lecoq, Wichura and others – and the defects of past experiments (trans. Bateson, 1909):

> Those who survey the work done in this department will arrive at the conviction that among all the numerous experiments made, not one has been carried out to such an extent and in such a way as to make it possible to determine the number of different forms under which the offspring of hybrids appear, or to arrange these forms with certainty according to their separate generations, or definitely to ascertain their statistical relations.

In selecting peas for his work, Mendel knew that they are usually self-fertilising and that different cultivated varieties differ from each other in a number of respects. First he tested a selection of stocks (34 in all) and, in a two-year trial, found a number to be true-breeding. This is one of the most important facets of Mendel's work. Then, after carefully removing the unopened stamens of selected flowers, he crossed pea plants that differed in a pair of contrasting characters. Using the useful terms devised by Bateson & Saunders (1902) for the first and second generations (F_1; F_2 etc.), we may take as an example the cross between unpigmented plants (white seeds, white flowers, stem in axils of leaves green) and pigmented plants (grey or brownish seeds – with or without violet spotting, flowers with violet standards and purple wings, stem in axils of leaves red). Here Mendel discovered that the first generation of hybrids, the F_1, were all pigmented plants: 'pigmented' Mendel spoke of as 'dominant', the character 'unpigmented' he termed 'recessive'. He obtained the same result when pigmented plants were seed or pollen parent – in other words reciprocal crosses gave the same results. Following natural self-fertilising, in the next F_2 generation he discovered that the recessive character (unpigmented) reappeared along with pigmented plants in a numerical ratio of 3 pigmented:1 unpigmented (Fig. 4.2). In the next, F_3, generation, Mendel discovered that unpigmented plants bred true, whereas only one-third of pigmented plants did so. On selfing, the other two-thirds of pigmented plants gave pigmented to unpigmented plants in a 3:1 ratio. The 3:1 ratio in the F_2 was in reality a ratio of 1 true-breeding pigmented:2 non-true-breeding pigmented:1 unpigmented.

Mendel obtained essentially similar results in crossing other peas differing in single characters:

Character	Dominant	Recessive
Stature	Tall	Dwarf
Seed shape	Round	Wrinkled
Cotyledon colour	Yellow	Green
Pod shape	Inflated	Constricted
Unripe pod colour	Green	Yellow
Flower position	Axillary	Terminal

Particulate inheritance

To explain his results Mendel postulated the existence of 'factors' as he called them. Using current nomenclature, the dominant character, in our example 'pigmented', may be denoted by a factor C, and the recessive 'unpigmented' by *c*. True-breeding parental stocks, CC and *cc*, produced C and *c* gametes respectively, which at fertilisation gave an F_1 of constitution *Cc*. These F_1 plants, in appearance pigmented, produced in equal numbers two sorts of gametes, C and *c*, which (mating events being at random) gave three kinds of plants in the F_2 generation in the proportion 1*CC*:2*Cc*:1*cc* – a ratio of 3 pigmented:1 unpigmented. Mendel realised that, owing to the operation of chance, an exact 3:1 ratio would not be achieved in practice. His results came close to expectation: in our example his F_2 consisted of 705 pigmented:224 unpigmented plants, giving a ratio of 3.15:1.

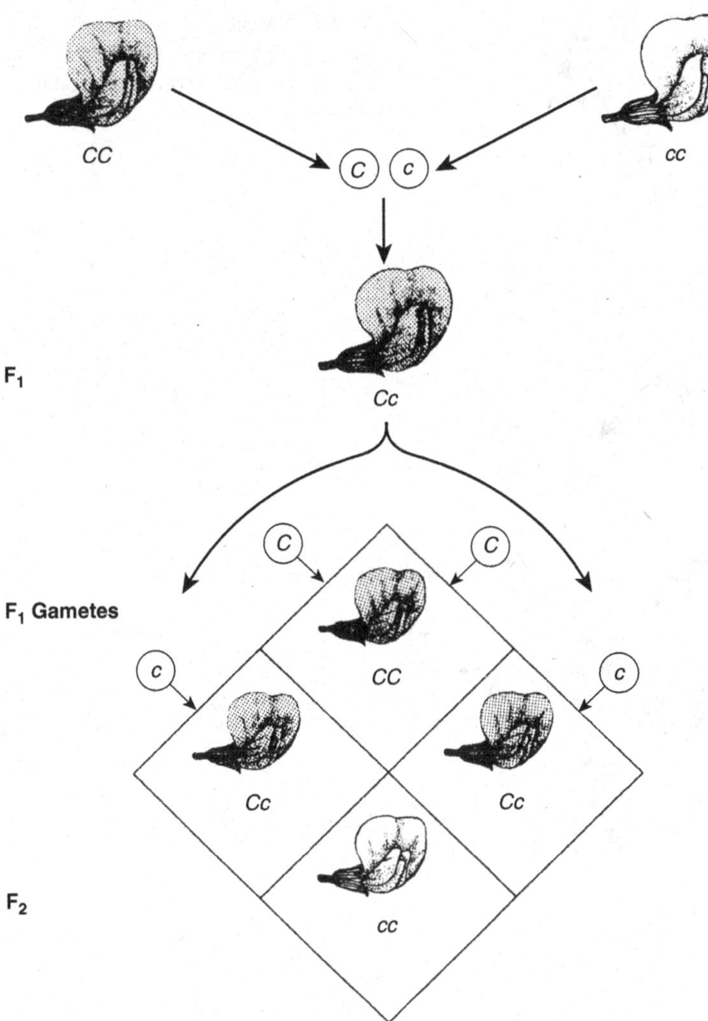

Fig. 4.2. One of Mendel's single-factor crosses, in which 'pigmented' allele (allele *C*) is dominant to 'unpigmented' (allele *c*). Note the 3:1 ratio of dominant to recessive phenotype in the F_2 progeny.

Mendel's hypothesis of factors which can coexist in an F_1 and which segregate intact at gamete formation was subject to a further test, that of backcrossing the F_1 (*Cc*) to the recessive parent (*cc*). As expected, the progeny were in the ratio 1 pigmented:1 unpigmented. These confirmatory results of Mendel vindicated his explanation of the earlier crosses.

Mendel's two-factor crosses

We must now examine what happened when Mendel made 'two-factor' crosses. One of his experiments, incorporating his theory of determinants, may be represented by the following outline (Fig. 4.3). Mendel confirmed the genetic constitution of each category of plants by examining the progeny of selfed F_2 individuals. The important principle he discovered in these experiments was that the F_1, besides producing *YR* and *yr* gametes, as did its parents, also produced gametes *Yr* and *yR* in numbers equal to those of the parental type. This equality of numbers of the four types of gametes established the *independent assortment* of the pairs of factors. Independent assortment was confirmed by Mendel when he crossed

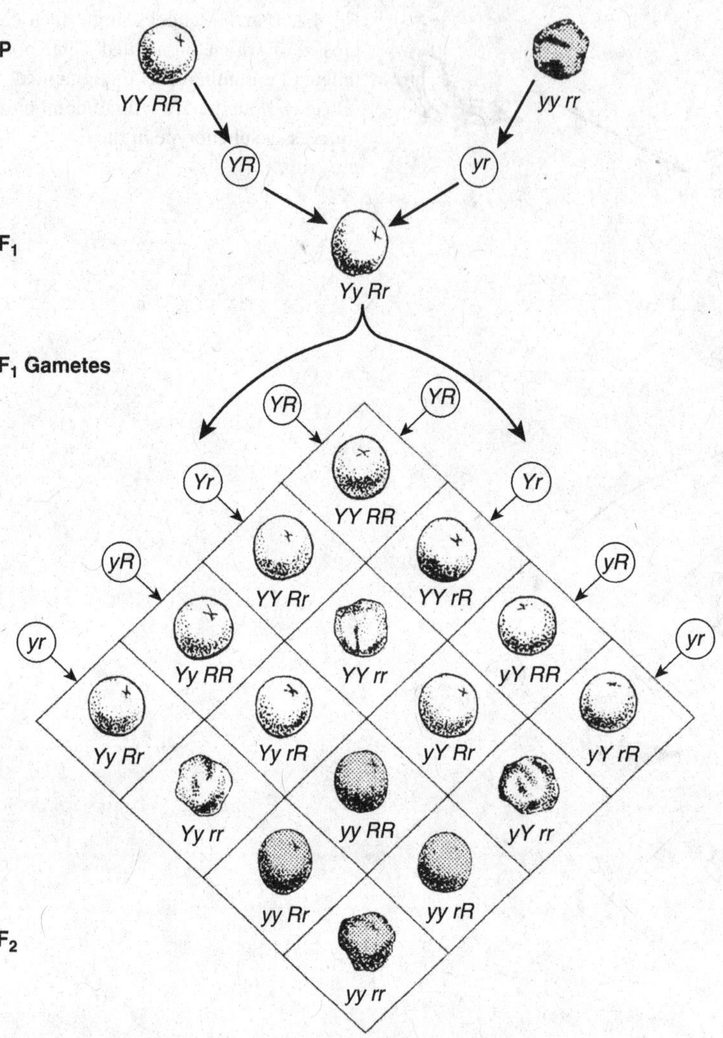

Fig. 4.3. A Mendelian two-factor cross: pure breeding yellow, round Peas (*YYRR*) × green, wrinkled (*yyrr*). The appearance of the F_2 progeny was:

A Mendelian two-factor cross: pure breeding yellow, round Peas (*YYRR*) × green, wrinkled (*yyrr*). The appearance of the F_2 progeny was:

	Yellow round	Yellow wrinkled	Green round	Green wrinkled
Theoretical phenotype	9	3	3	1
Mendel's result	315	101	108	32
Experimental ratio	9.8	3.2	3.4	1

the F_1 (*YyRr*) with the double recessive parent (*yyrr*). As he predicted from his earlier results, four classes of offspring were produced in a 1:1:1:1 ratio.

Further work, in three-factor crosses in peas, and crosses in French beans, is reported in Mendel's paper.

A series of letters from Mendel to Nägeli have survived (see Stern & Sherwood, 1966 for English translations). These reveal many facets of Mendel's work. First, Nägeli did not appreciate the significance of Mendel's findings, and he questioned some of the results, for instance, whether unpigmented plants produced in the F_2 were true-breeding. He thought that the progeny of hybrids must be variable and that in the case of unpigmented F_2 peas, repeated selfing would eventually lead to segregation. Secondly, it was fortunate that Mendel's experiments on peas were more or less complete before 1864, for Mendel writes that in the following summers infestations of pea beetle (*Bruchus pisi*) made cultivation of peas difficult and finally impossible in Brno. The letters also reveal that Mendel carried out hybridisation experiments on more than 20 other groups of plants, e.g. *Aquilegia, Cirsium, Dianthus, Mirabilis, Viola* and *Zea* (see Iltis, 1932).

Historians of the early history of genetics have critically examined what might be called 'the traditional account' of Mendel's work using many primary sources of information, including Mendel's published papers, his letters to Nägeli and annotations in the books in his library etc. (Orel & Matalová, 1983; Orel, 1984, 1996; Bowler, 1989). No potential source has been neglected. There is even a published illustration of Mendel's table-cloth with its embroidered monogram (Anon., 1965). First, scholars have critically examined the aim of Mendel's work. It has been argued that he was not only concerned with studying heredity but also whether hybrids were constant or variable, in the context of a major concern of his day, namely, the role of hybrids in the origin of new species (Olby, 1979). Callender (1988) has proposed that Mendel was an opponent of the theory of descent with modification. However, evidence suggests that he is unlikely to have known about Darwinism when he started his experiment. Later, Mendel obtained a German translation of the *Origin* (published in 1863) in which he made marginal notes, and had read Darwin's book before he presented his paper in 1865. These marginalia, and what little survives of Mendel's writings, do not provide enough to decide whether Mendel supported or opposed Darwinism (Fairbanks, 2008).

On the Internet and in various publications, it has often been said that 'Darwin had a reprint of Mendel's paper in his library, but that it was uncut and therefore unread' (Fairbanks, 2008). To clarify, we now know that Darwin 'owned Focke's (1881) *Die Pflanzen-Mischlinge*, which briefly refers to Mendel's paper, but the pages that refer to Mendel in Darwin's copy of Focke's book were uncut' (Fairbanks, 2008). So Darwin could not have read them.

In traditional accounts of Mendel's work Nägeli is 'blamed' for encouraging Mendel to study hybrids in the genus *Hieracium* (Mendel, 1869; English translation in Stern & Sherwood, 1966). It is now known that reproduction in this genus is aberrant (we shall see why in Chapter 7). However, Mendel had begun his studies of *Hieracium* before he started his correspondence with Nägeli (Iltis, 1932). If Mendel's primary purpose was to study different sorts of hybrids in relation to the development of new species, then it is quite understandable that he should study not only the segregating hybrids of pea etc. but also the 'constant' hybrids of the genus *Hieracium*. Thus, it has been suggested that Mendel's experiments on *Hieracium* were not undertaken in the expectation that they would yield the same results as those on peas (Callender, 1988). It is important to realise, therefore, that Mendel did not derive a generalised scheme of heredity for all organisms from his experimental results on peas. Indeed, it was only in the twentieth century that geneticists stated Mendel's findings as 'laws' applicable in plants and animals, including humans. He thought of his work as demonstrating the method by which the laws of heredity could be worked out. This point is clearly stated towards the end of his paper on peas when Mendel wrote: 'It must be the object of further experiments to ascertain whether the law of development discovered for *Pisum* applies also to the hybrids of other plants.'

Secondly, historians have discussed at length whether Mendel's notion of paired factors or elements was conceptually equivalent to the paired alleles of classical genetics (Brannigan, 1979; Olby, 1979, 1985).

Thirdly, there is another point of great interest: did Mendel falsify his experiments? These have been analysed in detail by Fisher (1936), Edwards (1986) and others. Mendel's data, taken as a whole, fit expected ratios far too well, and consistently do not deviate as much as would be expected by the operation of the laws of probability. Fisher argues cogently that Mendel probably knew what his results would be before he started his experiments and that in reality his experiments were a confirmation or demonstration of a theory he had already formulated. Maybe doubtful individuals were classified to fit expectation, perhaps by an assistant, or aberrant families may have been excluded from the final results (Sturtevant, 1965). The excessive goodness of fit of Mendel's results does not seem to be in dispute, but the conclusion that deliberate falsification was involved has not been accepted by Wright (1966) or by Fisher himself, who wrote that Mendel's 'report is to be taken entirely literally, and that his experiments were carried out in just the way and in much the order that they are recounted' (Fisher, 1936).

However, historians of science have continued to examine Mendel's experiments. A recent volume by Franklin *et al.* (2008) entitled (perhaps optimistically) *Ending the Mendel–Fisher Controversy* brings together the previous work on the subject by many authors, together with a translation of Mendel's papers and Fisher's (1936) analysis. Every facet of Mendel's work is exhaustively examined and subject to various statistical analyses. The authors evaluate historic opinion as well as recent investigations. For instance, soon after the finding of Mendel's paper, Bateson (1913) wrote that 'it is very unlikely that Mendel could have had seven pairs of varieties such that the members of each pair differed from each other in only one considerable character'. However, as a result of this new appraisal, Fairbanks (2008) agrees with Fisher that 'Mendel was not guilty of deliberate fraud in the presentation of his experimental results': that Mendel's data fit his expectations extraordinarily well is correct, but may be explained without invoking fraud; and that Fisher had great admiration for Mendel and his work and, as Franklin (2008) emphasises, he would have been 'quite unhappy with those who used his work to diminish Mendel's achievement'. Finally, Fairbanks (2008) takes the view that 'short of a miraculous discovery of Mendel's original notebooks, other questions will forever remain unresolved', and that 'it is time to end the controversy'.

Pangenesis

Perhaps we should now compare the ideas current at the end of the nineteenth century with those of Mendel that superseded them. Darwin, in his astonishingly productive later years, gave a great deal of thought to the problems of heredity, and (Darwin, 1868) in *The Variation of Plants and Animals under Domestication*, he put forward his theory of 'Pangenesis'. This theory, in many ways derived from Hippocratean ideas about the direct inheritance of characters, suggested that cells of plants and animals threw off minute granules or atoms (Darwin called them gemmules), which circulated freely within the organism. It was these gemmules that were transmitted from parent to offspring. Blending of gemmules occurred in the progeny. The phenomena of 'segregation' of a recessive plant in an F_2 or subsequent generation Darwin could account for only by suggesting that sometimes the gemmules were transmitted in a dormant state.

The theory of pangenesis, with its notion that gemmules came to the reproductive cells from all parts of the body, provided a mechanism for the inheritance of acquired characters, a possibility envisaged by Darwin. This view was challenged by Weismann (1883), who, in the words of Whitehouse (1959), disputed the idea that 'something from the substance of each organ was thought to be conveyed to the reproductive elements'.

It is not necessary to go farther into Darwin's ideas, as they received no support from experiments. Galton, searching for evidence of gemmules, intertransfused blood of different-coloured rabbits and studied the colour of their offspring. There was no evidence that the presence of 'foreign' blood in a female rabbit made any difference to the colour of her progeny (Darwin,

1871a; Galton, 1871). For a detailed account of this fascinating episode in the history of genetics, see Pearson's (1924) *The Life, Letters and Labours of Francis Galton* and Gillham's (2001) *A Life of Sir Francis Galton.* Galton himself had many ideas about heredity. Those he developed most forcibly were based upon a belief in blending inheritance. Unlike Mendel, he did not carry out any breeding experiments, but analysed records of human families in developing his 'law of ancestral heredity'. Taking height as an example, 'an individual inherits 1/4 of his characteristics from each parent, 1/16 from each grandparent, 1/64 from each great-grandparent, and so on. When the trait does not blend – as in the case of say, whether an individual would have blue or brown eyes – ancestral inheritance would have an effect on the ratio of traits in a population' (Sapp, 2003). Galton's 'law' was a statistical statement of general patterns in samples, rather than a genetic analysis (Galton, 1889).

Mendelian ratios in plants

There had only been a few references to Mendel's work before 1900 (Olby & Gautry, 1968), but this is perhaps to be expected as the paper was published in a obscure journal. Then, tradition has it that de Vries, Correns and von Tschermak, who had all been conducting breeding experiments, independently rediscovered Mendel's laws. In Fisher's (1936) words, Mendel's paper on peas 'had at this time the triple aspect of a confirmation, an anticipation, and an interpretation of their own researches'. However, the notion of independent rediscovery has recently been dismissed as a 'myth' (Corcos & Monaghan, 1990). Sapp (2003) provides a recent assessment of the evidence, which provides a good illustration of the complex human and psychological background to scientific advances. 'Tschermak, did not actually understand the significance of Mendel's work when he first referred to it. De Vries and Correns both insisted that they had read Mendel only after they had conducted their own experiments and reached their own interpretations. Thus, each was "anxious to protect [his] priority, and have his work regarded as independent of the work of Mendel and other rediscoverers" (Weinstein, 1977).' In fact, there was a widespread belief amongst commentators that de Vries at first intended to suppress any reference to Mendel, and that his plans were interrupted when he found that Correns was going to refer to the monk (Sturtevant, 1965). This inference is based on de Vries's failure to mention Mendel when he first announced his discovery in a short abstract before Correns's paper. De Vries mentioned Mendel's work only later in two longer papers, in which he remarked that it was *trop beau pour son temps* (Zevenhuizen, 2000). It has also been argued that 'Correns, realizing that he had lost priority to de Vries, referred to Mendel's work as a strategy to minimize his loss and effectively undermine the priority of de Vries's claim to the discovery'. Clearly, 'Mendel's revival in 1900 took place in the context of a priority dispute between Correns and de Vries': for further analysis of this dispute see Simunek, Hossfeld & Breidach (2012).

When Mendel's results became available in 1900, it was soon realised that his hypothesis of segregating factors, or 'genes' as they came to be called, could explain the results obtained for many plants and animals. Bateson (1909) gives a representative list of plants in which Mendelian inheritance was discovered. The characters involved range from those of the general growth habit, to details of the leaf, flower, fruit and seed. It is very interesting to see that variants known for many years were investigated. For instance, in experiments by de Vries, white flower colour in a variety of *Polemonium caeruleum* (described by John Ray in the seventeenth century) was shown to be recessive to the normal blue colour, Not only did morphological characters show Mendelian inheritance, but so did physiological traits. An example is disease resistance in Wheat (*Triticum*) infected with the fungus *Puccinia glumarum*, where susceptibility was shown by Biffen to be dominant (see Engledow, 1950).

Table 4.2. **Frequency distribution of corolla length in the cross *Nicotiana forgetiana* × *N. alata* var. *grandiflora*. (Data of East, 1913)**

	Length of corolla (mm)														
	20	25	30	35	40	45	50	55	60	65	70	75	80	85	90
N. forgetiana	9	133	28	–	–	–	–	–	–	–	–	–	–	–	–
N. alata var.	–	–	–	–	–	–	–	–	–	–	–	–	–	–	–
grandiflora	–	–	–	–	–	–		–	–	1	19	50	56	32	9
F_1	–	–	–	3	30	58	20	–	–		–	–	–	–	–
F_2	–	5	27	79	136	125	132	102	105	64	30	15	6	2	–

Cases of independent segregation in two-factor crosses were also discovered. For example, in crossing a white-flowered 'three-leaved' *Trifolium pratense* with a red-flowered, 'five-leaved' variant, de Vries (1905) obtained an approximate fit to an expected 9:3:3:1 ratio, the characters 'red-flowered' and 'five-leaved' being dominant.

Gradually, Mendelian explanations, for many single discontinuous variation patterns, were accepted by most botanists, and a number of useful terms were introduced. The alternative factors *A* and *a*, as, for example, tall and dwarf in peas, were spoken of as *alleles* (allelomorphs) of a gene by Bateson & Saunders (1902), who also introduced the term *heterozygous* (*Aa*) to describe a zygote or individual with two unlike alleles, and *homozygous* (*AA, aa*) for one with two alike.

Mendelism and continuous variation

Notwithstanding the success of Mendelian explanations of familiar patterns of variation, universal acceptance did not follow. The biometricians, led by Pearson, remained loyal to the 'law of ancestral inheritance' of Galton, which has been shown to be based upon blending inheritance (Porter, 2004).

Among the criticisms of Mendelism, one of great weight was that in certain crossing experiments no clear-cut segregation occurred in the F_2 generation. As an example, East's (1913) data for corolla length in F_1 and F_2 hybrids of *Nicotiana forgetiana* (female) × *N. alata* var. *grandiflora* (male) are given in Table 4.2. Here, a short-flowered plant was crossed with a long-flowered plant; the F_1 was of intermediate corolla length and the F_2, showing wider variation, did not segregate with Mendelian ratios. Is such a situation an example of blending inheritance? Pearson and his school of biometricians considered blending inheritance to be the general rule, Mendelian inheritance only applying in special circumstances. In the early years of the twentieth century, the problem of explaining continuous variation patterns was very urgent. An initial difficulty, in understanding continuous variation, was in estimating the environmental and genetic components of the variation pattern. This difficulty was largely removed by the work of Johannsen, to which we have already referred.

Yule (1902) was probably one of the first to suggest that many genes were involved in continuous variation. To show what he had in mind, we may take as an example human height, which follows a typical normal distribution, and, even though nutritional factors are highly important in determining the height of a person, the fact that Pearson and Galton showed a positive correlation (r about 0.5) between the height of parent and offspring provided a *prima facie* case of genetic control of height. Yule considered that a number of genes might be involved in determining continuous variation patterns, and in this case different genes might determine leg length, trunk length, neck length, etc. In order to make this hypothesis credible, it was necessary to demonstrate that the genetics of a single character could be controlled by at least two genes.

Such a situation was discovered in 1909 by Nilsson-Ehle, who studied hybrids between wheats with brown and white chaff (Fig. 4.4). In the F_1 of the cross, brown chaff was dominant. Inter-crossing of the F_1 gave an F_2 generation, not in the expected 3:1 ratio of brown:white, but in the ratio of 15 brown:1 white. This result was confirmed in a second experimental cross. Nilsson-Ehle considered that in this case two different genes were involved in chaff colour and that the 15:1 ratio was in reality a modified 9:3:3:1 ratio. The presence of a single dominant in an individual was sufficient to give brown chaff; only one-sixteenth of the progeny (of genotype ***aabb***) had

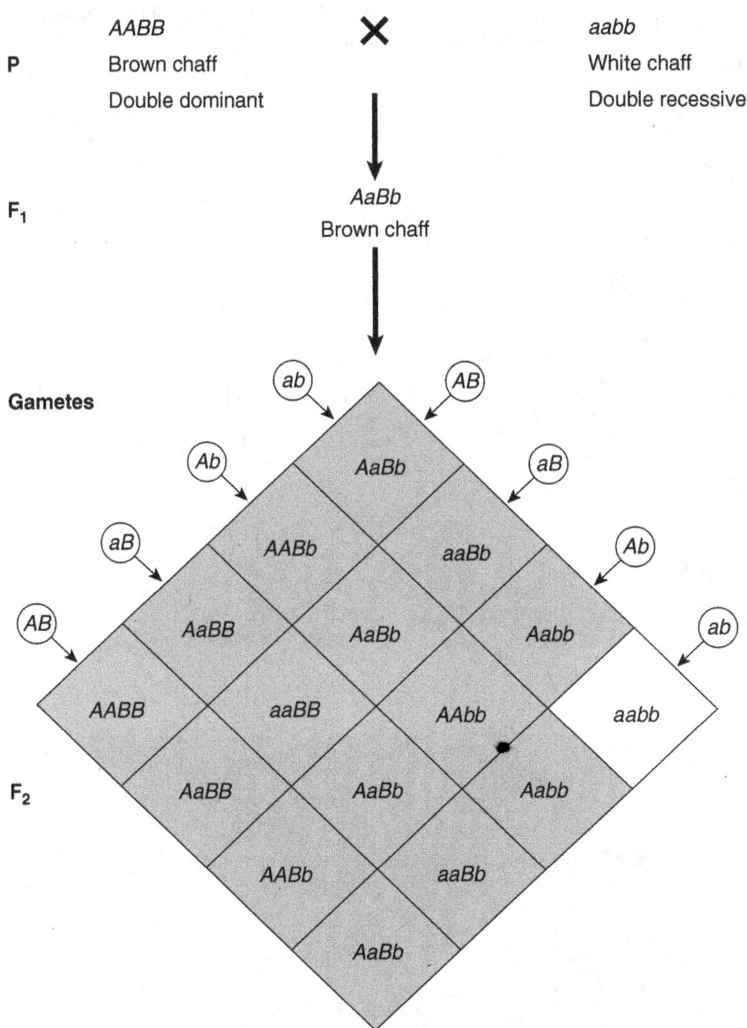

Fig. 4.4. Chaff colour in Wheat (*Triticum*). (From Nilsson-Ehle, 1909.)

white chaff. Here is a clear case of two genes affecting the same character.

These experiments of Nilsson-Ehle, which were paralleled by the independent work of East, provide the necessary basis for an understanding of the genetics of continuous variation. To demonstrate the principles we will examine a hypothetical case of flower colour (Fig. 4.5).

In this model we postulate that two different genes are involved: *A* and *B* being the dominant alleles determining red flower colour, alleles *a* and *b* determining white flower colour. In the example, we assume, however, that the effects of *A* and *B* are additive, the degree of red colour in the flower depending upon the number of *A* and *B* alleles present in an individual. Examination of the F_2 'chequer-board' shows that one-sixteenth of the progeny has four red alleles, four-sixteenths have three, six-sixteenths have two, four-sixteenths have one, and one-sixteenth has none. It should be noted that our

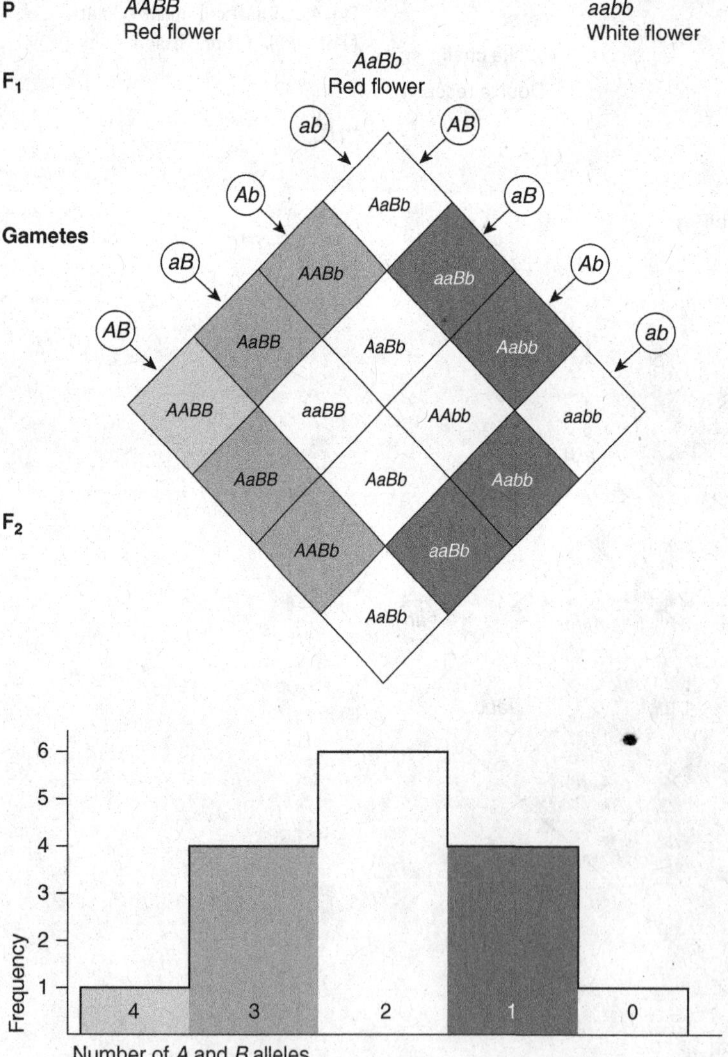

Fig. 4.5. Flower colour: a hypothetical case.

example still shows Mendelian segregation of 15 red:1 white on a broad classification, in detail with four different categories of red. Expressing the frequencies as a histogram, we obtain a distribution that bears a striking resemblance to a normal curve.

Consider now what might happen if a larger number of genes was involved. With six dominant genes, all additive in their effect for red colour, the F_2 would show very many categories of individuals, and a closer fit to a normal curve. Of great importance, too, the parental genotypes *AABBCCDDEEFF* (red) and *aabbccddeeff* (white) would be very infrequently segregated in the F_2. In fact only 1/4096 of the progeny would be *AABBCCDDEEFF* and, even more important, there would be a similar proportion of *aabbccddeeff*, which would be the only white phenotype. In actual practice, if the cross were made, even though Mendelian segregation had taken place at gamete formation in the F_1 plants, it is quite likely (especially if the F_2 is represented by a small number of plants) that no *aabbccddeeff* plants would be recovered at all. The F_2 progeny would then all be red-flowered, in different degrees giving a normal distribution curve.

Turning now to an actual experiment, the *Nicotiana* crosses of East, which we referred to earlier (Table 4.2), far from demonstrating blending inheritance, may more satisfactorily be interpreted on the basis of multiple factors affecting corolla length. The two variants of *Nicotiana* used differ in corolla length, and the F_1 from the cross is intermediate in length, indicating the absence or incompleteness of dominance. In the F_2 a wide array of corolla sizes is found, the frequency distribution approximating to a normal curve. Note that the extreme 'parental' corolla sizes are not represented in the data. East considered that there were probably four genes involved in the determination of corolla length.

Many investigations of continuous variation patterns in nature have given similar results to those of East, and elaborate genetic and statistical experiments since that time have demonstrated the general validity of the multiple-factor hypothesis. Such systems, in which the character is determined by several genes, are usually called *polygenic*.

Physical basis of Mendelian inheritance

So far we have not discussed the physical nature of Mendel's factors. In Mendel's day little or nothing was known about the physico-chemical basis of heredity but there was plenty of theoretical speculation. Nägeli, for instance, postulated a genetically active 'idioplasm'. By the time Mendel's work was rediscovered in 1900 the situation, however, was very different. The latter half of the nineteenth century had seen an enormous increase in interest in the microscopic study of plant and animal cells. Certain technical innovations such as the use of stained material (carmine was introduced in the 1850s, and haematoxylin and anilin dyes in the 1860s) and the perfecting of apochromatic lenses (by Abbé in 1886) enabled biologists to make a close study of all aspects of cell division and development (see Hughes, 1959). It is impossible in the space available to review the results of these studies in any detail, but the main conclusion was that the chromosomes discovered in cell division were clearly very important in heredity.

It was found that each species has a characteristic number, the diploid number, of chromosomes, visible in stained preparations of meristematic cells.
The account that follows applies to diploid organisms, i.e. those whose nuclei contain two like sets of chromosomes, one set from pollen and one from egg. There are, however, haploid organisms, e.g. certain fungi, whose nuclei contain only one set of chromosomes. In higher plants there is generally consistency of number, size and form of chromosomes of the meristematic cells of root-tip and shoot apex where chromosomes divide by *mitosis*. Essentially each chromosome divides into two daughter chromosomes and at the end of the process the two groups of daughter chromosomes are separated from each other by a new cell wall (Fig. 4.6).

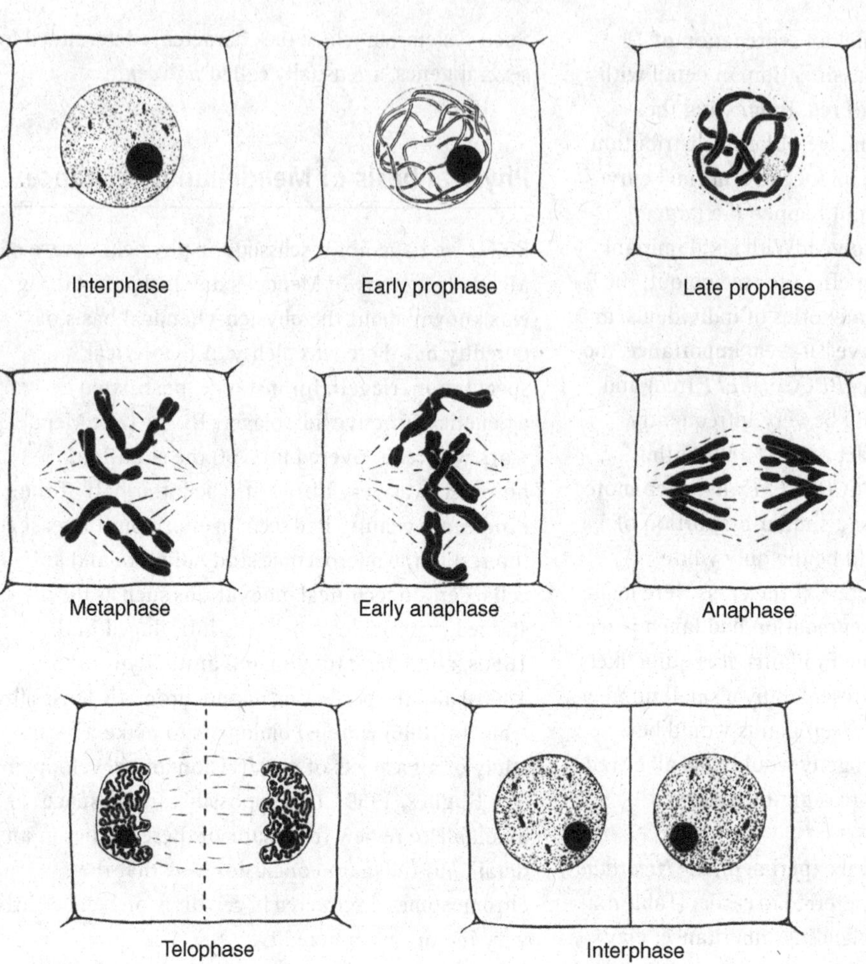

Fig. 4.6. The stages in mitosis in an organism with two pairs of chromosomes. (From McLeish & Snoad, 1962.)

Studies of the division of chromosomes in young anthers and in ovules revealed a different kind of nuclear division, the so-called *reduction division* or *meiosis* (Fig. 4.7). In this process the chromosome number is halved, the four derivatives having the haploid chromosome number. This halving compensates for the doubling in chromosome number following fertilisation of egg by sperm. Thus, in a diploid plant a haploid complement of chromosomes has come from each parent. Microscopic examination of favourable material establishes a most interesting fact: if maternal and paternal haploid complements are examined they are normally found to be exactly alike in appearance (except in the case of certain sex chromosomes). In the early stages of meiosis, homologous chromosomes, from maternal and paternal sources, pair together. Studies as early as those of Rückert (1892) suggested that in this paired state exchanges of chromosome material occurred.

This very brief outline of mitosis and meiosis gives some idea of the sort of knowledge about chromosomes that was available at the beginning of the nineteenth century. It was only a short time after the discovery of Mendel's work that various biologists, Boveri, Strasburger and Correns among them, saw a possible connection between Mendelian

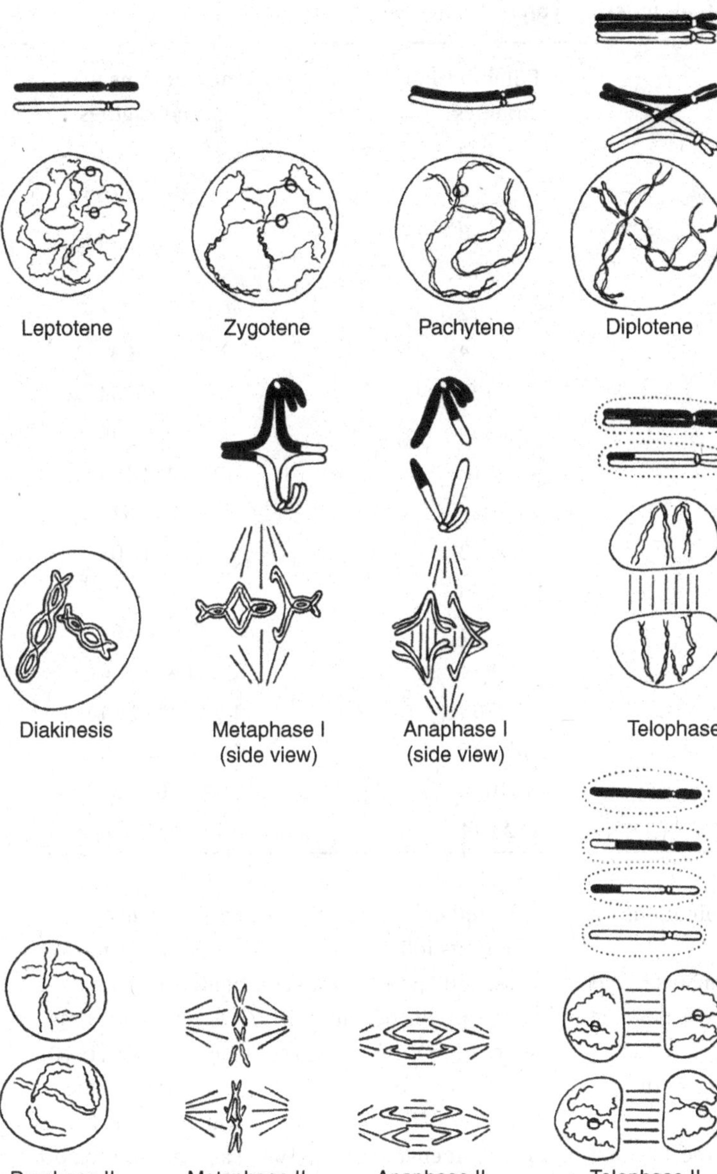

Fig. 4.7. The stages in meiosis in an organism with two pairs of chromosomes. The formation of one bivalent and its subsequent behaviour are shown diagrammatically above the appropriate stages. (From Whitehouse, 1965.)

segregation and chromosome disjunction. It was probably Sutton (1902, 1903), however, who first set out with clarity a cytological explanation of Mendel's findings. In his view the separation of maternal and paternal chromosomes of a homologous pair at the end of the first stage of meiosis resembled the postulated separation of factors, which Mendel suggested occurred at gamete formation. Further, if the orientation of pairs on the spindle is at random, a number of combinations of maternal and paternal chromosomes would be obtained in the gametes. If the chromosome number was very small the number of combinations would also be relatively small; on the other hand a diploid chromosome number as low as 16

Table 4.3. **Possible zygotic combinations (Sutton, 1903)**

Chromosome number		Combinations in gametes	Combinations in eventual zygotes
Diploid	Haploid		
2	1	2	4
4	2	4	16
6	3	8	64
8	4	16	256
10	5	32	1024
12	6	64	4096
14	7	128	16384
16	8	256	65 536
18	9	512	262 144
20	10	1024	1 048 576
22	11	2048	4 194 304
24	12	4096	16 777 216
26	13	8192	67 108 864
28	14	16 384	268 435 456
30	15	32 768	1 073 741 824
32	16	65 536	4 294 967 296
34	17	131072	17 179 869 184
36	18	262 144	68 719 476 736

would give 65,536 possible zygotic combinations (Table 4.3; Fig. 4.8) (see Sutton, 1903).

As many plants have chromosome numbers higher than this, a huge number of combinations is possible. We have here the beginnings of the chromosome theory of heredity, which is the basis of modern genetics.

Mendel postulated in his experiments the independent segregation of factors, and this view received support from the early geneticists. There were, however, increasing signs in the first decade of the twentieth century that not all genes segregate independently. Bateson, Saunders & Punnett in 1905, working with two-factor crosses in Sweet Peas (*Lathyrus odoratus*), did not get 9:3:3:1 ratios in F_2 families. Similar aberrant results were obtained from many organisms, amongst them the Fruit Fly (*Drosophila*) and the Garden Pea (*Pisum*). Many biologists followed Mendel in experimenting with peas, and up to 1917 an additional 25 character-pairs were examined (White, 1917). A very interesting series of crosses was made by de Vilmorin (1910, 1911) and subsequently by de Vilmorin & Bateson (1911) and Pellew (1913), working with 'Acacia' peas, a variant characterised by the absence of the normal leaf tendrils. The absence of tendrils was associated with wrinkled seed. The cross 'Acacia' × round seed and tendrilled leaf gave an F_1 with round seed and tendrilled leaves. The F_2, instead of segregating to give 9:3:3:1, gave the results in Table 4.4.

It is quite clear that the two factors are not segregating independently: the grandparental combinations of wrinkled/no tendril and round/tendril are being recovered with too high frequencies.

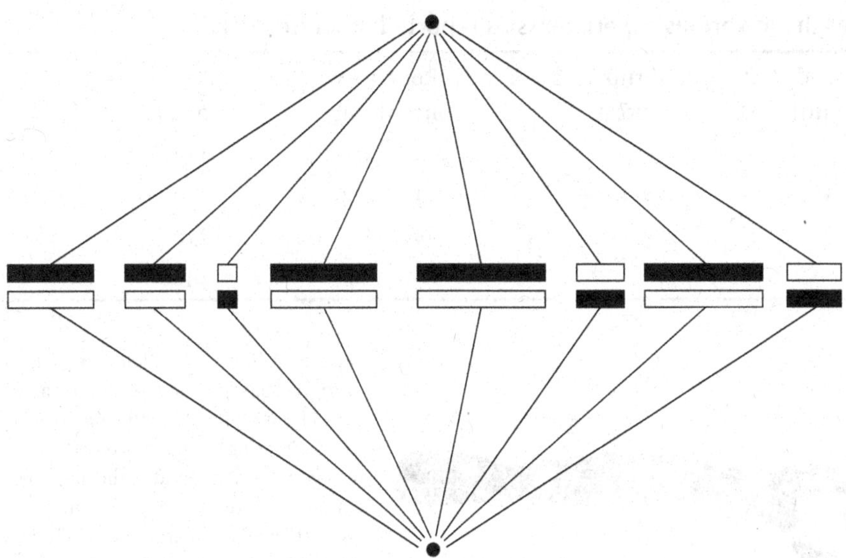

Fig. 4.8. The random orientation of bivalents at meiosis. At fertilisation a male and female gamete fuse, and each contributes a haploid set of chromosomes. Meiosis takes place in the diploid phase of the life cycle. Homologous chromosomes form bivalents (maternal chromosomes dark: paternal chromosomes, white), which orientate themselves at random about the equatorial region of the cell. The diagram represents one of the possible meiotic metaphase arrangements of the chromosome complement in a diploid cell with 16 chromosomes (8 pairs). As Table 4.3 shows, in a plant with $2n = 16$ there are 256 possible patterns of arrangement of the maternal and paternal chromosomes. As homologous chromosomes may carry different alleles, independent orientation means that many different combinations of maternal and paternal genes may be obtained in the gametes.

Various explanations were offered for this phenomenon, which came to be called 'partial linkage', later abbreviated to 'linkage'. Bateson & Punnett (1911) favoured an obscure 'reduplication' hypothesis; as time went by, however, the views of Morgan prevailed. He suggested that partially linked groups of factors were together on the same chromosome. In the Fruit Fly, *Drosophila melanogaster*, where $n = 4$, the extensive researches of Morgan and his colleagues established beyond doubt the existence of four such linkage groups. In pea, where $n = 7$, there are seven linkage groups. Historians of science are intrigued by the question: 'did Mendel detect but not mention linkage' in writing his famous experiments? After all Mendel studied 'seven traits in the pea plant' which has seven chromosomes (see Franklin *et al.*, 2008), and two of the seven traits studied by Mendel are reported to be linked (Lamprecht, 1961).

In the formation of a diploid organism the two gametes each carry one set of linkage groups: the haploid chromosome number. The appearance of occasional recombinants in small numbers in a cross such as that in Table 4.4 was accounted for by postulating an exchange of parts by homologous pairs in the first stage of maturation division. Evidence of such an exchange was seen in the chiasmata of prophase (Fig. 4.9).

The development of plant cytology

Early cytologists demonstrated that, normally, a particular diploid number of chromosomes was characteristic for a species. For example, the French cytologist Guignard (1891) made a clear drawing of the stages of mitosis in very young anthers of

Table 4.4. **'Acacia' Peas, results of various experiments as reported in White (1917)**

Source	Wrinkled seed, no tendril	Wrinkled seed, tendril	Round seed, no tendril	Round seed, tendril
de Vilmorin	70	5	2	113
de Vilmorin	99	4	1	170
Bateson	64	1	4	210
Pellew	564	15	20	1466

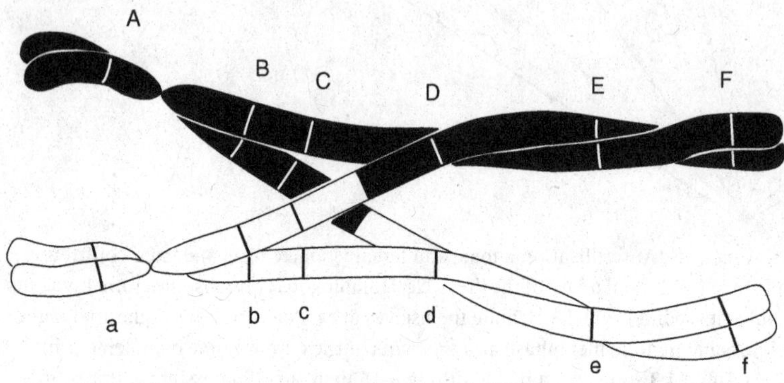

Fig. 4.9. A bivalent at diplotene with a single chiasma. The position of some genes (represented by letters) is indicated. Crossing over has occurred between one chromatid of the maternal chromosome (white), and one chromatid of the paternal chromosome (dark). (From McLeish & Snoad, 1962.)

Turk's-cap Lily (*Lilium martagon*), establishing the diploid number for that species as 24.

Descriptive cytology has made great strides from these early beginnings. A close study of individual chromosomes of the set reveals that they are distinct in such features as total length, length of arms and the presence of secondary constrictions. Chromosomes, in most plant species, but not all, have a localised distinct centromere or primary constriction that is involved with the spindle for the orderly movement of chromosomes in mitosis and meiosis. Other plant groups, for example members of the Cyperaceae/Juncaceae clade, have diffuse centromeres. Some chromosomes have terminal knobs on certain chromosomes, e.g. in maize (Heslop-Harrison, 2011). Also, the ends of chromosomes have distinctive regions called telomeres.

Given the results published by Guinard, it is instructive to compare his drawings with a diagrammatic representation of the haploid set of chromosomes of *Lilium martagon* prepared by Stewart (1947) (Fig. 4.10a; Fig. 4.10b) also shows that there are differences between *Lilium* species.

Such stylised diagrams are referred to as idiograms. The term karyotype is used for a more faithful representation of the chromosomes in drawings or photographs (see for example Fig. 4.11).

Chromosome number

An examination of *Lilium martagon* root-tip mitosis demonstrates that the diploid chromosome number is 24 and that this is the characteristic number for the species. There is a great deal of variability in chromosome number amongst plant species, e.g. *Haplopappus gracilis* has $2n = 4$ chromosomes, while the palm *Voanioala* has $2n =$ c.596 and the fern *Ophioglossum reticulatum* $2n =$ c.1200.

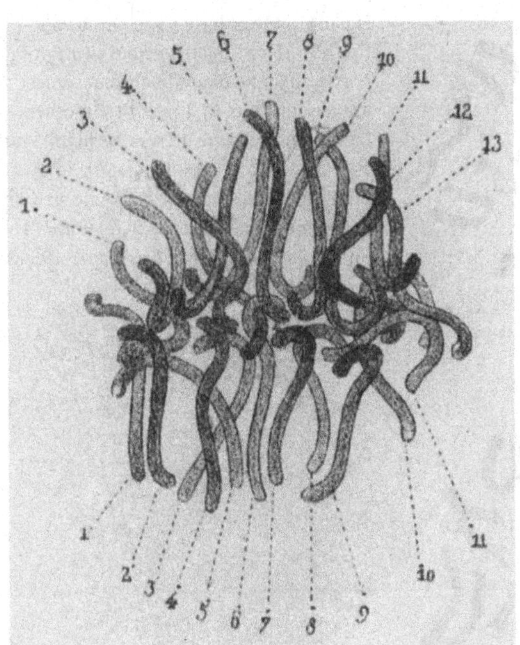

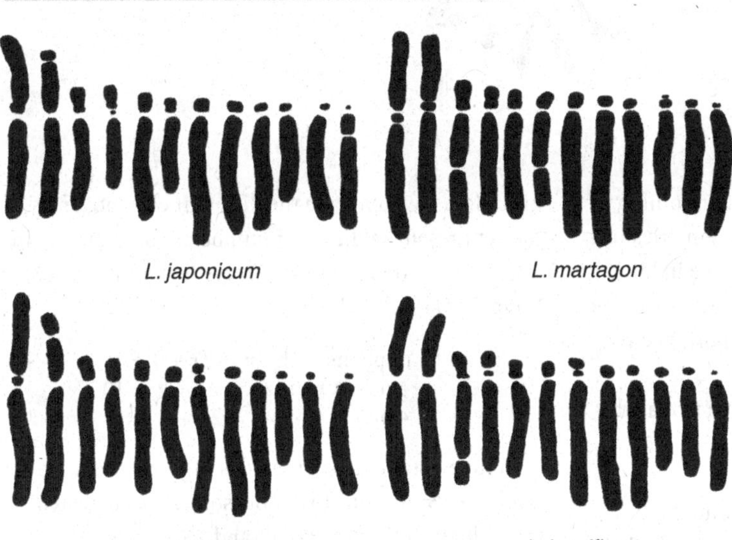

Fig. 4.10. Chromosomes of *Lilium*. (a) Mitosis in very young anthers of *Lilium martagon* ($2n = 24$) (×1500). (From Guignard, 1891.) (b) Idiograms of species of *Lilium* (×660). (From Stewart, 1947.)

In many genera, individual species form a polyploid series, in which high numbers are simple multiples of the lowest haploid number or the number which is taken to be the basic number of the genus (designated x; e.g. $x = 6$). Polyploids may arise from gametes that contain the unreduced number of chromosomes. Fusions between unreduced and 'normal' gametes (and between unreduced gametes) will produce in many cases viable plants with elevated chromosome numbers. For example, *Campanula rotundifolia* plants are divisible into three main groups, $2n = 34$, $2n = 68$ and $2n = 102$. In addition, changes in chromosome

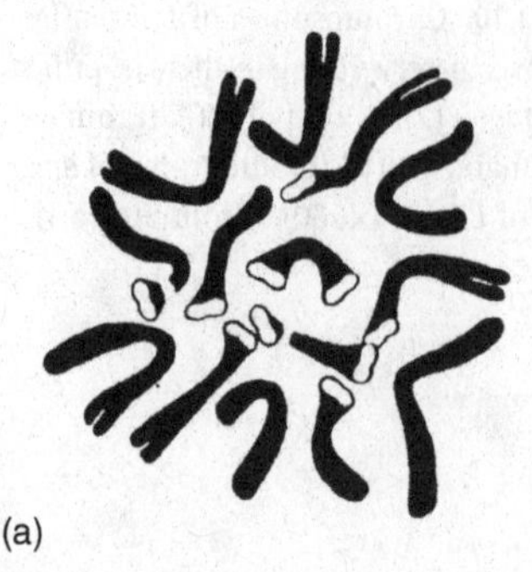

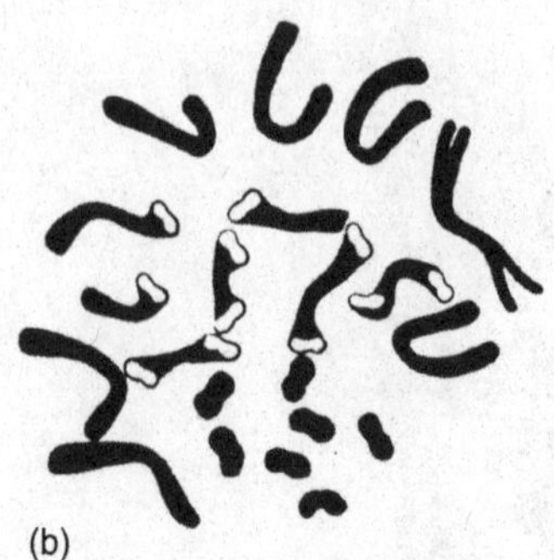

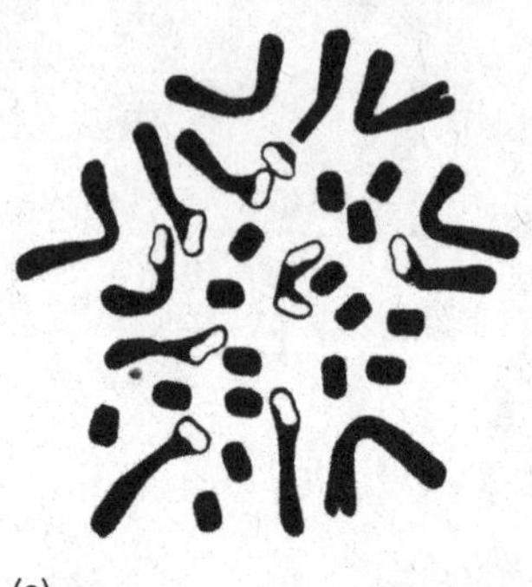

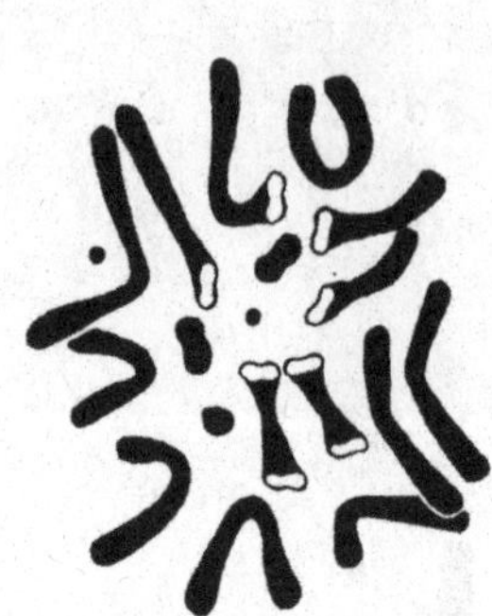

Fig. 4.11. Metaphase plates from root-tips of different plants of *Festuca pratensis*. In some plants B chromosomes are absent, as in (a) ($2n$ = 14). In others B chromosomes are present, as in (b) $2n$ = 14 + 7B and (c) ($2n$ = 14 + 16B). B chromosomes may be the same size, as in (c), or different sizes as in (d) ($2n$ = 14 + 5B) (All ×2700). (From Bosemark, 1954.)

number may arise following beakage and rejoining of chromosomes – so-called 'fission and fusion' (Heslop-Harrison & Schwarzacher, 2011). As we see in later chapters, polyploidy and other changes in chromosome number are of extreme importance in plant evolution.

Studies of many Linnaean species have revealed that there may be intraspecific variation in chromosome number. In diploid organisms, homologous pairs of chromosomes are found. Occasionally, however, mis-division of the paired chromosomes at meiosis may give gametes with more or fewer chromosomes than the haploid number. Chromosomes may be missing or represented more than once. Gametes with an incomplete haploid complement are usually defective, but those with one or more additional chromosomes may be fertilised and may develop into adult plants. Thus, in a large sample of plants there may occur individuals, called aneuploids, which have the different chromosomes of the set present in different numbers. For example, in a sample of 4000 plants of *Crepis tectorum* ($2n$ = 8) Navashin (1926) found:

10 plants with $2n = 2x + 1 = 9$
4 plants with $2n = 2x + 2 = 10$
4 plants with $2n = 2x + 3 = 11$

In investigations of progenies of *Taraxacum* plants ($2n$ = 24), Sørenson & Gudjónsson (1946) detected individuals with $2n$ = 22, 23 and 26.

In cytological investigations of individuals belonging to the same species, other differences in karyotype may be found. For example, in species with separate male and female individuals, the sex-determining mechanism may involve distinct sex chromosomes (Grant, 1975; Richards, 1986).

In addition to gross features of chromosome morphology, different parts of the chromosomes may show differential staining. A combination of cytological and genetic studies indicates regions of the chromosomes, called heterochromatic segments, different from 'normal' euchromatic regions. Heterochromatin has several distinctive cytological properties, being condensed when euchromatin is extended (Fincham, 1983). In some taxa the presence of heterochromatin may be accentuated by cold treatments before staining (Darlington, 1956). There may be a consistent pattern of distribution of heterochromatin, so-called constitutive heterochromatin, e.g. around the centromeres, but in many taxa there is variation in the pattern and a class of 'facultative' heterochromatin has been distinguished (Rees & Jones, 1977).

Also, chromosomes additional to the normal complement may be found (Fig. 4.11). Such accessory, supernumerary or 'B' chromosomes occur in at least 1,372 species of flowering plant species, with a degree of intraspecific variation in numbers (Jones, 1995). 'B' chromosomes are often, but not always, heterochromatic (Rees & Jones, 1977; Jones, Viegas & Houben, 2008). While they appear to be restricted to outbreeding species, overall the species with B chromosomes presents a highly heterogeneous grouping (Camancho, 2005). Evidence suggests that, at meiosis, they do not usually pair with the larger members of the set (the 'A' chromosomes). Indeed, they often behave as univalents at meiosis, not pairing with other chromosomes. Concerning the origin of B chromosomes, it has been suggested that they may be the result of interspecific hybridisation, or they have arisen intraspecifically from the A set.

A range of staining techniques may be used to study patterns of chromosome banding. For example, treatment with quinacrine produces 'Q-bands' visible under ultraviolet light in fluorescence microscopy. Preparations treated with giemsa dye give 'G-bands'. 'R-bands' may be revealed in heat-treated preparations, while 'C-bands' may become evident in chromosomes stained with giemsa followed by alkaline extraction.

Chromosome changes

Studies of meiosis may also reveal cytological differences between plants (Stebbins, 1966). Fig. 4.12 shows diagrammatically the main kinds of chromosomal change that result in genetic effects and visible cytological peculiarities at meiosis in the heterozygote. Such inversions, translocations, deletions and duplications of segments of chromosomes are to be found in naturally occurring individuals, as well as in experimental material. Clearly they provide a further important source of variation of great evolutionary significance.

Non-Mendelian inheritance

As we shall see in later chapters, molecular studies have transformed our understanding of many puzzling findings of pioneer geneticists. For example, Correns (1909), one of the rediscoverers of Mendel's work, showed that in the familiar American garden flower *Mirabilis jalapa* ('Four O'Clock' or 'Marvel of Peru') some variants with yellowish-green or variegated leaves showed normal Mendelian segregation, whilst a particular variant, *albomaculata*, with yellowish-white variegation, did not. Plants of the variant *albomaculata* produced occasional shoots that were wholly green and others that were white. Flowers on green shoots gave only green progeny whether pollinated from flowers on green, variegated or white shoots, and conversely flowers on white shoots, whatever the pollen parent, gave only white progeny (which died in the seedling stage). Variegated shoots gave all three kinds of progeny. In two respects this inheritance was clearly non-Mendelian: first, because the offspring resembled closely the female parent and the reciprocal crosses gave entirely different results; and secondly, because there was no regularity in the proportions of the phenotypes in the segregating families.

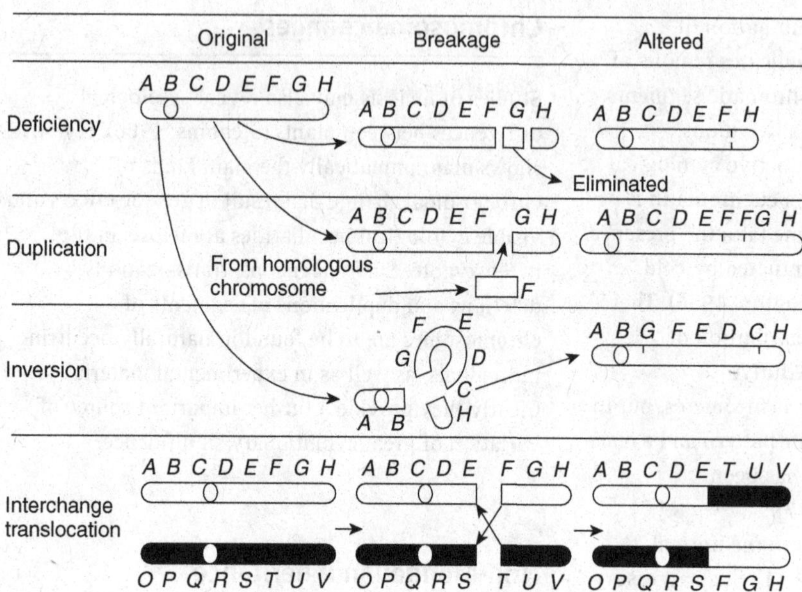

Fig. 4.12. Diagrams to show how chromosome breakage and reunion can give rise to the four principal changes that chromosomes may undergo. (After Stebbins, 1966.)

Over the years many other cases of non-Mendelian inheritance were discovered, e.g. in leaf variegation in the Garden Geranium (*Pelargonium*), the Evening Primrose (*Oenothera*) and many other genera (Kirk & Tilney-Basset, 1978; Evenari, 1989). Not all cytoplasmic inheritance concerns chlorophyll-containing plastids. It has been shown, by repeated backcrossing with species of Willow-herb (*Epilobium*), that 'alien' cytoplasm of one species can persist and give a variety of genetic effects with the nucleus of another species (Michaelis, 1954). Early geneticists were extremely puzzled by these findings. As we shall see in Chapter 6, our understanding of non-Mendelian phenomena has been transformed by molecular studies.

Patterns of variation

In previous chapters we have discussed the variation patterns found in samples taken from plant populations, and suggested that for descriptive purposes it may be useful to distinguish three broad types of individual variation: genetically determined, environmentally induced and developmental variation. This notion of three types of variation, however, gives an oversimplified picture of the nature of individual variation. There is strong evidence for the proposition that the phenotype and behaviour of the plant are determined by interactions between genotype and environment. As we shall see, different genotypes react differently to a given set of environmental conditions, and plants of identical genotype produce different phenotypes under contrasting environmental conditions. Moreover, the *interactions* between genotype and environment are further complicated by the complex sequence of changes that occurs as a plant develops from an embryo to the mature fruiting state.

Phenotypic variation

In the growth of an organism from fertilised zygote the genotype of a particular plant plays a vital role in determining the characteristics of the mature phenotype (perhaps a tree 30 m high) that is organised from raw materials drawn from outside the plant. There are complex close-knit interactions between genotype and environment at the level of the cell and of the whole plant. Concerning the genetic control of cellular processes, early ideas suggested that a gene provides information, which, in an appropriate environment, will contribute to a particular phenotype. However, accumulating evidence suggested that there was no certainty that a particular gene will always manifest itself. For example, in *Lotus corniculatus* certain plants known to possess alleles appropriate to the production of hydrogen cyanide when their foliage is crushed do not in fact produce the cyanogenic reaction in every circumstance (Dawson, 1941). Also, it was been found that a particular barley variant produces albino phenotypes out of doors, yet the same genotype, grown at higher temperatures in a glasshouse, has normal foliage (Collins, 1927).

Other complexities emerged in historic investigations. For example *Drosophila* geneticists have demonstrated conclusively that the manifestation of a gene can be modified, by altering the arrangement of genes in the chromosomes – the so-called position effect. Position effects in plants have been demonstrated by Catcheside (1939, 1947) studying *Oenothera*. Thus, two individuals with the same alleles may have different phenotypes.

Another complication displaces the early view that one gene determines only one characteristic. It was discovered that a gene may affect many quite different phenotypic characters – a phenomenon known as pleiotropy. A good example of this is provided by studies of *Aquilegia vulgaris* (Anderson & Abbe, 1933). In comparison with normal plants the mutant *compacta* is shorter, bushier and more branched, with erect, not drooping buds, shorter petals and less well developed sepals. Studies revealed that the manifold effects of this *compacta* mutation flowed directly or indirectly from the precocious secondary thickening of cell walls in many parts of the plant.

The extent of phenotypic variability in plants

In general, plants show greater phenotypic variability than is found in higher animals. In animals, individual variation is apparently held within very tight bounds by the early precision of mechanisms determining irreversibly the form and relationship of the main organs. To some extent this is true of plant structures, but there is very important difference between the plant and the animal, which resides in the fact that there is often a persistent meristem or growing-point tissue, on which a succession of organs of limited growth is initiated. The result of this difference is that the individual plant is open to much more environmentally induced variation over a much greater part of its life than is the animal.

Here, it may be helpful to consider the causes of phenotypic variation in plants. If material of a given genotype is divided into separate pieces (ramets) and grown in two or more different environments, different genotype–environment interactions may be produced (see Fig. 4.1). While the plant responds to the environment as a whole, it is possible, by appropriate experiments involving changes in particular factors, to deduce that certain elements of the environment are of particular importance. Test environments may differ in soil properties (e.g. water table, base status, level of toxic ions), and aspects of climate or microclimate (e.g. temperature extremes, wind exposure, rainfall). In their competition with one another in experiment or in natural communities, plants show many diverse interactions, for example to the effects of shading. Also of importance may be the influence of chemical substances (at present often hypothetical) produced by one taxon in suppressing the growth of others (so-called allelopathic effects: see Rice, 1984; Putnam & Tang, 1986; Rizvi & Rizvi, 1992; Reigosa, Pedrol & González, 2010).

The availability of organisms to form the natural symbioses characteristic of many plants is a critical factor in natural and experimental situations (Harley

& Harley, 1987), as are the presence and severity of grazing, pest attack and disease. Given the variety of different environments, a great diversity of genotype–environment interactions is possible in the growth and reproduction of individual plants. Indeed, we should note that the stresses of the environment may kill the plant, restrict its growth or perhaps prevent reproduction.

Developmental variation

Phenotypic variation should be viewed within the context of developmental variation. In most flowering plants the cotyledon stage of the seedling is very different from the adult – a fact which has, of course, excited the interests of botanists from early times, and which provided the basis for John Ray's inspired division of flowering plants into the monocotyledons and dicotyledons. How our views of this division have changed in the light of phylogenetic investigations is considered in Chapter 16. The developmental transition between the simple cotyledon and the often lobed or dissected mature leaf is generally rather abrupt, and provides the most familiar example of a phenomenon which Goebel (1897) called 'heteroblastic development', or the change from a juvenile to an adult phase accompanied by more or less abrupt changes in morphology. Strictly speaking, the cases which Goebel and others have mainly called 'heteroblastic' are those in which a juvenile leaf other than the cotyledons contrasts more or less clearly with an adult one, as, for example, in the case of Gorse (*Ulex europaeus*), in which the seedlings produce trifoliate leaves (of a type which is normal in related genera) before the simple ones (Fig. 4.13).

There seems, however, much to be said for extending the term to cover all ontogenetic phase changes (in leaf-shape, etc.), including the cotyledons, and including also changes between early and late phases which are more gradual.

If we investigate closely the detailed development of a plant with an adult leaf clearly different from the juvenile, the commonest situation is likely to be as shown in Fig. 4.14 for *Ipomoea purpurea* (*Ipomoea caerulea*), the Morning Glory, a common tropical climbing herb. In this illustration, taken from the work of Njoku (1956), the top line shows the shape of the first 10 leaves of a plant grown in the shade, and the second line shows the same series from a plant in full daylight. Here two things are evident: first, that the development of the adult three-lobed leaf-shape is gradual; and secondly, that the onset of the three-lobed leaf-shape in the developmental series is greatly modified by the environment, in this case by light.

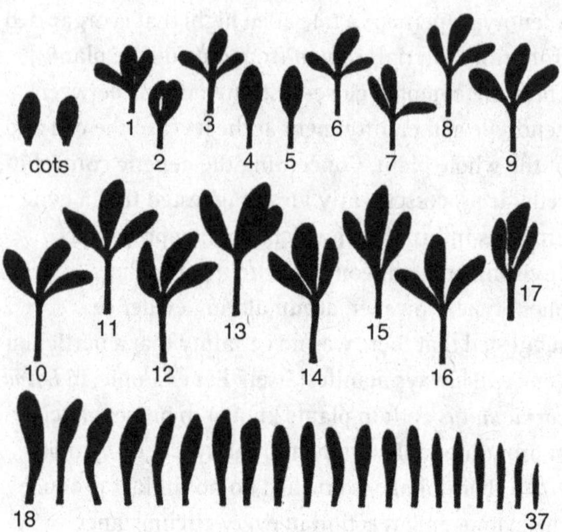

Fig. 4.13. Juvenile (top two rows) and adult (bottom row) foliage in *Ulex europaeus*. The leaves are numbered in sequence. (From Millener, 1961.)

Figure 4.14 also illustrates the effect of transferring the plant from shade to light at the stage of the unfolding of the second leaf, and also of transferring it from light to shade. In both cases there is a 'time-lag' in reaction which lasts until the sixth or seventh leaf, suggesting that the developmental processes which determine the form of the mature leaf are operating at an early stage in the differentiation of leaf primordia on the growing point, and that once these have reached a certain stage the effect of environmental factors is no longer operative. The important point to note is the irreversibility of change at a certain stage

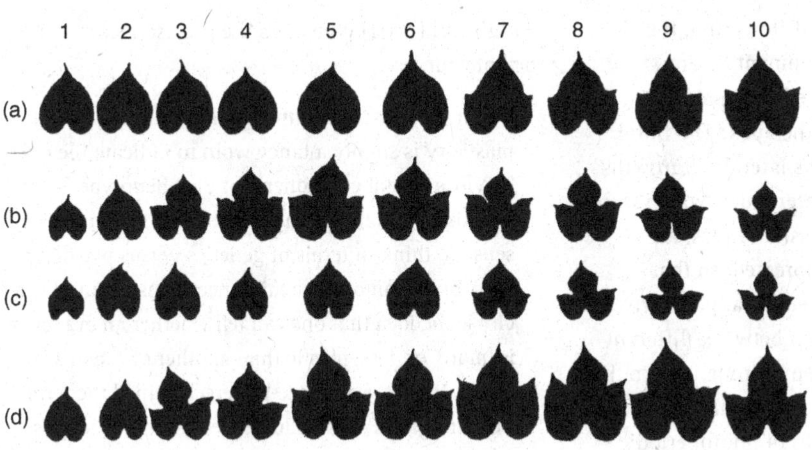

Fig. 4.14. Njoku's experiment with *Ipomoea purpurea* (*I. caerulea*) (×0.125). The first ten leaves are shown of: (a) a plant grown in the shade; (b) a plant grown in the light; (c) a plant transferred from shade to light as the second leaf unfolded; and (d) plant transferred from light to shade as the second leaf unfolded. (From Njoku, 1956.)

in the development of mature structures and its relation to developmental variation.

Another example of a situation where changes are generally irreversible is found in Ivy (*Hedera helix*). The wild plant, common in Britain and Atlantic Europe, shows a very marked heterophylly, the familiar lobed 'ivy-leaf' being produced exclusively on non-flowering shoots which are normally flattened and adapted for growth, attached to trees, or on the woodland floor. The flowering shoots, in contrast, are erect, branch more or less radially and bear simple leaves. Intermediate shoots and leaves are rare (Fig. 4.15).

Seedlings, as would be expected, produce lobed leaves and quickly assume the vegetative phase. If, however, portions of *either* vegetative or reproductive shoots are detached and rooted separately, the plants so produced normally continue to grow in the manner characteristic of the particular phase. This is apparently true even of intermediate shoots. In this way, whole plants of *Hedera* with a more or less erect habit and simple leaves can be propagated, apparently indefinitely, though 'reversion' to the juvenile vegetative phase can be induced, for example, by repeated cutting, by grafting on to the juvenile stock, or by spraying with the growth substance gibberellic acid (GA_3).

Fig. 4.15. Heterophylly in *Hedera helix* (×c.0.33). Note the simple leaves on flowering shoots. (From Ross-Craig, 1959.)

Turning to the phenomenon of flowering, the reproductive phase in the development of the individual plant is usually marked by an abrupt change in pattern of growth at the apex. The floral parts originate, like the leaves, as lateral outgrowths on the growing-point, but the internodes, which were very obvious between the successive leaves, are suddenly greatly reduced or suppressed, so that whorls or spirals of tightly packed floral parts are produced. There is great variation between different plants as to the influence of the environment upon the initiation of flowering, but there is usually some detectable effect, and in the cases of photoperiodic response the effect can be very great indeed. Certain species require exposure, often for very brief periods only, to a particular day-length before they will pass into the reproductive phase which, once initiated, can continue whether the particular day-length conditions are still present or not. This kind of adaptation has, of course, obvious importance in terms of wild populations; it may mean (as, for example, with certain plants of American origin in Europe) that the plant may be unable to flower and fruit when introduced into a country where the particular day-length conditions are not present, and in this way the spread of a species may be restricted. Other species may require cold treatment before flowering is initiated.

Phenotypic plasticity

Early experiments on the basis of variation established that cloned individuals of particular genotypes grown in different environments respond by producing different phenotypes. Such phenotypic plasticity was defined and explored by Woltereck as long ago as 1909. Historically, the significance and evolutionary importance of plasticity was emphasised by Schmalhausen (1949), Schlichting (1986) and Sultan (1987). For a more recent major review of the subject see Kaplan (2010).

Pigliucci (2001) considers the precise meaning of the concept.

> By far the most common misunderstanding is that plasticity is simply a fancy word to indicate the old 'environmental component' of the phenotype (Falconer, 1952), and that therefore it still makes sense to think in terms of genetics versus plasticity. Next in line among the misconceptions about plasticity is the idea that one can tell whether an organism is 'more' or 'less' plastic than another one, across the board. In fact, ever since Wolereck coined the term 'reaction norm', it should be clear that plasticity is a property of a genotype, and that it is specific to particular traits within a range of environments.

Some early experiments

Many investigations have studied the relative plasticity of different characters. Indeed, we have already referred to the question in Chapter 3. The classic studies of *Potentilla glandulosa* in California by Clausen and associates provide another example (see Bradshaw, 1965). As we shall see in Chapter 8, they discovered, in experiments involving clone transplants into a number of different environments, that there was considerable plasticity in vegetative parts, including plant height and the number of shoots, leaves and flowers. In contrast, there was less plasticity in shape and marginal serration of the leaves, in the shape of the inflorescence and in the size of parts of the flower (Clausen, Keck & Hiesey, 1940).

The variable extent of plasticity in related taxa is beautifully illustrated by the group of species of *Ranunculus* subgenus *Batrachium* that exhibit heterophylly (Fig. 4.16; Table 4.5).

Species growing on mud or in very shallow water produce only floating leaves, while taxa from deep or swiftly flowing water develop only finely divided submerged leaves. In contrast, species inhabiting

Table 4.5. **The occurrence of heterophylly in British species of *Ranunculus* subgen. *Batrachium*. (From Bradshaw, 1965)**

		Leaves	
Species	Habitat	Floating	Submerged
R. hederaceus	Mud or shallow water	Many	–
R. omiophyllus	Streams and muddy places	Many	–
R. tripartitus	Muddy ditches and shallow ponds	Some	Many
R. fluitans	Rapidly flowing rivers and streams	–	Many
R. circinatus	Ditches, streams, ponds and lakes	–	Many
R. trichophyllus	Ponds, ditches, slow streams	–	Many
R. aquatilis	Ponds, streams, ditches, rivers	Some	Many
R. peltatus	Lakes, ponds, slow streams	Many	Many
R. baudotii	Brackish streams, ditches, ponds	Some	Many

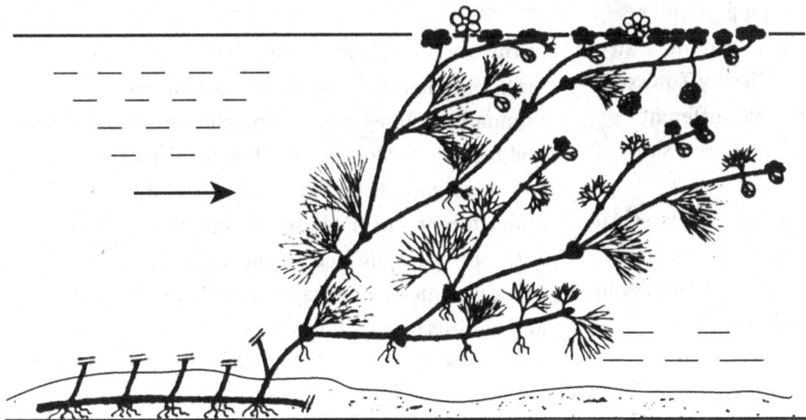

Fig. 4.16. Generalised diagram showing heterophylly in *Ranunculus* subgen. *Batrachium.* For convenience, only one shoot is illustrated. Close examination has revealed seasonal variation in leaf form, especially in the submerged leaves. (From Zander & Wiegleb, 1987.)

shallow water produce both types of leaves (see Cook, 1968; Zander & Wiegleb, 1987; Webster, 1988).

Cook & Johnson (1968) and Wells & Pigliucci (2000) have also studied phenotypic plasticity of other groups of heterophyllous semi-aquatic species.
In considering the responses of such species, Pigliucci makes the important point that the reactions of semi-aquatic species often yield spectacularly abruptly different phenotypes, a phenomenon sometimes referred to as polyphenism.

Experiments with *Persicaria amphibia* (*Polygonum amphibium*) have also yielded valuable information on phenotypic plasticity.

Plants growing on land and in water have very different phenotypes (Fig. 4.17). If ramets of cloned material are separately grown in conditions

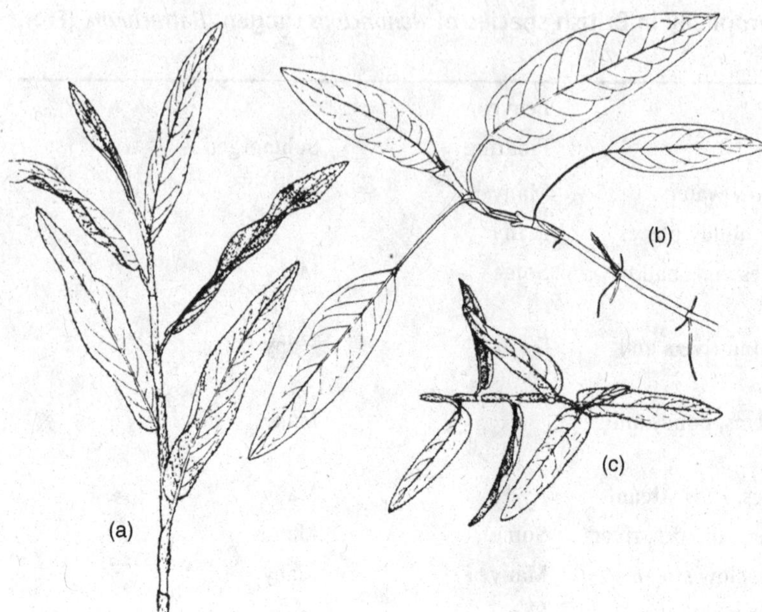

Fig. 4.17. *Persicaria amphibia* (*Polygonium amphibium*) (×0.5). (a) Terrestrial form, growing on land near water. (b) Aquatic form, with submerged stems and floating leaves. (c) Form growing in damp sand dunes. (From Massart, 1902.)

simulating land, waterlogged and submerged conditions, then the degree to which a particular individual may produce both the 'land' and the 'water' phenotype may be investigated. Studies by Turesson (1961) and Mitchell (1968) suggest that different individuals (presumably different genotypes) show different degrees of plasticity.

Morisset & Boutin (1984) consider that there is a clear tendency for pioneer species, in early successional habitats, to show greater plasticity than related taxa in later successional vegetation types. However, Schlichting & Levin (1984) point out that there is insufficient information to permit critical tests of hypotheses and, hence, difficulties in making generalisations. Clearly, many interacting factors must be considered including levels of heterozygosity, degree of relatedness of the taxa, as well as ecological factors.

Individual variation in plants

In this chapter a brief sketch has been provided outlining historic ideas. Building on these foundations, integrated approaches involving genetic and molecular biology, and biochemistry, have, in recent decades, completely transformed our understanding of the basis of individual variation. Major progress has been achieved with the intense investigation of model species such as *Arabidopsis* and various crop plants.

The next chapter considers how many aspects of Darwinism were challenged in the early twentieth century, and how, through early genetical and cytological investigations, these challenges were largely resolved.

5 Post-Darwinian ideas about evolution

For forty years after the publication of the *Origin*, Darwin's ideas were a source of tremendous public controversy. For this reason he never received any awards from the state, although he was awarded honorary degrees and decorations in plenty. Despite thousands of critical sermons, many accepted Darwin's views on evolution, but, as seen in this chapter, the concept of natural selection was not generally accepted, until the rise of genetics in the mid-twentieth century. The biological literature of the period is full of papers speculating about the adaptive significance of various structures, the probable course of evolution in the plant and animal kingdoms, and so-called 'evolutionary trees' showing phylogenetic relationships (Fig. 5.1).

Some of this work is of lasting interest, but there was a depressing tendency in the later years of the period for armchair biologists to produce highly speculative theories, and there was a lack of critical experiment with living material. Towards the end of the century, however, there were signs of an increasing interest in the possibility of using experiments for the investigation of evolutionary problems.

Experimental investigation of evolution

A good example of this change in climate is provided by the controversy that enlivened the pages of *Nature* and the editorials of *The Gardeners' Chronicle* in 1895. At a discussion meeting of the Royal Society, the Director of the Royal Botanic Gardens, Kew, W. T. Thiselton-Dyer, had shown specimens of the 'feral' type and cultivated variants of what he called *Cineraria cruenta* (now called *Pericallis cruenta*) – the gardener's Cineraria – and an extended account of his ideas was printed in *Nature* (1895). He suggested, as befitted an ardent and orthodox disciple of Darwin, that, as far as was known, the garden Cineraria was derived from *Senecio* (*Cineraria*) *cruentus* from the Canary Islands 'by the accumulation of small differences'. Bateson (1895a, b) responded to this view in a lengthy letter to the editor, questioning the assertions of Thiselton-Dyer. Bateson concluded, after a study of the literature, that modern Cinerarias arose from hybridisation between several distinct species, that selection was practised on variable hybrid progeny, and that 'sports' may have been important, as well as subsequent improvements as a consequence of the selection of small-scale variation.

The arguments in *Nature* continued back and forth with four letters from Thiselton-Dyer, three from Bateson and three from the biometrician Weldon. It became clear that argument could not settle the issue of the origin of the garden Cineraria.

The possibility that experimental hybridisation might shed light on the variation patterns occurred to Bateson, who enlisted the help of Lynch, Curator of the University Botanic Garden in Cambridge. Lynch raised stocks and made a number of artificial crosses, some of which were exhibited at a meeting of the Cambridge Philosophical Society in 1897. The report of the meeting (Bateson, 1897) says that the experiments 'were entirely consistent with the view that Cineraria was a hybrid between several species'. Lynch's experiments were published in detail in 1900. Here we have a clear case of speculation about evolution leading directly to experiment. It is interesting that a more recent review of the Cineraria problem (Barkley, 1966) reveals that it has received little attention since these early experiments.

In the period 1892–1910 some of the first experiments investigating natural selection were

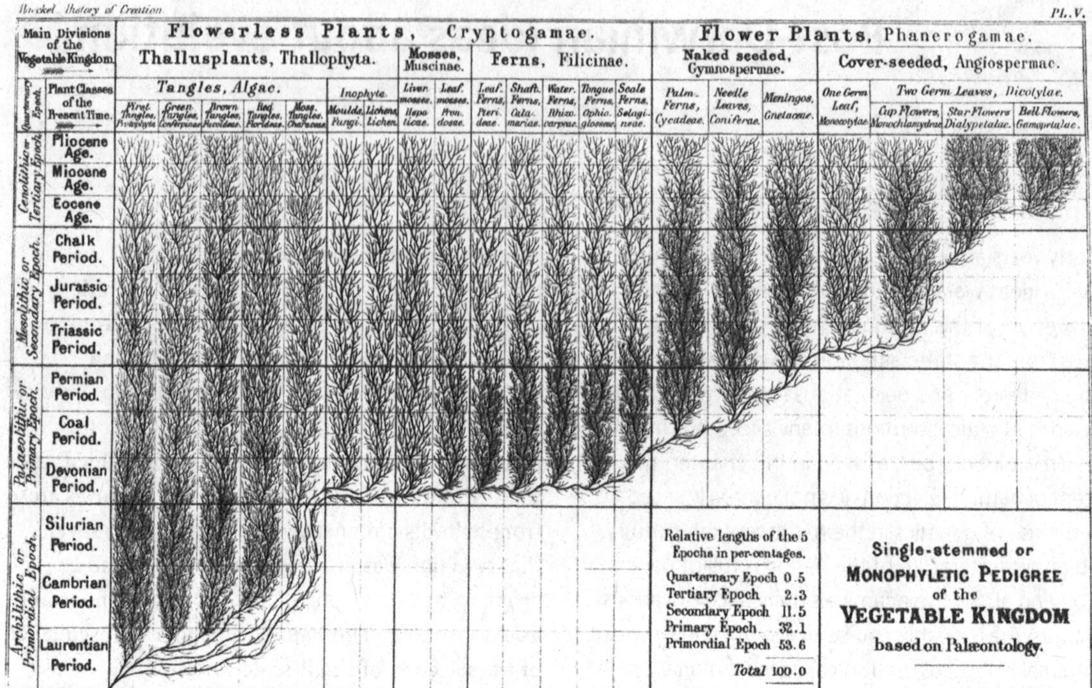

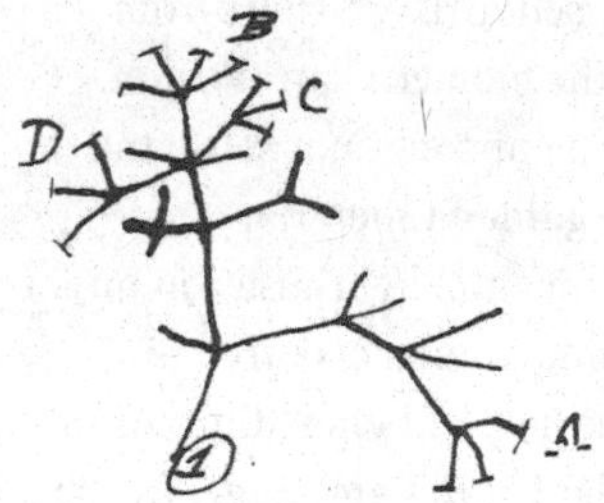

Fig. 5.1. Tentative sketch of a phylogenetic tree from Darwin's notebook (1837) (see De Beer, 1960–1), which contrasts with the baroque splendour of Haeckel's highly speculative 'monophyletic pedigree of the vegetable kingdom' (Haeckel, 1876).

carried out. Darwin had written in the *Origin* (chapter IV): 'Can we doubt (remembering that many more individuals are born than can possibly survive) that individuals having any advantage, however slight, over others, would have the best chance of surviving and of procreating their kind?' In 1895 Weldon wrote:

> The questions raised by the Darwinian hypothesis are purely statistical, and the statistical method is the only one at present obvious by which that hypothesis can be experimentally checked. In order to estimate the effect of small variations upon the chance of survival, in a given species, it is necessary to measure first, the percentage of young animals exhibiting this variation; secondly, the percentage of adults in which it is present. (Weldon, 1895b)

If the percentage of adults exhibiting the variation proved to be less than that in young animals, then some of the young animals must have been lost before reaching adulthood, and a measure of the advantage or disadvantage of the variation could be obtained. In putting these novel ideas to the test, Weldon (1898) investigated the variation in the crab *Carcinus maenas* at a site on Plymouth Sound. While it is inappropriate to give details of his results, we may note that his findings offered some support for the initial hypothesis. After several years' investigation (1892–8) he also concluded that the population was unstable. He deduced that changes were caused by the increasing amounts of china clay and sewage in the waters of Plymouth Sound, and carried out experiments investigating the death rate of captive crabs subjected to foul water. Crabs survived captivity in clean water, but only a portion of the variable population – those with small frontal breadth relative to their carapace size – survived in foul water. While it is true that these experiments may be criticised on grounds of sampling technique (and, indeed, in other ways), they do represent a major step forward in the design of investigations into natural selection and, as we shall see, are a model for some more recent botanical studies.

Weldon was also instrumental in encouraging some of the first field studies of selection. Di Cesnola (1904), a student of Weldon's at Oxford, noticed that in Italy there were green and brown variants of the Praying Mantis, *Mantis religiosa*. The green variant was found in grasses and the brown on vegetation burnt by the sun. In an experiment which lasted several days, individuals were tethered by silk threads as follows:

1. in a green grassy area 20 green and 45 brown individuals
2. in a brown area – 25 green and 20 brown individuals.

The 25 greens in the brown area and 35 of the browns in the green area were taken by birds or ants. It is significant that the individuals that matched the background vegetation were untouched. These studies were forerunners of many experiments by zoologists to test the supposed adaptive significance of protective coloration (see Cott, 1940). As in the case of the investigations on crabs, it is obvious that the experiment with *Mantis* is not beyond criticism, but it is based, nevertheless, on a novel approach to the study of the force of natural selection.

In plants too, studies of variation led to insights into natural selection. For example, in 1895, von Wettstein described the phenomenon of seasonal dimorphism in a number of hemiparasitic genera including *Euphrasia, Odontites* and *Rhinanthus.* As a result of careful investigation, many species appeared to have two subspecies: an early summer flowering variant (aestival) and a later variant (autumnal). He considered that the practice of haymaking in central Europe was important in the origin of the two types of subspecies. The maintenance of the annual habit was only possible if plants either fruited before midsummer hay-cutting, or elongated and matured after the crop had been taken. Plants flowering or in immature fruit at the time of haymaking would fail to reproduce, and selection would therefore favour both early and late flowering. This work provoked a good deal of controversy, especially about which subspecies was ancestral, and whether patterns were as simple as was suggested by von Wettstein. For full details of the historical studies in this area see Ter Borg (1972). Hay and pasture variants have now been studied in a range of species (see Briggs, 2009 for details)

Interesting results were also obtained from agricultural species. Brand & Waldron (1910) and Waldron (1912), working at Dickinson, North Dakota, USA, cultivated 68 samples of the important legume forage crop *Medicago sativa* (Lucerne or Alfalfa) collected from different parts of the world. Plants from Mongolia proved to be cold-resistant, whilst those from Arabia and Peru, on the other hand, were frost-sensitive. In the severe winter of 1908/9 the pattern of losses due to frost damage provided interesting information. For instance, in a strain from Utah, many plants died of frost damage, but out of the progeny of three specially resistant plants originating from this stock only 3.5% died. Brand & Waldron deduced that a frost-hardy strain could originate from extreme individuals of an otherwise frost-sensitive stock.

There was also interesting research on 'races' in hemiparasitic plants. Thus, three variants of *Viscum album* (Mistletoe) were described by von Tubeuf (1923) from broadleaved trees, Fir and Pine. Each was morphologically distinct in such characters as size and shape of leaves, and colour of berries. Some attempt was made (only partially successful) to test, by transfer of seed, whether there were three different physiological races of Mistletoe, each adapted to a different host.

The mutation theory of evolution

The experimental approaches employed by biologists at the turn of the century not only provided insights into biometry and natural selection, but also provoked Bateson, de Vries and others to propose a rival theory to that of Darwin – the mutation theory of evolution (Provine, 1986; Cock & Forsdyke, 2008). The theory stressed the importance of 'sports' and various other abruptly occurring new variants. While the theory was claimed to be new, it is clear, as we saw in Chapter 2, that it grew out of a long-standing interest in the subject of 'sports'.

Darwin argued that species were ever-changing entities, the products of natural selection; his thesis was descent with modification, involving continual and gradual change. De Vries and Bateson did not deny the existence of natural selection; in fact it still played a key role in their ideas of evolution. What was different was their view that new species arose abruptly by 'mutation'. They confined the significant changeability of species to distinct and probably short periods. They accepted the theory of descent with modification, but thought that the changes occurred abruptly, interspersed with periods of stability (Cock & Forsdyke, 2008).

What evidence could the 'mutationalists' find in support of their theory? First, they examined cases of the apparent abrupt evolution of new persistent variants. Most famous of these were plants of the genus *Oenothera* (Evening Primrose) studied by De Vries (1905). Secondly, they discussed problems of heredity. For the 'mutation theory' it was *discontinuous* patterns that were significant and, with the re-finding of Mendel's research in 1900, Bateson did not fail to point out that this provided a mechanism explaining discontinuous variation. In his view, continuous variation was the product of environmental factors.

A third piece of evidence was also forthcoming. In Johannsen's experiment, which we introduced in Chapter 4, selection had no effect upon mean seed weight. Try as he might, from a particular line he could not select a strain with larger or smaller beans. Some variations did occur, but Johannsen ascribed this to the effect of the environment. Bateson and others went further and argued that all 'fluctuating variations' found in nature were environmentally based. For the 'mutationalists' natural selection occurred only when the products of mutation were being sorted out.

By the beginning of the century a curious situation had developed. In opposing camps were the 'Mendelian-mutationalists' and the 'Darwinian-biometricians' (Crew, 1966; Waddington, 1966; Provine, 1971, 1986, 1987). The Darwinian-biometricians, for the most part, remained loyal to Darwin's theory of gradual change. It is remarkable that this group of mathematically minded scientists opposed Mendel's views, preferring instead Galton's law of ancestral inheritance. Indeed, the animosity between the forceful personalities involved contributed to a delay in the development of the subject. Furthermore, as has been pointed out by Fisher (1958), the lack of mathematical understanding in biologists possibly contributed to the neglect of Mendel's work at the time of its publication, yet, paradoxically, on its re-finding it was the mathematical biologists who opposed it.

The Evolutionary Synthesis

In the early decades of the twentieth century, many biologists thought that Darwinism had been eclipsed and was indeed dead (Huxley, 1942). Then, in the 1930s, through the work of Fisher (1929), Haldane (1932), Sewall Wright (see Provine, 1986) and many others, an integration of Mendelian genetics and Darwin's evolutionary theory took place. 'Neo-Darwinism' was born.

Historical researches make it plain that many biologists, from a range of disciplines, contributed to this 'Modern Synthesis' (see for example Mayr & Provine, 1980; Brush, 2009; Cain & Ruse, 2009 and references cited therein). The celebrated evolutionist G. Ledyard Stebbins highlights the role played by Dobzhansky (1935) in promoting the synthesis. This is clear from the research of Smocovitis (2006), who quotes from a draft manuscript of autobiographical reflections by Stebbins:

> nobody can deny that the leader of the mid-century storm of interest in evolutionary theory during the middle of the 20th century was Theodosius Dobzhansky. He was the only scientific evolutionist who combined a thorough knowledge of what was then modern genetics ... with a [sic] extensive knowledge of a deep interest in the forces of evolution that operate in nature. Dobzhansky was enormously persuasive; like all examples of Messianic promotion of a cause, his enthusiasm was captured captivating? Infectious? [sic] Furthermore, he had planned a campaign that would supplement his own writing with that of specialists in related fields like G. G. Simpson, Ernst Mayr and myself to produce well balanced synthesis of contemporary theories.

Major reviews of different aspects of evolution, were published by Dobzhansky (1935, 1937, 1941, 1951), Huxley (1940, 1942), Mayr (1942), Simpson (1944), Stebbins (1950) and Clausen (1951).
In considering these works, Smocovitis (2006) highlights the interrelationships of those involved with the Evolutionary Synthesis, 'exploring the multi-directional traffic of influence between Dobzhansky and Stebbins', but also the 'social and professional networks' that linked Stebbins, Edgar Anderson, Epling, the 'Carnegie team' of Clausen, Keck and Hiesey, the zoologists Lerner, Wright, Dunn and other 'architects of the synthesis', especially Mayr, Huxley and Simpson.

Elements of the New Synthesis

There was an acceptance that inheritance is particulate and that genetic material is borne on chromosomes. Segregation and recombination were widely demonstrated. Gradually, more and more evidence against the 'mutation theory of evolution' view was discovered. First, intensive genetic and cytological studies of many species, including species of *Oenothera*, were carried out, and it became obvious that the new persistent variants found in *Oenothera* were of several different kinds. Some were simple mutants; others were polyploid derivatives; and a further group was the result of complex interchanges of chromosome segments.

Other species did not give the same results, and the *Oenothera* situation was seen to be unique in its complexity.

Also, further studies were carried out to investigate Johannsen's ideas. Habitual self-fertilisation is characteristic of French Beans and, as we shall see later, this leads to genetic homozygosity.
The ineffectiveness of selection in his experiment is understandable considering the breeding system. However, it is only in the absence of genetic variability that selection is ineffective, a fact attested by many successful selection experiments with outbreeding organisms.

The idea of blending inheritance was finally demolished by research carried out at this time. Darwin's idea of the persistence of favoured variants under a regime of blending inheritance necessitated

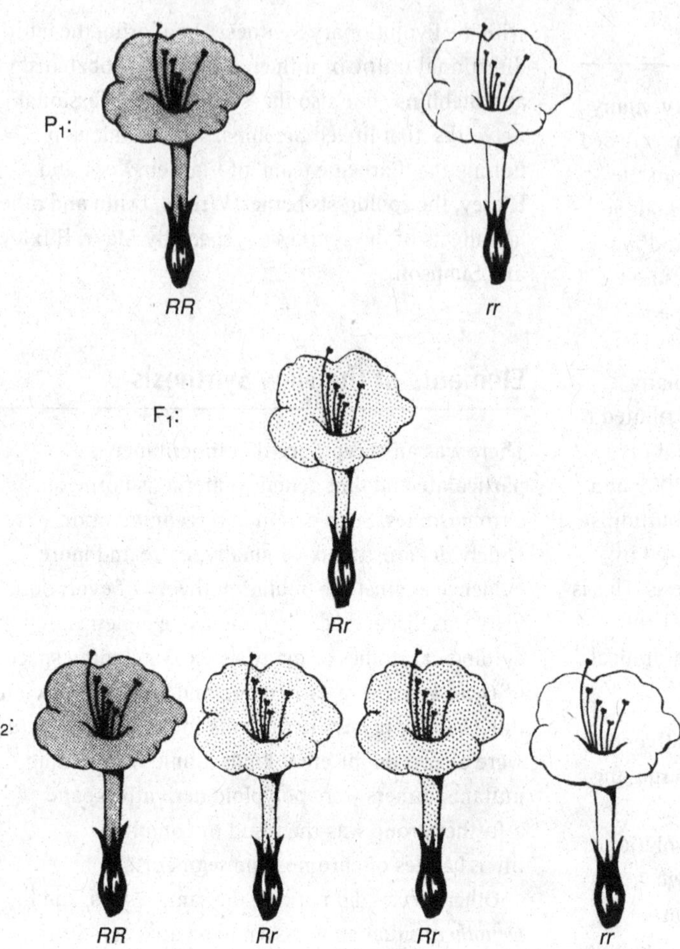

Fig. 5.2. *Mirabilis jalapa.* At first sight the production of pink-flowered offspring from red- and white-flowered parents looks like a case of blending inheritance, but, as the diagram shows, the situation can be explained in simple Mendelian terms if we assume the absence of dominance. Reprinted with permission of Macmillan Publishing Co. Inc. from Strickberger (1976) after Correns. © 1976 Monroe Strickberger.

a high mutation rate. Mathematical calculations, based on the probable mutation rate, showed that its incidence was likely to be much lower than that required to support the idea of blending (Fisher, 1929). Further, it became clear that cases of inheritance that were at first explained by blending were explicable in terms of Mendelian genetics – some as instances of systems with no dominance (Fig. 5.2), and others, as we saw in Chapter 4, as examples of inheritance controlled by many factors.

Concerning one of the major elements of the New Synthesis, Fisher came to important theoretical conclusions concerning the effectiveness of natural selection. Brush (2009) summarises Fisher's findings: if natural selection acts primarily on very small mutations, it might seem that it must be a very slow process. But Fisher showed that a rare gene with only a small selective advantage – a small mutation that produces no observable effect in the first generation – can spread rather quickly through a population if the same mutation is repeated (randomly but at a finite rate) in every generation. For example, if the selective advantage is 1 percent (and the mutation rate is one in one thousand million [per generation]), then 'in a species in which 1,000,000,000 come in each generation to maturity, a mutation rate of 1 in a thousand million will produce one mutant in every generation [on average], and thus establish the superiority of the new type in less than 250 generations, and quite probably in less than 10, from the first occurrence of the mutation, whereas, if the new mutation started with the more familiar rate of 1 in 1,000,000 the whole business would be settled, with a considerable margin to spare, in the first generation'.

In Fisher's calculations population sizes are very large. As we shall see in later chapters, Sewall Wright (1931) made major contributions to evolutionary theory when he examined the consequences of fragmentation and small population size on microevolutionary processes, especially the effect of chance events.

When the Evolutionary Synthesis was being established, paradoxically, direct evidence for natural selection was meagre. This is made very clear in a major critique by Robson & Richards (1936). They wrote: 'in Darwin's treatment of the subject no proof is adduced that a selection process has ever been detected *in nature*. Throughout the work such a process is *suggested* and assumed . . . in short, the proof is based on circumstantial rather than direct evidence'. They continue: 'on the evidence available at present natural selection has been accepted and its prestige created very largely on the desire for some such hypothesis'. Robson & Richards then consider the evidence available – eighteen case histories in animal groups – including the examples of crabs and *Mantis* cited at the beginning of this chapter. They conclude that 'there is a 'pathetic trust in observation *per se* . . . wholly inadequate data have sometimes been brought forward in support of the adaptive origin of certain examples of mimicry, protective coloration etc.' Overall they conclude that 'evidence for the occurrence of natural selection is very meager and carries little conviction'.

Coda

With the disappearance of the grounds for believing that speciation only occurred by 'mutation', and the demise of the theory of blending inheritance, a more unified science emerged. For those who wish to understand more of the historical context of the synthesis, Mayr & Provine (1980), Depew & Weber (1996) and Smocovitis (1996, 2006) should be consulted.

Since the establishment of the New Synthesis, our understanding of botanical evolution has been transformed, especially in the last few decades with the emergence of molecular approaches. At the time of the synthesis, a number of assumptions were made. According to Rose & Oakley (2007), biologists of the period saw an organism as having a 'well-organized library of genes', having 'single functions', that have been 'specifically honed by powerful natural selection'. Moreover, species were 'finely adjusted to their ecological circumstances'. Also 'the machinery of the cell' was considered to exhibit 'efficient design'. Furthermore, they envisaged simple 'trees of life' modelled on the basis of 'bifurcating graphs'. Throughout this book we will explore how new insights have emerged. For some the New Synthesis has been superseded: others write of an expanding or an extended synthesis (Rose & Oakley, 2007; Sapp, 2009; Pigliucci & Müller, 2010; Danchin *et al.*, 2011).

In the following chapters, we will examine how the patterns and processes of microevolution in plant populations have been explored, including investigations of the action of natural selection. Also, modern insights into plant speciation will be reviewed, by showing how models of gradual change and/or abrupt events have been tested. In addition, the ideas behind the Evolutionary Synthesis will be critically examined.

6 DNA: towards an understanding of heredity and molecular evolution

A revolution in biology is in progress – no less of a revolution than that caused by Darwin's theory of evolution by natural selection. The achievements of molecular biology have been so great and the progress so rapid that an enormous body of information now exists. In this chapter we focus on several key issues that are important in understanding later chapters.

- How the properties of nucleic acids provide the basis of heredity.
- Why molecular insights are changing our view of the mechanisms of evolution.
- How intensive molecular genetic studies of model flowering plant species (particularly *Arabidopsis thaliana* and a variety of crop plants) are revolutionising studies of many areas of plant biology.
- As the results of molecular investigations into these model and representative organisms for all the major groups are becoming available, do these point to a common origin for life on Earth? And what does the comparison of genomes of different groups indicate about evolutionary changes in gene order, chromosome number, the fate of duplicated DNA segments etc.?
- Finally, what are the key molecular techniques for studying pattern and processes in populations and species, and what properties of DNA have been exploited to allow, through sequencing and computational methods, spectacular insights into the evolution of the flowering plant branches of the Tree of Life?

Obviously, in this short book, we must be content to provide an outline of the ways in which this new knowledge of molecular biology illuminates these key questions. For comprehensive accounts of historical and recent developments in molecular biology, Clark (2005), Allison (2007), Craig *et al.* (2010) and Watson *et al.* (2014) should be consulted. For detailed accounts of our present understanding of the molecular biology of plants, see Smith *et al.* (2010), Jones *et al.* (2012) and Grotewold, Chappell & Kellogg (2015).

DNA: its structure and properties

The early geneticists visualised genes as 'beads threaded on a string'. In 1944 Schrödinger, in his fascinating book *What Is Life?*, suggested that genes could be complex organic molecules in which endless possible variations in detailed atomic structure could be responsible for codes specifying the stages of development. Since the1950s, spectacular progress has been made in our understanding of the structure of the hereditary materials and how they work. In the past it was considered that proteins might be the carriers of genetic information, but now it has been established that deoxyribonucleic acid (DNA) is the primary hereditary material.

The double helical structure of DNA was established by Watson & Crick (1953). For recollections of the circumstances leading up to the discovery, see Watson (1968). Other accounts consider in more detail the interaction of the scientists involved (Crick, 1988; Olby, 2009), including Wilkins and his colleagues at King's College, London, especially the role of Rosalind Franklin (Klug, 2004). Sayre (1975) and Maddox (2002) make the case that her personal contribution was, for a long time, not properly acknowledged.

Fig. 6.1a shows diagrammatically the sugar–phosphate 'backbone' of the two complementary

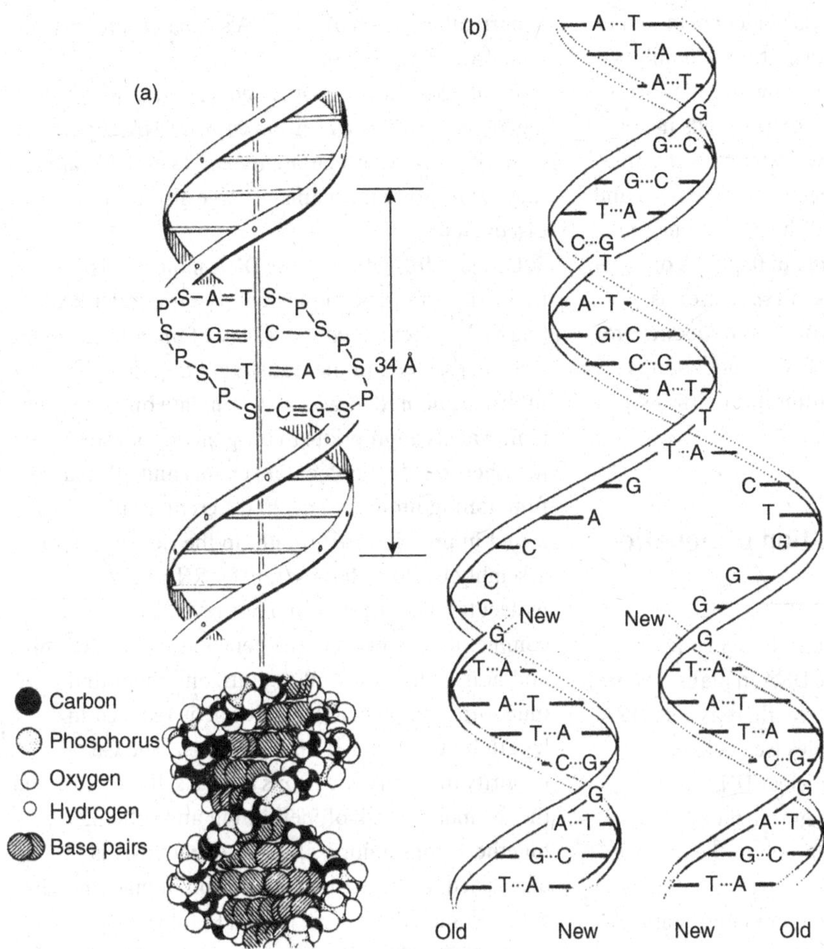

Fig. 6.1. The structure and mode of replication of DNA. (a) Structure of DNA, showing the complementary paired strands linked together by hydrogen bonds. Adenine (A) and thymine (T) always pair together, as do guanine (G) and cytosine (C). The backbone of the DNA molecule consists of the phosphate (P) and deoxyribose sugar groups (S). The DNA molecule is built up from many basic units called nucleotides, each of which consists of a base linked to a molecule of deoxyribose plus a phosphate group. (From Hayes, 1964; Berry, 1977.) (b) Replication of DNA as suggested by Watson & Crick. The complementary strands separate to produce a replication fork and each forms the template for the synthesis of a complementary daughter strand. (Reprinted by permission of Watson, J. D. *Molecular Biology of the Gene*, Menlo Park, California. The Benjamin Cummings Publishing Company, Inc., 1965.)

chains, which are held together by hydrogen bonding of nitrogenous bases. As indicated in the diagram, the pairing of these bases is highly specific. Four different bases are found in DNA: adenine pairs with thymine and cytosine with guanine.

DNA has a number of important properties, relating to its replication, transcription of genetic information, mutation, exchanges of DNA between molecules and DNA silencing. Each of these properties will be examined in turn (for recent reviews see Heslop-Harrison & Schmidt, 2007; Smith *et al.*, 2010). Historically, as DNA has been investigated, 'simple pictures' have been replaced by more nuanced understanding of the emerging complexities, and, at appropriate points in the book, further comments will be introduced concerning DNA and its role in cell functioning etc.

Replication of DNA

In the presence of the DNA polymerase enzymes, DNA is replicated without change, in a semi-conservative fashion, i.e. the bonds between the base pairs are broken at a replication fork (Fig. 6.1b). Each strand of the double helix acts as a template for synthesis of a complementary strand. There is evidence, in higher organisms, that replication begins at several points (replicons) in the long DNA molecule and that, from each point of initiation, replication proceeds in opposite directions along the two separated strands, until the replicons growing in each direction meet and fuse (Hardin, Bertoni & Kleinsmith, 2009). There may be as many as 35 000 such points in the DNA of *Vicia faba* (Lewin, 1990). The process of replication is carried out with great fidelity, with less than one error per 10^9 nucleotides added, and DNA replication is closely controlled occurring within the synthesis stage of cell growth and division.

Transcription and translation of genetic information

DNA, in its sequence of bases, provides a code of genetic information. Most of the DNA in plants resides in the nucleus, but, as we shall see, there is also DNA elsewhere in the cell, in organelles (see below). The coded information in the nuclear DNA, transcribed in the nucleus, is then translated into action outside the nuclear membrane in the production of a range of proteins by a process that assembles 20 amino acids in different combinations. In the transcription of information, DNA serves as a template for messenger RNA (mRNA), which differs from DNA in being usually single-stranded and in having ribose sugar in its nucleotides and the base uracil (U) instead of thymine (T).

Primary protein structure is specified by genes. At the molecular level a gene is a length of sequences of nucleotides of DNA. Starting from a fixed point, the code of the gene, represented by the base sequences, is read, with start and stop signals, as a series of non-overlapping triplets of nucleotides (codons). Almost all amino acids, the building blocks of protein, are specified by more than one triplet. Thus, there is 'redundancy' in the code (see Fig. 6.2). This figure also gives details of the three 'stop codons' that indicate the end of the amino acid sequence for which a 'particular stretch of the DNA sequence' provides the code (Aldridge, 1996).

In addition to the coding sequences, 'typically a gene is preceded by untranscribed DNA sequences, usually short, that regulate its expression' (Ayala, 2012), but other mechanisms have also been discovered.

Coded information in the DNA molecule is transcribed to the complementary messenger RNA (mRNA) molecule that moves out of the nucleus to the ribosomes. Here, in the translation of genetic information, mRNA acts as a template on which the amino acids appropriate to the genetic code are joined together to form polypeptide chains and, ultimately, functioning proteins. The 20 different amino acids found in proteins are brought to the ribosomes for assembly by 20 different transfer RNA (tRNA) molecules. Further research has revealed complexities, however. The gene has coding (exon) and non-coding (intron) regions, and the initial immature complete RNA molecule is reduced in length by the removal of the introns. 'The vast majority of eukaryotic exons are not self-splicing, and the formidable task of identifying and splicing together exons among all the intronic RNA is performed by large ribonucleo-protein machine, the spliceosome' (Craig *et al.*, 2010). Detailed studies have revealed an astonishing aspect of transcription: alternative splicing occurs. From an initial exon sequence, say ABCDEF, different mRNAs are produced – e.g. ABCD or ABCE – that specify different proteins. Considering the role of the introns, Aldridge (1996) stresses that 'by acting as spacers between exons, introns could have been the key to mixing and matching of exons to make new proteins with new functions. This exon shuffling is now thought to have been a major driving force in evolution.'

For a given cell or organism the total of all the RNA molecules produced is known as the transcriptome. But Adams (2008) notes the important point that 'in multicellular organisms nearly every cell contains the same genome and thus the same genes. However, not every gene is transcriptionally active in every cell – in

Second position

First position	U		C		A		G		Third position
U	UUU	Phe	UCU	Ser	UAU	Tyr	UGU	Cys	U
	UUC		UCC		UAC		UGC		C
	UUA	Leu	UCA		UAA	Stop	UGA	Stop	A
	UUG		UCG		UAG		UGG	Trp	G
C	CUU	Leu	CCU	Pro	CAU	His	CGU	Arg	U
	CUC		CCC		CAC		CGC		C
	CUA		CCA		CAA	Gln	CGA		A
	CUG		CCG		CAG		CGG		G
A	AUU	Ile	ACU	Thr	AAU	Asn	AGU	Ser	U
	AUC		ACC		AAC		AGC		C
	AUA		ACA		AAA	Lys	AGA	Arg	A
	AUG	Met	ACG		AAG		AGG		G
G	GUU	Val	GCU	Ala	GAU	Asp	GGU	Gly	U
	GUC		GCC		GAC		GGC		C
	GUA		GCA		GAA	Glu	GGA		A
	GUG		GCG		GAG		GGG		G

Fig. 6.2. The genetic code. The triplet codons given are those of the RNA, which provides the template for the production of the polypeptides from amino acids. The amino acids are given the standard three-letter abbreviations Met (Methionine), Val (Valine) etc. Note that almost all the amino acids are specified by more than one triplet. Stop codons mark the end of a sequence specifying a particular protein. (From Kendrew, 1994. Reproduced with permission of Blackwell Publishing.)

other words, different cells show different patterns of gene expression. These variations underlie the wide range of physical, biochemical, and developmental differences seen among various cells and tissues.' Control of the expression of genes is crucial as higher plants develop as multicellular organisms with specialist tissues, and such control, through mechanisms of gene activation and silencing, also provides the flexibility to adapt to impacts such as cell damage and disease, as well as providing the means whereby plants control the timing and amounts of gene products in response to changing signals from the environment.

Watson *et al.* (2014) give a detailed account of the molecular steps in the transcription process, and outline the various methods used to investigate the connection between genome and gene function, including making libraries of transcribed sequences, the use of microarrays of DNA probes etc.

Regarding RNA, it is important to note that there are several types of RNA molecule in the cell. There are noncoding RNAs of different length: long molecules; short RNAs (less than 200 nucleotides long); and microRNAs (18–24 nucleotides long). MicroRNAs (miRNA) are produced from longer RNA molecules by trimming, and these molecules act as repressors of gene expression, playing a key role in the fine-tuning of the activities of the cell through regulation of development, differentiation and communication, etc. (Wang *et al.* 2007). In addition, small interfering ribonucleic acids (siRNAs) are important, in many contexts, in gene silencing (Meyer 2007), for instance, in defence against infection. Balcombe (2011) notes that 'plants, like animals, have developed mechanisms to ward off disease and "remember" infections'. Thus,

> plant cells recognise the foreign genetic material of the virus, copy a section of the viral DNA into siRNA and use it as a 'specificity determinant'. The siRNA binds to the viral genetic material and, rather like hoisting a molecular flag to identify the marauder, causes a protein argonaute to bind and stop the virus from working. Just as our immune system can be primed by an infection so that we can fight it off quicker next time, plant cells retain the siRNA as a means of escalating the defence response next time the plant sees the invader.

Mutation of DNA

In the growth and development of the cells of a plant, each cell usually receives an exact copy of the hereditary material. This is achieved through a 'proof-reading' mechanism that operates to correct errors (Clark, 2005; Allison, 2007; Craig *et al.*, 2010). Where damage occurs this is recognised by enzymes and repaired. Undamaged sequences on homologous chromosomes or on the complementary strands of DNA are crucial for such repair. If damage is not repaired then the cell may die.

However, despite the mechanisms acting to perpetuate the DNA in an unchanged form, significant permanent change in base sequence and other elements of the DNA molecule may occur. (For a technical account of the changes at the molecular level, see Watson *et al.*, 2014.)

Such changes may be spontaneous or artificially induced by chemicals, ultraviolet radiation (nonionising radiation) and other sources of radiation. In a study of mutation in *Arabidopsis thaliana*, Ossowski *et al.* (2010) identified '99 base substitutions and 17 small and large insertions and deletions', which implies a 'spontaneous mutation rate of 7×10^{-9} base substitutions per site per generation'.

A number of different types of mutational change affecting the base sequences have been recognised (Fig. 6.3). In addition, mutation may occur through reciprocal chromosome exchange of segments, errors in replication and the insertion of 'mobile DNA elements' (see below). Mutations have several effects.

a Original sequence
.. ATG GTC CTC AGC ATA GCT TAT AGC ...
.. Met Val Leu Ser Ile Ala Tyr Ser ...

b Point mutation (missense)
.. ATG GTC **G**TC AGC ATA GCT TAT AGC ...
.. Met Val **Phe** Ser Ile Ala Tyr Ser ...

c Insertion leading to frameshift with premature termination
.. ATG GTG CTC AGC ATA GCT TA**T** TAG C ...
.. Met Val Leu Ser Ile Ala Tyr **STOP**

d Insertion leading to frameshift with altered amino-acid sequence and premature termination
.. ATG GTG C[**GA TAT CTC TGT GT**]T CAG CAT AGC TTA TAG C ...
.. Met Val **Arg Tyr Leu Cys Val Gln His Ser Leu STOP**

e Deletion leading to frameshift
.. ATG GT[G CTC AGC ATA G]CT TAT AGC ...
.. ATG **GTC TTA TAG C** ...
.. Met **Asp Leu STOP**

f Silent mutation no change
.. ATG GTG CT**A** AGC ATA GCT TAT AGC ...
.. Met Val Leu Ser Ile Ala Tyr Ser ...

Fig. 6.3. The diagram shows the effect of different types of mutation on an original sequence of triplets of DNA (top line), each coding for a different amino acid (second line). The amino acids are given the standard three-letter abbreviations Met (Methionine), Val (Valine) etc. (a) Original sequence. (b) Point mutation resulting from the substitution of one base by another. (c–e) Mutation in the reading frame of the triplets caused by the insertion or deletion of bases. (f) Because of the redundancy of the code, a base change may lead to a silent mutation as it does not result in a change in the amino acid sequence. (From Kendrew, 1994. Reproduced with permission of Blackwell Publishing.)

Many are deleterious, and a further group may be neutral or near neutral in effect and accumulate in the genome. Other mutations are, perhaps rarely, advantageous.

As we see below, some DNA in the genome of plants has no discernible function, and the effects of mutation, now outlined, apply to coding sequences etc. of the DNA.

With reference to an original sequence of bases, a point mutation may occur, resulting in an altered codon and the insertion of a different amino acid into a polypeptide chain. Also, as the coding bases are read as non-overlapping triplets, inversion, addition and deletion of bases will cause a mutation by altering the codons. However, given the redundancy of the code, the change of a nucleotide may sometimes produce a 'silent' mutation, and no change to the amino acid sequence results. While mutations occur randomly in the genome, there is evidence that they are not 'evenly distributed': there are 'mutation hot spots' (see Alexander, 2011).

Animal geneticists make a distinction between mutations in somatic cells and those that occur in the dedicated 'germline' that produces the reproductive cells. Germline mutations may occur and appear in the progeny of the next generation. In contrast, plants do not have a dedicated germ line. Somatic mutations occur in plants, and individuals may have phenotypically different mutated tissues. Somatic mutations may be more evident in very long-lived species, for example, in trees or clonally propagating individuals (Schaal, 1988). For such mutations to be passed to the next generation depends upon whether such mutations occur in the cells producing the gametes, but many other factors are important including the fitness of the progeny.

Exchanges of DNA segments

Another property of DNA, of crucial importance in evolution, is the existence of mechanisms to enable reciprocal exchanges of segments of nuclear DNA. There is evidence that the unreplicated chromosome of higher organisms contains a single very long DNA molecule, forming a complex structure with nuclear proteins, in particular histones. Thus, the DNA is wrapped around histones to form nucleosomes and, in arriving at its most contracted form in nuclear division, these structures are folded, looped and supercoiled (Smith *et al.*, 2010). As we have seen in an earlier chapter, mitosis is a process that allows orderly assignment of the identical products of DNA replication to daughter nuclei. In contrast, the process of meiosis permits the reciprocal exchange of DNA segments, as crossing over takes place between the homologous pairs of chromosomes. Evidence of such exchanges is provided by the chiasmata seen at the diplotene stage (see Fig. 4.9). It is beyond the scope of this book to examine the evidence for the molecular basis of crossing over: for full details of the most plausible model of the process – the junction model devised by Holliday – see Watson *et al.* (2014).

Between nuclear divisions the chromosomes de-condense; loops of chromatin become active in gene expression. Evidence suggests that the chromosomes take up non-random positions within the nucleus. In the fully condensed state, a chromosome may be of the order of c.2–15 μm. Astonishingly, if fully extended, the DNA is in the range c.7–300 mm long (Heslop-Harrison & Schmidt, 2007). How metres of DNA are packed *in vivo* within the nucleus is still unclear.

Gene silencing: epigenetic modification of DNA

Molecular studies have clarified another type of genetic change, namely gene silencing through epigenetic modification. These modifications are 'reversible', and do not alter the primary gene sequence. Support for a number of models of epigenetic silencing has come from a range of experiments. In brief, the accessibility of the gene may be 'blocked' by chemical changes to chromatin, which, as we have seen, is an intimate association of protein and DNA. Epigenetic silencing may also occur following methylation, which involves the addition of methyl groups to cytosine or adenine nucleotides in DNA (He, Chen & Zhu, 2011). These two mechanisms change the accessibility of the DNA to the molecular complexes that achieve transcription. In addition, post-transcriptional silencing may be achieved by another mechanism: the activities of small interfering ribonucleic acids (siRNA) that degrade the transcript (Meyer, 2007). These mechanisms of silencing and control are at the heart of the control of gene expression.

In some cases, there is evidence that epigenetic changes may pass down the generations. Such changes involve 'heritable inactivation of gene expression that does not involve any change in the deoxyribonucleic acid (DNA) sequence' (Meyer, 2007).

A particularly interesting example of epigenetic inheritance, involving methylation of DNA, relates to the peloric variant of Toadflax, *Linaria vulgaris*, described in Chapter 2. Jablonka & Lamb (2010) give the following details:

> Over 250 years ago, Carl Linnaeus described a morphological variant, Peloria, characterised by flowers that are radially rather than bilaterally symmetrical. Later generations of botanists assumed that Peloria was a mutant form. However, when Cubas and his colleagues (1999) studied *Lcyc*, the *Linaria* version of the *cycloidea* gene that in related species is known to control dorsoventral asymmetry, they found that the DNA sequences of the normal and peloric forms were identical. What is different is not the DNA, but the pattern of methylation: in the peloric variant, part of the gene is heavily methylated and transcriptionally silent. In other words, the Peloric phenotype is the result of an epimutation, not a mutation. Peloric strains are not totally stable, and occasionally branches with partially or even fully wild-type flowers develop on peloric plants, but it has been shown that the epigenetic marks on *Lcyc* are transmitted to progeny for at least two generations.

Cross-generational effects have been discovered in other situations (Weigel & Colot, 2012), for example, the epigenetic induction of plant defences. Agrawal, Laforsch & Tollrian (1999) report that when caged plants of Wild radish (*Raphanus raphanistrum*) are attacked by caterpillars of *Pieris rapae*, extra defensive mustard oil glycosides and more leaf hairs are produced in comparison with control plants not subject to such herbivory. Intriguingly, when seed is grown from plants attacked by caterpillars (with seed from control plants for comparison), these induced defences persisted in the next generation. In exploring alternative possible explanations, evidence suggested that persistent differences were not the result of changes in seed mass, or resources, relative to those in the control plants. Since these very interesting experiments, about a dozen cases of naturally occurring epialleles have been reported (Weigel & Colt, 2012); epialleles, in each of the cases, were genetically identical, but exhibited variability in gene expression as a consequence of the degree of DNA methylation.

Considering epigenetic phenomena, Meyer (2007), concludes that such 'pathways play a crucial role in the control of viruses and mobile elements', and we emphasise again that they have a significant role in plant development. In addition, Jablonka (2013) considers that epigenetic changes may be important in the generation of phenotypic variation.

This possibility was examined by Becker *et al.* (2011) and Schmitz *et al.* (2011) with inbred *Arabidopsis thaliana*, propagated by single seed descent for 30 generations in a greenhouse experiment. Commenting on these experiments Jablonka (2013) concluded that 'since the lines all had the same DNA (barring a few possible mutations and rare transpositions), and since there had been no change in the environment in which the plants lived, the results tell us about the frequency with which inherited variations in methylation occur'. Thus it was discovered that 'transgenerational epigenetic variation in DNA methylation may generate new allelic states that alter transcription, providing a mechanism for phenotypic diversity in the absence of genetic mutation'. Epigenetic changes, which are stably inherited through meiosis, could have important implications for phenotypic and developmental plasticity in the wild.

In the last few years, other aspects of epigenetics have been studied (Grant-Downton & Dickinson, 2006; Weigel & Colot, 2012). As we see in later chapters these investigations have extended our understanding of many aspects of variation and evolution in hybrid and polyploid plants (Lui & Wendel, 2003; Lui *et al.*, 2009; Schatlowski & Köhler, 2012; Hegarty *et al.*, 2013).

The plant cell: adaptive, neutral and junk DNA

Alvarez-Valin (2002) points out that 'until the late 1960s, all differences between species or among members of the same species were understood in terms of their contribution to the fitness of the organism or species bearing them. This view, known as the synthetic theory or neo-Darwinism, claimed that even the most minute features of organisms contribute to their adaptation. It seemed reasonable that the synthetic theory could easily be extended to the molecular level. In other words, if observable phenotypic differences were due to adaptation, differences at the molecular level should be as well.' However, reflecting on the early findings of molecular biologists, this interpretation was challenged by Kimura (1983) and others, who have proposed a neutral theory of evolution.

> The neutral theory asserts that the great majority of evolutionary changes at the molecular level as revealed by comparative studies of protein and DNA sequences, are caused not by Darwinian selection but by random drift of selectively neutral or nearly neutral mutants. The theory does not deny the role of natural selection in determining the course of adaptive evolution, but it assumes that only a minute fraction of DNA changes in evolution are adaptive in nature, while the great majority of phenotypically silent molecular substitutions exert no significant influence on survival and reproduction and drift randomly through the species. (Kimura, 1983)

Biologists are now content with the notion that random neutral genetic changes are part of the evolutionary picture (Cockburn, 1991; Alvarez-Valin, 2002), but it is difficult to determine the relative contribution of random drift and selective forces. Jablonka & Lamb (2010) review another crucial related topic. There is much debate about whether genes/molecules; cells; organisms and populations are all 'units of selection', or is one level paramount?

Regarding DNA in a great variety of organisms including plants, another important concept has been developed. Much of the DNA appears to be without discernible function – so-called junk DNA.
The background to this idea comes from studies of genome size. Thus, there is considerable variability in genome size or nuclear DNA content in different species (Heslop-Harrison & Schmidt, 2011). For instance, *Fritillaria assyriaca* has a very large genome size at *c.*130 000 Mbp (million base pairs), while *Genlisea* has 70 Mbp. These values were obtained by staining nuclei and estimating the DNA content, using a microdensitometer.

Further evidence is now becoming available.
In 2000, the first more or less complete genome for a plant was published, when a large team of researchers collaborated to determine the DNA sequence in *Arabidopsis thaliana*. Now, the genome of Rice has been sequenced and there is a wealth of information from other flowering plant (angiosperm) species. Heslop-Harrison & Schmidt (2007) list Alfalfa, Barley, Cabbage, *Citrus*, Grape, Lucerne, Maize, *Medicago*, Poplar, Potato, Rape/Canola, Sorghum, Soybean, Sugar Beet, Tomato and Wheat relatives. In 2011, the genome of the Potato was successfully sequenced (Potato Genome Sequencing Consortium, 2011), and other groups are also being

investigated. In addition, sequence data are accumulating for other groups, e.g. gymnosperms, bryophytes (Mosses & Liverworts) and pteridophytes (Ferns). (There are informative websites on the *Arabidopsis* genome (www.arabidopsis.org) and other genome projects (The Institute for Genomic Research, TIGR (www.tigr.org)).)

In celebrating the tenth anniversary of the completion of the *Arabidopsis* genome sequence, Feuillet *et al.* (2011) note that only for Rice has the 'genome sequence been finished to a quality level similar to that of the *Arabidopsis* sequence'. For other species the trend has been to produce 'draft genomes', and this 'could affect the ability of researchers to address biological questions of speciation and recent evolution or link sequence variation accurately to phenotypes'.

However, recent technical advances have resulted in the development of the so-called 'next generation sequencing methods'. These have further reduced the cost and accelerated the processes, making it potentially possible to undertake detailed investigations of a wide variety of plant species, even those with the largest genomes (Kelly & Leitch, 2011).

What have these genomic studies revealed? Sequencing the DNA in detail has confirmed that there is considerable variability in genome size and nuclear DNA content (Heslop-Harrison & Schmidt, 2011), and there is 'no simple relationship between the apparent complexity of an organism and the amount of DNA in its genome' – a phenomenon known as the C-value paradox (Aldridge, 1998). Thus, in plants, the genome of *Arabidopsis* is 157 Mbp with about 27 000 genes on five chromosome pairs. In contrast, the genome of Rice is 450 Mbp with 41 000 genes on 12 pairs of chromosomes (the predicted number of genes in both species has increased since they were first estimated: figures from Sterck *et al.*, 2007). These findings, together with the results of other molecular investigations, offer support for the notion of junk DNA in plant cells. However, Heslop-Harrison & Schmidt (2007) point out that these sequences, which might make up as much as 50–75% of the genome, have important structural roles in chromosome functioning (see further comment below).

DNA in the nucleus

About 80–90% of the DNA of the genome is located in the nucleus. In addition to the nuclear genomic elements, DNA is found in mitochondria and plastids. Also, it is very important to recognise that, typically, plant cells include other genomes – of virus, bacteria, fungi etc. These exist in a variety of relationships, including parasitism, infectiveness, symbiosis etc. (Dupré, 2010).

In characterising the DNA of the nucleus, Heslop-Harrison & Schwarzacher (2011) conclude that there are 'single- or low-copy coding sequences, introns, promoters and regulatory DNA sequences, but also various classes of repetitive DNA motifs that are present in hundreds or even thousands of copies in the genome'.

Some of these repetitive elements are in the form of tandemly arranged DNA motifs that occur adjacent to each other. Those whose repeats are in the region of 150–170 nucleotides are so-called satellite DNA. Those with much shorter repeats have been classified as microsatellites or simple sequence repeats (SSRs) (typically 1–5 bp) and minisatellites (10–50 bp). Microsatellites provide valuable genetic marker systems (see below).

As well as the tandemly repeated sequences, scattered repetitive elements occur throughout the genome and on all or most chromosomes. These are called transposons or mobile elements. Transposons and their degraded products may represent half or more of the entire DNA present in the genome.

Two broad classes of transposons have been recognised.

a) *Retrotransposons* (so-called because of their similarity to retroviruses). Adopting the metaphor of word processing, these elements may be visualised replicating by 'copy and paste' through an RNA intermediary to produce new DNA copies that can inset at new locations in the genome. They are only able to relocate following replication, by which means they increase in number.

b) *DNA transposons.* In contrast, this class of transposons can be visualised as 'cut and paste' elements. Excision from one site and insertion at another is accomplished by a specific enzyme (transposase), which is coded within the transposon itself. This group of transposons does not accumulate in such numbers as the copy and paste variant. Insertion may lead to a phenotypically visible mutation. If the transposon is subsequently excised, reversion may be observed.

An example of particular historical interest, of how a transposable element can cause a mutation, concerns the round/wrinkled peas investigated by Gregor Mendel (see Chapter 4). Graur & Li (2000) provide the following commentary on experiments by Bhattachayya *et al.* (1990) and Bhattachayya, Martin & Smith (1993). They report that

> the wrinkled variant arose through a mutation at the *rugosus* locus by the insertion of [a] transposon into the reading frame of a gene encoding a starch-branching enzyme. Due to the inactivation of the gene, the total amount of starch and the proportion of amylopectin (branched starch) are greatly reduced in homozygotes, while the amount of sucrose is increased. Increased sucrose in seeds causes a greater uptake of water, thereby increasing seed size. During seed desiccation, these seeds lose more water than seeds from plants possessing a functional starch-branching enzyme, resulting in a wrinkled pea.

Centromeres

These structures have distinctive staining properties, which do not de-condense between nuclear divisions (interphase), and provide attachments for the spindle proteins, as chromosomes move to opposite poles during mitosis and meiosis. Thus, they play a key role in nuclear and cell division. Associated with a particular protein, histone 3, centromeres are heterochromatic in staining and behaviour (see Chapter 5). Overall, there is considerable variability in DNA sequences in the centromere regions. While centromeres have some active genes, there are also long arrays of repetitive tandemly repeated non-coding DNA sequences (sometimes called satellite DNA) and large numbers of retrotransposons. Whether precise sequences of this repetitive DNA are necessary for the proper functioning of centromeres has long been an intriguing question. Now, there is increasing evidence that 'centromere position and function is not defined by the local DNA sequence context but rather by an epigenetic chromatin-based mechanism' (Valente, Silva & Jansen, 2012).

Telomeres

Typically in angiosperms, the ends of chromosomes are protected by hundreds to thousands of copies of a short repeat sequence ({TTTAGGG}n in *Arabidopsis* and most other flowering plants). Such structures play a vital role, as unprotected ends of DNA strands from different chromosomes could pair together leading to a tangled mass that could not dissociate properly at nuclear division. The terminal heterochromatic knob on certain chromosomes in maize has been shown to be composed of sequences of repetitive DNA (Laurie & Bennett, 1985).

B chromosomes

What follows draws on the excellent reviews of Camacho (2005), Jones, Viegas & Houben (2008) and Houben, Nasuda & Endo (2011). First reported in animals about a hundred years ago, in the 1920s, B chromosomes, sometimes referred to as supernumeraries, were detected in Rye and Maize. They are now reported from *c.*1500 species of plants, but they may not be found in all individuals of a species within a population. In a survey of 979 species, Bs were present in about 8% of monocots, but only 3% of eudicots. The survey also suggests that Bs might be more common in groups with larger genomes. Bs are not necessarily present in the same numbers in related groups. For example, B chromosomes are present in 27.2% of the species in the Commelinales, but only 4.3% of species of the sister group order Zingiberales have Bs. Where they are present Bs may have a significant impact on genome size. For instance, in Rye the chromosome set is 8114 Mbp. Where 4Bs occur, they contribute an additional 3200 Mbp.

B chromosomes arise from the A set, but do not pair with them. At meiosis, Bs often survive, and their numbers may increase. Various hypotheses have been advanced as to how Bs could originate from the A set. They could be formed from a single or successive events. They could arise from trisomics; from unequal translocations between chromosomes; or be the result of cytological events following hybridisation. In one recent molecular study, of *Brachycome dichromosomatica*, it was discovered that the Bs were not simply individual pieces of DNA from one source, but were 'built up' from tandem repeat sequences of different A chromosomes (Houben *et al.*, 2001).

Reflecting on present knowledge, Jones, Viegas & Houben (2008) conclude that 'despite many attempts ... there is as yet no convincing argument that Bs have any selective advantages except under certain severely drastic conditions' (Rees & Hutchinson 1973; Holmes & Bougourd, 1991). What is clear is that high numbers may be deleterious to fertility. It remains an open question whether Bs carry genes that are active and whether they are in some measure a handicap to the individuals that possess them. There are many classic studies of B chromosomes and Jones, Viegas & Houben (2008) emphasise that the time is ripe to revisit these case histories, with 'new ideas and new investigative tools'. Considering the relationship of B chromosomes to the A set, they conclude that, on present evidence,

> B chromosomes are the ultimate genome parasites, occupying the nucleus of their host and exploiting all the nuclear machinery needed for their replication and transmission although, in some cases, they have a different replicative chronology to that of the A chromosomes. They differ from transposons, and other forms of selfish DNA, in that they are autonomous elements, and can vary their transmission rate, their number within individual plants and their frequency in natural populations. They have the capacity to spread themselves, and to optimize their survival strategy, but within certain constraints imposed upon them by their hosts. (Jones, Viegas & Houben 2008)

The concept of junk DNA revisited

Important new research on the human genome has now been published. The notion that 98% of the DNA is 'junk' has, for some, been seriously undermined by the findings of the recent Encode project involving 440 researchers in 32 institutes across the world (www.genome.gov). Their results, published in more than 30 papers in prestigious scientific journals, provide evidence that 80% of human genomic non-coding sequences can now be assigned a role in biochemical functioning, especially in the regulation of the relatively small number (*c.*20 000) of protein-coding genes. Faced with this evidence, some have concluded that the concept of junk DNA should be abandoned, but others are not yet convinced that the issue is fully resolved (see for example Graur *et al.*, 2013 and references cited therein). In addition, more research is necessary to clarify the control of gene regulation in plants, especially to investigate the functional significance of very different genome sizes in related species. Eddy (2012) notes that: 'there are many examples of related species in the same genus that have haploid genome sizes that differ by three- to eight-fold; this is particularly common in plants, as seen in species of Rice (*Oryza*), *Sorghum*, or Onions (*Allium*). The Maize (*Zea mays*) genome expanded by about 50% in just 140,000 years since its divergence from *Zea luxurians* (and not merely by polyploidization)'.

DNA in plant cells: mitochondria and chloroplasts

In addition to the nuclear genetic apparatus of plant cells that we have just outlined, non-nuclear genetic elements must now be considered. Historically, it was supposed that cell organelles might have originated from nuclear genes by fragmentation. However, on the basis of electron microscopy, biochemical and genetic studies, and especially through molecular investigations, overwhelming evidence suggests a different origin: several of the key elements of the eukaryotic cell, including mitochondria and chloroplasts, originated through endosymbiosis. This idea, first suggested by Schimper in the 1880s (see Howe, 2007), was subsequently promoted by Mereschkowsky (1905) and Margulis (1971). They proposed that a proto-eukaryotic cell engulfed a free-living prokaryotic cell that was subsequently retained to mutual symbiotic benefit. Thus, mitochondria are the result of endosymbiosis, most likely, alpha-proteobacteria, whose closest-living relatives are members of the *Rickettsia* group. To put this event in historical context, it is envisaged that about two billion years ago there were three lineages of organisms: bacteria, Archaea (single-celled micro-organisms) and primitive eukaryotes, the latter having received 'a major influx of genes' through 'endosymbiosis' (Lake & Rivera, 1994).

Then, later, perhaps just over a billion years ago (Lake & Rivera, 1994), a secondary event occurred in which a photosynthetic bacterium (cynobacterium) was engulfed by a mitochondria-containing cell (the product of primary endosymbiosis), and, through endosymbiosis, became the plastid, that now occurs in a range of forms: chloroplasts, coloured chromoplasts etc. This photosynthetic ancestor produced lineages with different types of plastid: red algae and green algae/land plants (McFadden & van Dooren, 2004). Here, we are principally concerned with higher plants, but note that, in the evolution of other groups – euglenoids, dinoflagellates etc. – there is evidence for further, subsequent, endosymbiotic events.

Concerning the proposed monophyletic origin of plastids, Stiller, Reel & Johnson (2003) provide counter-arguments. They consider that the endosymbiosis that led to the evolution of chloroplasts might have occurred many times (i.e. as polyphyletic), with selection pressures resulting in convergent evolution in genetic, biochemical and ultra-structural traits. Howe (2007) and Veseg, Vacula & Krajcov (2009) provide critical overviews of the evidence, pro and con, with Howe questioning whether the research tools available permit the accurate reconstruction of events that occurred of the order of 10^9 years ago. However, the concept of the endosymbiotic origin of organelles is very strongly supported, especially by molecular studies, as organelle DNA is bacteria-like, and without the histone proteins of nuclear DNA (Vesteg, Vacula & Krajčovič, 2009; Smith *et al.*, 2010).

In the course of time the organelle genomes were fully integrated into the evolving cell biology by loss of genes – some were redundant, others were transferred. Thus, chloroplast genes have been transferred to the nucleus, some to mitochondria. In addition, there are large sections of the mitochondrial genome in the nucleus (Heslop-Harrison & Schmidt, 2007). As a consequence, plastids and mitochondria have become obligatory symbionts, incapable of independent existence outside the host.

Loss of chloroplast genes can arise in other circumstances, e.g. in the evolution of non-photosynthetic parasitic species such as *Epifagus* (De Pamphilis & Palmer, 1990).

Evolutionary changes in the chloroplast structure

The chloroplast genome is in the range 120–200 kbp. 'In many species, part of the molecule is repeated in an inverted configuration, with single copy regions of unequal size, referred to as the small and large single copy regions, separating the repeats' (Howe, 2007). Genome organisation of the chloroplast is highly conserved. However, there have been changes in structure in different groups of plants. 'These include expansion and contraction of the inverted repeats, rearrangements and sequence evolution at the nucleotide level.' Howe then emphasises a key point, that 'sequence evolution at the nucleotide level, including substitutions, insertions, deletions and alterations in microsatellites is often used in phylogenetic analysis' (for up-to-date information on chloroplast genomes see http://megasun.bch.umontreal.ca/ogmp/projects/other/cp_list.html).

Genetics of chloroplasts and mitochondria

From the early twentieth century onwards non-Mendelian and cytoplasmic inheritance, often involving leaf colour and variegation, have been reported (see Chapter 4). Molecular studies have greatly increased our understanding of the puzzling cases: they are explicable in relation to the behaviour of mitochondrial and chloroplast genomes. Self-replicating closed circles of DNA occur in the chloroplast (cpDNA) and in the mitochondria (mtDNA). The DNA present in these organelles expresses its genes. The genetics of these DNAs will depend upon how replicas, whose division is not linked with the cycle of cell division, are distributed as proto-organelles. In a cross, are they inherited both from the male and the female side? Investigations have revealed different behaviour. For instance, in conifers inheritance of plastid transmission is paternal (Schaal, O'Kane & Rogstad, 1991).

In flowering plants inheritance is often maternal, but, in some cases, it is paternal or biparental (see reviews by Harris & Ingram, 1991; Olmstead & Palmer, 1994; Raubeson & Jansen, 2005). Howe (2007: 582) stresses the important point that where chloroplast genes are maternally inherited, they 'are effectively in an asexual gene pool'.

Arabidopsis thaliana: its role as a model species

The sequencing of the *Arabidopsis* genome was a highly significant milestone in the history of plant biology (Arabidopsis Genome Initiative, 2000). Now, other plant genomes have been sequenced and many new insights into plant evolution are being discovered. Intensive studies of *Arabidopsis thaliana*, and certain crop species, have been undertaken using a wide range of molecular, genetic, biochemical and physiological techniques. This focus on model species has led to unprecedented advances in our understanding of plant development, variation and evolution. Here we briefly introduce a number of key advances relating to the major themes of the book.

Arabidopsis thaliana has a number of properties that make it a good choice as the first model plant (Koornneef & Schere, 2007). It is a small regularly self-fertilising member of the Brassicaceae (chromosome number $n = 5$) found in ruderal habitats in the Northern Hemisphere. Native in many areas, the plant is likely to have been introduced into North America and Japan, as well as Australia. This species can easily be grown in climate chambers, where it completes its life cycle in six to eight weeks, allowing six experimental generations per year. In contrast, in the 'wild', *Arabidopsis* typically (but not universally) reproduces by a single generation per year. This is because some stocks have seed dormancy: many also require vernalisation before they will flower (see below for further details).

The morphology and genetics of a great variety of different variants of *Arabidopsis* have been closely studied, and a wide range of experimental material from different geographical areas has been collected together in centres in the USA and UK (The Arabidopsis Information Resource: TAIR: www.arabidopsis.org) and Nottingham Arabidopsis Stock Centre, which holds more than 300 000 different stocks of seed representing over half a million genotypes (www.arabidopsis.info). The different lines selected for investigation in *Arabidopsis* are often referred to as 'ecotypes' (see Chapter 8). Weigel (2012) prefers the neutral term accession, as the word ecotype implies, perhaps wrongly, that each line 'has a unique ecology and is adapted to specific environments, as opposed to differing only in genotype from other varieties'.

Genetic maps of *Arabidopsis* have been prepared: e.g. in 1983 chromosome maps with 76 morphological markers were published. Many 'molecular' markers have since been added. The chromosomes are relatively small, but the extended pachytene chromosomes (in early meiosis) have been used for detailed cytogenetic studies.

While a small amount of outcrossing has been detected in plants visited by insects, *Arabidopsis* grown under controlled conditions is regularly self-fertilising, and, by selfing and 'single-seed descent', inbred lines can be generated to produce near isogenic lines (NILs) and, in crossing experiments, recombinant inbred lines (RILs) (Weigel, 2012). In addition, Weigel notes that recent advances have made it possible to bypass 'the lengthy inbreeding process' by producing 'doubled haploid plants from recombinant populations' (see Ravi & Chan, 2010).

In the study of morphological and functional genetics, induced mutations also play a pivotal role, and these have readily been produced in *Arabidopsis* – by irradiation, or chemical treatment of the developing flowers, such as ethyl methanesulfonate (EMS). In the next generation, selfing leads to the segregation of newly formed recessive mutants. The properties of mutants are assessed by screening under specific experimental conditions, following protocols developed in many laboratories (see Koornneef & Schere, 2007 for details); altered gene function can be linked to altered DNA sequence, and clones can be produced from specific DNA sequences of interest.

In essence, molecular cloning involves several steps. A DNA fragment is generated using enzymic digestion, and this is combined, using DNA-ligase, with sequences necessary for replication and identification, to produce a vector. Transfectation to a suitable active cultured cell system then follows. The DNA of interest is subsequently identified by the genetic markers built into the vector construct. For example, in many protocols the cell medium contains an antibiotic and only those cells containing the appropriate antibiotic resistance markers – from the vector – will grow. In other cases colour markers are employed. Confirmation of the success of the process is the next step, using PCR restriction fragment analysis and/or DNA sequencing. (For full details of molecular cloning in plants, see Green & Sambrook, 2012.)

As a model plant species *Arabidopsis* has another advantage: it can be genetically transformed by the introduction of novel or altered genes using a variety of techniques. The 'novel' gene construct contains appropriate sequences plus marker elements to enable the transformed products to be identified from amongst unaltered cells. The cell walls of plants represent an obstacle to the simple introduction of constructs. Transformation may be achieved by employing the natural virulence systems of the bacterium *Agrobacterium tumefaciens*, the agent of crown gall (for details see Hooykaas, 2007). In other species different techniques have been devised, involving the direct treatment of protoplast from which the cell walls have been enzymically degraded. Or genetic constructs can be introduced through cell walls damaged by micro-laceration, e.g. by agents such as helium gas, electric discharge, the vortexing of cells with silicon carbide fibres, or by particle guns. *Arabidopsis* transformation can also be achieved using vacuum treatments (Crouzet & Hohn, 2007).

Studying developmental processes: the role of model plants

Through integrated studies involving molecular genetics, biochemistry, physiology etc. major insights have been obtained that have an important bearing on plant variation and evolution. For example, investigations of variation in *Arabidopsis* have included studies of copper tolerance, disease resistance, drought responses, flowering time, seed dormancy, sodium accumulation, trichome density, zinc responses etc. Techniques have been devised to identify, map, and validate the causal genes/polymorphisms, and examine shared ancestry through the use of Genome-Wide Association Techniques (GWAs) (see Weigel, 2012; Colautti, Lee & Mitchell-Olds, 2012). In these investigations, comparison of inbred lines with and without various traits has led to the identification of candidate genes (Atwell *et al.*, 2010) and in some cases these genes have been cloned.

At appropriate points in the book, the power of these new integrative approaches will be illustrated in discussing breeding systems, regional variation in plants, abiotic and biotic environmental effects and phylogenetic investigations. In addition to examining how the cell works at the molecular level, biologists are now testing hypotheses that cut across what, in the past, have often been separate specialisms. They are working towards an integrated understanding of the molecular genetic basis of variation between the different major groups of organisms. They are also attempting to test models that investigate the molecular/genetic mechanisms underpinning differences in morphology, metabolism, physiology, developmental responses to the changing environment, pests/diseases, geographical, ecological, and phylogenetic/evolutionary relationships.

While the model plant species, selected for intensive study, have shed light on many fundamental biochemical, physiological and developmental processes, some fascinating developmental questions remain to be investigated, for example, the phenomenon of heteroblasty, where there is an abrupt change in form and function during plant development. Such heteroblastic behaviour is found in *Hedera helix* (see Fig. 4.15). Early investigations reported that if the hormone gibberellic acid was applied to the 'adult phase', reversal to 'juvenile morphology' occurred (Doorenbos, 1965). In an important review of the subject, Zotz, Wilhelm & Becker (2011) point out that none of the species so far selected as model species have the pronounced phase changes typical of heteroblasty and many of the species exhibiting the phenomenon, such as *Hedera helix* (a long-lived woody species), would not be ideal model species. Thus, the underlying genetic, hormonal and developmental control of heteroblasty remains to be investigated.

Another question has continued to intrigue plant biologists (Cooke, 2006), namely whether 'complex biological patterns' are 'governed by simple mathematical rules', such as the Fibonacci number sequence, discussed in Chapter 3. In his comprehensive review, Cooke considers the properties of the Fibonacci sequence (1,1,2,3,5,8,13 . . . where successive numbers are the sum of the two preceding numbers), evidence concerning the positioning of whorls of leaves, and whether structures follow the sequence in achieving 'optimum packing'. He concludes that 'phyllotactic whorls of leaf homologues are not positioned in Fibonacci patterns . . . the consensus starting to emerge from different subdisciplines in the phyllotaxis literature supports the alternative perspective that phyllotactic patterns arise from local inhibitory interactions amongst existing primordia already positioned at the shoot apex, as opposed to the imposition of a global imperative of optimal packing'. Research on this interesting question will no doubt continue.

Phylogenetic studies

Molecular tools have also revolutionised the study of the evolutionary relationships in other ways, giving insights at different timescales, including the distant past. Graur & Li (2000) set out the possibilities and purpose of modern phylogenetic studies.

> The DNA of every living organism is an accumulation of historical records . . . Admittedly, the information contained in these records is in a disorderly multi-layered state, scattered and at times fragmentary. Some of it is hidden or camouflaged beyond recognition, and parts of it are lost without trace. Charles Darwin's words on the imperfections of the [geological] evolutionary record sound as accurate and pertinent in the molecular era as they were in the last century. 'A history of the world, imperfectly kept and written in a changing dialect. Of this history we possess the last volume alone. Of this volume, only here and there a short chapter has been preserved: and of each page only here and there a few lines'.
> The purpose of molecular evolution is to unravel these historical records, fill in the missing gaps, put the information in order, and decipher its meaning. Since each evolutionary process leaves its distinctive marks on the genetic material, it is possible to use molecular data not only to reconstruct the chronology of evolution but also to identify the driving forces behind the evolutionary process.

In later chapters, we consider how far molecular approaches have provided insights into evolution through the devising of phylogenetic trees etc. Also, consideration will be given to the interrelationships of different fields of plant biology. Cronk, Bateman & Hawkins (2002) emphasise that 'recent technical advances mean that, at least in theory, the same homologous string of nucleotides' can be studied to test a very wide range of hypotheses in the study of variation. Through molecular studies many 'disparate phenomena, previously studied in isolation', are being 'cross-linked into a more unified discipline', encompassing the fields of systematics, genetics, morphology, development, population and ecological genetics.

Common origin to life on Earth

The major focus of this book is on the evolution of the flowering plants. But, here, it is appropriate to stress a most important conclusion concerning the relationships between the different groups of organisms. With regard to the origin of life itself, Darwin believed that: 'All the organic beings which have ever lived on this earth have descended from some one primordial form, into which life was first breathed' (Darwin, 1859). Recent advances, many based on the intensive study of a wide range of model species representative of many different groups of organisms found in the biosphere, have been extremely revealing of evolutionary relationships across all groups of living organisms. A wide range of evidences now point to a common evolutionary origin of 'Life on Earth'. There has been much debate about the genetic/biochemical and other characteristics of what has been called the Last Universal Ancestor (LUA) or the Last Universal Common Ancestor (LUCA) from which all current life has descended, and whether this 'root of the tree of life' can be identified (Maynard-Smith & Azathmáry, 1995; Theobald, 2010).

In the present context, two points are worth stressing. First, given that, with few exceptions, the genetic code provided by DNA is universal, we have one of the strongest indicators for the 'common origin ancestry of all life' (Aldridge, 1996). However, several lines of evidence suggest that, as RNA can store information and act as a catalyst in biochemical reactions, it seems possible that RNA was the 'original' hereditary material (Maynard-Smith & Azathmáry, 1995; Le Page, 2012), and that DNA subsequently evolved as a derived more stable storage molecule. Secondly, it is very interesting that molecular evidence suggests that around 3.5 billion years ago, the Last Universal Common Ancestor (LUCA) had about 100 genes for 'making RNA and proteins' and this 'core machinery is still found in life today' (Le Page, 2012).

Turning to the molecular genetic characteristics of flowering plants, Smith *et al.* (2010) highlight a number of key findings of the Arabidopsis Genome Initiative (2000).

- 'Approximately half of *Arabidopsis* genes have a clear counterpart in animals or fungi, while approximately 150 of the protein families ... are unique to plants'.
- There are '400 genes that are related to those of cyanobacteria ... probably inherited from the plant-specific endosymbiotic event that gave rise to the plastid'.
- Also, there is also a 'nearly complete copy of the mitochondrial genome, inserted near the centromere of chromosome 3'.
- 23% encode metabolic enzymes including those for secondary plant compounds and the characteristic chemical structures that make up the cell walls of plants.
- 17% are gene transcription factor families, of which '45% are unique to plants'. This reflects the separate evolutionary history of the animal and fungal lineages that diverged of the order of c.1.5 billion years ago in the Precambrian Era.
- However, some of these transcriptional factor families of genes have 'their counterparts in animals and fungi', e.g. the MADS family that are involved with the control of development processes.
- In the *Arabidopsis* genome there are a large number (about 4000) transposons 'inserted mostly near centromeres and in heterochromatic regions'.

Comparative study of other plant genomes makes it clear that there are considerable differences in size, partly due to the number of transposons. For instance, the Rice genome is about three times larger than *Arabidopsis thaliana*, first estimated at about 26 000 genes (Arabidopsis Genome Initiative, 2000). This figure was later increased to 27 000: in contrast, Rice has 41 000 predicted genes (Sterck *et al.*, 2007). While transposons account for at least 10% of the *Arabidopsis* genome, for Rice the figure is c.35%. Some of the differences between *Arabidopsis* and Rice in numbers of genes may reflect the different developmental pathways, morphology and physiology of the eudicot and monocot groups. The genome of Poplar has also been sequenced. This tree species differs from *Arabidopsis* in having many more genes that code for the biosynthesis of lignino-cellulose that forms the structural elements of the cell walls in trees.

Advances in comparative genomics

If the genomes of *Arabidopsis*, Rice and Poplar are compared, another highly significant pattern is revealed. Sections of the genomes of all three are duplicated, and almost identical segments appear at multiple sites. It is clear that the genome was duplicated in whole or in part at some stage in the past. The likely cause of this duplication is palaeopolyploidy, a process, in the distant past, whereby the genome was enlarged by the duplication of chromosome sets – the so-called Whole Genome Duplication (WGD).

Detailed analysis using comparative genomic methods reveals several significant palaeopolyploidy events in the evolution of the angiosperms (Fig. 6.4). Since the groundbreaking sequencing of the *Arabidopsis thaliana* genome in 2000, an increasing number of plant genomes have been sequenced and comparing genomes is very informative. For comparative purposes, different genomes can be represented in linear diagrams or concentric circles of sequences, often with additional information from genetic mapping. These are devised and studied using

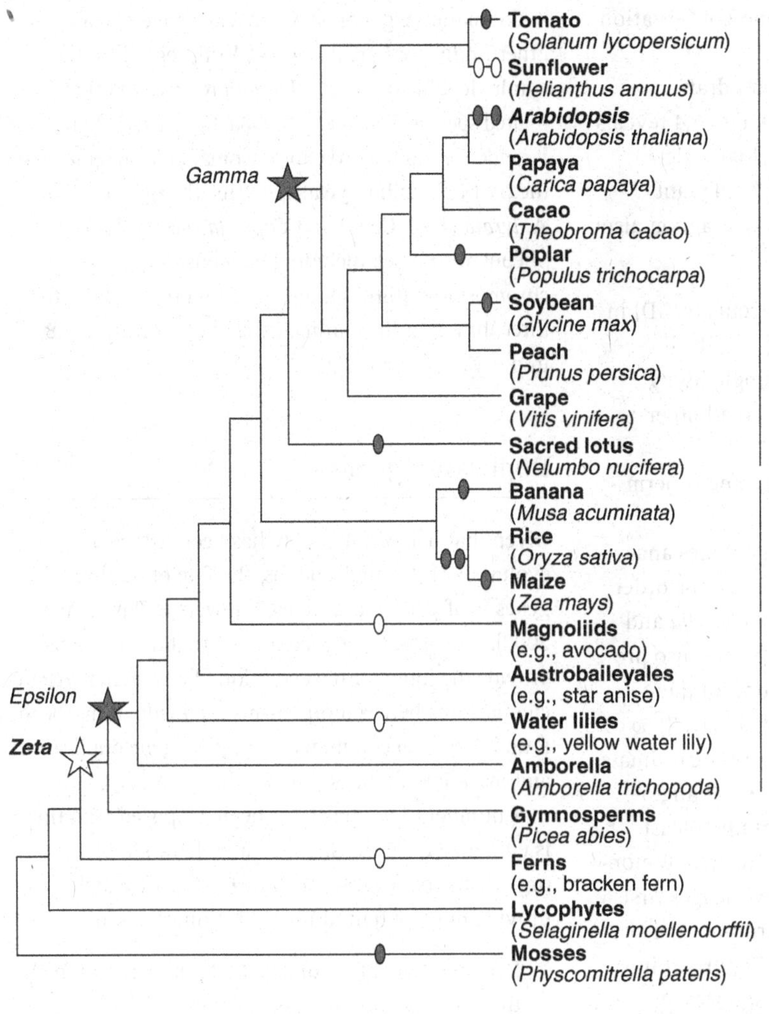

Fig. 6.4. An overview of land plant phylogeny, including the relationships among the major lineages of angiosperms. Representatives with sequenced genomes are shown for some lineages. But in the case of basal angiosperms, which lack genome sequences (except for *Amborella*) and non-flowering lineages, larger plant groupings are indicated. Hypothesised polyploid events (WGD) in land plant evolution are indicated by symbols. Epsilon star indicates the common ancestry of the angiosperms showing that *Amborella* is sister to all other extant angiosperms. The zeta and gamma HGD events are also shown. Additional polyploid events are indicated with ellipses. Events supported by genome-scale synteny analyses are filled, while those supported with frequency distribution of paralogous gene pairs or phylogenomic analyses are empty. (From *Amborella* Genome Project, 2013.)

a range of visualisation tools and/or statistical evaluations involving sequence alignment algorithms (Lyons & Freeling, 2008).

It is revealing to compare the genomes of closely and distantly related species to see the extent to which gene sequences are conserved. Because of the importance of grass species in agriculture, the genomes of Wheat, Barley, Maize, Rice, Sorghum, Millet and Sugar Cane were the first large group to be compared. While they have different chromosome numbers, Smith *et al.* (2010) note that 'if the chromosomes of each species are dissected into large fragments and realigned, a shared gene order emerges that was probably present in the common ancestor of grasses . . . this conservation of gene order, known as synteny, is weaker in more distantly related species, because genome rearrangements have had more time to reshuffle the ancestral gene order'. Thus, looking at other comparisons, there is a high degree of collinearity between *Arabidopsis*, *Carica* (Papaya) and *Populus*, but macrosynteny is not evident in the comparison of the genomes of the more distantly related *Lactuca* (Lettuce) and *Arabidopsis*, but still some local synteny is detectable, with gene conservation in small regions.

Taken from the account of the recent draft sequencing of the *Amborella* genome, Fig. 6.4 reveals our present understanding of the phylogenetic location of the various groups of the land plants (*Amborella* Genome Project, 2013). The diagram also indicates:

- Hypothesised ancient polyploidy events (WGD) in land plant phylogenies.
- That *Amborella trichopoda* is the single living representative of the sister lineage to all other extant flowering plants.
- The position of the monocots in the angiosperm lineages.
- Following extensive comparative genomes analyses, 'the remarkable conservation of gene order (synteny) among the genomes of *Amborella* and other angiosperms'. These investigations also provided evidence for 'the reconstruction of the ancestral gene arrangement in eudicots (c. 75% of all angiosperms)', with a more up-to-date estimate of numbers of genes. Thus, 'an ancestral angiosperm gene set was inferred to contain at least 14,000 protein-coding genes', and 'relative to non-angiosperm seed plants, 1179 gene lineages first appeared in association with the origin of the angiosperms. These include genes important in flowering, wood formation, and responses to environmental stress.'

Palaeopolyploidy and chromosomal changes

Returning to *Arabidiopsis thaliana*, genomic evidence has provided other insights. As the species has $n = 5$ chromosomes, historically it would not have been categorised as a polyploidy (Tate, Soltis & Soltis, 2005). Yet, as Fig. 6.4 shows, there is evidence for two chromosome doublings and one tripling event in its remote ancestry (Lysak *et al.*, 2009).

Comparative genomic approaches have revealed other findings of great interest. With regard to the cytological history of *A. thaliana* relative to other related species, Tang *et al.* (2008a, b) conclude that there is evidence for '9 to 10 chromosomal rearrangements in the past few million years since its divergence from *A. lyrata* (Rock Cress) and *Capsella rubella* (Pink Shepherd's Purse), including condensation of 6 chromosomes [through breakage and rejoining] into three, bringing the chromosome number from $n = 8$ to $n = 5$'.

Duplicated genes

Given that palaeopolyploidy has been detected in the evolution of the angiosperms, the fate of duplicated genes is of particular interest (Flowers & Purugganan, 2008). Increasing evidence reveals that duplications provide the route for the evolution of new genes. Where comparisons between organisms are made, it has been found that from common ancestry, by gene duplication and mutation, families of genes have evolved.

A number of models have been proposed, and there is evidence from the study of a wide range of organisms for several outcomes, some of which provide huge potential for evolutionary change.

1. Degenerative mutation(s) lead to 'non-functioning' through silencing or loss of genes.
2. Through mutational changes, duplicated genes come 'to specialize such that each performs a different subset of the original functions of the ancestral gene (subfunctionalisation)'.
3. Duplicated genes may come to acquire 'new functions (neofunctionalisation)'.
4. Duplicated genes may 'not change in function and simply remain redundant'.
5. 'Combination of these different process can occur'. (quoted from Taylor & Raes, 2005).

Concerning changes to genes in the course of evolution, it has been useful to distinguish orthologues (that diverge from the same ancestral gene after speciation) from paralogues (that arise from a duplication events in the genome).

In this chapter we have considered polyploidy, as it influences gene duplication in genomes. However, the process is a very important route to plant speciation. In later chapters, we consider mechanisms of abrupt speciation through duplication of genome sets in polyploidy, and also numerical changes in base numbers involving chromosome breakage and rejoining. Also, the fate of duplicated genes will be examined in newly formed polyploidy species.

Techniques for studying genetic variation

The biologist interested in plant variation and evolution is presented with an ever-increasing choice of techniques. In research first developed in the 1950s and 1960s, it proved possible to study proteins, particularly the genetics of enzymes, using gel electrophoresis (Weeden & Gottlieb, 1979; Gottlieb, 1981a, 1982, 1984; Crawford, 1990; Weeden & Wendel, 1990). Such studies investigated downstream products of the DNA. While investigations using DNA are now the preferred means of testing hypotheses, isozymes provided the valuable research tool for testing many hypotheses in plant systematics, microevolution etc., and very interesting case histories are to be found in the literature (see Soltis & Soltis, 1990; Hillis, Moritz & Mable, 1996; Weising *et al.*, 2005).

Recently, the emphasis has been on investigations of the DNA itself, through the development of many different techniques, including sequencing. Initially, sequencing was a manual, time-consuming, labour-intensive process. Now, automated DNA sequencing has been perfected in the development of 'next generation sequencing methods'. These have reduced the cost and accelerated the process, making it possible to undertake detailed investigations of non-model plant species, even those with the largest genomes (Kelly & Leitch, 2011).

Here, it is not possible to give detailed accounts of the techniques employed in molecular investigations. But it is important to outline a number of key properties of DNA that provide the means for advanced genetic investigations of many kinds.

Specific breakage of the DNA DNA exists as very long molecules in plant cells, and, in many techniques, specific breakage is a component. This is achieved by treating extracted DNA with restriction endonucleases from bacteria, whose 'natural' function is to degrade 'foreign' DNA. More than 100 restriction enzymes are known. The points of cleavage are, therefore, so precisely determined that restriction sites may be regarded as specific genetic markers.

Amplification of DNA: the polymerase chain reaction (PCR) For many investigations it is necessary to amplify the DNA of interest. This is achieved through the use of the PCR technique. The reaction involves cycles of precisely controlled heating and cooling. The PCR method has been automated, an advance made possible by the isolation of a heat-stable form of DNA polymerase (Taq) from *Thermus aquaticus*, a bacterium living in hotsprings (Tindall & Kunkel, 1988). Unlike the normal form of the enzyme, Taq is able to withstand the high temperatures at the denaturing part of the cycle without loss of activity. From a very small initial sample of DNA, huge numbers of copies of amplified sections of DNA can be made by the PCR method.

Separating enzymes and DNA fragments by gel electrophoresis In comparative studies of proteins (including enzymes), and DNA from different sources, electrophoresis is a key element in the experimental investigation. Samples are loaded onto an appropriate gel, across which an electric field is generated. The molecules of isozymes migrate in accordance with their size, and they move, depending on their net charge, to the positive or negative pole. In many DNA studies, fragments are separated by gel electrophoresis: smaller fragments 'travelling' further from the origin on the gel than larger fragments.

Visualisation of the products of electrophoresis In the studies of isozymes, to make the location of the

Table 6.1. **Characteristics of different methods of assessing genetic diversity. (From Frankham, Ballou & Briscoe, 2002. Reproduced with permission of Cambridge University Press)**

Method	Source	Non-invasive sampling	Cost	Development time[a]	Inheritance
Electrophoresis	Blood, kidney, liver, leaves	No	Low	None	Co-dominant
Microsatellites	DNA	Yes	Moderate	Considerable	Co-dominant
DNA fingerprints	DNA	No	Moderate	Limited	Dominant
RAPD[b]	DNA	Yes	Low–moderate	Limited	Dominant
AFLP	DNA	Yes	Moderate–high	Limited	Dominant
RFLP	DNA	No	Moderate	Limited	Co-dominant
SSCP	DNA	Yes	Moderate	Moderate	Co-dominant
DNA sequencing	DNA	Yes	High	None	Co-dominant
SNP	DNA	Yes	Moderate–high	Considerable	Co-dominant

[a] Indication of the time taken to develop the technique so that genotyping can be carried out in studies of threatened species.

[b] There are sometimes problems of repeatability with RAPDs (see text for further details). All other methods are highly repeatable.

separate molecules visible an appropriate substrate and stain are added to the gel. In the study of molecular profiles, the separated patterns of DNA are visualised by a variety of techniques, including staining, or the use of DNA probes labelled with fluorescent dyes.

Genetical interpretation of the results of electrophoresis The characteristics of different molecular methods for assessing genetic diversity are set out in Table 6.1 (Frankham, Ballou & Briscoe, 2002).

Different molecular techniques Here, a number of techniques are introduced that are listed below: others are mentioned, as appropriate, in later chapters.

Restriction Fragment Length Polymorphism (RFLP) A profiling technique once widely used, RFLP is now more or less obsolete. In essence, comparisons are made between homologous samples of DNA treated with one or other restriction enzymes. By examining fragment lengths in electrophoresis, RFLP identifies any differences in the location of restriction enzyme sites.

Random Amplified Polymorphic DNA (RAPD) This approach involves PCR amplification using arbitrarily chosen groups of nucleotides. RAPD techniques may be developed relatively quickly and at low cost (Frankham, Ballou & Briscoe, 2002). However, these advantages must be balanced against a concern about the reproducibility of RAPD techniques. In an investigation, subsets of DNA from two *Populus* × *euramericana* clones were investigated independently in nine laboratories using agreed protocols with the same batch of chemicals and primers (Jones *et al.*, 1998). It was concluded that (a) the RAPD was not sufficiently reproducible between laboratories; and (b) that reproducibility was not certain even in a single laboratory, if there were changes in chemical sources, equipment or personnel. Nonetheless, the low cost and short development time in the use of

RAPD has resulted in the publication of a number of case histories in the literature.

Amplified Fragment Length Polymorphism (AFLP) This technique involves the selective amplification, by PCR, of restriction fragments from a complex mixture. In international comparisons of RFLP/AFLP, mentioned above, it has been shown that AFLP techniques have high reproducibility (Jones *et al.*, 1997). This was the conclusion of research in seven laboratories, in which researchers used the same protocol and the same batch of chemicals to study the same DNA sample.

Mini- and microsatellites Patterns of repeated 10–60 base-pair (bp) units have been investigated in some studies – the so-called minisatellites (Heslop-Harrison & Schmidt, 2007). Also, microsatellites have been investigated, consisting of arrays of 1–5 bp sequences in short tandem repeats (STRs), or simple sequence repeats (SSRs) (Heslop-Harrison & Schmidt, 2007). How these repeat sequences, sometimes present in high numbers, have been employed in genetic fingerprinting for population genetics and conservation is reviewed in Bruford and Saccheri (1998) and Baker (2000).

Single nucleotide polymorphism (SNP) Comparison of sequence data reveals that sometimes two fragments of DNA from different plants differ in a single nucleotide polymorphism. This constitutes an allelic difference, and techniques have been devised to examine SNP differences between paired chromosomes and individuals within species.

Selecting marker systems to test hypotheses

A wide array of genetic markers is available and has contributed to unprecedented progress in the investigation of plant evolution, providing insights into breeding systems, gene flow and paternity, variation in geographical and ecological correlated variability, the past history of populations, patterns of hybridisation and speciation.

Reflecting on these studies, Holderegger, Kamm & Gugerli (2006) consider the implications of the genetics of the two principal types of nuclear DNA molecular markers, namely that they are either co-dominant or dominant (see Table 6.1).

- 'Co-dominant markers provide the possibility to score the identity of the two gene variants (alleles) that a diploid individual possesses at a given gene. Scorings may be aa, ab, cc, cd etc. Co-dominant markers include allozymes (proteins) and the highly variable microsatellites.'
- 'In contrast, dominant markers create banding patterns that resemble a barcode. For each individual, tens or hundreds of bands can be generated, resulting in a DNA fingerprint. Each band of this fingerprint refers to a locus with only two alleles. Let us assume that one of these alleles, a, is dominant over the other allele b. The genotypes carrying the dominant allele, i.e. aa and ab, thus show the same band. One cannot discriminate the heterozygote from the homozygote genotype. In contrast, the genotype bb does not show a band. As a consequence, dominant genetic markers are usually scored in a band presence/band absence manner. Corresponding marker types are RAPDs (Random Amplified Polymorphic DNA), ISSRs (Inter Simple Sequence Repeats) or AFLPs (Amplified Fragment Length Polymorphisms)'.
- The genomes of mitochondria (mtDNA) and chloroplasts (cpDNA) in plants are usually 'uniparentally transmitted'. Holderegger, Kamm & Gugerli (2006) stress that 'organelles have a single copy genome (haploid) with only one variant per gene. In fact, the whole organelle genome behaves as a single gene, since here is no recombination'.
- They also stress the very important point, that, while there are some exceptions, most of the 'molecular markers presently used in population genetics have one thing in common. They are essentially neutral and do not undergo selection.'
- For further details of these markers and techniques, see Frankham, Ballou & Briscoe (2002).

Molecular approaches have made it possible to study some very intractable areas of genetic variation. Two areas of study serve to illustrate this point.

Combining genetic and molecular approaches: quantitative trait loci By using simple crossing experiments and examination of segregation patterns the genetic inheritance of many traits can be deduced and mapped. But, many loci are involved in the inheritance of other traits, the so-called quantitative trait loci (QTLs), and these cannot simply be mapped by classical Mendelian approaches. This problematic situation becomes more tractable if DNA markers are closely linked to QTLs (Smith *et al.*, 2010). Polymorphic DNA markers are selected whose presence is correlated with the QTLs of interest. DNA markers and traits are then examined in progeny of crosses between different strains, inbred lines etc. The degree of correlation of QTLs and markers is then estimated in statistical tests, bearing in mind that some apparent correlations may be chance events. Mapping is based on inferences from these statistical tests (see Lynch & Walsh, 1997). Some very complex genetical situations have been detected, involving dominance, interaction of loci, and linkage. QTL analysis is not restricted to backcrosses, F_2s of inbred lines etc., but has also been employed in the study of differences between populations and species.

Studying DNA from herbarium specimens, sub-fossils and fossils PCR techniques allow the small amounts of DNA to be amplified and this has encouraged the study of plant samples, some of considerable age (Willerslev & Cooper, 2005). For instance, DNA has been regularly extracted from plant specimens in herbarium collections (Lookerman & Jansen, 1995; Andreasen, Manktelow & Razafimandimbison, 2009). Microsatellite genetic fingerprints have been studied in herbarium specimens of *Cypripedium calceolus* (Orchidaceae) (Fay & Cowan, 2001). Gugerli, Parducci & Petit (2005) review the prospects for the successful extraction and analysis of DNA from fossil plants. We return to the subject of ancient DNA (aDNA) in Chapter 17.

Concluding remarks

In a historical account of developments in molecular biology and related subjects, Sapp (2003) reviews the emergence of a more complex picture than was envisaged by those that developed the 'New Synthesis'. As we have seen in this chapter, it is surprising that there are few transcribed genes relative to the amount of DNA in the genome. A high percentage of the DNA present in the genome appears to be non-coding. But, as we have seen above, there has been a great deal of debate about whether this DNA (often called junk DNA) has any adaptive value. Important new evidence from the study of the human genome suggests that far from being 'junk', much of this non-coding DNA is involved in regulation and control. However, it remains to be seen whether these findings also apply to plants, where related species often have very different genome sizes.

The historic concept of genes as beads on a string, and the notion that one gene gives rise to one protein, must now be abandoned. This conclusion follows from the finding that eukaryotic genes contain coding exons and non-coding introns, and mRNA transcripts are edited by protein/RNA and spliceosomes are involved in the removal of the non-coding introns. Also, there is increasing evidence that pieces of RNA can be spliced in different combinations, as part of genetic regulation during development.

Turning to another issue, the central dogma of early molecular biology was that genetical information was transferred from DNA → RNA → amino acids → proteins. This concept has been challenged, in a number of ways. For instance, it was commonly assumed that protein shape was determined by amino acid sequences, but investigations have revealed that, in some cases, proteins do not fold properly unless guided by 'chaperone proteins', the so-called chaperonins. Also, animal studies have revealed exceptions to the notion that information transfer occurs only in the direction of DNA → proteins.

Prions, which contain neither DNA nor RNA, appear to be pure proteins. Sapp (2003) notes that

> when the prion invades the brain, it refolds a normal brain protein to match its own infective three-dimensional shape. Thus, no information is conferred from nucleic acid to protein: only its conformation – its three-dimensional shape – changes. The newly-folded protein then becomes infectious and acts on other normal proteins, setting up a chain of reactions that propagates the disease.

The discovery of epigenetic effects, outlined above, is transforming our understanding of plant development and genetics, with extremely important implications for theories of evolution, especially the Modern Synthesis developed in the 1930s. For example, as we have seen, some epigenetic phenomena may owe their origin to methylation of cytosines in DNA, and highly methylated DNA has been found to be transcriptionally inactive. Intriguingly, there is increasing evidence that DNA methylation patterns may, in some cases, be inherited from one generation to the next, through sexual reproduction. The implications of these findings are extremely important, should they provide a route for the transmission of acquired characteristics.
In addition, further evidence for the inheritance of acquired characters comes from the realisation that mitochondria and chloroplasts are most likely to owe their origin to endosymbiosis.

These findings have clear implications for evolutionary theory. The strictly orthodox neo-Darwinian theory of evolution, typified by the Modern Synthesis, had no place for the ideas of Lamarck. But an increasing body of evidence suggests that inheritance of acquired characters occurs in certain circumstances. However, Weigel & Colot (2012) introduce a note of caution: 'since the 1990s the molecular basis of hundreds of naturally occurring phenotypic variants has been identified in crop and wild species, and overwhelmingly, DNA sequence differences are involved'. They stress that 'the number of epialleles that we know of in plants is only about a dozen ... Thus, the extent to which epigenetic variation contributes to phenotypic variation is still not known with certainty'. Considering the wider picture, they call for 'an explicit theory of population epigenetics that describes the parameters under which epimutations could contribute to evolution'.

Turning to a final point, many historic studies of microevolution and systematics are reported in the literature. They often employed the 'cutting edge' methods of their day, methods that have now been largely superseded. In the light of our present knowledge, these earlier studies can be seen as variously defective or incomplete, but ignoring them would be short-sighted, for they are often worthy of further attention. This is because historic investigations have repeatedly provided inspiration for more critical experiments using the latest molecular tools. In this way, new insights can be built on knowledge from the past. For instance, the study of *Tragopogon* species and their hybrids begins with Linnaeus (Chapter 2). From the 1950s, investigations of hybridising Tragopogons, whose parents were introduced from Europe to the USA, have provided the material for many investigations using different research tools, resulting in a continuous series of revealing insights into the microevolutionary processes of hybridisation and speciation.

7 Breeding systems

The study of a range of plants in the wild or on display in a Botanic Garden reveals a bewildering array of floral types. Many books describe in some detail a selection of pollination mechanisms, often discussing in highly technical botanical language the variety of floral structure. Special terms have, quite properly, been devised by botanists to enable them to write concise, accurate plant descriptions. Although the botanical literature reports extensively on the structures involved in reproduction, in our opinion, it does not pay sufficient attention to the variety of breeding systems in plants, systems of which complex structures and pollination mechanisms are only a part.

We have noted in earlier chapters the role of the internal sources of genetic variation, namely mutation and recombination, and have seen that a vast number of gametic types is theoretically possible as a consequence of these factors. Which gametes are actually brought together to form the zygotes, however, depends, to a great extent, upon the breeding system of the plant concerned.

In this chapter, as a prelude to our discussions of variation within and between species, we consider three areas:

A. The different breeding systems found in flowering plants.
B. Breeding behaviour discovered in studies of wild populations.
C. The evolution of breeding systems.

Our account discusses how studies of breeding behaviour have developed, and emphasises that a knowledge of the different breeding systems provides an indispensable framework for understanding the complexities of patterns and processes found in nature.

A. The different breeding systems found in flowering plants

There are three basic breeding mechanisms, which we examine in turn.

Outbreeding

In many animal groups outbreeding – crossing between different individuals – is rendered likely by sexual differentiation. In higher plants, however, separation of the sexes is the exception. According to Richards (1979), only about 4% of the flowering plants are dioecious. A slightly higher figure is reported by Renner & Ricklefs (1995), who note that *c.*6% of angiosperm species are dioecious, these species being distributed amongst 7% of genera. The incidence of dioecy varies in different floras. While only *c.*3% of British flowering plants are dioecious, much larger percentages are reported, for example, from groups of distant oceanic islands such as Hawaii (28%) and New Zealand (13%). Tropical forests may also have high numbers of dioecious species (e.g. 23% of dioecious species were reported in a sample of woody plants from a lowland rainforest in Costa Rica (Bawa, Perry & Beach, 1985)).

Most higher plants are hermaphrodite and the typical angiosperm flower has a zone of pollen-bearing stamens (androecium) surrounding a gynoecium containing one or more ovules. Even in the case of simple unisexual flowers of the catkin-bearing woody plants (Amentiflorae), male and female flowers are usually found on the same individual. Such juxtaposition of stamens and ovules suggests that self-fertilisation would be the most likely mode of reproduction, and it is interesting to trace the historical development of ideas which force us to conclude that many plants are adapted to facilitate cross-fertilisation (i.e. crossing between different individuals) and to minimise or prevent self-fertilisation (Whitehouse, 1959; Richards, 1986; Barrett, 1988, 2010a, b).

The sexual function of flowers was established in the seventeenth and eighteenth centuries (Proctor & Yeo, 1973; Proctor, Yeo & Lack, 1996). In 1793 Sprengel published his classic book in which he produced excellent descriptions of the wind and insect pollination of plants, but he was apparently unaware that flowers are primarily adapted for cross-fertilisation. It was Darwin who concluded, after a detailed study of many orchid species, that the orchid flower was 'constructed so as to permit of, or to favour, or to necessitate cross-fertilisation'. Reflecting further on the orchid studies, Darwin wrote (1876):

> It often occurred to me that it would be advisable to try whether seedlings from cross-fertilised flowers were in any way superior to those from self-fertilised flowers. But as no instance was known with animals of any evil appearing in a single generation from the closest possible interbreeding, that is between brothers and sisters, I thought that the same rule would hold good with plants.

As early as 1866 Darwin was studying inheritance in *Linaria vulgaris,* and as part of his studies raised two large beds of seedlings of self-fertilised and cross-fertilised individuals. To his surprise the 'crossed' plants, when fully grown, were taller and more vigorous than the 'selfed' progeny. Darwin's interest was thoroughly aroused and he investigated, over many years, the effect of cross- and self-fertilisation in a number of species, e.g. *Ipomoea purpurea, Mimulus luteus, Digitalis purpurea, Zea mays.* He gave great attention to experimental design; for example, his basic comparative test was devised as follows. Seed from cross- and self-fertilised plants was germinated and a 'crossed' seedling was matched against a 'selfed' seedling, several such comparisons being made in each of a number of pots. A partition was placed between the two sets of seedlings but in such a way as to make sure that both sets of plants were equally illuminated. Other types of pot experiment and garden trial were also carried out and the effect of crossing and selfing in some cases was studied for a number of generations. With some exceptions, his results revealed that the progeny of cross-fertilised plants were superior in performance when compared with the progeny from self-fertilisations. (He examined one or a number of measures of performance such as height, weight or fertility.) An example of his results with Maize is set out in Table 7.1.

While we might accept Darwin's (1876) general conclusions from his experiments that progeny from cross-fertilised plants were generally taller etc. than progeny from self-fertilised plants, it is interesting to discover, as Darwin himself realised, whether the

Table 7.1. **Darwin's experiment with Maize (*Zea Mays*): height (in inches) of young plants raised from seeds obtained by cross- and self-fertilisation**

	Cross-fertilisation	Self-fertilisation
Pot I	$23\frac{4}{8}$	$17\frac{3}{8}$
	12	$20\frac{3}{8}$
	21	20
Pot II	22	20
	$19\frac{1}{8}$	$18\frac{3}{8}$
	$21\frac{4}{8}$	$18\frac{5}{8}$
Pot III	$22\frac{1}{8}$	$18\frac{5}{8}$
	$20\frac{3}{8}$	$15\frac{2}{8}$
	$18\frac{2}{8}$	$16\frac{4}{8}$
	$21\frac{5}{8}$	18
	$23\frac{2}{8}$	$16\frac{2}{8}$
Pot IV	21	18
	$22\frac{1}{8}$	$12\frac{6}{8}$
	23	$15\frac{4}{8}$
	12	18

results of a particular experiment were statistically significantly different. Darwin consulted his cousin Galton, who, employing the crude statistical techniques available at the end of the nineteenth century, had to conclude that the experiment was based on too few plants. It was not until 1935 that Fisher made a close analysis of Darwin's experiments. He concluded that the experimental design was fundamentally sound in comparing the growth of an 'outcrossed' with a 'selfed' seedling. (However, planting several plants in each pot must have led to competition and perhaps a better planting arrangement could have been devised.) Fisher used statistical tests first devised in the early years of the twentieth century for studying small samples, and discovered that the Maize result above was just statistically significant.

As a result of his botanical investigations – which led to the publication of three books (Darwin, 1862, 1876, 1877a) – Darwin was able to see how the enormous range of floral types and physiological differences in behaviour, such as different times of maturity of stamens and stigma on the same flower, could be viewed as adaptations to ensure cross-fertilisation (Barrett, 2010b). Why it might be beneficial for progeny to be cross-bred rather than the product of self-fertilisation, Darwin was unable to decide.

In his experiments on 'cross' and 'self' progenies, Darwin discovered that in some cases the attempt to produce progeny by self-fertilisation failed: certain plants were self-sterile. Examples of self-sterility had also been noted, by other botanists. Again Darwin was unable to account satisfactorily for this phenomenon.

Great progress has been made in our understanding of self-sterility, which may arise from a number of causes. In most species self-incompatibility is involved. A fertile hermaphrodite plant is incapable of producing zygotes following self-pollination. There is a mechanism operating at the pre-zygotic stage – involving both pollen and style or stigma tissue – preventing self-fertilisation.

The first, and inconclusive, studies of self-incompatibility were made by Correns (1913) in experiments with *Cardamine pratensis*. He suggested that a genetic mechanism was involved, but it was not until the experiments of East & Mangelsdorf (1925) working with *Nicotiana* that the genetics of self-incompatibility became clearer. The elements of the scheme they proposed – a gametophytic incompatibility system – have now been found to apply to very many, but not all, incompatible species (Takayama & Isogai, 2005; Karron *et al.*, 2012).

The stages are as follows:

1. Each plant is heterozygous for a gene S; e.g. S_1 S_2. (Many, perhaps scores of, alleles occur: say S_3, S_4, S_5, etc.) The style and stigma (pistil) in the flowers of a plant are maternal tissue containing nuclei with the diploid chromosome number and in consequence style and stigma contain S_1S_2 in a plant of genotype S_1S_2.
2. At meiosis prior to pollen formation, segregation occurs and pollen, which contains nuclei with a haploid chromosome number, receives one of the two S alleles. The pollen hereafter behaves in pollination in accordance with its S allele genotype.
3. Pollen arrives at the stigma, which may be specially adapted for pollination by wind, insect or other means. Often stigmas are 'wet' from the secretion of exudates, but in some cases they are 'dry', for example in the grasses.
4. The pollen grain hydrates and germinates to give a pollen tube that grows intracellularly through a special tract of transmitting tissue. In some species the tract is solid, while in others it lines the hollow style.
5. If the S allele present in the pollen is also found in the style, the growth of the pollen tube in the style is progressively slowed over several hours, the tip of the incompatible tube growing abnormally and becoming occluded with callose. The accumulation of callose in incompatible tubes provides a means of exploring the behaviour of pollen from different sources, as callose 'plugs' fluoresce in ultraviolet light, when stained with decolourised aniline blue. If pollen and stylar tissues have dissimilar S alleles,

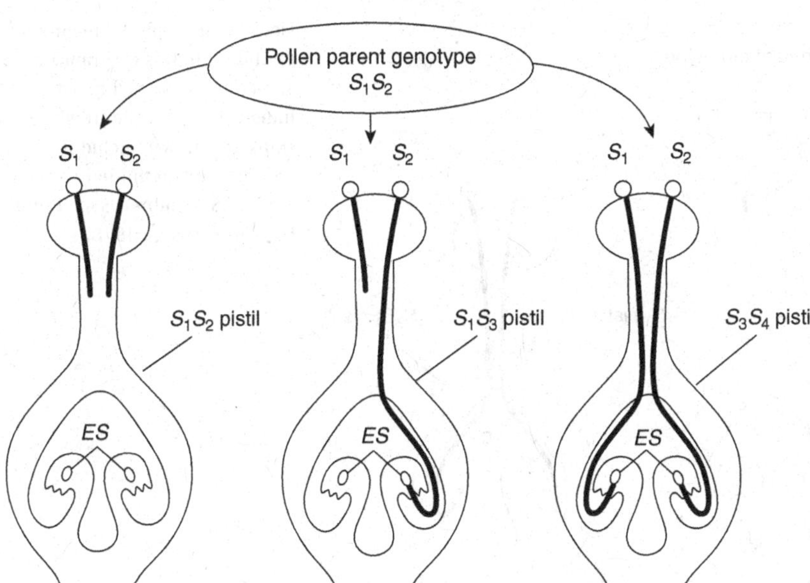

Fig. 7.1. Gametophytic self-incompatibility. A pollen parent of genotype S_1S_2 will be infertile, semi-fertile or fully fertile according to the genotype of the female plant. In most species with this system, incompatible pollen tubes are inhibited in the style. ES = embryo-sac. (From Heslop-Harrison, 1978.) Note that in some families gametophytic self-incompatibility is more complex genetically, being controlled by two or more loci (De Nettancourt, 1977).

growth of the pollen tube proceeds normally and fertilisation may occur. (In some plants, however, for example the grasses, the incompatibility reaction appears to occur at or near the stigmatic surface rather than in the style.)

Given such a genetic mechanism, it is now clear why certain self-fertilisations fail to produce offspring. The S alleles segregating in the pollen (say S_1S_2) are both represented in the stylar tissue of the same plant and all pollen is arrested in its growth down the style. The consequences of self-pollination and pollination by individuals of various 'S' genotypes are displayed in Fig. 7.1.

It is important to note that a given plant will not necessarily be fully fertile with all other *individuals* of the same species, even if they differ substantially genetically. The S alleles represented in pollen and style are decisive. The self-incompatibility reaction arises from an interaction between the gene products of pollen and style, preventing free access of pollen tubes to the ovule, permitting pollen tube growth in certain cross-fertilisations and preventing pollen tube growth on self-pollination. While most species with the gametophytic type of self-incompatibility are controlled by a single locus (the S-supergene; see below), it is important to stress that another mechanism has been discovered in grasses, where two unlinked loci (S and Z) are involved. A pollination will fail if the loci in the grain match those in the pistil (Baumann *et al.*, 2000; Klaas *et al.*, 2011).

In 1950, American botanists studying *Parthenium* (Gerstel, 1950) and *Crepis* (Hughes & Babcock, 1950) discovered what has come to be known as the sporophytic incompatibility system. The following are the important differences from the gametophytic system just outlined:

1. Individuals are heterozygous for the S gene, say S_1S_3, and, as before, segregation occurs at meiosis preceding pollen formation. The pollen, notwithstanding the 1:1 ratio of segregation of S alleles, behaves as if it had the genotype of the plant that produced it. Evidence suggests that proteins, amongst which are recognition components responsible for the incompatibility reaction, are produced by the tapetal cells of the anther pollen sacs. These substances are exported to the pollen wall during its development.

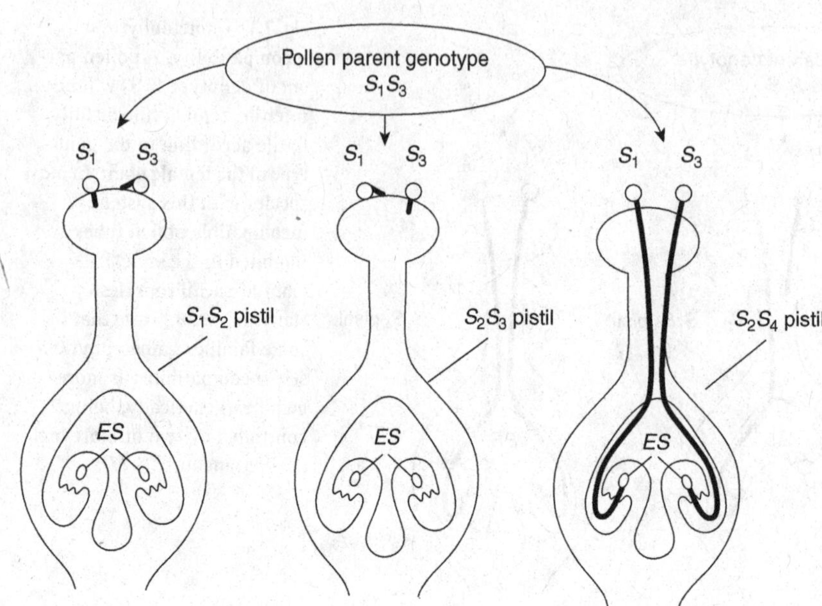

Fig. 7.2. Sporophytic incompatibility. In this diagram S alleles are presumed to act independently. Other relationships are known, including dominance and mutual weakening. ES = embryo-sac. (From Heslop-Harrison, 1978.)

2. In general, pollen is rejected if its pollen parent shares the same S allele as the stigma and style (pistil) of the female parent. (However, in some species there are more complex reactions including dominance of certain alleles. For example, see Allen *et al.* (2011), and Brennan *et al.* (2011), who studied dominance interactions and modifiers in *Senecio squalidus.*) The results of certain cross-pollinations are set out in Fig. 7.2.
3. The site of pollen tube failure is often different in sporophytic systems, occurring on or close to the stigmatic surface. As in gametophytic systems, the accumulation of plugs of callose indicates incompatible pollinations.

Darwin's list of self-sterile plants was quite short, but self-incompatibility is now thought to be widely distributed, in both the monocotyledons and dicotyledons (Charlesworth, 1985). Having carried out a thorough survey, Igic, Lande & Kohn (2008) found that 'at least 100 plant families reportedly contain self-incompatible species' but they note that 'this is probably a conservative estimate' as proper tests for self-incompatibility (and its control) 'are rarely performed or are infrequently reported'.

Molecular studies of self-incompatibility systems

The control of self-incompatibility and the identity of the substances involved in such reactions have been the subject of intense research (Charlesworth *et al.*, 2005; Takayama & Isogai, 2005; Igic, Lande & Kohn, 2008; Charlesworth, 2010). The self-incompatibility systems of a number of 'model' plant species have been studied in great detail. Undoubtedly much remains to be discovered, as only a small number of species have been examined. The gametophytic system, outlined above, has been investigated in members of the Solanaceae, Rosaceae, Papaveraceae and Scrophulariaceae (in particular, *Antirrhinum*). Species of the Brassicaceae have also been examined and these are found to have the sporophytic system.

Reviewing these studies of model plants so far investigated, Charlesworth *et al.* (2005) conclude that the S-locus region contains 'separate pistil and pollen protein genes' and that self-incompatibility is a 'lock-and-key' system rather than 'recognition through expression of the same proteins in pistil and pollen'.

Here, in outlining recent advances, we stress the very important point that several different mechanisms have been discovered (Takayama & Isogai, 2005). In the gametophytic self-incompatible system of Solanaceae, Rosaceae and *Antirrhinum*, 'the pistil S-RNase protein is expressed in stylar tissue'. This protein 'is taken up by growing pollen tubes, degrading RNAs within incompatible pollen tubes and causing arrest of growth in the style' (Charlesworth *et al.*, 2005).

A second different gametophytic process has been revealed in studies of *Papaver rhoeas* – the so-called S-glycoprotein mechanism. An extracellular protein is found on the stigmatic surface: the male component of the system has yet to be fully characterised. In incompatible pollinations, pollen tubes are arrested very quickly, involving an influx of calcium ions (Franklin-Tong, 2003; Thomas & Franklin-Tong, 2004).

In the sporophytic self-incompatible Brassicaceae system the 'pistil protein is a receptor kinase, *SRK* . . . anchored to the stigma membrane, presenting its extracellular S domain to the ligand molecules on the pollen surface' (Charlesworth *et al.*, 2005). 'The pollen component is a small cyteine-rich protein (*SCR*) present on the pollen coat. When the pollen touches the stigma cell and the *SRK* recognizes the matching *SCR* protein on the pollen coat, a signal-transduction cascade is triggered in the stigma cell, which prevents it from sustaining germination of the pollen grain' (Smith *et al.*, 2010). However, models of the mechanism(s) that arrests pollen germination have yet to be fully tested (see Takayama & Isogai, 2005; Smith *et al.* 2010).

Considering these findings, Charlesworth *et al.* (2005) emphasise that, 'despite their name, self-incompatibility systems do not involve a self-recognition system, but are based on genes encoding proteins with receptors and ligands that recognize one another'. Furthermore, 'the evolution of such a system in species that lack incompatibility is likely to be a very rare event' (Charlesworth, 2006). Reviewing recent progress, Charlesworth (2010) stresses that

> when the sequences of the S-locus genome region of some of these species were obtained, separate genes were found for the recognition proteins expressed by pollen grains and pistils (i.e. the genetic S-locus includes at least two incompatibility genes: one for the pistil incompatibility protein and one or perhaps more than one gene controlling the pollen incompatibility types). With two or more genes, correct combinations of alleles must be maintained to avoid combinations that allow self-compatibility, and obviously very close genetic linkage between pistil and pollen genes will prevent such disadvantageous combinations. However, it is mystifying how new combinations (alleles with new incompatibility types) can ever arise given that at least two genes must change appropriately –this is a puzzle that has not yet been solved.

Homomorphic and heteromorphic incompatibility systems

Surveys of self-incompatible groups have established the finding that plants having gametophytic incompatibility systems are all homomorphic: there are no structural differences associated with different S alleles. In contrast, some plants with sporophytic incompatibility systems are heteromorphic. (Heterostyly is reported from approximately 25 angiosperm families (Barrett, 1992).) The existence of different forms of flowers in the same species was the subject of some of the most famous studies made by Darwin.

Darwin carried out crossing experiments with heterostylous plants of both the distylic and tristylic type (Fig. 7.3). In the case of the Cowslip, *Primula veris*, he showed that there are two forms of the flower, 'pin' and 'thrum', each with a characteristic syndrome of characters (Fig. 7.4). The pollinations, pin × thrum and thrum × pin, yielded good seed set; selfing pins or thrums or crossing pin × pin and

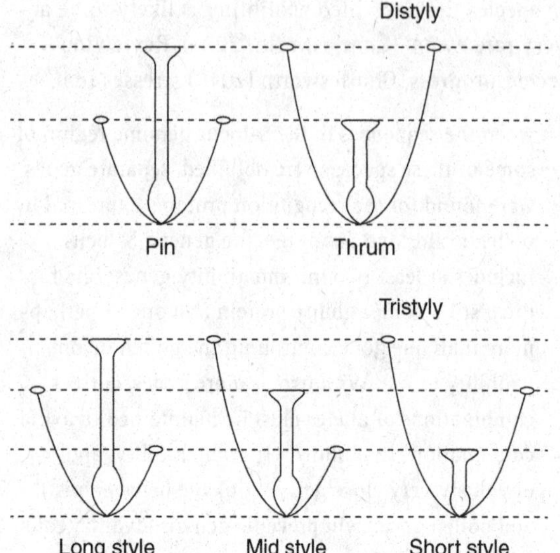

Fig. 7.3. Symbolic representation of distyly and tristyly. In each system the compatible pollinations only involve anthers and styles at the same level, and therefore the following are incompatible combinations: pin × pin, thrum × thrum, long × long, mid × mid and short × short. (From De Nettancourt, 1977). Darwin (1877a) listed 14 families in which heterostyly had been confirmed. This list has now been extended to c.25 families (Barrett, 1992).

thrum × thrum yielded very much less seed. In view of the comparative self-sterility of pin and thrum plants, Darwin concluded that distyly was a device favouring outcrossing. Pollinators visiting *P. veris* were likely to pick up pollen in a pattern related to anther position. Subsequent visits would transfer pollen to stigmas at 'an equivalent height'.

Investigations have revealed the genetics of the situation in *Primula*, and the following model has been proposed. Dimorphy and the incompatibility reaction are controlled by blocks of tightly linked genes, sometimes called a 'supergene'. Thrum plants are heterozygous GPA/gpa, while pin plants are homozygous gpa/gpa, where:

Controlling female characters: G: short style, short stigmatic papillae and thrum female incompatibility versus g: long style, long papillae and pin female incompatibility;

Controlling male characters: P: large pollen, thrum male incompatibility versus p: small pollen and pin male incompatibility;

Controlling anther position: A: high anthers (as in thrum) versus a: low anthers (as in pin).

It has been found that other heteromorphic sporophytic plants are di-allelic (Lewis, 1979). The genetic information in the 'supergene' is very tightly linked, but occasionally crossing over occurs to give various types of homostyle (Richards, 1997). We will consider the breeding behaviour of these variants below. Ganders (1979) has provided a valuable review of the characters in which heterostylous plants vary (style length/colour; stigma size/shape/papilla shape; stamen position/anther size/pollen size and number/pollen shape, size, sculpturing and food reserves; corolla size/pubescence).

Darwin also studied Purple Loosestrife, *Lythrum salicaria*, in which three kinds of flowers are found corresponding to different levels of anthers and stigmas in the flowers; eighteen different interpollinations were needed to investigate the effects of six kinds of anther and three kinds of style, and these Darwin carried out. The number of seeds in six pollinations between anthers and stigmas at the same height was much greater than those produced in other pollinations (Fig. 7.3).

Using molecular approaches, a number of geneticists are now actively testing the classic S-supergene model of heteromorphic self-incompatibility and its functioning (see McCubbin, 2008).

Late-acting self-incompatibility systems

By definition self-incompatibility mechanisms are pre-zygotic, acting as pollen lands on stigmatic surfaces or as pollen tubes grow down the style (Barrett, 1988). However, Seavey & Bawa (1986) suggest that other situations could be considered as manifestations of self-incompatibility, so-called late-acting reactions that occur in the ovary before or after

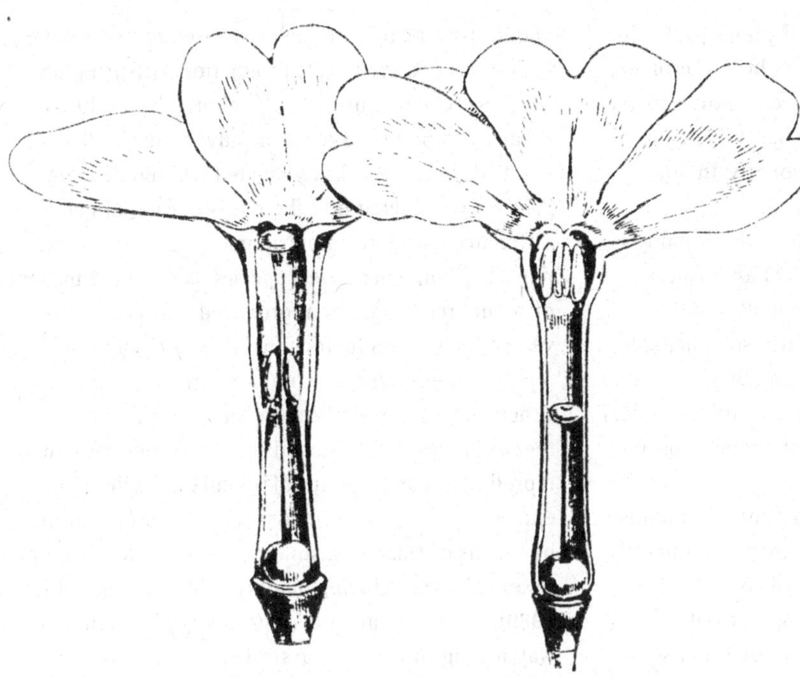

Fig. 7.4. Heterostyly in *Primula veris* (×3). Darwin (1877a) discovered that the long-styled (pin) form always had a much larger pistil with a globular, rough stigma standing high above the anthers. Pin pollen was oblong in shape. In contrast, the short-styled (thrum) plants always had a short pistil about half the length of the corolla, with a smooth stigma (depressed on the summit). Pollen in the anthers, which stood above the stigma, was spherical and larger than in the pin.

fertilisation, resulting in the abortion of the products of self-fertilisation over a short timeframe. A number of researchers discuss the possibility of late-acting self-incompatibility in various groups, for instance, Australian Myrtaceae (Beardsell *et al.*, 1993); *Spathodea campanulata* (Bittencourt, Gibbs & Semir, 2003); *Caesalpinia echinata* (Borges, Sobrinho & Lopes, 2009); *Cyrtanthus breviflorus* (Vaughton, Ramsey & Johnson, 2011).

However, others point out that embryo abortion could occur for a number of different reasons, inbreeding depression being the most likely cause (see below). Indeed, it is possible that both late self-incompatiblity reactions and early effects of inbreeding depression may occur. Such a possibility is suggested by experiments with Monkshood (*Aconitum kusnezoffii*) by Hao *et al.* (2012), who report that reduction in seed set in self-pollination experiments involves late-acting self-incompatibility. In addition, as some of the resulting embryos were aborted at different stages of development, it is suggested that early-acting inbreeding depression may be the cause.

Self-fertilisation

The second main breeding method involves self-fertilisation. In his account of 'cross'- and 'self'-fertilisation of plants, Darwin (1876) reported a list of species that yielded seed when covered by a net to exclude insect visitors. In 1877, he discussed the presence in some species of specialised flowers that never open, pollination occurring automatically in the closed bud, which gives rise directly to a fruiting structure. Such flowers, which are, of course, self-fertilising, are referred to as cleistogamous. Following a literature search and as a result of his own researches, Darwin listed 55 genera in which cleistogamous flowers had been found, and investigations by botanists over the years have added to the list (Uphof, 1938). In some cases cleistogamy appears to be more or less obligatory, as in several grass species (McLean & Ivimey-Cook, 1956), but, as we shall see below, cleistogamous flowers occur seasonally in some species.

A recent study has established that cleistogamy is quite widespread, being found in 693 angiosperm species belonging to 228 genera and 50 families (Culley & Klooster, 2007). The study also concluded that after close study more cleistogamous species may be found and that this breeding behaviour has almost certainly evolved independently many, perhaps 34–41, times.

Following observations and experiments, botanists have concluded that some species are predominantly self-fertilising, but in many cases this is a presumption based on the following types of evidence. It may be readily demonstrated that some species are self-compatible, and it can also be observed that some plants, visited by insects in summer, can flower and fruit in autumn, winter or spring, when insects are absent. Further, progenies of some species are remarkably uniform in appearance and it is assumed that such uniformity results from persistent inbreeding. As we shall see later, such evidence is circumstantial, and for definitive studies the use of genetic markers is essential in the study of breeding behaviour.

If some plants are predominantly or obligately self-fertilising, how is this to be equated with the demonstration by Darwin of general superiority of outbred plants? This paradoxical situation will be examined below.

Apomixis

In this third reproductive mode, reproduction is achieved without fertilisation, the sexual process being wholly or partly lost (the term and its definition are according to Winkler, 1908). Historically, two types of system were distinguished: agamospermy and vegetative apomixis. A plant reproducing by seed apomixis or agamospermy seed is set, but no sexual fusion has occurred in its production. Offspring have the genetic constitution of the plant that produces them. Such plants have all the advantages of the seed habit without the risks associated with pollination. Winkler considered that there is no essential genetic difference between simple agamospermy and asexual reproduction, where some species, lacking the means of sexual reproduction, reproduced entirely vegetatively. Thus he included these two types of reproduction under the common term 'apomixis'. Others have followed Winkler's approach by regarding vegetative spread in the absence of sexual reproduction as a variant of apomixis, while excluding those cases where vegetative propagation – by means of rhizomes, stolons, runners, etc. – occurs in sexually reproducing plants (Stebbins, 1950; Stace, 1989). The consequences of grouping plants in this way are significant. If the seed apomixis and vegetative reproduction are classified together under the term apomixis, then 60% of the British flora would be included (Richards, 2003). However, if the term is used only for taxa with asexual reproduction through seed, then a much smaller fraction of the British flora would be included.

Now, many botanists do not regard these vegetative systems as apomixis, and would restrict the use of the term apomixis to 'agamospermy'. Let us examine the argument. Those including vegetative reproduction under the blanket term apomixis have in mind a number of interesting situations. For instance, female plants of the dioecious Canadian pondweed *Elodea canadensis* were introduced into Europe in the nineteenth century and these introductions, even in the absence of male plants, spread widely by vegetative growth and fragmentation (Sauer, 1988). In other species, specialised structures, such as small, aerial, readily detachable bulbils, may be produced (Fig. 7.5).

Considering the contrary position, we may cite the view of Van Dijk (2009) who concludes that 'vegetative reproduction in plants is not apomixis'. He points to significant differences between

Fig. 7.5. Bulbils. (a) *Saxifraga cernua*: 1 (×1); 2. A cluster of bulbils; 3. Bulbils at various stages of development. (From Kerner, 1895.) (b) *Poa alpina*: 1. *P. alpina* with bulbils replacing its flowers (×1); 2. A portion of the inflorescence; 3. A miniature grass-plant developed between the glumes of a spikelet of *P. alpina.*

vegetative reproduction and seed apomixis. (a) 'Probably all perennial plants can reproduce vegetatively. If we include vegetative reproduction in apomixis, the term becomes meaningless.' (b) 'Although the maternal genotype is maintained with vegetative reproduction, there are good reasons not to include vegetative reproduction in apomixis. Meristems are multicellular, and therefore mutations result in chimeric tissues.' In contrast, 'apomicts *sensu stricto* go through a single cell stage, being equivalent to a zygote, which increases the chances of establishment of mutations, and restricts the possibility for transmission of viruses to the progeny'. (c) 'Vegetative daughter plants generally establish themselves close to the mother plant whereas seeds are often efficiently dispersed. Seeds are a dry phase, related to dormancy and survival of adverse conditions. Vegetative reproduction, in general, does not involve dormancy.'

Agamospermy

Turning now to the details of agamospermy, we discover that it was first described in 1841 by J. Smith in plants of the Australian species *Alchornea ilicifolia* growing at Kew Gardens. The deduction that seed development had occurred without fertilisation could safely be made, as the Kew collections consisted entirely of female plants. Embryological studies that revealed some of the underlying mechanisms of seed apomixis were first carried out by Murbeck and Strasburger on *Alchemilla* and *Antennaria*. Since those classical studies, a wealth of detailed examples has accumulated and several very important generalisations can now be made.

First, apomixis occurs very rarely in diploids: there is a well-established correlation between apomixis and polyploidy (Bengtsson, 2009). Only in one diploid species (*Boechera holboellii*, a close relative of *Arabidopsis thaliana*) is apomixis confirmed, with this breeding behaviour being associated with the presence of an 'extra' B chromosome (Sharbel *et al.*, 2005).

Secondly, seed apomixis is a modification of normal sexual seed development. 'All outer aspects of sexual reproduction may be retained both structurally and functionally, with the exception that the fertilised egg is replaced by a suitably prepared diploid cell from the mother' (Bengtsson, 2009).

Apomixis has been detected in ferns. Apomixis is rare in the gymnosperms (Van Dijk, 2009), but sometimes cloning or twinning of embryos occurs from a single sexual zygote – a process know as cleavage polyembryony (see Mogie, 1992). Turning to the angiosperms, apomixis has been detected in more than 40 flowering plant families in at least 140 genera (Whitton *et al.*, 2008). There are different forms of seed apomixis (see below). If these variants do not have a common ancestry, then apomixis has arisen independently many times in angiosperm evolution (Van Dijk, 2009).

Thirdly, there are certain flowering plant families that show a great deal of apomixis affecting several genera; the outstanding familiar examples in the Northern Temperate flora are in the Rosaceae, Poaceae (Graminae) and the Asteraceae (Compositae). Van Dijk (2009) emphasises that in Poaceae and Asteraceae there is a clear clustering of apomixis at the (sub-)tribal level, suggesting pre-adaptation for apomixis or common ancestry (Van Dijk & Vijverberg, 2005). Equally remarkable is the absence of apomictic behaviour in some very large plant families, such as the Leguminosae. Fourthly, there is in these families a rather obvious correlation between taxonomic difficulty and the occurrence of apomixis. Particularly in the nineteenth century, many 'critical' genera or species-groups were detected, in which the taxonomists found extreme difficulty in reconciling the points of view of the 'splitters' and the 'lumpers' as to specific delimitation. Many of these groups are now known to be agamospermic. Familiar examples are *Crataegus* (in North America), *Rubus* and *Sorbus* in the Rosaceae, and *Hieracium* and *Taraxacum* in the Asteraceae.

There is a wide literature on apomixis. Classic studies, begun in the nineteenth century, are reviewed in Gustafsson (1946, 1947a,b), Stebbins (1950), Grant (1981), Van Dijk & Vijverberg (2005), Asker & Jerling (1992) and Richards (1986). Recent developments in our understanding of apomixis are presented in Richards (2003), Hörandl *et al.* (2007), Whitton *et al.* (2008), Tucker & Koltunow (2009) and Van Dijk (2009).

Apomictic phenomena

Complete agamospermy is not difficult to detect. For example, in certain plants of the Common Dandelion (*Taraxacum officinale*), sterile emasculation of all the florets, if performed carefully, followed by immediate 'bagging' of the capitulum, will result in a perfect head of fruit. The hypothesis of partial apomixis, on the other hand, cannot be tested in this way. Even more difficult are the cases of *pseudogamy*, in which pollination is necessary for seed formation but nevertheless the embryo is not formed by sexual fusion. Indeed, the detection of pseudogamous situations is so difficult that we may well suspect them to be more common than we know at present. Maternal (matroclinous) inheritance is usually an indication of pseudogamy. If an apparent cross between two plants differing obviously in easily scored characters produces a rather uniform F_1 resembling very closely the female parent, one should look for pseudogamy in the details of embryo formation. This is how the phenomenon was first suspected in the case of *Ranunculus auricomus*. It has been demonstrated in 'crosses' in *Potentilla*. Now molecular markers are the preferred choice in such studies. If 'crosses' are made between plants with different genetic markers, then the hypothesis that progenies are maternal in their genotype can be critically tested (Asker & Jerling, 1992).

Embryology of apomixis

Within the restricted space available here it is not possible to give any detailed account of the embryological and cytological complexity of apomictic groups. Here, we outline some key information: for more comprehensive accounts see Gustafsson (1946, 1947a, b), Battaglia (1963), Nogler (1984), Asker & Jerling (1992) and Naumova (1993).

When we come to look at apomixis at the embryological level, we find that we are not dealing with a single, standard pattern of development, but with a whole range of situations having in common only one feature, namely that they involve abandonment of the fusion of gametes in the normal sexual process as a necessary preliminary to embryo and seed development. As in many biological topics, we find terminological difficulties: the terms used here generally follow Gustafsson. Where Battaglia differs substantially, we have indicated the alternative term in brackets.

Before apomictic situations can be understood, we must know the normal pattern for a sexually reproducing flowering plant. (In ferns, apomixis is necessarily somewhat different because the gametophyte is a free-living plant separate from the sporophyte; most fern apomixis is technically *apogamy*, in which vegetative cells of the gametophyte give rise to a new embryo directly, thus omitting the stage of gamete production.)

Figure 7.6a illustrates in diagrammatic form the essential features of the development and fertilisation of the ovule of a typical angiosperm. Note the following points:

1. The megaspore originates by meiotic division as one of four products of an initial megaspore mother-cell; and normally the other three nuclei degenerate early.
2. A mature ovule ready for fertilisation contains a single embryo-sac, which corresponds to the free-living gametophyte generation in the ferns (where it is called a 'prothallus').
3. The embryo-sac develops by means of three ordinary mitotic nuclear divisions from the megaspore. This embryo-sac contains eight nuclei, one of which is an egg-cell or female gamete. 'The gametophyte consists in most species of eight nuclei and seven cells – one cell is binuclear (fused nuclei: n+n), the central cell. One of the other cells is the egg cell' (Van Dijk, 2009).
4. Van Dijk continues: 'in sexual flowering plants, there is a so-called double fertilization: the pollen grain tube contains two nuclei (both n), of which one fertilizes the egg cell ($n + n = 2n$) and the other fertilizes the central cell of which the two nuclei fused ($(n + n) + n = 3n$). The fertilized egg cell forms the embryo and and the fertilized central cell forms the endosperm.' This is a temporary tissue that nourishes the developing embryo. He stresses the important point that the 'maternal to paternal genome ratio in endosperm is 2:1'.
5. The embryo divides mitotically, and 'together with the maternal seed coat the embryo and endosperm make the seed'. After germination, the mature diploid plant develops.
6. Considering the essence of normal sexual process, the meiotic division that occurs in the formation of the megaspore (and also a similar meiotic division in the formation of the microspore or pollen grain (Fig. 7.6a, b) results in the production of a cell with a single set of chromosomes – the so-called haploid state. Subsequent mitotic divisions replicate this haploid set, so that the gametophyte generation, and the male and female gametes produced are all haploid. The sexual fusion of egg-cell (female gamete) with pollen-tube generative nucleus (male gamete) restores the diploid state in the zygote. This cycle of haploid gametophyte generation succeeded by diploid sporophyte generation is of fundamental significance in the plant kingdom, and can be traced from the more complex algae

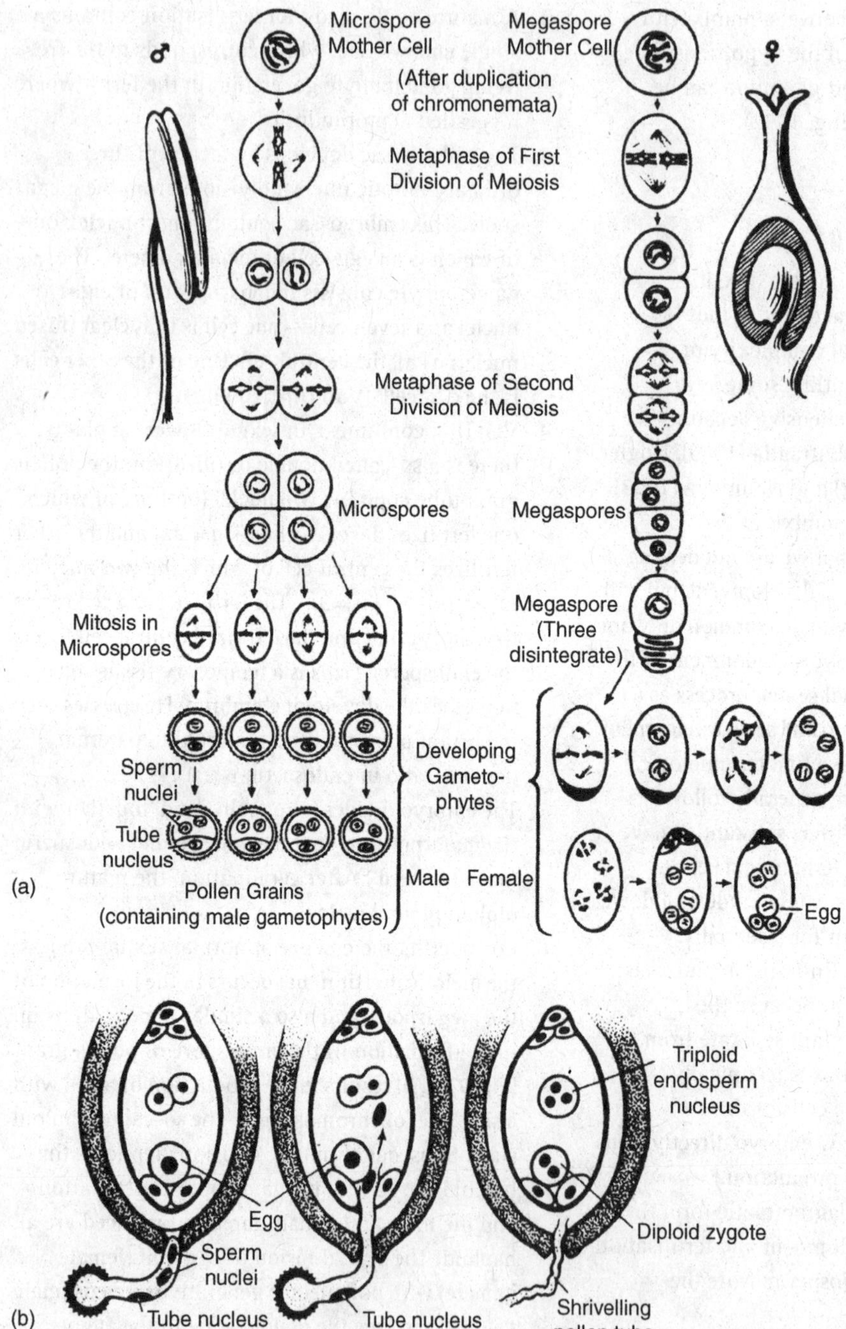

Fig. 7.6. (a) Development of pollen and ovules in sexual angiosperms. (b) Fertilisation in angiosperms. Two sperm nuclei unite with the egg and central nucleus respectively. (From Winchester, 1966.)

right through to the flowering plants. The apparent simplicity of the life cycle of the flowering plant, involving pollination, seed setting and dispersal, disguises a complex evolutionary history of suppression of the free-living gametophyte generation and the free-swimming gametes, which are still present in other members of the land flora such as the ferns.

Apomictic behaviour

Returning to apomictic flowering plants, we find that the great majority of apomictic deviations from the normal embryological pattern of development involve the production of an embryo-sac with the unreduced number of chromosomes, from which, an egg-cell develops directly, without fertilisation by a male nucleus. Thus, the diploid embryo-sac, so formed, requires no complement of chromosomes from a male gamete to restore the normal sporophyte number. In this way, the commonest kinds of apomixis cut out the meiotic stages from the life cycle, so, at first sight (but see below), the possibilities of variation generated by sexual reproduction are lost. It is for this reason that apomictic reproduction is genetically equivalent to vegetative propagation.

If the origin of the unreduced gametophyte is investigated, it is generally possible to distinguish between two situations. In the first, which is called *diplospory* (*gonial apospory*), the gametophyte arises from an unreduced megaspore.

Diplospory

Van Dijk (2009) notes that 'a normal reductional meiosis is replaced by a non-reductional division which can be mitotic-like' (e.g. in *Hieracium, Antennaria*) or meiosis may be interrupted at the first stages (*Taraxacum, Tripsacum*). 'In both types of diplospory two unreduced megaspores (2n) are produced, of which one degenerates and the other develops into an unreduced gametophyte with an unreduced egg cell.'

Apospory

In contrast, considering *apospory* (*somatic apospory*), Van Dijk (2009) stresses that 'in addition to the normal reduced megagametophye (n) a second but unreduced (2n) megagametophyte is produced from a non-spore cell (aposporous initial). Within a single ovule there is competition between two developing gametophytes. The 2n aposporous gametophyte forms an unreduced egg cell, that develops parthenogenetically into an embryo being genetically identical to the mother plant.' Thus, 'the reduced, fertilization-dependent gametophyte is suppressed and an apomictic seed is formed'. Such behaviour is found in a number of taxa including *Pilosella, Hypericum perforatum* and *Poa pratensis*. However, as we see below, a low percentage of functional sexual egg cells may be produced and some sexual offspring may arise in aposporous species.

Considering these two major routes to apomixis, it is apparent that the diplosporous condition involves a less radical departure from the normal sexual pattern than does the aposporous. Considering the interpretation of the embryology in the two cases, it is easier where the megaspore mother-cell is clearly differentiated in an early stage in the ovule (that is, the archaesporium is unicellular), as, for example, in the Asteraceae (Compositae). However, there are some groups (for example, Rosaceae) in which there is a multicellular archaesporium, and in such cases a decision as to whether a particular cell that has undergone apomictic development is or is not part of the archesporial tissue may be difficult or even almost arbitrary. Thus, the fact that both diplospory and apospory occur in *Potentilla neumanniana* (*Potentilla tabernaemontani*) is not indicative of any fundamental difference in this case (Smith, 1963a, b). Asker & Jerling (1992) may be consulted for details of the different variants of diplospory and apospory that have been detected in the embryology of the angiosperms.

Diplospory and apospory are the commonest apomictic situations, but we must briefly mention the range of possibilities that Battaglia calls aneuspory. In such cases the megaspore mother-cell undergoes a more or less irregular meiosis to form the megaspore. In apomictic *Taraxacum*, at the first division of meiosis, there is usually no chromosome pairing and, instead of producing two nuclei, a single restitution nucleus is produced. The second division then produces a dyad of unreduced cells (instead of the normal tetrad) and the lower one of these functions as a gametophyte initial, giving the normal eight-nucleate embryo-sac. However, in some cases chromosome pairing does occur in *Taraxacum*, and this may allow some crossing-over and reassortment of genetic material, which is not possible in the simple cases of diplospory and apospory. 'Sub-sexual' complexities of this kind may be more widespread, and more important in their effect on variation patterns, than has yet been established.

Embryological and other investigations have detected another apomictic mode of reproduction, namely adventitious embryony. Typically, after normal embryo-sac formation and fertilisation, additional somatic embryos develop from the cells of the nucellus or integuments. Thereafter, embryos, formed from sexual and asexual processes, compete for resources, resulting in different proportions of sexual and asexual products, e.g in *Citrus* the percentage of asexual seed varied between 33% for Eureka lemon and 100% in Darcy mandarin (Reuther, Batchelor & Webber, 1968). Adventitious embryony is the least well known of the modes of apomictic reproduction. Reports suggest that it is found in the Rutaceae, Celastraceae and Orchidaceae, and in other groups: indeed this variant of apomixis might be the most taxonomically widespread in the angiosperms.

One final question concerns the function of pollination in pseudogamous species. Whitton *et al.* (2008) review the evidence.

1. In most cases 'pollen is necessary for the proper development of the endosperm, with at least one of the pollen nuclei fusing with at least one of the polar nuclei'.
2. 'Adventitious embryony is usually pseudogamous ... pseudogamy is prevalent among aposporous apomicts, whereas anomalous endosperm formation is more common with diplospory' (Richards, 1986).
3. 'The requirement for fertilization of the endosperm selects for the maintenance of at least some viable pollen' (Noirot, Couvet & Hamon, 1997).
4. 'Apomicts with autonomous endosperm formation tend to produce less viable pollen, and, in some cases, are male sterile' (Meirmans, Den Nijs & Tienderen, 2006; Thompson & Whitton, 2006; Thompson *et al.*, 2008).
5. 'In rare cases apomicts have been shown to require pollination to stimulate seed development, even though neither the embryo nor endosperm is fertilized' (Bicknell *et al.*, 2003).

The role of pollen in the evolution of apomictic plants

Whitton *et al.* (2008) stress that early researchers on apomixis concentrated their attention on the development of the seed. However, it is clear that residual sexual function is retained in pseudogamous species. They may produce some meiotically reduced viable pollen. The possible contribution of pollen to the spread of apomixis in the hybridisation between apomictic (supplying male gametes) and sexually reproducing plants was not properly acknowledged until Maynard Smith (1978) and Mogie (1992) drew attention to its likely importance. We consider the evidence for such hybridisation and its consequences below.

Genetics of apomixis

Full analysis of genetical studies of apomicts is provided by the reviews of Ozias-Atkins (2006) and Whitton *et al.* (2008). After a detailed examination of the experimental evidence, Whitton and associates conclude that: 'Interesting trends emerge from the genetic data obtained thus far . . . most of the alleles controlling apomixis are dominant.' Here we examine important elements that need to be included in any model of the genetic control of apomixis.

1. Genetic crosses are only possible in certain circumstances. Most studies have examined F_1 ratios generated by crossing an apomictic plant as the pollen parent with a sexually reproducing female plant. Clearly, such crosses are restricted to those apomicts that produce viable pollen.
2. It has long been recognised that there is an association between apomixis and polyploidy. Dominant control of apomixis is not surprising. If, hypothetically, apomixis were to be controlled by recessive alleles, then there would need to be multiple copies of recessive alleles in polyploids.
3. It would appear that there are two distinct elements in the control of apomixis: the production of unreduced megagametophytes, originating through apospory or diplospory, and the subsequent development of the embryo through parthenogenesis. Whitton *et al.* (2008) are surely correct when they write: 'the chance that two mutations causing these two shifts would occur soon after each other within a small nascent apomictic population seems prohibitive, yet without the simultaneous emergence of the two traits, apomixis seems unlikely'. Asker & Jerling (1992) favour a model where a single mutation is responsible for both these changes. To date, studies of apomictic plants do not support such an idea. However, their analysis might have traction, if it could be shown that normal sexual diploids have alleles for parthenogenesis with little or no penetrance. Such alleles could then become important, if apomictic behaviour were introduced by a mutation, giving rise to the development of unreduced ovules.
4. There is yet another possibility. As we see in later chapters, many apomicts are of hybrid polyploid origin. Perhaps the two mutations needed for the functioning of apomixis might be separately present in polyploid parental stocks and brought together by hybridisation leading to a functional apomict.
5. Other elements might need to be incorporated into a model, for example, the genetic control pseudogamy, and in some apomicts the autonomous development of endosperm occurring without fertilisation.
6. Devising a universal model of apomixis may be unrealistic: clearly the phenomenon occurs in different forms and in many unrelated families, suggesting that apomixis has evolved many times in plant evolution.

Molecular nature and origin of apomixis

As part of the intense study of molecular development in model plants, investigations of reproductive pathways have been carried out, with particular attention to the behaviour of mutants. Here we are concerned with their overall conclusions. Tucker & Koltunow (2009) report the 'identification of novel apomixis-like mutants in sexual plants'. With regard to the behaviour of apomicts, they consider that 'the initiation of sexual reproduction preceding apomixis in most apomicts and the similar expression of reproductive marker genes in some sexual and apomictic species provide support for the theory that apomixis evolved and is manifested as a modified form of sexual reproduction, instead of an entirely novel reproductive pathway'. They have devised a model of the initiation of apomixis, involving the following steps.

a. Reproduction is almost certainly a necessary cue for the induction of apomixis.
b. 'The subsequent initiation of apomixis is likely to be dependent on positional information that varies between diplospory and apospory, and the precocious induction of a mitotic program that offers a competitive advantage over sex.'
c. Subsequent steps in the apomictic process until embryo-sac maturity make use of the basic sexual framework, at least in the case of apospory.
d. Thus, the current evidence suggests that 'apomixis evolved from the same molecular framework supporting sex'.

The research effort to understand the molecular genetics of apomixis is gathering pace. If genes for apomixis can be identified and introduced transgenically into crop plants, especially grasses, there is a huge potential for propagating and exploiting elite cultivars in agriculture (Bicknell & Koltunow, 2004; Ozias-Atkins, 2006; Barcaccia & Albertini, 2013).

Some dogmas about seed apomixis

Some generalisations emerging from early studies of apomixis have been widely circulated and often uncritically accepted amongst ecologists, taxonomists and others. These 'dogmas' have been identified (Gornall, 1999), and are based on the assumption that agamospermic taxa are obligately apomictic. For example, 'agamosperms are usually of hybrid origin and therefore highly heterozygous' (Gustafsson, 1946, 1947; Stebbins, 1950). They are 'often (always?) genetically invariant lineages' (Löve, 1960; Janzen 1977). And they 'are evolutionary dead-ends' (Darlington, 1939) and can produce 'new variation only on an old theme' (Stebbins, 1950). As we see in this and later chapters, all these generalisations are being examined, and often challenged, as hypotheses about variability and 'origins' are investigated by molecular methods.

B. Breeding behaviour discovered in studies of wild populations

Consequences of different reproductive modes

Having discussed the three main modes of reproduction, we may now examine the consequences of reproduction in each mode.

What happens as a result of repeated self-fertilisation is highly important in our understanding of breeding systems. (For an interesting review of the history of this subject, see Wright, 1977.) Studies of inbreeding in Maize by East and Shull at the beginning of the century provided an important model, which has been confirmed in many other studies of crop plants. In the heterozygous diploid the dominant allele often shelters recessive alleles that are deleterious in the homozygous state. Self-fertilisation quickly results in the segregation of lethal or sublethal types as homozygous recessives are produced. Unless specially looked for in the seedling stage these types, which may die at a very early stage of growth, may be undetected even in garden or glasshouse culture. (For a thorough discussion of the evolutionary consequences of deleterious mutations, see Charlesworth & Willis, 2009.)

Further selfings produce rapid separation of the material into uniform lines, often called pure lines, differing from each other in various vegetative and reproductive characteristics (Fig. 7.7). The continued selfing of uniform lines may be rendered impossible as some plants may become weak or sterile. Surviving lines may be characterised by plants of reduced vigour and fertility. If plants of pure lines originating from different parental stocks are crossed together, hybrid vigour (so-called heterosis) may be demonstrated. Such hybrid plants are characteristically of great vegetative vigour and high fertility (Table 7.2).

It is important to note that crossing genetically closely related plants from lines derived by repeated self-fertilisation from the same original parental stock

Table 7.2. **Hybrid vigour in Maize (*Zea*). Crossing inbred lines of Maize P_1 and P_2 yields F_1 plants showing hybrid vigour. Repeated self-fertilisation through several generations results in diminution in height and loss of yield. (From Jones, 1924)**

	Parents		Successive generations							
	P_1	P_2	F_1	F_2	F_3	F_4	F_5	F_6	F_7	F_8
Number of generations selfed	17	16	0	1	2	3	4	5	6	7
Mean height (inches)	67.9	58.3	94.6	82.0	77.6	76.8	67.4	63.1	59.6	58.8
Mean ear length (cm)	8.4	10.7	16.2	14.1	14.7	12.1	9.4	9.9	11.0	10.7
Mean yield (bushels per acre)	19.5	19.6	101.2	69.1	42.7	44.1	22.5	27.3	24.5	27.2

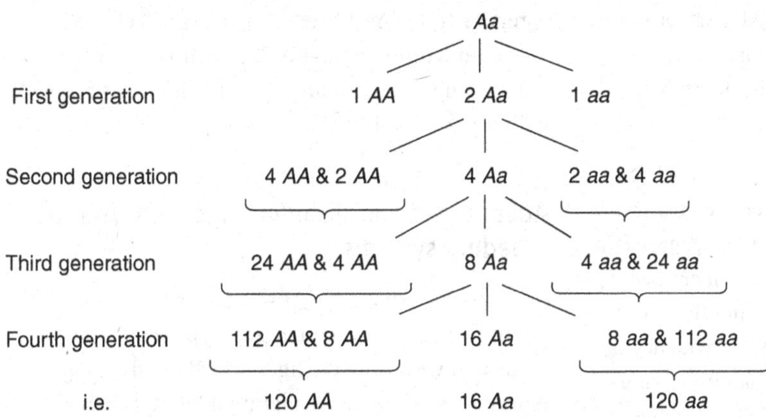

Fig. 7.7. Diagram showing the effect of selfing on a heterozygote. In the absence of selection, the proportion of heterozygous individuals rapidly declines in successive generations, and pure lines are established. (From Wilmott, 1949.)

will not give heterotic plants. It is clear from the model that repeated self-fertilisation will yield complete homozygosity in a few generations unless the heterozygous state is favoured by selection (Fig. 7.7). However, many plants are polyploids, having more than two representatives of a gene, and, as we shall see in a later chapter, a greater number of generations will be required to produce complete homozygosity in such plants.

The phenomenon of heterosis, so pronounced in experimental crosses with inbred lines, was not a new discovery, being often reported in the studies of early plant hybridists (Roberts, 1929). The underlying causes of loss of vigour or fertility on repeated selfing and the heterotic effects in products of crossing inbred lines have been the subject of intense study. The molecular biology of hybrid vigour is being actively investigated. Balcombe (2011) writes:

> plants use RNA silencing to shut down the selfish DNA in their genomes. We now realise that because different varieties of the same plant have different selfish DNA and different siRNAs to combat it, this could be one mechanism for explaining the mysteries of hybrid vigour . . . crossing two varieties results in a mix of siRNA that is different to that of either parent. In some cases, the new mix optimally turns the right combination of genes on and off and results in offspring that is better than either parent.

For more detailed accounts of this controversial subject the reader is referred to Charlesworth & Willis (2009) and Schnable (2013).

In considering generalisations that might be made on the effects of inbreeding, it is important to note that our ideas are based largely on results with a few crop-plant species. However, a few studies of wild plants have also been made (Charlesworth & Charlesworth, 1987; Charlesworth & Willis, 2009). Judged by seed set, germination rate, plant size, fertility or survival, there is often clear evidence of inbreeding depression, even for those species that are regularly self-fertilised. However, it should be noted that most of the data sets do not provide estimates of all components of fitness.

Different models have been proposed to account for inbreeding depression. Vergeer, Wagemaker & Ouborg (2012) characterise the current theories as follows:

> negative effects of inbreeding have until now been exclusively explained by classical genetic theories: the partial dominance hypothesis, i.e. the expression of deleterious recessive alleles owing to increased homozygosity in inbred individuals; and the over-dominance hypothesis, i.e. the reduced frequency of superior heterozygote genotypes. In addition, inter-acting effects between alleles at different loci (i.e. epistasis) have been suggested to contribute to inbreeding depression.

However, in recently published research (Vergeer, Wagemaker & Ouborg, 2012), there is evidence for 'an epigenetic role in inbreeding depression'. Vergeer *et al.* 'compared epigenetic markers of outbred and inbred offspring of the perennial plant *Scabiosa columbaria* and found that inbreeding increases DNA methylation', but 'inbreeding depression disappears when epigenetic variation is modified by treatment with a demethylation agent [5-azacytidine], linking inbreeding depression firmly to epigenetic variation'. While there may be a crucial role for epigenetics in inbreeding depression, further research is needed to see if demethylation effects persist in further generations.

Using isozyme and DNA markers, it has proved possible to test some of the predictions of models of breeding behaviour. Here we note that Gottlieb (1981a) compared the variation in outbreeding species with selfers. As predicted, selfers have statistically significantly less genetic variation, in the mean proportion of loci that are polymorphic, the mean number of alleles at polymorphic loci and the mean level of heterozygosity.

DNA fingerprint methods have also been employed to examine variation in species with different breeding systems (see Chapter 9). For example, Wolff, Rogstad & Schaal (1994) investigated population variation in *Plantago* species. *Plantago major,* which is regularly self-fertilised, showed relatively little variation within populations. In contrast, the outcrossing *P. lanceolata* exhibited high variation within populations.

Advantages and disadvantages of different breeding systems

One possible advantage of repeated self-fertilisation might be that well-adapted genotypes could be replicated with little change. A further advantage, especially in extreme or marginal habitats, where crossing between plants might be hazardous or fail altogether, is that self-fertilisation is an assured method of producing offspring (Lloyd, 1979a, b). An appreciation of the potential long-term disadvantages of inbreeding enables us to recognise the advantages of the outbreeding mode of reproduction. As we have seen, structural features or physiological mechanisms prevent or discourage self-fertilisation and lead to crossing between different individuals. The role of incompatibility mechanisms is very important in considering breeding within populations. Lewis (1979) has drawn attention to the important point that while some fruits (or seeds) from a given parent may be dispersed some considerable distance, many fall close to the parent, developing and flowering as a family group. This group, which may include parents and other relatives in plants with a long life cycle, is made up of genetically related individuals. Crossing between close relatives leads to inbreeding depression, although complete homozygosity is not achieved so swiftly by such matings. Lewis has shown that incompatibility systems will restrict crossing between close relatives, the degree of restriction depending on the genetic mechanism of the incompatibility system. Moreover, the outbreeding enforced by self-incompatibility allows new mutations arising in different individuals to be 'brought together' and, perhaps of equal importance a few exceptionally favourable progeny may be produced (Richards, 1986).

In general terms, in an ever-changing biotic environment, especially the presence of disease organisms (Burdon, 1987), obligate outbreeding would appear to have advantages associated with sexual reproduction, but there are costs to be borne. Compared with regularly self-fertilising species, a greater amount of biomass has to be employed in producing flowers, nectar etc. If only one genotype is present in an area, it may not be able to reproduce sexually. Reproduction may be rendered uncertain or unlikely by environmental factors influencing, for example, the variety and numbers of pollinating insects. Moreover, given that plants surviving to reproduce successfully in a habitat may be considered well adapted, outbreeding might seem to offer only the possibility of loss of such variants as each generation produces new variability. While this variability may include individuals of high fitness, there may be a considerable 'genetic load' of less fit progeny.

The third mode of reproduction – apomixis (either by vegetative (asexual) means or by agamospermy) – facilitates the reproduction in quantity of well-adapted plants of maternal genotype, with little or no genetic load. The agamospermic development of large quantities of identical progeny has been likened to the production line of the Model T Ford (Marshall & Brown, 1981). In some apomictic plants the pollen is defective and it has been suggested that the 'male costs' may be lower (Richards, 1986). Apomixis may also offer the possibility of reproduction by seed in plants with 'odd' or unbalanced chromosome numbers, such plants being unable to produce viable products at meiosis and likely to be totally or partially seed-sterile in sexual reproduction. Seed apomixis provides all the advantages of the seed habit (dispersal of propagules and a potential means of survival through unfavourable seasons). As we shall see later, apomictic plants are often of polyploid and hybrid origin. Therefore, this type of breeding system may be viewed as a means of conserving high heterozygosity (Asker & Jerling, 1992). Apomixis would also appear to be important at the edge of the range of many species, allowing populations to persist in territory in which various factors – including lack of insects – limit the extent, or exclude the possibility, of sexual reproduction (e.g. at high altitude or latitude).

While we might postulate various advantages of the apomictic mode of reproduction, it is clear that there are also some theoretical disadvantages. Mutations arising in different lineages cannot be brought together. The generation of 'new' variability would seem, at first sight, to be restricted or prevented in apomictic plants. Some have seen apomixis as an evolutionary 'blind alley' (Darlington, 1939). We return to this question later in the chapter.

Vegetative apomixis would appear to have limitations related to senescence and diseases. Whilst higher animals usually have a clearly defined life span, it is not clear whether there are ageing processes in perennial plants that would restrict or prevent natural asexual reproduction in the long term. However, observations on cultivated *Citrus* plants suggest that repeated vegetative propagation leads to senescence (Frost, 1938). In addition, reproduction by vegetative means carries with it the possibility that virus or disease might build up in the plant. A very good example (Richards, 1986) is provided by the sterile hybrid Primula (*P.* × *scapeosa*) produced in 1949 by crossing two Himalayan species. The plant was subsequently propagated by vegetative means and widely planted in gardens in different parts of the world. By 1982, it had ceased to be grown, as the plant had became debilitated by infection with cucumber mosaic virus. There is evidence that the sexual cycle provides a means of 'purging' the plant system of some disease organisms by providing a 'clean egg' (Richards, 1986). However, some viruses (Matthews, 1991) and other disease organisms are seed-borne.

An analysis of the three modes of reproduction reveals, therefore, that each has its advantages and disadvantages. While it might be important in certain open, early successional habitats to reproduce unchanged, well-adapted genotypes, by self-fertilisation or apomixis, in other habitats, with high spatial, temporal and biotic heterogeneity, plants capable of producing variable progeny would appear to be at a selective advantage.

Two commonly employed verbal models help us to see the force of this argument. 'The Tangled Bank' model – drawn from a famous quotation in Darwin's *Origin* about the interactions of plants – emphasises the heterogeneity of plant communities (Bell, 1982), whereas a second model considers the predator/prey and host/pathogen cycles at work in plant communities. Plants must produce the 'new' genetic variation necessary to 'stay in the game'. They have been likened to the Red Queen in *Alice* who ran to stay in the same place.

In addition to spatial, temporal and biotic heterogeneity, we must also consider vegetational and climate change over the longer timescale. There is evidence for such change from weather records, documents, plant remains, depictions of extreme winter conditions in works of art, etc. for fluctuations in climate in recent, as well as historical, times (Lamb, 1970). Intuitively, we can appreciate that a lineage lacking the capacity to produce variation might be at a serious selective disadvantage in competition with lineages capable of change. A lack of variation might prevent a lineage from withstanding the selection pressures associated with, say, migration during a period of global climate change. Important too, and perhaps less well known, are systematic changes, for example in climate, in the shorter term (Mather, 1966).

Reproductive assurance and the genetic 'quality' of progeny

Two interrelated issues are crucial in the evolution of breeding systems. Does the breeding system provide the assurance of reproductive success – with a high probability of progeny being produced and developing in the next generation? At the same time, given, for example, the likely deleterious effects of inbreeding, are 'progeny of high genetic quality' being produced and establishing themselves?

In pioneering work by Darlington, Mather, and Huxley the evolution of genetic systems was investigated, involving breeding systems, chromosome number structure and behaviour etc. (Barrett, 2010a). Considering the breeding system component of such systems in various groups – inbreeders, outbreeders, races, species, populations and individuals – these authors considered breeding behaviour in relation to the degree to which it delivered immediate fitness or flexibility; the extent to which the products displayed variance or invariance. Stebbins (1950) summarises the essentials of these pioneer efforts to understand plant breeding systems.

> All species populations of the more complex organisms should include individuals which are not perfectly adapted to the immediate contemporary environment of the species, but may be adapted to new environments to which the species will become exposed. As shown by Mather (1943), immediate selective advantage is gained by the *fitness* of particular individuals exposed to the selective forces, but the survival of the evolutionary line in a changing environment depends upon *flexibility*, expressed in terms of its ability to produce new gene combinations . . . in all evolutionary lines, a compromise is necessary between the conflicting requirements of fitness and flexibility . . . different groups have established compromises at very different levels, some in the direction of maximum stability and fitness at the expense of flexibility and others immediate fitness of all offspring sacrificed to the maintenance of flexibility.

As Stebbins (1950) notes, Darlington also developed the concept of 'anticipation' in genetic systems. 'Emphasis must be placed on the fact that the selective advantage of a particular genetic system benefits, not the individual in which it arises but its posterity. Whether an individual produces offspring of many different types or all similar makes no difference to the organism itself, but may have a profound effect on the survival of the race.'

Recently there has been criticism of these early approaches. First, in their ideas about pre-adaptation, but also as they imply a form of 'group selection' in ideas about the evolution of races, etc. Commenting on the studies of Darlington and others, Barrett (2010a) writes:

> unfortunately, influential work on the evolution of genetic systems during this period was based on the notion that the amount of genetic variation in a population was the main target of selection, with inbreeding and outbreeding promoting 'immediate fitness' versus 'long-term flexibility', respectively (Darlington, 1939). However, by the 1970s, the difficulty with this essentially 'group-selection' perspective was exposed largely through the work of David Lloyd (reviewed by Barrett & Charlesworth, 2007), and a return to models of individual selection

that characterised the investigation of R. A. Fisher and others. Thus, it is crucial to consider what happens when new breeding system variants appear in a population, either by mutation or migration. The fate, breeding behaviour and fitness of progeny of *individual* plants must be considered, as genetically variable populations encounter the forces of selection in microevolution. Charlesworth (2006) also emphasises the same point: 'selection acts on individuals in the short term, causing breeding system changes', therefore 'mating systems are properties of individuals not of populations'. Thus, modern studies of plant breeding systems focus on the testing of theories and models of individual selection.

Breeding systems in wild populations

Many different lines of evidence contribute to an understanding of breeding behaviour. Here, we examine some of the historic approaches to an understanding of breeding systems.

Inferring the breeding behaviour from flower structure or pollinator activities is unwise. For instance, the flowers of certain taxa of *Calyptridium* (Portulacaceae), which are regularly visited by insects, and therefore on logical grounds likely to be cross-pollinated, are in fact regularly self-pollinated by the insects that visit them (Hinton, 1976).

We may now examine, in outline, a number of different situations, showing how individuals within populations have different potential in breeding behaviour, and discuss a number of experiments which have shown how environmental factors provide a 'trigger', switching a plant from one mode of reproduction to another (Heslop-Harrison, 1964; Asker & Jerling, 1992).

Outbreeding combined with vegetative reproduction

Individuals of many self-incompatible species, producing variable progeny by outbreeding, are capable of considerable lateral vegetative spread. Decay of plant connections yields clonal patches of well-adapted genotypes often of considerable size and age, e.g. *Trifolium repens* (Harberd, 1963) and *Lysimachia nummularia* (Dahlgren, 1922).

Outbreeding in association with vivipary

Some species of genera such as *Agrostis, Allium, Deschampsia, Festuca, Poa* and *Saxifraga* have the capacity to reproduce not only by the sexual processes but also by vivipary, a condition in which tiny plantlets are produced in the inflorescence instead of, or mixed with, ordinary florets. In normal sexual reproduction, the generation of variation is possible, whilst the viviparous propagules reproduce the genotype of the plant that produces them (Fig. 7.5).

In experiments with *Poa bulbosa* (Youngner, 1960) it was shown that conditions of long day length and a short cold period followed by high temperatures yielded sexual inflorescences. In contrast, short day length and low temperatures yielded viviparous inflorescences, and mixed panicles of sexual and viviparous products resulted from long day/low temperature and short day/high temperature combinations.

Outbreeding combined with occasional self-fertilisation

It might be assumed that self-incompatibility mechanisms totally prevent selfing in nature. However, there are many cases known where largely self-incompatible species are capable of producing seed on selfing, for example the *Primula veris* stocks studied by Darwin. For a number of years experimental evidence has accumulated which suggests that self-fertilisation may occur in 'self-incompatible' species under certain conditions, for example, in material subjected to high temperatures, in situations where pollination of ripe stigmas is long delayed, or at the end of the flowering season (De Nettancourt, 1977). The rigidity, or otherwise, of incompatibility systems is being examined in considerable detail.

There have also been recent studies based on the analysis of endemic species of *Tolpis* from the Canary Islands using allozymes (Crawford *et al.*, 2010). In experiments on selfed seed set, almost all these endemic species tested were shown to be self-incompatible. However, in one species complex – *Tolpis laciniata* – self-incompatibility proved to be leaky to different degrees: some self-seed was set, albeit in low percentage. There was some variability between populations: the percentage of selfed seed ranged from 2% to 17.8%.

Very revealing studies have been obtained using molecular tools. For instance, Mena-Ali & Stephenson (2007) have examined the self-incompatibility system in the weedy invasive species *Solanum carolinense* in which the 'strength' of the self-incompatibility response 'is a plastic trait'. PCR methods have been used to identify the S alleles present and a two-generation crossing experiment has examined selfed and cross progenies. Mena-Ali & Stephenson discovered that one allele 'S_9 sets significantly more self seed than other S alleles in the population sampled'. Considering the mechanisms involved they suggest that variations in self-fertility may be 'due to factors that directly influence the expression of self-incompatibility by altering the translation, turnover or activity of the S-RNase'.

It is clear that molecular approaches are broadening our understanding of the functioning and possible leakiness of self-incompatibility systems. Also, accumulating evidence suggests that wide generalisations about the breeding behaviour of species/populations may very well be highly misleading: for example, some species, labelled as self-incompatible, may have populations containing individuals with mixed-mating strategies.

An important point to stress here is that 'leaky' behaviour has been discovered in other breeding systems, for example in so-called sub-dioecious species (Richards, 1997). Because a species is classified as having separate sexes in a flora, it does not follow that all individuals are strictly dioecious. The intriguing area of 'inconsistent males and labile sex expression' is reviewed by Ehlers & Bataillon (2007) and Renner (2014). Considering an example, *Withania aristata* (Solanaceae) is a dioecious endemic species from the Canary Islands (Anderson *et al.*, 2006). It shows 'weak gender plasticity in that a few male plants sometimes produce a very few fruits'. It has been suggested that 'leaky dioecy may facilitate the establishment after long-distance dispersal of colonizers' (Baker & Cox, 1984; Percy & Cronk, 1997).

However, it should not be assumed that the behaviour in *Withania* is typical or the norm. Studies of two species of *Dombeya* from La Réunion island have revealed that both are 'cryptically dioecious, each sex retaining non-functional morphological structures of the other sex' (Humeau, Pailler &

Thompson, 1999). 'One species is strictly dioecious, whilst the other shows leaky dioecy, with 8/10 males producing fruit during two years of study'. In the present context it is important to note that males set fruit and did so 'following outcrossing but not after self-pollination despite the growth of pollen tubes in the style. This suggests the occurrence of a late acting self-incompatibility system.'

Outbreeding combined with regular self-fertilisation

In some species (e.g. *Viola*) the spring-formed insect-pollinated flowers allow the possibility of outbreeding (Fig. 7.8), but in the summer, cleistogamous flowers are produced in which self-fertilisation is automatic.

Day-length is critical in the regulation of flowers. Borgström (1939) has shown that plants grown under 13–15 hours light per day produce normal flowers, whilst longer days (>17 hours), typical of early summer, induce the formation of cleistogamous flowers.

Cleistogamy, as we saw earlier, has been regarded as a rather rare phenomenon, but the capacity to produce cleistogamous flowers may be more widespread than hitherto realised (Richards, 1997). In some species, such as *Cardamine chenopodifolia*, there are normal flowers above ground, while cleistogamous flowers are produced below ground.

In considering what is widely seen as a 'fail-safe' reproductive strategy, Diaz & Macnair (1998) point out that the 'advantage of cleistogamy is its cost efficiency in seed production'. In cleistogamous flowers the 'structures for attracting pollinators are reduced to a minimum . . . with less resource allocation to attractive and male functions'. Also, cleistogamous species combine the capacity to produce 'cheap' flowers with an 'ability to detect environmental cues and respond by expressing the most effective breeding system for the conditions'. Thus, cleistogamous species may continue to flower and set seed under conditions of shade, temperature and light that are inadequate for the production of normal flowers.

Some of these ideas have been investigated experimentally. For instance, in *Impatiens capensis* the energetic costs of producing cleistogamous flowers appears to be only two-thirds those of normal flowers (Waller, 1979). However, seed from normal flowers – produced with the possibility of outcrossing – had better germination, survivorship and competitive success when compared with the obligatorily selfed progenies of cleistogamous flowers (Waller, 1984). Further investigation of cleistogamy would be valuable. For instance, it should not be assumed that seed from normal flowers is always produced by outcrossing. For example, in *Mimulus nasutus* both cleistogamous and normal flowers are autogamous, and so the seed of the normal flowers 'has no genetic advantage' (Diaz & Macnair, 1998).

In many plant species, male-sterile individuals occur as rare mutants but in some species, perhaps in a greater number than previously acknowledged, populations have a high proportion of female plants together with hermaphrodite individuals. Such species, which are particularly frequent in the Lamiaceae (Labiatae), are referred to as gynodioecious, a term coined by Darwin (1877a). Darwin's early studies of *Thymus* (Fig. 7.9) are particularly interesting. Gynodioecy permits the generation of variation in the crossing of hermaphrodite and female plants, whilst allowing the possibility of selfing in hermaphrodites, although it should be noted that some gynodioecious plants are self-incompatible, e.g. *Plantago lanceolata* (Hooglander, Lumaret & Bos, 1993).

While the gynodioecious breeding system is quite common, only a few cases of androdioecy have been reported (Charlesworth, 1985). In this breeding system, male and hermaphrodite individuals coexist, e.g. *Datisca glomerata* (Datiscaceae) (Fritsch & Rieseberg, 1992); *Phillyrea latifolia* (Oleaceae) (Aronne & Wilcock, 1994); and *Mercurialis annua* (Euphorbiaceae) (Pannell, 2008). In the case of *M. annua*, Pannell established that functional androdioecy occurs. While the species has male and co-sexual plants,

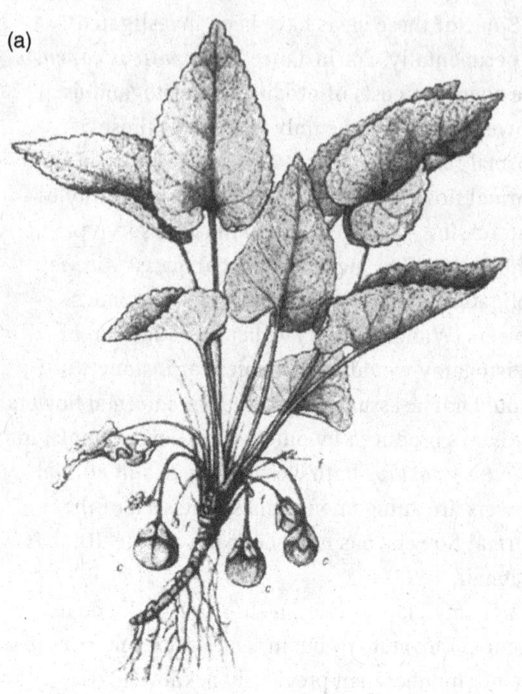

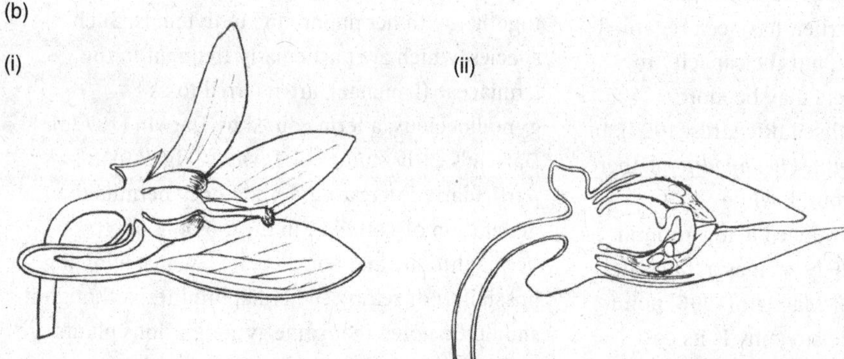

Fig. 7.8. Cleistogamy in *Viola*. (a) *Viola hirta:* plant with cleistogamous flowers, f, and developing capsules, c (×0.66). (b) *Viola riviniana* open (i) and cleistogamous (ii) flowers in longitudinal section, showing in the latter the crumpled style in contact with the developing anthers (×2.0). (From McLean & Ivimey-Cook, 1956.)

the co-sexuals have separate male and female flowers on the same plant.

Mixed reproduction: selfing and outcrossing in different proportions

There is a wide range of behaviour and floral morphology in self-compatible species, and while the percentage outcrossing in some species is very small, in others it may exceed 50%. Many factors may influence the rate of outcrossing, including floral structure, the behaviour of pollinators and environmental conditions.

In a very valuable reappraisal, Lloyd (1979a) makes the very important point that 'self-pollination is not a single unvarying process'. He distinguishes between models of:

a) 'prior' self-fertilisation, where the ovules are spontaneously fertilised without any opportunities for crossing, with

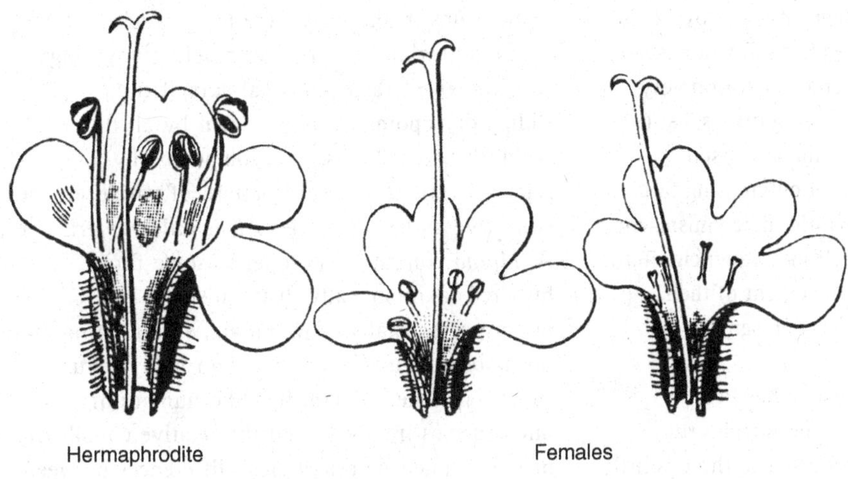

Fig. 7.9. Gynodioecy in *Thymus vulgaris* from sketches sent to Darwin from Mentone. The larger flowered hermaphrodite is figured left and the smaller flowered male sterile 'female' flower is shown on the right. (From Darwin, 1877a.)

Darwin raised plants from purchased seeds and discovered that the male sterile plants produced a greater quantity of seed than the hermaphrodites. He suggested that this extra seed production was 'compensation' resulting from the reallocation of resources in the production of male sterile plants. Studies by Assouad *et al.* (1978) confirm that male sterile plants produce more seeds than hermaphrodites in this species. However, seed production is not the only factor to consider. Male sterile plants outcross with the hermaphrodite plants. Hermaphrodite plants, on the other hand, may reproduce by self-fertilisation with the possibility of inbreeding depression (Charlesworth & Charlesworth, 1978).

Dommée, Assouad & Valdeyron (1978) discovered an interesting pattern of distribution of the two flower types in *Thymus vulgaris* in southern France. In unstable conditions – grasslands and old fields – the percentage of male steriles was high, allowing maximum seed production and outcrossing. In contrast, in more stable areas of rocky outcrops, the frequency of male steriles was much lower and there may be a greater level of autogamy. It is suggested that the frequency of male steriles in populations is subject to natural selection.

The genetics of gynodioecy has proved to be complex. In those species most extensively studied, sex determination has been found to be determined by nuclear and cytoplasmic genes (see text). For a review of our developing understanding of the biology and genetics of gynodioecious plants, see Lewis & Crowe (1956), Lloyd (1975), Widén (1992), Maurice *et al.* (1993), Ehlers, Maurice & Bataillon (2005) and McCauley & Bailey (2009). For a review of flower size dimorphism in plants see Delph (1996).

b) 'competition' between self and outcrossed products following the visits of pollinating insects and
c) 'delayed' self-fertilisation where there is an opportunity for cross-fertilisation and, should this fail, selfing will occur.

This analysis raises some very important issues concerning the relative competitive abilities of self- and cross-pollen. Darwin (1876) studied the progenies from self-compatible plants grown close to plants of a different variety or amongst plants of the same variety. It might be expected that, under such conditions, only selfed progenies would be produced. However, on the basis of character combinations, or vigour in the progeny, Darwin identified outcrossed plants in the progenies of seed produced by particular mother plants. This led him to postulate that the outcrossed pollen was 'prepotent' on the stigmas of the plants – in competition with 'self' pollen 'outcrossing' pollen 'won'. Support for this hypothesis has come from a few but not all experimental tests (see

Lloyd, 1992 for details). However, given Lloyd's reappraisal of self-compatible plants, it would seem important to re-examine pollen competition very carefully for 'prepotency can cause outcrossing to prevail whenever it is possible and allow self-fertilisation by default whenever outcrossing fails ... Altogether, the best-of-both-worlds mechanisms may be the cause of intermediate selfing frequencies in the angiosperms as a whole, but the extent of their occurrence is largely unproven at present' (Lloyd, 1992).

Another mixed breeding system has been described. In some self-compatible species the flowering sequence is 'synchronised so there is little or no overlap between staminate and pistillate phases of the individual plant' (Cruden & Hermann-Parker, 1977). Thus, there is, in effect, temporal dioecism and, as different plants are likely to be at different phases, outcrossing is encouraged. However, if flowers are unpollinated, selfing can take place.

Facultative and obligatory apomixis

In some genera, apomixis seems to have replaced completely the sexual processes in the great majority of species. In the Lady's Mantles (*Alchemilla*), for example, plants of the common northern European species-group to which Linnaeus gave the general name *A. vulgaris* show defective pollen, often degenerating in the tetrad stage, and precociously ripening fruit – sure indications that pollination is not necessary for seed formation. Indeed, with the exception of a very dwarf alpine species *A. pentaphyllea* and a very few alpine taxa belonging to another subsection of the genus, so far as is known, all *Alchemilla* species in Europe are apomictic. It is interesting that Robert Buser, the Swiss expert on *Alchemilla* who achieved an unrivalled knowledge of the plants in field, in herbarium and in cultivation, had rightly suspected from field evidence that certain puzzling intermediate populations in the Alps were hybrids of *A. pentaphyllea* and other species before anything was known of their genetical complexity. Obligatory apomixis, as is shown by all the 'vulgaris' Alchemillas, is accompanied by a relatively straightforward pattern of variation; the collective Linnaean species *Alchemilla vulgaris* and *A. alpina* consist (in Europe) of some three hundred taxonomically distinguishable microspecies, many of which are wide-ranging and no more difficult to identify than many sexual species in other genera. It is the number of microspecies involved, and the relative complexity of the detailed morphological differences between them, which make such critical groups the concern of so few botanists.

Studies of apomixis are increasingly confirming an extremely important conclusion: total or obligatory apomixis is much less common than partial or facultative apomixis. Bengtsson (2009) stresses two important points: 'asex is almost always associated with some sex'; and 'most asexuals are genetically variable'. Indeed, since we can never be certain that sexual reproduction is quite ruled out, even in cases such as *Alchemilla*, it may be that strict obligate agamospermy does not occur. In facultatively apomictic plants, amongst them the taxonomically difficult genera of *Rubus* and *Potentilla* in the Rosaceae and *Pilosella* in the Asteraceae (Compositae), apomictic embryos develop automatically without sexual fusion, but in addition some egg-cells are produced with the reduced number of chromosomes. Such eggs may be fertilised, giving rise to sexually produced embryos. It is predicted that obligate apomixis may yield little variation. In contrast, facultative apomixis combines a capacity to reproduce successful genotypes unchanged, with a mechanism allowing the generation of variation on an occasional or regular basis.

As we saw earlier, facultative apomixis is well illustrated by the adventitious embryony in the genus *Citrus*. According to Asker & Jerling (1992),

> usually pollination is followed by the double fertilisation of a reduced sexual embryo sac. The embryo and endosperm start to develop and the stimulation of embryo development . . . results in the growth of further adventive [apomictic] . . . embryos in the nucellus . . .Typically one of the apomictic embryos invades the embryo sac, out-competing the other apomictic and sexual embryos, and utilises the sexually produced endosperm. In other cases, the sexual embryo and one (or more) of the apomictic embryos coexist, sharing the same endosperm.

Thus sexual and apomictic progenies can be produced.

A good example of how occasional sexual crossing might yield a complex pattern is provided by the British representatives of the Series Aureae of *Potentilla*, which are, so far as is known, all pseudogamous apomicts. Most British plants can be classified as either *P. neumanniana (P. tabernaemontani)*, a rhizomatous, mat-forming perennial of chalk and limestone mainly in the lowlands, or *P. crantzii*, a non-rhizomatous perennial with an unbranched woody stock, typically found on calcareous cliffs in upland or mountain areas in Britain. In some parts of upland Britain, especially in northern England, however, puzzling intermediate plants occur, which obscure the otherwise fairly clear distinction between the two species. These plants mostly have higher chromosome numbers than normal specimens of either species. During a detailed study Smith (1963a, b, 1971) showed that, if large numbers of progeny were raised from 'crosses', occasional aberrant individuals could be detected because they differed from the normal offspring that resembled closely the female parent. One such individual, raised from the 'cross' between a *Potentilla neumanniana* (*P. tabernaemontani*) with $2n = 49$ from Fleam Dyke, Cambridgeshire, as a female parent and *Potentilla crantzii* with $2n = 42$ from Ben Lawers, Scotland, as pollen parent, was found to have 70 chromosomes. There can be little doubt that, in this case, an unreduced egg-cell with 49 chromosomes had been fertilised by a normal reduced pollen grain with 21 chromosomes, to give a zygote, and eventually a mature sporophyte plant, with $2n = 70$. A hybrid apomict had been synthesised. Such a plant can reproduce both vegetatively (to a limited extent) and agamospermously, and might have established a more or less uniform population in nature.

Moving to a more general model, sexual reproduction may not be rare in facultative apomictic plants and hybridisation between different variants could take place many times, giving populations differing subtly from each other according to the exact genetic constitution of their parents. Such events are particularly likely in pseudogamous species, where the pollen is largely normal. Furthermore, pollen from an obligate apomict may take part in sexual crosses in which the female parent is a sexual or facultatively apomictic plant (Asker & Jerling, 1992). Even the very limited sexuality of *Alchemilla* may for this reason be of far greater significance than we think.

Environmental control of facultative apomixis

It has been found in experimental studies of *Dichanthium aristatum*, for example, that environmental factors are important in determining the pattern of facultative apomixis. Thus, day-length is important in determining the balance between apomictic and sexual reproduction. Under continuous short days, up to 79% of embryos produced were apomictic; under long days, after floral induction in short days, only about 47% of embryos were aposporously produced (Knox & Heslop-Harrison, 1963). Similar results have been discovered for a number of other species (see Quarin, 1986). However, in controlled experimental studies of progenies of two variants of pentaploid Buffelgrass (*Pennisetum ciliare*), grown at 8, 12, and 16 hrs photoperiods, there was no relationship between the numbers of sexual embryo-sacs produced and photoperiod. Yet, under field conditions the number of sexual embryo-sacs was highest in spring and summer and lowest in late summer and autumn (Hussey *et al.*, 1991).

C. Evolution of breeding systems

Having discussed that different breeding systems are found in higher plants, it is appropriate at this point to make some preliminary observations on the evolution of the different systems. At the outset it is important to stress that we do not know the breeding system of the early flowering plants. However, a suggestion has been made by Whitehouse (1950), based on the widespread occurrence of self-incompatibility mechanisms in many flowering plant families. He proposed that this condition was ancestral, and postulated that the significance of the closed carpel 'lies in the protection of the ovules, not from desiccation or the attack of animals, but from fertilisation by the individual's own pollen, without appreciably restricting cross-fertilisation'. De Nettancourt (1977) supports Whitehouse's view and discusses the implications of the corollary proposition, namely that self-compatibility is a secondary, derived condition in modern angiosperms. He agrees that different breeding systems are adaptive under different conditions. This would imply that the initial diversification through the Cretaceous period was possible because of the efficiency of the outbreeding mechanism, but in later periods other factors were selectively more important, which favoured other breeding systems.

A great deal has now been published on the evolution of flowering plant breeding systems (see for example Barrett & Harder, 1996; Charlesworth, 2006; Barrett, 2010a, b). Here we note a number of the hypotheses concerning major trends.

1. For many years botanists have considered that gametophytic self-incompatibility (multi-allelic) is ancestral and that sporophytic systems are derived from it (see for example Richards, 1986). In a recent review of breeding system evolution, two conclusions emerge from a variety of evidence (Igic *et al.*, 2008). Firstly, given the distribution of gametophytic self-incompatibility systems, and phylogenetic considerations, the view that the ancestors of most eudicots (*c*.90–100 million years) were self-compatible is rejected. Secondly, it seems likely that 'the common ancestor of c.75% of dicots' had gametophytic self-incompatibility and that sporophytic self-incompatibity 'arose independently at least 10 times within the higher eudicots'.

With regard to the evolution of heterostyly, it seems possible that the ancestral stocks from which they arose were homomorphic (Charlesworth & Charlesworth, 1979), and that heterostyly arose as a mechanism favouring precision of pollination by increasing the chance of a stigma receiving compatible pollen (Yeo, 1975; Barrettt, 1992). Lloyd & Webb (1992), however, have examined the case for considering that at least some of the features associated with heterostyly could have evolved prior to self-incompatibility. Barrett & Shore (2008) have reviewed the evidence for the evolutionary pathways leading to the evolution of distyly. Summarising our present knowledge, Barrett (2010b) concludes that 'the evolutionary origins of heterostyly remain poorly understood . . . we are still some way from obtaining a clear picture of the stages in the build up of the polymorphism, especially the order in which the morphological polymorphisms and heteromorphic incompatibility are assembled'. Furthermore 'the molecular basis of heterostyly has not yet been determined for any species'.

2. Charlesworth (2006) reviews a second major evolutionary trend. 'The change from outcrossing to inbreeding is a repeated evolutionary transition'. As we have seen, such changes may occur in response to the 'low availability of conspecific potential mates or low availability of pollinators to ensure cross fertilization . . . self fertilization provides potential advantages of freedom from need for conspecifics, and from pollinating animals'. The effects of a 'selfing mutation can be very large' and 'long-term effects cannot prevent its spread throughout the population'.

In the last decade, molecular studies have provided revealing information on the evolution of self-compatibility. Investigations of the Mediterranean species-pair *Capsella rubella* (self-compatible) and *C. grandiflora* (self-incompatible) are consistent with the model that *C. rubella* separated from *C. grandiflora* recently (c.30 000–50 000 years ago) through the breakdown of self-incompatibility (Guo *et al.*, 2011). The limited variability of *rubella* suggests that it arose 'through an extreme population bottleneck . . . by a single selfing individual, most likely living in Greece'.

Arabidopsis thaliana is a self-compatible species in the Brassicaceae, a family in which self-incompatibility is widespread. Molecular investigations have revealed 'that the S-locus of *A. thaliana* harbored considerable diversity, which is an apparent remnant of polymorphism in the outcrossing ancestor' (Tang *et al.*, 2007).

These findings show that the fixation of a single inactivated S-locus allele cannot have been a 'key step in the transition to selfing'. *A. thaliana* has 'remnants of multiple old alleles of the S-locus', a situation that is 'inconsistent with a selective sweep involving a single S-locus mutation causing a transition to SC (self-compatibility) in this species'(Bomblies & Weigel, 2010). Reviewing the evidence on genetic variability at the S-locus, Tang *et al.* (2007) suggest that 'selfing most likely evolved roughly a million years ago or more'. However, in considering hitherto overlooked fossil evidence, Beilstein *et al.* (2010) consider that the split between *Arabidopsis lyrata* and *A. thaliana* occurred about 13 Mya.

In a further exploration of the evolution of self-compatibility in *A. thaliana*, Tsuchimatsu *et al.* (2010) consider the question, did selfing arise 'through mutation in the female specificity gene (S-receptor kinase, SRK), male specificity gene (S-locus cysteine-rich protein; SRC . . .) or modifier genes'? Their molecular evidence points to the widespread presence of a 'disruptive 213-base-pair (bp) inversion in the SRC gene' detected 'in 95% of the European accessions' studied. Furthermore in interspecific crosses between *A. thaliana* and *A. halleri* (as pollen donor), there is evidence that for some accessions of *A. thaliana* the female components, including SRK, are still functional. In considering these findings, Tsuchimatsu and co-workers point out that 'under conditions that favour selfing, mutations disabling the male recognition component are predicted to enjoy a relative advantage over those disabling the female component, because male mutations would increase through both pollen and seeds whereas female mutations would increase only through seeds'.

Experiments have also examined what happens when S-locus genes are transferred, using biotechnological methods, from self-incompatible variants of *A. lyrata*, to *A. thaliana*. It was discovered that self-incompatibility could be restored in *A. thaliana*, but not in all genotypes tested (Nasrallah *et al.*, 2004). This finding suggests that *Arabidopsis thaliana* populations may have 'independent mutations that caused or enforced the switch to self-fertility'. Further investigations, studying cpDNA and microsatallite variation, have revealed that not all *A. lyrata* populations are self-incompatible: there are varying degrees of self-compatibility, in populations in the Great Lakes region of Eastern North America, 'suggesting that selfing either evolved multiple times or has spread to multiple genetic backgrounds'. Overall the results 'do not suggest a single transition to selfing in this system, as has been suggested for some other species in the Brassicaceae' (Foxe *et al.*, 2010).

Many outbreeding species have inbreeding relatives (see for example *Leavenworthia* (Brassicaceae) (Lloyd, 1965); *Stephanomeria* (Asteraceae) (Brauner & Gottlieb, 1987); *Mimulus* (Phrymaceae) (Macnair, Macnair & Martin, 1989); and *Arenaria* (Caryophyllaceae) (Wyatt, Evans & Sorenson, 1992)). The evolution of autogamy in many groups is associated with changes in floral characters.

For instance, diversification in the genus *Phlox* in relation to pollen vectors has been studied by Grant & Grant (Fig. 7.10). They conclude that the group coevolved with different pollinators – bats, bees, hawk moths, hummingbirds etc. Six lines of specialised radiation were postulated, and, in the present context, it is significant that each group has its independently evolved small-flowered self-fertilising autogamous species. Self-compatible

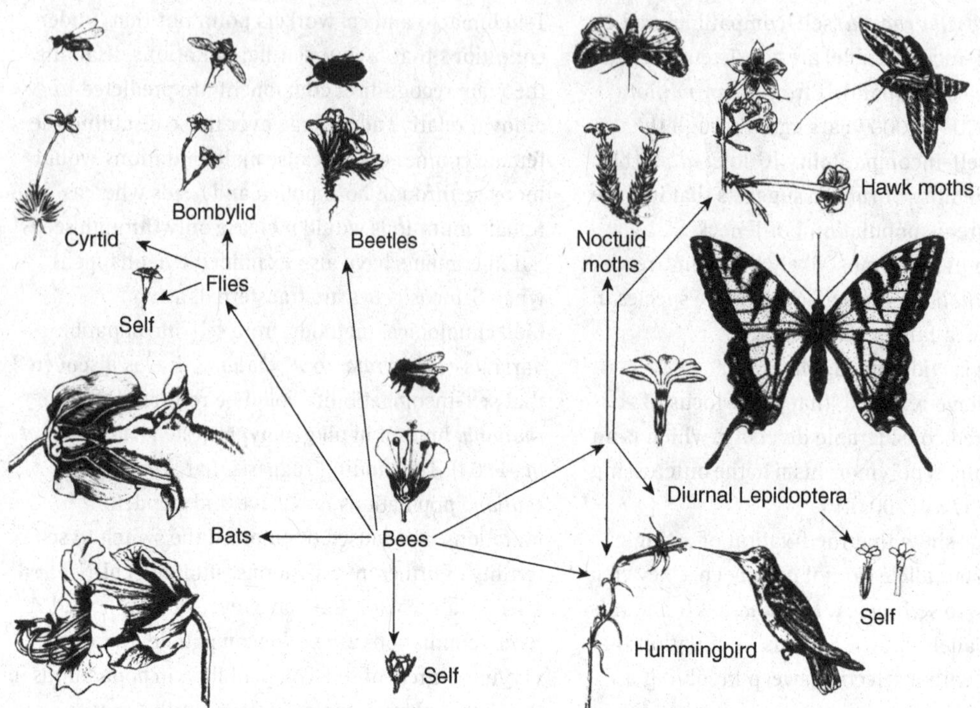

Fig. 7.10. It seems probable that, as a consequence of natural selection, the ancestors of a single group may through evolutionary divergence become adapted to a range of environments. A splendid example of presumed adaptive radiation for pollination by different pollen vectors is illustrated by present-day pollinators and flowers of the Phlox family (Polemoniaceae). The following species of pollinators and flowers are illustrated as representative of each group: bees, *Polemonium reptans* and *Bombus americanorum*; bats, *Cobaea scandens* and *Leptonycteris nivalis*; cyrtid flies, *Linanthus androsaceus croceus* and *Eulonchus smaragdinus*; bombylid flies, *Gilia tenuiflora* and *Bombylius lancifer*; beetles, *Linanthus parryae* and *Trichochrous* sp. (Melyridae); noctuid moths, *Phlox caespitosa* and *Euxoa messoria*; hawk moths (Sphingidae), *Ipomopsis tenuituba* and *Celerio lineata*; diurnal Lepidoptera, *Leptodactylon californicum* and *Papilio philenor*; birds, *Ipomopsis aggregata* and *Stellula calliope*; self (autogamous), *Polemonium micranthum* (bottom), *Gilia splendens*, desert form (left), *Phlox gracilis* (lower right). (From Stebbins, 1974, redrawn from Grant & Grant, 1965.) This branching diagram has some elements of a tree. Now the relationship between groups is most often explored using modern phylogenetic methods of analysing DNA sequences (see Chapter 16).

species may also show other character changes (Ornduff, 1969). Twenty-one such changes have been detected in detailed studies of a number of genera (Table 7.3).

However, self-compatible species are not always small-flowered. Adaptive radiation – in relation to such vectors as hummingbird, butterfly and moth – have been studied in the Orchidaceae (van der Pilj & Dodson, 1966) and, again, there are many instances of reversions to autogamy, but the flowers may not be reduced in size. How is this paradoxical situation explained? As we have seen above, many 'mixed' breeding systems occur. Such systems can provide the advantages of outcrossing, while allowing the reproductive assurance of self-fertilisation. It should not be assumed, therefore, that self-compatible species are on an evolutionary trajectory towards complete autogamy.

Concerning the steps in the development of self-compatible variants, a number of intensive studies of

Table 7.3. **List of 21 morphological and phenological character changes that are often associated with the evolution of autogamy. (Modified from Ornduff, 1969, by Wyatt, 1988)**

Outcrossing progenitors	Autogamous derivatives
Flowers many	Flowers fewer
Pedicels or peduncles long	Pedicels or peduncles shorter
Sepals large	Sepals smaller
Corollas rotate	Corollas funnelform, cylindric, or closed
Petals large	Petals smaller
Petals emarginate	Petals less emarginate
Floral colour pattern contrasting	Floral colour pattern less contrasting
Nectaries present	Nectaries reduced or absent
Flowers scented	Flower scentless
Nectar guides present	Nectar guides absent
Anthers long	Anthers shorter
Anthers extrorse	Anthers introrse
Anthers distant from stigma	Anthers adjacent to stigma
Pollen grains many	Pollen grains fewer
Pollen presented	Pollen not presented
Pistil long	Pistil shorter
Stamens longer or shorter than pistil	Stamens equal to pistil
Style exserted	Style included
Stigmatic area well defined, pubescent	Stigmatic area poorly defined, less pubescent
Stigma receptivity and anther dehiscence asynchronous	Stigma receptivity and anther dehiscence synchronous
Many ovules per flower	Fewer ovules per flower

distylous plants have been very revealing, e.g. *Primula* (Ernst, 1955), *Armeria* (Baker, 1966), and *Turnera ulmifolia* complex (Barrett, 1985). These have shown how crossing over in the supergene, which controls the distylous breeding system, has given rise to self-fertile homostyle plants.

Another example of the evolution of self-compatibility is provided by the study of tristyly in *Eichhornia paniculata* (Barrett, 1980a, b, 1985; Barrett *et al.*, 1989a, b; Barrett, Dorken & Case, 2001; Barrett, Ness & Vallejo-Marín, 2009). In Brazil and the Caribbean, large populations of this tristylous species are self-incompatible (with short-styled-S; mid-styled-M; and long-styled-L variants). In contrast, small populations are often dimorphic (some plants show self-compatibility) or monomorphic (all plants fully self-compatible). Barrett, Dorken & Case (2001) give the following details.

> The shift from outcrossing to selfing in *E. paniculata* occurs in two successive stages. First, the S-morph is lost from small populations through drift resulting in dimorphic (L, M) populations. Secondly, selfing variants of the M-morph spread to fixation . . . resulting

> in the loss of the L-morph. Selfing variants are favoured in dimorphic populations because the transmission bias of selfing genes is not countered by strong inbreeding depression, or pollen discounting. Selfing variants also benefit from reproductive assurance when pollinator service is insufficient and episodes of colonization and extinction of populations are frequent.

All things considered, it seems likely that self-compatible variants are at a selective advantage in areas of drought and human disturbance. Recent research involving morphological, DNA sequences and genetic analysis has extended these findings.

> The primary pathway from outcrossing to selfing involves the stochastic loss of the short-styled (S-morph) from trimorphic populations, followed by the spread of selfing variants of the mid-styled (M-morph). However, the discovery of selfing variants of the long-styled (L-morph) in Central America indicates a secondary pathway and distinct origin for selfing. Comparisons of multi-locus nucleotide sequences from 27 populations sampled throughout that geographical range suggest multiple transitions to selfing. (Barrett, Ness & Vallejo-Marín, 2009)

Concerning the long-term evolutionary consequences of self-fertilisation, Stebbins (1957) proposed that this 'irreversible' transition may be an 'evolutionary dead end'. In reviewing present evidence, Wright, Kalisz & Slotte (2013) accept that 'while some aspects of the hypothesis of selfing as a dead end are supported by theory and empirical results, the evolutionary and ecological mechanisms remain unclear' and more investigations are needed. For example, Karron *et al.* (2012) point out that much needs to be clarified about changes in physiological and floral traits that are part of the transition to selfing in many groups. They ask the question: 'do traits such as small flower size, reduced separation of anthers and stigma and decreased investment in male allocation evolve simultaneously with self-fertility, or sequentially? If sequentially, is there a predictable order? Surprisingly little is known about such questions.' Also, 'a diversity of environmental factors may influence transitions in the mating system . . . interactions with natural enemies or variation in the abiotic environment can alter the expression of inbreeding depression'. In addition, further investigations should be carried out into the 'evolutionary stability of mixed mating systems' and the role of phenotypic plasticity.

Finally, in considering the techniques available to study these problems Karron and associates highlight the potential of next generation sequencing for providing the means to study hitherto intractable problems.

> 'With next-generation sequencing technologies, researchers are increasingly able to genotype hundreds of markers within populations of non-model organisms, and probe the genetic architecture of population and individual traits critical to answering important questions about plant mating systems' (Karron *et al.*, 2012) and for 'fast-tracking' the discovery of candidate genes in investigations of self-incompatibility systems, sex determination, and 'studying both maternity and paternity in dispersed seeds and juvenile plants in natural populations'.

The role that polyploidy might play in the evolution of the transition from outbreeding to inbreeding will be considered in Chapter 14.

3. A third set of evolutionary trends link the evolution of self-compatibility and dioecy. Baker (1955, 1967) has suggested that, with self-compatible individuals, a single propagule is sufficient to start a sexually reproducing colony on an island or in a new habitat. In contrast, a single individual of a self-incompatible species would be unable to found a new colony reproducing by seed. There is support for this theory (elevated to a law by Stebbins, 1957), for example, in a study of plants of the Galápagos Islands: out of the 52 species studied, all those for which conclusive evidence was available proved to be self-compatible (McMullen, 1987). However, in parenthesis we should note that information is available for only about 20 per cent of the native

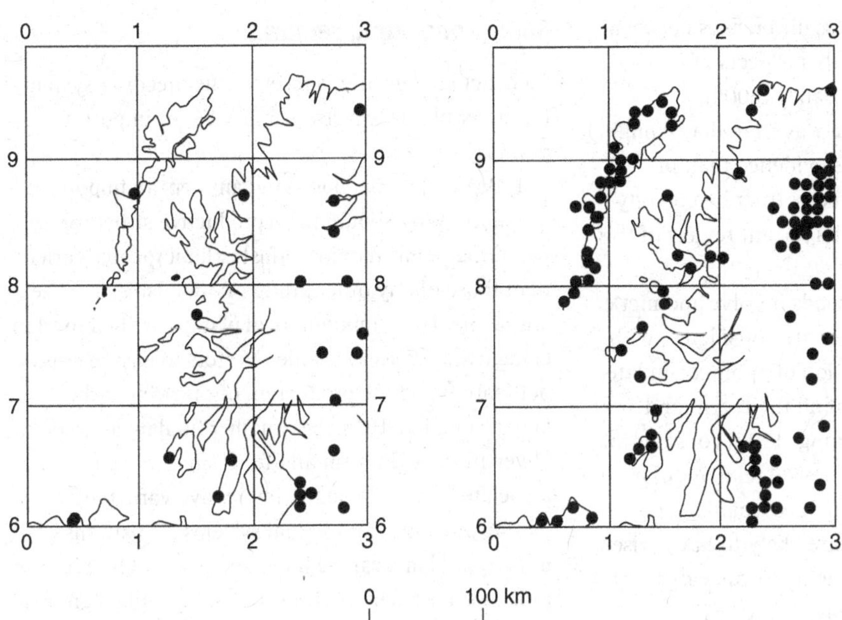

Fig. 7.11. Distribution in western Scotland of the self-incompatible Poppy *Papaver rhoeas* (left) and the self-compatible *P. dubium* (right). *P. dubium* has colonised many more of the Scottish Isles. (From Richards, 1986. Maps reproduced with permission from *Atlas of the British Flora*, Perring & Walters, 1976.)

species (Chomorro *et al.*, 2012). Support for Baker's view has also come from studies of *Nigella* in the Eastern Mediterranean (Strid, 1970) and *Papaver* (see Fig. 7.11) on Scottish islands (Richards, 1986). Baker's Law will be examined in more detail below.

At the beginning of this chapter, we noted that dioecious plants are disproportionately common on certain islands, e.g. New Zealand and Hawaii. Perhaps long-range dispersal has brought both sexes to the islands in two or more separate colonising events or perhaps, on a single occasion a many-seeded propagule arrived. However, another possibility must be considered. Self-compatible colonising populations may suffer inbreeding depression, and it has been suggested that selection may favour mutations in sex expression to yield, perhaps via gynodioecy, a fully dioecious system with its obligate outbreeding system. Charlesworth (2006) notes that at least two successive steps are involved. First, 'evolving females or males – generating gynodioecious or androdioecious populations respectively – and then changing the co-sexuals into males or females in one or more steps'. This model receives support from studies of the genetics of gynodioecy. To recap, populations consist of hermaphrodites and female plants. In a comprehensive review of all aspects of the subject, Ehlers, Maurice & Bataillon (2005) provide the following details:

> for most gynodioecious species, a combination of cytoplasmic genes causing male sterility (hereafter CMS) and nuclear restorer alleles determine the sex of an individual plant ... female plants carry a CMS gene blocking the development of functional anthers, while hermaphrodites either do not have such a CMS gene or carry at least one nuclear restorer allele that restores male function ... CMS genes are located in the mitochondria, and thus are usually maternally inherited ... a CMS gene can invade a population of hermaphrodites provided that females who carry it have even a slight female fitness advantage compared to the female fitness of hermaphrodites... higher seed set in females than hermaphrodites is indeed often observed.

For further details of the fitness differences between female:hermaphrodites and other aspects of gynodioecy, see McCauley & Bailey (2009).

This is not the only possible way that dioecy might arise; Ornduff (1966) provides evidence that, in *Nymphoides*, dioecy arose directly from heterostyly (see also Lloyd, 1979b for an important review of this area).

4. Other breeding system trends involve apomictic taxa. Here, a few general issues are considered, with further reference to the evolution of apomixis in later chapters. After a full review of all the recent and historic evidence, Asker & Jerling (1992) consider the commonly held belief that the sexual relatives of most agamospermous plants are self-incompatible and, therefore, apomictic variants are likely to have arisen from such plants. In many genera this appears to be the case; for instance in *Limonium*, the self-incompatible sexual relatives are heterostylous (Ingrouille & Stace, 1985). Berry, Tobe & Gómez (1991) suggest that in the genus *Erythroxylum* (Erythroxylaceae) agamospermy is derived from distyly. However, the tendency for near relatives to be self-incompatible is not an exclusive trend. In *Antennaria media* (*A. alpina* group) apomixis is found in a dioecious plant, the female being the only variant present over much of its range (Bayer, Ritland & Purdy, 1990). Furthermore, the relatives of apomictic *Citrus, Potentilla* and *Aphanes* are self-compatible.

Considering the evolutionary significance of apomixis, Bengtsson (2009) reviews current evidence and draws several important conclusions. 'Asexuality appears constantly amongst sexual eukaryotes', but there is no evidence for 'reverse evolution – from full asexuality to sexuality . . . The relative disadvantage of asexuality is directly seen in the eukaryotic [phylogenetic] tree in all its immensity, where sexuality dominates all major branches.' While an 'asexual lineage that monopolizes a suitable niche' may exist for a long time, 'asexual lineages, in general, are more short lived than sexual ones . . . the lack of recombination gives asexual lineages a long-term disadvantage'.

Some concluding remarks

To conclude our brief survey of the breeding systems in higher plants, we stress a number of important points.

1. Barrett (2010a) poses and answers an important question: 'Why should the reproductive structures of flowering plants (angiosperms) exhibit greater variety than those of any other group of organisms? . . . The answer lies in the immobility of plants and their need to engage the services of pollen vectors to ensure cross-pollination and the production of offspring of high genetic quality.' He makes another fundamental point: 'diversification in form and function of flowers is associated with an equally impressive variety of mating strategies and sexual systems . . . patterns of mating in plants can be both complex and highly promiscuous in comparison with many animal groups.'

Thus, for example, we see, in later chapters, how hybridisation is widespread in many angiosperm groups, leading to complex reticulate patterns of speciation.

2. In his writing about evolution in addition to natural selection Darwin (1859, 1871a) defined and discussed sexual selection in animals. Willson (1979) outlines what is involved: 'competition amongst members of one sex for the privilege of mating with individuals of the other sex, and some preference for, or at least advantage to, members of the second sex for mates that have characteristics permitting them to win the intrasexual competition'. Here briefly we face the question, does sexual selection occur in plants? Willson (1979, 1983) and Queller (1987) have both reviewed the subject; both make valuable suggestions about the significance of pollen/ovule ratios, pollination by pollen clumps etc. but struggle with the issues of choice and competition in the reproduction of plants. Queller provides an important insight when he concludes that 'due to their lack of mobility . . . female plants best exercise choice not by limiting matings but by mating promiscuously and choosing after pollination . . . such choice can be for traits of pollen tubes, endosperms or embryos'.

3. A major theme of this chapter is, quite rightly, the way 'reproductive features regulate the degree of inbreeding' (Charnov, 1988). For completeness, however, it is essential to stress again, briefly, an often overlooked and highly significant component of breeding behaviour and its evolution. Charnov stresses that 'both pollen and ovules contribute equal numbers of chromosomal genes to seeds'. Therefore, the 'male' (pollen) side of reproductive biology must be considered. It may be postulated that natural selection acts on both 'male' and 'female' reproduction. Does success in the 'female' mode reduce success in the 'male'? Does evolution by natural selection result in compromises in 'male' and 'female' success? How is this translated into use of resources in the plant/flower in producing various structures in the diverse groups of higher plants? All these fascinating questions are reviewed by Lovett Doust & Lovett Doust (1988) and more recently by Hodgens & Barrett (2008). A key issue that presents considerable practical difficulty is how to estimate 'female' and 'male' reproductive investment and success, especially in the field. Also, there are problems of deciding which parts of a hermaphrodite flower are acting as part of 'male' function and which are properly 'female'.

D. In the past, morphological studies, crossing experiments etc. have provided many insights into plant breeding systems. Molecular tools have revolutionised many fields of plant biology, and as we see below, a wide range of microevolutionary hypotheses and models have now been critically tested. With the wider use of next generation sequencing methods there is the exciting prospect of an increased understanding of evolutionary changes in breeding systems in higher plants. For example, progress is being made on the molecular identification of genes involved in the 'selfing syndome'. This involves the comparative study of 'sex allocation in male v female function and flower morphology, in particular flower (mainly petal) size and the distance between anthers and stigma' (Sicard & Lenhard, 2011).

Looking to the future, Charlesworth (2006) notes that it is important to increase our understanding of breeding system functioning, resource allocation and inbreeding depression in functioning ecosystems. We have seen that self-incompatibility systems are maintained over long evolutionary times. Further information is required on the timeframe for changes to selfing. Given the apparent weaknesses of regular inbreeding, is there a deficit of old selfing lineages and, most importantly, how far is it possible for incompatibility systems to re-evolve in selfing taxa? Some dioecious plants have sex chromosomes, the correct functioning of which depends upon the maintenance of many linked loci. As selection eliminates recombination between X and Y chromosomes, will Y chromosomes accumulate deleterious mutations (Charlesworth, 2002)?

In a wide-ranging review of breeding systems, Barrett (2010a) emphasises that the way ahead must involve combined approaches. 'Molecular and developmental studies can tell us how sexual diversity may arise but they cannot tell us why.' In combination with molecular studies, 'imaginative field experiments' will be needed including 'manipulations of the abiotic and biotic features of populations' to determine 'the selective mechanisms responsible for evolution and maintenance of reproductive traits'. 'Fortunately, the sedentary nature of plants and the fact that they can be easily cultured, crossed and cloned makes them ideal organisms for manipulative field experiments'. Such experiments have the potential to increase our understanding of plant breeding systems, and the intricacies of plant–insect/plant–bird interactions, for example, the evolution of plant–insect mutualisms (Bronstein, Alarcón & Geber (2006); the mechanisms and evolution of deceptive pollination in, for instance, the orchids (Jersáková, 2006); the dual role of floral traits: in pollinator attraction and plant defence (Irwin, Adler & Brody (2004); the evolution of bird-pollinated flowers (Cronk & Ojeda, 2008); and the evolutionary consequences of pollen theft by a range of animal species (Hargreaves, Harder & Johnson, 2009). In addition, we will also gain valuable insights into contemporary issues of major concern, namely how plant breeding systems are likely to change in species that are introduced into new territories; experience climate change; or become endangered and face extinction.

Our discussions of variation have brought us to the point where we have explored the *potential* variation generated by a variety of plant breeding systems. We now turn our attention first to a review of historic studies of factors influencing variation between and within populations (Chapter 8) and then, in Chapters 9–11, we survey our current understanding of microevolutionary pattern and process in plant populations, with further consideration of breeding systems. Later chapters will examine the role of hybridisation and polyploidy in microevolution, including the influence of these processes on the evolution of breeding systems.

8 Intraspecific variation and the ecotype concept

We saw in Chapter 7 how different breeding systems can be expected to produce different patterns of variation. If we are to understand the variation patterns actually found in nature and the processes that give rise to these patterns, we must discover how the potential variation in seeds relates to the variation of reproductively mature plants. Historically, the first advances in this field were made by means of comparisons between plants belonging to the same species but from different populations. Taxonomists, biometricians and, later, geneticists became interested in genetic variation in the wild, and many of their studies converged at one point, namely, the controversy over the 'reality' of the intraspecific groups, which could be distinguished in nature, and whether they were the subspecies or varieties of the taxonomist or the 'local races' of the biometrician. A new look at this old question was provided by the famous researches of Turesson published in the early 1920s.

Turesson's pioneer studies and other experiments

At the time of Turesson's experiments, the question of the reality of 'local races' was combined with another controversial issue, namely how much of the observed variation in natural populations was the result of the direct modification of plants subjected to severe environmental stresses. By the end of the nineteenth century, many botanists reasoned that distinctive intraspecific variants were merely 'habitat modifications'. Turesson, however, pointed out that in all previous cases known to him, only a partial test of the 'habitat modification' hypothesis had been carried out. For example, he considered the studies of *Lathyrus japonicus* (*L. maritimus*) undertaken by Schmidt (1899). Baltic populations of this plant have dorsiventral leaves, whilst on the North Sea coast of Denmark the plant has isolateral leaves. Schmidt showed, by experiment, that watering the Baltic variant with sodium chloride solutions induced a leaf structure typical of Danish plants. Given that the North Sea has a higher percentage content of salt than Baltic waters, Schmidt deduced that the leaf structure of the plants on the North Sea coast of Denmark was merely a habitat modification.

The logic of this type of deduction did not satisfy Turesson. His approach to the problem was to grow samples of several variants of a species in a standard garden, to see if 'distinctiveness' was retained or lost. He collected living plants (and in certain cases seeds) of many common species from a variety of natural habitats in southern Sweden, and grew them in experimental gardens first at Malmö (1916–18) and subsequently at the Institute of Genetics at Åkarp. In this way he studied, for example, shade variants, dwarf lowland plants from coastal habitats, and succulent variants, in most cases growing these plants alongside collections of the same species collected from ordinary inland habitats (Turesson, 1922a, b).

In some cases the distinctness of the variants was lost in cultivation in an inland garden, but usually the distinctive plants originating from extreme habitats retained their characteristics in cultivation, even in the absence of shading, salting, etc. These observations were clearly at odds with the notion that extreme variants were nothing more than habitat modifications, and the persistence of distinct variants under standard conditions suggested to Turesson that the variation had a genetic basis.

Many of Turesson's early experiments were carried out on the Composite *Hieracium umbellatum*. This plant is common in southern Sweden where its principal habitats – woodland, sandy fields, dunes and cliff tops – may all be found. In each of these habitats a distinctive plant was discovered in the field. By careful sampling and cultivation, Turesson found that, with few exceptions (for example certain prostrate plants from sandy fields), distinctive variants retained their characteristics in cultivation. The results of these experiments were consistent with those obtained in studies of other species, and again Turesson considered that patterns of residual difference had a genetic basis.

Hieracium umbellatum is a common plant in southern Sweden and Turesson was able to collect many samples from each habitat type. A close study of his extensive collections, after a number of years of cultivation, suggested to him the exciting possibility that habitat-correlated patterns of genetic variation were present, that is to say, in a particular habitat of *H. umbellatum* a certain race of characteristic morphology was invariably present. In the appropriate habitat there was to be found a dune race, a woodland race, etc. Turesson called these local races 'ecotypes' and described five, as follows (note that in these descriptions he considered anatomical and physiological traits (e.g. flowering times) as well as morphological features):

1. *An ecotype from shifting dunes*
   Narrow leaves and slender, less erect, sometimes more or less prostrate stems. Marked power of shoot regeneration in autumn. Leaves tough and thick with three to four layers of palisade cells. Fruiting in early September.
2. *An ecotype from sandy fields and stationary dunes*
   As 1, but power of shoot regeneration in autumn weak or lacking. Extremely prostrate in growth habit.
3. *An ecotype from western sea cliffs*
   Broad leaves, and more or less prostrate stems. Growth form contracted and bushy. Cells of leaves more or less distended. Fruiting late September to early October.
4. *An ecotype from eastern sea cliffs*
   As 3, but plants tall and almost as erect as in 5.
5. *An ecotype from open woodland*
   Stout, erect plants with lanceolate leaves of intermediate width. Leaves thinner with two or, at most, three palisade layers. Fruiting in September.

Turesson notes that additional ecotypes might be discovered in future studies. *Hieracium umbellatum* is a member of a genus famed for its apomictic reproduction. In considering Turesson's results it seems essential, therefore, to take into account the breeding behaviour of the plant. In a partial examination of the breeding system of his material, Turesson performed castration experiments, removing the upper half of unopened flower-heads with a razor. No fruits developed. This evidence supports the view that reproduction is sexual and not obligately apomictic. Plants of *H. umbellatum* proved in fact to be self-incompatible, and artificial crosses between plants of the dune ecotype and between plants of the cliff ecotype produced progenies in which the ecotypic characteristics of each were perpetuated, confirming the genetic basis of the discovered differences. Lövkvist (1962) has re-sampled at many of Turesson's *H. umbellatum* sites, and found broadly similar patterns of variation in cultivation trials. He also re-examined the breeding system of southern Swedish material of *H. umbellatum*, and found no evidence of apomixis. However, apomixis has been reported in this species (e.g. Bergman, 1935, 1941, and references cited therein) and may influence patterns of variation elsewhere.

Considering the origin of ecotypes, Turesson made two important deductions. He concluded, first, that the finding of widespread habitat-correlated genetic variation does not support the view that the variation patterns are largely governed by chance; rather the evidence suggests that natural selection operates in natural populations, well-adapted genotypes being selected in each habitat. This idea is expressed many times in Turesson's writings; for example, he says (1925) 'Ecotypes . . . do not originate through sporadic variation preserved by chance isolation; they are, on the contrary, to be considered as products arising through the sorting and controlling effect of habitat factors upon the heterogeneous species-population.' Turesson further concluded that a close study of the variation within and between ecotypes of *H. umbellatum* revealed patterns of leaf morphology that suggested a 'local' origin for coastal ecotypes from the widespread inland populations. It was possible that an appropriate ecotype could be produced many times, that is to say polytopically, and it was not necessary to postulate the invasion of Sweden by fully formed standard ecotypes after the last glaciation.

In a series of long papers published from 1922 onwards, Turesson eventually described ecotypes in more than 50 common European species. His first papers were about the plants of southern Sweden, but later (1925, 1930) he experimented with material collected from distant localities in all parts of Europe and also showed physiological differences between some of his stocks (1927a, b). Analysis of the behaviour of his extensive collections in cultivation enabled him eventually to distinguish two kinds of ecotypes, namely edaphic and climatic ecotypes, where the most important environmental effects were soil type (as in the case of *Hieracium umbellatum* in southern Sweden) and the climatic influences, respectively.

As early as the beginning of the eighteenth century there was a considerable amount of observational evidence that common species did not flower at the same time in different localities. For example, Linnaeus (1737) noted the different flowering times of Marsh Marigold (*Caltha palustris*) (March in the Netherlands, April to May in different parts of Sweden, June in Lapland). Quetelet (1846), having studied the dates of first flowering of Lilac (*Syringa vulgaris*) in different parts of Europe, came to the conclusion that there was a retardation of 34 days for each advance of 10° northwards in latitude. He also compared flowering at different altitudes above sea level, and discovered a retardation of 5 days for every 100 m increase in elevation. The important environmental factor, controlling flowering, was thought to be temperature. Turesson, studying the behaviour in cultivation of a large number of spring-flowering species, clearly demonstrated the importance of persisting genetic differences between plants originating from different climatic regions. Southern plants of such species flowered earlier in Turesson's experimental garden than plants of the same species collected from northern latitudes. He suggested that this group of plants is adapted to flower in the period immediately preceding the leafing-out of trees, a phenomenon which occurs earlier in the year in southern latitudes than in northern Europe.

In the botanical literature of the nineteenth century there are scattered reports that alpine plants flower earlier than lowland ones when both are cultivated in lowland gardens. Turesson's extensive experiments with species such as *Campanula rotundifolia* (Table 8.1) and *Geum rivale* enabled him to demonstrate that alpine ecotypes were smaller and retained their early-flowering habit in cultivation. He also carried out researches upon summer-flowering plants, showing that northern ecotypes were early-flowering and of moderate height, while southern plants were late-flowering and tall.

Western Europe was characterised by late-flowering plants of low growth; from Eastern Europe, on the other hand came taller early-flowering ecotypes. Turesson's contribution to our understanding of the patterns of variation within species is of very great importance: he demonstrated clearly the widespread occurrence of intraspecific habitat-correlated genetic variation. Adaptation to the environment was sometimes by plastic responses, but more frequently it had a genetic basis. Such studies were grouped together under the name of 'genecology' and the work was the model for many studies by other botanists. The work of Stapledon

Table 8.1. **Geographic variation in *Campanula rotundifolia***

(a) Results of transplant experiments from Turesson (1925). (Means of five measurements given)

		Field no.	Transplanted from	Length of stems (mm)	Width of middle-stem-leaves (mm)	Number of flowers on stems	Length of corolla (mm)	Width of corolla in the middle (mm)	Width of corolla at mouth (mm)	Length of corolla lobes (mm)	Length of calyx lobes (mm)	Power of regeneration of basal rosette-leaves	Year of collection	No. of plants
Norway and Sweden		99	Vitemölla	547.75	2.18	23.25	18.63	16.50	22.45	7.33	6.33	none–weak	1920	8
		206	Åhus	650.54	2.16	27.49	19.93	16.45	22.82	7.65	7.29	none–weak	1922	13
		270	Ulriksdal	334.30	1.86	20.33	17.13	14.99	20.2	7.02	5.63	none–weak	1921	14
	M	298	Åre	308.43	2.97	11.5	22.12	21.0	25.91	9.54	5.88	mostly strong	1921	17
		349	Bergen	378.67	2.73	9.19	20.56	20.53	25.67	8.41	6.36	weak–strong	1922	7
		240	Trondhjem	336	2.03	15.97	21.0	18.34	25.06	8.39	5.80	weak–strong	1922	14
Central Europe	M	19–25	Abisko (seeds)	250.10	1.99	13.97	24.47	20.48	27.68	9.32	7.90	strong	1921	seeds
		770	Freiburg	278.56	2.12	19.86	20.44	18.89	24.54	8.5	6.56	none–weak	1923	16
	M	796	Feldberg	224.66	4.29	6.88	23.45	21.82	25.32	8.76	7.89	strong	1923	14

(b) Progeny trial, from Turesson (1930)

Field no.		Source	No. of plants	Height (cm)			Earliness of flowering[a]		
				Mean	σ	m±	Mean	σ	m±
770		Freiburg	20	68.9	5.89	1.32	1.60	0.35	0.29
796	M	Feldberg	20	29.5	2.41	0.54	5.00	0.00	0.00
270		Ulriksdal	20	47.1	5.47	1.22	2.80	0.44	0.10
298	M	Åre	20	33.4	3.75	0.84	5.00	0.00	0.00

[a] note that a large mean corresponds to earlier flowering
M = montane localitites

(1928) is of special interest. Using the common pasture grass *Dactylis glomerata*, he studied the influence of hay cutting and animal grazing, and described a third class of ecotype, namely the 'biotic ecotype'. His work is summarised in Table 8.2.

Scandinavian botanists have made many notable contributions to genecology, and it is appropriate at this point to give an example of the important experiments of Bøcher (1963). He used the Turessonian technique of cultivation in a standard garden to examine the variation and flowering behaviour of collections of many European plants, and carried the analysis of variation into an important new area, namely the study of the timing of flowering in relation to the life history of the plant. For example, he discovered in cultivation experiments with *Prunella vulgaris* (Bøcher, 1949) that there were two main growth types in Europe, namely plants with a short vegetative phase, flowering in their first year, and plants with a longer vegetative phase, flowering in their second year. This latter group was further subdivided into plants that were short-lived and perennial types. The distribution of the two main types – first- and second-year flowerers – proved most interesting; for example, in Mediterranean regions subject to summer drought, only short-lived annual plants were found, whilst in areas with different climatic conditions biennial or perennial types were characteristic. Such patterns are likely to be the result of natural selection: only those plants whose life history 'fits' the growing season of a particular area will survive in the long term.

Experiments by American botanists

Some of the most famous experiments on ecotypes were published by Clausen, Keck & Hiesey (1940) on different species of plants collected on a 200-mile transect across central California, from a 'Mediterranean' climate in the west to an 'alpine' climate in the east. Turesson's method of studying ecotypes was to grow all his collections in a lowland garden. Such a method has the limitation that it may not allow certain traits to be revealed (e.g. tolerance or sensitivity to frost, or drought).

Craig (2005) provides full background details of the experiments. Under the leadership of Hall, with assistants Keck and Hiesey hired in 1926, three transplant gardens at the same latitude were established: at Stanford (30 m above sea level), Mather (1400 m) and Timberline (3050 m). By the time Clausen joined the group, in about 1930, many of the transplant experiments had been established. Hall died in 1932, and Clausen became the leader of the group. The results of the experiments were published from 1940 onwards.

To illustrate the very different conditions in the gardens, Fig. 8.1 gives climatic details for sites near Stanford and Timberline. Of especial importance are the extremes of temperature and the differences in the length of the growing season. In each garden, plants were grown spaced out in weed-free plots protected from grazing. The experimental plantings consisted, in the main, of clonally propagated stocks, each individual being grown and divided, and a ramet of each planted in each garden. Thus the growth and performance of each individual from samples collected from a range of different sites could be studied in a 'Mediterranean', an intermediate and an 'alpine' garden. Climatic ecotypes were studied in many species, particular attention being paid to *Potentilla glandulosa*, a species found from the coastal hills near the west coast of California to high altitudes in the Sierra Nevada. Their experiments made it possible to test the behaviour of diverse stocks in very different standard gardens. For example, they discovered that most lowland stocks died in the harsh climate of the alpine garden, and, at the Stanford garden, plants originating from high altitude remained winter-dormant under conditions that stimulated growth of lowland samples. Clausen and his associates (1940) decided that there were four distinct climatic ecotypes in *Potentilla glandulosa*, corresponding to the taxa subspecies *typica* (lowland), subspecies *reflexa* and subspecies *hanseni*

Table 8.2. **Biotic ecotypes in *Dactylis glomerata*. Stapledon (1928) discovered that grassland use determined the type of *Dactylis* present in a particular area**

		Per cent growth type					Per cent flowering behaviour			
		Hay	'Cup'	Tussock	Pasture	Per cent over 100 cm	Early 1	2	3	Late 4
Commercial hay stocks	A	59	36	2	3	78	40	50	9	1
	B	66	31	1	2	78	61	32	6	1
Old pastures		15	23	6	56	15	11	35	38	16
Hedgerows and thickets		26	35	25	14	31	17	35	34	14

Hay types, with their taller early-flowering plants were distinct from the shorter, late-flowering plants characteristic of grazed pasture. Pasture types had many more tillers than hay types, and a smaller percentage of tillers produced inflorescences. Plants from hedgerows had a wide range of variants. Even though this experiment did not reveal a discontinuous pattern of variation, Stapledon was content to interpret his results in terms of 'biotic ecotypes' (see Warwick & Briggs, 1978a, b, for a partial review of 'hay' and 'pasture' ecotypes).

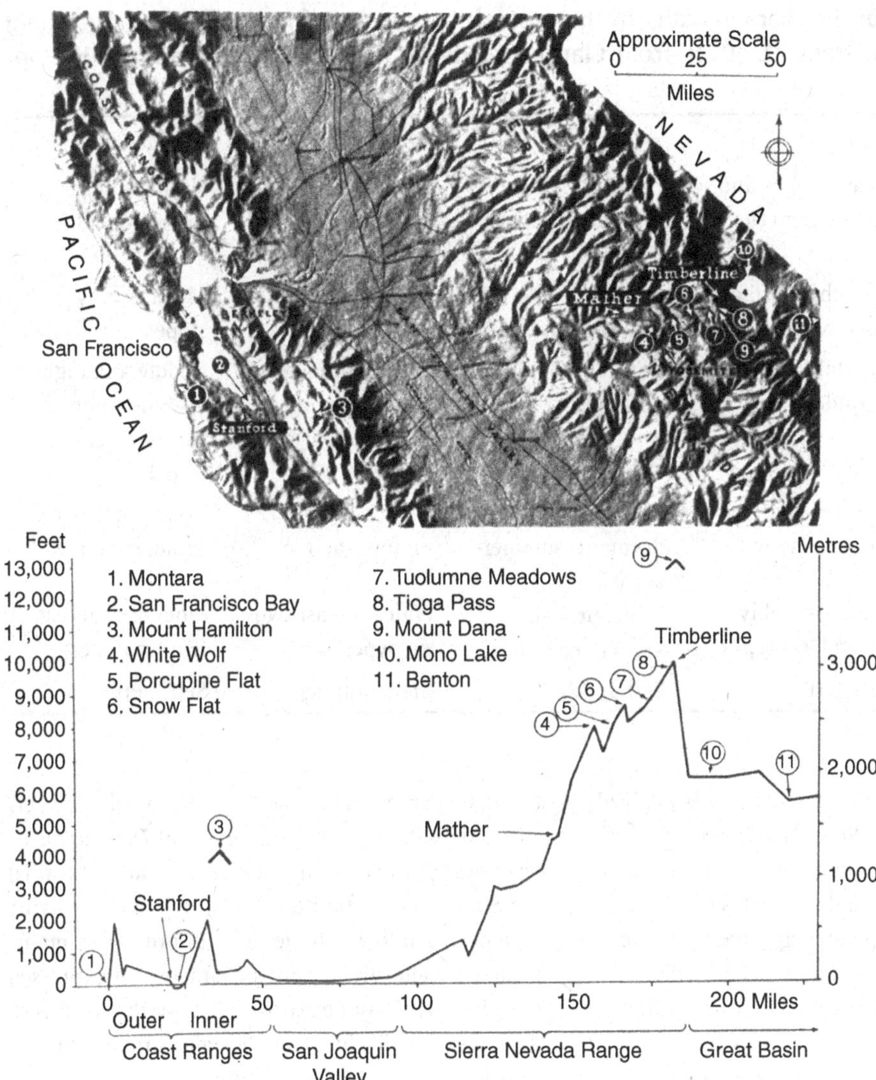

Fig. 8.1. Map and details for Stanford and Timberline sites in central California, used in the famous transplant experiments of Clausen, Keck & Hiesey. Annual variation in temperature and precipitation (US Weather Bureau data for 1925–35 inclusive) for Stanford and Timberline reveals that the lowland site has a 'Mediterranean-type' climate (Stanford), where active growth is possible throughout the year, whereas at Timberline (c.3000 m in the mountains) the active growth period is restricted to July to August.

(intermediate altitudes), and subspecies *nevadensis* (alpine) (Table 8.3).

In later writing, e.g. Clausen & Hiesey (1958), it was suggested that each subspecies was in fact made up of two or more ecotypes. The hypothesis that ecotypic variants of *Potentilla* differed genetically received support from a comprehensive series of crossing experiments carried out by Clausen & Hiesey (1958). When the very extensive results of their experiments on *Potentilla* were examined by Clausen, Keck and

Table 8.3. **A summary of the characteristics of the ecotypic subspecies of *Potentilla glandulosa* along the Central Californian transect. (Data from Clausen & Hiesey, 1958, as summarised by Heslop-Harrison, 1964)**

	typica	*reflexa*	*hanseni*	*nevadensis*
Distribution	Coast ranges and lower Sierra Nevada	Low and middle altitudes of Sierra Nevada	Meadows, mid-altitudes of Sierra Nevada	High altitudes of Sierra Nevada
Habitat	Soft chaparral and open woods	Dryish, open timbered slopes	Moist meadows	Moist, sunny slopes
Climatic tolerances as experimentally determined	Coastal to middle altitudes	Coastal to middle altitudes	Middle and high altitudes (poor survival near coast)	Middle and high altitudes (poor survival near coast)
Seasonal periodicity at Stanford (alt. 30 m)	Winter- and summer-active	Winter-active or -dormant; summer-active	Winter-dormant, summer-active	Winter-dormant, summer-active
Internal variation	Wide, probably several 'ecotypes'	Wide, probably several 'ecotypes'	Wide, at least two 'ecotypes'	Moderate, at least two 'ecotypes'
Self-compatibility	Self-fertile	Self-fertile	Undetermined	Self-sterile

Hiesey, they concluded that altitudinal races evolved through natural selection. Núñez-Farfán & Schlichting (2005) have reanalysed some of the data sets published by Clausen and associates (in particular, 107 F_2 clones planted at three elevations). This more complete statistical analysis provides extremely strong support for the action of natural selection in race formation.

Other American botanists made studies of ecotypes using the transplant stations at Stanford, Mather and Timberline. Lawrence (1945), for example, studied ecotypes of *Deschampsia cespitosa* (*Deschampsia caespitosa*), discovering differences in survival in different stations. Of especial interest were his studies of reproduction in the different transplants; although all individuals survived at Timberline, only the stocks native to that area were able to produce seeds in the short growing season. Such a finding, which is of crucial importance in understanding the genecology of the species, could not have been revealed in a lowland garden. A further point of general interest is revealed by their results with plants of *Deschampsia cespitosa* (*D. caespitosa*) from Finland (latitude 60°N) and South Sweden (latitude 56°N). When these plants were grown at low altitudes at Stanford (38°N) many of them became viviparous, a character not expressed in their native habitats. Growth in a garden with very different climatic characteristics may provoke an unusual response from the plants.

Experiments with several gardens separated by great distances are expensive to maintain, and botanists have devised ways of investigating ecotypes by varying the conditions in a single garden or laboratory. Turesson's experiments were carried out in a lowland garden on fertile soil, and in describing edaphic ecotypes he inferred the importance of soil differences in the wild. A more direct approach to the study of patterns of variation in relation to edaphic factors was made by Kruckeberg (1951, 1954). In one experiment, fruits of

Achillea borealis were collected from serpentine and non-serpentine sites in California. (Serpentine is a rock type which gives rise to soil with high levels of magnesium and low levels of calcium.) Two tons each, of a serpentine and a fertile soil, were collected and transported to the University of California Botanical Gardens, and stocks were grown from seed in soil bins, or pots, of the two soil types. Stocks raised from seed of plants native to serpentine soils grew well on the serpentine test soil, but, in contrast, plants from other soil types (shales, basalt, etc.) generally (though not always) grew badly or died. Kruckeberg's results on *Achillea borealis* and other species are consistent with the idea that a common species found on different soil types may be made up of a number of edaphic ecotypes.

A second example of the way in which diverse stocks may be presented with different environments in one garden or laboratory is provided by the use of glasshouses, growth chambers, etc., in which day-length, temperature and other factors may be varied. Samples may be tested in a variety of artificially controlled environments, in which, for instance, the responses of different stocks may be monitored under different day-lengths. In the first experiments studying the effect of different day-lengths, plants were grown on movable trucks. After a period of natural daylight, plants were moved into light-proof structures where they could be either in total darkness or given supplementary light from artificial sources. A good example of this type of experiment is provided by Larsen (1947), who studied *Andropogon scoparius*, a widespread and important forage grass in North America. Plants were collected from twelve localities from 28°15′N in Texas to 47°10′N in North Dakota. The grasses were given constant day-lengths of 13, 14 and 15 hours of light. None of the twelve samples flowered at 13 hours. Plants from the southern USA required a 14-hour photoperiod for floral induction, but a photoperiod of 15 hours was necessary for flowering in many northern plants.

Figure 8.2 illustrates the relation between latitude and day-length at different times of year. *Andropogon* plants growing in the southern USA naturally come

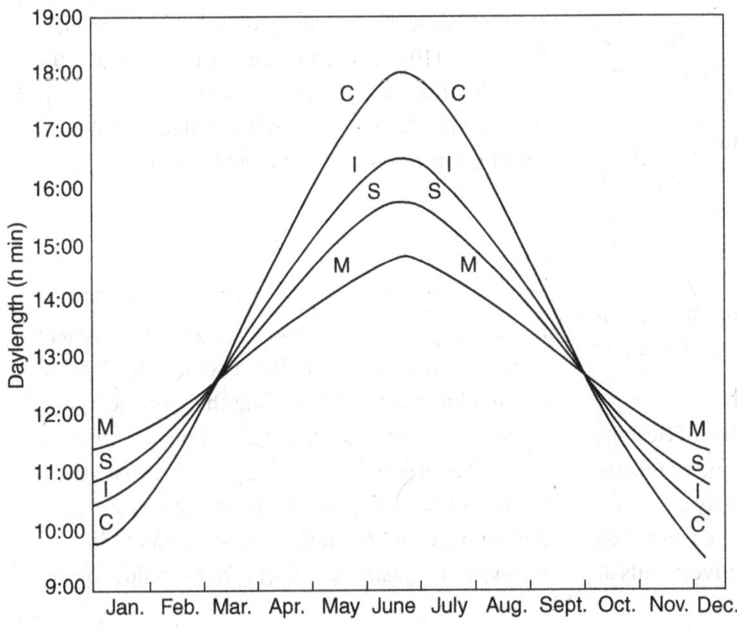

Fig. 8.2. Relation between latitude and day-length at different times of year. M = Miami, FL latitude c.26°N; S = San Francisco, CA, c.37°N; I = Ithaca, NY, c.42°N; C = southern Canada, 50°N. (From Curtis & Clark, 1950.)

into flower after receiving a photoperiod of 14 hours. Northern plants, with longer summer days, need a 15-hour day to come into flower.

As more sophisticated equipment became available, growth chambers were constructed in which many environmental factors (e.g. temperature, day-length) could be controlled. Adjacent chambers could be used to subject plants to different conditions. Splendid examples of photoperiodic responses of plants from different geographic areas are provided by the experiments of Mooney & Billings (1961) who studied *Oxyria digyna* collected from sites between 38°N and 76°N in North America, and the investigations of McMillan (1970, 1971) on *Xanthium strumarium*. These pioneer growth facilities were the forerunners of the sophisticated growth chambers now used widely in laboratory studies of *Arabidopsis* and other model plant species. Advances have also been made in the study of physiological processes in the field. In order to examine the effects of climate change, perturbation studies have employed open-top devices to investigate the effects of raising temperatures (Marion *et al.*, 1997) and elevated carbon dioxide (Gwynn-Jones *et al.*, 1997). Other factors have been manipulated:

A. Water and/or fertiliser additions 'to simulate increases in summer precipitation' (Wookey *et al.*, 1993).
B. Effects of warming on soil carbon storage and decomposition (Kirschbaum, 1995).
C. Manipulation of snow cover (Walsh *et al.*, 1997).
D. Various effects of warming on experimental ecosystems (Rustad *et al.*, 2001).
E. Enhancement of UV-B radiation using fluorescent tubes (Gwynn-Jones, Lee & Callaghan, 1997).

Physiological studies are advancing our understanding of ecotypes, and the reviews of Heslop-Harrison (1964), Hiesey & Milner (1965) and Bannister (1976) may be consulted for details of early studies. In recent years, the field of physiological ecology has expanded greatly and specialist reviews give details of researches in different fields (see Larcher, 2003; Lambers, Chapin & Pons, 2008).

The widespread occurrence of ecotypes

As a result of experiments in which plants have been grown in gardens or under controlled conditions, ecotypes have been described in hundreds of species. There is evidence that ecotypes occur not only in outbreeding species but also in species apparently predominantly inbreeding. There are also numerous studies of facultatively apomictic plants in which ecotypic patterns have been described, for example *Poa pratensis* (Smith, Nielsen & Ahlgren, 1946) and *Potentilla gracilis* (Clausen, Keck & Hiesey, 1940).

Of special interest is the finding of genetic heterogeneity in plants that are apparently obligately apomictic. Turesson (1943) discovered, within collections of European *Alchemilla glabra*, *A. monticola* (*A. pastoralis*) and *A. filicaulis*, that plants from Lapland and montane areas were earlier-flowering in cultivation than lowland stocks. The patterns of variation appeared to be ecotypic, but Turesson called the variants 'agamotypes' in recognition of the breeding system of *Alchemilla*. Bradshaw (1963a, b, 1964) and Walters (1970, 1986a) have described dwarf variants of an ecotypic nature in *Alchemilla*, the origin of which is plausibly due to selection in response to grazing by sheep.

Clines

In the experiments outlined above, the researchers were content to describe their material in terms of distinct local races, often using the term 'ecotype'. However, the ecotype concept was not without its critics. Langlet (1934), for example, pointed out that the most important habitat factors, such as temperature and rainfall, commonly varied in a continuous fashion, and thus one would expect

Table 8.4. **Results of soil analysis (air dry samples) and cultivation experiments with *Plantago maritima* (Gregor, 1946)**

Habitat	Mean scape length (cm)	Habit: grades (Percentage of sample in each grade) 1	2	3	4	5
Waterlogged mud zone (salt concentration 2.5%)	23.0 ± 0.58	74.5	21.6	3.9	–	–
Intermediate habitats with intermediate salt concentrations	38.6 ± 0.57	10.8	20.6	66.7	2.0	–
Fertile coastal meadow above high tide mark (salt concentration 0.25%)	48.9 ± 0.54	–	2.0	61.6	35.4	1.0

graded variation in many widespread species rather than discontinuous variation.

Support for this view was provided by Gregor (1930, 1938, 1946) who made an intensive study of *Plantago maritima* in northern Britain. Representative seed collections were made, and plants were grown in an experimental garden of the Scottish Society for Research in Plant Breeding. Table 8.4 gives an example of the sort of results obtained by those studies.

In this case all three sample zones are from the Forth estuary in eastern Scotland. If collections of *Plantago maritima* taken from different sites along a gradient from high to low salt concentration are compared, there was a difference in growth habit, with a progressive increase in scape height. In a similar fashion there are increases in scape volume and thickness; in leaf length, breadth and spread; and in seed length.

In 1938, Huxley, after surveying the literature, coined the useful term 'cline' for character variations in relation to environmental gradients (Huxley, 1938). Thus, a graded pattern associated with ecological gradients is referred to as an ecocline (a good example of this is Gregor's findings in his study of *Plantago maritima*). If the pattern is correlated with geographical factors, the term topocline can be employed. Clinal variation has been described in a large number of species, and a small selection of examples is given in Table 8.5.

How far are intraspecific patterns of variation explicable in terms of ecotypes and clines? Experiments, for example, by Bradshaw (1959a, b, c, 1960) on *Agrostis capillaris* (*A. tenuis*) have shown that much more complex patterns may be found in nature. Careful collections of living specimens of this grass were made mostly from localities in Wales. The stocks were grown, and then cloned material was planted into a number of experimental plots in North and Mid Wales, with an altitudinal range from sea level to about 800 m. A wide range of different responses was demonstrated by these experiments. Not only were plants different morphologically, but there were also physiological differences. For example, certain plants grew well on soils containing lead and other heavy metal residues; others, indistinguishable from them morphologically, died on this type of soil (we shall return to this interesting phenomenon of tolerance of heavy metal ions in Chapter 9). At this point it is important to note that Bradshaw could not delimit ecotypes in *Agrostis capillaris* (*A. tenuis*). This was not because extreme variants were not found in extreme habitats. On the contrary, many very distinctive plants were discovered: for instance, dense cushion plants from the exposed Atlantic cliffs at West Dale, South Wales. The problem was that, even though habitat-correlated variation could be demonstrated, the fact

Table 8.5. **Some historic studies of clinal variation**

Species	Variation	Reference
Allium schoenoprasum	Longitudinal cline in chromosome banding pattern in eastern North America	Tardif & Morisset (1991)
Anthoxanthum odoratum	Clines for various characters at mine/ pasture boundary	Antonovics & Bradshaw (1970)
Asclepias tuberosa	Clines for flower colour and leaf-shape in North America	Woodson (1964)
Blandfordia grandiflora	Morphological and reproductive characters in Australia	Ramsey, Cairns & Vaughton (1994)
Dactylis glomerata complex	Clinal variation in European populations in glutamate oxaloacetate transaminase gene frequencies (GOT I Locus)	Lumaret (1984)
Eschscholzia californica	Clines in California for various features	Cook (1962)
Eucalyptus spp.	Graded patterns of leaf glaucousness with extreme 'waxy' types in exposed habitats	Thomas & Barber (1974)
Geranium robertianum	Clines for hairiness	Baker (1954)
Geranium sanguineum	Decrease in leaf-lobe breadth west to east in Europe	Bøcher & Lewis (1962)
Holcus lanatus	First-year flowering in Southeast Europe. Second-year flowering in northern Europe	Bocher & Larsen (1958)
Juniperus virginiana	Clines in terpenoid content northeast Texas to Washington DC	Flake, von Rudloff & Turner (1969)
Lotus corniculatus	Flower colour variation in North England. Dark-keeled variant rare in west, increasing in frequency eastwards.	Crawford & Jones (1986)
Pinus strobus	Decrease in leaf length and number of stomata, increase in number of resin ducts, with increasing latitude in North America	Mergen (1963)
Silene latifolia	Clinal variation in seed morphology	Prentice (1986)
Viola riviniana	Clines in plant size	Valentine (1941)

that all kinds of intermediate plants were discovered made it utterly impossible to decide where one 'ecotype' ended and another began.

Does the concept of clines help in this situation? Bradshaw studied his material closely with this idea in mind. In many areas, even though clines might be described, he decided that the environmental gradients and the associated variation were too complex.

What, then, determines the patterns of intraspecific variation found in the wild? How can one reconcile the distinct ecotypes of Turesson and Clausen with the complex variation found by Bradshaw and many other researchers?

Factors influencing the variation pattern

Of first importance is the type of sampling technique used. Turesson and many other botanists collected widely spaced samples, whereas Gregor & Bradshaw carried out intensive sampling in small areas. Widely spaced samples taken from extreme habitats may exhibit a pattern of distinct 'ecotypes'. In contrast, samples taken from along smooth, regular gradients of soil or altitude may well give a pattern of clinal variation in the experimental garden. If, however, sampling is carried out in small areas, the plants being collected at random rather than along particular gradients, then experiments might reveal very complex patterns. Thus, in a very real sense, the mode of sampling largely determines the patterns 'discovered' in cultivation experiments.

Another aspect of sampling is important. An experimenter can choose either to collect a representative seed sample or to dig up mature plants. If both types of sampling are carried out on a single population, different patterns of variation might well be found. This is because mature plants have survived the rigours of natural selection. Seed collections, on the other hand, give an estimate of potential rather than actual variation. If several adjacent populations in different environments are examined, in a case where pollen can be transported from one population to another, sampling of mature individuals might well reveal a pattern of more or less distinct 'ecotypes'. On the other hand, because of gene flow between populations, seed samples will seem to reveal a more complex pattern in the same case.

Ecological, historical and geographical factors also influence the patterns discovered in experiments. If a species is found as small, non-contiguous populations, or if it has populations inhabiting two or more very different types of habitat, then the pattern of variation in the wild is more likely to be that of distinct 'ecotypes'. In contrast, common species, which throughout their geographical range are more or less continuously distributed over many habitats, will in all probability exhibit complex patterns of continuous variation. Also, the mode of pollination is important. Small populations of insect-pollinated species often exhibit ecotypic discontinuities, but these are less likely to occur in widespread wind-pollinated species.

Since Turesson's time there has clearly been a change of outlook. Ecotypes are now regarded as nothing more than prominent reference points in an array of less distinct ecotypic populations (Gregor, 1944). Some experimenters have been reluctant to designate ecotypes: they have instead carefully recorded the patterns of 'ecotypic differentiation' found in particular experiments (see, for example, Quinn, 1978). However, despite the difficulties of defining the word 'ecotype', online searches reveal that it is still being used for local and regional variants. Also, some of the regional, biochemical and developmental variants of *Arabidopsis* are referred to as 'ecotypes' in British, European and US stock lists and in publications.

With hindsight one can see in Turesson's own results the possibility that, in common species, variation patterns were more complex than the ecotype concept implied. For instance, where sandy fields and dunes were found as adjacent habitats, a considerable number of intermediate *Hieracium umbellatum* plants were found linking the two ecotypes. Similarly, in *Scorzoneroides autumnalis* (*Leontodon autumnalis*), Turesson (1922b) found a complex situation where meadows and pastureland ran down to the sea.

The refining of genecological experiments

Early cultivation experiments were often very crude; a few plants were dug up in the wild and planted in a garden. As we have just seen, the pattern of sampling will to a very large extent determine the outcome of an experiment. Furthermore, while a simple garden technique may serve to study major differences between population samples, the study of fine-scale variation has resulted in the devising of improved cultivation and other experiments. Thus, as genecology has developed, the methods of sampling and cultivation have been refined to enable statistical analysis of finer and finer differences between samples of plants.

Sampling populations

Much time and effort may be spent in growing and measuring plants and analysing results, but very little attention may have been given to sampling strategies; indeed, the word 'strategy' may be entirely inappropriate for samples of seed snatched at brief roadside stops on car journeys or obtained from Botanic Garden seed lists.

If statistical analysis is to be performed on the results, then ideally a random sample of plants must be collected. Ward (1974) has described a simple way in which two people may collect such a sample. Having decided on the area to be sampled, the recorders count the number of individuals in the area (or subsection of an area if the plant population is very large). A decision is then made on the size of the sample, say 25 plants out of 250. Using a table of random numbers (as found, for instance, in Fisher & Yates, 1963) or numbers 'drawn out of a hat', 25 numbers within the range 1–250 are 'selected' and placed in ascending order: say 5, 8, 14, 27, etc. On traversing the sample area again, one person calls out the number of each individual, 1–250, while the second person labels the individuals to be sampled, the 5th, 8th, etc., as determined by the random numbers. This random sample is then used for experimental investigation. Other methods of random sampling are discussed by Yates (1960), Cochran (1963), Greig-Smith (1964) and Green (1979). There are some theoretical and practical difficulties to be faced in undertaking such a sampling procedure, which will now be considered.

If the experimenter is studying apparent hybridisation, a random sample might not include all the 'interesting' plants of an area. A deliberate sampling of the plants of the area might be more appropriate in such circumstances. Should the study involve the investigation of variation across a vegetational discontinuity, e.g. woodland to grassland, it might be more informative to collect plants from a transect (sampling at, say, metre intervals) across the ecotone rather than collect a random sample. All will depend on the hypothesis being tested. There are many habitats, where the collection of random samples is very difficult (e.g. tropical rainforests, aquatic and wetland habitats, cliffs). However, where the collection of a random sample is a practical possibility it should be seriously considered.

Since populations often contain individuals at all stages of growth and development from seeds and seedlings to adult plants, a truly random sample should perhaps contain individuals in several different age classes. In practice, a subset of the population is often sampled. The following might usefully be distinguished:

1. 'Individuals' present as ungerminated seed in the soil ('seed bank').
2. Seedlings, a transitory stage in many habitats, but more important in some plant communities; for instance, in tropical rainforest many tree species growing in deep shade have a long seedling stage: only if disturbance in the canopy causes greater illumination of the ground flora do the seedling trees develop into adults.
3. Immature individuals.
4. Mature individuals.
5. Seeds attached to 4.
6. Diseased and damaged plants. Sometimes, as in the case of the 'choke' disease of grasses caused by fungus, the plants are vegetatively vigorous but the fungal infestation suppresses the formation of inflorescences (Bradshaw, 1959c).

Subsets 4 and 5 are most commonly sampled by experimenters. Different subsets may reveal quite different spectra of variation; we shall see examples in Chapter 9 when we consider attempts to study the effects of natural selection by comparing the variation of different subsets in cultivation.

One of the biggest difficulties in sampling populations concerns the definition of an individual. In open vegetation it is usually possible to define individuals in annual plants and to see patches of individual perennial plants. In closed swards, however, the problem is more difficult. Sometimes the presence of 'marker' genes (e.g. leaf marks in *Trifolium repens*: Davies, 1963) might reveal the extent of particular individuals: but such markers are rare. Theoretically it might be possible to trace root systems in an attempt to establish the extent of individuals, but the practical difficulties are enormous. Furthermore, in some plants, e.g. certain forest trees, root-grafts occur which unite the root systems of several different individuals (Graham & Bormann, 1966; Böhm, 1979). Studies of patterns of allozymes in Strangler Figs, which form a woody sheath around many tropical trees, provide evidence that apparent individuals are in reality genetic mosaics, caused by root fusions of a number of plants (Thomson *et al.*, 1991).

The problem of defining the individual is further complicated by clone formation, in which the vegetative continuity of an individual breaks down, producing a clonal patch of several individuals, which, baring somatic mutation, have identical genotype (Harper, 1978). Evidence for clonal populations is usually circumstantial, but direct evidence is available in some cases. Investigations using molecular methods have greatly increased our understanding of populations of clonally propagating species. For example, in a study of a population of Bracken (*Pteridium aquilinum*) in Virgina using 6 polymorphic isozyme loci, as many as 45 genotypes were detected in the study area (Parks & Werth, 1993). Some of these clones were very extensive. In an investigation of the same species in North Wales, an extensive triploid clone was detected and this was mapped using isozyme markers (Sheffield *et al.*, 1993). Extensive clones, some of great age, occur in some habitats (Table 8.6).

Why is knowledge of the extent of individual genotypes important in sampling? Suppose we collect two population samples A and B. Fortuitously, sample A could consist of 25 pieces of a widespread clone, whilst sample B could consist of material of 25 genetically different individuals. A comparison of the two 'populations' in a cultivation trial is likely to show that they are different, but interpreting this difference as a real population difference could be misleading. Perhaps population A is largely composed of the clonally propagated individuals of one genotype, while B is variable. On the other hand, populations A and B might both be variable, and the multiple sampling of one clone in population A might be merely the consequence of poor sampling technique. Harberd and others, who have made a special study of the problem (Harberd, 1957, 1958; Wilkins, 1959, 1960; Ward, 1974), recommend that spaced samples be collected from populations. From all the evidence available the probable maximum extent of clonal patches is estimated. Sampling at points separated by distances greater than this estimated clonal patch size is then carried out. Widely spaced samples are to be recommended to counteract another problem that arises on studying plant populations. Fruits and seeds are often shed very close to the plant which produced them and 'family groups of close relatives', perhaps involving several generations, may be found (see, for example, Linhart *et al.*, 1981). Distorted comparisons can arise if a sample containing a group of closely related plants is matched against a set whose members are totally unrelated.

It must be noted, however, that wide spacing of samples is somewhat at odds with the aim of studying small systems in detail. Such studies as those of Smith (1965, 1972) and Harper and associates (Harper, 1983) reveal enormous variation within sites. There seems to be no easy solution to the problems raised by clonal propagation: the experimentalist must make the best judgement possible in each situation in relation to the hypothesis under consideration.

Table 8.6. **Some examples of historic studies of clones (Lines of evidence: F. = field observations; C. = cultivation trials; H. = hybridisations; I. = studies of isozymes; M. = DNA fingerprinting)**

Evidence	Example
C.	*Anemone nemorosa* (von Bothmer *et al.*, 1971): large number of clonal patches, of limited size, in Swedish habitats.
C.F.I.	*Betula glandulosa* (Hermanutz, Innes & Weis, 1989): clones mapped on Baffin Island, at northern limit of species.
F.I.	*Decondon verticillatus* (Eckert & Barrett, 1993): a survey of this tristylous species reveals that at the northern margin of its range in eastern North America, populations may consist of only one of the three style variants and reproduction is exclusively by clonal propagation.
C.H.	*Festuca rubra* (Harberd, 1961): evidence of many genetically different individuals in a study of an area of south Scotland. One particular variant occurred at points *c.*220 m apart. If this area was achieved by radial growth then the clone must be *c.*400–1000 years old. However, perhaps the present distribution has been achieved by dispersal of fragments by animals or other causes, or as a consequence of vivipary, which has been recorded in this species (Smith, 1965). Widespread clones also found in *Festuca ovina* (Harberd, 1962).
F.I.	*Larrea tridentata* (Sternberg, 1976; Vasek, 1980): extensive clonal patches, visible on aerial photographs, in the Mojave Desert, California. By radiocarbon dating oldest clone may be 11 700 years old. Isozyme studies reveal that parts of apparent clones are indeed isoclonal.
C.H.	*Lysimachia nummularia* (Dahlgren, 1922; Bittrich & Kadereit, 1988): self-sterile clones found in many parts of North and Central Europe; presumably sexual reproduction only takes place in populations where individuals with different *S*-alleles occur together.
F.M.	*Phragmites australis* (Neuhaus *et al.*, 1993): large and small clones 'mapped' in Berlin and north-east Germany using DNA fingerprinting *a.* by digestion using restriction enzymes *Alu*I or *Dra*I, with the oligonucleotide $[GATA]_4$ used as a probe in hybridisation; or *b.* by RAPD reactions followed by separation of the amplification products on agarose gels and staining with ethidium bromide).
F.M.	*Populus tremuloides* (Rogstad, Nybom & Schaal, 1991): clones of various sizes mapped using DNA fingerprinting; (digestion with restriction enzymes *Dra*I, *Hae*III *or Hinf*I and hybridisation with the M13 probe).
F.I.	*Solidago altissima* (Maddox *et al.*, 1989): using isozyme markers, clones were mapped in sites at different stages of old field succession near Ithaca, New York.
F.	*Ulmus* spp. (Rackham, 1975): by studying in British woodlands patterns of morphological variation together with incidence of fungal diseases and timing of coming into leaf and leaf fall, evidence of very extensive clonal patches was discovered.

Another question to be resolved before sampling is undertaken concerns the number of sites and samples within sites. Suppose we study a single site with two different soil types, A and B. Patterns of variation may be revealed in samples drawn from the two subsites A and B, and at the end of an experiment some differences related to soil type may be found in plants originating from the two subsites. The experimenter must then decide whether the differences are ecotypic or whether they owe their origin to random variation. With one A/B comparison it is difficult to rule out random events (Wilkins, 1959). A more penetrating study of the patterns of variation might be made, by studying several areas, where subsites of type A and B are juxtaposed. Furthermore, in collecting from the wild, a bulk seed sample may be made to represent each of the subsites A and B, or the seeds from a random collection of mature individuals may be separately collected and packeted at each subsite. Family lines may then be grown, patterns of variation within lines offering some insights into the breeding system of the plants under study. This type of sampling – a hierarchical or nested pattern – has much to recommend it, allowing not only a number of A/B comparisons to be made, but also providing some information on variation within subsites. For instance, the plants under study might be obligate apomicts; while the progenies of different 'seed parents' might differ, there might be little or no variation within progenies. In this circumstance, the cultivation of plants from bulked seed samples would fail to reveal an important strand in the variation pattern.

Cultivation experiments

A study of variation usually requires cultivation of plants. This is true not only of field collections brought into a common environment to investigate the nature of variation patterns, but also of many sophisticated genetic and physiological studies. In many cases the experimenter wishes to grow material from diverse sources under the same conditions. Thus, if population samples are collected in the wild, and if there are interesting phenotypic differences between populations, a Turessonian cultivation experiment might be carried out, to see if differences between populations persist in cultivation.

At first sight a requirement to grow material 'under the same conditions' appears to present little difficulty. A moment's reflection, however, is sufficient to remind the reader of the variation in soil fertility, drainage, pests and diseases within even the most uniform experimental plot in garden or field. The notion that glasshouses provide a uniform environment is quickly dispelled by studying investigations of yields of vegetable crops on benches in different parts of experimental glasshouses (see, for example, the little-known experiments of Lawrence, 1950).

In designing genecological experiments, the botanist has much to learn from the agricultural scientist. Farmers wish to grow high-yielding varieties of crop plants and, since the middle of the nineteenth century, research workers have struggled to perfect experiments designed to study yield. In this short book we cannot provide a complete review of this interesting subject and will confine our attention to a few important general issues. Notable advances in the design of field experiments came with the work of Fisher, who studied the famous long-term Broadbalk Wheat experiment at Rothamsted Research Station in south England. Box (1978) provides a useful historical review of field experimentation and explores in detail Fisher's many contributions to the subject.

The basic ideas behind the design of cultivation trials are as follows:

1. Experiments must be designed with sufficient replications of the varieties, populations, treatments, etc. Thus, in a simple experiment on yield in, say, spring Wheat, several plots of each variety must be grown.
2. Soil fertility and other edaphic factors often vary across garden plots and fields, but it is commonly found that adjacent sites have similar fertility, etc. Thus, Fisher (1935) recommended that the ground available be divided into uniform blocks (not necessarily square). Each block should contain a full complement of the material under study. Within blocks the small plots of each variety should be *randomly* arranged. In early experiments in agriculture and forestry it was hoped that, by careful husbandry, varieties could be given the same conditions. But a critical approach to experimentation suggests that this is a forlorn hope: it is impossible to ignore the variability

induced by environmental factors. With a proper layout of experiments differences between blocks can be *measured* to give an estimate of the random element of variation introduced into the experiment.

3. Another important factor in the design of field experiments is the effect of position. If plants are growing in blocks, those in the centre of the block will be surrounded by neighbouring individuals; in contrast, plants on the margins of blocks are likely to be adjacent to bare soil and subject to very different amounts of root and shoot competition. Thus, it is recommended that 'guard rows' of similar plants be planted around the blocks, to provide uniform conditions for the experimental material. Guard rows, usually of the same species as the plants under study, are discarded at final harvesting of the experiment.

It is clear that these ideas can with profit be incorporated into the design of genecological experiments, and indeed advanced field trial techniques were employed in the famous genecological experiments of Gregor and his associates in studies of variation in *Plantago maritima*, to which we have already referred (Gregor, 1930, 1939; Gregor, Davey & Lang, 1936; Gregor & Lang, 1950).

In a simple genecological experiment each individual, say of plants A, B, C and D, may be clonally propagated, the experimental garden may be divided into small blocks and a ramet of each individual A, B, C and D planted in a weed-free plot surrounded by guard rows of the same species. The position of each ramet within blocks is determined by random numbers.

The fundamental ideas influencing the layout of simple field trials may also be incorporated into the design of more complex genecological experiments such as population trials, family lines and experiments involving populations given various treatments. Several excellent books with fully worked examples of various designs are now available for the biologist. Especially suitable for beginners are Salmon & Hanson (1964); Bishop (1971); Parker (1973); and Clarke (1980). More advanced treatment will be found in Campbell (1974); Ridgman (1975); Snedecor & Cochran (1980); Sokal & Rohlf (1969, 1981); Yates (1981); Stuart (1984); and Mead (1988).

Studies of agricultural crops have resulted in other important insights into the design of field experiments. At first sight it would seem reasonable to suppose that repeated experiments with the same varieties (or genotypes) would 'give the same results'. In practice there are considerable differences from year to year in the results of experiments estimating yield in cultivated stocks. The principal causes of variability are differences in weather and changes in the incidence and severity of various pests and diseases (which are themselves probably correlated with past or present weather conditions). An experiment by Nelson (1967) emphasises the importance of year-by-year differences in a genecological experiment. He was studying variation in *Prunella vulgaris* collected from many sites in the USA, by growing material at Berkeley, California. Usually there is no winter frost in this area, but, exceptionally, a very cold period occurred from 20 to 24 January 1962, providing him with a unique experiment, which revealed that some of his plants were frost sensitive.

In designing genecological experiments other factors must be taken into account:

1. In experiments begun with samples of seeds, Roach & Wulff (1987) point out that there may be maternal effects, i.e. there may be a 'contribution of the maternal parent to the phenotype of its offspring beyond the equal chromosome contribution expected from each parent'. Such contributions may (a) be cytoplasmic; (b) flow from the greater contribution of maternal genes ($2n$) than male genes (n) to the ($3n$) endosperm; or (c) result from the fact that maternal tissues contribute to developing fruits and seeds. Thus, maternal effects are often manifest in differences in seed size and

mineral composition. Roach & Wulff (1987) review the various techniques available to measure maternal effects in cultivation trials. Evidence for paternal effects – via 'male' cytoplasm – is also reviewed in the same paper. Maternal and paternal effects are examples of what are sometimes called 'carry-over' effects. An excellent investigation into the effects of parental environment on the next generation has been carried out by Elwell *et al.* (2011) using *Arabidopsis thaliana.* 'A single lot of seed was planted in six environmental chambers and grown to maturity. The seed produced was mechanically sieved into small and large class sizes then grown in a common environment ... Analysis of variance demonstrated that seed size effects were particularly significant early in development, affecting primary root growth and gravitropism, but also flowering time. Parental environment affected progeny germination time, flowering and weight of seed the progeny produced ... these data indicate that life history circumstances of the parental generation can affect growth and development throughout the life cycle of the next generation to an extent that should be considered when performing genetic studies.'

2. 'Carry-over' effects are also possible in experiments begun with vegetative material, such as clone transplants. Thus, the length of an experiment may be crucial if the investigation involves material dug up from the wild and transplanted into a garden for, as Turesson (1961) discovered, an extended period of adjustment may be necessary before plants may be said to have outgrown the effects of their original habitats. Indeed, it may be difficult to convince a sceptic that a complete adjustment is ever made, especially in the case of woody plants. Experiments with herbaceous plants have also been revealing. For instance, from a 43-year-old pasture in Canada, Evans & Turkington (1988) grew samples of *Trifolium repens* collected from patches dominated by different species of grasses. At the end of a field trial, lasting for four months, significant differences were detected between the samples for a number of characters. Then a second trial was begun with the same samples, using material produced by vegetative propagation. It is of very great interest that there were no significant differences between the samples after 27 months. This experiment is a clear indication of the importance of 'carry-over' effects in relation to the duration of experiments. A second example makes some further important points. 'Carry-over' effects could arise from the use of unequal-sized pieces of material used to begin clone experiments. In a study of many facets of garden trials, Davies & Snaydon (1989) examined the effect of tiller size – small versus large – in *Anthoxanthum odoratum* on a number of measures of performance in a garden trial. They discovered no evidence for a major problem with 'carry-over' effects in this case, as there were no differences in survival, height or date of flowering. However, large tillers produced slightly larger plants.
3. The pre-treatment of seeds and seedlings prior to the experiment is very important. There will be differences in the speed of development of plants between those sown as seed and those set out in the field as young plants. The timing of the experiment in relation to such seasonal factors as cold periods may also be crucial. Thus, some plants will not flower unless subjected to cold treatments, and spring and autumn sowing will yield different results.
4. The treatment of plants during the experiment has a profound effect upon their growth and performance. The experimenter must decide whether to water plants in dry weather, apply fertilisers, etc.
5. The incidence of pests and diseases causes considerable problems. In particular, experiments in glasshouses often turn into a struggle to control various insect and fungal pests, and the liberal use of pesticides may be the only means of 'preserving' the experiment. It is important to realise that 'spot-treatments' of badly infected individual plants may seriously affect the randomised design of the

experiment. Therefore, in the design of garden and field trials the decision is often taken to allow non-catastrophic invasions of pests and diseases to take their toll of the experimental material. In this way it may be possible to see if any individuals or populations are resistant to fungal or insect attack. Studies of the effects of non-fatal pests and diseases may add a further dimension to our knowledge of population variation.

6. Agricultural experiments are often designed to be left until final harvest, when estimates of yield are made on fruiting material, and in other cases the experiment is so constructed as to permit regular intermediate harvests at selected periods between sowing and final harvest. Such experiments may be poor models for experiments in the ecological genetics of plants, in which a great deal of information may be gathered by 'non-destructive scoring' of the plants over weeks or months. For instance, given adequate spacing between plants, plant height at different times could be measured, and the timing of flowering and fruiting could be studied. Also, samples of leaves could be removed for study, provided that all the material in the experiment is treated alike. Thus, a good deal of quantitative information might be obtained by repeated scoring of an experiment. Sometimes it is unnecessary to make measurements: the stages of development or incidence of damage by pests may be recorded by classifying the material into a small number of 'character states'.

The designed experiment

So far we have discussed a number of important factors in the design of genecological experiments. For both the experimenter and the botanist who wishes to interpret the scientific literature it is crucial to take proper account of the problems and possibilities of sampling and cultivation. These are elements in a larger canvas, however. Many authors have stressed that genecologists should aim at a *designed experiment* in which hypothesis, sampling, cultivation, analysis and interpretation all take their proper places.

The generation of germinal ideas is a mysterious process. Armed with a knowledge of the literature, provoked by the observations and comments of others, the botanist notices something of interest in the patterns of variation. From this initial interest an idea emerges for an experiment. The process by which ideas occur to experimenters is not to be seen as a mechanical process, but as a creative act much as is required for practice of the arts. Next, the investigator formulates a hypothesis (a conjecture) leading to an experiment, the results of which are used to consider whether, as a logical consequence, the hypothesis is confirmed or rejected. Experimenters must be aware that they might have a preference for a particular outcome, and that it is a crucial error to avoid/ignore results that do not fit such expectations. Turning briefly to an area of some complexity, it is important to note that classical approaches (and indeed popular belief in the present day) consider that experiments can prove or partially confirm a hypothesis. However, an alternative viewpoint has been championed by the philosopher Karl Popper (1963) in his book *Conjectures and Refutations* (Magee, 1973; Medawar, 1984, 1991). He stressed that an hypothesis (and experimental design) should be framed in such a way that it is falsifiable, disprovable. In the present context, these ideas lead to devising and statistical testing of a 'null hypothesis' (see below). Popper's views form the basis of modern scientific experimentation, but they are not universally accepted and some have pondered whether the 'falsification' approach is how scientists carry out experiments day to day. Here we consider the importance of designed experiment in genecology, the results of which are almost invariably subject to statistical tests.

The best way to appreciate the different elements in the designed experiment is to study an example. We have chosen to present the results of a simple study on *Plantago major* (Warwick & Briggs, 1979). Our account should be seen as a simplified (certainly oversimplified) introduction to a central concern of science, namely how to devise, execute and interpret experiments. We hope that biologists reading our account will be encouraged to study the many excellent introductory books (which we have noted above) on the design and statistical analysis of experiments.

An experiment to study the variation in *Plantago major* growing on droves (grassy tracks) at Wicken Fen Nature Reserve, Cambridgeshire, England

Many thousands of visitors visit the famous Wicken Fen Nature Reserve each year and the droves (grassy tracks) that cross the fen are subject to severe trampling pressure. *Plantago major* occurs in the heavily trampled areas (as a small prostrate plant) and also in the adjacent grassy sward (in which it is a larger, erect plant).

Ecotypic differentiation has been reported in *P. major* (Turesson, 1925; Groot & Boschuizen, 1970; Mølgaard, 1976) and there is evidence from a number of genecological studies which suggests that differentiation might occur over short distances, despite gene flow. Therefore the possibility exists that dwarf prostrate variants might be selected on the pathway, while taller plants would be at a premium in the adjacent grassy sward. Thus, we could formulate the hypothesis that samples taken from the wild might retain their distinctness in cultivation. As the differences involved are those of size, our hypothesis is not very precise in its present form. We cannot make any definite prediction as to the degree of difference to be retained; indeed as we are dealing with quantitative differences it is not at first sight clear how one can make a prediction as to the degree of difference that 'needs' to be retained in order to accept the hypothesis. So far our hypothesis is too vague. However, a precise hypothesis is possible in this case, namely that on cultivation we expect *no* difference between groups of *Plantago* after cultivation. Such a hypothesis is known as a 'null hypothesis'. The concept of the null hypothesis is widely used in biology and such a hypothesis, that zero difference is expected between two sample groups, should always be formulated as part of a designed experiment, for a precise initial hypothesis is likely to lead to a well-designed investigation.

Unbiased samples, 10 from the trampled area and 10 from adjacent grassy swards, were collected in the autumn of 1974. *P. major* is not a clonally propagating species (although it may be cloned in gardens: Marsden-Jones & Turrill, 1945), but spaced samples were taken at least 10 m apart. Plant material was potted up in John Innes No. 1 compost and the pots, which were randomly arranged, were plunged to the rims in the sand of an outdoor plunge bed. Spacing between pots was very generous and guard rows were not necessary.

In order to allow us to examine the null hypothesis, a statistical test is necessary to enable us to compare the two groups of samples. The test should allow us to compare the variation between and within groups. Clearly variation between groups (from trampled path versus adjacent grassy sward) is only likely to be significant if it can be shown to be significantly greater than variation within groups. We shall use for our test the analysis of variance technique, which works by estimating the significance of variation between groups by comparing it with variation within groups. The variation in some measurable trait of twenty plants of *Plantago major* is, by this test, partitioned in such a way as to enable us to see the variation due to subsites at Wicken, while at the same time giving us an estimate of the variation within groups.

The steps in the analysis of variance are a simple extension of those used in Chapter 3. To recapitulate, we showed that:

$$\text{variance}(s^2) = \frac{\sum (x - \bar{x})^2}{n - 1}$$

Table 8.7. ***Plantago major*: length of longest leaf after *c.*10 months cultivation in the Botanic Garden, Cambridge University**

	Wicken Fen: trampled areas on droves	Wicken Fen: grassy swards adjacent to droves
	30.5	33.3
	33.4	28.0
	25.5	21.9
	34.2	26.0
	27.4	24.0
	26.5	28.4
	31.5	32.2
	29.3	27.0
	24.8	26.3
	28.0	26.0
Mean	29.110	27.310
Total	291.100	273.100

Grand total = 564.200

$$\text{Correction factor} = \frac{564.200^2}{20} = 15916.082$$

$$\text{Sum of squares (total)} = (30.5^2 + 33.4^2 + 25.5^2 \ldots 26.0^2) - C$$
$$= 16133.080 - 15916.082 = 216.998$$

Having calculated the total sum of squares we now calculate the variations between and within groups. Between groups is estimated by

$$\frac{291.100^2}{10} + \frac{273.100^2}{10} - C = 15932.282 - 15916.082 = 16.200$$

Within groups is estimated by subtracting 16.200 from the total sum of squares. Within groups sum of squares = 216.998 – 16.200 = 200.798. Subdivision of the sum of squares into its two parts has been accomplished and the degrees of freedom (19 in all: one less than the number of observations) may now be determined for each component. Between groups: 2 groups, therefore 1 degree of freedom. Within groups: 10 observations per group, each loses 1 degree of freedom, total 18.

The analysis of variance may now be set out in a table showing the sources of variation, the divisions of degrees of freedom and sum of squares. Mean squares (variances) are now calculated. The between-groups mean square gives the variance of the two groups about the grand mean, while the within-groups variance gives the variance of individual values about the two sample means.

Source of variation	Degrees of freedom	Sum of squares	Mean square (variance)	Variance ratio (F)	Probability
Between groups	1	16.200	16.200	1.452	>0.05
Within groups	18	200.798	11.155		
Total	19	216.998			

The sum of the squares of the deviations from the mean could be calculated by subtraction of each value from the mean, squaring the difference and summing the resulting squared deviations. Alternatively we suggested that the sum of (deviations from mean)^2 (sum of squares) could more readily be calculated by employing the formula

$$\text{sum of squares} = \sum x^2 - \frac{(\sum x)^2}{n}$$

where $\frac{(\sum x)^2}{n}$ is known as the Correction Factor or Term, C.

We may now examine (Table 8.7) the steps in the calculation of simple analysis of variance on the *Plantago major* experiment. The *null hypothesis* is that there is no difference in leaf length between plants grown from trampled drove and from grassy sward.

If this null hypothesis is to be confirmed then there should be little or no difference between the variances between and within groups. To estimate the relative size of these two variances we calculate the variance ratio (the *F* value – in honour of R. A. Fisher who developed analysis of variance). If, however, there is a real difference between groups, we would expect variation between groups to exceed that of the variance within groups. Tables of probabilities appropriate to different values of *F* are available. In the case of the *Plantago* experiment it is clear that a good deal of the variation is *within* groups and that the difference between groups is small. The mean values are very similar. Indeed, leaf length of plants grown from the small plants of the trampled area slightly exceeds that for the samples from the tall sward. The null hypothesis, that there is no statistically significant difference between the two groups of plants, is supported by our results. On the strength of present evidence, we have no reason to suppose that 'ecotypic' differentiation has occurred in the trampled and tall sward subsites.

The *Plantago major* investigation was part of a more extensive study of this species in various grasslands (Warwick & Briggs, 1979, 1980b). Table 8.8 sets out another comparison. Small phenotypes were found not only in trampled areas (as on the droves at Wicken), but also in closely mown lawns. Samples of plants from the Botanic Garden lawn in Cambridge and from Wicken droves (trampled areas) were compared. The variation between groups in this case is statistically significantly greater than the variation within groups. Therefore, the null hypothesis, namely that samples do not differ in leaf length, receives no support from the experiment. There would appear to be a real difference in leaf length between the two samples. In Warwick & Briggs (1979, 1980b) details are given of the highly distinctive plants of *Plantago major* discovered in the lawns of Cambridge colleges and gardens.

Our examples of analysis of variance are of a very simple kind, with division of the variation into two parts. Much more elaborate experiments may be devised, and the 'overall variation' discovered in experiments may be divided into many parts estimating, where appropriate, the variation due to blocks, population differences, family lines within populations, interacting factors, random events, etc. Analysis of variance is also the basis of tests of variation within and between groups in molecular investigations. By looking at the relative magnitude of different segments of the variation, very considerable insights into population variation may be obtained.

Analysis of variance is a most elegant technique, which must, however, be used with care. It should only be employed in analyses where the results are 'normally distributed' and in which the variances of the contributing population samples, treatment values, etc., are equal or approximately so. Various tests have been devised to study the 'properties' of arrays of figures to see if they are appropriate for analysis of variance (e.g. for details of Bartlett's test see Salmon & Hanson, 1964; Sokal & Rohlf, 1981). Sometimes it is possible to 'transform' the results to produce equality of variances. For instance, the

Table 8.8. ***Plantago major*: the effect of *c.*10 months cultivation on samples of small phenotype from Wicken droves and Botanic Garden lawns; length of longest leaf (cm)**

	Wicken: trampled areas on droves	Botanic Garden lawns
	30.5	11.8
	33.4	20.7
	25.5	8.9
	34.2	22.6
	27.4	24.0
	26.5	14.1
	31.5	13.1
	29.3	16.0
	24.8	12.5
	28.0	12.0
Mean	29.110	15.570
Total	291.100	155.700

Grand total = 446.800

$$\text{Correction factor} = \frac{446.800^2}{20} = 9981.512$$

$$\text{Sum of squares (total)} = 30.5^2 + 33.4^2 + 25.5^2 \ldots 12.0^2 - C$$
$$= 11228.860 - 9981.512 = 1247.348$$

$$\text{Between groups sum of squares} = \frac{291.100^2}{10} + \frac{155.700^2}{10} - C$$
$$= 916.658$$

Within groups sum of squares = 1247.348–916.658 = 330.690

Source of variation	Degrees of freedom	Sum of squares	Mean square (variance)	Variance ratio (F)	Probability
Between groups	1	916.658	916.658	49.895	< 0.001
Within groups	18	330.690	18.372		
Total	19	1247.348			

unsatisfactory raw data may be converted to square roots or to logarithms. If the results cannot be satisfactorily transformed, then other statistical tests – so called non-parametric tests – may be applied (Sokal & Rohlf, 1981). Such tests do not make any assumptions that the values from the experiment are normally distributed or have equal variances. Non-parametric tests should be more widely used in biology, for the results of many experiments and observations show enormous departures from normality.

The interpretation of experiments

Whatever the results of particular experiments, there are usually grounds for a cautious interpretation of genecological studies.

However many plants are grown, or studied in experiments, the size of samples that can conveniently be grown is often minute relative to the size of wild or semi-natural populations. For instance, according to the estimates of Barling (1955), populations of *Ranunculus bulbosus* may reach 257 000 per acre in the English Cotswolds, and continuous populations in adjacent fields of pasture were estimated to contain 14 million plants. Such figures are by no means exceptional.

Many experiments are carried out in conditions remote from those in nature. For example, studies of metal tolerance in plants involve measurements of root growth in very simple culture solutions (see Chapter 9). In the wild, plants grow in soils where conditions are quite different. The attempt to simplify situations in order to study individual factors is clearly justified, but the investigator must not make too facile an extrapolation from simple laboratory tests to the natural situation.

As in the case of metal tolerance many experimentalists isolate individual factors of presumed importance and make special studies of the tolerances of population samples. A fascination with the study of critical or limiting factors should not blind the student of evolution to the fact that the concept of a factor is an abstraction. Often particular factors are chosen for study largely because the means to control or vary them in precise ways are available in laboratories. It is often forgotten that plants respond to their environments as a functioning whole. Realising this difficulty, a number of botanists are becoming interested in the experimental studies of the adaptive significance of variation in plants by carrying out experiments in the field. The garden trial with its weed-free, spaced plants is not entirely satisfactory as a means of studying 'adaptation', for the competitive interaction between plants is absent. Thus, there is continuing interest in the reciprocal clone-transplant experiment, in which cloned material of diverse origin is transplanted into swards subject to different treatments. By close mapping and labelling of plants the survival and growth of transplants may be studied (see Chapter 9). Care in the laying-out and recording of such experiments may overcome the difficulties that cast doubt on the historic studies of Bonnier & Clements (see Chapter 4), By studying the way plants behave in such experiments, the experimentalist may have a very direct insight into the responses of plants to 'whole' environments. There is obviously a place for both tolerance tests and reciprocal transplant investigations in the repertoire of techniques available to the genecologist.

A final problem facing the experimentalist is that of deciding the causes of the underlying patterns of variation under study. Even after long and complex cultivation experiments, it is not possible to conclude with certainty that residual variation, in say a garden trial, is 'genetic' in origin: breeding experiments are necessary to see if characteristics are transmitted by seed. Such experiments are impractical for most species, especially if they are long-lived. As we shall see in the next chapter the advent of molecular methods has utterly transformed the study of the genetics of population variation, and has shed new light on the processes of microevolution.

9 Pattern and process in plant populations

The word 'population', like so many other familiar terms, seems to present little difficulty until we have to define it. In statistics, the concept of population is an abstraction signifying a theoretically large assemblage of individuals from which a particular group under consideration is a sample. Most biological uses of the term, however, imply the total of organisms belonging to a particular taxonomic group (or 'taxon') that are found in a particular place at a particular time. Scientific research involves the testing of models. The population model of outstanding significance, in the study of variation, is quite specific, being the local 'interbreeding group of individuals sharing a common gene pool' (Dobzansky, 1935 in Rieger, Michaelis & Green, 1976). In earlier editions of this book, the term 'gamodeme' was used for such populations. The deme terminology was devised by Gilmour and associates (Gilmour & Gregor, 1939; Gilmour & Heslop-Harrison, 1954; Gilmour & Walters, 1963; Walters, 1989a, b, c). Details of the system are given in the glossary. It is with some regret that we abandon the use of this term, but must face the fact that the '-deme' terminology has not been accepted or is used in ways not intended by its authors (Briggs & Block, 1981).

Population geneticists have coined other names – Mendelian or panmictic populations – for groups essentially similar to gamodemes. All these units represent idealised 'model systems' (see Maynard Smith, 1989; Silvertown & Lovett Doust, 1993; Silvertown & Charlesworth, 2001).

The model underlying the idealised population is the Hardy–Weinberg Law. In a population with two alleles A and a, which are selectively neutral and in which mating is at random, the expected frequencies of A and a (p and q respectively) are

$$(p + q)^2 = p^2 + 2pq + q^2 = 1$$

This represents the 'null' model (Cockburn, 1991), for a large population of constant number in a diploid organism, with sexual reproduction, non-overlapping generations, the same allele frequencies in male and female, no gene flow or migration and no selection operating. In such a situation the relative frequencies of the alleles will not change from generation to generation.

Clearly, in natural populations 'in the wild', many factors will cause deviations from the Hardy–Weinberg equilibrium, and these have been modelled by population geneticists (Silvertown & Lovett Doust, 1993; Silvertown & Charlesworth, 2001). As we shall see, it is not possible to determine the exact limits of Mendelian populations. The best that can be achieved is to use statistical models to determine the probable limits of gene flow. For example, mutation rates are usually very low and do not influence populations in the short term. In many plant species there is considerable overlapping of generations. Concerning the 'linkage' of genes, we have already considered, for example, the breeding system in *Primula* where a 'supergene' of closely linked genes controls distyly. The account of the various breeding systems found in plants (Chapter 7) provides many excellent examples of situations, including self-incompatibility, autogamy and apomixis, where the Hardy–Weinberg 'null' position is vitiated by various kinds of non-random mating. Pollinator behaviour may also produce non-random (so-called assortative) mating. For instance, in *Raphanus raphanistrum*, which is sometimes polymorphic for flower colour (white/yellow), pollinators may show strong preferences for one flower colour variant (Kay, 1978). Also, such important factors as gene flow, chance events and natural selection must be considered. As we shall see, the mathematical model systems of the population geneticist provide a very valuable framework for considering these elements of microevolution, but, as we are concerned, in the main, with what happens 'in the wild', we will tend to use mostly verbal models, at the same time giving details of published sources containing mathematical treatments.

Variation within and between populations

Population geneticists have devised a number of measures of genetic variation within and between populations (Wright, 1950; see Hartl & Clark 2004 for full details). The development of techniques for studying plant isozymes provided information about genetic variation in a wide number of plants (Hamrick, Linhart & Mitton, 1979; Brown, 1979; Gottlieb, 1981a; Hamrick & Loveless, 1989; Hamrick & Godt, 1989; Cole, 2003).

There is clear evidence that populations of species with different breeding system, geographic range, life-form and seed dispersal have different levels of variation (Table 9.1). In considering these historic data, some caution is necessary, as there are differences in how components of genetic variability have been calculated (Bossdorf *et al.*, 2005). Also, there are debates about how far such values correlate with quantitative variation (Reed & Frankham, 2001), and, given the diverse breeding behaviour of plants, there is some scepticism about the validity of broad generalisations about such groups as selfers.

Essentially, these early estimates provided a valuable 'snapshot' of genetic variation within and between populations, as revealed at a specific date, by a particular sampling method and set of markers, and they have provoked a great deal of detailed investigation on the nature of the processes underlying pattern in populations.

Now population variation is generally investigated using an array of molecular methods. Such a very large number of population studies have been published that comprehensive summary is not possible. Here, we stress again a point made in Chapter 6: most of the genetic markers used in population genetics are 'essentially neutral' and do not reflect the 'evolutionary potential' for adaptation in populations and species (Holderegger, Kamm & Gugerli, 2006).

The key processes to examine in elucidating population variation are natural selection, chance events and gene flow (the dispersal of gametes and zygotes within and between populations). In this chapter we examine case histories that focused principally on natural selection in plant populations. Then, in Chapter 10, we consider investigations that have studied how chance influences gene flow, consider what happens when new populations are formed from small numbers of individuals drawn from a larger population (founder effects) and the consequences of severe reductions in population numbers (bottleneck effects). In discussing these different factors, we stress that these 'mechanisms of evolution do not act in isolation' (Andrews, 2010). Then in Chapter 11, we examine the factors that determine whether populations flourish or decline to extinction.

Selection in populations

Examining first the evidence for natural selection in populations, we note that the processes that Darwin grouped together under this blanket term are now classified into three main types (Fig. 9.1).

The first type, so-called stabilising or normalising selection, is a process tending to produce conformity and stability. At first sight it is surprising to find selection acting in this way, as many people equate natural selection with change. Imagine, however, a population, which is well adapted to its environment. The environment may, of course, change climatically and biotically with the seasons,

Table 9.1. **The amount of allozyme variation for different groups of higher plants. (From Brown & Schoen, 1992, using information from Hamrick & Godt, 1989.) The measures of variation are: A = allelic diversity, as number of alleles per locus; h = a measure of gene diversity; G_{ST} = the extent of population divergence, as the proportion of total diversity found among populations ($G_{ST} = 1 - h_p/h_s$); and N = number of species. Subscripts s and p refer to measures within species and populations, respectively**

Many comparisons may be made between different groups of species. For instance, note that in comparison with outcrosses species, selfers show less diversity within species and populations but greater divergence between populations. Also, that endemic and narrowly distributed species, in this sample set, are less variable than widespread species.

	Species			Population		Divergence
Categories	N	A^a_s	h_s	A_P	h_P	G_{ST}
			Breeding System – Pollination			
Selfing	78	1.69	0.12	1.31	0.07	0.51
Mixed						
– animal	60	1.68	0.12	1.43	0.09	0.22
– wind	9	2.18	0.19	1.99	0.20	0.10
Outcrossing						
– animal	124	1.99	0.17	1.54	0.12	0.20
– wind	102	2.40	0.16	1.80	0.15	0.10
			Geographic Range			
Endemic	52	1.80	0.10	1.39	0.06	0.25
Narrow	82	1.83	0.14	1.45	0.11	0.24
Regional	180	1.94	0.15	1.55	0.12	0.22
Widespread	85	2.29	0.20	1.72	0.16	0.21
			Life Form			
Annual	146	2.07	0.16	1.48	0.11	0.36
Perennial herbs	119	1.70	0.12	1.40	0.10	0.23
Long-lived woody	110	2.19	0.18	1.79	0.15	0.08
			Seed Dispersal			
Gravity	164	1.81	0.14	1.45	0.10	0.28
Attached	52	2.96	0.20	1.68	0.14	0.26
Explosive	23	1.48	0.09	1.25	0.06	0.24
Ingested	39	1.69	0.18	1.48	0.13	0.22
Wind	105	2.10	0.14	1.70	0.12	0.14

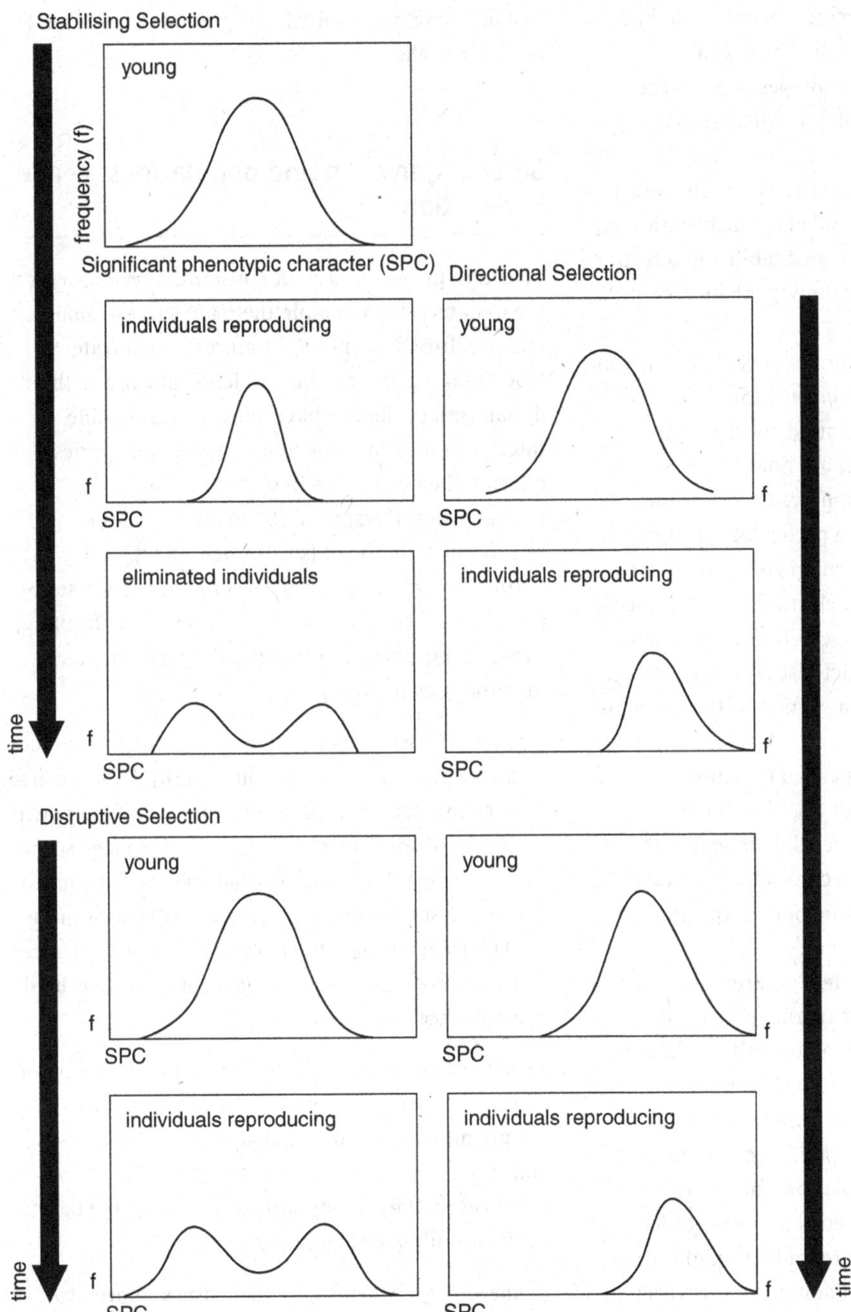

Fig. 9.1. The three types of natural selection. (From Hardin, 1966.)

but we assume that, sometimes, it is not changing directionally and fundamentally. By sexual reproduction an array of phenotypes is produced, a sample of which may exhibit a typical normal distribution for some character; most of the individuals in the array depart little from the mean, but some segregants in each tail of the distribution are markedly distinct from the mean. Stabilising selection has the effect of eliminating individuals that depart significantly from the mean, giving a bimodal distribution of eliminants. For instance, in a study of flowering in *Tephroseris integrifolia* (*Senecio integrifolius*), Widén (1991) found evidence for stabilising selection in some, but not all, years.

The second type of selection occurs where the environment is changing in a particular direction, for instance, where a site is becoming contaminated with a heavy metal such as zinc. Selection may then occur producing a directional change in the mean values for significant phenotypic characteristics. Again, as a consequence of selection, a portion of the young is eliminated.

The third and final type is called disruptive selection. In this case individuals, of different genotype, may be at a selective advantage in different places within the total area occupied by a population, resulting in a pattern of genetic polymorphism in a mosaic or patchy environment.

These models of natural selection provide guidelines for the design of experiments aimed at detecting selection in action in populations. Many experimental studies of natural selection make comparisons of adult material from contrasting sites. Other investigations have examined the young and mature individuals in a population, for, as we have seen from the models described above, selection often involves the maturation and reproduction of only a portion of the population – adults are less variable (in significant phenotypic traits) than the young from which they develop. Thus, if the variation in young and adult is compared, it should be possible to detect selection at work. This idea is of course not a new one. As we saw in Chapter 5, as long ago as the 1890s, the zoologist Weldon studied variation in young and adults in crabs.

Selection in changing populations: r- and K-selection

Selection pressures are not constant, as ecosystems are dynamic. For example there are successional changes from bare ground to forest, open water to woodland etc. In devising models that capture this dynamism, ecologists have shown considerable interest in how intraspecific variants and species allocate their resources in open and closed communities in response to the three modes of selection just outlined (see Harper, 1977).

On bare ground derived from natural processes of erosion, as well as human activities such as farming, overgrazing, rubbish dumping, Harper characterises the processes involved.

> During phases of population increase after a crash or after a new invasion, individuals are by definition free from interference from neighbours and of their own species. During the phase of population increase the pre-eminent force selecting individuals that contribute most descendents to subsequent generations is their fecundity and the effect of fecundity is enhanced by precocity (provided that generation follows hard on the heels of generation).

As a consequence there is 'selection for allocation of a large fraction of resources to seed production', and 'no premium on attributes that confer competitive ability'.

However, Harper emphasises that when the habitat begins to fill up with plants:

> the struggle for existence that ensues contributes a special element to the process of natural selection, for the chance of an individual leaving descendents now depends on the share of needed resources that can be captured from competitors. Precocity and high fecundity are now relatively unimportant compared

to the factors that confer aggressiveness, e.g. height, a perennial or quickly renewable canopy, larger but fewer seeds. In dense communities both inter- and intraspecific encounters favour individuals with longer life, longer juvenile periods spent gaining competitive ascendancy.

Fitness

It is important at this point to define more carefully what is meant by 'selective advantage' by introducing the concept of 'fitness'. Fitness is relative and may be defined by the *number* of descendants left by an individual compared with those of other individuals. What is clearly important in microevolutionary change is the *relative* contribution of offspring made to the next generations.

With regard to a particular environment:

relative fitness may be estimated = performance of alien genotypes/performance of native genotypes.

In some cases where extreme habitats have been studied the calculations of fitness are simplicity itself. For instance, only heavy-metal-tolerant variants survive on highly contaminated soil, and seedlings unable to tolerate heavy metals in the rooting medium may almost all die on germination, making no contribution to the next generation. In most habitats, however, conditions are less harsh, and individuals of different tolerance survive and reproduce with varying degrees of success. To estimate fitness directly in such situations is an enormous challenge and may not be feasible. But it may be possible to estimate relative fitness based on the comparison of the survival, growth and reproduction of plants in various experiments (Levin, 1984; Silvertown & Charlesworth, 2001). In situations where several estimates of 'relative fitness' have been made from different measurements of vegetative and reproductive characters of the same material, very different results have been obtained (see Warwick & Briggs, 1980a, b, c). Wherever possible, estimates of reproductive success should be obtained.

In considering relative fitness, Harper (1977) stresses a most important point.

> Natural selection operates on individuals in a population and selectively favours those individuals and those genotypes that contribute most descendants to subsequent generations . . . The critical word in this argument is descendants – the contribution of parents to subsequent generations. *It does not follow that an organism that produces a large number of progeny will also leave a large number of descendants.*

Finally, we stress an important general conclusion that has emerged from studies of selection. In his famous mathematical studies of selection outlined in Chapter 5, Fisher (1929) showed that selection would be effective even if there was only a very small selective advantage between individuals in populations. It is now clear from the study of extreme habitats, however, that very large selective advantages must sometimes be operating in nature, and we might therefore conclude that sometimes selection may act very swiftly indeed.

Developments in the investigation of populations

Early studies in genecology, now more often called ecological or molecular genetics, often involved the study of widely spaced sites and the examination of gross differences in phenotype, say as large as those found in Mendel's stocks of peas – tall (c.2 m or more) versus dwarf (c.0.5 m). Almost any garden trial, however crude, could hardly fail to 'reveal' such differences in height. However, with a growth of interest in studying small sites as systems has come a fascination with small-scale variation in its ecological context. Investigations of this type require well-considered sampling techniques, combined with proper attention to design, layout, scoring and above all to statistical analyses of experiments. Amongst the repertoire of techniques for the study of selection, the transplant experiment is proving a powerful tool in analysing population variation. To provide a framework in which population variability and the speed of evolutionary change may be judged, knowledge of the history of study areas is very important.

In times past, many ecologists chose to study what they thought, often erroneously, were natural communities (Heslop-Harrison, 1964; Briggs, 2009). However, as we see in this chapter many insights into microevolution have come not only from investigation of what are taken to be 'wild' plants, but also from the study of weedy, invasive and introduced species in human-disturbed habitats (Bishop & Cook, 1981; Bradshaw & McNeilly, 1981; Taylor, Pitelka & Clegg, 1991; Davies, 1993; Briggs, 2009). In such areas, extreme habitats are often juxtaposed; for instance, areas with heavy-metal pollution can occur as islands in a sea of pasture land; grasslands managed for grazing may be found close to hayfields, or arable land; lawns are surrounded by flowerbeds; and improved grasslands, subject to fertiliser and pesticide treatments, are located next to untreated grasslands. In studies of plant variation in such heterogeneous habitats very strong selection pressures have often been identified.

Selection: the study of single factors

Natural selection is most readily investigated by studying intraspecific variation at grossly dissimilar sites differing in toxicity, exposure, temperature, light, moisture, salinity, etc. Such *extremely different sites* are sometimes distant from each other, sometimes side by side. Plants are collected and cultivated and subjected to a tolerance test. In such tests a major habitat factor, in which experimental sites are presumed to differ, is investigated under garden or laboratory conditions.

Experiments of this type reveal many patterns explicable in terms of selection (see Table 9.2 for some examples).

In judging tolerance testing it is important to assess the completeness or otherwise of the evidence. The assumption that the habitats differ in critical factors should be investigated fully by quantitative methods. For instance, if it is supposed that two sites differ in copper contamination, measurement of copper in the soil should be made and the availability of the copper *to the plant* should be examined. Bearing in mind the difficulties of extrapolation from laboratory to the field situation, tolerance tests should be as 'natural' as possible. In a large number of cases the investigation has reached the point where, for example, plants from sites A and B have been shown to be subject to different levels of factor X in the field. In a tolerance test, manipulation of factor X reveals a difference in sensitivity to X, which is consistent with the level of exposure to X each receives in the wild. In relation to a specific factor each group of plants seems to be best able to survive, grow and reproduce in its native conditions. In such circumstances the biologist often suggests that the pattern is the result of natural selection. But commonly the evidence for this view is circumstantial, especially if the genetic basis of the difference has not been examined.

Studies of several interacting factors: *Lotus* and *Trifolium*

In cases where a genetic polymorphism has been investigated in detail, it has become apparent that the effect of many factors must be considered. This important point is beautifully illustrated by studies of cyanogenesis (the production of hydrogen cyanide: HCN) in which we now know that many factors interact. It is worth describing this phenomenon in some detail.

In the Sudan campaign of 1896–1900, a number of British transport animals were poisoned by eating a local species of *Lotus* (Dunstan & Henry, 1901). Chemists, interested in the losses, discovered that certain species of the genus contain cyanogenic

Table 9.2. **Examples of historic studies of ecotypic differentiation selected to indicate the variety of tolerance tests devised by genecologists and ecologists. (See reviews by Clements, Martin & Long, 1950; Heslop-Harrison, 1964; Antonovics, Bradshaw & Turner, 1971; and Crawford, 1989)**

1. *Soil*
(*i*) *Use of natural and other soils*
Limestone soil; *Teucrium scorodonia* (Hutchinson, 1967)
Colliery waste; *Agrostis capillaris* (*A. tenuis*) (Chadwick & Salt, 1969)
(*ii*) *Natural soil plus additions*
Mine soil plus garden soil; test used widely by Liverpool group studying heavy-metal tolerance (Bradshaw & McNeilly, 1981)
(*iii*) *Water culture experiments*
Effects of chromium, nickel, magnesium; *Agrostis* spp. from serpentine (Proctor, 1971a, b)
Various heavy metals; e.g. arsenic (Pollard, 1980), cadmium (Coughtrey & Martin, 1978), zinc (Al-Hiyaly, McNeilly & Bradshaw, 1988), and see review of heavy-metal tolerance by Macnair (1993)
Effect of aluminium (Chadwick & Salt, 1969; Davies & Snaydon, 1973b)
Effects of sodium chloride (e.g. Ab-Shukor *et al.*, 1988) in *Trifolium repens*
(*iv*) *Soil water stress*
Pots allowed to dry out; *Pseudotsuga taxifolia* (Pharis & Ferrell, 1966)
(*v*) *Flooding*
Pots with different water regimes; *Veronica peregrina* (Linhart & Baker, 1973)
2. *Light*
Variation in light regimes using shade tubes; *Teucrium scorodonia* (Hutchinson, 1967)
3. *Exposure*
Plants growns in pots in exposed sites; *Agrostis stolonifera* (Aston & Bradshaw, 1966)
4. *Effects of loss of foliage*
Effects of grazing animals; various taxa (see Watson, 1969; Jones, Keymer & Ellis, 1978)
Effects of mowing; *Plantago major* (Warwick & Briggs, 1980b)
5. *Air pollution*
Gas chambers, etc. (Taylor & Murdy, 1975; Horsman, Roberts & Bradshaw, 1979; Taylor, Pitelka & Clegg, 1991)
6. *Herbicides*
Various tests on many taxa (see Lebaron & Gressel, 1982; Warwick, 1991)

glycosides. If leaves are bruised, the glucoside is broken down and hydrogen cyanide is liberated. Further studies revealed that, while some plants of *Lotus corniculatus* are cyanogenic, others are acyanogenic (Armstrong, Armstrong & Horton, 1912). Dawson (1941) studied this polymorphism in the south of England. Using sodium picrate papers, which redden in the presence of hydrogen cyanide, he tested samples from different populations. By crossing cyanogenic and acyanogenic plants, Dawson was able to show that it was likely that the presence of glucoside is dominant to its absence. The genetics of the situation is complicated, however, as *Lotus corniculatus* is a tetraploid.

In *Trifolium repens*, another species polymorphic for cyanide production, Corkhill (1942) and Atwood & Sullivan (1943) demonstrated that, in this species, cyanogenic glycoside presence (allele *Ac*) is dominant to glucoside absence (allele *ac*). The glycosides involved – linamarin and lotaustralin – are found in the vacuoles of the cells of the leaf and stem. The enzyme hydrolising the glycoside to produce the cyanide is determined at an independent locus. Presence of this enzyme (Li) is dominant to its absence (li). The hydrolysing enzyme is found in the cell wall. In cyanogenic plants, subject to herbivory, 'cell rupture from tissue damage brings the enzyme into contact with the substrate and generates the cyanogenic response' (Olsen, Sutherland & Small, 2007). The extent to which this cyanogenesis polymorphism occurs in other *Trifolium* species has been investigated by Olsen, Kooyers & Small (2014).

Kooyers & Olsen (2012) summarise the molecular basis of cyanogenesis in *Trifolium repens*, as follows: '*Ac*/*ac* corresponds to a gene presence/absence polymorphism at the locus CYP79D15, which encodes the P450 protein catalysing the first step in cyanogenic glycoside biosynthesis pathway (Olsen, Hsu & Small, 2008); Li/li corresponds to an unlinked gene presence/absence at Li, which encodes the linamarase glycoprotein precursor' (Olsen, Sutherland & Small, 2007). Further molecular analysis has recently been published (Olsen, Kooyers & Small, 2013).

Distribution of cyanogenic variants

Returning to historic studies, the distribution of the variants in *T. repens* was examined by Daday (1954a, b), who showed that cyanogenic plants were present with high frequency in south-west Europe. In contrast, acyanogenic plants predominate in north-east Europe (Fig. 9.2). At intermediate sample stations different proportions of the two variants were found; there is in fact a 'ratio-cline' across Europe. Interesting observations were made also upon the frequency of cyanogenic plants at different altitudes in the Alps, and again a ratio-cline was discovered, high frequencies of cyanogenic plants being reported from low altitudes. This frequency declined with increasing elevation, until at high altitudes all the plants in the sample proved to be acyanogenic (Fig. 9.3).

Trifolium repens was introduced into North America within the last 500 years, and similar clines in cyanogenesis have been reported by Daday (1958), who examined samples collected in localities from southern Louisiana to Alaska. Clinal patterns have also been detected in Canada by Ganders (1990). Recently, Kooyers & Olsen (2012) reported a latitudinal cline in samples drawn from a 1650 km transect from New Orleans (86% cyanogenic) to Wausau in Wisconsin (only 11% cyanogenic).

In interpreting his findings on the distribution of the different variants in Europe, Daday showed that there was a correlation between cyanogenesis and January mean temperatures, a decrease in temperature being associated with an increase in frequency of the acyanogenic variant. It appeared likely from these investigations that winter temperatures played some direct role, through natural selection, upon the frequency of cyanogenic plants of *Trifolium repens*, or that the locus concerned with glycoside production is genetically linked to genes involved with fitness responses at different temperatures (see Daday, 1965).

The role of herbivory

Studies by Jones (1962, 1966) shed new light on the frequency of cyanogenic plants of *Lotus corniculatus* in different English localities. He observed that while such plants were relatively free from damage by small invertebrates, many acyanogenic plants showed signs of having been grazed by slugs and snails. Following these observations, he carried out some simple experiments in which various species of slugs and snails were confined with cyanogenic and acyanogenic plants of *Lotus*. The experiments were

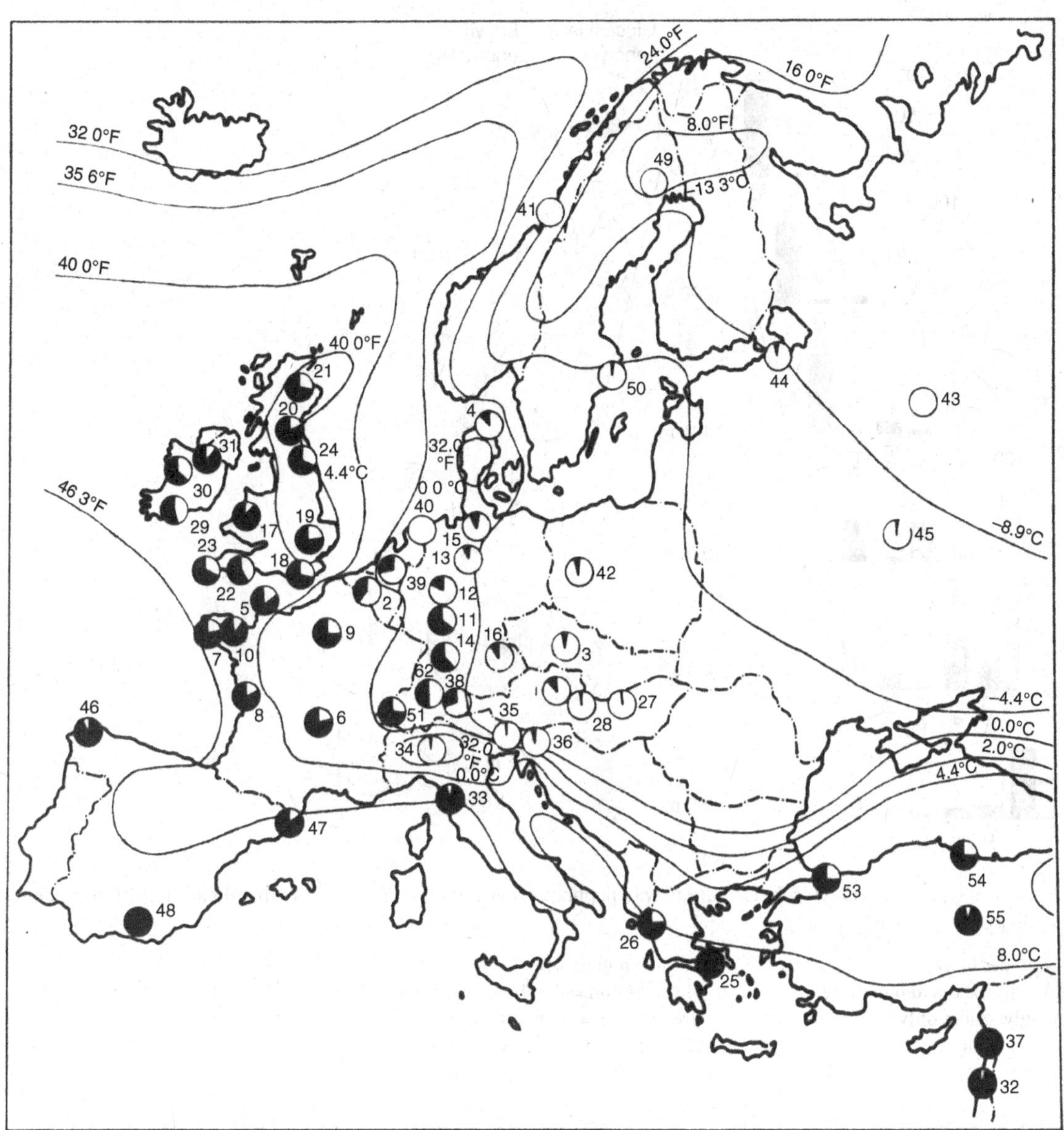

Fig. 9.2. Distribution and frequency of the cyanogenic variant of *Trifolium repens*. Black section: frequency of cyanogenic variant. White section: frequency of acyanogenic variant. ________ January isotherms. (From Jones, 1973; after Daday, 1954a.)

repeated many times and Jones obtained good evidence that two snails, *Arianta arbustorum* and *Helix aspersa*, and two slugs, *Arion ater* and *Agriolimax reticulatus*, showed selective eating of the acyanogenic plants when offered both variants. This experiment proved of great interest to ecologists and many have repeated the tests on both *Lotus corniculatus* and *Trifolium repens*. For example, in 'food choice' experiments under laboratory conditions as well as in outdoor enclosures, Saucy *et al.* (1999) and Viette, Tettamanti & Saucy (2000) presented cyanogenic and acyanogenic varieties of clover to the

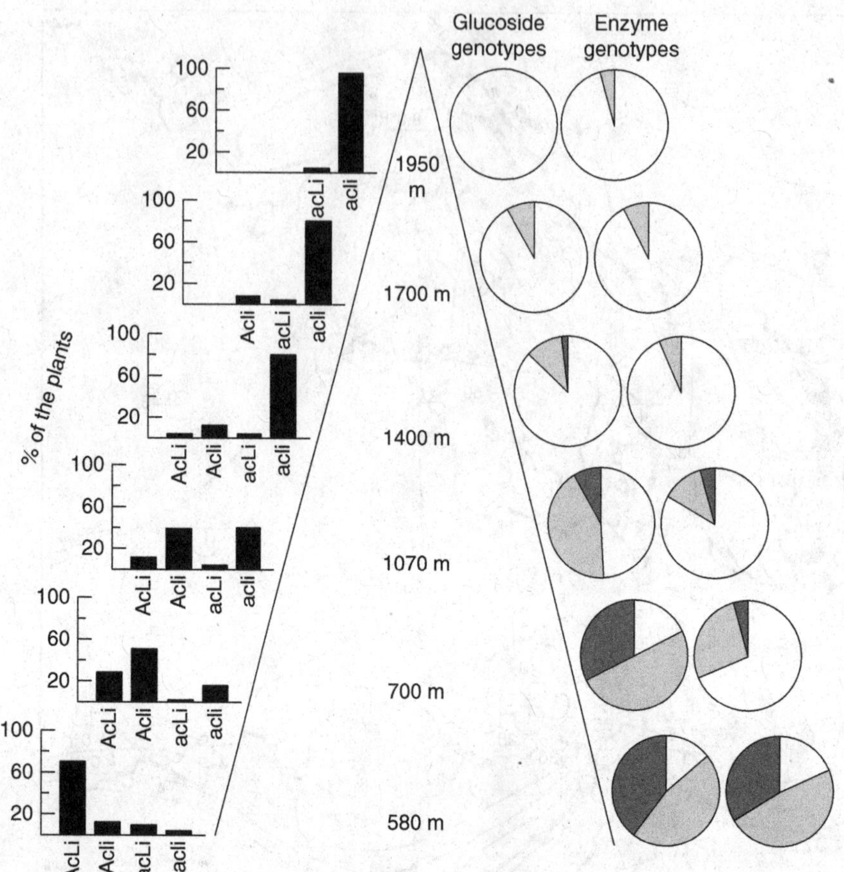

Fig. 9.3. Phenotypic and genotypic frequencies in wild populations of *Trifolium repens* from different altitudes. (From Jones, 1973, after Daday, 1954b.)

Phenotypes (left):
AcLi – glycosides and enzyme
Acli –glycosides only
acLi – enzyme only
acli – neither glucosides nor Enzyme

Estimated genotypes (right):
Black section = dominant homozygotes
Grey section = heterozygotes
White section =recessive homozygotes

vole *Arvicola terrestris*. They discovered that the vole had a preference for the acyanogenic variant.

However, not all the experiments produced such clear-cut results (see, for example, those of Bishop & Korn, 1969). It has been argued that the particular conditions used in the test influence the results. Important factors to take into account are the food materials and how they are made available to the animals, whether they are hungry or not, and the variability in animals. To quote the opinion of Jones (1972): 'in the same way that some men like beer, and others do not, individual molluscs have different palates'. The balance of evidence seems to favour the view that cyanogenic glycosides provide a defence mechanism against certain small invertebrates. In *Lotus*, however, this defence is by no means absolute (Crawford-Sidebotham, 1971). For example, Lane (1962) has shown that the larvae of the Common

Blue Butterfly (*Polyommatus icarus*) show no preference for acyanogenic plants of *Lotus*; in fact, they produce an enzyme, rhodanese, which converts cyanide into harmless thiocyanate. Hughes (1991) has reviewed the evidence for the defensive role of cyanogenesis against herbivores in *Trifolium repens*. While cyanide production protects against grazing by some molluscs, leatherjackets, weevil larvae and some aphids, in contrast other herbivores – field crickets, grasshoppers, two species of aphid and some other mollusc species – were unable to discriminate between cyanogenic and acyanogenic plants.

How can one explain the patterns of distribution of acyanogenic and cyanogenic *Trifolium* plants discovered by Daday in the light of Jones' findings? It seems most likely that the distributions of animals that selectively eat acyanogenic plants are correlated with climatic factors. In Atlantic regions of Europe with mild winter temperatures and at low altitude, a great number of small invertebrates likely to eat plants may be found. It is clear that one can postulate a selective advantage of the cyanogenic plants in western Europe. What is the selective advantage of the acyanogenic condition? To explain the pattern of high frequency of the acyanogenic variant at high altitude and in continental conditions it has been suggested that frost may be important. Conditions of extreme cold will freeze the cells of plants, releasing the enzymes that break down any glycosides present. The production of cyanide through its inhibitory effect upon plant metabolism could then place the plant at a strong selective disadvantage relative to the acyanogenic variant. However, after reviewing the evidence then available, Hughes (1991) could find no support for the notion of a differential effect of frost damage.

Recent laboratory studies by Olsen & Ungerer (2008) have investigated the hypothesis of frost-induced cyanide autotoxicity in cyanogenic variants by subjecting cold acclimated plants of different cyanogenic genotype to freezing temperatures. They discovered that 'cyanogenic genotypes exhibited lower freezing tolerance in species-wide sample'. While this would seem to provide a definitive test, comparisons of 'paired' cyanogenic and acyanogenic plants drawn from the polymorphic population showed no similar trend. Therefore, Olsen & Ungerer (2008) point out that a definitive study would require larger samples.

Reviewing all the accumulated evidence of clinal variation in *Trifolium*, Olsen, Sutherland & Small (2007) conclude that there is strong evidence for 'rapid adaptive evolution in these introduced populations'. However, Kooyers & Olsen (2012) reflect on important issues.

1. 'The clover polymorphism arises through two independently segregating simple Mendelian genes ... Plants that produce one cyanogenic component without the other may be maladaptive because they bear some of the energetic costs of cyanogensis but without the herbivore defence benefits ... some polymorphic mid-cline populations in Europe have been reported to show strong linkage disequilibrium between the two unlinked cyanogenesis genes, with a statistical excess of AcLi and acli cyanotypes ... this pattern potentially suggests persuasive epistatic selection against Acli and acLi plants'. However, in their recent studies of clines in central North America, Kooyers & Olsen (2012) found no evidence of such epistatic selection. This raises as yet unanswered questions about the maintenance of the polymorphism in the native and introduced range.
2. Also, Kooyers & Olsen point to a potentially complicating factor influencing patterns of cyanogenesis. They write: 'Ennos (1982) notes that some mollusks possess β-glucosidase enzymes in their guts that could function in the same way as linamarase in hydrolysing cyanogenetic glycosides and inducing a deterrent effect. Thus, clover populations in a community rich in mollusks may be equally well protected whether they are Acli or AcLi cyanotypes. Under such conditions, no epistatic selection would be expected.'

Small-scale influences on patterns and process in cyanogenic species

So far our account has concentrated on the interaction of climatic and biotic factors on the large scale. Small-scale experiments have also been carried out. For example, Ennos (1982) sowed seeds of White Clover 'polymorphic for the presence/absence of both cyanogenic glycosides and the hydrolyzing enzyme linamarse were introduced into three natural habitats', where they were subject to 'natural herbivory'. The experiments were frequently examined and it was discovered that selection was very important at the seedling stage. 'Estimated selection coefficients against plants lacking linamarase were in the region 0.3'.

While the study of local variation suggests that grazing by small invertebrates is important, other factors may also be involved in *Lotus*. It is not possible to summarise all the results discovered in experiments, but they suggest that a number of interacting factors may influence patterns of cyanogenesis, for there is evidence that water stress, mammal grazing, trampling, insect damage and salt spray may all be important in influencing small-scale pattern (Ellis Keymer & Jones, 1978a, b; Keymer & Ellis, 1978). Recent research has confirmed that drought stress is an important factor in the evolution of cyanogenesis clines in *Trifolium repens* (Kooyers *et al.*, 2014).

The classic experiments of Harper and his associates, in a small 'superficially dull' pasture at Henfaes in North Wales, have added greatly to our understanding of cyanogenesis and other genetic variation in *Trifolium repens*. Burdon (1980) collected many samples from the field and, as a result of cultivation tests, concluded that the population was very variable indeed. Some of this variability has been shown to be under genetic control: e.g. *T. repens* is genetically polymorphic for cyanogenesis and leaf markings. Genetic polymorphism is also suspected for other characters; for instance, some plants were more resistant than others to the fungi *Pseudopeziza* and *Cymadotheca*. Variability in the population was confirmed when Trathan, using isozyme markers, discovered that there are *c.*50 genotypes per square metre. Many previous studies of cyanogenesis have been carried out under laboratory conditions, and, realising the possible limitations of this approach, Dirzo & Harper (1982a) investigated the distribution of different variants of the cyanogenesis polymorphism in relation to the distribution of active molluscs in the field. Cyanogenic plants predominated in areas of very high and high mollusc density, while the frequency of acyanogenic plants was greatest in the areas with very low mollusc grazing. This association is statistically significant, and consistent with the hypothesis that cyanogenetic plants are at a selective advantage in certain areas of the field but not in others. While cyanogenesis confers advantage in certain parts of the field, transplant studies at Henfaes by Dirzo and Harper (1982b) revealed that the cyanogenic morph was more susceptible than the acyanogenic to the pathogenic rust *Uromyces trifolii*. The grazing preferences of sheep were also examined, particularly in relation to the genetic polymorphism associated with leaf markings (Cahn & Harper, 1976a, b). Evidence suggested that sheep were selective. They ate more of the unmarked leaf variants and more of the commoner morphs than the rarest. There was no evidence of a distasteful chemical or that cyanogenesis is associated with lack of markings. This could be an example of apostatic selection, in which rare morphs are at a selective advantage relative to the common variants. Harper (1983) reflecting upon the patterns detected in the Henfaes field, points out the weather varies from year to year, leading to differences in the severity of attack by fungal pathogens, differences in sward height and grazing intensity. Thus, some of the evident pattern in this clonal species will reflect recent, as well as past, selective forces. Clearly, interactions between many factors must be considered in order to obtain a balanced picture of microevolution.

For an insight into the way the study of cyanogenesis is leading to a more complex picture, the reviews of Jones, Keymer & Ellis (1978), Hughes (1991) and Olsen, Sutherland & Small (2007) should be consulted.

Reciprocal transplant experiments

By employing reciprocal transplant methods additional evidence about patterns of variation may be obtained. Suppose that two extremely different habitats (a and b) occur in the same region or side by side. Preliminary evidence of tolerance tests suggests that for the species under study each of two morphs (A and B) is best suited to its own native habitat (A in a and B in b). Clearly such a hypothesis might be put to the test by growing an appropriate number of replicates of A and B in each habitat, using either cloned or genetically uniform material for the purpose. Usually, before planting out, A and B are first grown for a period in a uniform garden, to minimise the possibility of carry-over effects.

As we have already seen in earlier chapters, this type of experiment has had a long history. The early experiments by Bonnier and Clements were technical failures, but recently there has been a revival of interest in reciprocal transplant studies. These investigations are characterised by the great care taken in their execution. Davies & Snaydon (1976) marked their experimental plants with wire, and in the studies by Warwick & Briggs (1980a, b) experimental plants were grown in random array at spaced intervals in a carefully mapped area. In many other cases 'native' stocks perform best in their own habitats, and alien stocks do less well (Levin, 1984).

The experiments of van Tienderen and van der Toorn (1991a, b) provide an excellent example of the use of reciprocal transplants. They studied local adaptation in *Plantago lanceolata* in the contrasted habitats of pasture and hay meadow. Some background information is necessary to appreciate the context of their experiments. Evidence suggests that selection has favoured different variants in areas of different land use, in particular hay meadows and pasture (see van Groenendal, 1986 and references cited therein). With their erect, tall growth habit and early fruiting to produce heavy seeds, plants from hay meadows are adapted to succeed in establishing and growing in tall swards and reproducing before the date of hay cropping. In contrast, plants from pasture are smaller, with short-leaved decumbent rosettes, bearing many small inflorescences producing light seed. This variant represents a different syndrome of adaptive traits, related to survival and reproduction under grazing by domestic and other animals. From these experiments, it was deduced that each variant had a different set of co-adapted traits, life-history strategies and reproductive 'tactics' related to the habitat of origin. This hypothesis was put to the test in a reciprocal transplant experiment (van Tienderen & van der Toorn, 1991a, b), which provided very clear evidence of the selective disadvantage suffered by the alien transplants (Table 9.3). This experiment is noteworthy for making several assessments of relative fitness. It is clear that, while vegetative survival of alien plants is relatively high, such plants were presented with a series of challenges as the experiment progressed and relative fitness declined during the flowering and fruiting phases.

Reciprocal transplant experiments have very clear attractions for testing certain hypotheses; they permit examination of responses to the *totality* of the environment, including competition with the native flora at each site. Such investigations subject the plants to more natural conditions than those employed in many tolerance tests.

Experimental evidence for disruptive selection

Tolerance tests and reciprocal transplant experiments with a number of species suggest that genetic polymorphism occurs and that different morphs may be at a selective advantage in different parts of

Table 9.3. **Relative performance under field conditions, of clonal transplants of *Plantago lanceolata*, from early mown hay meadow on clay soil and grazed pasture on sandy soil, transplanted into hay and pasture. (From van Tienderen & van der Toorn, 1991a.) For comparative purposes the performance of plants transplanted into their 'native' site is set at 100%. 'Alien' transplants perform less well than plants in their native site. The comparative reproductive failure of the pasture variant transplanted into the hayfield is particularly significant. It is late flowering and fruiting, and is damaged by early season haymaking before it reaches reproductive maturity**

	HABITATS			
	Hay cut as usual		Pasture grazed by cattle	
Source of cloned material	pasture	hay	pasture	hay
Vegetative survival	70	100	100	90
Success in flowering	52	100	100	60
Success in producing at least one ripe fruiting spike	50	100	100	30
Total seed yield	3	100	100	40

a mosaic or patchy environment. Developing the ideas on disruptive selection outlined earlier, we may envisage a model system as follows. Across the line of contact of two dissimilar habitats gene flow may occur between the polymorphic variants of an outcrossing species, 'well-adapted' adults on each side of the divide being cross-pollinated by pollen from the other morph. As a result of gene flow at the seed dispersal stage, seed of different genotypes will be scattered round the site. The variation in young plants, as they germinate and develop, will be subjected to disruptive selection and only the appropriate morph will survive in each area of the patchy environment. The 'potential' population represented by the young plants will be put through the sieve of selection and only well-adapted plants will survive to reproductive maturity. As this area is one of the most interesting in ecological genetics, we consider a study that tests this model.

The effect of artificial fertilisers on crop plants was an important issue in the mid-nineteenth century, and in 1856 Lawes & Gilbert laid out the famous Park Grass Experiment at Rothamsted, UK, which was designed to compare the yield of hay in plots treated either with artificial or natural fertilisers. For our purposes it is not necessary to give details of treatment and control plots (for details see Brenchley & Warington, 1969; Snaydon, 1970; Thurston, Dyke & Williams, 1976). It quickly became apparent that grassland productivity could be increased by artificial fertiliser treatments, but fortunately the experiment has continued with an early summer hay crop to the present day. One effect of fertiliser treatments, especially in plots treated with ammonium sulphate, is the lowering of the pH. Gradually plots became acid and in 1903 (after some tentative applications of lime in the 1880s and 1890s) regular liming of the southern half of each plot was undertaken. In 1965, further subdivision of some plots was made to give four subplots, each with a different lime treatment. Looking at the experiment today one is struck by the crispness of the boundaries between

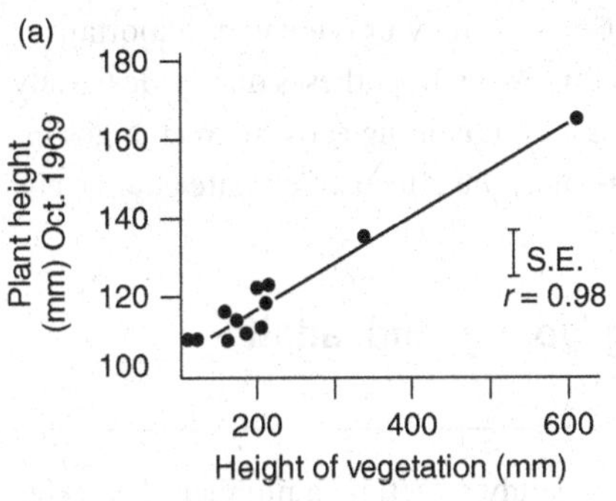

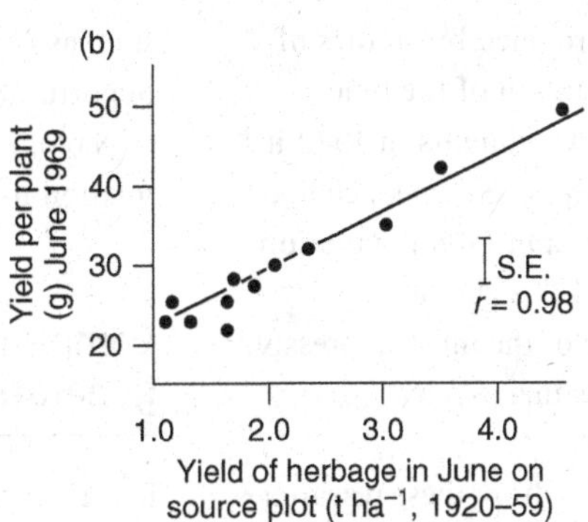

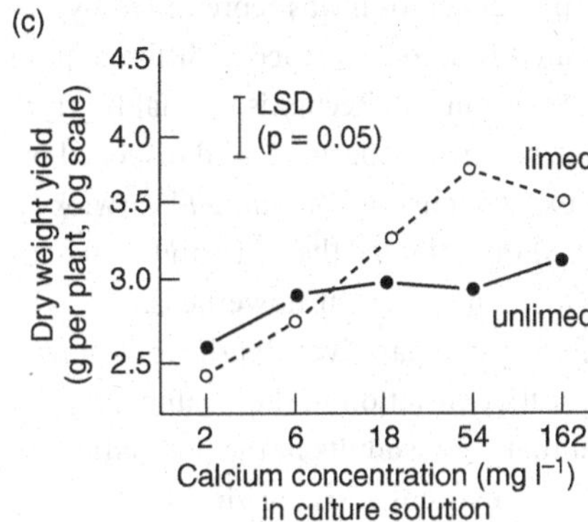

Fig. 9.4. Graphs summarising some results of studies of *Anthoxanthum odoratum* collected from the Rothamsted Park Grass Experiment and grown in a special trial. (From Snaydon, 1976.) (a) Relation between plant height and height of vegetation in source plots. (b) Relation between plant weight and yield of herbage in source plots. (c) Effect of calcium on dry weight of plants from limed and unlimed source plots. (Snaydon, 1970, 1976; Snaydon & Davies, 1972; Davies & Snaydon, 1973a. See also Davies, 1975 – response to potassium and magnesium; Davies & Snaydon, 1974 – response to phosphate.)

plots. Clearly there is little or no sideways movement of nutrients on this more or less level site, and plots differ markedly in species composition, vegetation height and productivity. The plots were laid out on a pre-existing grassland said to have been grassland for several centuries (although there are faint traces of plough marks in part of the area). From what was likely to have been uniform grassland, the experimental regime has produced such marked differences in adjacent plots that the patterns can be seen from the air. Snaydon and Davies realised that, because several species occurred in many of the plots, the Park Grass Experiment offered a unique opportunity to test Darwinian views on selection. Were differences due to fertiliser and lime treatment detectable within one of these common species? How quickly could changes take place?

Studying the grass *Anthoxanthum odoratum*, Snaydon and Davies discovered that, when material was collected from various plots and cultivated in a garden trial, the plants from tall vegetation plots were generally taller and heavier than plants from short vegetation plots (Fig. 9.4). Furthermore, plants showed ecotypic differentiation in relation to edaphic factors. For example, plants from acid unlimed plots, where the soil was deficient in calcium, needed much less calcium than plants from limed plots. Studies of gene flow were made and reciprocal transplant experiments carried out from acid to limed plots and *vice versa*, revealing that plants grew best on their native plots (Davies & Snaydon, 1976). Relative fitness of the alien plants was only 0.59 on the limed subplots and 0.75 on the unlimed subplots.

The evidence from all these experiments suggests that population differentiation has taken place in response to fertiliser and lime treatments, as a consequence of natural selection acting over the last 120 years. Indications that important changes might be possible in an even shorter time are provided by studies of vegetation following the alteration of the lime addition regime in 1965. In experiments in 1972 it was discovered that, after only seven years, changes could be detected in the *Anthoxanthum* populations. The highly imaginative use of this classical experiment has provided one of the most impressive studies of pattern and process in a patchy environment (Davies, 1993).

An interesting question arises from these researches. There is evidence for genetic differentiation in *Anthoxanthum*, but have similar processes occurred in other species in the Park Grass experiment? Until recently, there was only limited information for other species (see Silvertown 2001; Briggs, 2009). However, investigations of the short-lived species *Tragopogon pratensis* have proved informative (Dodd *et al.*, 1995). Records reveal that the species was not present in the area in the 1930s, but from 1945 to 1960 it began to appear in a number of plots, though later it retreated to 'core area' on plots 19, 20 and 2.

As a hypothesis for future study, Silvertown (2001) suggests that, while the seeds of *Tragopogon*, with their very large pappus, have the potential to colonise the plots of the Park Grass Experiment, this species lacks the necessary genetic diversity on which selection can act. Support for this view comes from a number of genetical studies. In the Park Grass area *Tragopogon* is monomorphic for certain allozymes loci (PGI, PGM). Studies elsewhere, employing inter simple sequence repeat (ISSR) DNA markers, discovered no detectable genetic variability at all within or between 10 families of 12 progeny in 12 loci (Zietkiewicz *et al.*, 1994). Also, investigations of American material revealed that the species was monomorphic for allozyme and DNA loci (Roose & Gottlieb, 1976; Soltis *et al.*, 1995). Thus, in the Park Grass experiment *Tragopogon pratensis* has a relatively small core area, and behaves as a short-lived 'outbreak' species across other plots, where local colonisation is followed by extinction (Silvertown, 2001).

Silvertown *et al.* (2010) make an extremely strong case for the continuation of long-term experiments such as Park Grass, as they provide very important opportunities for testing hypotheses on a wide variety of ecological issues, including ecosystem changes in relation to land use, the effects of climate change etc.

Comparing 'young' and 'adult' generations

In an attempt to study selection, a number of workers have compared young and adult generations. In most cases, the 'young' generation was represented by collections of mature seed developed in different parts of the patchy environment. Recognising that this type of study neglects certain aspects of seed dispersal by gene flow, in experiments on *Poa annua* Warwick & Briggs (1978a, b) investigated the adult and seed bank flora in both lawn and adjacent flowerbeds. The results of these comparative experiments show that the spectrum of variation in the 'young' is different from that of the adults in the population. This evidence is consistent with the view that disruptive selection is a potent force in certain situations.

These experiments should not, however, be accepted as providing a complete picture of natural situations, but rather should be seen as first attempts to come to terms with the complex patterns and processes found in natural populations. The following complications should be considered in the interpretation of past and future experiments.

The model system predicts that in a particular generation the spectrum of variation in the 'young' may exceed that of the adults. Most of the comparisons so far attempted have compared adults with 'young' of the next generation. This comparison raises a number of problems. First, it is not clear whether the habitats under study are at equilibrium and whether the forces of selection are operating at the same intensity year by year. Furthermore, it is difficult to assess the importance of various demographic factors. Many plants are potentially long-lived, and in some of the study areas it is not known how frequently individuals succeed in establishing themselves. Secondly, the model system envisages two entirely different habitats separated by an abrupt boundary. Realistically, models should take account of the evident heterogeneity within habitats in nature and the existence of transitional conditions – the so-called ecotone – between habitats. In a remarkable study of a mine/pasture transition, Antonovics & Bradshaw (1970) discovered very interesting clinal patterns in *Anthoxanthum odoratum*. Thirdly, the complications raised by the existence of a seed bank in the soil have yet to be properly assimilated in experiments on disruptive selection (see below). Here, we note that a source of potential variation may be present in the soil seed bank that is constantly being eroded by loss of seed viability and by predation, whilst at the same time 'new variation' is being added by current seed production and dispersal (Harper, 1977).

Co-selection in swards

We have already discussed some of the results of the detailed study of *Trifolium repens* in the field at Henfaes, North Wales (Harper, 1983). Another aspect of the work, involving reciprocal transplants, has yielded very significant results. Different parts of the field are dominated by one of four different species of grass and, using reciprocal transplant and other experiments, Turkington & Harper (1979a, b) and Turkington (1989) found support for the extremely interesting hypothesis that clovers are not distributed at random over the field, specific clones being associated with each grass species. In a further series of experiments on plants from a different site, several *Lolium* genotypes were collected, together with their associated clover plants. *Trifolium repens* has nitrogen-fixing root nodules and the causative organism, *Rhizobium*, was extracted from each clover plant. When the *Lolium perenne*, clover and *Rhizobium* were grown together in different combinations, Chanway, Holl & Turkington (1989) discovered that the best performance was achieved by the *Rhizobium*, *Trifolium* and *Lolium* combinations that had existed together in the 'wild'. These findings support the view that there is a complex pattern of co-selection at work and, clearly, it will be of very great interest to see if the phenomenon occurs in interactions of other plant species.

The speed of microevolutionary change: agricultural experiments

Some of the most convincing examples of the power and speed of selection come from studies of crop plants. Artificial directional selection has yielded spectacular results in agricultural crops (e.g. Fig. 9.5). Another example is provided by artificial selection over nine generations for high oil content in oats (*Avena sativa*) (Bone & Farres, 2001).

Other experiments have studied natural selection, principally to discover the speed of change and which varieties yield best at a given site. For instance, in an investigation by Harlan & Martini (1938), 11 varieties of Barley (*Hordeum*) were mixed together in such proportions that an equal number of plants of each variety might be expected to grow. Seed samples were sent out to ten experimental stations in different parts of the USA: all stations received the same spectrum of 'potential variation'. At each site the Barley was harvested and seed saved to sow a plot in the following season. A sample of seed was sent each year (1925, 1936) to Washington, and the proportional representations of the 11 varieties were determined in a field trial. The results of the annual census for 1930 are shown in Fig. 9.6. Different varieties predominated in different areas, and there was rapid reduction (or even elimination) of the less well-adapted varieties. In all areas the variety that would eventually dominate the plot in a particular site was quickly evident.

Experiments of this type, where a 'standard' seed mixture is planted out in different sites, have been attempted with a number of crop plants and have

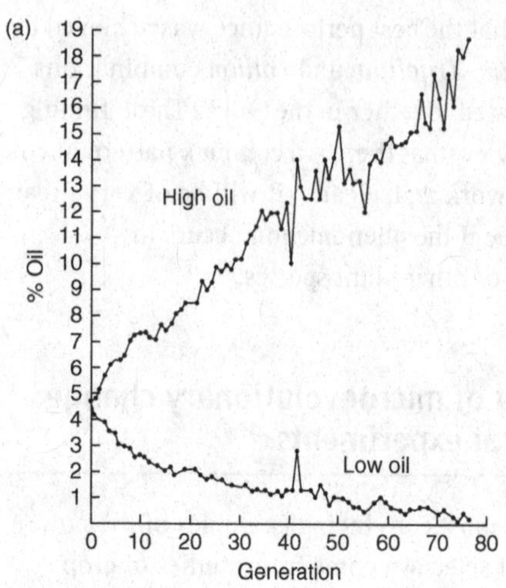

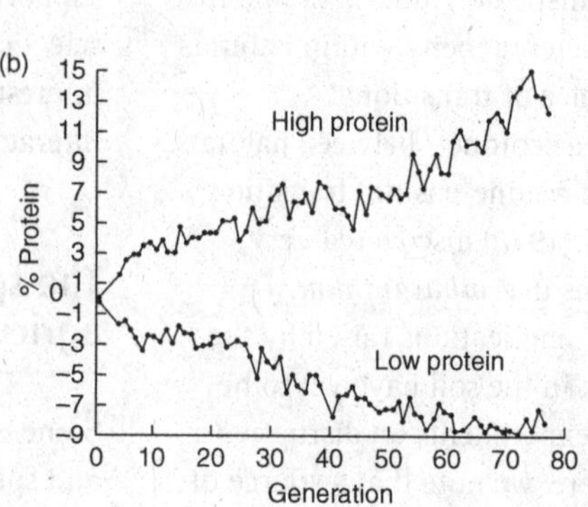

Fig. 9.5. Directional selection for high and low oil and protein content in Maize (*Zea*) after 76 generations. (From Solbrig & Solbrig, 1979, after Dudley.)

yielded broadly similar results (Snaydon, 1978). Such experiments provide clear evidence that selection acts quickly and may be of great intensity.

An investigation by Brougham & Harris (1967) is particularly revealing. Two varieties of Rye Grass (*Lolium*) were sown (together with White Clover) in an experimental plot. After establishment the sward was subjected to lax grazing for six months. The plot was then divided into a number of subplots, which were subjected to lax, moderate or continuous grazing regimes. Here, we are not concerned with the detailed results of this experiment; it is sufficient for our purposes to note that major changes in population composition were detected within four months of the application of the grazing regimes. The implications of this and many other experiments on agricultural stocks is that strong selective forces may act very quickly to change the composition of populations.

Rapid change in polluted sites

Heavy metal contaminated areas are found in many parts of the world by the weathering of naturally occurring ores (Ernst, 1998a). Contaminated sites also occur as a result of human activities in mining, refining, utilising metals and their compounds in a wide range of circumstances and ultimately recycling metals or by discarding them in household and industrial waste. Some instances of toxicity have required investigation to determine the source: for instance, irrigation of fields with what turned out to be 'mine' water, and applications of sewage sludge to agricultural land. Also, contamination, at levels toxic to some organisms, may result from the use of fungicides that contain copper in apple, hop orchards and vineyards.

Some plants have evolved variants able to grow on contaminated sites. A full account of heavy metal tolerance is beyond the scope of this book. Antonovics, Bradshaw & Turner (1971), Macnair (1993) and Ernst (1998a) provide important reviews on which we base our comments.

Heavy metal tolerance has been described in a number of species. However, formal genetic analysis of its control has been attempted in only a few cases. Tolerance has generally been found to be a dominant trait. For instance, single major genes – subject to the influence of modifying genes – appear to control copper tolerance in *Mimulus guttatus* (Macnair, 1983) and arsenate tolerance in *Holcus lanatus* (Macnair, Cumbes & Meharg, 1992). Contaminated sites differ in the suite of heavy metals contaminating the soil. Some studies have suggested that tolerance to one

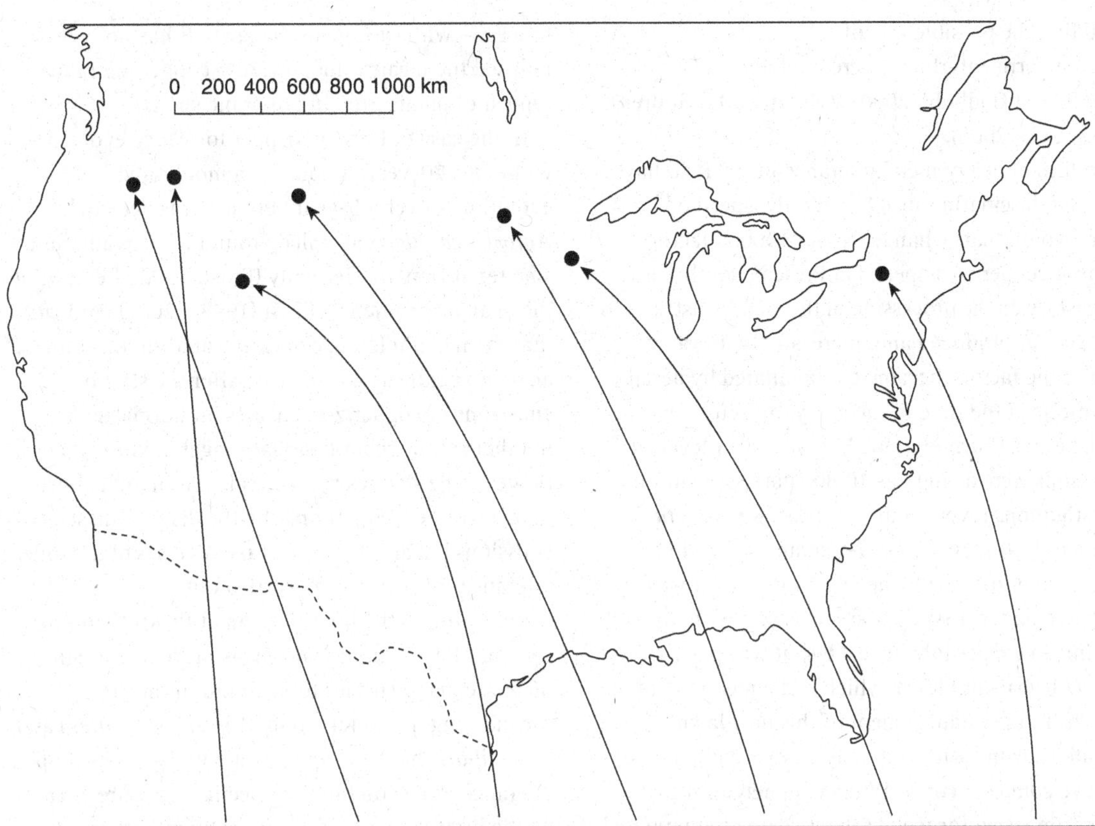

Varieties	Pullman Washington	Moro Oregon	Aberdeen Idaho	Moccasin Montana	Fargo North Dakota	St Paul Minnesota	Ithaca New York
Coast & Trebi	150	125	159	102	156	121	75
Gatami	1	3	20	73	20	16	46
Smooth Awn	5	10	6	54	23	37	47
Lion	3	3	21	44	14	34	44
Meloy	6	3	9	12	0	5	0
White Smyrna	276	276	119	89	17	5	1
Hannchen	30	48	109	55	152	215	17
Svanhals	23	26	33	31	80	57	8
Deficiens	5	0	7	2	1	0	0
Manchuria	1	6	17	38	37	10	262
Number of plants of each variety in a sample of 500 – figures for 1930							

Fig. 9.6. The results of an experiment with Barley (*Hordeum*), in the USA, showing rapid selection of the variety suitable to different climatic conditions. (From Harlan & Martini, 1938.)

heavy metal confers tolerance to other metals. In contrast, other investigations point to the independent evolution of tolerances at individual sites. Further studies of co-tolerance to different heavy metals are required. While some progress has been made on studying the physiology of metal tolerance (and hyperaccumulation; see below), much has still to be learnt of the mechanisms involved, including the possible role of root–mycorrhizal–rhizosphere interactions (see Jentschke & Godbold, 2000; Meharg, 2003; Göhre & Paszkowski, 2006).

Studies of heavy metal tolerance in the field have provided fascinating insights into the speed of microevolutionary change. About 1900, a factory refining copper was opened at Prescot, south-west Lancashire. In the processing of the metal, dust, rich in copper, was produced and, increasingly, the area around the factory became contaminated by aerial pollution. At the time of the study, by Wu, Bradshaw & Thurman (1975), total copper levels in lawn soils were as high as 10 800 parts per million. Investigations revealed that *Agrostis stolonifera* plants in lawns were copper-tolerant to varying degrees and that, while the vegetation cover was complete on the 15-year-old lawns laid with normal turf (mean copper tolerance 42%), it was patchy on a new (8-year-old) lawn, which had a mean tolerance of 32%. In the establishment of this new lawn, repeated sowing with commercial seed had failed to achieve complete cover. When commercial seed stocks are tested for metal tolerance on contaminated soil, most seedlings die, but a few reveal their metal tolerance by growing normally (Bradshaw & McNeilly, 1981). From this study it was deduced that selection was taking place each time seed was sown on the contaminated new lawn area. Only copper-tolerant varieties survived, and such was the toxicity of the ground that only a few survivors were likely from any seed batch. The possibility that all the survivors were of a single genotype was examined by studying plants in cultivation and by isozyme analysis, and it was discovered that several tolerant genotypes were present in the contaminated lawns. As well as the lawns, the boundary area developed over four years from rough grassland was also examined. It had a mean copper tolerance of 21%. This splendid example of the detailed study of an anthropogenically disturbed site suggests that the different grasslands can be seen as a time series, the copper tolerance increasing – as a multi-stage process – with the age of the lawn. It has not yet reached the mean value of c.70% copper tolerance typical of plants growing on mine spoil.

In the case of Prescot, copper tolerance evolved within a c.70-year period. Even more rapid evolution – over a 5-year period – was detected in *Agrostis capillaris* sampled around a zinc/cadmium smelter at Datteln, Germany (Ernst, 1990). Reviewing the available evidence, Ernst (1998a, 2006) confirms that the metal tolerance may, in some circumstances, arise in as little as 4 or 5 years after the start of emissions of copper/zinc in sites on normal soils. In other cases, the process was longer, taking decades. However, he stresses that the calculation of 'exposure time' is influenced not only by the date of the start of emissions but also when the first analyses of the soils/tolerances of plants were carried out.

Returning to the investigation at Prescot, another finding of great significance was made. Although there are many species to be found in the areas surrounding the factory, only *Agrostis stolonifera* and *A. capillaris* have the appropriate variation in copper tolerance to respond to the selection pressure exerted by the highest levels of copper contamination from aerial fall-out (Wu, 1990).

The notion that selection might be constrained because of lack of the appropriate variation has received support from other experiments on metal tolerance. Electricity pylons are coated with zinc compounds to protect them against corrosion. Over a period of time, zinc leaches from the metal structures (Fig. 9.7) and, in areas of acid soils, becomes a serious source of local contamination (mean total zinc 1250–6500 $\mu g\ g^{-1}$ under pylons versus 170–320 $\mu g\ g^{-1}$ from control sites 10–50 m from the pylon).

Pylons represent a series of replicated situations in space. Is the genetic outcome the same at each site? Tests reveal that zinc-tolerant *Agrostis capillaris* plants occurred at a number of sites, but not all those tested (Al-Hiyaly *et al.*, 1990, 1993). Samples of seed were collected from several pylon sites. In the cases where tolerance had been detected in adults collected under the pylon, the local seed population responded

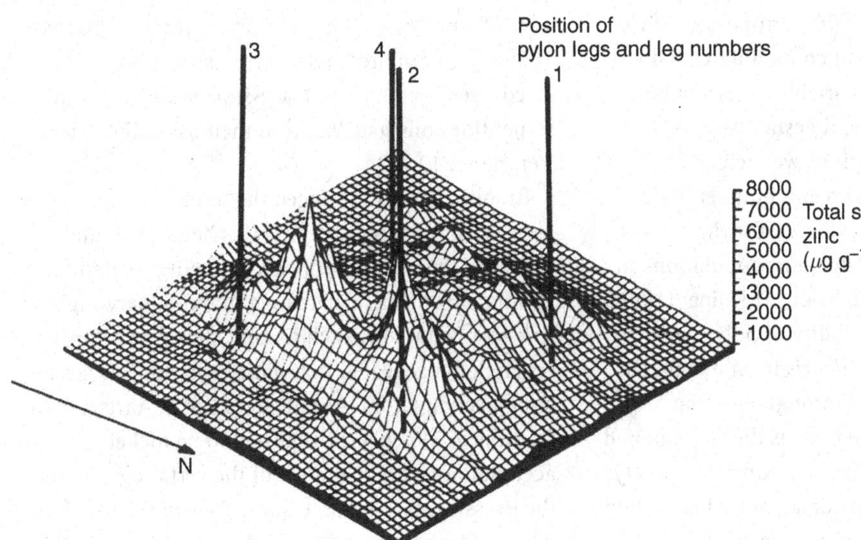

Fig. 9.7. Total soil zinc concentration in 399 soil samples from an electricity pylon in North Wales. Data smoothing by extending sampling scale ×2.5; intermediate values obtained by averaging. (From Al-Hiyaly, McNeilly & Bradshaw, 1988.)

positively to selection for zinc tolerance. However, at a site where no zinc-tolerant plants occurred under the pylon, no response to selection was detected in seedlings raised from local seed. This confirms that the absence of zinc-tolerant plants under the pylon was due to the lack of the appropriate genetic variation in the vicinity. This finding supports the view that lack of appropriate variation may limit evolution, and indicates 'the stochastic nature of the evolutionary process because of the randomness in the occurrence of the necessary variability' (Al-Hiyaly *et al.*, 1993).

A study by Meharg, Cumbes & Macnair (1993) of populations of *Holcus lanatus* provides a different perspective on the availability of 'tolerance genes' in uncontaminated areas. On arsenate-contaminated mines on Dartmoor, south-west Britain, plants were all arsenate tolerant. However, on uncontaminated sites distant from mines there was a high frequency of arsenate-tolerant plants. The significance of this finding is, as yet, unclear.

Origin of metal-tolerant populations

A number of possibilities must be considered: heavy-metal-tolerant variants found on mines etc. could have come from plants growing on 'natural' outcrops of rock. Also, it is possible that human activities might transfer tolerant variants from mine to mine, mine to refinery etc. (see Ernst, 1990). One such example is the finding of the New World copper moss *Scopelophila cataractae* on the slag heaps of smelters in South Wales.

Another hypothesis, for which there is increasing evidence, is the '*de novo* selection of metal-tolerant ecotypes from the surrounding vegetation' (Ernst, 1998a) and that these can be found in their sites of origin. Evidence with which to judge this hypothesis has come from the studies of Wu *et al.* (1975), Gartside & McNeilly (1974) and Walley (1974). They discovered that heavy metal variants are present in small number in 'normal'populations. In experimental tests, most seedlings died on heavy metal soils, but there were some survivors that grew effectively.

The possibility of recurrent origins of heavy metal variants has also been examined. For example, Bush & Barrett (1993) investigated populations of heavy-metal-tolerant plants of *Deschampsia cespitosa* growing on mine debris at Sudbury and Cobalt in Canada. Samples from non-contaminated sites were also sampled. Population variability was examined using isozymes (9 enzyme systems 19 putative loci).

They concluded that: 'the results corroborated the prediction that colonization of contaminated habitats reduces the level of genetic variability, particularly where populations have recently established'. Also: 'Cobalt and Sudbury populations were clearly differentiated by unique alleles at a number of enzyme systems, providing evidence for the independent origin of metal-tolerant populations in the two mining areas.' Schat, Vooijs & Kuiper (1996) have also examined the possibility of polytopic origins in studies of *Silene vulgaris* from metal-rich sites in Germany and Ireland through crossing experiments. They concluded that, as the German and Irish plants had common genes, tolerance was likely to have resulted 'from independent parallel evolution in local non-tolerant ancestral populations'.

In another investigation, Vekemans & Lefèbvre (1997) have examined variation in allozymes in heavy-metal-tolerant variants of *Armeria maritima.* Their results suggested that 'multiple independent evolutionary origins of heavy-metal tolerant populations have occurred in the absence of strong genetic bottlenecks'.

Several hypotheses have been suggested to explain the absence of bottleneck effects in the evolution of heavy-metal-tolerant populations (Mengoni *et al.*, 2001), namely successive colonisation events, a high number of tolerant plants in the primary populations, pollen flow from the neighbouring populations, environmental heterogeneity and human disturbance.

Hyperaccumulation

So far we have been considering heavy metal contamination in human-influenced habitats. Populations of plants have also been examined at more 'natural' contaminated sites, albeit ones where past (and present) human activities are still apparent.

In 1865, plants of *Thlaspi calaminare* (now *Noccaea caerulescens*, formerly *T. caerulescens*), growing on zinc-rich soil, were found to contain an astonishing 17% of zinc dry weight in their ash (Risse in Sachs, 1865). Furthermore, investigations in 1948, discovered that plants of *Alyssum bertolonii* from serpentine soils had 1% Ni in their ash (Miguzzi & Vergnano, 1948).

Brooks (1977) introduced the term hyperaccumulators to define this behaviour, and reviewed the biology of plants growing on natural and human-influenced sites containing very high levels of metal ions such as zinc, nickel etc. Further research has revealed that more than 400 species are hyperaccumulators (Assunçáo, Schat & Aarts, 2003). Of these, *c.*317 species, reported to be nickel accumulators, are members of the certain genera of the Brassicaceae, e.g. *Alyssum, Thlaspi.* In addition, hyperaccumulation of nickel has independently evolved in several other genera such as *Phyllanthus* (Euphorbiaceae), *Hybanthus* (Violaceae) etc. About 15 species are zinc accumulators, including *Thlaspi caerulescens* (now in the genus *Noccaea*) and *Arabidopsis halleri* (formerly *Cardaminopsis halleri*). There are a few Zn accumulators in other families. Some Zn-hyperaccumulators, including *Thlaspi caerulescens* and *Arabidopsis halleri,* also accumulate cadmium.

Concerning the origin of populations, Pauwels *et al.* (2005) investigated metallophyte (M) and non-metallophyte (NM) populations of *Arabidopsis halleri* using PCR, RFLP and cpDNA markers. They discovered that 'geographically isolated groups of M were more genetically related to their closest NM populations than each other'. The evidence suggested that 'M populations have been founded separately from distinct NM populations without suffering founding events and that evolution towards increased tolerance observed in the distinct M population groups occurred independently'.

There has been a good deal of interest in the adaptive significance of metal accumulation. Experimental evidence of various kinds suggests that high Ni provides protection against bacteria, fungi and insects, while accumulation of zinc in the tissues provides a deterrent against insect herbivory (see Pollard &

Baker, 1997; Assunçáo, Schat & Aarts, 2003). Metal accumulation in this context has been interpreted in relation to trade-offs, which propose that protection by zinc/nickel comes at less cost to the plant than defensive secondary metabolites. However, reflecting on the results of a study of the palatability of the Zn hyperaccumulator *Thlaspi caerulescens* to snails (*Helix aspersa*), Noret *et al.* (2005) concluded that the design of ecologically realistic palatability experiments is very demanding, as they must test not only the reaction of herbivores to difference in metal content, but must also take into account the degree to which other chemical defences are active in the test material.

Given the worldwide occurrence of an estimated 52 million hectares of polluted land, including mines, industrial plants etc., there is considerable commercial interest in species capable of hyperaccumulation, for harnessing the process would allow what has become known as phytoremediation (Nascimento & Xing, 2006; Memon & Schröder, 2009; Vangronsveld *et al.*, 2009).

> Compared with conventional methods of soil remediation, the use of plants provides several striking advantages. It is cheap: after planting only marginal costs apply for harvesting and field management . . . It is a carbon-dioxide neutral technology: if the harvesting biomass is burned, no additional carbon dioxide is released into the atmosphere beyond what was originally assimilated by the plants during growth. (Peuke & Rennenberg, 2005)

The model species being intensively studied are close relatives of *Arabidopsis thaliana.* On average, *A. halleri* shares c.90–95% of coding regions with *A. thaliana*, and *Thlaspi caerulescens* has c.88% DNA identity. Because they share coding regions, studies of hyperaccumulation can take advantage of the techniques and expanding understanding of the molecular genetics of *A.thaliana*.

Studies of hyperaccumulation are investigating the physiological basis of the processes of accumulation, chelation and the role of antioxidants (Krämer, 2010). However, there are some difficulties using these species for direct removal of contamination. For instance, *T. caerulescens* grows very slowly, has a long vernalisation period before it flowers and is susceptible to fungal and insect attack when grown on 'normal' soils. If the underlying mechanism and genetics of hyperaccumulation can be fully understood, perhaps key traits can be successfully transferred, by genetic manipulation techniques, to fast-growing plants for the more efficient removal and recycling of zinc, nickel, cadmium etc. through phytoremediation.

Serpentine ecotypes

Building on the classic investigations of Kruckeberg (1951, 1954), reported in Chapter 8, there have been many investigations of plants growing on serpentine. Our account is drawn from a major review of serpentine floras (Brady, Kruckeberg & Bradshaw, 2005).

As we have seen, soils on this rock type are characterised by having low calcium/magnesium ratios, and frequently nickel, iron, chromium, cobalt salts etc. are present in sufficient quantities to render the soils toxic to most plants. However, there are other significant soil properties that present a challenge to plant colonisation. Soils are generally deficient in key nutrients (nitrogen and potassium compounds and phosphates). In addition, soils are often shallow, subject to erosion and higher ambient temperatures.

Brady, Kruckeberg & Bradshaw (2005) have made several observations highly relevant to our concerns in this chapter. The majority of serpentine-tolerant taxa are not hyperaccumulators. Ecotypes in this habitat type exhibit a 'syndrome' of characters. For example, they are not only metal tolerant but also drought tolerant etc. Little is known about the genetics of adaptation to serpentine, and the study of the physiological adaptations 'is not well understood' (Brady, Kruckeberg & Bradshaw, 2005).

The linked concepts of cross-tolerance and preadaptation are widely discussed in the literature on serpentine ecotypes. In terms of soil fertility, susceptibility to erosion etc., the contaminated spoil heaps surrounding abandoned metal mines share many of the habitat characteristics of serpentine outcrops, which are geological features of considerable age. As many 'serpentine' species are also found on mines, it has been suggested that such plants, are 'preadapted' to invade mining sites many of which are very recent in geological time (Macnair, 1987). Preadaptation has also been discussed in other contexts. Procter (1971) showed that plants of *Armeria maritima* from coastal areas grow comparatively well on serpentine soils. Likewise, coastal ecotypes of *Silene maritima* show comparable growth to serpentine-tolerant races, when both are grown on serpentine soils. However, serpentine-tolerant ecotypes do not grow so well on normal soils as non-serpentine-tolerant races (Kruckeberg 1951, 1954). But in other experiments, it has been discovered that *Silene dioica* populations are tolerant of serpentine soil, regardless of whether their native habitat was serpentine or non-serpentine (Westerbergh, 1975). These observations suggest that tolerance in this case is 'constitutive' rather than site specific.

There is evidence that serpentine tolerance has arisen independently in different serpentine outcrops. For instance, evidence comes from the study of cpDNA in samples collected in Italian populations of *Alyssum bertolonii* from four regions (Mengoni *et al.*, 2003). Population-specific marker patterns provide evidence for independent recurrent origin of populations in this species. Evidence has also been discovered for parallel evolution and site-specific selection for serpentine tolerance in *Cerastium alpinum* during the post-glacial colonisation of Scandinavia (Nyberg Berglund, Dalgren & Westerbergh, 2003).

Sulphur dioxide tolerance

Turning to another set of case histories, studies of sites of known history have provided circumstantial evidence for rapid change in populations under natural selection in response to atmospheric pollution. Thus, a factory producing smokeless fuel from coal was opened in a country district in West Yorkshire in 1926, and a good deal of sulphur dioxide pollution was produced from the works. Ayazloo & Bell (1981) discovered that plants of the grasses *Dactylis glomerata, Festuca rubra, Holcus lanatus* and *Lolium perenne*, growing in the vicinity of the works, were significantly more tolerant of sulphur dioxide than samples of the species further from the factory.

The rapidity with which selection pressures change is revealed in another experiment on air pollution. In 1975, the Sports Turf Research Institute of Britain set out plots of *Lolium perenne* in a polluted area of Manchester. Wilson & Bell (1985) raised plants from the original seed mixture and collected living samples each year from the plots. Then the sulphur dioxide tolerance of plants raised from the original seed – an unselected population – was compared separately with the tolerance of those collected each year (1976–82) in the polluted site. They discovered that the samples taken from the urban plot in 1979 and 1980 – which suffered significant pollution with sulphur dioxide – were statistically significantly more tolerant of sulphur dioxide than the 'original' material. However, the samples for 1981 and 1982 were not significantly different in their sulphur dioxide tolerance from the controls. As sulphur dioxide concentration in the areas declined in the early eighties, due to air pollution legislation and industrial recession, it seems clear that the selection pressure favouring sulphur dioxide tolerance was relaxed during this period.

Ozone tolerance

Studies of the effects of another atmospheric pollutant have yielded interesting results. Ozone is an important component of air pollution, especially in country districts downwind of conurbations. It is formed when oxides of nitrogen, resulting from vehicle exhaust emissions etc. react with oxygen in sunlight. *Plantago major* is a very common weed and it is exposed to different levels of ozone in different parts of Britain. In a standard test, populations from southern Britain exposed to high levels of ozone have proved to be much more resistant than those from more northerly areas (Reiling & Davison, 1992). Moreover, there is evidence that resistance patterns changed. Two of the populations showed increased ozone resistance over a five-year period that included two summers with high ozone levels (Davison & Reiling, 1995).

Evolution in arable areas

So far in this chapter, we have examined grasslands of various sorts – hay, pasture and lawns. Investigations of arable weeds have also produced some very interesting insights into the action of natural selection (Baker, 1965, 1974, 1991; McNeil, 1976; Warwick, 1990a, b; Briggs, 2009).

Barrett (1983) has reviewed the phenomenon of crop mimicry by weeds.

> The selective forces imposed by agricultural practices have resulted in the evolution of agricultural races of weeds ... Such associations can involve a system of mimicry, whereby the weed resembles the crop at specific stages of its life history and, as a result of mistaken identity, evades eradication. Mimetic forms of weeds are most likely to be selected by hand-weeding of seedlings, or by harvesting and seed cleaning procedures.

In the past, before thrashing and seed cleaning were improved, harvested crops were regularly contaminated, because the seed of mimetic weeds had sufficient resemblance in key physical attributes of the crop seed that they were not efficiently separated. By this means, such species as *Agrostemma githago* were a regular component of the weed flora before the widespread use of seed cleaning and herbicides. Barrett draws attention to a number of other important examples. For instance, *Echinochloa crus-galli* is a weed complex with many different variants, including mimics of Rice. In a comparison of rice mimics and other weedy variants the precision of the mimicry became evident: for example, anthesis in the mimetic variant coincides exactly with the flowering of Rice, presumably because genotypes flowering earlier than the crop would be visible and, therefore, weeded out. Rice has now become an important crop in California, where modern methods of agriculture are being used, rather than the labour-intensive methods of the Far East. It will be interesting to study the effect of mechanised farming on the crop mimic. Barrett draws attention to other strategies found in arable weeds: for example, *Aethusa cynapium* has tall and dwarf genetic variants (Weimark, 1945). The dwarf variant of the corn field, which survives below the level of cutting at harvest, flowers and fruits after harvesting, a good deal later than the tall variant, which is found in field margins and on waste ground.

Weed control is a characteristic of human-influenced habitats, and there is increasing evidence that weeding may act as a selection pressure, favouring variants of common weeds that are capable of precocious development. Thus, *Senecio vulgaris* from Botanic Gardens and other well-weeded sites have a quicker rate of development than samples taken from poorly weeded or non-weeded sites (Kadereit & Briggs, 1985; Theaker & Briggs, 1993). There has recently been much interest in the evolution of life histories. It is perhaps significant that these studies were particularly in vogue in the 1980s and employed 'monetarist' language to model and measure investments/costs/benefits/trade-offs etc. Stearns (1992) defines the key concept: 'A trade-off exists where a benefit realized through a change in one trait is linked to a cost paid out through a change in another.' In the case of *Senecio vulgaris* from Botanic Gardens, the 'benefit' of precocious reproduction, in the face of frequent regular weeding of the flowerbeds, incurs costs. In comparison with material from poorly weeded or non-weeded habitats, such early reproduction is associated with shorter stature, fewer leaves and earlier onset of senescence.

Herbicide resistance

The development of herbicide-resistant variants of common weeds provides another example of the evolution of weeds in relation to agricultural practices. Resistance has been confirmed in many more than 100 weed species (see Warwick, 1991; Cousens & Mortimer, 1995; Busi *et al.*, 2013 and the website www.weedscience.org).

It is of great significance to agriculture that resistance to glyphosate has been detected. Glyphosate is the world's most important and widely used herbicide, and resistance to glyphosate has now been detected in 15 weed species in 14 different countries (see Powles, 2008; Heap, 2009).

In many population studies, tolerance to a single herbicide has been established. However, as many herbicides are often used it is not surprising that multiple resistances have been detected, e.g. triple-resistant variant of *Chenopodium album* from Hungary (Solymosi & Lehoczki, 1989a, b).

In reviewing herbicide resistance mechanisms in detail, Powles & Yu (2010) point out that 'the great majority of herbicides inhibit specific plant enzymes (target site) that are essential in plant metabolism' and make a distinction between target-site versus non-target-site resistance. 'Evolved target-site resistance exists when herbicide(s) reach the target site at a lethal dose.' In contrast, 'evolved non-target site resistance involves mechanisms that minimize the amount of active herbicide reaching the target site', and 'include decreased herbicide penetration into the plant, decreased rates of herbicide translocation, and increased rate of herbicide sequestration/metabolism'.

Yu *et al.* (2009) have examined in considerable detail the multiple resistance found in two Australian populations of *Lolium rigidum*. They discovered that these populations had 'multiple resistances to glyphosate, acetyl-co-enzyme A carboxylase (ACCase) and acetolactate synthase (ALS)-inhibiting herbicides'. Resistance in all three categories of herbicide is due to the 'presence of distinct non-target site resistance mechanisms . . . glyphosate resistance is due to reduced rates of glyphosate translocation, and resistance to ACCase and ALS herbicides is likely to be due to enhanced herbicide metabolism'.

Simple nuclear-encoded Mendelian control of resistance has been discovered in some situations (see Cousens & Mortimer, 1995); for example, in *Conyza bonariensis* a single dominant allele is involved in resistance to paraquat. In other species, genetic control of herbicide resistance involves a semi-dominant gene (for instance, diclofop resistance in *Lolium multiflorum*; and paraquat resistance in *Hordeum glaucum* and *H. leporinum*); or one recessive gene (trifluralin resistance in *Setaria viridis*). In other species there is quantitative inheritance of herbicide resistance (e.g. resistance to barban in *Avena fatua*; and glyphosate resistance in *Convolvulus arvensis*).

In other species, herbicide resistance is determined by genes of the chloroplast genome. For example, many studies have been made on the triazine group of herbicides (Warwick, 1991). Triazines powerfully inhibit photosynthesis, and resistance is usually, but not always, maternally inherited. Molecular studies reveal that resistance in *Poa annua* is due to the loss of the herbicide-binding site, as a result of a point mutation in the chloroplast gene (Barros & Dyer, 1988).

Resistance arises after a few years' use of the triazine, at doses of this long-acting group of herbicides that achieve weed control for the whole growing season (without using other herbicides), and especially in areas where the same crop is grown year after year. Studies of variation in resistant populations, using isozymes, have revealed that initially, at least, they are not as variable as normal populations. It seems possible that the initial selection pressure exerted by the herbicide is so great that the population goes through an extreme bottleneck effect maybe of a single resistant individual, and chance may play a part in the spread of the resistant plants to new areas (see below).

There have been a number of other studies of the speed of selection for resistance from the date of introduction of a 'new' herbicide (see Maxwell & Mortimer, 1994 for details). Thus, while resistance to diclofop methyl was detected in *Avena fatua* after only 4–6 years, the development of resistance to paraquat/diquat in *Hordeum leporinum* was recognised only after 25 years.

Another important question has been investigated by a number of researchers. It is clear that plants resistant to a particular herbicide are at a selective advantage (relative to herbicide-sensitive individuals of the same species) where they encounter the herbicide in question. But do herbicide-resistant plants pay a cost of resistance, being less fit than sensitive variants where this herbicide is not being used. Because of the economic importance of weed control, there have been many investigations to see whether there are 'costs of resistance' and, if so, to estimate the magnitude of such costs. In an important review, Vila-Aiub, Neve & Powles (2009) conclude that measurement of costs is far from simple and that many studies in the literature have 'incorrectly defined resistance or used inappropriate plant material and methods to measure fitness'.
A framework for evaluating costs is proposed, involving control of the genetic background in stocks tested, characterisation of the biochemical basis of resistance, appropriate assessment of life-history traits, and the examination of costs under a range of situations including the effects of competition and ecologically stressful conditions. Taking account of all the evidence available, they conclude that there is 'unquestionable evidence that some herbicide resistance alleles are associated with pleiotropic effects that result in plant fitness costs . . . however these resistance fitness costs are not universal'.
Pleiotropic effects involve the production by one particular mutant gene of apparently unrelated multiple (or manifold) effects at the phenotypic level (Rieger, Michaelis & Green, 1976).

The regular application of herbicide does not necessarily lead to populations of herbicide-resistant plants, as was discovered in researches by Holliday & Putwain (1977, 1980). When seed stocks of *Senecio vulgaris* were screened for resistance to simazine – one of the triazine group – by sowing seed on to soil containing the herbicide, almost all the seedlings died, a surprising result considering that some of the sites had been treated with simazine for a number of years. An analysis of the ecological situation revealed an explanation. First, simazine applications were made once in the year and, after a time, the herbicide degrades in the soil. Moreover, some simazine-resistant *S. vulgaris* plants arising in the populations may be killed, because other herbicides were also used on the plots to keep down the weeds. Furthermore, there appeared to be a bank of simazine-sensitive seed in the soil, which germinated when brought to the soil surface, and could grow and reproduce in the autumn when herbicide activity was low or absent. Thus, the population dynamics of sites treated with herbicides may be interestingly complex, and intermittent or seasonal factors will not necessarily lead to directional or disruptive selection.

Taking into account the pattern of distribution of resistance – western USA, eastern Canada, various parts of Europe etc. – it has been suggested that intraspecific triazine variants of several weed species have arisen independently (that is to say, polytopically) in different places. Considering the local distribution of triazine resistance in car parks, along railways, and pathways etc., it is likely that herbicide resistance is also spread by anthropogenic seed dispersal.

There is a growing body of evidence for the polytopic origin of herbicide resistance in several species. For example, in 1987, resistance to the herbicide sulfonylurea appeared more or less simultaneously in populations of *Kochia scoparia* in six different states of the USA, and in Canada. In the 1980s, strains of the weed *Alopecurus myosuroides* resistant to urea herbicides appeared in Germany, the UK and Spain. Detailed studies have also been carried out in France, and given the localised patterns of farm management, it is concluded that polytopic origin of resistant mutants is likely. Random dispersal of these variants then occurs through extensive use of the machinery used in modern agriculture (see Colbach & Sache, 2001; Menchari *et al.*, 2006).

Weed evolution

Weed floras have changed with the increasing use of herbicides, which have imposed 'new selection pressures' (Murphy & Lemerle, 2006). A number of important trends have been detected (Briggs, 2009).

1. Where crops are grown at 'higher' densities: climbing weeds such as *Galium aparine* and *Fallopia convolvulus* (*Polygonum convolvulus*) are at a selective advantage (Håkansson, 1983).
2. Where herbicides are regularly used: loss of diversity in the weed flora and higher proportion of perennial weeds.
3. Where 2,5-D is used to control dicotyledenous weeds in cereal crops: there is an increase in grass weeds *Avena* species and *Alopecurus myosuroides* (Fryer & Chancellor, 1979).
4. Difficult to control weeds of the same or related genera by herbicides. For example, annual bolting weedy variants of *Beta vulgaris* grow in many of the fields of cultivars of the same species – the cultivated sugar beet. This biennial plant is harvested for the sugar content of its swollen underground parts, which do not develop in the annual weedy variants (Soukup & Holec, 2004; van Dijk, 2004).
5. Conservation tillage (no-till or reduced frequency tillage) and minimum cultivation: changes in weed floras resulting from shallower burial of seeds of weeds, increase in non-annual species such as *Conyza canadensis, Cirsium vulgare, Achillea millefolium* etc. (Murphy *et al.*, 2006).

In considering these trends, it is important not to draw simplistic conclusions. Clearly herbicide use is important in the evolution of weed floras, but other site-specific factors must be taken into account, e.g. changes in crop rotation and soil amendments, the extent of soil tillage and changes in mechanised harvesting etc.

Ecotypic variation in response to seasonal or irregular extreme habitat factors

In many of the examples we have been studying, habitats remain extreme throughout the year; for instance, in a soil contaminated by copper, metal ions may leach out into the soil solution all the year round. It is clear, however, that some habitats are subject to severe conditions only at certain critical times. What patterns are found at such sites?

Microevolution in a golf course In California, there is typically a period of rainfall from November to March, with almost no rain in the period May to September. In an imaginative study of a golf course in Davis, Wu, Till-Bottraud & Torres (1987) studied heritable temperature-controlled germination behaviour (namely the responses to high temperatures of 25°C) in populations of the unsown weed *Poa annua* sampled from three ecologically different areas.

The rough areas of the golf course are occasionally mown. In response to high temperatures, germination of samples from the rough was low: less than 30% with a mean of 5%. It is highly likely that seedlings germinating in the summer would be unable to survive, as there was no little or no natural rainfall and no irrigation of the roughs. Germination was promoted by the lower temperatures typical of the autumn rainy season, when there was a higher chance of the plants completing their life cycle.

In contrast, the greens on the course were watered daily to produce a year-long emerald-green playing sward. Thus, *Poa annua* seedlings on the green are provided with adequate moisture regimes all the year round. More than 60% of samples germinated at high temperatures. It is not surprising that selection has favoured variants capable of high-temperature germination, but it is unclear why this figure is not higher. Perhaps it is related to other selection pressures on the greens, namely very close mowing, intense interspecific competition, trampling, herbicide and fertiliser use.

The fairways of the golf course were also sampled. These areas were mown twice weekly and watered two or three times per week. A wide range of intermediate germination behaviour was detected in these areas. Given the management of the course and the constant human activity facilitating gene flow around the site, such a finding is not unexpected.

Two points are worth stressing. Given that the golf course at Davis was only 20 years old, it seems very likely that differentiation can occur rapidly on newly established courses. It would be interesting to examine the microevolutionary patterns of differentiation on golf courses of different ages, and, as the greens are, to some degree, separate islands within the playing area, to study the pattern of differentiation within the courses.

Seasonal changes in water table in natural habitats One of the most familiar concerns a variable water table, the habitat being dry in some parts of the growing season and wet or flooded in others. For example, Linhart & Baker (1973) investigated samples of annual *Veronica peregrina* growing in and around temporary vernal pools in California. Winter rains fill these small pools, which then begin to dry out in the hot summer months. Genetically determined differences were detected in experiments comparing samples growing in extremely different conditions, namely the periphery, which dries out first and where there is competition with other plants, and the centre of pools, which dry out later and where plants face more intraspecific competition (Linhart, 1988).

In tolerance tests, samples from the two subsites were compared. Seeds from the centre were characterised by simultaneous germination, yielding large seedlings with greater tolerance of intense intraspecific competition, and plants reached maturity earlier, producing larger seeds. In contrast, plants from the periphery had smaller seeds and exhibited later and prolonged germination. Seedlings were smaller but grew into larger later-flowering plants, with greater tolerance to interspecific completion (especially to grasses) and environmental extremes. In summary, there was evidence of genetic differentiation to local conditions at ecologically extremely different subsites that are only a few metres apart.

Aquatic and terrestrial conditions In another study, Cook & Johnson (1968) examined intraspecific variation in the heterophyllous species *Ranunculus flammula*, which has lanceolate aerial leaves on and linear leaves under water. They collected plants from a number of habitats in Oregon and grew them under aquatic and terrestrial conditions. They discovered that certain populations, which were likely to experience the most unpredictable regimes in the wild, could produce the extreme heterophyllous leaf types in the appropriate conditions. In contrast, plants from habitats that were less frequently flooded showed very little heterophylly in the experiments.

Phenotypic modification and genetic differentiation

This investigation raises a number of questions about the modification of the phenotype. In Chapter 4, we discussed the phenomenon of phenotypic plasticity, and showed that the phenotype is the product of the interaction of genotype and environment, and that any given genotype will produce different phenotypes in different environments. Each genotype is likely to have a characteristic breadth of phenotypic plasticity, itself under genetic control. At extreme sites, plants with a narrow range of responses may be selected, whereas at less extreme or variable sites plants with a wider spectrum of responses might be at a selective advantage. Let us examine why this might be so. Imagine plants growing on an exposed sea cliff. Plant A is genetically programmed to produce a tall phenotype, while B is of genotype appropriate to a dwarf plant. In exposed conditions 'B' is dwarf, but so too is 'A', and 'A' can be said to be a 'phenocopy' of 'B'. In the unusual event of a period with little wind, plant 'A' may grow tall, only to be severely damaged in the next storms. Plant 'B', however, is constrained genetically within a narrow range of phenotypic possibilities in height, and under unusual conditions it does not grow 'too tall' for the habitat. It is easy to see how selection might operate to favour plants in genotype B in such circumstances. However, given unpredictable or highly heterogeneous habitats, it is apparent that a plant with a wide phenotypic plasticity could be at a selective advantage.

The use of model plants in the study of microevolution

As we saw in Chapter 8 in our historical review, elucidating the presumed genetic basis of ecotypic patterns has always presented a significant challenge. Now, many areas of microevolutionary significance are being investigated with the benefit of a rapidly expanding base of molecular genetical information, especially in model species, such as *Arabidopsis thaliana* (Shindo, Bernasconi & Hardtke, 2007).

At first sight many naturalists will be surprised that *Arabidopsis* – for many a boring, undistinguished species – provides such a splendid opportunity for studying microevolution. However, given the wealth of molecular genetic studies in this species, Shindo, Bernasconi & Hardtke (2007) see many advantages. Despite the fact that *A. thaliana* is almost completely self-fertilising, the species is variable – eight subspecies have been described – with a wide native range across western Eurasia, in habitats from sea level to *c.*4000 m. Evidence suggests that it has been widely introduced into the Southern Hemisphere, North America and East Asia. Worldwide stocks of *Arabidopsis* are available from culture collections. As we saw in Chapter 6, such collections are often called ecotypes, a designation that suggests local adaptation of the plants to the site of collection. It is a point worth stressing that there is a case for using the neutral term 'accessions', to avoid any assumptions about the ecological status of the plants (Shindo, Bernasconi & Hardtke, 2007).

Bakker *et al.* (2006) note further advantages: there are nine species in the genus, making it possible to study interspecific relationships between *A. thaliana* and its near and more distant relatives in the family, which, of course, share a common evolutionary history. For example, as we saw in Chapter 7, comparative studies of the self-incompatible species *A. lyrata* have shed light on the evolution of self-compatibility in *A. thaliana.*

Here, to illustrate the range of microevolutionary experiments on model species, we focus on three key areas: (a) studies of the genomic DNA; (b) investigations involving cultivation experiments; and (c) integrated studies of key life cycle stages.

Detecting the signature of selection from genomic studies

Ehrenreich & Purugganan (2006) point out that 'adaptations are shaped by selection, which can leave a distinctive imprint on the levels and patterns of nucleotide variation in an organism's genome', and there are 'numerous statistical tests that use molecular variation data to identify genes that bear the signature of selection . . . The null hypothesis for these tests is often that the observed genetic variation is consistent with selective neutrality at the locus of interest (Kimura, 1983) and that significant departures from this neutral expectation may be indicative of the action of selection'. Walsh (2008) considers 'the nature of the signal that selection would leave'. He points out that when selection 'rapidly increases the frequency of an allele, linked sites also hitch-hike along for the ride' (Maynard Smith & Haigh, 1974). Thus, 'when selection (natural or artificial) favors a particular allele, sites linked to that allele are also dragged along to fixation, resulting in a region around the selected site showing reduced variation relative to the rest of the genome' following 'a *selective sweep*'. Furthermore, 'the more rapid the fixation, the more reduced the level of variation around the favored site and the larger the size of the region influenced by the sweep'. Statistical tests have been developed for identifying these signatures of selection (see Ehrenreich & Purugganan, 2006), including the Hudson–Kreitman–Aguard (Hudson *et al.*, 1987) and McDonald–Kreitman tests (McDonald & Kreitman, 1991). Walsh (2008) has written a very useful review of the models used in the study of signatures, and outlines important constraints and some unanswered questions in their use. In another important contribution, Siol, Wright & Barrett (2010) stress in detail the implications of some important assumptions in the neutral model, 'namely no population structure, a constant population size and random mating'. It follows that 'frequent and severe bottlenecks or extensive population subdivision are likely to strongly influence the power to detect selection, and our understanding of these influences on testing for selection is still rather limited'.

In an analysis of microsatellite variation in wild and weedy populations of *Helianthus annuus*, Kane & Rieseberg (2008) found evidence of 'selective sweeps'. Patterns of genetic variability suggested that 'between 1 and 6% of genes were significant outliers with reduced variation in weedy populations, implying a small but not insignificant fraction of the genome was under selection and involved in the adaptation of weedy sunflowers'. Moreover, weedy populations were often more closely related to nearby wild populations than to each other, suggesting that they may have evolved multiple times, 'although a single origin followed by gene flow cannot be ruled out'.

Signatures of other types of selection have also been investigated (e.g. in adaptation to climate in *Medicago truncatula*: Yoder, *et al.*, 2014; also see Kujala & Savolainen, 2012). Sometimes a genetic polymorphism is long maintained by selection, in situations where there is 'natural selection through heterozygote advantage, frequency-dependent selection or spatial–temporal selection for alternative alleles' (Tian *et al.*, 2002). Here, in contrast to the situation examined above, 'when a polymorphism maintained by selection is old, it will have an island of enhanced sequence variability surrounding it, which represents a detectable signature of selection'. Tian and co-workers have examined such a site in *A. thaliana*, containing the disease-resistance gene *RP55*. Roux *et al.* (2012) have investigated recent and ancient signatures of balancing selection around the S-locus in *Arabidopsis halleri* and *A. lyrata*. As we saw in Chapter 7, balancing selection maintains multiple alleles at the S-locus in populations of these self-incompatible species.

In Chapter 7 we discussed the evolution of selfing from outcrossing ancestors. Genomic studies of the selfing species *Leavenworthia alabamica* have identified demographic signatures leading to the conclusion that in one race selfing was likely to be 'relatively old' and linked to a severe population bottleneck event, while selfing arose more recently in other races of the species.

Studies of local adaptation involving cultivation experiments

Fournier-Level *et al.* (2011) grew a geographically diverse set of *Arabidopsis* accessions, maintained as inbred lines from natural populations in replicated common gardens at sites in Oulu (Finland), Halle (Germany), Norwich (UK) and Valencia (Spain). These sites represented climatic extremes in the native range of the species. In measuring adaptive traits, survival and fruit production were examined in detail, and convincing evidence of local adaptation was obtained. This information was examined in relation to information about the genomic variability of the accessions, and the authors 'identified candidate loci for local adaptation from a genome-wide association study of lifetime fitness in geographically diverse accessions'. Moreover, 'fitness associated loci exhibited both geographic and climatic signatures of local adaptation'. This elegant case study highlights the insights that can flow from the combined use of molecular genetics with classical genecological approaches involving replicate common garden trials in regions with different climate.

A second case study involves investigations of selection in the model species, Bread Wheat (*Triticum aestivum*) (Rhoné *et al.*, 2010). A composite stock of wheat was developed by crossing parental inbred lines with diverse contrasting characteristics for fitness-related traits such as height, flowering time, resistance levels etc. Subsamples of this composite stock were grown in three sites in France with different climatic conditions – near Vervins, northern France; Le Moulon, near Paris; and Toulouse, southern France. They discovered that 'in Northern populations, where winter is long and cold, plants flowering too early in the spring may be exposed to frost damage, whereas in Southern populations plants flowering later may suffer from heat and drought stress in early summer'.
The experiment, carefully designed to avoid inadvertent artificial selection, was continued for 12 generations. At each generation bulk harvests were collected at each site and a random selection of seed (at least 5000 seeds) was sown to provide the next generation. There was strong evidence for the action of natural selection on flowering time. Genomic studies of the stocks grown in the experiment identified six candidate genes that were presumed to be 'involved in trait expression'.

Combined studies of crucial life cycle traits

One critical transition in life-history evolution is the change from the vegetative to the reproductive state. The timing of this change in relation to the 'growing season' is highly significant. If a plant is to have high biological fitness it is crucial that reproduction is successful within a window of equitable growing conditions. As we have seen in the last chapter, there is plenty of evidence for ecotypic and clinal variation in life-history traits. If a plant is to have a high biological fitness, it is crucial that reproduction is successful within this window. Remarkable insights have been obtained in studies of *Arabidopsis* (Donohue, 2002; Mitchell-Olds & Schmidt, 2006). Many of the strains are winter annuals: seed germinates in the autumn, and flowering occurs in the spring induced by vernalisation – extended periods of exposure to low temperatures. However, other strains of *Arabidopsis* come quickly into flower with no requirement for cold period vernalisation. Molecular genetical analysis and physiological studies have established the basis of these differences between these two extreme types. The following account is drawn from Giakountis *et al.* (2010).

1. 'The semi dominant locus FLOWERING LOCUS C (*FLC*) and dominant locus FRIGIDA (*FRI*) are present in the winter annual and are required for the vernalization response'.
2. In its activity *FLC* represses flowering, its activity being promoted by *FRI*.
3. 'Exposure of plants to several weeks of cold periods causes *FLC* transcript levels to fall, due to changes in the chromatin structure of the *FLC* locus'. The mechanism involves an 'epigenetic switch that represses the expression of *FLC* itself through chromatin remodeling' (Chaing *et al.*, 2009).
4. Quick-flowering 'summer annuals carry alleles of the genes *FLC* or *FRI* that reduce the activity of one or both genes' (Giakountis *et al.*, 2010).
5. It seems likely from a number of lines of evidence that 'the *FRI* allele from the late-flowering ecotypes is the ancestral form of the gene, with early flowering evolving independently at least twice from the late flowering ecotypes through deletion events leading to loss of *FRI* function' (Johanson *et al.*, 2000).
6. 'The independent appearance of early-flowering ecotypes in the evolution of *Arabidopsis* suggests that there has been strong selection in some

environments for ecotypes that do not require vernalization' (Johanson *et al.*, 2000).

7. In a study of the distribution of ecotypes Johanson *et al.* (2000) draw the following conclusions. 'The majority of late-flowering ecotypes are from northern latitudes, whereas most of the early-flowering ecotypes were collected in central and eastern Europe ... Flowering early without vernalization may be an advantage where severe winter weather prevents germination ... in climates that would support more than one generation per year, or where there is a selective advantage in escaping agricultural harvesting (Toomajian *et al.*, 2006), succession or summer drought'. Human disturbance has been a major factor in the microevolution of *Arabidopsis*, exposing the plants to 'novel environments' with strong selective pressure for adaptive mutations in relation to flowering behaviour (Johanson *et al.*, 2000). In a recent very intensive genomic investigation of *Arabidopsis*, Hu *et al.* (2006) extend these findings, concluding that 'selection has acted to increase the frequency of early flowering alleles at the vernalization requirement locus *FRIGIDA*'. Moreover, 'selection seems to have occurred during the last several thousand years, probably as a response to the spread of agriculture'.
8. It should not be supposed that *Arabidopsis* populations contain only one of the two major variants: those requiring a standard length of vernalisation and those that do not. Populations in the wild contain genotypes with different vernalisation requirements, as adaptations to different climatic regions (Jones 1971a, b, c; Mendez-Vigo *et al.*, 2011). In addition, investigations have revealed that individuals requiring vernalisation will generally flower even if a cold period does not occur. In such circumstances, flowering is often delayed.
9. Here we have examined the importance of vernalisation in the flowering behaviour of *Arabidopsis*. But other factors are also crucial. Photoperiodic effects on flowering (and other life-history events such as date of bud opening etc.) have been discussed in earlier chapters. To recap, as day length varies with latitude, species with a wide geographical distribution often show differences in photoperiodic responses in flowering behaviour in different parts of their range. Also, gibberellin hormonal pathways also play major roles in the flowering. Intensive studies into all these phenomena in *Arabidopsis* have made it possible to outline a Flowering Time Gene Network – a ' large genetic network of genes representing a variety of different light sensing, hormone signaling and developmental pathways' (Flowers *et al.*, 2009) (Fig. 9.8). (For more recent reviews of this rapidly developing area of research see Hemming & Trevaskis, 2011; Turnbull, 2011.)
10. Another life cycle trait of great significance and subject to strong selection is the timing of seed germination (Chiang *et al.*, 2009). Evidence suggests that FLOWERING LOCUS C (*FLC*) acting within a network of other genes is associated with temperature-controlled seed germination.

A number of examples of the investigation of model plants and their relatives in the elucidation of microevolutionary patterns and processes are outlined in Table 9.4.

Arabidopsis: experiments on phenotypic plasticity

Not only have investigations of model species contributed to our understanding of natural selection, they have also presented important opportunities to examine the comparatively neglected phenomena of phenotypic and developmental plasticity, concepts introduced in earlier chapters. Figure 9.9 gives the results of an elegant experiment on plasticity using stocks of *Arabidopsis* (reported in Pigliucci, 2010).

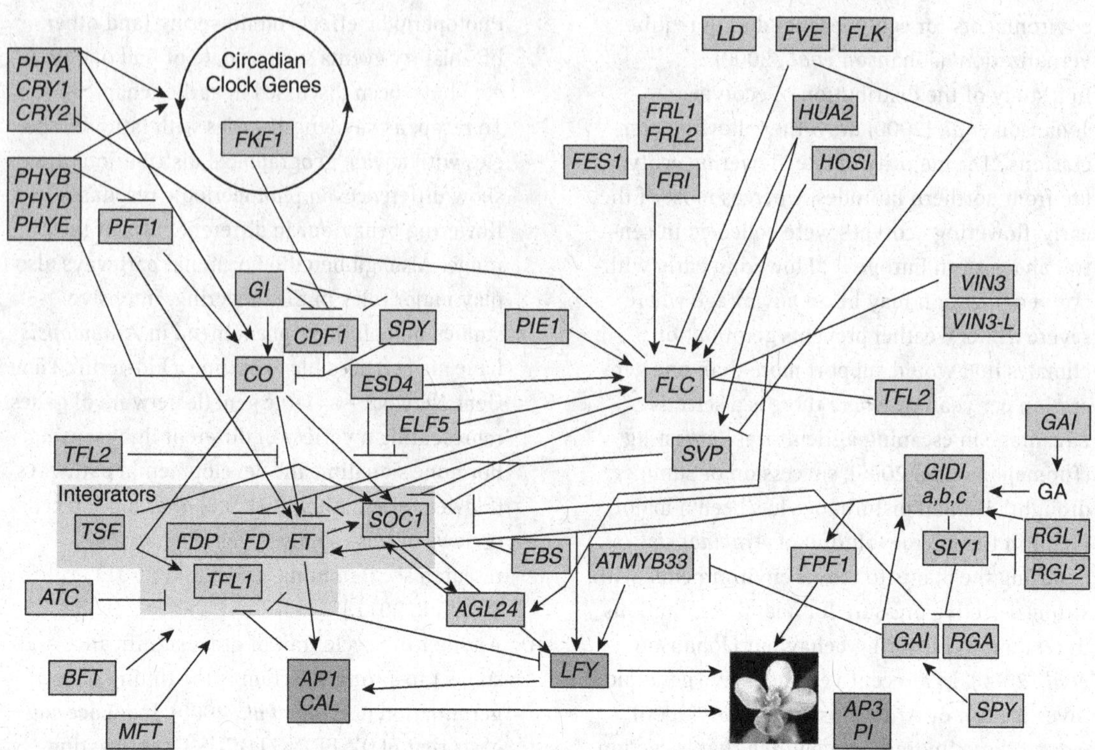

Fig. 9.8. The flowering time gene network in *Arabidopsis thaliana* compiled from published genetic studies. (From Flowers *et al.*, 2009.)

Five lines of *Arabidopsis thaliana* were collected from different parts of Europe: Spain, southern France, northern France, Netherlands and Sweden. As this model species is automatically self-fertilising, the breeding system leads to pure lines. All five provenances were represented in three growth chambers, each of which was set up to simulate a different photoperiod representing the environments of southern, central and northern latitudes in Europe.

The number of fruits formed was chosen as the measure of phenotypic response. Pigliucci (2010) provides the following commentary on the experiment. 'Each line represents a distinct genotype's reaction norm. The diagram illustrates the properties of genetic variation (different heights of the various norms), environmental variation (when individual norms have a slope different from zero) and genotype-by-environment interaction (when different norms are not parallel to each other)'.

In the past not enough attention has been paid to examining the evolutionary significance of phenotypic plasticity in plants. Now researchers are considering the phenomenon more closely, to discover how it might be initiated, maintained and evolve in the longer term. They are also facing the key questions: what are the limits of the process and to what extent are the responses adaptive? (see Kaplan, 2010).

Also physiological plasticity is being examined. An intriguing example of this phenomenon is the variability in the expression in cyanogenic plants of *T. repens* (Till, 1987; Hughes, 1991). Some cyanogenic plants of *Lotus corniculatus* are of stable phenotype, whilst others are cyanogenic only at certain times of the year and under some conditions. Concerning *Lotus*, Ellis, Keymer & Jones (1977b) are inclined to see this

Table 9.4. ***Arabidopsis thaliana*: some examples of the use of this model plant and its relatives in the study of adaptively significant traits**

Carbon dioxide levels: In experiments, highly relevant to an understanding of the effects on plants of past and future climate change, responses to CO_2 as a selective agent were examined. Using genotypes from different locations and altitudes, random crosses were used to produce hybrid stocks subjected to five generations of selection in four CO_2 controlled growth chambers: two with low CO_2 (20 Pa) and two at high CO_2 (70 Pa). '*Arabidopsis* showed significant positive responses to selection for high seed number at both 20 and 70 Pa CO_2' (Ward *et al.*, 2000).
Climate-adaptive genetic loci and pathways: candidate genes identified with genome-wide scans (Hancock *et al.*, 2011).
Clinal variation: 189 genotypes from 17 populations grown under controlled conditions, clinal variation along climatic gradients in Spain (Montesinos-Navarro *et al.*, 2011).
Defence against herbivores: Glucosinolates are important in defence against herbivores in the Brassicaceae. Variable MAM genomic region controls the types of glucosinolates produced. 25 ecotypes studied – genotypic variation patterns in MAM suggest the possibility of balancing selection acting on one of the genes (MAM2) (Kroymann *et al.*, 2003).
Flooding tolerance: 86 accessions studied: significant variation in flooding submergence tolerance detected (Vashisht *et al.*, 2011).
Freezing tolerance: 2 ecotypes studied; evidence that CBF2 is a candidate gene (Alonso-Blanco, Mendez-Vigo & Koornneef, 2005).
Latitudinal flowering time clines in North America: where *A. thaliana* was introduced 150–200 years ago. Investigations of flowering time genes FRI, FLC & PHYC carried out (Samis *et al.*, 2012).
Nitrogen supply: ecotypes tested in normal and low N supply; significant differences detected in the extent of growth reduction in limiting conditions (North *et al.*, 2009).
Pathogen resistance: Role of R-genes in pathogen resistance. 'R-proteins monitor the plant cell for the presence of pathogen-secreted proteins. Upon their detection, they then rapidly activate robust plant defence pathways, including local cell death, cell wall reinforcement, production of secondary compounds and systemic acquired resistance' (Shindo, Bernasconi & Hardtke, 2007).
Salt tolerance: common candidate single gene on chromosome 5 in the 350 accessions tested (Katori *et al.*, 2010).
Seed dormancy: strong evidence for QTLs identified (Huang *et al.*, 2010).
Investigations with other *Arabidopsis* species
Metal tolerance: Pauwels *et al.* (2008). Investigations with *A. halleri* of the accumulating molecular evidence for the genetic basis of zinc tolerance.
Herbivory: *Arabidopsis lyrata*. 'Strong positive correlations between leaf trichome [leaf hair] density and resistance to leaf herbivory, demonstrating that the production of leaf trichomes increases resistance to damage' by the moth *Plutella xylostella* (Sletvold *et al.*, 2010). In *Arabidopsis thaliana* several genes have been identified that affect trichome density, e.g. lack of function glabrous variants at the locus GLABRA1 (GL1) (Weigel, 2012).
Serpentine soils: are high in heavy metal content and with low Ca/Mg ratios. Such areas present an extreme environment in which ecotypic differentiation has been detected. Comparative studies of *A. lyrata* from granite and serpentine soils employed 'a tiling array [from *A. thaliana*] that had 2.85 million probes' – evidence that calcium-exchanger 7 is an excellent candidate gene for adaptation to low Ca/Mg ratios in this species (Turner *et al.*, 2008).

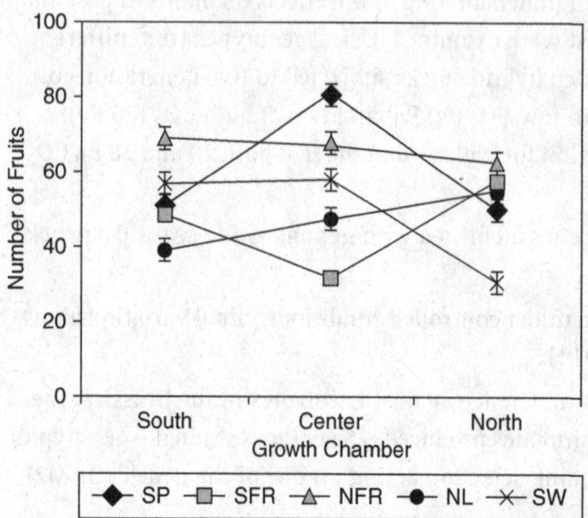

Fig. 9.9. *Arabidopsis thaliana*: a simple set of norms of reaction from a plasticity experiment. (Data from Josh Banta of the Pigliucci laboratory.) The horizontal axis represents environment (simulated photoperiods of northern, central and southern latitudes in Europe) and the vertical axis is a measure of phenotype. Each line represents a distinct genotype's reaction norm. The diagram illustrates the properties of genetic variation (different heights of the various norms), environmental variation (when individual norms have a slope different from zero) and genotype-by-environment interaction (when different norms are not parallel to each other). Origin of provenances tested: SP = Spain; SFR = Southern France; NFR = Northern France; NL = Netherlands; SW = Sweden. (From Pigliucci, 2010.) © 2010 Massachusetts Institute of Technology, by permission of The MIT Press.

physiological flexibility as adaptively significant, the cyanogenic mechanisms effective against herbivory being 'switched off' at just those times when grazing pressure is likely to be low, and the risk of damage to the plant by other factors is at its highest.

This behaviour could be controlled by the epigenetic mechanisms outlined in Chapter 6. Many of the studies of epigenetics in plants have been carried out in controlled environments. Epigenetic phenomena in relation to stress factors in natural and human-influenced ecosystems have yet to be properly examined. Reviewing existing information, Jablonka (2013) concludes, 'non-genetic information can be transmitted to the next generation', and 'the stress-induced mechanisms that lead to wide-ranging genetic and epigenetic modifications are of obvious importance in both adaptive evolution and in speciation; environmentally induced changes often begin as matched to conditions in which they may prove advantageous, unlike mutations, which begin as random with respect to environment'.

Adaptive and non-adaptive characters

In this chapter we have examined the adaptive significance of various morphological traits and physiological characteristics. We have to acknowledge, however, that in many cases we do not know whether particular characters are adaptive or not.

The case of the weed *Spergula arvensis* reveals the necessity of critical investigation. New (1958, 1959) showed that a ratio-cline existed in the British Isles, populations of *Spergula* in the north and north-west having significantly higher proportions of the variant with smooth seed-coat than those in the south and east, where papillate seed-coats contributed quite high frequencies (Fig. 9.10).

Examination of herbarium material for Europe as a whole confirmed the general tendency for the smooth variant to be characteristic of higher latitudes. Twenty years later, New (1978) was able to show that the ratio-cline for seed-coat pattern had not significantly changed in Britain (though there *was* evidence for a significant change in proportion of strongly glandular-hairy plants in the populations). Cultivation experiments showed conclusively that the variant with smooth seeds was less tolerant of high temperatures and low humidity than the variant with papillate seeds. New concluded that the apparently non-adaptive seed-coat character, which she had shown to be determined by a single gene without dominance, must be correlated with genotypic differences in physiology, which were themselves clearly of selective importance in different climates. Indeed, further studies revealed that papillate seeds germinate more easily than smooth ones under dry

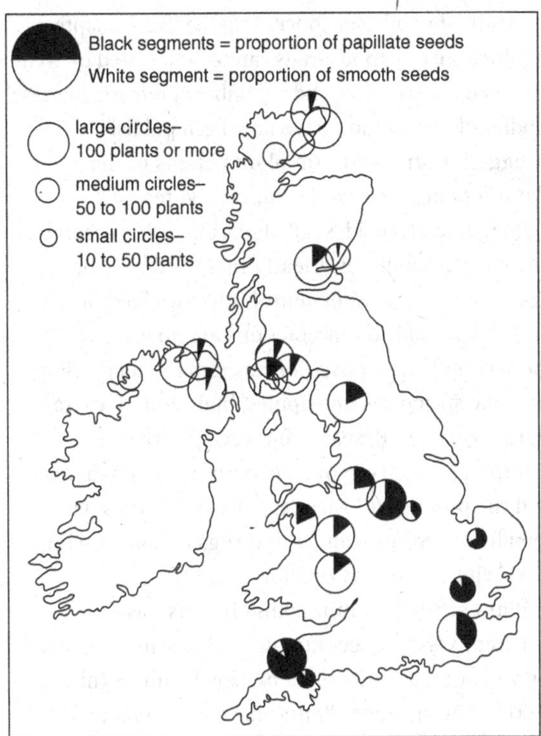

Fig. 9.10. Distribution in the British Isles of variants of *Spergula arvensis* with smooth and papillate seed-coats. (From New, 1958.)

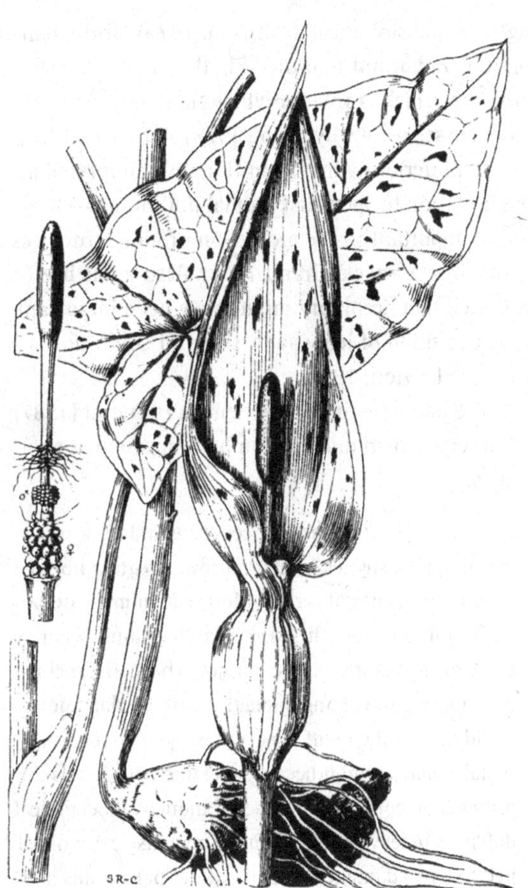

Fig. 9.11. *Arum maculatum*, spotted variant (×0.5). (From Ross-Craig, 1973.)

conditions, so that an apparently non-adaptive character has, after all, been shown to have direct adaptive significance (New & Herriott, 1981). Further investigations of this species are recommended, for despite close examination of the performance of experimental mixtures of the two seed variants, Silvertown (1992) did not discover how the polymorphism in this species is maintained.

Turning to a second example, Lords and Ladies (*Arum maculatum*) shows remarkable polymorphism for the spotted versus unspotted leaves, and for purple versus yellow spadix (Fig. 9.11). Attempts to demonstrate that the spadix colour affects the quantity or quality of small insects trapped by the remarkable pollination mechanism have been unsuccessful. And it is not clear whether the presence of anthocyanin spots in the leaves is of adaptive significance (Prime, 1960). Whether these traits are adaptive or non-adaptive has still to be determined.

Biologists have puzzled over the adaptive significance of other intraspecific polymorphisms, for example, leaves of different shapes, dissection, leaf geometry, and as we have seen in *Arum* surface markings. Perhaps in some of these cases the polymorphism is not itself adaptively significant, but it is 'linked' to adaptive physiological traits concerned with water relations, gas exchange, photosynthesis etc. (Givnish & Vermeij, 1976; Givnish, 1979). Also, many species are polymorphic for hairiness, some individuals being glabrous, others obviously hairy. Again the adaptive significance of such variation might be 'physiological' (Johnson, 1975), or the hairs might be important in protecting the plant against herbivore attack (a role for which there is strong evidence; see review by Levin, 1973; Harborne, 1993). Even the inferior ovary – traditionally interpreted as an adaptation to specialist animal pollination – may also be important in the protection of vital structures against insect attack (Grant, 1950). Also it has been suggested that herbivory might have been important in the evolution of leaf shape, size and coloration (Brown & Lawton, 1991).

On the basis of studies by Niemela & Tuomi (1987) and others, Brown and Lawton have speculated as follows:

> Evidence that some insects recognize and respond to leaf shape . . . suggest(s) that narrow, irregular hollows on the leaf blades of certain Moraceae mimic caterpillar feeding damage. These irregular 'incisions' occur on some of the plant's leaves, but not others, enhancing resemblance to feeding damage . . . [f]alse damage could protect the plant in several ways. Ovipositing females may avoid it because real damage signals the presence of competing larvae, or induced biochemical defences in leaves; false damage may also be avoided because real damage serves to attract parasitoids and predators . . . A wide range of plants outside the Moraceae also have leaves with herbivore-like indentations and holes . . . The hypothesis that 'pseudo-damage' is mimetic and protective against herbivores (including vertebrates) deserves more attention, along with the related suggestion that variegated colour patterns on certain leaves may serve a similar function, for example by mimicking leaf-mines.

Plant mimicry is now seen as adaptively significant in reducing predation by animals (Wickler, 1968; Wiens, 1978; Barrett, 1987; Stowe, 1988). For instance, members of the genus *Lithops* are so well camouflaged in their stony background in African deserts that they are referred to as living stones, and Australian mistletoes of the genus *Amyema* are practically invisible on their hosts as they mimic its foliage (Barlow & Wiens, 1977).

Clearly, hypotheses concerning the likely adaptive significance of various traits can only be tested by well-designed experiments, and a number of remarkable case studies of coevolution have now been published.

Egg-like structures on plants Plants of the genus *Passiflora* are protected from herbivores by cyanogenic glycosides, alkaloids and other secondary chemical compounds. Clearly these defences have been breached as *Heliconius* butterflies feed on the plants. Coevolution interactions are now seen to be more complex, as *Passiflora* species produce yellow egg-like structures on stipules, buds and on extra-floral nectaries. Research has revealed that caterpillars of larvae of *Heliconius* are cannabalistic, and the presence of these egg-like structures deters female butterflies from laying eggs on plants with 'fake' eggs (Williams & Gilbert, 1981).

Plant structures mimicking insects Cases of apparent visual insect mimicry, of possible adaptive significance, are described by Lev-Yadun & Inbar (2002). For instance, *Xanthium strumarium* and *Arisarum vulgare* have epidermal spots on the stems/petioles that resemble ants; the dark anthers of *Paspalum paspaloides* look like aphids; and immature pods of three species of wild legumes have conspicuous reddish spots on their seed pods that appear to mimic lepidopteran caterpillars.

Associations with ants As we saw earlier, cyanogenic glycosides offer protection against herbivory, as tissues damaged by herbivores produce 'defensive' HCN. *Acacia cornigera* lacks the glycosides found in other related species, but has large spines, which are adaptively significant in deterring grazing vertebrates. However, these spines are adaptively significant in another way. Some are hollow, and occupied by colonies of ants. Intensive studies have revealed a mutualistic association between *A. cornigera* and ants of genus *Pseudomyrmex*, which are attracted to the plant by proteinaceous secretions produced at the tips of leaflets (Janzen, 1966). On *A. cornigera* individuals with ant colonies, caterpillars and other small herbivores are predated by the ants, and they also remove climbing vines from the plant and its immediate surroundings. In considering the fitness advantage of the presence and activities of ants, it is highly significant that individuals of this species lacking ants are severely attacked by herbivores and soon become infested with climbing vines.

So far we have considered morphological features. Investigations of the properties of many secondary chemical compounds in contemporary plants, provide evidence that such chemicals have adaptive value as a defence mechanism against herbivores. A number of models of coevolutionary interactions have been developed. For instance, some have considered the interaction to be an 'arms race' as successive changes in the chemical defences of plants under 'attack' are followed by counter-adaptation by herbivores (Spencer, 1988). We now recognise that the chemical defences of plants are at least as complex as the more traditional structural defences such as spines and prickles, and several case studies are considered in later chapters (Harborne, 1993). The coevolution of insects and plants has produced other very complex situations, as, for example, in the orchids of the genus *Ophrys*, where the plants have evolved to provide false clues to attract male insect pollinators, not only by producing mimetic flowers resembling the female of the species but also specific female sex hormones to lure males without providing any reward (see Dettner & Liepert, 1994).

Concluding remarks

As we have seen in this chapter, there is increasing evidence for natural selection in plant populations. Where there is information concerning the history of sites, there is convincing evidence that change may be rapid (Thompson, 1998), and that in extreme habitats selection pressures may be high. Insights have come from a range of situations, many of them in human-influenced habitats, where wild and weed species of plant populations encounter extreme/contrasting habitat conditions, in hay meadows/pasture; mines spoil/pasture; hay meadow given different fertilizer treatments; flowerbeds/lawns etc.

In addition, there is also growing evidence for selection in natural habitats, e.g. on serpentine and heavy-metal-containing soils etc. While the study of extreme habitats has provided invaluable evidence of natural selection, there is every reason to believe that natural selection, at different intensities, occurs in all populations.

Recent studies have provided strong evidence for recurrent evolution. The possibility of recurrent origin of ecotypes is mentioned in the early literature, e.g. Turesson (1922). In an important review Levin (2001a) draws together evidence for the recurrent origin(s) of ecotypic variants. First he makes the point that it could be 'tacitly assumed that the key ecological and genetic transitions ... occur once, and that the new taxon spreads to suitable sites'. However, there is no reason why distinct ecotypic variants 'cannot arise from separate lineages within an ancestral species at different times and different places ... Directional selection may act in similar ways across sites and times, and thus provides a mechanism for recurrent evolution.' He considers that full tests of origin are necessarily 'rare' with 'almost no unassailable examples'. However, as we have seen in this chapter, there is a growing body of circumstantial evidence, from the use of molecular markers of various kinds, for the independent recurrent origins of single-factor variants in heavy-metal-contaminated sites, serpentine outcrops, and herbicide resistance in weeds growing in field crops etc. But, 'while the many documented cases of herbicide resistance do in fact provide firm evidence that parallel genetic changes cause convergent phenotypes, parallelism at the genetic level is not always the case – there are a number of examples wherein the mechanism underlying resistance to the same herbicide differs among and even within species' (see Baucom, 2016, who reviews current research on underlying molecular genetics and physiology of the remarkable 'repeated evolution of herbicide resistance'). There is also evidence for the polytopic origins of multi-factor variants. For instance, the investigations of Kane & Rieseberg (2008) strongly suggest that weedy populations of Common Sunflower (*Helianthus annuus*) may have evolved recurrently from nearby wild populations. Evidence for multiple origins also comes from study of the evolution of self-compatible variants from self-incompatible lineages. As we saw in Chapter 7, self-compatibility has frequently evolved and is often associated with a whole syndrome of characters related to an evolutionary shift in resource allocation, such as smaller flower size, lower pollen/ovule ratios etc. The concept of recurrent origins – sometimes called polytopy – is very valuable for the study of other aspects of microevolution.

In this chapter we have reviewed the enormous advances in our understanding that have come from the intensive study of model species, particularly *Arabidopsis thaliana*. As we have seen, the move to study not only laboratory stocks but also naturally occurring populations of this species has provided major new insights. One especially promising avenue of research is to incorporate reciprocal transplant experiments into 'genomic' research (Ågren & Schemske, 2012). Such approaches provide 'the long-standing gold standard to test the hypothesis that populations are locally adapted' (Lowry, 2012).

Many advances have also been achieved in investigations of other well-chosen model species. As some of these species are related to *Arabidopsis thaliana*, it is possible to build on the molecular genetic knowledge of this species etc. in the devising of critical experiments on other species.

Siol, Wright & Barrett (2010) reflect on recent advances in genomic studies of plant populations, including the detection of signatures of selection.

> Exciting as the past decade has been in giving us new insights into the genomic structure of plant populations, the advent of so-called 'next generation' sequencing holds even more promise. These new techniques generate quantities of data that are orders of magnitude greater than classic sequencing methods . . . evolutionary analysis of genomic data is still in its infancy and many formidable challenges face the field of evolutionary bioinformatics . . . the first involves the sheer number of sequences that must be dealt with . . . for the first time in population genetics history, the limiting factor is the availability of methods and models, and not the data on which to address evolutionary questions . . . the parallel improvement of next-generation sequencing techniques and computational and analytical tools should allow large-scale interspecific comparisons of the historical and contemporary context in which selection acts at the molecular level. These approaches will yield important insights into the interactions between demography and adaptive evolution in plant populations.

Some of the most convincing examples of the action of natural selection acting in the wild have come from studies of extreme habitats, where plants are subjected to 'stresses' such as herbivory, metal pollution, herbicides etc. Further important insights are emerging too from molecular studies of clines (e.g. Chen *et al.* (2012, 2014) have identified candidate genes for adaptive traits in *Picea abies*).

Recent genomic studies have revealed that besides 'signatures of selection', other genetic changes may be detected, which may have profound implications for the study of adaptive responses in populations of plants. For example, transposable elements may be 'activated' (Kilian *et al.*, 2007). Also, heritable stress-induced methylation epigenetic change patterns may be induced, as was discovered in experiments with apomictic *Taraxacum* species (Verhoeven *et al.*, 2010). Some important questions have still to be investigated: do such changes occur in sexual species, and what are the stresses that might precipitate changes?

In this chapter, we have focused mainly on natural selection, with some reference also to chance effects and gene flow. In the next chapters, further consideration of these key microevolutionary processes will examine recent advances in our understanding.

10 Pattern and process: factors interacting with natural selection

Chance has profound effects

It is important first to differentiate between chance events and those that involve natural selection. Ridley (1996) provides an elegant example. He imagines a long line of loaded packhorses toiling, in single file, up a long precipitous mountain track. Clearly, elements of natural selection might be important, as some animals might be inherently more nimble and survive the journey in greater numbers than clumsy horses that fall to their death. However, in his example, the hazards of the alpine journey also include massive falling rocks that sweep horses to their death into the ravine below. Animals are not genetically predisposed to avoid the sudden chance arrival of massive falling rocks. The important point to emphasise is that horses swept off the track by such chance events are likely to be a random sample from the original population that started out on this hazardous journey. Furthermore, a new mutant horse – more sure footed than the average – might by chance be carried away with others of lesser fitness. The potential of this mutation is, therefore, not realised. Chance determines which, of a group of horses, survive the falling rocks.

Chance effects – so-called stochastic events – play a major role in all evolutionary processes, not only in chromosomal and DNA changes, but also in hybridisation, speciation and extinction. In population biology, for instance, chance plays a huge role in which flowers are pollinated by animals/wind, and which seeds/fruits are dispersed to 'safe sites' allowing successful germination. Indeed, the effect of chance is highly significant in all the processes of population biology: dispersal, establishment, growth and reproduction.

Random genetic change in populations is known as genetic drift or in the older literature as the 'Sewall Wright' effect. 'Alleles may be fixed or lost, especially from small populations, because of random sampling errors and without regard to their adaptive values' (King, Stansfield & Mulligan, 2006). Thus, chance effects are particularly important if the population is reduced by a bottleneck effect to a very small size (Fig. 10.1). Wright (1931) was the first to point out that, in such small populations, irregular random fluctuations in gene frequency occur that may result in the fixation or loss of one or more alleles. Such chance effects may be very important also in the declining populations of endangered species.

As we see below, insights into the effects of chance have also been revealed by the use of molecular markers in the close study of 'new' populations that may be founded by one or a very few individuals. Such derivative populations are subject to the so-called founder effects, having only a random subset of the variation of the larger ancestral population from which they are drawn.

Concerning the interaction of key processes, some progress has been made in computer simulations that model with a 'reasonable level of reality and complexity' the effects of 'drift, selection, migration and population sub-division' (see Frankham *et al.*, 2002). However, it has to be accepted that, in many circumstances, it is not possible definitively to separate the effects of chance and selection in the wild.

Gene flow: population variability and structure

Because of the basic biology of plants, chance events play a key role in the functioning of plant populations. With many obvious exceptions

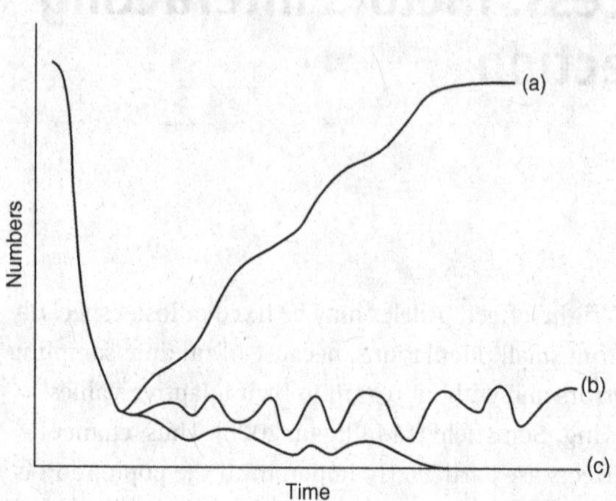

Fig. 10.1. The effect of a severe reduction in numbers of individuals in a population – the so-called bottleneck effect. (a) Recovery in numbers: the new population may not differ phenotypically from the original one but some genetic variability may be lost. (b) No recovery in numbers. As an effect of chance only, a small random selection of the population may survive but becomes genetically less variable, with irregular fluctuations in gene frequency and fixation or loss of one or more alleles without regard to their adaptive value. (c) Extinction of the population.

(e.g. certain species in aquatic ecosystems) adult plants are generally stationary (Ashley, 2010). Dispersal occurs through the movement of gametes and offspring (as pollen and seed dispersed by gravity, wind, water or animals). If we wish to understand the recruitment of new genotypes or species to a given population, the extent, distance and direction of such dispersal is vitally important. Moreover, it is clear that gene flow is a key evolutionary process. It has a major influence, at wider spatial scales and longer timeframes, on patterns of species colonisation, migration, persistence and the speciation process itself. Practical concerns are also important, for example, as agriculturalists attempt to propagate 'pure' cultivars of crop plants, and estimate the risk of gene flow between genetically modified (GM) and other non-GM plants (involving not only the crops themselves but also wild relatives) (Ashley, 2010).

In order to determine which plants are interbreeding it is necessary to study patterns of gene flow, i.e. the source and ultimate destination of fertilising pollen and the pattern of dispersal of fruits and seeds. In the past, where genetic markers were available in crop plants, the first direct measures of gene flow were carried out. But, for wild plants, historically only circumstantial evidence was available, based on a study of the movement of pollen and seed.

To appreciate both the limitations of pioneering investigations and the magnitude of the recent advances made possible through the use of molecular tools, it is important to provide some historical background. Then, we review the modern molecular approaches that have provided some more definitive measurements of 'actual' gene flow and paternity in wild plants.

Gene flow: early ideas

Wright's *F*-statistics model is the basis for estimating the variation within and between populations (see Table 9.1).

As we saw above, the quantity F_{ST} estimates the amounts of variability amongst populations. Holderegger, Kamm & Gugerli (2006) point out that if this value is unity it means that different alleles are found in the different populations. If F_{ST} is zero it means that the same alleles are found in the different populations at the same frequency. Intermediate values indicate the degree of similarity between populations. Given that extensive gene flow tends to homogenise populations, values for F_{ST} provide some indication of the extent of this phenomenon. However, it is important to stress that these estimates may reflect what has happened over long timeframes rather than indicating contemporary gene flow.

Turning to more field-based studies, until the 1970s, models of gene flow in natural populations were not clearly formulated. But, according to Ehrlich & Raven (1969), many biologists thought that gene flow could be extensive (see, for example, Merrell,

1962; Mayr, 1963). This view was supported by the following observations of naturalists and others:

1. Pollen, present in enormous quantities in air at the appropriate times of year, is widely dispersed (even being detected, for example, 50 km out to sea in the Gulf of Bothnia, an area of the Baltic Sea between Finland and Sweden: Hesselman, 1919).
2. The various mechanisms for ensuring insect and wind pollination were assumed to be widely effective.
3. The assumption was made that fruits and seeds were well adapted to wide dispersal in nature.

These general impressions led naturally to the view that the members of a widespread species may be united in the same gene pool by gene flow. The species was viewed as the important evolutionary unit, gene flow amongst its members making it a breeding unit. Speciation was seen as the breakdown of the cohesive gene pool within a species to produce daughter species through the evolution of isolating mechanisms.

However, in the 1970s a new perspective on gene flow came from various lines of study (Levin & Kerster, 1974; Richards & Ibrahim, 1978; Levin, 1988; Falk & Holsinger, 1991). On the basis of comprehensive review of all the evidence and new experiments then available, the view that gene flow is widespread and 'effective', and that interbreeding populations are therefore large, was subject to radical reappraisal (Ehrlich & Raven, 1969; Levin, 1978a). It was now contended that gene flow in nature may be much more restricted than previously thought. It seemed likely that species existed in nature as a multitude of small sub-populations, rather than single or a few large ones.

Gene flow: agricultural experiments

Some of the most interesting historical information on gene flow has come from studies of crop plants (Levin & Kerster, 1974), forest trees (e.g. Wright, 1953) and weeds (Cousens & Mortimer, 1995). The development of superior cultivars by plant breeding requires the production of large quantities of 'pure' seed, and there has, therefore, been a good deal of interest in estimating the minimum distance required to prevent crossing between two different cultivars. Table 10.1 shows the isolation requirement for a number of crop species.

It is clear that only a relatively small distance is required in many cases to prevent contamination. Experiments with mixtures of cultivars differing in genetic markers have also provided important insights. Thus, the incidence of hybridisation between mixtures of two crops depends upon planting arrangements in the field. The degree of crossing may also be influenced by the presence and disposition of taller plants. For example, in an experiment studying gene flow in Cotton (*Gossypium*), which is insect-pollinated, the degree of intercrossing of two cultivars of Cotton was very much influenced by growing barriers of Maize (*Zea*), c.2–3 m high, between the rows of plants (Pope, Simpson & Duncan, 1944). In the agricultural landscape the presence of hedgerows and plantations has been known to act as a barrier to crossing (Jensen & Bogh, 1941; Jones & Brooks, 1952).

Gene flow: historic insights from the movement of pollen

Govindaraju (1988) reviewed the evidence then available on pollen movement in 115 species. On average, gene flow in wind-pollinated species was greater than that of those pollinated by animals.

Wind pollination By sampling the pollen content of the air, using sticky slides or pollen traps, it was discovered that from an isolated source, pollen is distributed in a leptokurtic fashion, much pollen being detected relatively close to the parent plant, with only a small portion travelling some distance from the parent. In interpreting these various studies it was stressed that many factors influenced dispersal,

Table 10.1. **Isolation requirements for seed crops (After Kernick, 1961; from Levin & Kerster, 1974)**

Species	Breeding system[a]	Pollination agent[b]	Isolation requirement (m)
Gossypium spp.	S	I	200
Linum usitatissimum	S	I	100–300
Camellia sinensis	S	I	800–3000
Lactuca sativa	S	I	30–60
Avena sativa	S	W	180
Hordeum vulgare	S	W	180
Oryza sativa	S	W	15–30
Sorghum vulgare	S	W	190–270
Triticum aestivum	S	W	1.5–3.0
Cajanus cajan	SC	I	180–360
Citrullus vulgaris	SC	I	900
Apium graveolens	SC	I	1100
Carthamus tinctorius	SC	I	180–270
Papaver somniferum	SC	I	360
Vicia faba	SC	I	90–180
Pastinaca sativa	SC	I	500
Voandzeia subterránea	SC	I	180–360
Nicotiana tabacum	SC	I	400
Coffea arábica	SC	IW	500
Hevea brasiliensis	C	I	2000
Helianthus annuus	C	I	800
Brassica campestris	C	I	900
Daucus carota	C	I	900
Lycopersicum esculentum	C	I	30–60
Anethum graveolens	C	I	300
Brassica oleracea	C	I	600
Allium cepa	C	I	900
Raphanus sativus	C	I	270–300
Brassica oleracea	C	I	970
Carica papaya	C	I	100–1600
Zea mays	C	W	180
Chenopodium ambrosoides	C	W	180–360
Cannabis sativa	C	W	500
Secale cereale	C	W	180
Beta vulgaris	C	IW	3200

[a] S, species which are predominantly self-fertilising; SC, species in which self- and cross-fertilisation are of similar importance; C, species in which there is predominant or exclusive cross-fertilisation.

[b] I, insect pollination; W, wind pollination.

including wind speed and direction, as well as the effect of height presentation of pollen, pollen grain sizes and the influence of environmental heterogeneity in biotic and topographical factors.
It was also important to study the actual pollination mechanism in the field. *Plantago lanceolata*, a plant thought on grounds of morphology to be wholly wind-pollinated, is in fact visited by insect pollinators (Stelleman, 1978).

Animal pollination While many plants are apparently well adapted for insect (or bird/bat) pollination, it does not follow that those plants are all effectively pollinated. Competition for pollinators occurs (Dafni, 1992), or there may be too few plants in the population for all to be effectively pollinated (Jennersten, 1988). Also, entomologists were increasingly inclined to study insect behaviour in terms of cost–benefit analysis. Foraging for food (pollen or nectar) is an energy-consuming activity. 'Optimal foraging' may be undertaken by insects: they visit the flowers offering the best return for the energy expended in searching. The flowers of a particular species under study must now be viewed as an array of 'floral offerings' presented to a wide variety of insects, and many factors must be taken into account, such as the degree of faithfulness of an insect to particular plant species, as well as patch size, spatial pattern and density effects, heterogeneity and seasonal changes. Given such complexities, the only course of action open to the student of evolution is to study populations individually. Insects and birds must be observed to see which plants are actually visited. So that patterns of dispersal may be discovered, pollen was stained with dyes (Simpson, 1954; Sindu & Singh 1961), or made radioactive (Schlising & Turpin, 1971), and its physical movement followed by setting pollen traps (Greenwood, 1986; Caron & Leblanc, 1992).
The movement of pollinators has been closely monitored (Levin & Kerster, 1974; Walther-Hellwig & Frankl, 2003). Sometimes intraspecific differences in size of pollen grains have enabled observations of gene flow to be undertaken (e.g. Richards & Ibrahim, 1978). Dafni (1992) provides a critical review of these early methods available for studying pollen movement. Clearly, trapping may collect pollen from several sources. These methods provide only indirect evidence of pollen flow. As we have seen, what is important in microevolution is 'actual patterns of fertilization and gene flow' (Ashley, 2010).
In summary, as with wind dispersal of pollen, leptokurtic distribution of pollen is likely as a consequence of the activities of pollen vectors.

Gene flow: historic studies of seed dispersal

Wind dispersal. Early investigations of the distribution of propagules was studied by catching seeds or fruits on sticky tapes or traps or by looking for seedlings around individual plants. Ashley (2010) reviews the expanding range of ingenious methods of marking/tagging seeds, e.g. attaching threads (Wenny, 2000; Jansen, Bongers & Hemerick, 2004); wire tin tags (Li & Zhang, 2003, 2007); magnets (Sork, 1984); spraying seeds with radioactive tracers (Carlo *et al.*, 2009); or marking them with fluorescent dyes and looking for these in faecal samples of birds (Levey & Sargent, 2000). Sometimes the distribution of marker genes from a carrier parent has been examined (see, for example, Bannister, 1965, for a study of the distribution of various markers in plants of progeny found in a study of *Pinus radiata*). On the basis of these studies, dispersal of medium to large seeds probably conforms to a leptokurtic distribution, though the 'tail' may be very long. A major problem in these investigations is that typically there is very low recovery of marked seeds, and the means of study, tagging etc., may influence the dispersal patterns (Xiao, Jansen & Zhang, 2006).

In reviewing historic studies and risking a generalisation, Levin & Kerster (1974) consider that even in species with fruits and seeds apparently well adapted with wings or plumes, most fruits or seeds travel relatively short distances from the parent plant.

However, they encouraged the experimental approach to study what happened in the field, to displace the simple notion that, if a plant has a plumed fruit or seed, *ipso facto* its progeny must be widely scattered over large areas at each generation. They emphasised that detailed studies were essential to determine what happens in particular species. For instance, can it simply be assumed that all the seeds produced by a plant behave in the same way? Cheplick & Quinn (1982, 1983) have discovered a very interesting reproductive strategy in the grass *Amphicarpum purshii*. Plants produce not only normal spikelets of 'far dispersed' aerial fruits, but also larger and heavier subterranean 'near dispersed' fruits. In many Asteraceae (Compositae), dimorphic or polymorphic achenes are found. For example, in *Heterotheca latifolia* the ray florets produce a thin-walled short-lived fruit with a pappus, facilitating dispersal by the wind. In contrast, the ray florets give rise to thick, fibrous achenes with no pappus, and these achenes form a seed bank in the soil (Venable & Levin, 1985). These multiple dispersal strategies have been considered by Schoen & Lloyd (1984) and Cox (1988). Seed and germination heteromorphism have been reviewed by Silvertown (1984) and by Olivieri (2001).

Animal dispersal Reviewing early studies it had to be concluded that very little is known about the primary distribution of propagules by animals (Harper, 1977). Cousens & Mortimer (1995) provide some excellent examples of animal dispersal of weeds, both by adhesion to fur and by the ingestion of seeds, subsequently defecated in a viable condition. However, these studies indicated that much remained to be discovered about primary dispersal. Also, secondary dispersal may complicate seed flow in some species. For instance, in some species of the genus *Viola* primary dispersal is by means of explosive release and wind, but each seed has a protein-rich elaiosome on its surface. This is a food source for ants, which carry seeds back to their nests, giving a different ultimate pattern of seed dispersal (Beattie, 1978; Huxley, 1991; Huxley & Cutler, 1991).

'Neighbourhoods' in wild populations

To determine the limits of Mendelian populations it is necessary to have accurate knowledge of gene flow. Given the complexities of populations in the 'wild', biologists have had to settle for *estimating* the size of 'breeding groups'. To make a judgement of the number of individuals randomly mating and the area they occupy Wright (1943, 1946) devised a 'neighbourhood' model. 'A neighbourhood is defined as an area from which about 86% of the parents of some central individual may be treated as if drawn at random' (Levin, 1988). Discussion of the detailed assumptions, background and equations used is beyond the scope of this book, but full details are presented in Levin & Kerster (1974) and Crawford (1984). It is clear that Wright's equations, first formulated for bisexual mobile animal populations, are not ideally suited to the study of sessile and often hermaphrodite and clonal plants (Crawford, 1984). However, neighbourhood models have influenced botanical studies, as attempts have been made, by increasingly sophisticated means, to investigate population structure in relation to gene flow.

Employing historic information, several difficulties were encountered in the calculation of neighbourhoods in plants.

a. Pollinators. In estimates, based upon pollinator flight distances, it is unclear whether pollen carry-over occurs beyond to a second flower, and to a third, etc. It is also difficult to take full account of insect behaviour, as there may or may not be some directionality in their flight paths.
b. Seed dispersal. It often proves difficult to study seed dispersal in the wild, and estimates involving spaced plants have sometimes been used (see, for example, Gliddon & Saleem (1985) who studied *Trifolium repens*).
c. Dispersal distances. In the determination of both pollen and seed dispersal it is impossible to make

Table 10.2. **Neighbourhoods in various herbs. (Published estimates, details in Levin, 1988)**

Species	N_a (m^2)	N_a diameter (m)	N_e
Phlox pilosa	41	3.61	533
Liatris cylindracea	66	4.58	633
Liatris aspera	35	3.34	175
Viola pedata	48	3.91	432
Viola rostrata‡	25	2.82	167
Primula veris	30	3.09	7.4
Plantago lanceolata	10	1.78	17
Avena barbata	4	1.13	140

‡ Estimates do not take into account selfing by cleistogamic flowers.

N_a = Neighbourhood Area

N_e = Neighbourhood Size

an accurate assessment of the 'tail' of the distribution in natural vegetation.

d. Choice of study area. Technical difficulties of watching pollinators and studying seed dispersal were likely to have encouraged the experimenter to choose a relatively 'simple' area with a low number of plants within an open area. It is unclear whether the results in Table 10.2 are typical of the species in question. Clearly, the neighbourhood size is greatly influenced by the density of the species being studied and the height of the vegetation in which it is growing.

e. Estimates of neighbourhoods do not represent constants for particular populations. Year-by-year differences in genetic neighbourhoods are likely in pollinator activity etc. Populations of plants with overlapping generations also present a complicating factor. For instance, biennial plants often have first-year non-flowering rosette individuals along with flowering and fruiting second-year plants. At first sight it would appear that there are two separate gene pools in biennial plants. However, studies (e.g. of *Senecio jacobaea*: see Harper, 1977) have shown that second-year plants, prevented from flowering by insect or other damage, may flower in their third or later years.

Historically, there are many problems in estimating neighbourhoods. Those who have studied the subject in detail are most aware of the difficulties. Thus, Levin (1988) states 'the neighbourhood values that I and others have calculated are only rough approximations'.

Summarising these early studies undertaken from the 1970s onwards, the evidence from the few species studied in detail suggests that the dispersal of fertilising pollen and seeds *might* be restricted; that neighbouring plants are likely to mate together, and that adjacent plants in a population are often likely to be genetically related. How far have these historic ideas been confirmed by recent studies employing molecular markers?

Gene flow: studies using molecular tools

Clearly, historic investigations of gene flow in wild populations had obvious limitations. The use of molecular tools is revolutionising our understanding of plant gene flow. For an individual or group of plants, the aim is to determine the exact sources of pollen and seeds, and investigate the dispersal curves associated with one or more vectors contributing to gene flow (Dennis, Green & Schupp, 2007). In these challenging studies that estimate so-called 'dispersal kernels', statistical approaches are often employed, for example maximum-likelihood procedures (Rubledo-Arununcio & Garcia, 2007; Ree & Smith, 2008).

The earliest 'molecular' investigations of gene flow examined allozyme markers. If there are some unique alleles in a population, then gene flow between it and other populations can be estimated (for details, see Hamrick, 1990). For instance, Muller (1977) examined the distribution of a unique allele (of the enzyme leucine aminopeptidase: LAP) in a population of *Pinus sylvestris* and discovered the allele in progenies of trees 80 m from the source. Researchers have also studied gene flow in artificial populations using allozymes (Smyth & Hamrick, 1987). For example, Schaal (1980) used such a marker to study the movement of pollen in populations of *Lupinus texensis*. She found that estimates of gene flow obtained from the distribution of allozymes in progenies was greater than that obtained by the study of pollen flow.

Polymorphic allozyme markers have also been used in studies of paternity. In a pioneering study of the dioecious species *Chamaelirium luteum* in a forest in North Carolina, the genotypes of many plants were established using 11 electrophoretic markers (Meacher & Thompson, 1987). From this detailed knowledge it was possible to determine the realised gene flow patterns and 'most-likely' pollen parent of progenies of various plants (Fig. 10.2).

Studies of paternity have also been made in populations of Wild Radish (*Raphanus sativus*) (Ellstrand & Marshall, 1985b). Typically, one to four (mode two) pollen parents have fertilised the different ovules on each maternal parent.

Studies using polymorphic allozymes led to higher estimates of gene flow than those discovered by studying the movement of pollen and seeds. For example, Ellstrand & Marshall (1985a) concluded that 8–18% of the gene flow in *Raphanus sativus* may occur between populations separated by as much as a kilometre. Furthermore, Friedman & Adams (1985) discovered that 40% of the pollen fertilising an orchard of *Pinus taeda* came from at least 400 m away.

Studies of isozymes have also provided important new information on long-distance dispersal to isolated oceanic islands. For example, the fern *Asplenium adiantum-nigrum* is the allotetraploid derivative of two strictly European diploid species. The tetraploid occurs on the extremely isolated islands of the Hawaiian archipelago. Samples of the species were collected and variability in isozymes studied (Ranker, Floyd & Trapp, 1994). There is evidence that the species is highly inbred and that different collections were genetically different. The pattern of difference could not simply be accounted for by mutation. Since the parental diploids are not known from the islands, the differences between samples are likely to be the result of multiple colonisation events (a minimum of three and a possible maximum of 17) by natural long-distance dispersal of the dust-like spores of this fern.

It is of course important to recognise that some multiple introductions have resulted from human activities. For example, the medicinal and ornamental plant *Bryonia alba* was introduced to the western United States from Europe in the nineteenth century. Studies of the allozymes suggest that there were two or possibly three separate introductions (Novak & Mack, 1995). Other examples of multiple introductions will be examined below.

Gene flow: insights from the use of microsatellite markers

There are limitations of the use of allozyme markers for studies of gene flow, as they 'generally lack the levels of variability required for categorical assignment of individual parents for most seeds sampled in a study' (Chakraborty, Meagher & Smouse, 1988). In contrast, microsatellite markers (also known as SSRs) are widely distributed in plant genomes,

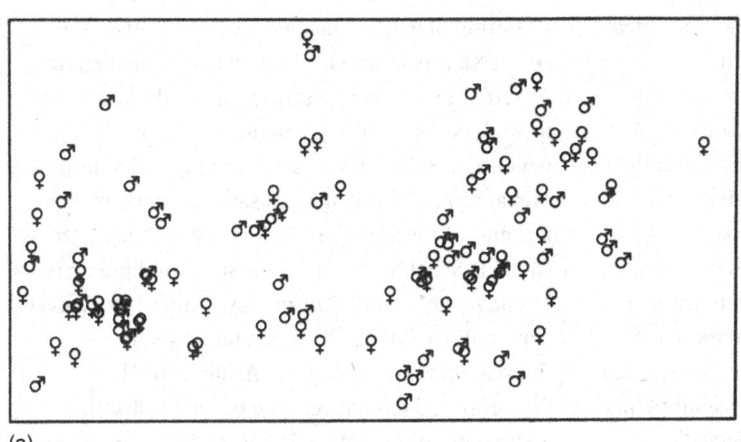
(a)

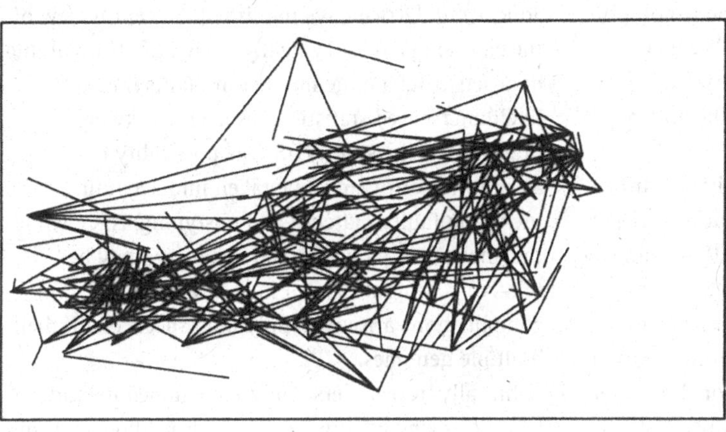
(b)

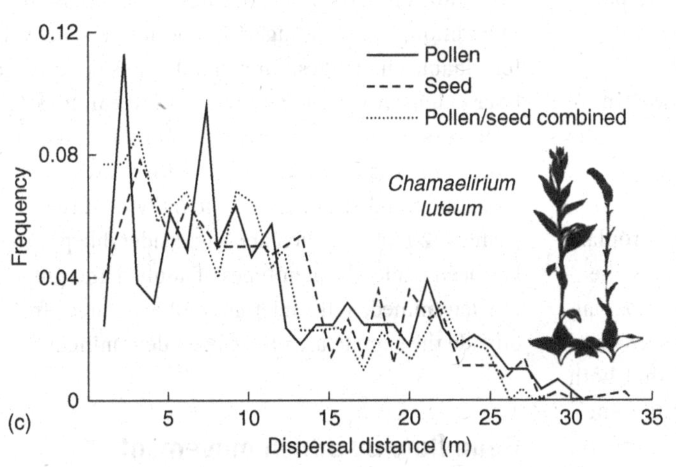

(c)

Fig. 10.2. Map of a population of *Chamaelirium luteum* showing (a) the distribution of males and females; (b) lines connecting females with their mates in 381 natural crosses; and (c) the dispersal of pollen and seed within the population. (Meagher, 1986, and Meagher & Thompson, 1987, from Silvertown & Lovett Doust, 1993.)

offering valuable neutral genetic markers for critical study of gene flow and mating behaviour.

What follows draws on the excellent account by Ashley (2010). Typically, these markers are of 2–6 base pair sequences repeated 12 to many times: their variability arises from mutations involving the number of repeats. These markers allow the identification of pollen source and, in some cases, permit the identification of seed parents in trees or shrub species. By studying the markers in the leaves or cambium of the bark it is possible to characterise, for 5–10 microsatellite loci, the genotype of adult plants. By investigating seed collected from selected individuals, it is possible to determine the genotypes for these loci at each of these microsatellite loci, bearing in mind that a seed would have two alleles, one from the seed parent and one from the pollen parent.

Before we consider the results it is important to examine some requirements and limitations of the study of microsatellites. Ashley notes that many species have hermaphrodite flowers and it is 'generally impossible to determine whether an assigned parent is the seed parent or pollen parent when using nuclear microsatellites'. For dioecious species, assignments are possible 'because the search will be for a compatible male/female pair', and maternal and paternal origins can be determined using uniparentally inherited markers – mtDNA and cpDNA. For example, in a study of gene flow in studying populations of the dioecious species *Silene latifolia* (*S. alba*), McCauley (1994) writes: 'With maternal inheritance the genetic structure of the chloroplast DNA should reflect seed movement, whereas the genetic structure of the nuclear-encoded allozyme loci should reflect the movement of both seeds and pollen'. The results of the study revealed that both seeds and pollen contribute significantly to gene flow.

As part of investigations, various maternally inherited seed tissues have been studied (technically endocarp, pericarp or tissue of the seed wings in flowering plants), or megagametophytic tissue (in conifers). It is important to note that while certain maternal tissues are present in the seed, they are lost to study as the seedling emerges and grows. It is beyond the scope of this account to give details of the preparation and use of microsatellite markers. For an account of the selective cloning involved etc., see Glenn & Schable (2005). As the study of these markers has advanced, scoring of microsatellites has evolved from radioactive labelling, to staining and fluorescence techniques (see Ashley, 2010). Concerning the interpretation of the results, the assignment of parentage is not straightforward and Jones *et al.* (2010) have published a recent review of the emerging issues. Typically, assignments involving data sets from a large number of plants now use likelihood-based statistical tests with software programs such as CERVUS. The possibility of assignment errors must be taken into account, including faulty matches and mistyping. Also, while the analysis of paternity and gene flow may be relatively straightforward in diploid species, complications arise with polyploid species with their multiple genomes.

Initially, researchers might have hoped to study situations where there are hundreds or thousands of individuals in large tracts of forest, grasslands etc. In addition, investigating some species presents very formidable challenges, for example, those plants that have extensive clonal growth. So far, researchers have concentrated on comparatively simple situations in nature, such as forest fragments with a low to moderate number of identifiable individual trees (Ashley, 2010). For such areas, considerable progress has been made: the genotypes of adult(s) and progeny are determined, and sometimes pollen sources from outside the sample area have been determined.

Gene flow by pollen movement

In a literature search, Ashley (2010) discovered 53 papers on microsatellite parentage studies. This is a new field of research: only 4 papers were published in the 1990s. Many investigations have tracked pollen movements in wind-pollinated trees and shrubs in the genera *Fagus, Fraxinus, Juglans, Pinus, Populus, Quercus* etc. Early results from the study of *Quercus* species confirmed the leptokurtic distribution of pollen from a source, but pollination was not just from near neighbours with low pollen immigration rates. Indeed, it was discovered that often more than half the pollinations from selected adults were from sources outside the stand. Furthermore, as more results from wind-pollinated species emerged, the level of gene flow suggested that wind pollination in trees in many cases is effectively panmictic [i.e. mating at random] not only over large spatial scales, but even for small fragments and remnant populations. However, there were exceptions in the case of extremely isolated stands of trees. But even here there was some immigrant pollen. For example, in a study of *Pinus sylvestris* in central Spain, evidence suggested that 4.3% of the immigrant pollen had travelled at least 30 km from the nearest conspecific source (Robledo-Aruncio & Gil, 2005). Thus, considering this and other examples cited by Ashley, geographically isolated groups of plants are not necessarily reproductively isolated. Also, there was no strong evidence of directionality in pollen dispersed by wind. Despite these advances in our understanding, Ashley concludes that the many studies of gene flow involving microsatellites have not been helpful in determining the length and size of the tail of the leptokurtic distribution of pollen from a source. In the words of Ashley, it may be a 'fat tail' rather than a 'thin tail'.

Animal-pollinated trees and shrubs

Various species have been examined using microsatellites, namely insect-pollinated trees of the genera *Aesculus, Dipterocarpus, Eucalyptus, Ficus, Magnolia, Malus, Prunus, Sorbus, Swietenia* etc.; bird-pollinated species of the genera *Calothamnus* and *Symponia*; bat-pollinated members of the Hymenaea; bee-pollinated *Rhododendron*; and the cactus *Polaskia* (pollinated both by insects and birds).

Surprisingly high pollination distances were detected in some cases, for instance, in a study of the tree *Dinizia excelsa* in the Amazon (Dick, 2001). Janzen (2001) has postulated that isolated trees remaining in an agricultural landscape after forest clearance were effectively the 'living dead', having little or no reproductive potential and, therefore, of questionable conservation value. However, in testing these ideas, Dick discovered that even isolated trees 'received pollen, from multiple pollen donors covering distances of hundreds of meters, thanks to exotic African honey bees that were more common pollinators than native bees in disturbed habitats' (Ashley, 2010). Similar results were obtained for fragmentary populations of *Swietenia humilis* in Honduras (White, Boshier & Powell, 2002). Investigations of the bird-pollinated shrub *Calothamnus quadrifidus* found that, in fragmentary populations in the agricultural landscapes of Australia, 43% of pollinations were from distant populations up to 5 km away. Studies of the African fig tree *Ficus sycomorus* in Namibia revealed mean pollination distance was 88 km, with an extreme value of 160 km (Ahmed *et al.*, 2009). Ashley notes that this species is pollinated by a 'small host-specific fig wasp . . . that lives for only 48 hours and must be carried on the wind. Unlike the findings for wind-pollinated trees, *F. sycomorus* shows a marked east-to-west directionality of pollination flow matching the predominantly easterly winds.'

Studies have also revealed that even in the most extremely isolated populations of *Prunus mahaleb* (Hoobee *et al.*, 2007) and *Sorbus torminalis* (Hoebee *et al.*, 2007) there were still low levels of pollination from distant stands. Therefore, in the absence of proper testing, it should not be assumed that lone individuals and extremely small stands are always effectively reproductively isolated.

Seed/fruit dispersal

As seed/fruit dispersal, germination and seedling establishment are all complex processes, it is not surprising that there have been fewer attempts to track gene flow by seed dispersal in wild plants using microsatellites. Gene flow and establishment by seeds/fruits must be seen as a process with several potential steps in a seed dispersal cycle, involving primary and secondary dispersal. There may be several primary seed dispersers to consider, e.g. birds and rodents, and sometimes two or more sequential phases in dispersal. For instance, viable seed may be distributed in dung (Schupp, Jordano & Gomaz, 2010). Also, seed and seedling predation, and seedling competition must be considered. As a consequence, there is a 'jumbling [of] the match between initial seed deposition and the resulting distribution of seedling recruits' to a population (Ashley, 2010). To undertake complete investigations of gene flow, by seed dispersal, is a formidable challenge indeed (Beckman & Rogers, 2013).

Considering population structure, some studies have discovered low-level seed immigration into study populations (Ashley, 2010). But in others, such as *Fraxinus excelsior*, with its impressive winged wind-dispersed seeds, high numbers of immigrant seeds – *c*.50% – were found in a study area in Scotland (Bacles *et al.*, 2006). Moreover, investigations in Panama by Hardesty *et al.* (2006), of *Simarouba amara* (an insect-pollinated tree species with vertebrate dispersed fruits), revealed that 'germinated seedlings were seldom the offspring of the nearest or nearby reproductive adults' (Ashley, 2010).

As we have seen above, seed dispersal is particularly complex where there are a number of different groups of animal dispersers. For example, there is evidence that the seed dispersal kernel in *Prunus mahaleb* has at least two components: a short-distance component (small herbivores) and a long-distance element (birds and carnivorous mammals).

Recent insights into gene flow from the study of transgenic crop plants

Transgenic crop plants are now widely grown in North America and elsewhere and the possibility of gene flow has been examined in several situations: (1) Crossing between GM and non-GM variants of the same crop. (2) Hybridisation between GM cultivars and wild and weedy taxa. (3) The escape of transgenic plants as seed into non-agricultural habitats.

Transgenes involved in possible gene flow may be identified by numerous methods (Chandler & Dunwell, 2008). For instance, gene flow may be tracked if the inserted transgenic construct contains genetic information for the production of a fluorescent protein. In some cases, herbicide resistance has been introduced in transgenic constructs, and the presence of resistance genes can be confirmed by painting leaves or testing adults/seedlings with the appropriate herbicide. Other methods involve the identification of the construct using PCR-based tests.

Gene flow involving transgenic plants may be examined in deliberately designed trials and experiments involving blocks of GM plants with non-GM plots alongside and at different distances. In other investigations, conspecific 'sentinel' or 'bait' individuals have been grown various distances from the GM crop. Or experimenters have investigated, opportunistically, situations where GM individuals interact with non-GM plants in agricultural, urban and also semi-natural ecosystems.

Chandler & Dunwell (2008) have reviewed the literature on pollen dispersal distances for various crop plants. While Barley, Wheat, Tomato, Soybean, Common Bean and Rice have relatively small pollen dispersal distances (of the order of 12 m or less) for others pollen dispersal distances are greater, e.g. *Brassica napus* (up to 500 m). For the turf grass *Agrostis stolonifera* dispersal distances were primarily under 2 km, but the GM construct was detected in sentinel plants 20 km distant. And in studies of Sugar

Beet, pollen dispersal was identified up to 300 m from source using 'bait plants'.

As a result of these investigations the isolation distances recommended on the basis of historic studies (Table 10.1) have been confirmed or modified for different crops. For instance, for *Brassica napus* an isolation distance of 150 m is now recommended, while for Maize, 750 m is the necessary distance between stocks for complete isolation. But a lesser distance is required to meet some less stringent statutory requirements of the European Union and other regulatory authorities. To minimise, or largely prevent, pollen transfer, it is also recommended that buffer zones be established around GM crops, especially where they may come into contact with non-GM conspecifics. What constitutes a safe and effective isolation distance is a crucial question for many agriculturalists, especially those propagating stocks for the 'seed trade'. It is also a very live issue for farmers growing 'organic' non-GM crops. For their crops to meet 'organic' standards, there should be extremely low levels of, or no, contamination from GM cultivars.

There have been many investigations of gene flow by dispersal of GM seed. GM grasses are being developed for golf courses and other uses (Ge *et al.*, 2007). For instance, GM glyphosate-tolerant Creeping Bent Grass (*Agrostis stolonifera*) was planted out in 2002 in a designated control area of 162 ha in Oregon, USA (Reichman *et al.*, 2006; Snow, 2012). 'Despite efforts to restrict gene flow, wind-dispersed pollen carried transgenes to local *A. stolonifera* and *A. gigantea* as far as 14 km away, and to sentinel plants placed as far as 21 km away' (Watrud *et al.*, 2004). The attempt to kill all the escapees failed and in another sampling of the area, 62% of 585 plants were found to have the transgene (Snow, 2012). Moreover, Zapiola & Mallory-Smith (2012) also detected an intergeneric hybrid between *A. stolonifera* (female parent) and Rabbitfoot grass (*Polypogon monspeliensis* as pollen parent). It is not clear how long these GM plants might survive 'in the wild', but the findings of Zapiola *et al.* (2008) 'highlight the potential for transgenic escape and gene flow at a landscape level'.

The case of GM oilseed rape/canola provides another excellent example. Studies have revealed that GM seed may be dispersed by the movement of machinery in agricultural landscapes (Claessen *et al.*, 2005a, b; Warwick, Beckie & Hall, 2009), leading to the development of GM canola seed banks in soils, where wheat/rape are grown in rotation (Gruber & Claupein, 2007), and the establishment of GM volunteers Rape plants in fields and along roadsides (Garnier *et al.*, 2008; Pivard, *et al.*, 2008; Knispel & McLachlan, 2010).

A survey of 16 populations of 'escaped Canola populations' in southern Manitoba from 2004 to 2006 reveals another aspect of the consequences of gene flow from GM herbicide-resistant (HR) crops (Knispel *et al.*, 2008). 'Glyphosate resistance was found in 14 (88%) of these populations, glufosinate resistance in 13 (81%) and imidazolinone resistance in 5 (62%) of the tested populations.' Furthermore, gene flow amongst different HR canola varieties has led to the development of multiple HR Canola plants, which were detected at 10 sites (62%). Also, it is important to recognise that very long-distance dispersal of GM seed takes place as seed is exported for food production (von der Lippe & Kowarik, 2007; Bailleul *et al.*, 2012). Investigations in Japan have revealed that plants of Canola have grown from spillage of HR GM seed (both glyphosate- and glufosinate-tolerant) imported for food production from as far away as Canada. Ferals have been found not only around Japanese harbours, but also near processing plants (Kawata, Murakami & Ishikawa, 2009). Some of these feral GM plants have now established, as perennials, in some roadside habitats in Japan where they reproduce freely.

Considering the question of long-distance dispersal of GM crops, this has occurred not only through the adventitious presence of GM seed in supposedly pure seed stocks, but also through unregulated trade and illegal activities in a number of countries (Ho, Zhao & Xue, 2009; Warwick, Beckie & Hall, 2009). Smuggling of GMs has also been reported (Gealy *et al.*, 2007).

There has also been a concern as to whether transgenes might spread into wild crops. Gilbert (2013) considers the situation in Oaxaca, Mexico, where Quist & Chapela (2001) discovered that 'locally produced Maize contained a segment of the DNA' present in the transgenic 'glyphosate-tolerant and insect-resistant Maize'. This finding caused an outcry 'for contaminating Maize at its historic origin – a place where the crop was considered sacred'. But this paper was criticised by some for 'technical' deficiencies and withdrawn. Later research did not detect the GM sequences (Ortiz-Garcia *et al.*, 2005). However, in samples taken from the Oaxaca area, Piñeyro-Nelson *et al.* (2009) located the same transgenes as Quist. Other studies have also detected the transgene sequences across Mexico (Dyer *et al.*, 2009) and within local communities (Mercer & Wainwright, 2008). Snow *et al.* (2003) conclude that 'it seems inevitable that there will be movement of transgenes into local Maize crops . . . but it is difficult to say how common it is or what are the consequences'. How far transgenes in crop plants will 'escape', through gene flow, into non-GM crop variants and wild relatives, and the consequences of such flow are much debated and more research is needed (for reviews of the evidence see Warwick, Beckie & Hall, 2009; Nicolia *et al.*, 2013).

Other concerns have also surfaced. It is clear that plants can be engineered to produce pharmaceutical and industrial chemicals (Davies, 2010). The natural choice for such transgenic stocks would be to employ cultivars of crop plants used for food for humans and domesticated animals, as genetic transformation of these species is well studied. However, even if containment and separate processing were to be organised, there is a risk of accidental or deliberate cross-contamination between food plants and those that produce organic chemicals. (For a discussion of these risks and their possible resolution see Chandler & Dunwell, 2008.)

Genetic modification of other groups of plants is in progress, including ornamental plants, forest and plantation trees, sports turf and pasture grasses and biofuel species. Many biologists are considering the impact of growing GMs of cultivated, wild and weedy species. Planting of GM stocks in the wild is also being contemplated. For instance, populations of American Chestnut (*Castanea dentata*) were almost completely eliminated from North America forest ecosystems when the fungus *Cryphonectria parasitica* was accidentally introduced on imported chestnuts (perhaps from Japan) in the early 1900s. Reintroduction of GM-disease resistant plants of this heritage species into forests is being considered (see Chandler & Dunwell, 2008, and the website of the American Chestnut Restoration Project: www.fs.fed .us/r8/chestnut/qa.php).

Knowledge of gene flow is critical in plant conservation

Because of habitat loss and other factors, many plant species are threatened with extinction in human-influenced ecosystems. The remaining populations are often small, and it is supposed that gene flow between the extant fragmentary populations is limited or impossible. Two examples illustrate conservationists' attempt to estimate gene flow, to provide baseline information on which to devise effective conservation strategies. As we have seen above, there are problems in applying Wright's neighbourhood model in study of gene flow in plants (see Epperson, 2007 for critical theoretical analysis). However, in the absence of a better model, some botanists, including those providing the two case histories, continue to employ the neighbourhood model in their research.

- Star Cactus (*Astrophytum asterias*) is a self-incompatible federally listed endangered species in southern Texas and northern Mexico that is highly restricted in its distribution (Blair & Williamson, 2010). Seeds fall beneath the parent plants, and, therefore, the principal potential means of gene flow is the movement of pollen by pollinating bees.

Gene flow by pollen was studied in a 1.9 ha patch by adding fluorescent powder to a source plant and detecting the presence of the dye in the flowers within the study area. About 80% of all recipient plants were located within 30 m of the source plant, with 'the longest dispersal event' being 142 m. Neighbourhoods were estimated to contain 42 individuals with a neighbourhood area of 0.094 ha. This result indicates the potential for population subdivision within the larger patch, as a consequence of 'restricted pollen dispersal'.

- The critically endangered tropical New Caledonian conifer *Araucaria nemorosa* has been studied using microsatellite techniques, to estimate 'a maximal dispersal envelope around the extant populations' (Kettle *et al.*, 2012). Estimates of Wright's genetic neighbourhood ranged from 22 to 876 trees, with seed dispersal in the range 11–84 m. Seed dispersal of the order of <100 m is judged to be too small 'to allow *Araucaria nemorosa* to disperse to new more hospitable sites within an ecological relevant time scale'. Such information provides important information on which to base decisions about conservation management and species restoration.

Gene flow: future directions of research

Reflecting on recent investigations with microsatellites, Ashley (2010) notes that 'the major surprise is that distance explains only a portion of the variation in mating patterns, often a relatively modest portion, and this holds for both wind- and animal-pollinated plants. The traditional view of pollination biology, and the assumption of all plant dispersal models, is that distance is paramount; it is usually the only parameter considered.' In her review she suggests 'that models of seed and pollen dispersal based simply on distance will provide poor predictions of plant gene flow and dispersal patterns for many plants. While parentage studies have certainly made the dispersal scene more complex, the emerging paradigm of dispersal is much richer and multi-faceted.'

Our views on population structures, sub-structuring and functioning must take account of these new findings. For example, if in some cases gene flow is much more extensive and complex than previously thought, estimating the size and number of individuals in genetic neighbourhoods is likely to be even more problematic, as many of the actual pollen and seed parents of a population may be outside any area that it is practical to choose as a study site.

Attempting to access the direction of future research, Ashley (2010) stresses the need to combine molecular and ecological approaches in the study of gene flow. Thus, further investigations should take account of flowering behaviour, breeding systems, the physics of pollen dispersal in wind columns, the detailed biology of pollen including its maturation (and denaturing) under different conditions, mate choice in plants, and the behaviour of pollinators and seed dispersers, with regard to primary and secondary seed dispersal.

Gene flow is at the heart of models of population functioning, sub-structuring and speciation.
In contemplating current knowledge, evolutionists face a problem common in science. How far do case histories provide a firm basis for generalising about gene flow in plants? Or are such investigations highly method-bound, and only provide information that is site/time/observer/species specific? Given our current knowledge, it is clear that some caution is necessary in interpreting experimental results in this rapidly advancing field, and the reader should be suspicious of sweeping generalisations about gene flow in plants.

11 Populations: origins and extinctions

Populations do not exist in isolation: they owe their origin to pre-existing populations. Thus, populations establish, flourish, languish, decline and revive, and finally, for all manner of reasons, become extinct (Harper, 1977).

Metapopulations

A family of metapopulation models linking all these stages has proved important in the study of animal populations – from kangaroo rats to monk seals, from mountain sheep to butterflies. The concept has also influenced the study of plants.

Models take as their starting point that species occur in patches. At any one time some but not all suitable patches are likely to be occupied. Unoccupied patches are colonised by dispersal of seeds or fruits from an occupied patch. In time, while populations in newly occupied sites may expand, eventually numbers may decline and the population becomes extinct. Each population, therefore, has a finite life. Thus, the success of the species regionally will depend upon there being healthy colonies from which recolonisation can repopulate empty patches, where the species has become extinct. Brussard (1997) describes three different types. (i) 'Classical metapopulations consist of several small extinction-prone local populations connected by a moderate amount of migration'. (ii) Mainland populations or large habitat blocks that supply colonists to small outlying island populations, including islands of land in the sea or lakes; 'island' water bodies scattered across a landscape; and isolated mountain tops in a lowland plain etc. (iii) Metapopulation systems may also consist of populations that 'are declining to extinction because dispersal is too infrequent' or does not occur to facilitate re-establishment.

The metapopulation concept has been employed in interpreting patterns of behaviour in plant populations. For example, *Pedicularis furbishiae* has been studied at sites along the banks of the St John River in Maine, USA. A total of 28 colonies were mapped over a four-year period, and the patterns of colonisation/extinction were consistent with the metapopulation model (Menges, 1990, 1991).

In another case study employing metapopulation models, populations of the short-lived perennial species *Silene latifolia* (*S. alba*) were examined over a number of years in roadside habitats in Virginia, USA (see Altizer, Thrall & Antonovics 1998; Thrall *et al.*, 1998). Overall, 7500 habitat patches were detected, of which some 400–500 were occupied in any one year. The extinction rate of populations was of the order of 14–22% per annum. Extinction risk was increased by the presence of the anther-smut disease *Ustilago violacea* in 16–19% of the populations. Infected plants produce fungal spores in the anthers, rather than pollen, leading to the sterilisation of diseased individuals.

The metapopulation concept is employed in the interpretation of other situations, island populations of *Silene dioica* in Sweden (Giles & Goudet, 1997; Giles, Lundqvist & Goudet, 1998; Ingvarsson & Giles, 1999) and a number of species in the scrub-lands of south-central Florida, USA (Quintana-Ascencio & Menges, 1996).

However, some investigators stress that it is important not to try to 'fit' all populations into the metapopulation concept. For instance, in studies of *Silene tatarica* in Finland, investigations of riverside populations of this endangered species found that recolonisation did not occur in discrete patches, but occurred in areas wherever there was suitable habitat (Jäkälämiemi *et al.*, 2005). Therefore, the process did not conform to the metapopulation model of discrete patches linked in colonisation/extinction/ recolonisation cycles.

There is a lively debate amongst plant population biologists about the metapopulation concept (see Freckleton & Watkinson, 2003; Ouborg & Eriksson, 2004; Honnay *et al.* 2005). Given a long-enough timescale do all plant populations behave as metapopulations? As some species have seed banks in soils and/or persistent underground structures, can the researcher be sure that a population is extinct? Furthermore, as many natural and human influences impinge on habitats, can it be assumed that once-occupied patches remain suitable habitat for a species? Indeed, the notion of habitats being either suitable or unsuitable presents some difficulties in dealing with phenotypically plastic and physiologically adaptive plants (Honnay *et al.*, 2005). And, as many plant species are very long-lived as individuals or as ramets of clones, the postulated extinction–recolonisation cycles may be beyond the period of research projects or indeed human lifetimes.

In what follows, different aspects of population structuring and functioning will be critically discussed in exploring key aspects of colonisation (especially founder effects), population growth, persistence (clonal growth habit and seed banks), and extinction processes. Our understanding of these processes has been greatly increased by investigations of invasive species (Simberloff, 2013) and endangered species.

Founding events and bottleneck effects

A new population may be founded by one or a very few individuals, and such derivative populations will, by chance, have only a random reduced subset of the variation of the larger ancestral population from which it is drawn. Moreover, a new sexually reproducing population may result from the progeny of a single immigrant of a self-compatible species (Baker, 1955, 1967). However, a single individual of a self-incompatible species, introduced to a new area, will be unable to reproduce sexually (see Chapter 7). Cheptou (2012) provides a critical review of current evidence for this simple model, proposed more than 50 years ago, that 'links two traits of importance in ecology, namely dispersal and mating system'. While, at first sight, these ideas seem 'self-evident', the accumulated evidence has 'not established a general pattern of association'. Cheptou considers that Baker's Law 'encompasses a variety of ecological scenarios, which cannot be considered *a priori* as equivalent'.

Turning to another issue, there are genetic implications if existing populations are by chance greatly reduced in numbers and subsequently revive beyond a bottleneck. Dlugosch & Parker (2008a) note that

> rare alleles that persist through a bottleneck have the opportunity to become more common, and in general, large shifts in allele frequencies are predicted. For molecular markers, we expect most of these shifts to have no effect on fitness. For other types of Mendelian traits, however, the evolutionary importance in shifts in allele frequencies and losses of rare alleles are likely to be highly idiosyncratic. While many rare alleles are deleterious, a few particularly those under frequency dependent selection, may have important fitness consequences

(e.g. self-incompatibility alleles: Elam *et al.*, 2007). Thus, in a self-incompatible species, sexual reproduction may be prevented or significantly reduced if 'S' alleles are lost in a bottleneck event.

While founding events may arise from the arrival of one or a very few individual plants/propagules, it is important to recognise that multiple founding events may occur. These have the potential to increase the 'trait diversity in founding populations' and influence the breeding behaviour as different genotypes perhaps from 'differentiated source populations' are brought together (Dlugosch & Parker, 2008a).

Designing experiments

Founder effects are particularly likely in the establishment of new populations following *natural* long-range dispersal to oceanic (Franks, 2010) and other habitats effectively isolated by ecological and geographical factors, such as lakes and isolated mountain peaks. Such effects are important too where *human-mediated* long-distance dispersal of plants/seeds/fruits is deliberately or accidentally transferred to new areas. Molecular investigations provide invaluable evidence for considering hypotheses concerning founder and bottleneck effects/single or multiple introductions etc., especially if these investigations are combined with taxonomic and historical studies. In comparing introduced populations with the 'native' populations from which they might have originated, properly designed sampling strategies are very important.

In considering the evidence for such events, it is important to recognise that there are inherent uncertainties in the reconstruction of past events. Herbarium specimens and plant records are often available, but these do not necessarily indicate the date of first arrival of a species into new territory, or its continuing presence in a new area.

Given the widespread human influences on plants, there are also uncertainties in identifying natural and derivative ranges of species. For many widespread species there is also the issue of how to design an adequate sampling strategy for comparative studies of species that are widespread in both their natural and invasive ranges.

In addition, if populations of introduced plants are being examined some time after their introduction, genetic drift and/or selective forces and human activities may have influenced the pattern of variation. Also, if the population remains small for many generations, mating between close relatives may occur, leading to individual weakness and infertility through the effects of inbreeding (Frankham *et al.*, 2002). Moreover, it may be difficult to distinguish an initial founding effect in the distant past from genetic depauperisation in species that have recently become endangered through human activities (Habel & Zachos, 2012).

Case histories of founder events

Morphological traits Founder effects have been examined in plant populations in cases where selectively neutral floral traits are being considered. For instance, Rafinski (1979) investigated stigma colour polymorphism (white versus orange) in *Crocus scepusiensis* found growing in the Gorce Mountains of Poland. Populations differed widely in frequency of the different morphs, and Rafinski considered that a major determinant was the harvesting of hay. By chance, small isolated populations of *Crocus* arise by founder events from seed falling from the hay as it was being carried along forest trails. In considering this example, it is important to acknowledge that our understanding of floral colour polymorphisms is rarely complete: the assumption that stigma colour is a neutral trait may not be justified. The polymorphism may be adaptively significant in pollination, in ways we do not yet understand.

Allozymes Founder effects have been proposed to account for the lower level of genetic variation in introduced populations. For instance, Schwaegerle & Schaal (1979) studied variation in isozymes in 11 populations of *Sarracenia purpurea* (Pitcher plant) from eastern North America. A population from Ohio, growing on the 17-acre island in an artificial lake, proved particularly interesting, as a population in excess of 100 000 plants was present at the time of sampling. Even though the population was very large, it was genetically depauperate relative to some of the others studied. 'Only one polymorphic locus was found, where the mean number for all populations of *S. purpurea* in this study is 2.5 loci. Similarly, average heterozygosity per individual for all loci at Cranberry Island, 0.042, is 50% below the species mean of 0.089'.

This is most likely to be the result of a 'founder effect', as the population of pitcher plants on the island was established by the planting of a single individual in 1912 by a student of Ohio State University, Freda Detmers. When the genetics of this population was investigated, the island colony had developed through perhaps 8–15 generations from first establishment.

There is also evidence for chance effects in populations of *Sarracenia* introduced from Canada to Ireland in 1906. The initial introduction of seeds and rootstock was made at Termonbarry and from this population several new populations were founded in different parts of central Ireland. By examining 25 enzyme systems, Taggart, McNally & Sharp (1990) studied the variation in the six extant populations, and discovered that the number of polymorphic loci was reduced in the derivative populations, each of which had themselves passed through a severe founder event, only a small number of plants having been transferred to start each new population. However, if the initial populations had been small for a number of years, it is also possible that genetic drift occurred.

Sarracenia purpurea has also been introduced into Switzerland. As a result of detailed genetical studies of two populations using RAPD markers, Parisod, Trippi & Galland (2005) consider that both founder effects and selective forces are also likely to be important determinants of population differences.

Founder effects in weedy and ornamental species

Weedy species are often unwittingly introduced to other parts of the world, and in some cases the populations of newly arrived species lack genetic variation, as measured by genetic markers, and this has been attributed to founder effects, e.g. Australian populations of *Avena barbata* and *Bromus mollis* (Brown & Marshall, 1981), *Chondrilla juncea* (Burdon, Marshall & Groves, 1980), *Echinochloa microstachya* (Barrett & Richardson, 1986) and *Emex spinosa* (Marshall & Weiss, 1982); Jamaican populations of *Eichhornia paniculata* (see Barrett & Shore, 1990), populations of *Striga asiatica* in the United States (Werth, Riopel & Gillespie, 1984) and introductions of *Hypericum canariense* from the Canary Islands to North America and Hawaii (Dlugosch & Parker, 2008b).

A cautionary note is necessary here. The low level of allozyme variation found in many invasive species may reflect variability in 'neutral' markers, and does not necessarily indicate a lack of variability in potentially adaptive characteristics. For instance, five weed species introduced to Canada all had low levels of allozyme variability; however, they exhibited substantial between- and within-population variation in morphology and flowering behaviour in a garden experiment (Warwick, 1990a). In the case of *Sorghum halepense* and *Panicum miliaceum*, genetic exchanges between the weeds and cultivated taxa may have contributed to the variation.

Founder effects may occur as ornamental plants are introduced into gardens from overseas. These species often later escape into semi-natural ecosystems, for instance the pestilential weed Japanese Knotweed (*Fallopia japonica*) is of major importance in the UK. Sixteen populations were examined using RAPD markers: all proved to be identical (Hollingsworth & Bailey, 2000). While only one aspect of the genetic variability of the stocks was examined, it suggests that plants came from a common source.
The complete absence of male plants in the introduced range is also a strong indicator of a founder effect in the introduction of this species, which now propagates itself as a weed by clonal vegetative reproduction. However, recent studies have introduced a new element of complexity to this simple interpretation of the situation in Europe. Bailey, Bímová & Mandák (2009) report that 'a significant proportion of the Japanese Knotweed *s.l.* is not *F. japonica* var. *japonica*, but the hybrid between it and *F. sachalinensis* (another species introduced into Europe). 'This hybrid is able to backcross to either parent with the potential to replace the missing male

F. japonica; by the same process, the hybrid is generating the genetic diversity so conspicuously lacking in *F. japonica*' (for a full discussion of the interactions between introduced *Fallopia* species, their taxonomy, ecology, cytological variability and their control as invasive species in both Europe and North America, see Bailey, Bímová & Mandák, 2009).

Successive founder events The native range of *Rubus alceifolius* stretches from Vietnam to Java. The variability of populations introduced into Indian Ocean islands and Australia has been examined using AFLP markers (Amsellen *et al.*, 2000). Evidence suggests that the plant 'was first introduced into Madagascar, perhaps on multiple occasions', and that the 'Madagascan individuals were the immediate source of plants that colonised other areas on introduction. Successive nested founder effects appear to have resulted in a cumulative reduction in genetic diversity.' While dispersal by birds may have occurred, it seems more likely that human activities were involved in the spread of this weed. Further investigations revealed that in the native range the species reproduces sexually, but in the introduced range there is a switch towards apomixis (Amsellem, Noyer & Hossaert-McKey, 2001; Amsellem, Chevalier & Hossaert-McKey, 2001).

Evidence for multiple introductions In several case histories, the newly introduced populations do not exhibit lower genetic variation relative to the proposed source or ancestral populations. In these cases, introduction to the new territory may have occurred many times, or perhaps a single variable population was somehow introduced. Nineteen case histories are considered in the review by Dlugosch & Parker (2008a) including the following examples.

Introductions from Europe: To Australia: *Echium plantagineum* (Brown & Burdon, 1983). To Canada: *Apera spica-venti* (Warwick, Thompson & Black, 1987). To North America: *Alliaria petiolata* (Durka *et al.*, 2005); *Bromus tectorum* (Novak, Mack & Soltis, 1993; Novak & Mack, 1993); *Phalaris arundinacea* (Lavergen & Molofsky, 2007).

Introductions from the Mediterranean: To California: *Trifolium hirtum* (Jain & Martins, 1979). To North America: *Avena barbata* (Garcia *et al.*, 1989).

Introductions from southern Europe: To British Isles: *Hirschfeldia incana* (Lee *et al.*, 2004).

Introduction from Spain: To Ireland: *Rhododendron ponticum* (Erfmeier & Bruelheide, 2011).

Introductions from North America: To Europe: *Ambrosia artemisiifolia* (Genton *et al.*, 2005).

Tracing the origin and spread of new populations

At first sight it might be thought impossible to determine with any degree of accuracy from which part of the native range of a species any particular newly introduced population might have originated, but some remarkable insights have come through molecular studies.

Identification of sources of introduced taxa

Studies of variation in neutral allozymes in populations of Reed Canarygrass (*Phalaris arundinacea*) in its native and introduced range in North America have provided important information on multiple introductions of the species (Lavergne & Molofsky, 2007; Fig. 11.1).

Turning to further examples, the native ranges of *Bromus tectorum* stretch from Europe through to southwest Asia. The species has been introduced into temperate grasslands worldwide. Studies of allozyme variation have revealed that emigrants have come from at least two sources in Europe. The GOT-4c multilocus genotype, detected in southern Germany and the Czech Republic, has been found in North America, Argentina, Hawaii and New Zealand. The Pgi-2b genotype, found in the native range in

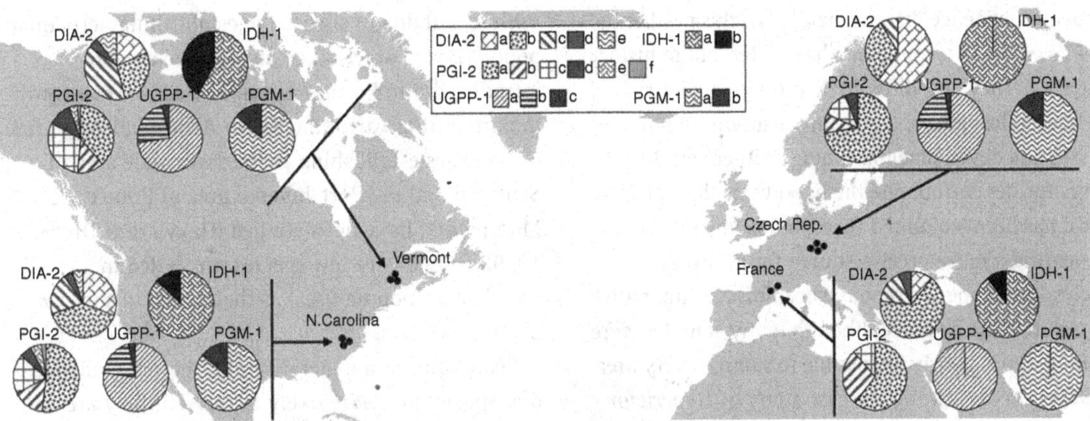

Fig. 11.1. Geographic distribution of neutral genetic diversity of Reed Canarygrass (*Phalaris arundinacea*) for five highly variable allozyme loci. Pie charts display allele frequencies within central v. southern regions of occurrence in the native range (Czech Republic v. France) and invasive range (Vermont vs. North Carolina). Note that the alleles unique to southern France (DIA-2d and IDH-1b) and the Czech Republic (PGI-2d, UGPP-1c, and PGM-1b) co-occur within the invasive regions of the Reed Canarygrass. Lavergne & Molofsky (2007) report that multiple introductions of European strains have occurred in North America since the mid-nineteenth century. This has resulted in the continental-wide genetic diversity of Reed Canarygrass in its European range being redistributed and recombined in the introduced North American populations, yielding a number of novel genotypes with high evolutionary potential. Thus, multiple immigration events may trigger future adaptation and geographic spread of this invasive species by preventing genetic bottlenecks and generating genetic novelties through recombination. (From Lavergne & Molofsky, 2007.) Copyright (2007) National Academy of Sciences, USA.

France, Spain and Morocco, now occurs in Nevada, California, the Canary Islands and Chile. In western North America evidence suggests that there may have been six independent founding events (Novak & Mack, 2001).

Avena barbata is native to the Mediterranean basin, and south-west Asia through to Nepal. It has been introduced very widely into Mediterranean-type climatic zones. Intensive studies of the Californian invasive populations revealed that the genetic diversity in native and introduced ranges was similar, with no severe genetic founding effects. The similarity of the south-west Spanish and Californian populations in allozyme markers indicates the origin of the invasive populations on ships sailing from Spain to the New World (Garcia *et al.*, 1989).

The genetic variability in 26 populations of the introduced *Alliaria petiolata* was examined using microsatellite markers (Durka *et al.*, 2005). Representative samples were collected from Tennessee to Ontario and Wisconsin to New York. To examine variability in the 'native range' of the species, 27 populations were sampled across a wide area: from the British Isles to the Czech Republic and from Sweden to Italy. Overall the introduced North American populations were less diverse than those in the native range, and there was no evidence of bottlenecks. Patterns of microsatellite variation suggested that the British Isles and northern and central Europe were the most likely regions from which the introduced variants came.

More advanced molecular approaches have now been employed in the study of the variability and origins of invasive plants. For instance, Guggisberg, Huang & Rieseberg (2013) have examined European native populations and North American introduced material of *Cirsium arvense* in garden trials and by transcriptome techniques (allowing investigation of loci involved in stress responses etc.). The results of their work are too extensive to discuss here, where we note that they discovered evidence that the material of this pestilential weed from the two continents has 'different life history strategies'. They also differ in 'R-protein mediated defence against pathogens [and] sensitivity to abiotic stresses and developmental timing'. Moreover, their results further corroborate the hypothesis that the New World has been 'colonised twice, independently'.

Locating a more precise source for an invasive species. In the species we have been discussing, native ranges are large and, even though many samples were examined, it is clearly impossible to sample every area. However, in one significant case study of *Hypericum canariense* var. *canariense*, introduced populations have originated from a very restricted native range, and therefore it has been possible to carry out satisfactory replicate sampling of the whole native range and introduced populations. This plant is native and restricted to three of the Canary Islands (Tenerife, Gomera and Hierro). The species has been introduced to California and Hawaii, where it has become invasive. In a study of the genetic variation in native and three introduced populations (Kula on the island of Maui, Hawaii, and Point Loma, San Diego and coastal San Mateo, California), sequence data and AFLP profiles were examined and the source of the introductions was identified as the island of Tenerife (Dlugosch & Parker 2007). 'The genetic similarity of the invasions to one another suggested that they shared a common source (i.e. a single horticultural collection).' Tenerife is only 2034 km^2 in area, providing an unusually precise source region for these invasions'. Dlugosch & Parker (2007) discovered that 'the invasions of *Hypericum canariense* show genome-wide losses of variation ... each invasion has lost ~45% of the expected heterozygosity in populations from Tenerife, indicating a strong bottleneck'.

Once introduced, some species fail to establish

While many introduced species successfully establish and flourish in new territories, others may grow for a while, but do not succeed in establishing permanent populations. For instance, many species were introduced into Europe as viable seeds on raw wool imported from Australia, South Africa and Argentina etc. Species establishing in the Tweedside area of Scotland and the Port Juvenal area of France, near Montpellier, have been studied (Hayward & Druce, 1919). Only a very few species succeeded in establishing populations, as they failed to survive Scottish winters.

Searching for a generalisation, there is evidence that species are more likely to thrive if they are introduced into areas where climatic conditions are similar to those in their native areas. Thus, Mediterranean species have been successfully introduced into areas with a similar climate – Chile, California, South Africa, Australia – where they encounter hot summers and cool wet winters.

However, some species brought to Europe in wool have eventually succeeded in establishing themselves. Ernst (1998b) gives the following details of the perennial introduced species *Senecio inaequidens* in the Netherlands, where there have been at least two independent introductions. The plant arrived 'at Tilburg with sheep's wool from South Africa in 1939, where it failed to establish a permanent population'. However, 'in 1942, a new colony was established at Liege. The species, which is self fertile, has now dispersed widely along railway lines and roads.'

Development of populations: the lag phase

In the development of new populations the number of individuals may increase slowly at first. Crooks & Soule (1999) examine mathematical models of these so-called lag effects. There are many reasons why populations may become extinct or fail to increase at first. All the genotypes necessary for the full functioning of the breeding system may not arrive in the new population. For instance, three flower morphs are found in *Lythrum salicaria* (see Chapter 7), and in newly formed colonies in North America only one or two of the morphs may be present initially, as a result of founder effects (Eckert, Manicacci & Barrett, 1996).

Another example is provided by *Senecio squalidus* introduced into Oxford Botanic Garden about 300 years ago. There was a marked delay in the spread of this species, which eventually escaped onto the railway systems and colonised new areas, especially in sites in London damaged by bombing in the Second World War. We return to this example and its lag phase in a later chapter.

In other cases, the cause(s) of the lag phase are unknown. For instance, in Florida two introduced woody species (*Melaleuca quinquenervia* and *Schinus terebinthifolius*) are colonising the Everglades National Park. It was several years before these species became invasive.

Also, many edaphic, and biotic factors might influence the establishment phase of introduced species (Ewel, 1986; Mack *et al.*, 2000). Some species become invasive as human activities cause major habitat disturbance or provide conduits for dispersal, e.g. road, railway and canal construction etc. Some have suggested that, in populations of new colonists, favourable mutations have to occur before species are able to exploit the habitat available (Crooks & Soule, 1999).

Changes in populations following introduction

A number of case studies have investigated population variation after introduction, and a number of explanations have been devised to account for the changes detected: (a) plasticity and adaptability; (b) sorting out introduced variants; or (c) adaptive change. We examine each in turn. It is important to note that the different mechanisms are not mutually exclusive: all three may have operated, sequentially or collectively in some or all of the localities, since their introduction to new territories.

Phenotypic plasticity and developmental adaptability

In 1965, Baker (1965) formulated an influential concept: general-purpose genotypes were the key to the colonisation process. New colonies of introduced plants owed their success to their capacity for phenotypic plasticity and developmental adaptability. A number of researchers have employed the concept in interpreting their experiments. For instance, Parker, Rodriguez & Loik (2002) studied *Verbascum thapsus*, a species introduced both as a medicinal and ornamental garden plant into North America from Europe. By 1880, the plant had reached California and it has become invasive on roadsides and waste ground as well as in natural habitats such as Yosemite National Park. Samples of seed from ten populations were grown in a garden trial and in growth chambers. In this investigation there was evidence of considerable plasticity and it was concluded that colonisation was 'not driven by rapid adaptation'. Rather, Baker's notion of general-purpose genotypes was favoured. Further experimentation would be very interesting, especially a reciprocal transplant experiment in the wild. However, there is likely to be considerable resistance to planting out such trials of an invasive species.

Turning to another excellent example, across western and central Europe, *Buddleja davidii* (variant spelling *Buddleia davidii*) is a widely distributed *invasive* ornamental shrub originating from central China. By means of common garden experiments and reciprocal transplant experiments, Ebling *et al.* (2011) investigated population variation in 20 populations. They found no evidence of clinal variation and concluded that there was a 'continent wide shuffling of cultivars' originating from the horticultural trade and that 'invasive spread has been facilitated by phenotypic plasticity rather than adaptation to climate'.

Baker's concept of general-purpose genotypes has also been employed to interpret the variability in invasive populations of *Taraxacum* and *Chondrilla*. Both these taxa have been widely introduced in North America from Europe, where they exist as a mixture of sexual and widely dispersed apomictic variants. Considering variation in these genera, Van Dijk (2003) concluded that 'allozymes and DNA markers indicate that apomictic populations are highly polyclonal ... In *Taraxacum* clonal diversity can be generated by rare hybridization between sexuals and apomicts, the latter acting as pollen donors ... some clones are geographically widespread and probably represent phenotypically plastic general-purpose genotypes'. Overall,

> the long-term evolutionary success of apomictic clones may be limited by lack of adaptive potential and the accumulation of deleterious mutations. Although apomictic clones may be considered as 'evolutionary dead ends', the genes controlling apomixis can escape from degeneration and extinction via pollen in crosses between sexuals and apomicts. In this way, apomixis genes are transferred to a new genetic background, potentially adaptive and cleansed from linked deleterious mutations. Consequently, apomixis genes may be older than the clones in which they are currently expressed.

Natural selection acting on the variants introduced from multiple introductions *Capsella bursa-pastoris* has been introduced into North America from Europe. (Neuffer & Hurka, 1999; Neufffer & Linde, 1999). Investigations using a range of genetic markers reveal a number of important conclusions. (a) The invasive variants have a smaller number of genotypes than in their native range in Europe. (b) There may have been as many as 20 different independent introductions. (c) Variants in Iberia and California share some of the same markers, indicating that the plant was introduced with Spanish colonisation of California. These early-flowering genotypes of the weed are likely to have been introduced as contamination of grain. (d) There is evidence for another broad wave of colonisation from introductions of later-flowering genotypes into eastern North America from more northern parts of Europe. As waves of human immigrants moved from east to west across America these variants were introduced to new territory. The gold rush of 1848 was probably responsible for the first arrival of *Capsella* from more easterly regions to northern California. Contrasting ecotypes now occur in California: late-flowering in the mountains and early-flowering in the summer-dry areas. In interpreting present-day patterns of variation it has been proposed that 'the European gene pool' was introduced 'without major genetic changes' (Linde, Diel & Neuffer, 2001) and that selection has acted on *pre-adapted genotypes*. Such an interpretation fits with what is known of the biology and the predominantly selfing breeding system of the species, which offers limited opportunities for hybridisation between early- and later-flowering variants.

Evidence consistent with a rapid evolution of flowering time has been obtained in controlled growth chamber investigations of the Asian annual grass *Microstegium vimineum* that is now invasive in 20 states of the USA (Novy, Flory & Hartman, 2013). While multiple introductions from appropriate latitudes in Asia are considered as an alternative explanation, herbarium and other records suggest that the plant, first noticed in about 1910, has made steady range expansion northwards in recent years. Taking present evidence into account suggests that there has been rapid evolution of flowering time and biomass.

Species have also been introduced from North America to Europe. Weber & Schmidt (1998) studied *Solidago altissima* and *S. gigantea* introduced into Europe about 250 years ago. Population samples were collected along a transect across Europe – from Sweden to northern Italy. Cloned material was grown in a garden trial, and clinal variation in flowering times was evident with northern populations flowering earlier than those from southern locations. This cline, in flowering time, is similar to that found in North America. There are two possible explanations of these results. (a) Pre-existing ecotypic variants from North America were 'sorted out', by natural selection. However, at present, there is no evidence that the species was introduced into Europe many times. (b) It is also possible that there has been rapid evolution, by newly arising mutants or genetic recombinants, of ecotypic variants from the introduced plants through perhaps 10–20 generations.

Clines in flowering time have also been described for European populations of the introduced invasive *Impatiens glandulifera* (Kollmann & Bañuelos, 2004). Whether these are recently evolved or the result of multiple variable introductions has still to be determined.

Life history evolution: case histories Returning to the investigations of *Hypericum canariense* referred to above, samples were collected from the Canary Islands (Hierro, Gomera and Tenerife) and from the three introduced populations (on Maui, Hawaii, and two sites in California – San Mateo and San Diego). Growth and reproductive characteristics were examined in a common garden experiment at Santa Cruz, California (Dlugosch & Parker, 2008a).

> A latitudinal cline in flowering time has developed among the invasions ... This cline is particularly exciting in that it represents rapid local adaptation among introductions from the same source, where that source is a region with minimal latitudinal range itself ... Thus, novel differentiation can arise even in a case where it would seem highly unlikely, and our studies provide an illustration of local adaptation proceeding despite strong founder effects.

First introduced to the Oxford University Botanic Garden in 1794 from Mount Etna, Sicily, *Senecio squalidus* eventually escaped and colonised scattered localities. Then from c.1850s, the plant spread rapidly along railway tracks, reaching as far as southern Scotland by about 1955. In the study of a range of samples (from Oxford (latitude 51°N) and Edinburgh (55°N), with three samples from intermediate latitudes), common garden trials and greenhouse experiments were set up by Allan & Pannell (2009) in Oxford and Edinburgh. They examined growth, drought tolerance, temperature sensitivity and flowering behaviour. Their results 'are largely consistent with the hypothesis of rapid adaptive divergence of populations of the species within the introduced range, with genotypes typically showing home-site advantage ... southern genotypes were more tolerant of dry conditions and high temperatures and flowered later than northern genotypes'.

As we have seen in earlier chapters, tristylous *Lythrum salicaria*, introduced c.200 years ago in the eastern territories of North America, has now become a widespread invasive (Thompson *et al.* 1987). In a study of variation, material was collected from 25 populations along a 1400 km transect across the latitudinal range of the species (see Barrett, Colautti & Eckert, 2008). The growing season extremes covered by this sampling ranged from 138 to 256 days, 'based on an 8°C degree growth threshold' for the species. Cultivation trials were set up using open-pollinated seeds (technically a set of half-sib seed families were examined). Field studies and the result of cultivation experiments revealed a cline in flowering time, plants from higher latitudes being earlier flowering. However, a cautious interpretation of the results is presented. Perhaps populations were introduced into North America from different parts of Europe and there has been 'ecological sorting' in the introduced range. This hypothesis is unlikely given the history of the introduction and spread of the species in North America. Moreover, *Lythrum salicaria* is spreading northwards in Canada, and again it is unlikely that plants 'pre-adapted' to these northernmost conditions have been introduced directly. Further grounds for caution are the limitations of common garden experiments. Evidence from such studies alone 'do not constitute definitive evidence of adaptation' (see Chapter 8 for details of why this should be so). In this example and *indeed in other case studies considered here*, reciprocal transplant experiments of cloned material would provide a more critical test of presumed local adaptation.

As invasive species successfully extend their range, adaptation to new circumstances involves many characteristics. In further studies of this field trial with *Lythrum*, other potentially coevolving characteristics were examined: seedling growth rate, time to reproductive maturity and mature height etc. (Colautii, Maron & Barrett, 2009; Colautti, Eckert & Barrett, 2010; Colautti & Barrett, 2011). These painstaking investigations and analyses provide an important model for other investigations.

The breeding behaviour of invasive species is another potential arena for microevolutionary change. Thus, Colautti, White & Barrett (2010) studied the breeding system of *Lythrum* in a range of populations by self- and cross-pollinations, to investigate whether plants on the expanding front of the northwards invasion exhibit higher levels of self-compatibility (SC). They discovered that

> just over one-quarter of plants set at least one fruit after self-pollination, and there was a significant effect of style morph on variation in SC, with the mid-style morph most compatible. Although variation in SC was detected in 11 of the 12 populations, there was no evidence the SC increased towards the northern range limits as a result of the invasion process.

What factors contribute to the success of new populations?

When new populations are established by migration from an existing population many factors are involved in their success or failure (Harper, 1977). Introduced species in new territories are a special case of such migration and establishment of new populations, but such studies may provide important pointers to migration success/failure in general. Species invasions are particularly interesting as they provide a means of studying current or recent microevolutionary processes acting in contemporary ecosystems (Keller & Taylor, 2008).

Some of the introduced species have become so successful that they are now considered to be invasive. This success could be related to one or more of the following factors (see Hierro, Maron & Callaway, 2005).

a) A species may arrive in new territory released from its natural enemies.
b) There may be an empty niche in the new territory.
c) Major habitat disturbance (often by human activities) may provide an ideal habitat for the new arrival.
d) Success in establishment may follow if there is sustained propagule pressure.
e) In the native range, introduced plants may produce allochemicals that are relatively ineffective against their customary well-adapted neighbours. In new territory the allochemicals produced by the introduced plants may give the newcomers greater 'fitness', as the members of the recipient ecosystem may not have encountered these chemicals before.
f) Many invasive plant species were introduced to new territories as ornamentals grown in gardens and glasshouses. The increased vigour detected in some introduced species (Keller & Taylor, 2008) may in part be the result of artificial selection for size and other characteristics, as species were introduced into the flower trade before escaping to the wild (Mack & Lonsdale, 2001). This hypothesis is supported by investigations of the invasive shrub *Ardisia crenata* introduced into Florida from Japan (Kitajima *et al.*, 2006).
g) Potentially selection for agronomic traits may also increase the performance of invasive species. This possibility has been examined in studies of Reed Canarygrass (*Phalaris arundinacea*) by Jakubowski, Casier & Jackson (2011). Their evidence suggests that 'breeding efforts are not responsible for wetland invasion by reed canarygrass'. However, there is a concern that the active breeding programmes now being established with Reed Canarygrass, and other species, to select cultivars as sources of biofuels, may offer the

opportunity for future escapes into the wild of variants with the potential for increased invasiveness (Calsbeek *et al.*, 2011).

h) There is evidence of mutual co-adaptation of grasses and grazing animals, and of weeds and arable crops in Europe. In the colonisation of new territories, many animals, plants, pests and diseases that had coevolved together in the Old World, were all brought to new territories by human activities. Crosby (1986) discusses their impact on pre-existing ecosystems overseas. 'The Old World quadrupeds, when transported to America, Australia and New Zealand, stripped away the local grasses and forbs, and these, which in most cases had been subjected to light grazing before, were often slow to recover. In the meantime, the Old World weeds, particularly those from Europe and nearby parts of Asia and Africa, swept in and occupied the bare ground. They were tolerant of open sunlight, bare soil, and close cropping and of being constantly trod upon and they possessed a number of means of propagation and spread . . . the success of the portmanteau biota and its dominant member, the European human, was *a team effort by organisms that had evolved in conflict and co-operation over a long time'* [emphasis added].

Herro, Maron & Callaway (2005) consider ways in which these untested, or partially explored, hypotheses can be properly evaluated. Two examples indicate the potential of exploring this interesting area.

Changes in toxicity Lankau *et al.* (2009) have investigated the chemical defences in *Alliaria petiolata*, a Eurasian species invasive in the understory vegetation in North American forests. They note that

> like all members of the Brassicaceae, *A. petiolata* does not form connections with mycorrhizal fungi and it produces glucosinolates, a class of secondary compounds that break down into products toxic to herbivores, fungi and other plants . . . If these allochemicals weaken the mycorrhizal associates of native plants enough to reduce their ability to compete with the invader, this could explain *A. petiolata*'s greater abundance in its introduced versus its native range.

However, the investigation of Lankau and associates reveals a marked decline in phytotoxin production in older populations of the invasive. Several explanations of this finding are possible. Perhaps, as colonies of the invasive increase in size, intraspecific competition became important and the production of allochemicals 'entailed a cost but accrued no benefit'. Or perhaps 'the benefit of allopathic traits may also decrease over time if native plants or soil microbes evolve resistance'. Taking the results of this investigation as a whole, Lankau and associates consider that patterns of reduction in phytotoxic production are the result of natural selection rather than founder effects. Lankau (2012) has extended investigations of this species by carrying out reciprocal transplant experiments. These support the view that coevolution between the invasive plant and native plants is driven by chemical competition and soil biota.

It should not be assumed that phytochemical changes are always in the same direction. Cultivated Parsnip (*Pastinaca sativa*) brought by settlers from Europe to Virginia, USA, by 1609, escaped cultivation and has now become a widespread invasive plant. In its invasive range, few of the insects native here attack the plant, which contains toxic furanocoumarins (Zangerl & Berenbaum, 2005). In 1869, the principal coevolved herbivore of *P. sativa* in Europe – the Parsnip Webworm (*Depressaria pastinacella*) – was accidentally introduced into North America and it is now widespread. Chemical analyses of a series of dated herbarium specimens have established that, in the earliest period of establishment (1850–69), *P.sativa* had lower levels of furanocoumarins than European stocks of the same age. Since the Webworm has increased its range in North America 'widespread shifts' have taken place, leading to an 'increase in noxiousness' of *P. sativa.* Clearly there are some limitations in the use of herbarium for historical studies, but these findings strongly suggest the action of natural selection. 'It is likely that the high cost of furanocoumarin defense led to selection for reduced investment before reassociation with the Parsnip Webworm.' This case history has implications for biological control programmes where coevolved insects are introduced to control an invasive plant species in its introduced range. 'If the genetic variation exists in weed populations to respond to selection pressure from their coevolved associate, [then] enhanced chemically-based resistance may evolve.'

Population persistence and stability: clonal growth

As we saw in Chapter 8, some plants are capable of great longevity through clonal reproductive growth. Populations may be composed of plants of different genotype (genets), reproducing vegetatively by the production of vegetative structures (ramets). Crucially, clones have different growth characteristics. Harper (1977) characterises two extreme variants of a range of growth forms. Some plant clones grow out to form a recognisable patch with an expanding front – the phalanx mode. From the point of initial colonisation, the plant exploits 'an area of habitat space' that increases with time. At the opposite extreme, other clonal species produce 'exploratory structures' (stolons etc.). This 'guerilla mode' often results in fragmented clonal growth, as the plant infiltrates the vegetation, to occupy under-exploited microsites.

Indirect methods of detecting clones and estimating longevity have been devised: counting of annual rings in wood/growth increments/bud scars; and examining morphological/biochemical markers (known or suspected of being genetically determined). Sometimes, other information such as radiocarbon dating and aerial photography can be revealing.

Historically, the study of certain variable self-incompatible species provided a means of estimating the size and longevity of clones. Variability was studied in garden trials of population samples, and the material classified into different individuals on the basis of morphology, phenology, susceptibility to pests and diseases, etc. The behaviour of different plants in crossing experiments was then studied to test the following model. In crosses between dissimilar-looking plants of a self-incompatible species, a 'full seed-set' suggests that the individuals have different S alleles and are therefore different genotypically. Conversely, crosses between plants that were morphologically indistinguishable, yielding little (or no) seed, were likely to share the same S alleles, and therefore likely to be of the same genotype or 'isoclonal'. This method was used to study variation in populations of *Festuca rubra* and other species and revealed the presence of extensive clones (see Chapter 8). Some caution is necessary in interpreting experiments of this type, as the method depends upon a thorough knowledge of the type of incompatibility mechanism involved – a requirement never completely satisfied with wild species.

Recent investigations of size and age of clones employing molecular tools (including allozymes and DNA markers) provide the means of making more critical tests of hypotheses concerning clone size and longevity. The advantage of these more direct methods of analysis is that they provide a large number of genetic markers at low cost, allowing extensive sampling with minimal impact on the populations sampled. However, in reviewing progress in this area, De Witte & Stöcklin (2010) draw attention to issues of ambiguity and 'mis-scoring of fingerprints'. Also, 'repeated samples' from the same genet, 'but from different ramets', do not always yield identical profiles, which may 'result from somatic mutations, from contamination in the laboratory, or from scoring errors' (Douhovnikoff & Dodd, 2003; Bonin *et al.*, 2007). Therefore, repeatability tests and statistical tools are necessary, in these, as in other, molecular analyses (Lasso, 2008).

Size and longevity of clones: new insights using molecular markers

Many tree species exhibit clonal growth in arctic and subarctic regions, but is this the prevalent means of reproduction? This question was examined by Viktora, Savidge & Rajora (2011) in a study of the genetic structure of Black Spruce (*Picea mariana*) populations in the Yukon using microsatellite DNA markers. They discovered that three populations of the species 'growing on flat terrain at relatively low elevation had unique multilocus genotypes, indicating an absence of clonal structure'. However, on the north slopes of Mount Hansen, 'the majority of sampled individuals belonged to eight genetically distinct clones (genets). Clone size differed by altitude, the dominant genet being nearest the timberline–tundra ecotone.' They concluded that in Black Spruce 'reproduction is variable and adaptive, being primarily sexual' in areas previously subject to fires, but reproduction was 'mixed vegetative–sexual in the anthropogenically undisturbed subalpine population'.

In other studies there is considerable evidence for extensive old clones. For instance, on the basis of morphological distinctness, and patterns of growth visible on aerial photographs, apparent clones of *Populus tremuloides* have been estimated at 10 000 years old. One population covered 81 ha with 47 000 stems (Kemperman & Barnes, 1976). Analysis of microsatellite divergence based on mutation accumulation has now extended estimates of age of large clonal patches in this species to 12 000 years (Ally *et al.*, 2008).

Many non-tree species also form clonal patches, and studies with DNA markers reveal that the oldest genets may be hundreds or thousands of years old (deWitte & Stöcklein, 2010). For instance, the dwarf shrub *Rhododendron ferrugineum* has been estimated to be 300 years old (Escaravage *et al.*, 1998), while a genet of the alpine species *Carex curvata* was estimated at 2000 years (Steiger *et al.*, 1996).

Clonal reproduction is very important in aquatic species, through the production of turions, winter buds and shoot fragments. Very large clones – one up to 15 km – were detected in a study of microsatellites in samples of the Seagrass *Posidonia oceanica* from 1544 sites across 40 locations in the Mediterranean (Arnaud-Haond *et al.*, 2012). Such enormous clones are estimated to be hundreds to thousands of years old, 'suggesting the evolution of general-purpose genotypes with large phenotypic plasticity'.

Clones are important also in crop plants. For example, since the early 1800s, Vanilla (*Vanilla planifolia*) has been grown on islands in the Indian Ocean. Evidence suggests that almost all the accessions have probably arisen from one very old genet. The clonal propagation of grape varieties is very common. The famous Swiss cultivar 'Rouge du Pays', which is still being cultivated, is mentioned in a manuscript dating from 1313 (Vouillamoz *et al.*, 2003).

Organisms reproducing over long periods by constant repeated mitosis are likely to accumulate somatic mutations, and recent investigations have studied how this somatic variation can be exploited to study size and age of clones. Investigations of an Elm genet (*Ulmus procera*) date the origin of this tree to Roman times. Ramets were propagated vegetatively, and, by human introduction, can now be found growing at sites as far apart as Spain and Britain (Gill *et al.*, 2004). *Ranunculus repens* forms large clones. The possession of extra petals in the flowers of old clones is indicative of somatic mutations, and the antiquity of the meadow habitats in which these variants grow (Warren, 2009).

Implications of clonal growth in populations

From the microevolutionary and ecological standpoints, clonal propagation of a genotype offers 'a number of advantages' (Chapter 7). To recap, it provides reproductive assurance at economical cost in areas where sexual reproduction is limited, fails or is impossible, often at the limits of the distribution of the species, altitudinally and/or geographically. Such behaviour is likely to be adaptively significant in extreme habitats, where there are few pollinators, where the necessary genotypic variability for sexual reproduction is not present, or where plants may be lone individuals of an obligatory outbreeding species. A capacity for clonal growth may permit the survival and reproduction of triploid or chromosomally unbalanced plants that 'pass the test of mitosis', but not the rigours of sexual reproduction. Ecologically, plants with long-lived clonal growth habit may provide elements of stability and continuity in ecosystems, by the presence of many different genets, or, in some cases, by the domination of the habitat by very large clones that may limit the establishment of new individuals arising by sexual reproduction.

As we have seen, clonal reproduction is very important in many invasive species. For example, the European flowering rush (*Butomus umbellatus*) is seriously invasive in the wetlands of northern USA and southern Canada. Barrett (2011) gives the following details. 'Diploid populations reproduce by seed and asexually through pea-sized vegetative bulbils and rhizome fragmentation, while triploid populations are sterile, propagating exclusively through clonal reproduction ... the diploid cytotype is far more abundant with a much broader geographical range.' However, reproduction by the diploid appears to be largely through asexual means. 'This is reflected in very low levels of genetic diversity throughout North America compared to Europe where diversity is much higher ... In contrast, North American triploid populations do not produce bulbils and can only propagate via rhizome fragmentation. This is a slower process, limiting rates of clonal propagation and the spread of the triploid cytotype.'

Are there some possible limitations of clonal habit? In an important review, Vallejo-Marin, Dorken & Barrett (2010) consider the implications of clonal growth for the functioning and evolution of breeding systems and draw attention to issues that have yet to be critically examined. They point out that 'sexual and asexual reproduction usually occur simultaneously, and this can lead to allocation trade-offs, and antagonism between reproductive modes'. Moreover, they outline key issues that should be addressed when studying any particular species: population structure, architecture and flowering behaviour of clones. Considering the flowering behaviour, genet diversity and clone size and form (guerilla, phalanx or intermediate) are important, as there may be a tendency for ramets of the same clone to flower at the same time. A critical question is how far proximity of flowers of the same genotype leads to selfing rather than outcrossing in self-compatible species, and the degree to which such intra-genet selfing might lead to inbreeding depression. By means of pollination manipulations and the study of genetic markers it is possible to estimate the degree of selfing, an approach used in the study of populations of the clonal species *Decodon verticillatus* by Eckert (2000). He discovered that c.30% of seed produced resulted from self-fertilisation and 28% of the seed came from between ramets selfing (so-called geitonogamous self-fertilisation). Repeated selfing in self-compatible species has fitness consequences. Vallejo-Marin, Dorken & Barrett (2010) discuss the important concept of pollen discounting where there is 'a loss of outcrossed siring success caused by self pollination'. To give a more complete picture of breeding behaviour in self-compatible clones, Vallejo-Marin, Dorken & Barrett (2010) suggest that it would be of considerable interest to investigate whether peripheral parts of a clone produce more outcrossed progeny than flowers at the centre.

In self-incompatible species all male gametes transferred between ramets of the same genotype are 'effectively lost from the pollination process', causing complete pollen discounting. However, many self-incompatible plants have clonal growth habit. As we have seen, such behaviour is adaptively significant in circumstances unfavourable to mating, as it provides reproductive assurance.

Vallejo-Marin, Dorken & Barrett (2010) call for carefully designed experiments to investigate the implications of clonal growth for the balance between asexual and sexual reproduction, and for studies of the evolution of floral mechanisms to reduce self-fertilisation (technically geitonogamy) in clonal plants. Such adaptive changes promoting outcrossing could involve the relative timing of female and male functions, and transitions from hermaphrodite flowers to dioecy etc.

Considering all the evidence available, in some populations of a species reproduction may be exclusively asexual through clonal propagation. But, as we have seen, investigations using molecular markers have detected genetic variability in many cases, indicating the possibility of some 'successful' sexual reproduction.

Is it possible to distinguish the products of asexual and sexual reproduction, particularly with molecular markers? Examining this possibility, Mes (1998) writes: 'the relative contribution of different modes of reproduction to genetic variation in natural populations is most commonly determined by isozymes. Asexual reproduction is indicated by a lack of segregation (within loci), a lack of recombination (between loci), or both'. As we have seen earlier,

> most DNA markers available for genotyping individuals, such as Random Amplified Polymorphic DNAs (RAPDs) and Amplified Fragment Length Polymorphisms (AFLPs) are dominant. Therefore, in studies of natural populations heterozygotes cannot be distinguished from homozygotes and characteristics of segregation, such as fixed heterozygosity, the absence of certain segregating genotypes, and other deviations from Hardy–Weinberg equilibrium, cannot be estimated by dominant markers.

In examining possible solutions to the problem, Mes (1998) has employed very detailed comparisons of molecular data sets and his paper may be examined to see how 'character compatibility tests' are employed to distinguish asexual and sexual reproduction.

Turning to another issue, while observation suggests that there is no obvious penalty for extreme age in clonal growth, perhaps future studies might detect some form of senescence in old clones. One route to mortality could be the accumulation of deleterious somatic mutations, attack by virus infections or other pathogens.

While long-lived clones are found in many species, it is important to stress that changing ecological situations may influence the balance between asexual and sexual reproduction. This is clearly indicated in studies of Aspen in Yellowstone National Park, USA. As we have seen above, this is a long-lived clonal species and early research revealed that reproduction by seed was very rare.

In the early management of Yellowstone Park, predator animals (wolves, mountain lions, coyotes, lynx, bobcat, foxes, otters, martens, weasels etc.) were controlled to encourage game animals (particularly elk). Wolves were eventually exterminated (Budiansky, 1995). This policy led to increasing numbers of herbivores and overgrazing of many areas of the park. Exclusion experiments (with stock-proof fencing) revealed the very strong grazing pressures on the Aspen; unprotected stands showed a steady decline and deterioration. Another National Park policy – fire suppression – was also practised in Yellowstone. Natural and human-set fires were extinguished as soon as possible. This resulted in a huge build-up of flammable material on the forest floor. Then, in 1988, enormous fires destroyed the vegetation over large sections of the park. After the fires, there was a major recruitment of new individuals of Aspen. Samples of seedlings were excavated and their annual rings counted. Genetic variation in these samples was examined using RAPDs markers. These combined studies revealed that 92% of these new plants were genetically distinct – the result of recruitment following sexual reproduction – while only 8% were ramets of existing plants (Romme *et al.*, 2005). In addition, to these fire-induced changes, grazing pressures on Aspens and other plants are now being modified by the predation of elk by wolves, newly reintroduced in the Park in the 1990s. Kauffman, Brodie & Jules (2010) conclude that elk numbers are still high despite the arrival of wolves in the ecosystem, and 'aspen recovery in Yellowstone will only occur if wolves (in combination with other predators and climatic factors) further reduce the elk population'. In addition, drought and conifer invasion are other important factors influencing Aspen regeneration. Romme *et al.* (2005) make two additional points of considerable importance. First, it seems highly likely that in the past, large forest fires allowed periodic episodes of seedling recruitment of Aspen. Also, it is possible that the demographic and genetic composition of the present populations will change as competition between plants increases. This and other selection pressures make it likely that few of the many genetically distinct plants in this seedling cohort will survive in the longer term. The case history of Aspen in Yellowstone provides a fine example of differences in genetic variability in plants of different age classes related to forest history and herbivory.

Genetic sub-structuring in populations of plants may have other causes, including historical factors such as forest clearance (Young & Merriam, 1994). Historical factors resulting in fragmentation may have long-standing influence on patterns of genetic variation through their influence on gene flow, and fruit/seed dispersal etc. Vekemans & Hardy (2004) review the use of genetic markers to examine the non-random distribution of genotypes in plant population studies, using data from genetic markers analysed by spatial autocorrelation techniques.

Another major factor influencing population persistence: seed banks

Persistent seed banks in the soil are characteristic of many plant species that live in habitats subject to unpredictable spatial or temporal disturbance regimes (Honnay *et al.*, 2008). Population survival through unfavourable periods is facilitated by the persistence in the soil of dormant viable seed. Aggregated seed banks may remain viable for decades or perhaps hundreds of years, sometimes as high as c.10 000 seeds m^{-2}. Soil seed banks have been detected in a wide range of different habitats. Many complex interacting factors influence whether such banks are ephemeral or persistent, for instance, seed dormancy, seed predation and attack by pathogens, and with what frequency the necessary climatic/ecological conditions occur for successful germination.

A number of botanists have compared the genetic variability of adult plants with that represented in the seed bank. For instance, Falahati-Anbaran *et al.* (2011) compared the seed bank and above-ground variation in Norwegian populations of *Arabidopsis lyrata* ssp. *petraea* by examining samples from populations and variability in seedlings emerging from soil core samples. They discovered that the genetic composition of the seed bank and above-ground cohorts were highly similar, with little genetic difference between them. However, there were more unique (so-called private) alleles in the above-ground samples.

Studying the variability in the threatened aquatic species *Nymphoides peltata* in Japan, Uesugi *et al.* (2007) sampled adults and seedlings (that emerged from seed banks) in populations from Lake Kasumigaura. Genetic variability was investigated by studying 10 satellite markers. Low genetic diversity was detected in the18 genets found as clonal adult plants. Also, 430 seedlings were examined. These were plants that emerged from the seed bank at six locations along the natural shoreline and at three restoration sites. 'The seedlings showed genetic variation different from the adults.' Seven alleles, presumed lost in the adults, were detected in the seedlings, and it was concluded that the seed bank had some potential to help restore diversity in this species.

Other studies have revealed different patterns. Wild Rice is an endangered species subject to restoration conservation in China. The genetic variability represented in the seed bank was compared with that of adult plants by studying SSR DNA markers (Lui *et al.*, 2006). Of significance for the conservation management of Wild Rice is the finding that, compared with adult plants, the seed bank contained reduced levels of variation (about 72%).

Turning from these individual case studies, what general statement can be made about the relative variability of adults versus seedlings from soil seed banks? Reviewing 13 published investigations Honnay *et al.* (2008) discovered no evidence 'of high levels of genetic diversity accumulating in the soil seed bank. If genetic differences are present between the standing crop and the seed bank, they are very likely the result of local selection acting directly or indirectly as a filter on the alleles present in the seed bank.' They stress that it is 'not very fruitful to continue comparing seed bank genetic diversity with above ground plant genetic diversity, unless this is performed under different selection regimes'.

While some plant species have seed banks below ground, others have above-ground banks. *Banksia spinulosa* is one such species. After a fire that consumes the above-ground biomass, plants resprout from lignotubers in the soil. *Banksia* plants also invest heavily in sexual outcrossing. Seeds are contained in cone-like structures (technically woody follicles) held for several years in the canopy. Fire stimulates the release of the seed from the cones. In a study of an extensive population in southern New South Wales, Australia, adults were collected and samples of cones representing different years' reproductive effort (Ayre *et al.*, 2010). Genetic variability was examined using microsatellite markers. It was discovered that genetic variability increases rapidly within the aerial seed bank, and estimated that, over a period, cones come to have '100% of the adult diversity'.

Processes involved in the extinction of populations

Having considered various microevolutionary factors – chance, selection etc. – that influence population variability, it is important to consider factors that drive populations to extinction. At the outset, we stress that, over different timeframes, all species – rare or common –tend to decline and become extinct.

Metapopulation models predict that where many populations occur in a territory the loss of some populations can be temporary, and new populations are established following gene flow. But conservationists are concerned that many rare and endangered species are in serious decline, with no large populations nearby from which extinct populations could be reinstated through natural gene flow. In such cases, appropriate management is recommended to prevent extinction, through reintroduction and other measures (see Chapter 20).

The effect of several interacting factors that can drive a population towards incipient extinction has been modelled as a vortex (Gilpin & Soulé, 1986). Once a population of a species has become small by a bottleneck effect (often caused by human activities in damaging and changing habitats resulting in habitat loss), problems in seedling establishment and/or reproduction may occur (Fig. 11.2).

Decline in numbers in response to these factors is exacerbated by population subdivision and consequent curtailment of gene flow: populations often become totally isolated genetically. Reduction in population size then opens the door, in the longer term, to inbreeding depression and genetic drift that may result in genetic depauperisation and an inability to respond to further changes in the habitat. Thus, there is a tendency for small populations to be driven to extinction.

As yet all the interlocking effects of the vortex model have not been investigated in any particular species, including those that are endangered (Meffe & Carroll, 1994). However, some facets of the model have been examined in different investigations.

Demographic stochasticity

In a healthy population, birth rate is often, but not always, more or less sufficient to 'compensate' for the

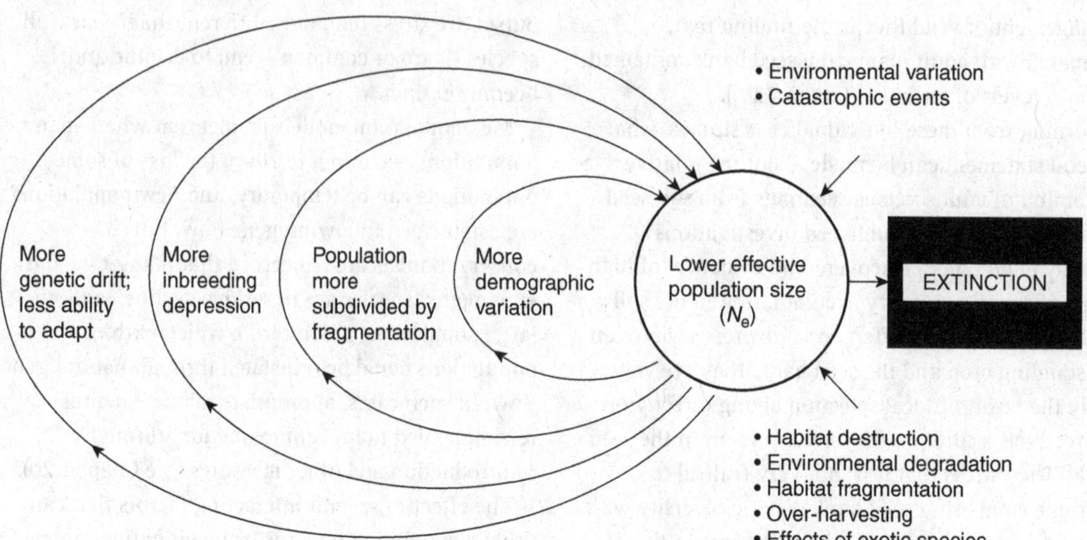

Fig. 11.2. Extinction vortices progressively lower population sizes, leading to local extinction of species. Once a species enters a vortex, its population size becomes progressively lower, which in turn enhances the negative effects of the vortex. (Adapted from Guerrant, 1992, and Gilpin & Soulé, 1986, from Primack, 1993.)

death rate (Silvertown & Lovett Doust, 1993). But small fragmented populations may encounter demographic problems, and support for this hypothesis is offered by a number of investigations. For example, in a study of *Gentiana pneumonanthe* in the Netherlands, Oostermeijer *et al.* (1992) discovered that some populations were dynamic, with a high number of seedlings and low percentage of adult plants. However, others, subject to increased nutrient status, groundwater depletion and inappropriate management practices, were at a greater risk of extinction, for they were static or senile consisting mainly of flowering individuals and hardly any seedlings. Clearly, a census of the number of adult flowering individuals may not be a good guide to the health of a population in the longer term.

To understand the status of populations it is important to carry out demographic monitoring of all life cycle stages (Davy & Jeffries, 1981). Such studies have revealed some important complications in monitoring certain rare plants. For example, studies of *Gentiana* (Oostermeijer *et al.*, 1992) and certain orchid species, e.g. *Ophrys sphegodes* (Hutchings, 1989), have revealed that individuals may remain underground in a dormant state for one or more years.

In the study of the demography of plant populations the presence of persistent banks of seed in the soil is a complication found in some species. Thus, the above-ground population may become very small or even disappear, but because there is a persistent bank of seed in the soil, the population may reappear should the right conditions prevail. For instance, the Fen Violet (*Viola persicifolia*), lost to Wicken Fen, Cambridgeshire, for the last 60 years, was 'rediscovered' in 1980 when a seedling was produced in a sample of fen soil taken from under scrub and placed in an unheated greenhouse (Rowell, Walters & Harvey, 1982). Subsequent investigations revealed the presence of a large population in a different part of the fen, where it had not been detected in earlier visits. It is significant that scrub had recently been removed from this area and the surface soil disturbed (Rowell, 1984). It seems very likely that this population also arose from seed buried in the soil. In the past, when the Fen violet was more abundant, large areas of Wicken Fen were dug for peat. Rowell argues that the key to the conservation of the species is to ensure that the soil is periodically disturbed. Two important points emerge from this example. It is clear that a species may be declared extinct locally when invisible populations of seed may still be present. The converse is also true. Conservationists may cling to the erroneous belief that, where a rare species has disappeared, viable seed of the species may still be found in the soil.

Pollen limitation causes an Allee Effect

In small and declining populations of plants and animals there is also evidence that the reproduction may be curtailed by failure to find mates – the so-called Allee Effect. In plants this failure often results from pollen limitation (Knight *et al.*, 2005). For instance, Jennersten (1988) studied two populations of *Dianthus deltoides* in an area of southern Sweden. Site A contained a large, more or less continuous population in a 1 ha meadow surrounded by grasslands and forest. Site B on the other hand contained a much smaller population in two habitat fragments in a sea of arable fields. Site B had fewer flowering individuals and seed set was lower. A thorough investigation discovered that the number of ovules per flower did not differ between the A and B sites and at neither site did development of seeds appear to be resource-limited. However, fewer insects visited the plants in site B, and, as hand pollination of flowers increased the level of seed set in site B but not site A, it was concluded that low seed set was explained by pollination limitation in the small fragmented population.

Studies of the rare perennial herb *Tephroseris integrifolia* (*Senecio integrifolius*) in Sweden also discovered that in small populations reproduction fails through lack of insect visits, but the investigation also suggested that population size is not the only factor: the density of flowering individuals is also important (Widén, 1993).

Baker, Barrett & Thompson (2000) studied the possibility of pollination limitation in *Narcissus assoanus*, a species that flowers in the very early spring, when pollinators are often infrequent. In studies of this kind, often only one population is examined and only in one season. Here, five populations of this self-sterile, insect-pollinated species were examined in the South of France over a 2–3 year period. Supplementary pollinations with outcross pollen increased seed set by up to 19%. Four of the five populations exhibited a degree of pollen limitation. From the microevolutionary perspective it is of considerable interest that two of the populations were pollen limited in one year but not another.

It might be assumed the pollen of wind-pollinated species is ubiquitous so pollen limitation would not occur. This hypothesis was tested by Davis *et al.* (2004) in experiments on populations of *Spartina alterniflora*, at Willapa Bay, Washington State. They carried out hand pollination together with exclusion experiments in which the inflorescences were enclosed in clear plastic bags capped with bridal veil. Unmanipulated controls were also included. Nine-fold more pollen was detected on stigmas in areas of high density compared with low-density plants at the leading edge of the colony, where 'pollen impoverishment' was discovered. It seems likely that this pollen limitation persists until the vegetative growth produces coalescence of the sward into continuous meadow. Davis and co-workers consider that pollen limitation could contribute to a lag effect between the introduction and rapid spread of the plant.

Ghazoul (2005) considers a number of cases where pollen limitation has been detected and others where no effect has been discovered.

Pollination disruptions and declines Insects are key pollinators for many species of plants. Over the past decade, concerns have been expressed about pollinator declines and disruptions both in wild and cultivated ecosystems, leading to the setting-up of the

Convention on Biological Diversity (International Pollinator Initiative, www.cbd.int/decision/cop/?id=Fallee.). The following brief account is drawn from Potts *et al.* (2010), who have reviewed our current understanding of this important issue, which has huge economic implications for the efficient pollination of many crop plants, as well as the functioning of wild species. While Ghazoul (2005) questions whether there is a global crisis, recent research has shown that there is 'clear evidence of recent declines in both wild and domesticated pollinators, and parallel declines in the plants that rely on them' (Potts *et al.*, 2010). 'Most wild plants (80%) are directly dependent on insect pollination for fruit and seed set, and many (62–73%) of plant populations investigated showed pollination limitation, at least some of the time (Burd, 1994; Ashman *et al.*, 2004), although this may vary markedly between sites and seasons.' However, 'those with the most specialized pollination requirements might be expected to be most at risk, but there is little evidence for this' (Aguilar *et al.*, 2006).

Declines in the domestic honeybee stocks have been dramatic in some areas. In the USA, there has been a 59% loss of colonies between 1947 and 2005, while in central Europe there has been a 25% loss between 1985 and 2005. No single driver has been identified for insect declines, including the threats to the honeybee. Many potentially simultaneously interacting factors are under investigation: (a) habitat fragmentation and degradation (fire, intensive grazing, mechanisation of agriculture with the intensive use of agrochemicals); (b) ecosystem changes in the pollination of native species following the introduction of alien species; (c) increase in the number of pathogens attacking bees, e.g. Varroa mite on honeybees; and (d) the effects of climate change. These multiple interacting stresses are being investigated in many parts of the world, including the recently reported colony collapse disorder of honeybee leading to losses in pollinator services (Cox-Foster *et al.*, 2007; Anderson & East, 2008). Concerning insects, Potts *et al.* (2010) conclude that 'threats to pollinators are diverse, and might interact: the current challenge is to better quantify the relative importance of a range of drivers and in particular their synergistic effects'.

Declines and disruptions have also been reported in other pollinators. Elmqvist (2000) provides a 'frightening picture of pollinator disruption that emerges from oceanic islands' with 'reciprocal extinction of entire pollinator sets, and plant guilds that support them . . . entire pollinator sets, including indigenous birds, bats and insects are disappearing. In some cases, introduced organisms begin to fill the role of pollinators, but all too often pollinator niches for indigenous plants (but not invasive species) remain empty.'

Effects of fragmentation

The vortex model suggests the great importance of population fragmentation in the extinction of species. A reduction in area may lead to serious population decline and this has the potential to disrupt patterns of gene flow, and demographic functioning, together with genetic changes. A critical consideration for individual species within an ecosystem is the availability of suitable habitat patches. The fate of such patches is intimately bound up with processes that bring about fragmentation at the habitat level.

To illustrate key points that are of general significance in considering the processes of fragmentation, we draw on the review by Honnay *et al.* (2005) who have examined the effects of fragmentation on the herbaceous plants of the forest floor. Here, through considering one particular vegetation type, we establish some general principles relating to continuity, change and extinction of patches and fragments.

A. Ecosystems develop through serial stages, and they are profoundly influenced by the impact of extreme events such as damaging winds, serious floods, major fires, etc. Biotic influences are also important, for instance, pathogen or insect attacks. These natural factors often result in habitat fragmentation. Smaller events are also of great significance at the 'patch level', such as the opening of forest clearings following individual tree fall.

B. In addition to these natural events, a multitude of different human activities result in fragmentation at many scales, leading, in some cases, to total loss or reduction in fragments and different degrees of isolation between fragments and patches of different size. Such changes are of great importance in considering the conservation of rare and endangered species (see Chapter 20).

C. In Britain, the boundaries of some forest fragments were largely set centuries ago, and human management has led to a degree of continuity of forest cover (Rackham, 2008). Different sites have individual histories of exploitation, through activities such as coppicing, wood pasture etc. Others have been cleared for agriculture, but later, as in the eastern USA, forest has re-established on abandoned crop-lands (Whitney, 1994).

D. Fragment size and shape determine the length of the patch edge, where interactions with adjacent ecosystems occur, including potentially negative influences. In the case of forest fragments surrounded by agricultural lands, these include changed microclimate with higher temperature, light, and lower humidity typically reaching 20–50 m into the patch (Honnay *et al.*, 2005). In addition, edge effects between forests and agricultural lands include the impact of agrochemicals, together with an increased opportunity for invasive organisms to penetrate the forest.

E. In reviewing the biology of forest herbs in relation to fragmentation and population viability, Honnay *et al.* (2005) conclude that many are long-lived perennials typically with clonal growth. Indeed, 85% of European herb plants have some degree of clonal growth. Often seed production is low and there is limited seedling recruitment. 'Colonization [between fragments] was more successful in high-connectivity landscapes (c. 50% forest cover) than in low-connectivity landscapes (c. 5% forest cover).' In addition, evidence suggests that long-range dispersal is rare, and Honnay *et al.* (2005) consider that fragments of forest 'more than 2km from an ancient source will probably never completely recover in terms of herbaceous plant species composition'.

F. Typically small fragments of forests have fewer herbaceous species and generally contain a less heterogeneous selection of habitats than larger fragments or more or less continuous tracts of forest. There is increasing evidence that species may be lost from newly created fragments and patches. The timescale of such losses will depend upon the longevity of the plants concerned. Therefore, in addition to short-term changes, there may an extinction debt that is only paid in the future. For instance, Turner *et al.* (1994) have studied plant extinctions in the Republic of Singapore. As a result of forest clearance in the last century, only forest fragments remain today. Many plants are long-lived and population turnover in these species is very slow. It may take decades or centuries for extinction to occur in long-lived species in remnant forest patches. In contrast, short-lived plants have become extinct quite quickly. For example, the epiphytic flora has declined with the reduction in the numbers of big old trees and the drier microclimate that accompanies fragmentation. Also, for certain epiphytic orchids, already under pressure from habitat change, the *coup de grâce* has sometimes been delivered by plant hunters.

Genetics of small populations

The vortex model predicts that small populations may become genetically depauperate. A number of studies of rare and endangered plants have examined variation in isozymes: for instance, no variation was detected in electrophoretic patterns of isozymes in populations of endemic streamside species *Pedicularis furbishiae* of the St John Valley, Maine, USA (Waller, O'Malley & Gawler, 1987). DNA markers have also been employed to study variation in rare species: for example, low variation was detected in the endangered island endemic *Malacothamnus fasciculatus* var. *nesioticus* of Santa Cruz Island, California (Swensen *et al.*, 1995), and in Swedish populations of the rare species *Vicia pisiformis* (Gustafsson & Gustafsson, 1994).

With regard to the vortex model, the most informative of the studies compare variation in small and large populations to test the hypothesis that genetic erosion occurs in populations with small numbers of individuals. Evidence consistent with this hypothesis has been detected in studies of variation in isozymes in the New Zealand endemic *Halocarpus bidwillii* (Billington, 1991); *Salvia pratensis* and *Scabiosa columbaria* in the Netherlands (van Teuren *et al.*, 1991); *Eucalyptus albens* in south-western Australia (Prober & Brown, 1994); and, to some degree, in studies of *Gentiana pneumonanthe* in the Netherlands (Raijmann *et al.*, 1994). However, Ellstrand & Elam (1993) note that in some cases small populations had the same level of genetic variation as large populations. In these cases, historical factors may have been more important; in particular the number of years since the bottleneck effect, in relation to the length of the life cycle of the plants.

In interpreting these findings, it must be stressed that much of the variability in isozymes is likely to be 'neutral' and non-adaptive. Very few studies have considered the question, is there evidence for the erosion of 'adaptive and potentially adaptive' morphological traits in small populations? In pioneer studies of this question, Ouborg & van Treuren (1995) collected seeds from the two smallest and the two largest populations of *Salvia pratensis* in the Netherlands. These were grown in a common garden experiment. They studied a wide range of characters and discovered differences between populations, none of which were correlated with population size. They concluded that while there was evidence for a decrease in isozyme diversity in the smaller populations, the genetic variation underlying morphological traits of consequence to fitness had not been affected and the small populations were likely to be at an early stage of genetic erosion. Similar conclusions were drawn from studies of large and small populations of *Scabiosa columbaria* (van Teuren *et al.*, 1993) and *Silene flos-cuculi* (*Lychnis flos-cuculi*) (Hauser & Loeschcke, 1994).

Evidence for loss of variability of adaptive significance in small populations has also come from studies of breeding systems. For example, Demauro (1993, 1994) discovered that in the Great Lakes endemic *Hymenoxys acaulis* var. *glabra*, a remnant population in Illinois produced no seed. Crossing experiments with plants from other sites revealed that the plant is self-incompatible and that the Illinois population was all of the same mating type and, thus, prevented from reproducing by seed. The author concludes: 'small populations of self-incompatible species are vulnerable to extinction if the number of self-incompatible alleles, either as a result of a bottleneck or of genetic drift, falls below the number needed for the breeding system to function'.

However, investigations of populations of *Aster furcatus*, a rare primarily self-incompatible species from Wisconsin, USA, have revealed that another outcome is possible (Reinartz & Les, 1994). Small populations of this species were found to have a reduced amount of variation at isozyme loci and seed set proved to be limited by low numbers of S-alleles. But, in some populations, self-compatible individuals were detected and, as they are able to reproduce by seed, it is very likely that such plants will be at a selective advantage and their numbers increase in small populations.

Because the level of inbreeding increases as the number of reproducing individuals declines (Falconer, 1981; Charlesworth a & Charlesworth, 1987), inbreeding depression is another genetic factor to be considered. A number of investigations have recently attempted to study the difficult question of the degree and effects of inbreeding in populations in apparent decline. For instance van Treuren *et al.*

(1993) grew seed samples from *Scabiosa columbaria* from small and large populations in Holland, and examined progenies obtained by three different manipulations: self-fertilisation, within-population crosses and between-population crosses.
The performance of these progenies was followed throughout their life cycle. On average, the within-population progeny showed a four-fold, and the between-population crosses almost a ten-fold, advantage over selfed progeny, indicating that the species is highly susceptible to inbreeding. Three points emerge from these findings. First, the enhanced fitness of between-population progenies points to the effect of factors that limit or prevent gene flow between small isolated populations. Secondly, crosses within small populations may be between close relatives and such crosses may lead to inbreeding depression, albeit less severe than repeated selfing. This may explain the difference between the within-population and the between-population progenies. Thirdly, it has been suggested that inbreeding for several generations in small populations might purge the genetic load and the difference between selfed and outcrossed progenies might then be less in small compared to large populations. The fact that van Treuren *et al.* (1993) found no clear relationship between population size and level of inbreeding depression suggests that the genetic load has not been substantially reduced in the smaller populations. Thus, it seems possible that the small *Scabiosa* populations studied have only comparatively recently been reduced in size. Clearly, the history of the population will affect the level of inbreeding depression detected in experiment. Ouborg & van Treuren (1995) and Honnay *et al.* (2005) may be consulted for details of experiments on other species with small populations.

Minimum viable populations

While there is support for the vortex model from isolated pieces of research on different species, a thorough study has yet to be made of an endangered species that investigated all the interlocking factors contributing to the endangerment and potential extinction of its populations. Even though our understanding remains incomplete, it is clear from these studies, and those of many animal species, that, if the future of endangered species is to be secured, conservationists must pay attention to the size of populations of such species, in particular the number of individuals actually reproducing: the so-called effective population size (Ne), a concept developed by Sewall Wright (1931, 1938). This is defined as the number of individuals in a population contributing offspring to the next generation.
In practice, effective population sizes are usually lower than ecologically observed population sizes, which may include juvenile, non-flowering and diseased individuals. In a population with a large number of non-breeding individuals, Ne is likely to be smaller, perhaps a great deal smaller, than the census population number N. As we have seen, very small populations are vulnerable to genetic drift (for details of the history, assumptions and calculation of the effective population size Ne, see Kliman, Sheehy & Schultz, 2008; Charlesworth, 2009).

How large should the effective population be to ensure population survival? This question was first considered by zoologists, who formulated the concept of the minimum viable population (MVP). 'The MVP to retain evolutionary potential in perpetuity is the equilibrium population size where loss of quantitative genetic variation due to small population size (genetic drift) is matched by gains through mutation. Franklin (1980) estimated this to be a genetically effective size (Ne) of 500 individuals (50 to avoid inbreeding)' (Traill *et al.*, 2010). This estimation became known as the 50/500 rule (Franklin, 1980; Soulé, 1980). Frankham (1995) stressed that Ne is often smaller than the census population. Thus, the effective Ne population size of 500 might require a census population as high as 5000 adults, or perhaps higher (Lande, 1988; Franklin & Frankham, 1998). Such figures greatly exceed the census population size of many threatened and endangered species.

The general applicability of the 50/500 rule has been questioned (Lande, 1988), for it is clear that species differ radically. Instead of applying what amounts to a 'rule of thumb' in conservation management, separate individual assessments should be obtained through population viability analyses (PVAs). Traill *et al.* (2010) note that: 'typically, PVAs are stochastic system models which project changes in population abundance over time and account for demographic and environmental variation, catastrophic events, density dependence and inbreeding depression (Gilpin & Soulé, 1986). [The] likelihood of success is measured on a probability scale and projections into the future can be scaled to years or generations'.

Thus, '[a] minimum viable population for a given species in a given habitat is the smallest isolated population having a 99% chance of survival for 1000 years' (Shaffer, 1981). Survival probabilities may be adjusted in the model to give different probabilities and different timeframes, say 90% and 500 years (Primack, 1993). PVAs have been investigated for a number of plant species (see Traill, Bradshaw & Brook, 2007 for a meta-analysis of procedures and published estimates for a wide range of organisms – vertebrates, fish, plants, insects). They note that such approaches require full census information collected in wild populations and repeated observations to provide detailed quantitative information on the transition between life cycle changes. Using this data, together with information on environmental variables, simulation programmes are then employed using software such as RAMAS Metapop (Akçakaya, 2002).

Having arrived at an estimated MVP, zoologists, who are often dealing with mobile organisms, have also calculated a Minimum Critical (or Dynamic) Area necessary for the survival of such a population. As plants are generally static organisms, which differ widely in density and distribution within the habitat, botanists are not so concerned with trying to estimate MVAs.

The MVP estimates, generated by PVA, have been employed in a number of conservation management programmes designed to counter the extinction threats experienced by populations of a threatened species (Briggs, 2009).

Concluding comments

As invasive and endangered species are of practical, economic and social significance, research funding has often been made available, and as a consequence there have been considerable advances in our understanding of microevolutionary processes – establishment, expansion, decline and extinction – especially through the application of molecular approaches.

Considering these studies, it is clear that these microevolutionary insights often come from the study of extreme situations. For example, in Chapter 9 we discussed the evidence for adaptive responses to single factors such as pollution, herbicides etc. Invasives and endangered species present another series of extreme situations and exciting work is in progress revealing the much more complicated multifactorial interactions in expanding and declining populations in relation to a range of different timescales from a few years to decades or longer periods. As we have seen, many important insights are being obtained into the fundamental processes of microevolution: natural selection, founder effects, gene flow, and the many factors that determine whether populations are expanding or contracting to extinction.

The study of invasive and endangered species has also provided an opportunity to study another important aspect of evolution, namely to what extent does evolution follows predictable paths? Thus, we have seen the current excitements as researchers examine whether invasive species will establish the same patterns of adaptive variation in their introduced range that are found in their native range. Also, critical to our understanding of microevolution processes, to what extent do different populations of endangered species react differently to factors underlying extinction risk?

With regard to extinction processes, there have been recent advances and further progress is expected in our understanding of fragmentation processes, the possible decline of pollinators, the genetics of small and declining populations, and the effects on inbreeding in small populations etc. Attempts to calculate minimum viable populations of plants through population viability analysis have provided important information for conservationists. However, models consider only a limited range of important factors.

In a very important review of the future of ancient woodlands, Rackham (2008) identifies many environmental issues, some of which have the potential to produce profound changes to components of forest ecosystems. Thus, he considers in detail the threats posed by habitat destruction and fragmentation, the depletion of forest resources, effects of pollution and eutrophication, influence of fire or lack of fire, results of excessive shade, excessive number of large herbivores (such as deer and other ungulates), the effects of invasive species, infilling with trees those areas formerly managed as grassland within wood pasture, climate change, the impact of plant diseases (many global) and the interactions between all these threats. Clearly it is important to work towards modelling the fullest range of threats impacting on populations and species of interest.

While studies of endangered species have provided many insights, it could be argued that these are atypical, as humans often intervene decisively to try to prevent the demise of such species through conservation measures (Chapter 20). Likewise, in the case of invasive plants, human interventions also occur, as drastic measures are taken to try to control such species that exhibit invasive behaviour. Management may involve, for instance, physical removal, herbicide treatments or the employment of biological control measures. Whilst biological control has a chequered history, there have been some notable successes in the control of invasive plants (Caltagirone, 1981) and others are being attempted.

- In the 1920s, infestations of Prickly Pear (*Opuntia* spp.) in Australia were controlled by the importation and release of the pyralide moth *Cactoblastis cactorum* from Argentina.
- Successful biological control of Klamath Weed (*Hypericum perforatum*) has been achieved. This was brought to North America by the first European settlers, and by the 1940s more than 400 000 acres of grazing land in California was invaded by this species. This infestation – toxic to livestock – was reduced to less than 1% of its original size following the importation and release of the beetles *Chrysolina hyperici* and *C. quadrigemina* from Australia.
- Japanese Knotweed (*Fallopia japonica*) was introduced to the UK as an ornamental garden plant in the early nineteenth century. Now the plant is a seriously invasive species that is difficult to control by physical removal of its extensive underground root systems. Herbicide treatments are not always appropriate, as adjacent plants may be damaged. Releases of a sap-sucking insect (*Aphalara itadori*) and a leaf spot fungus (*Mycosphaerella polygoni-cuspidati*) have the potential to control the plant, and experiments are in progress to assess the safety of widespread release (www.environment-agency.gov.uk).

Clearly, in the case of endangered and invasive species, human interventions occur, and it could be argued that such populations are 'unnatural' and do not reflect what happens in the wild in 'natural populations'. However, increasingly, it has to be recognised that through habitat loss, exploitation, destruction and management, anthropogenic influences are increasing across the globe in almost all ecosytems (Briggs, 2009).

12 Species and speciation: concepts and models

Since the time of John Ray, whose own attempt at a definition of species we discussed in Chapter 2, there has been no universally agreed definition of 'species'; different definitions have been devised by biologists working in different specialist fields. Thus, the word 'species' has different meanings for different biologists. Here, we examine five influential definitions that focus on different aspects of pattern and process in evolution. (For thorough reviews of species concepts, see Stuessy, 2009; Wilkins, 2009).

The morphological species concept

Historically, the naming, description and classification of species have been based largely upon morphological details of herbarium specimens, and to a lesser extent living material collected from wild or cultivated sources. This is supplemented by geographical and sometimes ecological information. The aim of the taxonomist is to provide a convenient general-purpose classification of the material, a classification that will serve the needs of biologists in diverse fields.

It is quite obvious that in order to communicate experimental findings to others, by word of mouth, in the literature and through databases, the experimentalist, like any other botanist, must be able to name plants unambiguously. To this end, an International Code of Botanical Nomenclature has been agreed. The development of this Code has a fascinating history (Smith, 1957). By 1900, four rival codes of practice were employed in different herbaria. Discussions of the problem occupied taxonomic sessions at International Botanical Congresses in Vienna (1905), Cambridge (1930) and Amsterdam (1935), and the successive Congresses, now at approximately 5-yearly intervals, are the occasion for continued revision of the Code. The international agreements leading to a unified Code must be recognised as a major achievement.

One meaning of the word 'species' is now clarified. We may say that species are convenient classificatory units defined by trained biologists using all the information available. Clearly there is a subjective element in their work, and we must therefore face the fact that there will sometimes be disagreements between taxonomists about the delimitation of particular species, but there is a very large measure of agreement, for all except 'critical groups', in regions where the flora has been studied for many years. In the taxonomic process, a type specimen is designated and a diagnosis provided that specifies the important distinguishing characteristics of the 'new' species from others of the same group (Stace, 1980; Stuessy, 1990, 2009).

Museum taxonomists recognise species on the basis of discontinuities and patterns of correlated variation in morphology. Reproductive behaviour is a key element in other species concepts (see below), but generally, taxonomists do not have enough direct information to take this factor into account in their work. However, Stuessy (2009) notes that: 'it is likely that the morphological discontinuities recognised formally' in species designation 'do reflect biological limits of isolation and commonality of interbreeding and genetic divergence'. Where experimental evidence is available – cytogenetic, molecular, chemical, anatomical, ecological, geographical etc. – taxonomists will generally consider this in their work.

Species definitions: taking into account pattern and process

Acknowledging the Darwinian insights into the evolution of life on Earth, other definitions of species attempt to relate pattern to process in the evolution of species.

The Evolutionary Species Concept

In order to deal with both modern and fossil lineages, Simpson (1961) proposed an evolutionary species concept. 'An evolutionary species is a lineage (an ancestor–descendent sequence of populations), evolving separately from others and with its own unitary evolutionary role and tendencies.'

Phylogenetic (cladistic) species concept

In the post-Darwinian period, biologists have been fascinated by the idea of determining the phylogenetic history of groups of organisms. Now, the armchair speculations of the nineteenth century have given way to modern phylogenetic investigations. The spectacular advances in molecular phylogenetic analysis have generated further interest in species concepts (see Davis, 1995; Luckow, 1995). In this context, Cracraft (1983) considers that: 'A species is the smallest diagnosable cluster of individual organisms within which there is a parental pattern of ancestry and descent.'

Ecological species concept

In an attempt to include ecological elements in a species definition, van Valen (1976) considered species to be: 'A lineage (or closely related set of lineages) which occupies an adaptive zone minimally different from that of any other lineage in its range and which evolves separately from all lineages outside its range.'

The definitions are considered to be important in the specialisms in which they developed. However, our fifth definition, the Biological Species Concept, has proved highly influential across many fields of biology.

The Biological Species Concept

This is a most important concept that informs the investigations of many biologists (Coyne, 1994). During the 1920s and 1930s, the botanists Turesson, Clausen and others suggested a number of new terms for units at or about the level of species based upon breeding behaviour, an idea with a long history as we have seen. The zoologist Dobzhansky also played a crucial role, considering species definitions in the light of the Evolutionary Synthesis. In his highly influential book, *Genetics and the Origin of Species*, Dobzhansky (1937) quoted his own definition drawn from a paper he wrote in 1935. He proposed 'to define species as that stage of evolutionary progress . . . at which the once actually or potentially interbreeding array of forms become segregated in two or more separate arrays which are physiologically incapable of interbreeding'. He continues:

> the definition of species just quoted differs from those hitherto proposed in that it lays emphasis on the dynamic nature of the species concept. Species is a stage in a process, not a static unit. This difference is important, for it frees the definition of the logical difficulties inherent in any static one. At the same time, our definition can not pretend to offer a systematist a fixed yardstick with the aid of which he could decide in any case whether two or more groups of forms have or have not reached the species rank. This drawback is unavoidable. A systematist is forced to describe the changing patterns of life in terms of abstract static conceptions.

Building on Dobzhansky's approach the eminent zoologist Mayr, whose special interest was in birds (Haffer, 2007), developed the Biological Species Concept (BSC). Biological species are defined as 'groups of actually or potentially interbreeding natural populations which are reproductively isolated from other such groups' (Mayr, 1942). This definition has proved highly influential in both zoological and botanical researches. Mayr (1963) set out the strengths of the BSC approach.

> The typological species concept treats species as random aggregates of individuals that have in common 'the essential properties of the type of the species' and that 'agree with the diagnosis' . . . This static concept ignores the fact that species are reproductive communities. The species is also an ecological unit that . . . interacts with other species with which it shares the environment. The species, finally, is a genetic unit consisting of a large, intercommunicating gene pool.

In a later chapter we consider current views of this model. Here, we note that the concept has been much criticised. A number of biologists pointed out that the concept of 'potential' interbreeding was very difficult and, later, Mayr (1969) removed the words 'actually or potentially' from his definition. Then, in response to further criticisms, Mayr (1982) presented yet another version. 'A species is a reproductive community of populations (reproductively isolated from others) that occupies a specific niche in nature.' This definition too has caused argument, principally concerning the difficulties of defining the concepts of 'niche' and 'reproductive community'.

Mayr's Biological Species Concept, in its different forms, makes it clear that a key step in speciation is the acquisition of reproductive isolation. Thus, groups of related plants, which are distinct at the level of biological species, do not interbreed when growing in the same area in nature. They are said to pass the test of sympatry, that is, of growing together without losing their identity through hybridisation.

The mechanisms that keep biological species separate have been closely studied for many years, and will be examined in detail below, but, in general, as Table 12.1 shows, isolating factors fall into three groups.

Table 12.1. **A classification of isolating mechanisms in plants (Levin, 1978b)**

Premating	
Spatial	
1. Ecological	
Reproductive	
2. Temporal divergence	
(a) Seasonal	
(b) Diurnal	
3. Floral divergence	
(a) Ethological	
(b) Mechanical	
Postmating	
4. Reproductive mode	
5. Cross-incompatibility	
(a) Pollen–pistil	*Pre-zygotic*
(b) Seed	*Post-zygotic*
6. Hybrid inviability or weakness	
7. Hybrid floral isolation	
8. Hybrid sterility	
9. Hybrid breakdown	

In some cases, it is proposed that pollination may be prevented; for instance, biological species found in the same area may grow in slightly different habitats, or flower at different times of day or in different seasons, and/or, for reasons of flower structure or pollinator behaviour, cross-pollination may not be successfully achieved. Even if cross-pollination occurs, pollen may fail to grow down the style. It is also possible, if plants are regularly and automatically self-pollinating, that cross-pollination may be prevented or its frequency may be greatly reduced.

The model also highlights the importance of a group of so-called 'post-zygotic' mechanisms. The seed from a cross between two biological species may fail to develop properly as a consequence of incompatibility between embryo, endosperm and maternal tissues

(Valentine, 1956). In other cases, where hybrids are produced, they may show various signs of defective development or reduced fertility. More pronounced difficulties have been detected in other situations, where hybrids may be viable but sterile, or they may be weak as well as sterile. Finally, defective progeny, in a cross between two putative biological species, may be discovered only in F_2 or later generations.

In an important review of isolating mechanisms, Levin (1978b) suggests that entities distinct as biological species are not generally separated by only one isolating mechanism and, furthermore, that barriers to crossing are not encountered simultaneously, but may be seen as a series of resistances which have to be overcome, if crossing is to be effected. Thus, it is proposed that, potentially, a number of barriers – both pre- and post-zygotic – may be discovered between biological species.

The notion of 'biological species' began to catch the imagination of experimentalists. As we noted in Chapter 5, the concept was a major element of the New Synthesis developed from various perspectives in a series of highly influential books including: Dobzhansky (1937, 1941, 1951), Huxley (1940, 1942), Mayr (1942), Simpson (1944), Stebbins (1950), Clausen (1951) and Heslop-Harrison (1953). As we see in later chapters, from this ferment of discussion the following influential ideas emerged for critical analysis and testing:

1. As the definition of biological species (and more or less equivalent groupings described in the 1920s and 1930s) involved a test of breeding behaviour, experimentalists considered that entities defined thereby were more objective than the 'species' of the herbarium taxonomist (cf. Gregor, 1931; Müntzing, Tedin & Turesson, 1931; Clausen, Keck & Hiesey, 1939).
2. The species and classifications produced by taxonomists – the so-called 'alpha taxonomy' – should be modified in the light of experiments to give a more perfect system eventually leading to an 'omega taxonomy' in which all the knowledge of biologists reached proper synthesis (Turrill, 1938, 1940).
3. Crossing experiments and other information, such as chromosome numbers, might reveal something of the phylogeny of groups, and the existing classificatory systems, which are a mixture of convenient arrangement and phylogenetic speculation, could be modified to allow classification to reveal the evolutionary pathways leading to present-day patterns (Darlington, 1956, 1963).

These ideas were attractive to some botanists and provided a stimulus for researches of various kinds. To give the correct historical perspective, however, we shall see that they were not universally accepted, and a great deal of argument ensued. While some biologists were concerned to try to apply the Biological Species Concept in taxonomy, others saw the concept as a model, useful for thinking about the processes involved in reproductive isolation, rather than a practical tool. We will examine the present views of the relation of experimental and taxonomic categories in Chapter 15, after presenting some of the ideas and results produced by the work of experimentalists who came to be known as 'experimental taxonomists' or 'biosystematists' (Camp & Gilly, 1943).

Origins of species

We may now turn our attention to the origins of species. Here we introduce a number of models. How these have been tested and developed will be considered in the following chapters. With regard to speciation the use of the plural 'origins' is important, as biologists now accept that there are a number of modes of speciation, which may conveniently be grouped under two heads, namely 'gradual, sometimes called geographic, speciation' and 'abrupt speciation'.

Gradual (Geographic) Speciation

Population geneticists point to four groups of processes of importance in the separate evolution of different geographically isolated 'sub-populations' derived initially from a single original local interbreeding population.

Mutations As genetic mutations are the result of random events, it is likely that, over time, patterns of genetic change will be different in the derived isolated populations.

Other effects of chance As we have seen, chance may influence the variation in populations. Some of the derived populations, dispersing to new territory, may be subject to founder effects and, if the numbers of individuals remains or becomes small, then genetic drift may occur (Wright, 1931) and alleles might be completely lost by accidents of sampling, or in other cases rare alleles might by chance become more frequent, without regard to their adaptive value.

Selection. As derived populations will be subject to different climatic, edaphic and biotic factors, the action of selection is likely to lead to genetic differentiation between the derived populations.

Migration. The extent and direction of change could be greatly influenced by the degree to which derived populations are effectively isolated from gene flow. By virtue of the independent small cumulative changes possible in different isolated sub-populations of an original interbreeding population, it seems probable that, given a long enough period of geographical and genetic isolation, genetic differentiation in a group of derived populations of common origin might proceed first through an ecotypic phase and then, later, with the gradual evolution of genetic isolating mechanisms, derivatives at the rank of biological species might be formed. Historically, such gradual speciation is thought to be highly important in the evolution of species, especially in the light of recent advances in our understanding of the geological and climatological history of the Earth. However, models of gradual speciation raise a number of issues. With regard to the postulated genetic changes in this type of speciation, are they small and gradually accumulated, or are some abrupt events involved?

Abrupt speciation

As we saw in Chapter 2, one of the controversies stimulated by Darwin's work concerned the possibility of 'saltations', or abrupt changes in the course of evolution. Darwin himself saw evolution as a continuous, gradual process, with no place for sudden 'large' events, but those who argued for abrupt change, in some circumstances, have been vindicated. There is now overwhelming evidence for abrupt speciation in plants. Thus, chromosome repatterning and changes in chromosome number may occur. Genetic events may change the breeding system. These abruptly arising changes in some individuals of a population may lead, as we shall see in Chapter 14, to the sympatric origin of new species.

Polyploidy

In many cases, polyploidy is involved in abrupt speciation, but other chromosome changes are possible and likely to be very important. As this chapter is designed to provide an introduction to a more thorough treatment in later chapters, at this point we confine our attention to polyploidy, and will consider other models of abrupt speciation later.

According to Rieger, Michaelis & Green (1976), the term 'polyploid', for a plant containing more than the normal number (two) of sets of chromosomes, seems to have been first defined by Strasburger (1910). Winkler (1916), in an important paper, describes what is probably the first clear case of experimental production of polyploids in the Tomato (*Lycopersicum esculentum*) and the related Nightshade (*Solanum nigrum*). Before this, a great deal of interest had centred on the work of De Vries with the Evening Primrose (*Oenothera*) (to which we have already alluded in Chapter 5) and, in particular, upon a 'mutation' that was called '*gigas*' because it was generally larger than the parent *Oenothera erythrosepala* (*O. lamarckiana*). This '*gigas*' mutant had been shown to possess twice the normal somatic chromosome number of 14, and there was argument between De Vries and Gates as to the significance of this difference, Gates (1909) holding the view that the chromosome doubling was itself a cause of the differences in morphology between '*gigas*' and the normal plant. Winkler's demonstration that his experimentally produced polyploid Tomatoes, with double the normal chromosome complement, differed also from the 'parent' diploid in similar ways to the '*gigas*' mutant strongly supported Gates' interpretation, and subsequent work showed that artificial polyploids generally differed in the larger size of all their parts, from the mean cell size to the size of the whole plant.

Soon after Winkler's paper, Winge (1917) made an important contribution in distinguishing between the simple doubling of the chromosome number in a single individual (autopolyploidy), and a more complicated situation where polyploidy followed hybridisation (allopolyploidy). (The terms auto- and allopolyploidy were coined by Kihara & Ono in 1926.) A number of early tests of Winge's hypothesis were made.

- Clausen and Goodspeed (1925) crossed a diploid species (*Nicotiana glutinosa* – $2n = 24$) with a tetraploid (*N. tabacum* – $2n = 48$) and produced a fertile hexaploid ($2n = 72$).
- Following the production of hybrids between the Radish (*Raphanus sativus* – $2n = 18$) and the Cabbage (*Brassica oleracea* – $2n = 18$), spontaneous chromosome doubling occurred to give a fertile tetraploid *Raphanobrassica* ($2n = 36$) (Karpechenko, 1927, 1928).

Autopolyploidy can be explained as follows. A diploid plant receives a haploid set of chromosomes (a genome) from each parent. Its constitution can be represented as AA. If the plant is subject, for example, to temperature shocks, the regular process of mitosis may be disturbed and, instead of two cells each with the diploid number of chromosomes, a single diploid cell with four times the haploid number may be formed (AAAA):

$$\mathrm{AA} \xrightarrow{\text{doubling}} \mathrm{AAAA}$$

In this way polyploid cells arise and may give rise to polyploid branches on diploid plants. Experimentally, polyploid cells can be produced with the drug colchicine, which acts as a spindle inhibitor preventing regular disjunction of chromosomes. Thus, chromosome replication in colchicine-treated material is not combined with the proper division of the products into two daughter nuclei, and the formation of a nuclear membrane around all the replicated chromosomes yields a polyploid cell, which may be involved in the production of autopolyploid seeds (AAAA). An autopolyploid plant may also be produced by different means, namely by the fusion of two unreduced gametes (AA + AA).

Let us now consider the origin of a new species through allopolyploidy. In essence, this mode of speciation 'involves the reuniting of divergent genes and genomes through sexual hybridization' (Rieseberg & Willis, 2007).

Two related diploid species that have diverged by gradual speciation from a common ancestor may be different both chromosomally and genetically, and may be represented as AA and BB. A hybrid between the two species, AB, may very well be highly infertile, as there is insufficient homology between the A and B genomes for proper pairing at meiosis. The hybrid plant may produce a polyploid branch, of genetic constitution AABB, by the process outlined above. However, in the AB hybrid a very small but significant percentage of unreduced (AB) eggs and pollen may be produced, together with unbalanced haploid meiotic products. On fusion of unreduced gametes, a plant with the constitution AABB is produced, in which the chromosome number has been effectively doubled:

AA × BB -----> AB ------> AABB

In this simple case we are dealing with a tetraploid with twice the normal diploid number. Such plants are sometimes referred to as 'amphidiploids' or 'amphiploids'. Other kinds of polyploids with extra genome sets are described with the appropriate term: 'triploid', 'hexaploid', etc. The level of 'ploidy' can be represented as the multiple of the 'basic number' x, which is the haploid number of the presumed original diploid or diploids; thus a triploid can be represented by $3x$, a tetraploid by $4x$, etc. (If this notation is used, it is then possible to retain n and $2n$ to indicate the functional haploid and diploid numbers, as distinct from the presumed polyploid relationships within a whole genus or group of species.)

Meiosis in the new allopolyploid is more normal than in the diploid hybrid (AB), as genomic pairing – A with A and B with B – can occur. If, however, there is still a high degree of homology between A and B genomes (they were perhaps derived from a common ancestor by gradual speciation), then more complex pairing of the chromosomes may occur and groups of three and four chromosomes may be found (Fig. 12.1).

Allopolyploid derivatives are reproductively isolated from their parents, as can be seen by examining what happens when an allopolyploid AABB (with gametes AB) is crossed with one of its 'parental' species, AA (with gametes A). Triploid individuals of constitution AAB are produced. Even though A genomes may pair at meiosis, there is no pairing partner for the B genome, and a highly irregular meiosis occurs, which leads to infertility in the hybrid. An isolating mechanism now exists between diploids and their derived allopolyploid

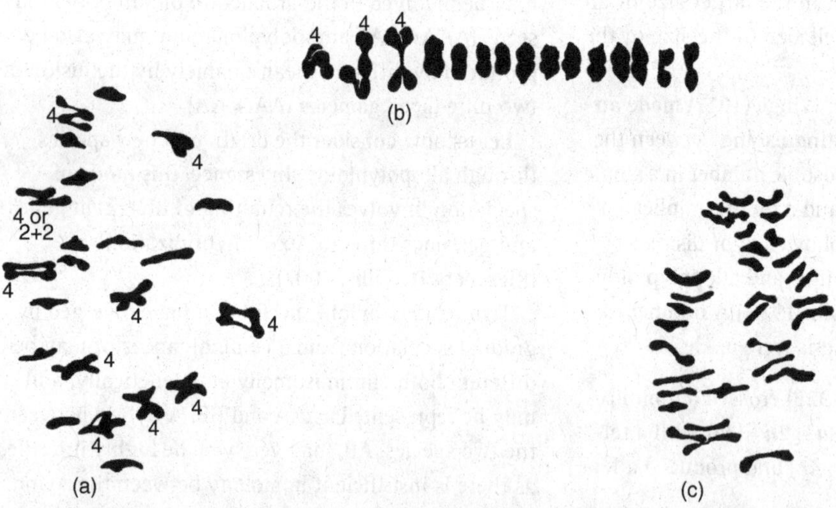

Fig. 12.1. (a) Meiosis (metaphase 1) in autopolyploid Watercress ($2n = 4x = 64$) prepared by colchicine from *Rorippa nasturtium-aquaticum* (*Nasturtium officinale*) ($2n = 2x = 32$).

(b) Meiosis (metaphase 1) in *Primula kewensis* ($2n = 4x = 36$) (× 1600). Note the three quadrivalents.

(c) Meiosis (metaphase 1) in wild tetraploid Watercress *Rorippa microphylla* (*N. microphyllum*). ($2n = 4x = 64$) (× 2500). ((a) and (c) from Manton, 1950; (b) from Upcott, 1940.)

In tetraploids, a range of different cytological behaviour is found. In autotetraploids (of type AAAA), quadrivalents are frequently found, as in (a). In allotetraploids (of type AABB), where each chromosome has a pairing partner, normal bivalent pairing is found, as in (c). Sometimes a mixture of quadrivalents and bivalents is discovered as in (b). Present insights into chromosome behaviour have revealed difficulties with these simple historic models (see text).

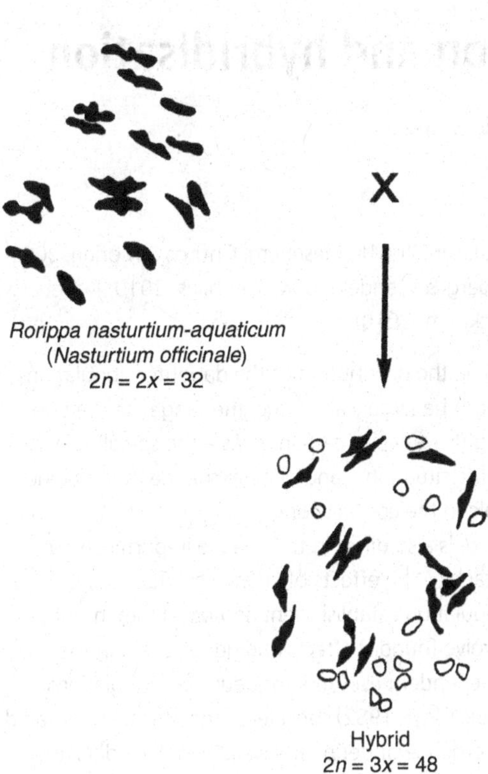

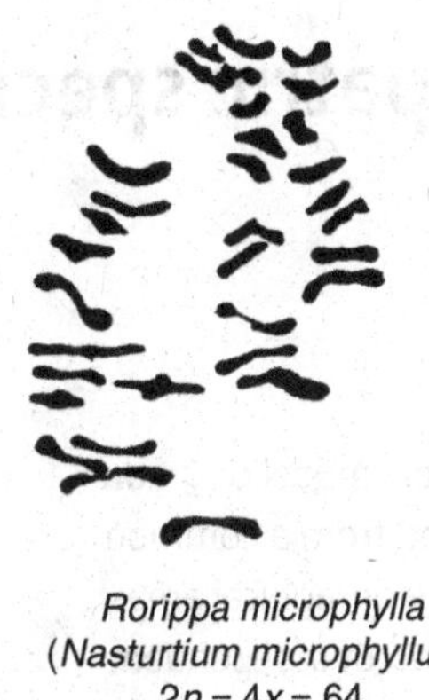

Fig. 12.2. Meiosis in a triploid hybrid (×1600). Note the mixture of bivalents (black) and univalent (white) at metaphase 1 of meiosis in the triploid. (From Manton, 1950.)

(Fig. 12.2). It is by the abrupt origin of an isolating mechanism in this way that new biological species arise by the process of polyploidy. New species may originate where the two parental species occur together, that is, sympatrically. Moreover, as new polyploids are produced by single, abrupt events, here we have a mechanism whereby, within a human lifetime, new, fertile species may arise.

In the chapters that follow, these two contrasting models of speciation and others are examined in detail, and their relative importance in evolution assessed.

13 Allopatric speciation and hybridisation

In Chapter 12 we presented a simple model of gradual speciation. Two populations derived from a common ancestor and occupying different geographical areas (i.e. allopatric) pass through a period of independent change yielding derivatives that are reproductively isolated from each other. In such cases the existence of isolating mechanisms is revealed if the taxa come to occupy the same area (i.e. become sympatric).

Allopatry may arise in many different ways. A comprehensive review of evolutionary plant geography, in relation to geological, climatological and historical factors, is presented in Chapter 17. Here, we note a few of the many possibilities, including vicariance and dispersal. 'Vicariance is the appearance of a barrier that allows fragmentation of the distribution of an ancestral species, after which the descendent species may evolve in isolation' (Morrone, 2009). For example, geographical isolation of daughter populations may result from the destruction of land bridges (e.g. the opening of the Irish and North Seas following post-glacial sea-level changes). In the longer geological perspective, vicariance events may be the result of major geological processes, such as continental drift and associated mountain-building.

New isolated populations may also result from long-range dispersal of propagules to new territories, including 'islands' of different sorts, whether they be oceanic islands, isolated mountain peaks, landlocked lakes or areas associated with specialised rock types (e.g. serpentine).

To take account of the complexities of different situations likely to be important in nature, a group of different models of allopatric speciation have been devised incorporating a range of assumptions (Grant, 1971; Levin, 2001b; Rieseberg, Church & Morjan, 2003; Rieseberg & Wendel, 2004; Bomblies, 2010; Rieseberg & Blackman, 2010).

1. While the distribution of the daughter populations might be largely allopatric, the ranges of the two might overlap in certain areas – the so-called parapatric situation – and gene exchange might be possible in the contact zones.
2. Models also differ in the relative importance they attach to the effects of chance events. For instance, population establishment and development may involve founder effects and genetic drift in the independent evolution of daughter populations. Thus, Mayr (1982) considered the possibility of 'rapid divergence of peripheral isolates or founders' (so-called peripatric speciation) (Baldwin, 2006).
3. However, Rieseberg & Wendel (2004) note 'the declining influence of models of speciation based on population bottlenecks and genetic drift' with attention being focused on 'divergent natural selection as a cause of divergence'. In this family of models 'reproductive isolation evolves as a byproduct of adaptation to different habitats'. Significantly, Rieseberg, Church & Morjan (2003) detect another change of emphasis. 'Early models emphasized the importance of geographical isolation, because gene flow was thought to prevent geographically proximal populations from diverging through drift and selection (Mayr, 1942). However, studies over the past 40 years have shown time and again that speciation may occur despite gene flow between populations (McNeilly & Antonovics, 1968a, b; Caisse & Antonovics, 1978; Church & Taylor, 2007)'. They continue: 'evidence from experimental studies of selection,

quantitative genetic studies of species' differences and the molecular evolution of "isolation" genes, all agree that directional selection is the primary cause of speciation, as initially proposed by Darwin'.

4. Levin (2003) considers the ecological transitions involved in speciation. He points out that the process involves two steps. 'First there is the establishment of ill-adapted populations where ecological opportunity allows. This is followed by genetic refinement of populations, which permits them to be integrated into novel communities and habitats. These steps are more readily accomplished in unsaturated floras, where competition is less intense. Ecological transitions in saturated floras may be facilitated by disturbance.' In view of these considerations, Levin concludes that important information on the early stages of speciation may come from the investigation of invasive species adapting to their new territories.
5. In constructing model systems, different assumptions might be made about differentiation. Thus, morphological differentiation in daughter populations might proceed in tandem with the sort of genetic changes that yield eventual isolating mechanisms or, alternatively, morphological change and 'reproductive isolation' may evolve at different rates.

Evidence for gradual speciation

In considering evidence for various models of gradual speciation, the extended timescale of perhaps hundreds of generations presents an immediate difficulty. It seems highly likely that the precise determination of the processes of change from a single ancestral population to two biological species cannot with certainty be fully reconstructed. However, considerable advances in our understanding have been achieved through the use of a variety of experimental approaches, including the use of molecular genetic techniques.

An historical perspective to studies of speciation is an excellent way to introduce the latest insights. A highly influential approach was championed, for example, by Clausen (1951). If groups of different related taxa are examined, they may be at different stages in speciation. Thus, differentiation might be at the ecotypic stage, as, for example, in *Potentilla glandulosa* in California (see Chapter 8), whilst other populations, perhaps given sub-specific, semi-specific or specific rank by taxonomists, may be at a later stage in speciation. By examining a range of different types of situation, from ecotypes to island endemics, from local races to vicariads (two similar taxa occupying different geographical areas), a composite picture of allopatric speciation might be built up.

To test the ideas of Clausen, it is necessary to discover the degree of reproductive isolation between collections of different taxa – taxa which, on account of their distribution, morphology, etc., are likely to share, either closely or remotely, a common ancestor. Plants from different areas, whose cytotaxonomic background is well understood, are brought together and, by appropriate crossing experiments, the viability and fertility of any resulting hybrids is determined. In this way, the genetic basis of ecotypic differentiation in *Potentilla glandulosa* was investigated in crosses between alpine and coastal plants, and between plants from sub-alpine habitats and those from the foothills across a transect in central California. Clausen & Hiesey (1958) detected considerable genetic differences between the ecotypes. Also, working with several different genera, Clausen and associates carried out extensive and elaborate crossing programmes between a range of species and subspecies, the results of which are set out in crossing polygons, in which the fertility of F_1 (and sometimes F_2) hybrids is revealed in diagrammatic fashion as a geometric figure

(Clausen, 1951). Radford *et al.* (1974) illustrate a further range of examples.

Crossing experiment with species of *Layia*

The experiments of Clausen and associates on a range of groups provide classic examples of pioneering investigations of speciation (Clausen, 1951). For example, they studied the results of crossing different subspecies of Californian Tarweeds (*Layia glandulosa*). Evidence of partial genetic barriers is provided by some of these crosses (Fig. 13.1).

They also examined the cytogenetic and crossing relations of a range of species of the genus *Layia* by studying chromosome pairing in the F_1 hybrids (Fig. 13.2). The species fell into two major groups: those with a chromosome number $n = 7$ and those with $n = 8$. The degree to which crossing was possible between the different species was examined (Clausen, Keck & Hiesey, 1941). Reflecting on these experiments, Clausen (1951) concluded that '*Layia* is an excellent example of a genus in various stages in evolutionary differentiation'. Between the races or ecotypes there are 'moderate barriers to interbreeding . . . At an even more advanced stage, the species have become so distinct that the hybrid is much weaker than its parent species, or the species are unable to produce a viable hybrid.'

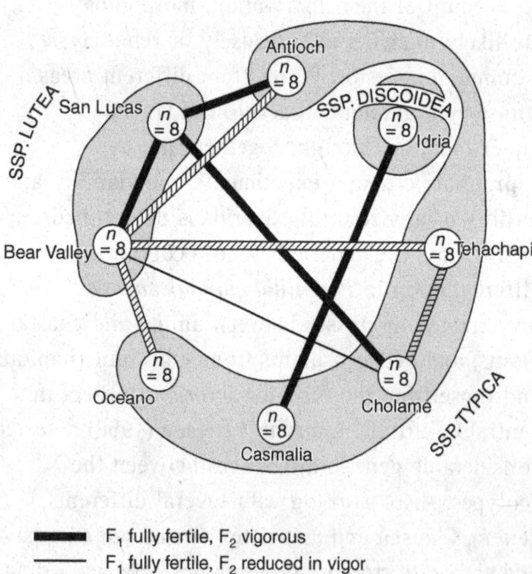

Fig. 13.1. Interspecific crossing within *Layia glandulosa*. (Reproduced with permission of the publisher, Cornell University Press, from Clausen (1951), *Stages in the Evolution of Plant Species*. © 1951 Cornell University Press.)

The interpretation of crossing experiments

First, there are important questions to be faced about the 'reality' or otherwise of ecotypes (Lowry, 2012). As we saw in Chapter 8, experiments have revealed that some species appear to be made up of distinct ecotypes: others have more continuous patterns of variation. Many factors can influence the pattern, as revealed in investigations; for example, the distribution, ecology, habitat preferences and breeding systems of the species under study. Also of crucial importance are the sampling strategies employed in different investigations, the cultivation conditions (one or several common garden sites with weeded or unweeded plots) and the experimental design (clone transplant or seedling stocks grown, etc.).

Turning to the crossing experiments, the choice of only one or two individuals, to represent what is taken to be a local race or ecotype, is problematic. Which samples are selected will influence the results. Furthermore, while crossing experiments provide valuable information, they require cautious interpretation. Also, a large amount of garden space may be needed to raise F_2 families, and, therefore, crossing programmes are often halted at the F_1 stage. There are other grounds for caution. Glasshouses are frequently used in breeding experiments. Insect pests and fungal diseases may interfere in crossing experiments, and success or failure of crossing may sometimes be influenced by these factors. The effects

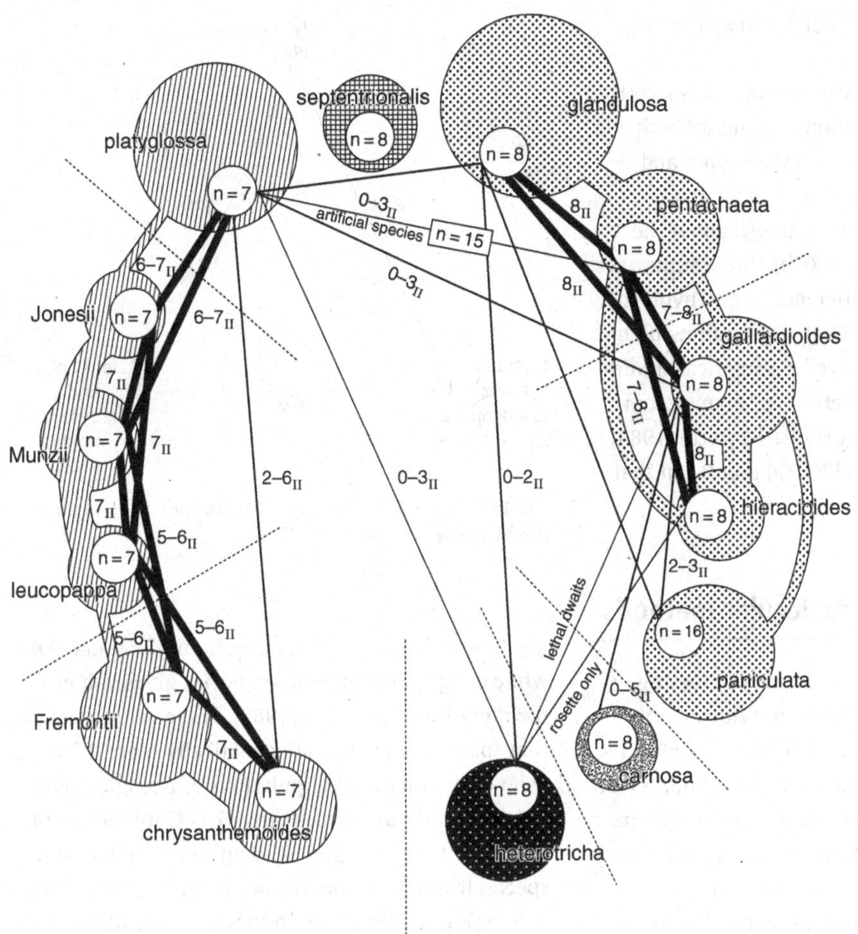

Fig. 13.2. Crossing polygon of *Layia* showing cytogenetic relationships. The black lines connecting species indicate successful hybridisations, and the width of the lines, the approximate degree of pairing between the parental chromosomes in the hybrid. The width of shaded connections between species indicates degrees of gene exchange within the genus. (From Clausen, Keck & Hiesey, 1941.)

of bagging flowers must also be considered. Geiger (1965) reviews the relevant literature, which indicates that temperature inside pollen/insect-proof bags might be up to 15°C higher than ambient temperatures in the daytime and −2°C lower at night. As pollen sterility and other effects may be induced at high temperatures, failure in crossing may be due to these external factors, rather than to intrinsic differences. Thus, pollen viability may decline quite rapidly, both inside and outside bags, and therefore it may be vital to use fresh pollen in crossing experiments (see Stone, Thompson & Dent-Acosta, 1995). Furthermore, the 'fertility' of pollen in presumed parental taxa and hybrids is commonly estimated, not by direct study of germinability or in a crossing test, but by staining with acetocarmine or other stain. Fully formed grains with nuclear staining are assessed as 'good' pollen, and misshapen, undersized, inadequately stained grains are judged to be 'bad'. The reliability of staining as an indicator of fertility is rarely, if ever, put to the test. In assessing crosses, the possible complicating factor of genetic incompatibility must also be considered (Chapter 7). Also, artificial crossing experiments do not often assess possible pre-zygotic isolating factors, but concentrate on the results of experiments that often involve crude surgery. Will such experiments reveal what would happen if isolated populations became sympatric in nature and only the normal agencies of pollen transfer were to operate?

Despite these caveats, the early work of Clausen & Hiesey (1958) provided important evidence of heritable character differences between races, ecotypes and species that underpinned speciation. Later, in the 1980s, considerable differences emerged about the genetics of speciation (Levin, 2001b). Some authorities argued that major genetic differences were involved (see, for example, Gottlieb, 1984). Others, drawing on theoretical considerations as well as empirical studies, proposed that the interactive effects of many minor genes were important (Barton & Charlesworth, 1984; Coyne & Lande, 1985). Levin (2001b) points out that both types of change may occur.

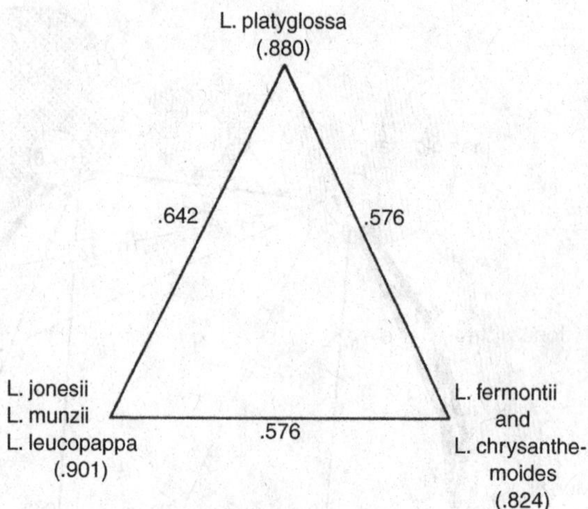

Fig. 13.3. Mean identity values for six species of *Layia*. (From Warwick & Gottlieb, 1985.)

Studies of *Layia* using molecular methods

Given some of the potential problems of crossing experiments, molecular tools now provide an important opportunity for a critical assessment of the models of the genetics of allopatric speciation. As we see in the following chapters, many of the taxa, first examined using cytotaxonomic methods, have now been re-examined with molecular techniques.

Further evidence concerning the evolution of the genus *Layia* has come from the studies of a number of populations of each of the six species that have the chromosome numbers $n = 7$ (Warwick & Gottlieb, 1985). Thirteen enzyme systems were analysed by means of electrophoresis of leaf extracts.

Before looking at these results we must consider the general question of how to estimate the similarity or difference between and within species. From the results of studying electrophoretic variability in enzymes, a statistic devised by Nei (1972) – the genetic identity – may be employed (Avise, 1994, gives details of the calculations). Genetic distance values range from 1.0 to 0.0. If two populations are identical, then the genetic identity value will be 1.0. If there are no alleles in common, the value will be 0.0. In comparisons of populations of the same species, mean genetic identities are usually in the range 0.90, while comparing congeneric species gives values in the range 0.67 (Crawford, 1989). These values indicate that species of a genus are genetically more different from each other than populations of a single species.

In the study of *Layia*, Warwick & Gottlieb (1985) discovered that the genetic identity for each of the species (estimated from the populations sampled) was 0.96 or higher. However, in the case of *L. platyglossa*, where the populations were more variable, the species had a genetic identity value of 0.88. (This finding of greater variability in *L. platyglossa* confirms the findings of Clausen, 1951.) The mean genetic identity values produced by comparing the six species are given in Fig. 13.3.

The genetic distance of *L. platyglossa* from the other species is similar to that reported in other comparisons of congeneric species. Concerning the high genetic identity between the three endemic species (*L. jonesii*, *L. munzii* and *L. leucopappa*), Warwick & Gottlieb suggest that they may be 'more closely related to each other than to other *Layia* [species] and that their divergence was relatively recent, presumably a consequence of recent (Pleistocene) isolation of different newly exposed soils'. Considering all their findings they conclude 'the genetic identity values are consistent with those expected on a model that the species diverged gradually as they adapted to geographically separate habitats'.

Phylogenetic studies of the Californian Tarweeds

The investigations of Warwick & Gottlieb have been greatly extended by the studies of Baldwin (2006), who examined phylogenetic trees obtained from a study of nuclear ribosomal DNA (rDNA) internal transcribed spacer (ITS) sequences. His investigations permit the critical testing of a number of hypotheses emerging from the historic studies.

He reports that with regard to the $n = 7$ members, the 'tree topology and branch lengths correspond well to Clausen *et al.*'s (1941) model of gradual allopatric diversification'. In contrast, the $n = 8$ group sister lineage that resolves into two subgroups reveals several examples that 'likely arose rapidly in peripheral or otherwise isolated, ecologically distinct settings'. Baldwin offers the following commentary on his results.

The endemic species *L. discoidea* grows on serpentine barrens, and lacks both ray florets and involucral bracts found in its near-relative *L. glandulosa*, which has white or yellow florets and grows 'on coarse sandy soils including sand dunes'. Considering the origin of *L. discoidea*, Clausen (1951) concluded that '*Layia discoidea* was probably an edaphic race of *L. glandulosa* adapted to serpentine soil, and possibly an ancient relict'. In contrast, Gottlieb, Warwick & Ford (1985), studying allozyme variation, pointed to the high genetic similarity between the two species, with values almost as high as conspecific populations. They concluded that *L. discoidea* was a 'peripheral isolate that diverged rapidly and recently in an ecologically marginal setting'. The phylogenetic studies of Baldwin offer support for the hypothesis of recent descent of *L. discoidea* from a sub-lineage of *L. glandulosa*, most probably, as Gottlieb suggests, from a variant with yellow flowers. Considering the differentiation of *L. discoidea* from ancestral *L. glandulosa* populations, evidence suggests that this 'did not involve major genotypic effects', as there is complete interfertilty between the two species. In his important review of more than 30 examples of proposed progenitor and derivative species, Gottlieb (2003) concludes that what Verne Grant called Quantum Speciation is an important mode of speciation, in which 'a derived species is budded off and acquires new traits, while the parental species continues more or less as before'. Such ecological selection, without radical genetic change to the progenitor species, is likely to be an important aspect of speciation in many groups, with accelerated evolution likely in 'peripheral populations' (Mayr, 1982; Barton & Charlesworth, 1984; Coyne, 1994). A recent issue the *Philosophical Transactions of the Royal Society of London* reviews Gottlieb's insights into this form of speciation, and celebrates his many contributions to other areas of plant evolution (Crawford *et al.*, 2014).

Considering *L. gaillardioides* and its relatives, the second group with $n = 8$, Baldwin makes two points. In contrast to the *L. glandulosa/L. discoidea* group, several of these taxa are intersterile. Also, while most *Layia* species are self-incompatible, two species of the *gaillardioides* group – *L. carnosa* and *L. hieracioides* – are self-fertile. The third taxon, *L. septentrionalis*, is a self-incompatible species.

Considering all the evidence now available, Baldwin (2006) considers that 'the 'evolutionary processes in *Layia* have been more dynamic' than was suggested by the biosystematic results obtained by the methods available to Clausen *et al.* (1941).

Speciation genes

Recently, major advances have been achieved in the study of speciation in some animal groups through the use of molecular approaches, e.g. *Drosophila*. Now, botanists are turning their attention to the investigation of speciation genes and genomic changes that underlie the speciation process in plants.

Rieseberg & Blackman (2010) and Bomblies (2010) provide important reviews of plant speciation genes. These are genes that 'contribute to a cessation of gene flow between populations' and, thereby, contribute to the establishment of reproductive isolation in the speciation process (Rieseberg & Blackman, 2010). In a number of cases it has proved possible to clone and characterise such genes providing validation of the genetics and regulation of candidate speciation genes, with clarification of the mutational events involving changes in coding, regulatory control, copy number, and microchromosomal rearrangements. A number of genes have been investigated in crop and other species, but their potential role in speciation, in the wild, has yet to be properly investigated. Here, it is only possible to give a brief outline of some of the case histories reported in these reviews.

First, to recap, isolating mechanisms act in different ways, some before the formation of the zygote; others are post-zygotic.

Pre-pollination mechanisms

Closely related species often have different flower colours, and colour shifts are clearly of great importance in plant speciation. Bomblies (2010) distinguishes two basic situations, namely: (a) 'alterations to the regulation of pigment biosynthesis' and (b) 'mutations in the biosynthetic enzymes themselves'.

Alteration of regulation Drawing on the research of Des Marais & Rauscher (2010), Bomblies notes that: 'in *Ipomoea* red variants of three species have been independently derived from ancestral blue or purple. All have reduced activity of the enzyme flavonoid 3′-hydroxylase (F3′H), which results in a shift towards a different branch of the anthocyanin pathway. This is a regulatory change: F3′H function is not lost in these species, rather, an expression change reduces F3′H in flowers but not in stems.' Another example is provided by investigations of *Petunia axillaris*/*P. integrifolia*, by Quattrocchio *et al.* (1999) and Hoballah *et al.* (2007). Loss-of-function mutations in coding sequences were detected that regulate anthocyanin floral colour production (see Rieseberg & Blackman, 2010). Recently, another case study has been published. Two ecotypes in *Mimulus aurantiacus* (Phrymaceae) have been studied in Southern California: an anthocyanic red-flowered morph in the west and an anthocyanin-lacking, yellow-flowered morph in the east, in which there are 'almost complete differences in pollinator preference' (Streisfeld, Young & Sobel, 2013). Investigating the molecular basis of flower colour difference, they discovered that 'a *cis*-regulatory mutation in an R2R3-MYB transcription factor results in differential regulation of enzymes in the anthocyanin biosynthetic pathway and is a major contributor to differences in floral pigmentation'. These findings provide evidence that 'divergent selection has driven fixation of alternative alleles of this gene between ecotypes'.

Mutations in pathways Turning to the second possibility – mutations in key enzymes – Bomblies (2010) notes that 'in angiosperm evolution, transitions from blue to red, and pigmented to unpigmented, flowers are more common than the reverse. This probably reflects the fact that such shifts often involve loss-of-function mutations in pigment biosynthesis pathway genes that are difficult to reverse.'

Post-pollination barriers

As we saw in Chapter 7, the *S* locus is crucial in self-incompatibility (SI) reactions: crossing experiments have revealed the gene sometimes contributes to interspecific incompatibilities. For example, Hancock *et al.* (2003) experimented with *Nicotiana* species, including genetic transformation studies. Rieseberg & Blackman (2010) summarise their findings as follows: 'the functional products of the *S* locus are S-RNases'. And 'studies have been performed to show that S-RNase genes contribute to interspecific rejection'.

It was discovered that 'species with SI reject pollen from other species, whereas species that are self-compatible (SC) will accept pollen from both SI and SC species'. In their review, Rieseberg & Blackman (2010) consider evidence for other genes that contribute to post-pollination pre-zygotic barriers in tomato and in crucifers, including *Arabidopsis.*

Genes and mutations that contribute to hybrid inviability Rieseberg & Blackman (2010) consider a number of examples in which hybrid inviabilities and sterility may evolve, and this important review should be consulted for technical details. Here, several areas of importance are noted.

1. A model that has informed many studies of speciation is the classic Dobzhansky–Muller (D–M) concept. 'As adaptive or nearly neutral substitutions accumulate in diverging lineages, substitutions may be fixed in one lineage that are incompatible with substitutions fixed in the other lineage', and as a consequence 'hybrid dysfunction results when these incompatible alleles are brought together in hybrid progeny' (Rieseberg & Blackman, 2010).
2. Sterility may arise as a consequence of chromosomal rearrangements. Such changes have been examined in very considerable detail in studies of quantitative trait loci (QTL) mapping of Sunflower (*Helianthus*) species (Lai *et al.*, 2005). Sterility genes have been cloned in Rice (Chen *et al.*, 2008) and these contribute to reproductive barriers and compatibility in *Oryza indica-japonica* hybrids.
3. Rieseberg & Blackman (2010) note that disease resistance genes might play an important role in the evolution of hybrid inviability. Four sets of observations support this idea developed by Bomblies & Weigel (2007). Rieseberg & Blackman (2010) provide a clear summary on this complex area of investigation.

i) Inter- and intraspecific hybrids often exhibit tumours, as well as necrosis or weakness.
ii) Symptoms of hybrid necrosis are similar to necrotic symptoms typically associated with environmental stress and pathogen attack.
iii) Hybrid necrosis usually results from the interactions of complementary genes, similar to classic D–M incompatibilities.
iv) Genetic characterisation of hybrid necrosis in crosses between tomato species and between *Arabidopsis thaliana* ecotypes has revealed that incompatibilities among complementary disease resistance genes are indeed the cause of necrosis (Bomblies *et al.*, 2007). At least one example is known where necrosis prevents interspecific gene flow in the wild (McNaughton & Harper, 1960).

Cytoplasmic male sterility: its possible role in speciation in plants

In brief, typically, in flowering plants the zygote contains nuclear genes transmitted from both the male and female, but the cytoplasm of the zygote (containing the maternally inherited chloroplast and mitochondrial genomes) is contributed by the seed parent. CMS results from changes in the mitochondrial genome, leading to the possibility of cytonuclear incompatibilities (CNI). Bomblies points out the possible implication of this for speciation. She notes that sometimes 'the cytoplasmic organelles from one parent cannot function together with the nuclear genome from the other parent ... CMS is a common type of CNI known from many plant species ... CMS aberrations arise frequently in hybrids implying that coevolution between mitochondrial genomes and interacting nuclear factors is crucial.'

CMS is characteristic of gynodioecious breeding systems (see Chapter 7). Active research is in progress to see if cytoplasmic male sterility (CMS) might also play a role in the evolution of isolating mechanisms (see Bomblies, 2010; Rieseberg & Blackman, 2010).

Greiner *et al.* (2011) have recently reviewed the evidence for the role of plastids in plant speciation. They conclude that 'plastome-genome incompatibility can establish hybridisation barriers, comparable to the Dobzhansky–Muller mode of speciation processes'.

Genomic changes involved in speciation

While a number of researchers have investigated the molecular genetics of speciation in plants by focusing on speciation genes, others are examining the nature of changes that occur at the genome level (Nosil & Feder, 2012). As we have seen in earlier chapters, DNA sequencing has revealed a great deal about the architecture of the genome, with progress being made in identifying 'signatures of selection'. Through the use of next generation methods, sequencing has becomes a great deal quicker and less expensive, and it will become increasingly possible to investigate what is happening at the genome level in the speciation process.

As genomes diverge, what DNA sequence patterns are likely to be indicative of speciation processes? Nosil & Feder (2012) envisage 'genomic islands of divergence', i.e. 'regions of the genome of any size, whose divergence exceeds neutral background'. They consider that 'speciation most probably involves the growth of such islands by divergence hitchhiking'. 'This is a process by which physical linkage to a divergently selected gene(s) increases genomic divergence for regions adjacent to a selected site on a chromosome.'

In a wide-ranging exploration of these emerging concepts, Strasburg *et al.* (2012) have reviewed 26 published genomic scans from a range of plant species. They report that 'the genome scan approach appears well suited for identifying genomic regions or even candidate genes that underlie adaptive divergence and/or reproductive barriers'. However, in the published studies 'genomic regions of high divergence generally appear quite small in comparisons of both closely and distantly related populations and for the most part these differentiated regions are spread throughout the genome rather than strongly clustered'.

Recently, Alcázar *et al.* (2012) have demonstrated the importance of combining genomic studies with crossing experiments in the exploration of signals of speciation within species. Taking advantage of the wealth of information available for the model species *Arabidopsis thaliana*, 80 accessions were sequenced and an intraspecific crossing programme was carried out. Post-zygotic incompatibilities were detected between different geographical accessions that conformed to the Dobzhansky–Muller model introduced above: i.e. 'novel alleles arise in isolated populations and when these are combined in crosses', they 'lead to lethality or sterility' in hybrids.
In judging the potential of these approaches, for studying other species and groups, Alcázar and associates consider that: 'many of the reproductive isolation processes observed in *Arabidopsis*' are likely to occur in other plant species.

Future prospects for the study of speciation genes and genomic architecture

The eventual aim is to understand – from beginning to end – the genomic patterns and processes involved in selection and speciation.

Studies of a number of genera reveal patterns of genetic divergence consistent with a model of allopatric gradual speciation, e.g. *Limnanthes* (McNeill & Jain, 1983); *Lisianthius* (Sytsma & Schaal, 1985); *Hosta* (Chung *et al.*, 1991); and *Streptanthus* (Mayer, Soltis & Soltis, 1994). The results of these historic investigations, together with those of more recent studies, raise a number of issues concerning future investigations of allopatric speciation (Bomblies, 2010; Rieseberg & Blackman, 2010).

First, it should not be assumed, without investigation, that members of a genus or other plant group have had the same evolutionary history. Moreover, historic and modern investigations confirm the existence of cryptic species in some genera.
In selecting experimental material for genomic and

genetic investigations, it is important to recognise that the speciation process does not necessarily produce visibly distinct taxa.

Secondly, while there will always be uncertainty about the magnitude and sequence of past events, advances in our understanding of the genetics of populations suggest that a range of *abrupt* events – founder, drift and bottleneck effects, genetic, chromosomal mutations – crucially influence what is sometimes envisaged to be a gradual speciation process. However, if abrupt events have small adaptive effects, and many such changes are 'needed' for speciation, then models of gradual accumulation of significant adaptive differences have validity.

Considerable progress has been made in modelling population structure, and gene flow in relation to speciation processes (see Rieseberg, Church & Morjan, 2003). They point out that Maynard Smith (1966) revealed the power of modelling to explore microevolutionary situations, when he demonstrated 'that divergence will occur when the strength of selection at a given locus (s) is greater than the migration rate (m)'. They also review recent investigations where a variety of situations relevant to speciation have been explored by computer simulation, including (a) the role of spatial structure and gene flow in relation to the development of D–M type incompatibilities; (b) situations where, in a series of isolated populations, migration only occurs between neighbours, and (c) attempts to incorporate metapopulation dynamics into models of divergence.

Turning to another issue, considerable advances in modelling and computational capacity will be necessary to make the most effective use of the increasing volumes of DNA sequence data. As the aim is to understand the development of isolating mechanisms as they operate in nature, looking to the future, it is clear that to enhance our understanding of plant speciation, combined investigations of crossing behaviour and genomic studies are to be preferred. Also, both chance events and selection must be considered. Furthermore there is a growing realisation that ecological and natural history aspects of speciation are very important, and that these are best studied in concert with molecular approaches.

Even with the use of sophisticated molecular tools, in conjunction with advanced and computational studies, Bomblies (2010) highlights perhaps the most intractable problem in studying allopatric speciation. As yet there is no way of determining the sequence of events in speciation: what were the initial and subsequent stages? Moreover, it is unclear the extent to which speciation involves not only *novel* mutational and other microevolutionary events, but also the sorting out of standing variation already accumulated over time in isolated populations.

Turning to another issue of importance, it has been established that symbiotic relationships are of great adaptive significance in many plant and animal groups. Intensive investigations are exploring the interactions between the host and microbiome associations of microorganisms. Increasingly it is being recognised that, in nature, many species exist as co-adapted 'hologenomes': beneficial associations of symbiotic organisms. This raises an important question: given the evidence for 'co-adapted associations', how do new species evolve? Studies of the Gall Wasp genus *Nasonia* have revealed that beneficial gut bacterial assemblages are species specific and 'cause lethality in interspecific hybrids' (Brucker & Bordenstein, 2013). These findings suggest that perhaps the breakdown of complex hologenomic relationships following hybridisation between divergent populations of plants may be an important isolating mechanism active in speciation.

Allopatric speciation and the taxonomist

As we have seen, evidence supports the view that allopatric speciation is one means by which plants evolve, and the different patterns of variation produced are of especial interest to taxonomists, who have named various subspecies or species according to the degree of morphological distinctness exhibited by plants. It is interesting, however, that crossing polygons do not reveal any consistent linkage between the degree of morphological differentiation and crossing behaviour. In some cases there is a suggestion that different entities recognised as taxonomic species are reproductively isolated from other such groups, but it is abundantly clear that there is no necessary correlation between the presence of sterility barriers and morphological differences. Two classic studies illustrate this point. Thus, in *Elymus glaucus*, which taxonomists treat as a single species (Snyder, 1950, 1951), there would appear to be sterility barriers between different collections (Fig. 13.4). On the other hand, certain taxonomic species of *Mimulus* (Vickery, 1964) appear to have fully fertile F_1 hybrids (Fig. 13.5). Clearly, the factors controlling morphological differentiation and post-zygotic isolating mechanisms may not necessarily evolve in step.

Natural hybridisation

So far we have been examining experiments that have involved artificial hybridisations. These investigations prompt a number of questions concerning natural hybridisation.

What are hybrids? Maunder *et al.* (2004) recognise two usages of the word. For geneticists a hybrid is the product of crossing genetically distinct lines. In contrast, for taxonomists the term refers to the offspring of crossing between two taxonomic species. We return to the vexed question of species definitions in later chapters.

How frequent is interspecific hybridisation? Two reviews indicate the widespread occurrence of hybrids. Knobloch (1971) catalogued 23 675 putative interspecific and intergeneric hybrids. Extrapolating from the incidence of hybrids in the British flora, Stace (1975) estimated that there could be *c.*78 000 naturally occurring hybrids in the flowering plants. However, as taxonomists disagree about species limits, estimates of the incidence of hybridisation depend upon taxonomic assumptions and practice (Maunder *et al.*, 2004). Also, these estimates of the incidence of hybrids may be misleading as they are based on studies of well-investigated temperate floras. For example, it is not clear how much hybridisation occurs in tropical regions.

Does the incidence of hybridisation differ in different taxonomic groups? Taxonomic studies have revealed that there is a high incidence of hybrids in certain plants groups (Ellstrand, Whitkus & Rieseberg, 1996); for example, willowherbs (Onagraceae), orchids, pines, members of the Rosaceae family, willows, Scrophulariaceae, grasses, composites and sedges.

Natural hybridisation in the wild: classic studies

Experimental investigation of natural hybridisation between different taxonomic species has also contributed to our understanding of speciation. The genus *Geum*, widespread in the temperate regions of the world, will serve to illustrate a number of important points.

The two most widespread European species are *Geum rivale* and *G. urbanum*. The latter, which has a somewhat 'weedy' tendency to which we shall refer later, has also become widely naturalised in North America. *Geum rivale* is a typical 'bee' flower and species of Bumble Bee (*Bombus*) are recorded as the commonest visitors (Fig. 13.6). The purplish colour and the somewhat concealed entrance to the hanging flower are features shown by many 'bee' flowers. Contrast with this the smaller, open, erect, yellow flower of *Geum urbanum*, which shows no specialisation for the visits of particular insects and seems to be frequently self-pollinated.

Over much of Europe these two species are sympatric. They are, however, usually effectively separated by ecological differences. *Geum rivale* often grows in damp shady places in southern and central Europe, and is more or less confined to the upland regions. It is absent from much of the Mediterranean

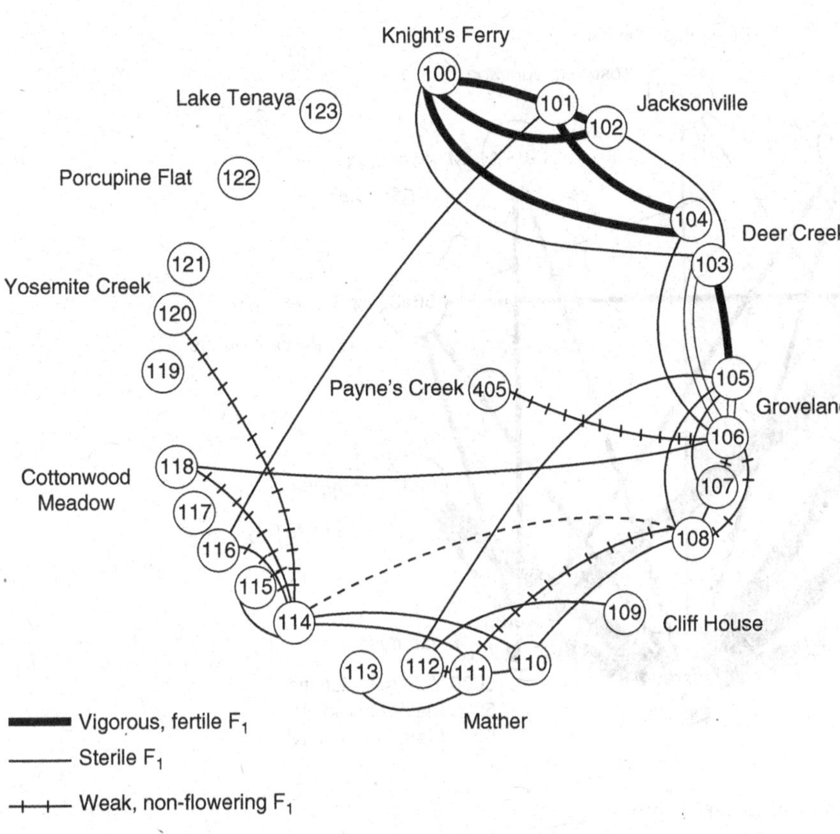

Fig. 13.4. A summary of hybridisation between populations of *Elymus glaucus* collected along a 75-mile transect in the Sierra Nevada. The diagram shows developmental behaviour and pollen fertility of the hybrids. Snyder concludes from the cytological behaviour of the hybrids at meiosis that much of the sterility is caused by small structural differences in the chromosomes and by specific genes. He also suggests that hybridisation in nature between *Elymus glaucus* and species of the related genera *Argopyron, Hordeum* and *Sitanion* might be responsible for much of the variability. (From Snyder, 1950, 1951.)

region, but it occurs in Iceland, from which *G. urbanum* is absent. Over most of lowland Europe, however, *G. urbanum* is the common plant, growing particularly in hedgerow and woodland communities affected by human activities. It has long been known that plants with somewhat intermediate characters sometimes occur in abundance in woods and scrub where the two species meet; such obviously hybrid plants were called *Geum intermedium* by Ehrhart as early as 1791. These hybrids have attracted much attention, mainly because the two parent species look so different, and in some places hybrid swarms are found in which there is a remarkable range of variation. Such a hybrid population formed the basis of a detailed genetic study by Marsden-Jones (1930), who was able to discover to some extent the inheritance along Mendelian lines of several of the characters determining the differences between the two species. This work was followed, and greatly enlarged, by the Polish botanist Gajewski, whose study of the genus *Geum*, published in 1957, is the fruit of many years' experimental study.

The situation described in detail by Marsden-Jones for a wet wood of Alder (*Alnus glutinosa*) at Bradfield, Berkshire, could be paralleled in a good many places in lowland England. On the other hand, Gajewski, who studied *Geum* mainly in Poland, emphasises that large hybrid populations are rare, and that even where the two species are growing close together, there are often very few intermediate plants. What is the cause of this apparent difference in efficiency of isolation between England and Poland?

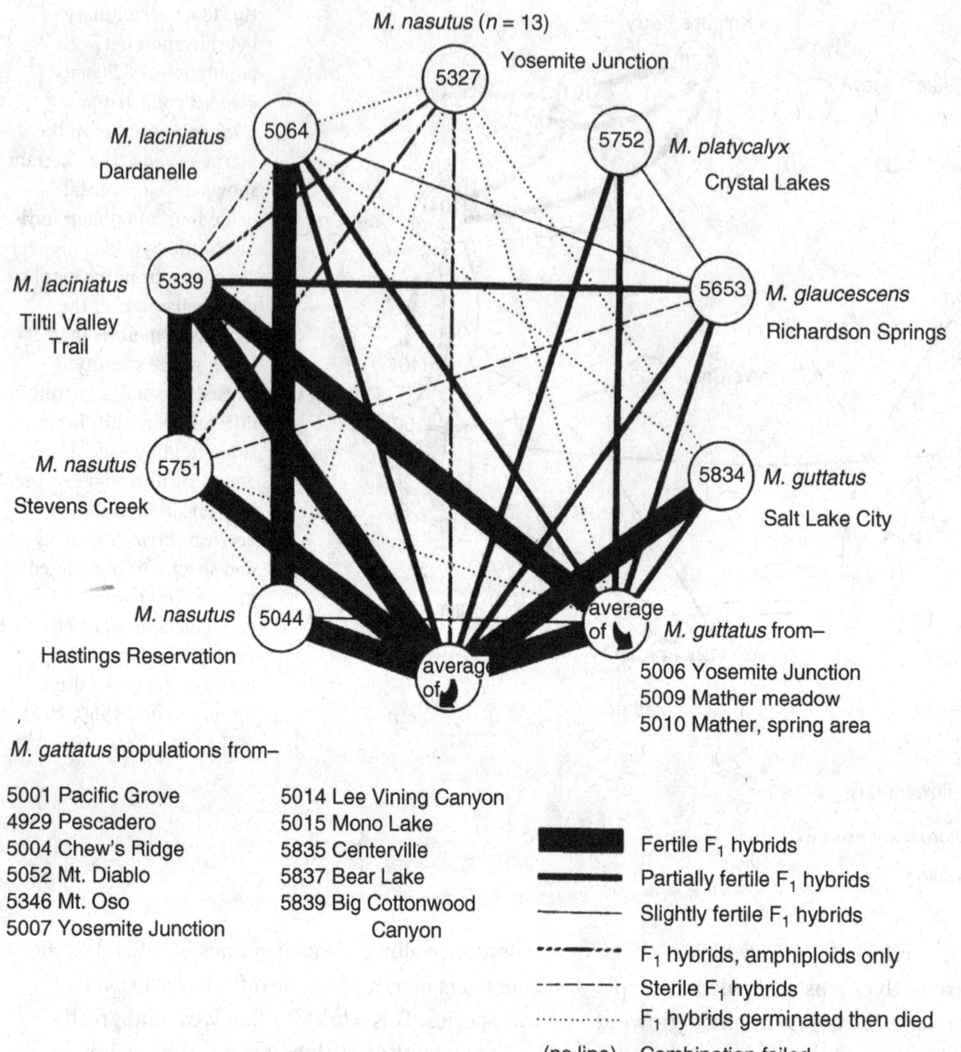

Fig. 13.5. Crossing polygon indicating fertility of F_1 hybrids between different populations of the *Mimulus guttatus* complex in North America. (From Vickery, 1964.)

The first relevant point is that artificial F_1 hybrids between the two species can be made, though not with ease, and that such plants are highly fertile, the F_2 showing a range of segregates as might be expected if (as Marsden-Jones' detailed genetic experiments bore out) many genes are involved in the specific differences. Judged purely in terms of the theoretically possible gene flow, therefore, *Geum rivale* and *G. urbanum* would fall within a single biological species. As a matter of fact, Gajewski's work demonstrated that all 25 species in the subgenus *Geum* (to which our two species belong) will hybridise with each other, and that most of these hybrids are at least partially fertile. In practice, however, a good many species-pairs or species-groups are allopatric and hybridisation does not take place in nature.

We have already seen that ecological preference will normally separate, at least partially, mixed

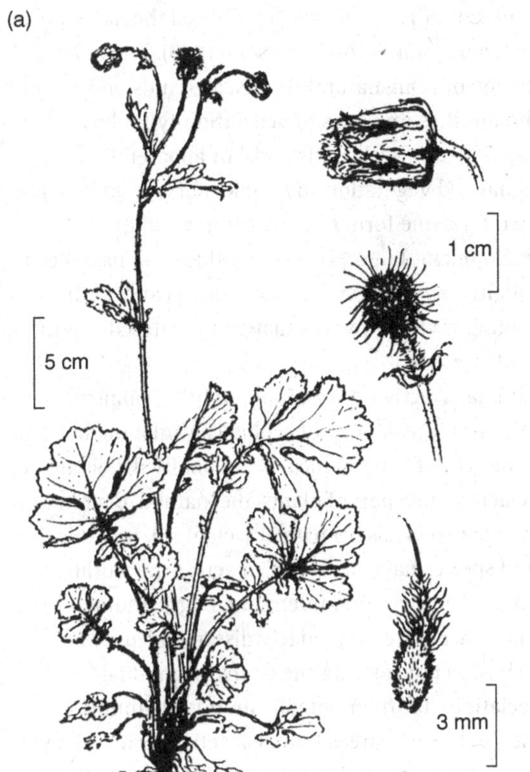

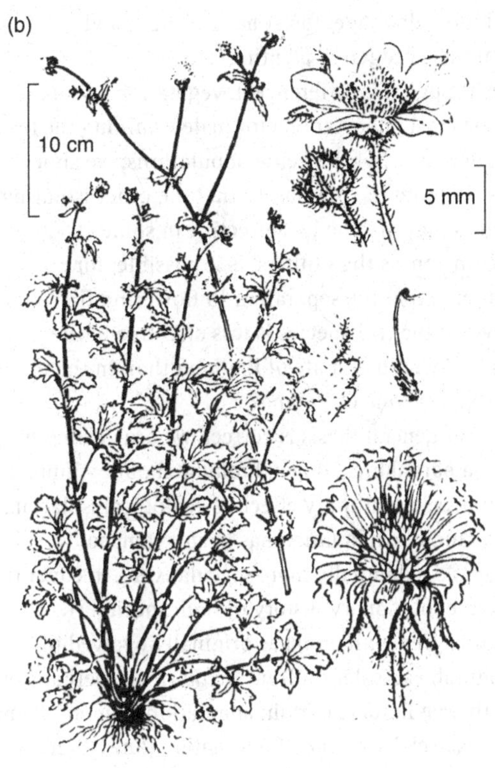

Fig. 13.6. (a) *Geum rivale*, (b) *G.urbanum*. (From Roles, 1960.)

populations of the two species. Moreover, the difference in structure and colour of flower is likely to attract different insect pollinators. To this difference should be added a rather obvious difference in the times of the beginning of flowering, which in Britain differs by 3 or 4 weeks, *G. rivale* being the earlier. This would mean that, even in mixed populations, seed set on the early flowers would be necessarily 'pure' in the case of *G. rivale*. Such considerations point the way to at least a tentative answer to our question. Gajewski records that he grew seeds taken from plants of each species growing in a mixed stand in a Polish locality where hybrid plants were rare, and found that the progeny were 'pure' with no detectable sign of hybridisation. Clearly factors such as the preferences of insect visitors, coupled with differences in flowering-time, effectively prevent more than a minimum of gene flow between the species in Poland. What is different about the English conditions?

The most important difference lies probably in the complex history of human interference with the vegetation. Gajewski's observations were made, partly at any rate, in the great forest nature reserve of Białowieza, in eastern Poland, where the forest is as little affected by human activity as anywhere in lowland Europe. Here such hybrids as he recorded were single individuals on roadsides and in forest tracks, where *Geum urbanum* was probably introduced by human activities. *Geum rivale* behaves as the original, native species. In England, on the other hand, most woodland is present as small 'islands' in a sea of agricultural land, and the disturbed marginal habitats suitable for *Geum urbanum* have clearly been enormously extended over the centuries by the anthropogenic interventions including drainage, forest management and clearance, hedgerow planting, etc.

Although this difference in vegetation history may be the most important factor in determining the local frequency of hybrid *Geum* populations, we should bear in mind the possibility that the other isolating factors may also be less effective in some circumstances than others. Is it possible, for example, that the separation in flowering time between the two species is less effective in the relatively mild climate of England than in the more continental one of Poland?

If our general thesis is correct, we are dealing here with a partial breakdown, brought about by humans' activities, of naturally effective ecological isolation. This kind of explanation has been extended by Gajewski, admittedly more speculatively, to cover the recent evolutionary history of both species. He pictures *Geum urbanum* as originally evolved in geographical isolation from *Geum rivale*, perhaps in south-east Europe. Certain adaptations, among them the unspecialised type of pollination (and often self-pollination) and the efficient, small, animal-dispersed fruits, made *Geum urbanum* an effective 'weed' of marginal woodland habitats created by human activities. In this way, the species became sympatric with *G. rivale* over much of Europe. In this new situation, the advantage lies with the 'weedy' species, for most vegetational change brought about by humans will favour it rather than its relative.

We have dealt at some length with the *Geum* example, because it provides an excellent paradigm of what we take to be a general phenomenon. Humans' activities in the exploitation and management of various areas have led to a variable degree of breakdown of ecological isolation. In areas long settled, there is hardly a square inch that has not in some way or other been influenced by human activity. Burning, forest clearance, drainage, grazing and other agricultural practices, mining, and all aspects of urban industrialisation from road building to atmospheric pollution – all these activities have contributed to changes in vegetation (Briggs, 2009).

Human influences have 'hybridised the habitats', in the famous phrase of Anderson (1949). Thus, the acreage of semi-natural forests, wetlands, mountain communities and coastal vegetation types has gradually diminished. 'Islands' of apparently unchanged vegetation sometimes survive on land less fitted for some form of agriculture or other development, but on inspection these too have been exploited by humans for food, fuel (peat, wood), building material (wood, thatching material), sport or in other ways.

Humans' activities result not only in complex patterns of breakdown of ecological isolation, but also in changes of geographical isolation. By deliberate or accidental transport of plants, the 'natural' distribution of some plants has been greatly changed. In particular, weed species have been carried across the world, and plants of horticultural interest and agricultural importance have been widely disseminated.

Thus, in considering the course of gradual speciation, a further complication is beginning to emerge. In some areas humans' activities are likely to have influenced, perhaps decisively, not only the past and present distributions of plants, but also patterns and processes of microevolution at different sites. Therefore, our models of gradual speciation must take account of the fact that in relatively recent times (geologically speaking) human influences have become the dominant factor in landscape change, greatly influencing plant distribution and ecology throughout the world. The breakdown of geographical and ecological isolation may lead, as in the *Geum* example, to hybridisation, and the consequences of such hybridisation are likely to differ in different circumstances. What consequences might we expect in various circumstances?

The consequences of hybridisation: some theoretical considerations

Consider the case of two taxa, A and B, which have become sympatric after a period of allopatry. They may have changed genetically in isolation to such an extent that, in sympatry, they are unable to cross freely, hybridisation being a rare event leading to infertile products (both pre- and post-zygotic factors may be important). The two populations are in effect behaving as two biological species. It may be, on the other hand, that the two populations have come to differ in morphology and ecological requirements, but crossing experiments reveal only partial, or no, major barriers to interbreeding. As the daughter populations A and B differ in ecological preferences, the fate of any hybrids produced in nature is likely to be influenced decisively by ecological factors.

If no intermediate habitats are found between those preferred by taxa A and B, then hybrids between A and B are likely to be at a selective disadvantage. In comparison with crosses A × A and B × B, the hybrids derived from the crosses A × B and B × A may yield fewer (or no) viable offspring. Selection, in favouring the progeny of A × A and B × B, will act against the products of hybridisation. Any partial isolating mechanism minimising A × B and B × A will be subject to selection. The perfecting of partial isolating mechanisms under the impact of selection may then take place, such a change being referred to as 'reinforcement', or in older literature the 'Wallace Effect' after Alfred Russel Wallace, the co-founder with Darwin of the theory of evolution by natural selection, who speculated on this subject (Grant, 1966).

If habitats intermediate between those preferred by A and B occur, then in such habitats hybrids between A and B may be 'fitter' (i.e. contribute a greater number of offspring to future generations) than the progeny of A × A and B × B. Any partial isolating factors developed in the once allopatric populations are likely to be overcome and the distinctness of the two populations may be lost, locally or regionally, in a mass of hybrids and backcrosses.

These two models, which are in reality extremes of a spectrum of possibilities, will now be examined in further detail. It is clearly very difficult to study natural populations to discover the validity, or otherwise, of the Wallace effect, but there is a certain amount of circumstantial evidence which supports the model.

The first line of evidence comes from studies of 'experimental' populations. In the Fruit Fly *Drosophila*, studies have shown that intensification of reproductive isolation may result from artificial selection. Rice & Hostert (1993) have reviewed 40 years of laboratory experiments on such speciation in animals.

A most elegant botanical example was studied by Paterniani (1969), who grew two varieties of Maize (*Zea mays*) together in an experimental garden. To explain Paterniani's experiment, some details of the stocks used are necessary. One variety had white, flint cob characters (genetically: *yy SuSu*): the other had yellow, sweet cob characters (*YY susu*). Thus, each stock had one dominant and one recessive marker gene. Pollen of the white flint stock had the genotype y Su, while pollen of yellow sweet had the genotype *Y su*. Any cross between the two stocks would yield kernels with yellow colour in the white flint parent and flint (often called starchy) kernels in the yellow sweet parent. Because these markers give pronounced effects visible in the developing cob, it is possible to determine the degree of hybridisation between varieties without raising progeny, enabling the experiment to be carried out in a relatively short time. Using the experimental stocks, planted in such a way as to maximise the opportunity for hybridisation between varieties, Paterniani investigated what happened when selection for reproductive isolation was carried out. This was achieved by selecting at each of a number of generations the cobs that showed the *least* hybridisation. From such cobs grains of phenotype white flint and yellow sweet were used to provide seedlings for the next generation. At the start of the experiment the degrees of outcrossing were 35.8% and 46.7% for white flint and yellow sweet stocks, respectively. After six generations of selection the levels of intercrossing were 4.9% for white flint and 3.4% for yellow sweet. The factors involved in this isolation were examined, and it was discovered that the number of days from sowing to flowering was probably the most important factor. The original stocks flowered on average at the same time, but by the end of the experiment the mean flowering time for white flint was 5 days earlier, while the mean for yellow sweet was 2 days later. By selecting against hybridisation the mean flowering times of the two stocks had been separated by about 7 days (Fig. 13.7). This experiment suggests that similar effects on flowering time might be expected in wild populations, where hybrids are at a selective disadvantage.

Empirical studies of reinforcement

Disruptive selection in polymorphic populations
The situation where genetically polymorphic species exist in a patchy or mosaic habitat would seem to offer the possibility of speciation (Thoday, 1972). Several studies of plants have been made on mine debris containing heavy-metal residues, sites that often exist as 'islands' in a 'sea' of pasture.

Some species occur both on the mine and in the adjacent pasture. There is evidence for gene flow between pasture and mine plants, followed by disruptive selection (Fig. 13.8). Furthermore, it is clear that pasture plants cannot grow on the mine debris and mine plants grow less well in the pasture (Bradshaw, 1976; Bradshaw & McNeilly, 1981). It seems that crosses within habitat type (mine × mine and pasture × pasture) will produce fitter progeny than the crosses mine × pasture and pasture × mine. Therefore, it has been argued that selection would favour any variants arising in the population that would restrict 'gene flow' by promoting within-habitat crosses. Experiments by Antonovics (1968) and McNeilly & Antonovics (1968a, b) working with *Agrostis capillaris* (*A. tenuis*) and *Anthoxanthum odoratum* showed that, although there was some overlap in flowering times between pasture and mine plants, those at the mine edge flowered about a week earlier, a difference maintained in cultivation. Further, plants of both species growing on mine debris, though normally self-incompatible, were found to be capable of a degree of self-fertility.

Both these traits may be seen as barriers to free gene flow and could be interpreted as the first steps in the speciation process. It is unclear, however, how much selfing actually takes place in wild populations. The studies of Lefèbvre (1973) suggest that, in *Armeria maritima* growing on mine debris, the potential for self-fertilisation exists, but in a study of a particular mine population, it was discovered that outbreeding was the rule.

These situations take us beyond the simple model of gradual allopatric speciation, with which we opened this chapter, for, as far as we know, the different polymorphic variants of each species have not developed in geographic isolation, but in closely adjacent areas. More studies are required to determine whether reproductive isolation – as a result of flowering time differences or changes in breeding behaviour – is the product of many genetic changes over a long period, or one or a few changes occurring abruptly. A model involving the abrupt origin of reproductive isolation through a simple genetic mechanism has received support from the studies of crosses between copper-tolerant plants of *Mimulus guttatus* from Copperopolis, California, and British non-tolerant material (Macnair & Christie, 1983). Crosses revealed the existence of a gene associated with copper tolerance, either closely linked or pleiotropic, which interacted with genes in the non-tolerant material to give F_1 inviability.

Wright (2013) has further investigated the situation. By employing high-resolution genome mapping techniques he discovered that 'copper tolerance and hybrid lethality are not caused by the same gene but are in fact separately controlled by two tightly linked loci . . . selection on the copper tolerance locus indirectly caused hybrid incompatibility allele to go to high frequency in the

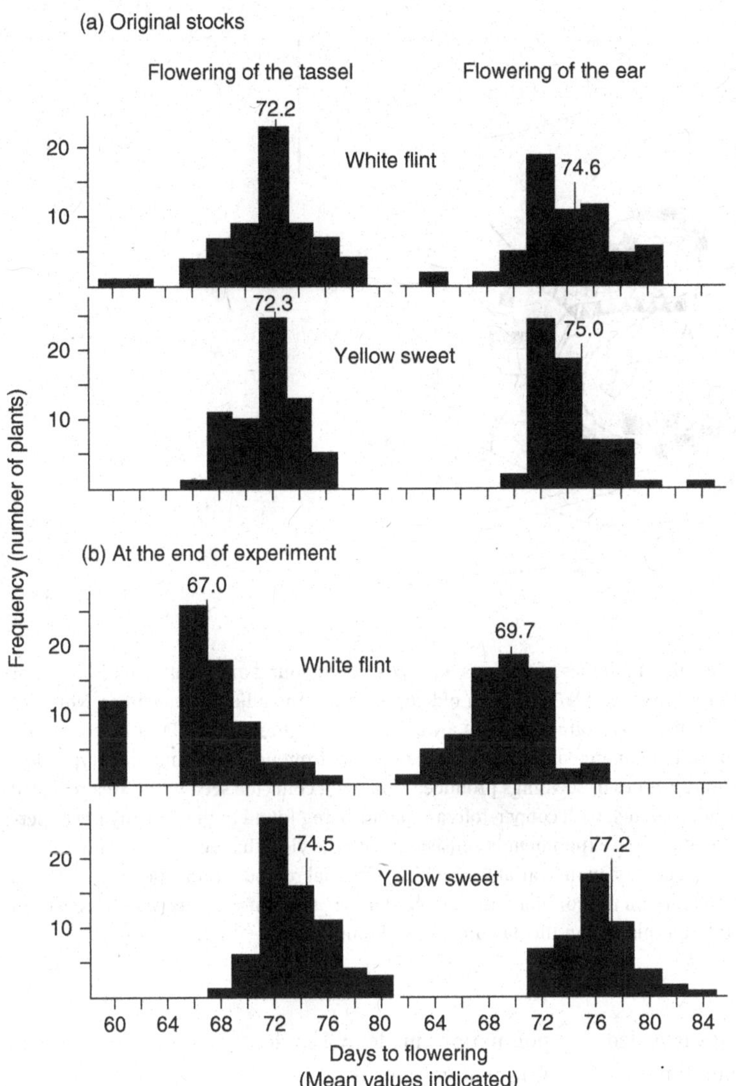

Fig. 13.7. Histograms showing the number of days to flowering for the original two varieties of Maize (*Zea*), and for the stocks after the selection experiment had run for six generations. Maize has separate male and female flowers on the same plant: hence, the separate recording of tassel (male) and ear (female). For details of the experiment, see the text. (From Paterniani, 1969.)

copper mine population' as a consequence of 'genetic hitchhiking'.

Reinforcement: crossing experiments. Other crossing experiments have yielded interesting evidence. If isolating mechanisms are initiated in allopatric populations but perfected in sympatric situations, then it follows that crossing within groups of sympatric and allopatric taxa might be revealing. Such studies have been carried out by Grant (1966), who studied nine species of *Gilia* (Fig. 13.9). Five of the species were sympatric, having overlapping distributions in the foothills and valleys of West California. The other four species were allopatric in maritime habitats in North and South America.

Table 13.1 shows that crossing between sympatric taxa yields fewer seeds per flower than that between allopatric taxa. A similar study of other *Gilia* species gave the same sort of result. However, crosses between

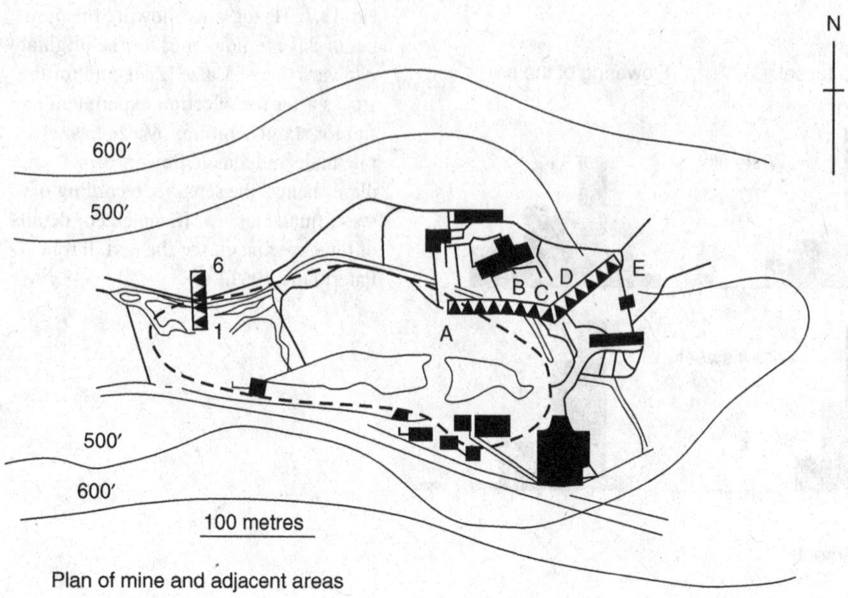

Fig. 13.8. In an attempt to study disruptive selection, a number of workers have compared young and adult generations. For example, McNeilly (1968) collected *Agrostis capillaris* (*A. tenuis*) from an old copper mine and adjacent pasture in North Wales (two transects 1–6 and A–E). Copper tolerance was examined using a water culture technique. Adults from contaminated mine spoil (up to 2700 ppm copper) proved to be more copper tolerant than plants from non-contaminated (typically 52–156 ppm copper) pasture adjacent to the mine. Studies of seedlings produced from wild collected seed in the area revealed a wider spectrum of copper tolerance than the adult plants, with copper-tolerant plants being found in the progeny of copper-sensitive plants collected from downwind of the mine. The experiment is consistent with the view that gene flow occurs in this wind-pollinated grass species, and strong selection occurs on the variable products of sexual reproduction. The only seedlings to survive to adulthood are likely to be copper tolerant on the contaminated areas, and non-tolerant variants (which have been shown to be better competitors than copper-tolerant plants) on the pasture areas. (From McNeilly, 1968.)

the *Gilia splendens* and *G. australis* groups revealed the opposite pattern, with sympatric populations giving greater seed yield, when crossed *inter se*, than allopatric populations similarly crossed. Thus, two of the three situations examined experimentally provided some support for a Wallace effect.

Reinforcement: flower colour. Population studies of wild plants provide a third area of evidence concerning the Wallace effect. A modified population, containing *Phlox glaberrima* (mean pollen size 55 μm), and *Phlox pilosa* (mean pollen size 30 μm), was studied by Levin & Kerster (1967). In order to investigate the effect of flower colour on interspecific pollinations made by Lepidoptera, transplants of red-flowered variants of *P. pilosa* were added to the naturally occurring white-flowered variants of this species at a site in Cook County, Illinois. Thus, two colour variants of *P. pilosa* were available for interspecific pollinations with the red-flowered *P. glaberrima*. Evidence of the grains on a sample of *P. pilosa* stigmas suggested that, while 30% of the red flowers of *P. pilosa* received alien pollen, only 12% of the white ones bore such pollen. It would appear that the flower colour differences in the unmodified population of *Phlox glaberrima* (red) and *P. pilosa* (white) might act as aids to pollinator discrimination

Table 13.1. **Comparative crossability of species with different geographical relations in the Leafy-Stemmed Gilias (Grant, 1966)**

Geographical relation of parental species	No. of combinations of parental species	Mean number of seeds per flower		Mean of means
Foothill species	9	0.0	0.1	0.2
inter se *(sympatric)*		0.0	0.1	
		0.0	0.4	
		0.0	1.2	
		0.0		
Maritime species	5	7.7		18.1
inter se *(allopatric)*		16.7		
		19.6		
		21.9		
		24.8		

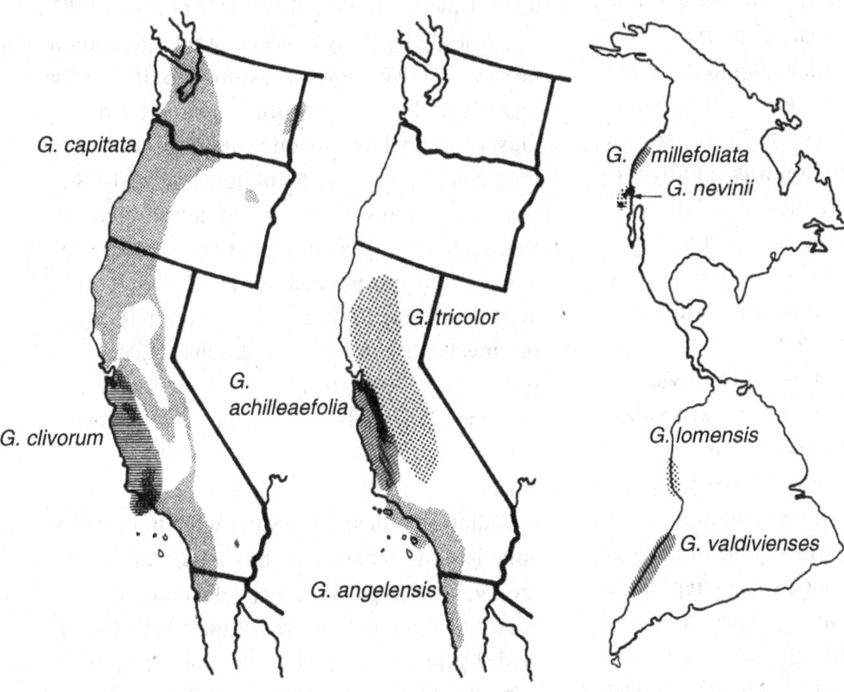

Fig. 13.9. Geographical distribution of the nine species of *Gilia* studied by Grant. (From Grant, 1966.)

and thus reduce interspecific pollinations. It is possible to devise models of selection for reproductive isolation between two daughter populations that have come to differ in flower colour, one population (A) being polymorphic for flower colour, while population B has only one colour phase, the flower colour of B being one of the colours possible in A. If pollinator activities result in cross-pollination, A × B, between similar colour variants and this leads to less fit progeny, then the balance of colour variation in population A could change. Ultimately A and B in sympatric situations could come to be characterised by different flower colour. These cases provide some support for the Wallace effect. Levin (1985) has carried out further experiments on character displacement in *Phlox*, and his paper makes reference to other examples where 'the presence of a related species may have been the stimulus for divergence in flower colour'.

Recent studies of reinforcement Experimental studies on animal species and computer simulation experiments offer support for the model of reinforcement (van Dijk & Bijlsma, 1994). However, Silvertown *et al.* (2005) note that 'empirical evidence of reinforcement has rarely been complete enough in any one case to exclude alternative interpretation', although recent studies of animal populations – flycatcher, salmon and insect – have offered strong support. But 'even these studies fall short of showing that a single ancestral population was split by the evolution of pre-mating reproductive isolation *in situ*. Indeed, it has been said that such evidence might even be almost impossible to obtain because the history of populations is usually not known (Barton, 2000).'

However, in a few special cases the history of land use is well known. Taking advantage of the wealth of earlier investigations of *Anthoxanthum odoratum* growing in the Park Grass Experiment (see Chapter 9), Silvertown *et al.* (2005) have investigated the possibility of reinforcement, where plots of hay were given different fertiliser treatments. They report that reinforcement 'has occurred at least once in populations of the grass *Anthoxanthum* ... where flowering time has shifted at the boundaries between plots. As a consequence, gene flow via pollen has been severely limited and adjacent populations that had a common origin at the start of the experiment in 1856 have now diverged at neutral marker loci'.

The emergence of the concept of introgressive hybridisation

Having considered cases where hybrids are postulated to be at a selective disadvantage, we can now turn our attention to what happens where hybrids might be at a selective advantage. In 1949 the American geneticist Anderson published a book entitled *Introgressive Hybridisation* in which he postulated that in some cases when species hybridise and the environment has been disturbed by humans' activities, incorporation of the genes of one species into another might occur, as a consequence of hybridisation and repeated backcrossing. This process, for which the shorter term 'introgression' is now used, Anderson claimed was much more widespread and important in evolution than had been previously thought (Fig. 13.10). He also emphasised that introgression might occur as a consequence of natural events (Anderson, 1953).

We might logically expect that introgression would be equally effective when nature herself does the upsetting. Floods, fires, tornadoes and hurricanes must certainly have operated upon natural vegetation long before the evolution of humankind. These phenomena too alter conditions catastrophically, leading to a breakdown in barriers between species, and provide unusual habitats in which hybrid derivatives may for a time find a foothold, thus serving as a bridge by which groups of genes from one species can invade the germplasm of another. Anderson (1949) also considered that 'the raw material for evolution brought about by introgression must greatly exceed the new genes produced directly by mutation'. However, a significant number of botanists have challenged this suggestion. In their review, Rieseberg & Wendel (1993) quote sceptics who are of the opinion that the 'ultimate contributions made by hybrids must be very small or negligible' and that introgression is a kind of 'evolutionary noise', being a primarily local phenomenon with only transient effects. Now, many argue that hybridisation is an important source of variability in species and populations (Abbott *et al.*, 2012).

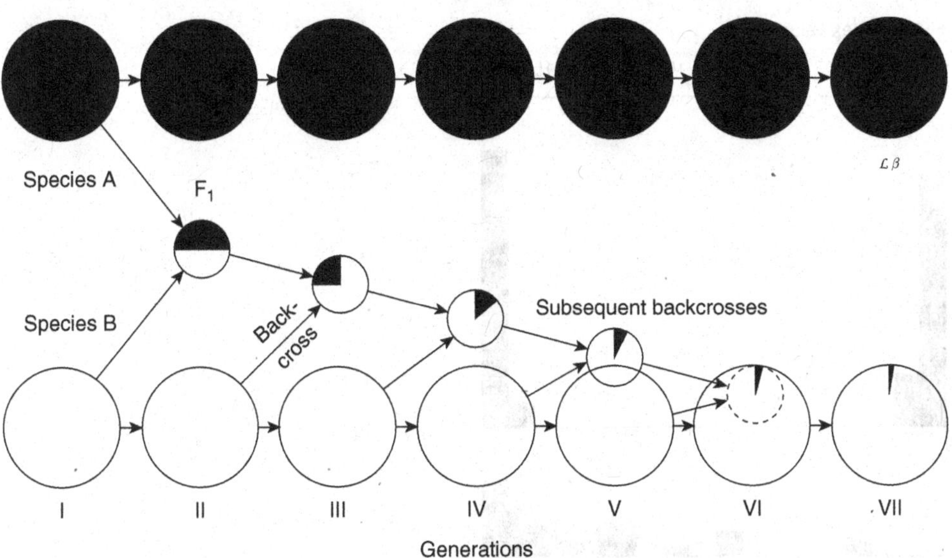

Fig. 13.10. Diagram illustrating introgression between two species. Backcrossing of the F_1 hybrid and subsequent backcrosses to species B ultimately results in the transfer of some genes from A into at least some individuals of species B. Reproduced by permission of John Wiley & Sons, Inc. from Benson (1962), *Plant Taxonomy*. © 1962.

Introgression: classic approaches championed by Anderson

Anderson invented several simple methods of displaying variation in hybridising populations; his hybrid index method is illustrated in Fig. 13.11. By devising a suitable scale of numerical values an investigator can make a rapid survey of field collections.

In the interpretation of complex populations, interspecific hybridisation is often assumed to be the cause of the pattern. But it would seem more intellectually honest to see the hybrid index method as a means of describing the degree of separation of plants of different morphology. As we shall see below, other evidence is necessary to assess whether interspecific hybridisation is actually involved. It is essential to have some clear idea of the variation to be expected in the supposed parental species. To this end samples should be collected from sites where the two 'pure' parents are not in contact with each other. (We may note in parenthesis that the collection of 'pure' parents is often a highly subjective and difficult task, especially in cases where the 'parental' taxa are broadly sympatric.) The hybrid index method yields a scale of variation, with pure parental colonies gaining the highest and lowest scores, and plants of intermediate morphology having intermediate scores.

The method has the disadvantage that the variation of individual plants is effectively 'lost' and, in the display of the results, plants with the same hybrid index score may differ phenotypically. Anderson's pictorialised scatter diagram technique, illustrated in Fig. 13.12, provides an attractive method of display that overcomes this problem to some extent. Material from complex and putative pure parental populations is scored for a number of features, and two quantitative characters are selected to generate an ordinary scatter diagram. The figures for each plant serve to determine the location of a spot on the diagram. Spots may be of different shape or colour, and are 'decorated' with appropriate arms to show the qualitative and quantitative characteristics of each specimen.

In the excellent example of studies of *Primula vulgaris* and *P. veris* populations, Woodell (1965) explored the variation in 'parental' populations at

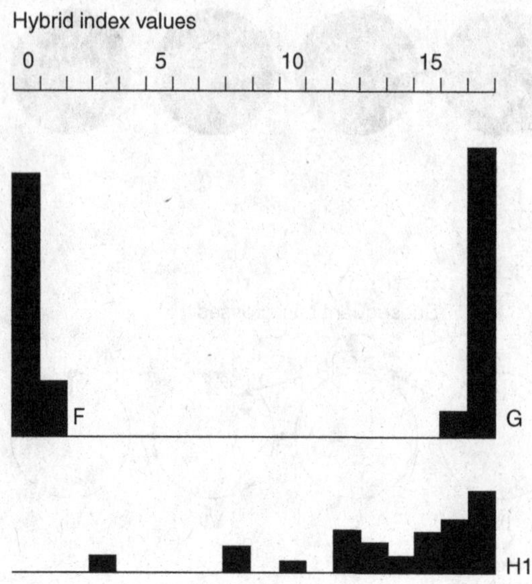

	Tube colour	Sepal blade colour	Sepal length	Petal shape	Exertion of stamens	Stylar appendage	Crest
Like *I. fulva* score	0	0	0	0	0	0	0
Intermediates score	1	1,2 or 3	1,2	1	1	1	1
Like var. *giganticaerulea* score	2	4	3	2	2	2	2

Fig. 13.11. Hybridisation between *Iris fulva* and *I. hexagona* var. *giganticaerulea* in Louisiana, USA, as an example of the use of hybrid index. First a list is compiled of the differences between the two taxa. Next, one species is arbitrarily chosen to be at the low end of the index and the other at the upper end. Specimens from natural populations were then scored character by character and given an appropriate score as outlined in the following scale:

Plants exactly like *I. fulva* score 0 for each character, giving a grand total of 0. Total score for plants like *I. hexagona* is 17. Intermediate plants score from 1–16. Riley's results shown here are for three populations (sample size 23). Colonies F and G are more or less pure parental species. Colony H1 contains apparent hybrid plants that in the main resembled *I. hexagona* rather than *I. fulva*. (From Riley, 1938.) A number of botanists (e.g. Stebbins, 1950; Grant, 1971) have considered this example to be a clear case of introgressive hybridisation. Others were more sceptical (e.g. Nelson & Plaisted, 1967).

Marley Wood and Dickleburgh (Fig. 13.12a) and in a complex population at Boarstall Wood (Fig. 13.12b: see the original paper for full details of the characters studied). This investigation raises a point of general importance. In scoring the material some characters (corolla diameter and calyx tooth length) have been given greater weight than others in the calculation of the hybrid index. For a discussion of weighting and its pitfalls, see Gay (1960) and Hathaway (1962).

Another method of displaying the variation of individual plants is provided by the polygonal graph method of Hutchinson (1936), later elaborated by Davidson (1947). An excellent example is the study of *Viola* hybrids by Moore (1959) (Fig. 13.13).

Various multivariate methods of studying field collections have also been devised, which are most conveniently carried out using computer packages. Fig. 13.14, for example, shows the pattern of variation

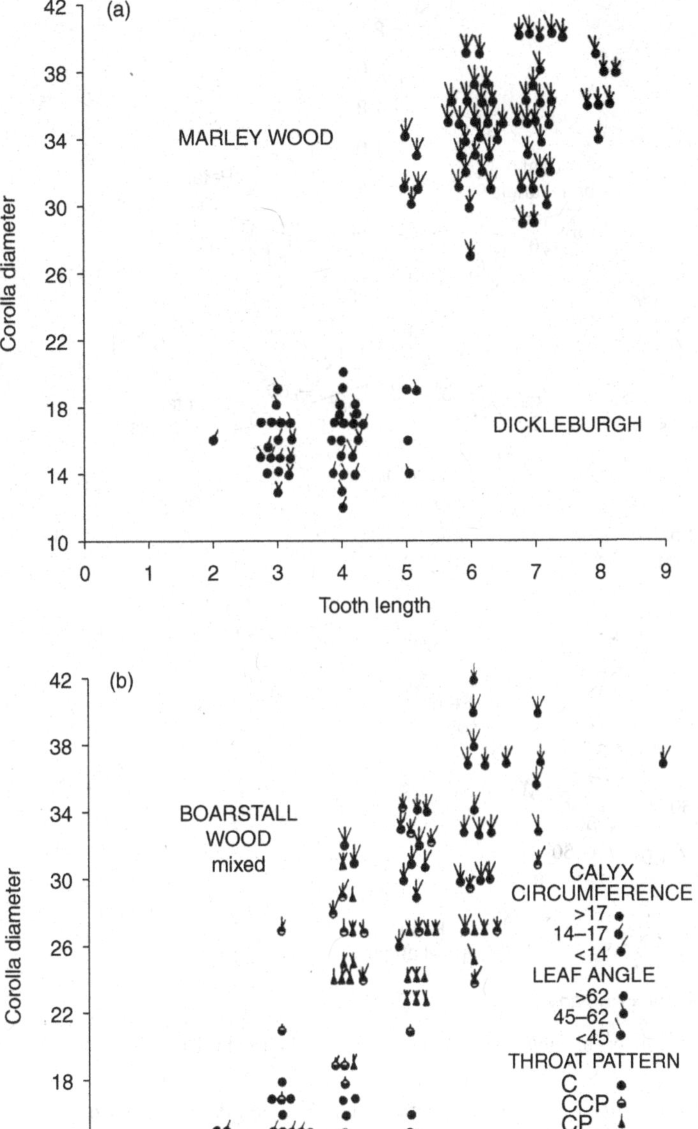

Fig. 13.12. Pictorilised scatter diagrams representing populations of *Primula vulgaris, P. veris* and hybrids in English woodland and meadow sites. (From Woodell, 1965.) (a) Scatter diagram of 'pure'populations of *P.vulgaris* from Marley Wood, Berkshire and *P. veris* from Dickleburgh, Norfolk. (b) Scatter diagram of population from Boarstall Wood, Buckinghamshire (with key to symbols used to prepare both scatter diagrams).

revealed by principal component analysis of populations of *Quercus robur, Q. petraea* and a complex population thought to contain hybrids (Rushton, 1978, 1979).

A character that is often scored in field samples, and sometimes in herbarium studies, is pollen stainability, as judged by staining with acetocarmine or other stains. As we have mentioned earlier, pollen

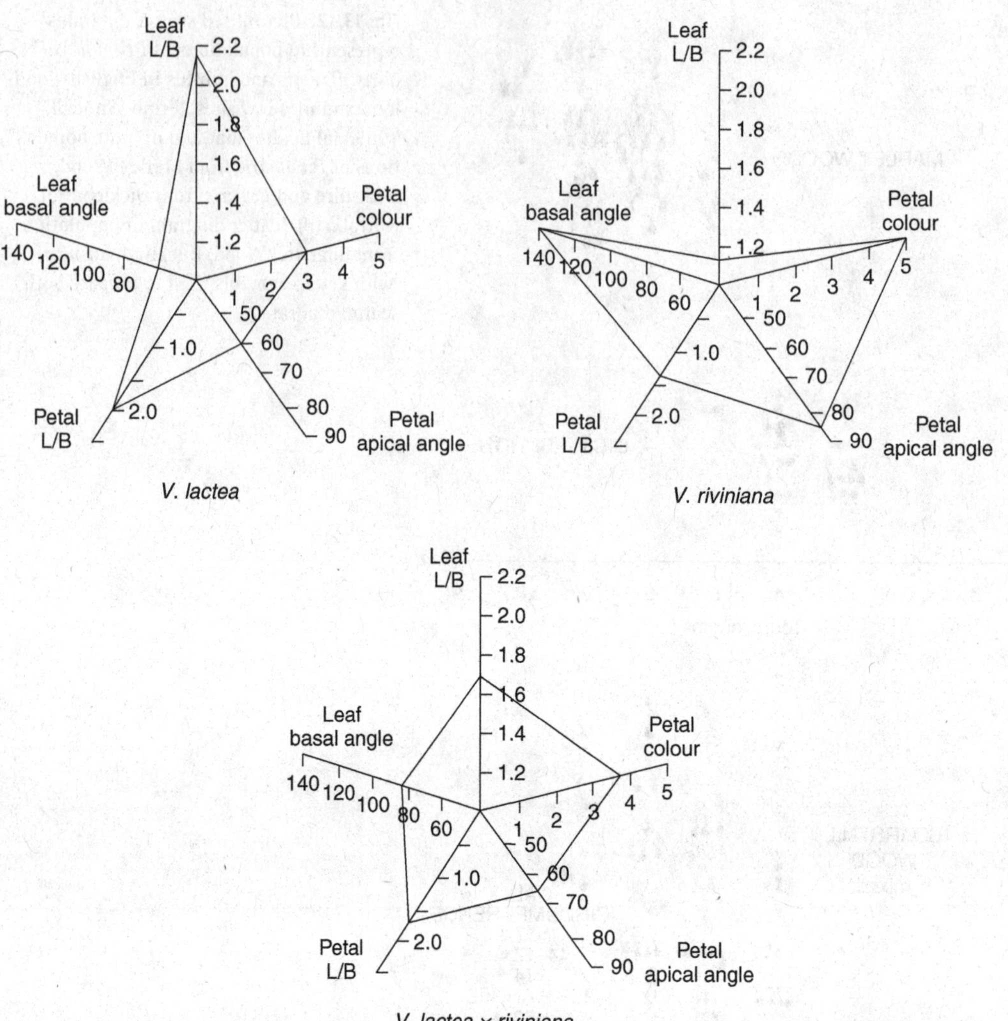

Fig. 13.13. Polygonal graphs of five quantitative characters of *Viola lactea, V. riviniana* and their hybrid. (From Moore, 1959.)

stainability sometimes masquerades as pollen fertility in the report of investigations. Since the relation of pollen stainability to fertility is rarely investigated, it would seem necessary to exercise caution in interpretation. The rationale behind the study of pollen stainability, in the context of natural hybridisation, is as follows. Two populations, as a consequence of genetic differences developed in their period of allopatric isolation, may produce primary hybrids of low fertility, and, in association with morphological intermediacy, low pollen stainability may therefore be taken as evidence of hybridisation.

Genetic investigations of hybridisation

Interpretation of the variation displayed in Andersonian pictorialised scatter diagrams and hybrid index histograms in terms of F_1, F_2 and backcross derivatives was often attempted in early studies. However, Baker (1947, 1951) insisted that intermediacy

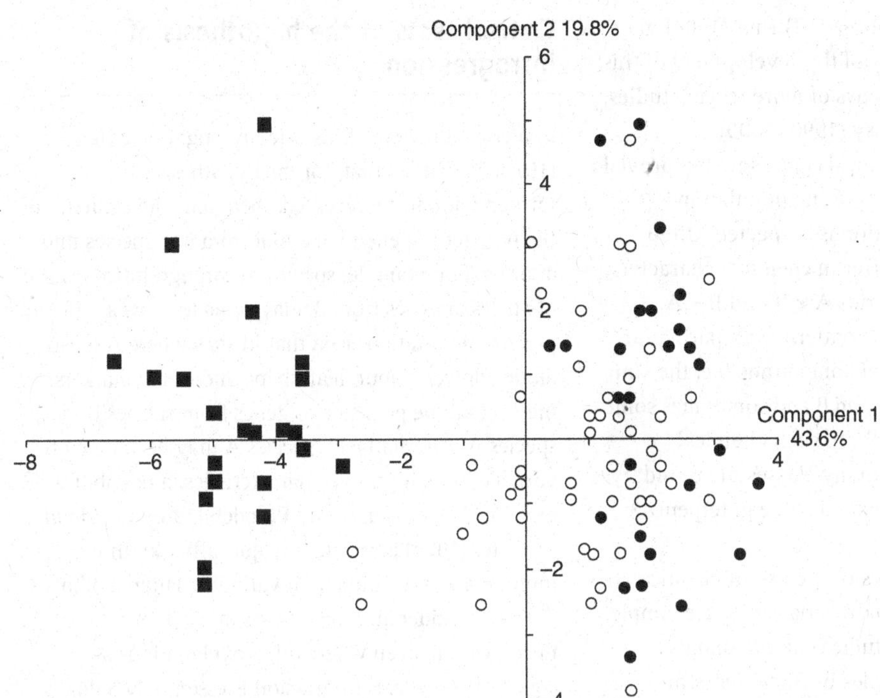

Fig. 13.14. Hybridisation between two species of Oak, *Quercus robur* and *Q. petraea*, in Britain (see Morris & Perring, 1974). This scatter diagram illustrates the use of principal component analysis in the separation of *Q. robur* (dots) from *Q. petraea* ((squares), and the intermediate nature of a putative hybrid population (open circles). (From Rushton, 1978, where further details of the investigation can be found.)

does not necessarily indicate hybridity and that the hypothesis of introgression should be put to the test. Historically, the route then available was to study the genetics of different traits by making artificial F_1, F_2 and backcrosses and comparing these synthesised plants with material collected from the wild.

Thus, in the case of the *Primula* investigation already discussed, the existence of earlier experimental hybridisation studies makes the interpretation more secure. Artificial F_1 hybrids have been made between *P. veris* and *P. vulgaris*, but only with *P. veris* as female parent, the reciprocal cross giving empty or imperfect seeds. Artificial F_1 hybrids are vigorous with pollen stainability *c.*30% of that of parental types (some cytological irregularities have been discovered in meiotic studies of hybrids) and backcrosses have been made (Valentine, 1975b, and references to earlier studies cited).

Sometimes, as part of genetic studies, progeny trials are carried out using seed collected in experiments and in the wild (see Heiser, 1949a, for examples). As the pollen parent is almost always unknown in seed stocks collected from nature, care must be exercised in interpretation (Baker, 1947).

If field collections are the sole evidence available, as they must sometimes be the case, it is now clear that only provisional hypotheses can be safely made, making it completely clear how far the situation has been explored, and in particular whether artificial hybrids have or have not been made experimentally.

Chemotaxonomy: historic investigations of hybridisation

In the 1960s, with the development of chromatography and other biochemical techniques, new sources of evidence became available for interpreting cases of interspecific hybridisation. Chemical investigations of plant variation have provided important insights at many levels. The reviews by Alston & Turner (1963a), Smith (1976),

Ferguson (1980) and Harborne & Turner (1984) may be consulted for the history of the development of this important subject. For reviews of more recent studies, see Stace (1989) and Stuessy (1990, 2009).

The examination of chemical characteristics proved particularly helpful in interpreting population variation where hybridisation is suspected. Often species A and species B differ in chemical characters, and, as a general rule, hybrids A × B and B × A, generally have an 'additive' pattern for a particular class of secondary chemical compounds (i.e. the sum of chemical constituents A and B and sometimes some 'hybrid' compounds as well). In such chemical taxonomic investigations many classes of secondary plant products have been examined, e.g. terpenes, alkaloids, phenolics, etc.

There are many examples of the use of chemical methods in studying complex variation. For example, Alston & Turner (1963b) studied the flavonoids present in population samples of *Baptisia*, in an attempt to resolve patterns of natural hybridisation. Another elegant example is provided by the investigations of Fröst & Ising (1968), who used chemical markers in their study of hybridisation between the widespread northern species of *Vaccinium, V. myrtillus* and *V. vitis-idaea*. The sterile F_1 hybrid between these two species has long been known as *V.* × *intermedium*. A study of the phenolic compounds of leaf extracts, by two-dimensional chromatography, was undertaken by Fröst & Ising using Scandinavian material. They discovered differences between *V. vitis-idaea* and *V. myrtillus* in two localities. While the variation within *V. vitis-idaea* was small, *V. myrtillus* showed considerable differences between sites. Generally speaking *V.* × *intermedium* (which may or may not have been produced from the particular individuals of the parents studied chromatographically) had the 'spots' of both parental taxa, but the patterns were not identical at the two sites. As the genotypes of parental stocks may differ, it is obvious that F_1 variation is to be expected, although Ritchie (1955a, b) found F_1 plants in British populations to be homogeneous morphologically.

Critical tests of the hypothesis of introgression

In a critical review of historic investigations, Heiser (1973) concluded that, for the hypothesis of introgression to be accepted, there must be evidence of the transfer of genetic material from one species into another. For example, species A, through introgression of alleles or genes from B, may come to show a different pattern of variation from that in sites where A exists alone. Flower colour, leaf-shape and other 'markers' may signal the presence of genes from species B in species A. Alternatively, species A may become more variable in quantitative characteristics, a possibility examined, for instance, by Woodell (1965) in *Primula*.

Heiser (1973) points to a major difficulty in the interpretation of patterns of variation. Often, botanists did not consider any hypotheses other than introgression, even where other explanations are equally plausible. Introgression is essentially a down-grade process, in which populations developed in isolation come together with local or regional blurring of pattern. Various other explanations for the variability of taxon A in the direction of taxon B might be devised, which do not necessarily involve present or recent introgression. For instance, as taxa A and B are likely to have developed from a common stock, mutations in A, which are independent of any involvement with B, might appear to be introgressants. Other patterns of variation may be primary (up-grade) situations of great complexity. Dobzhansky (1941) suggested that the 'remnants of an ancestral population from which two species had differentiated might have the appearance of hybrids'. Other possibilities include segregation in polyploid species (Gottlieb, 1972).

Studies of introgression using molecular tools

Having outlined the difficulty of deciding between conflicting hypotheses, Heiser (1973) concludes his review with the prediction that 'with the new tools

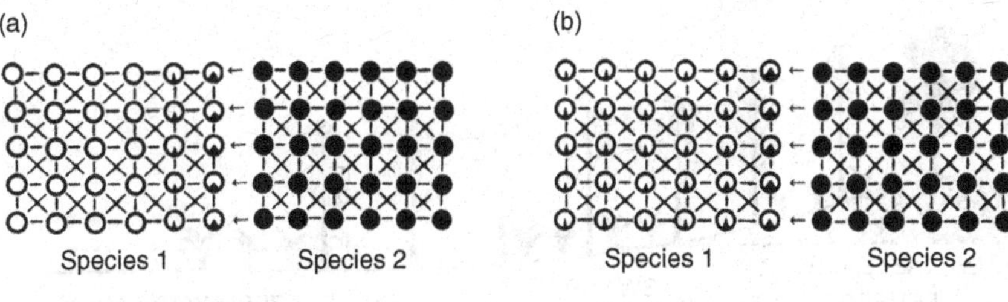

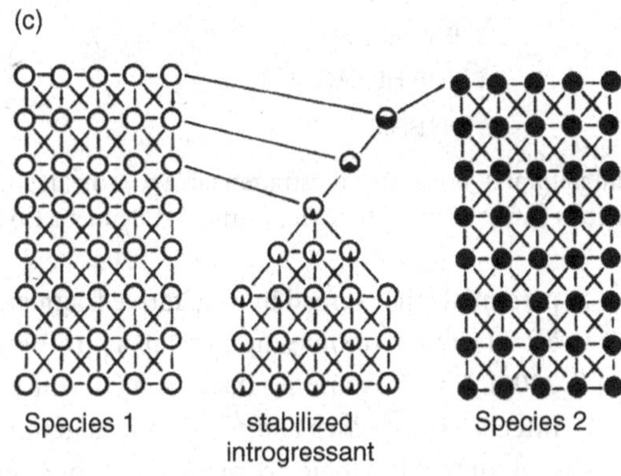

Fig. 13.15. Localised introgression, dispersed introgression and the origin of a stabilised introgressant. Open circles = populations of species 1; closed circles = populations of species 2; black lines = crosses between populations; and arrows = direction of introgression. (a) Unidirectional localised introgression from species 2 into species 1. (b) Unidirectional dispersed introgression. (c) Origin of a stabilised introgressant. (From Rieseberg & Wendel, 1993.)

from biochemistry and genetics now available . . . we may expect some contributions towards the solution of the problem in the future'. This statement can now be seen as prophetic, the use of molecular markers is transforming our understanding of introgression (Rieseberg & Wendel, 1993).

First, they consider some of the difficulties in basing an analysis of introgression on morphological characters alone. 'Morphological characters typically have an unknown, but presumably complicated genetic basis' and 'a non-heritable component that is difficult to estimate'. Furthermore, 'there are often few morphological characters differentiating hybridising taxa and these characters are often functionally or developmentally correlated'. For progress to be made in the critical testing of hypotheses concerning introgression, it is essential to have genetic markers. With the development of molecular tools, for the first time we have 'large numbers of independent molecular markers that allow the detection and quantification of even rare introgression' (Rieseberg & Wendel, 1993). Progress has been made in analysing not only local and dispersed introgression but also where introgression between two species has produced a stabilised introgressant (Fig. 13.15).

One hundred and sixty-five proposed cases of introgression are listed by Rieseberg & Wendel. Their list was not meant to be exhaustive, but concentrated on well-documented cases. Of the examples listed, 85% are from the dicotyledons, and nearly all the different growth form types are represented. More than 90% of the examples come from temperate zones, with 25% of the total being from California. In 65 of the 165 cases the authors consider that the evidence for introgression is strong.

In their ground-breaking review there are 37 examples of the study of introgression using chloroplast DNA, and, of these, 29 cases provide 'robust' demonstrations. In some cases, complex results are obtained, introgression being indicated by these cytoplasmic markers, but not confirmed by nuclear markers (isozymes, ribosomal DNA) studied in the same investigation. In an increasing number of cases, critical tests of hypotheses involving interspecific hybridisation

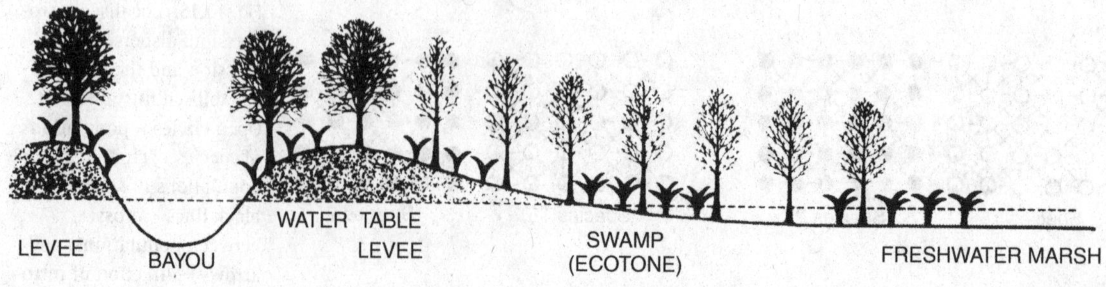

Fig. 13.16. Habitat associations for *Iris fulva, I. hexagona* and natural hybrids. This illustration advances the hypothesis that natural hybrids occupy an ecotone between the parental habitats (After Viosca, 1935, from Arnold & Bennett, 1993).

have been made using molecular tools e.g. some of the first genera to be critically examined include *Brassica* (Palmer, 1988; Song, Osborn & Williams, 1988), *Carduus* (Warwick *et al.*, 1989), and *Gossypium* and *Helianthus* (see Rieseberg & Wendel, 1993 for details).

Introgression in Louisiana Irises

To illustrate the power of the new molecular techniques, we examine in some detail introgression in *Iris*. Hybrids have long been known in *Iris* in the state of Louisiana, N. America. Indeed as we see below (Fig. 13.16), Anderson (1949) took, as one of his major examples of introgression, the patterns of hybridisation found along the banks of the Mississippi river in populations of two perennial clonally spreading *Iris* species.

- *Iris fulva* ($2n = 42$) with brick red flowers found in semi-shaded under-storey areas on wet clay soils along the banks of the water channels, the so-called bayou) and
- *Iris hexagona* ($2n = 44$) with violet flowers growing in full sun on open marshes and swamps subject to brackish water).

The two species have overlapping flowering times and common pollinators (Arnold & Bennett, 1993). Iris plants and their hybrids are long-lived clonal perennials that spread by vegetative fragmentation to establish new colonies up to 10 m apart (Bouck *et al.*, 2005). Notwithstanding the lowered fertility of interspecific crosses and reduced seed germination in their progenies, there are many plants of intermediate phenotype, especially in areas where human activities have hybridised the habitat (Riley, 1938).

Anderson (1949) provided a historical perspective to the ecology of the area. French settlers established farms that were at right angles to the river with long narrow holdings including fields, pasture and woodland. *Iris* hybrids grow in a mosaic of habitats of second-growth woodland, grazing lands, some of which were over-grazed with bare soil and cattle wallows.

Whether this pattern of variation is caused by introgressive hybridisation was at first disputed (for example, see Randolph, Nelson & Plaisted, 1967). However, using a range of molecular tools and a wide variety of experimental methods, Arnold and his associates (references below) have confirmed the introgressive nature of the hybridisation between the two species. In a long series of elegant experimental studies, many elements of the microevolution of *Iris* populations have been elucidated

Introgression investigated In experiments on interspecific crosses between *Iris fulva* and *I. hexagona*, pollen competition was examined. Conspecific/alien pollen mixtures were added to stigmas. Alien pollen proved to be at a competitive disadvantage, as hybrids were not produced in proportion to the two types of pollen (Carney *et al.*, 1994).

Species-specific genetic markers for the presumed parental taxa – *I. fulva* and *I. hexagona* – have been examined and the distribution of these markers in plants of different phenotype investigated (Arnold & Bennett, 1993).

Fig. 13.17 indicates the proportion of *I. fulva* and *I. hexagona* markers discovered in a sample of 42 plants. Introgression is elegantly confirmed. In the swamp cypress area, between the bayou and the road, there is a mixture of variants and a few *I. hexagona* plants. Beyond the road, in the marsh area, a range of variants was found. While there were no 'pure *I. fulva*' plants, the frequency of '*fulva*' markers increased in samples taken further and further out into the marsh.

In a further study, of many of the same samples, it was discovered that *I. fulva* and *I. hexagona* have different chloroplast DNA profiles (Fig. 13.18). As chloroplast DNA is maternally inherited, it is apparent that many of the hybrids (with a mixture of nuclear genes) found between the bayou and the road have *I. hexagona* (and its hybrid derivatives) as female parent. Given the distribution of the cpDNA markers, these plants have originated, not from seed dispersal, but by gene flow of *I. fulva* pollen carried by bumble-bees (Arnold, 1992).

Besides facilitating the analysis of localised introgression in *Iris*, analysis of molecular markers has made it possible to confirm that dispersed introgression also occurs. Thus, Arnold & Bennett (1993) report that marker genes for *I. fulva* have been found in *I. hexagona*, 10 km from the nearest *I. fulva* population. Likewise, *I. hexagona* markers have been detected in *I. fulva* plants, 25 km from the nearest colony of *I. hexagona*.

Investigating hybrid fitness. In the case of the two *Iris* species, tolerance tests, using different salt concentrations, revealed a fitness hierarchy: *I. hexagona* > hybrids > *I. fulva* reflecting the habitat preferences of the parental species and the intermediacy of the hybrids. In other investigations, studying shade tolerance, *I. fulva* proved to be more tolerant than *I. hexagona*, with hybrids showing intermediate tolerance.

Reciprocal transplant experiments revealed that some hybrids had higher fitness than parental taxa under 'intermediate' ecological conditions (Emms & Arnold, 1997).

Markers from another Iris species Molecular investigations have also established that some hybrid Irises contain not only markers for *I. fulva* and *I. hexagona* but also markers from a third species *I. brevicaulis* which grows in drier habitats (Arnold, 1993; Bouck *et al.*, 2005).

A new stable hybrid species Studies of hybridisation in *Iris* have shed light on yet another interesting question: can a stable hybrid species be produced by introgressive hybridisation? Arnold and associates, using the evidence of species-specific molecular markers, have discovered that *I. nelsonii* is a fixed derivative of the hybridisation of three *Iris* species *I. fulva*, *I. hexagona* and *I. brevicaulis* (Arnold, 1993). *I. nelsonii* grows in heavy shade in deep water and is, therefore, ecologically separated from its parental species. New stabilised introgressants have also been detected in *Helianthus* etc. (see Rieseberg & Wendel, 1993). In the next chapter, we will discuss the possibility that they may have had an abrupt origin.

Studies of introgression in other genera Close investigation of introgression in *Iris* has provided many insights. By using a range of techniques – some traditional, others involving advanced molecular techniques and new analytical computer tools – researchers have been able to focus on a number of important aspects of introgression.

Asymmetric introgression

Considering two species A and B, introgression may occur equally in both directions or it may be asymmetric, say predominantly or exclusively from A to B. Asymmetric introgression have been described in a number of investigations, e.g. *Rorippa palustris*/*R. amphibia* in human-influenced habitats (Bleeker & Hurka, 2001); *Morus alba*/*M. rubra* (Burgess *et al.*, 2005); *Silene dioica*/*S. latifolia* (Minder,

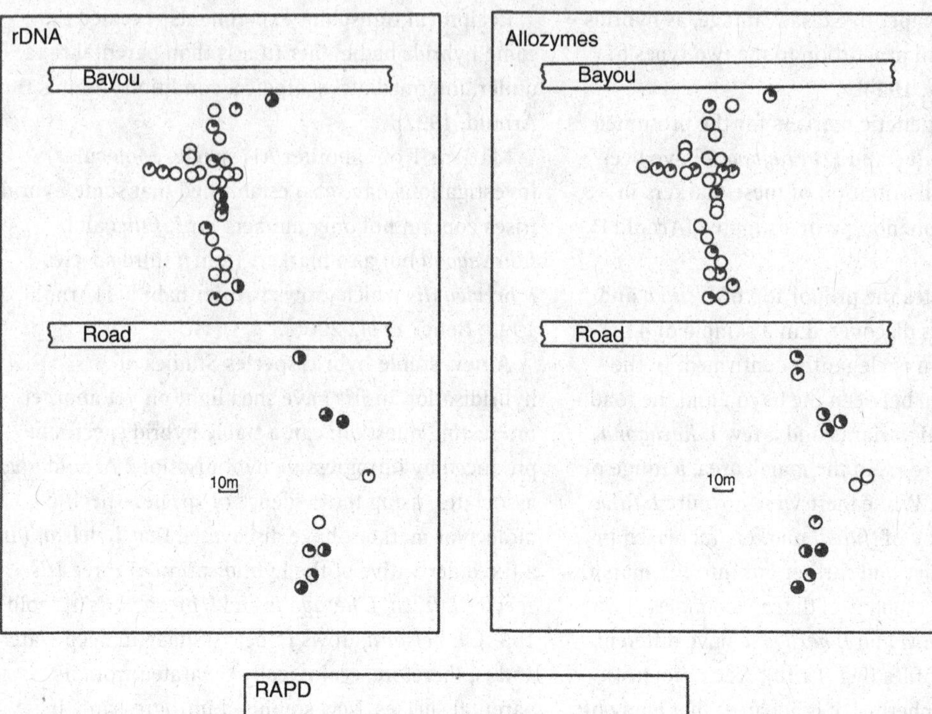

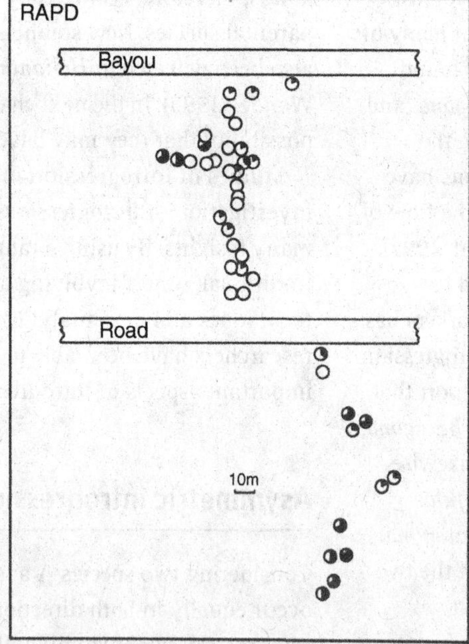

Fig. 13.17. Genetic variation in a sample of 42 individuals from the Bayou L'Ourse population. Each circle represents an individual plant. The filled and open portions of each circle represent the proportion of *Iris fulva* and *I. hexagona* markers respectively. The top left panel, top right panel and bottom panel illustrate the rDNA, allozyme and RAPD variations, respectively. There are three missing data points (plants) in the rDNA analysis. (From Arnold & Bennett, 1993.)

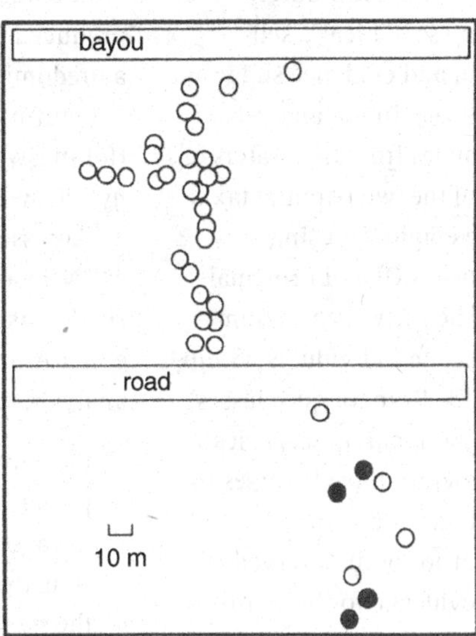

Fig. 13.18. Asymmetrical introgression between *Iris fulva* and *I. hexagona* is likely to result from pollen flow. (After Arnold, 1992, from Avise, 1994.) Each circle represents a single plant. Left: relative proportion of *Iris fulva* (shaded) and *I. hexagona* (unshaded) nuclear markers. Right: similar representation for maternally transmitted chloroplast DNA markers. Note in particular the population between the road and the bayou, where multilocus nuclear genotypes suggest the presence of advanced generation hybrids or backcrosses, despite the apparent absence of seed dispersal that would be registered by chloroplast DNA from *I. fulva* in this area. (After Arnold, 1992, from Avise, 1994.)

Rothenbuehler & Widmer, 2007; Karrenberg & Favre, 2008); *Ulmus pumila/U. rubra* (Zalapa, Brunet & Guries, 2009); *Helianthus annuus/H. debilis* (Scascitelli *et al.*, 2010); *Cypripedium tibeticum/ C. yunnanense* (Hu *et al.*, 2011); and *Eucalyptus aggregata/E. rubida.* (Field *et al.*, 2010).

Of factors that might promote asymmetric introgression, the relative frequency of the two species is thought to be most important, with introgression generally from the less frequent to the more frequent species.

Theoretically, differences in the breeding systems may also promote asymmetry in species introgression, but this area has been little explored. As we saw earlier in the chapter, hybrid swarms *Geum urbanum* × *G. rivale* were examined in detail by Marsden-Jones (1930). Building on these investigations the hypothesis of asymmetrical hybridisation in this case has been tested by examining the early evolution of a hybrid swarm from a disturbed area by a disused railway at Waters of Leith, Edinburgh (Ruhsam *et al.*, 2010; Ruhsam, Hollingsworth & Ennos, 2011). Samples of putative pure parental plants from other sites throughout the UK were included in the investigations.

The investigation employed the morphological characters used by Marsden-Jones (1930), together with molecular markers – chloroplast DNA (maternally inherited) – and nuclear markers (AFLP techniques). In addition, crossing experiments were carried out to explore parental breeding systems. Also, estimates of pollen fertility (as revealed by staining with 2% acetocarmine) were carried out, together with observations of the flowering periods of parental and hybrid taxa, and an examination of the ability of plants to set seed in the absence of pollinators (using both bagged and unbagged flowers).

Studies of breeding systems revealed that *G. urbanum* is self-compatible with a low outcrossing rate. In contrast, *G. rivale* possesses a leaky self-incompatible breeding system, and evidence suggests that outcrossing is high (70–90%). In the analysis of morphological markers of samples from the Waters of Leith (together with material of the two parental taxa), multivariate approaches were employed using Principal Coordinates Analysis (PCO) and Principal Component Analysis (PCA). The software program NEWHYBRIDS was used to 'assign individuals within the hybrid swarm to the most likely genotypic classes'. The hybrid swarm comprised both parental species together with F_1s and first generation backcrosses to *G. rivale* alone. Chloroplast data suggested that *G. rivale* was the pollen parent for both observed hybrid classes. There was no evidence for backcrosses to *G. urbanum*. Pollen fertility of the F_1 was only 30% lower than that of the parental taxa and was fully restored in the backcross hybrids. No genotypes resulting from the selfing of the F_1s were detected in the field, even though, in experiment, all the F_1s were self-compatible and had high pollen fertility (63%). Controlled crosses with F_1s can be made in either direction, but it is likely that, in the wild, *G. urbanum* is the maternal parent of the F_1s in the hybrid swarm. The viability of the seed was much lower when *G. rivale* was the mother plant. Overall, the results support the hypothesis that introgression is likely to be asymmetrical from the inbreeder (*G. urbanum*) to the outbreeder (*G. rivale*).

Cytoplasmic capture

The investigation of the two *Geum* species also considered another phenomenon associated with introgression, namely cytoplasmic capture.

Consider asymmetrical introgression involving two species A and B, with F_1 and backcrosses being produced. If B is the pollen parent of the F_1 with species A, and if A is the seed parent for the initial and subsequent backcrossing, then because of the maternal inheritance of cytoplasm and its organelles (mitochondria and chloroplasts), hybrids will be produced with the cytoplasm of A but with a predominance of nuclear genes from species B.

Returning to the *Geum* investigation, Ruhsam, Hollingsworth & Ennos (2011) concluded that: in the hybrid swarm *G. urbanum*, F_1 hybrids and backcrosses to *G. rivale* possessed haplotype A, the cpDNA variant typical of *G.urbanum*. In contrast *G. rivale* was polymorphic, with individuals having either A or another variant B. Considering the AFLP analysis, they note that:

> the most striking finding from our study is the high prevalence of backcross genotypes to *G. rivale* and the absence of backcrosses to *G. urbanum*. The data from the cpDNA analysis, which show no evidence of the species-limited marker from *G. rivale* in the backcrosses to *G. rivale*, are consistent with the F_1 being the female parent in these backcrosses. These findings are as predicted on the basis of the relative male fitness of the parental and hybrid genotypes and effectively result in asymmetric introgression of cpDNA and nuclear genes from *G. urbanum* into *G. rivale*. If this pattern of fertilisation of maternal hybrids by *G. rivale* pollen continues in future generations, individuals with nuclear genomes that are predominantly from *G. rivale*, but possessing the cpDNA from *G. urbanum* will be formed.

Effectively, the 'chloroplast haplotype of the inbreeder will be captured by the outcrossing species, but not *vice versa*'.

Earlier, a number of hybridising situations of 'cytoplasmic capture' have been identified (Rieseberg & Soltis, 1991). 'Plastid capture' was recently investigated in *Silene dioica* and *S. latifolia*. Minder, Rothenbuehler & Widmer (2007) found evidence of extensive introgressive hybridisation between the two species and evidence of 'unidirectional plastid introgression' from *S. latifolia* into *S. dioica*.

Cytoplasmic capture has been closely investigated in the genus *Quercus* (Petit *et al.*, 2003).

Chloroplast capture: another route

There is compelling evidence that introgression between different species may lead to chloroplast capture. However, this is not the only route. Recently, experimental evidence has been published (see below) that strongly suggests the possibility of horizontal transfers of chloroplast genomes through natural grafting, even between species that are sexually incompatible.

Transgressive hybridisation

How far are hybrids intermediate in character expression? A review of historic studies reveals that hybrids are often a 'mosaic of both parental and intermediate morphological characters rather than just intermediate ones, and that a large proportion of first (64%) and later generation hybrids (89%) exhibit extreme or novel characters' (Rieseberg & Ellstrand, 1993). There is considerable interest also in the evolutionary potential and significance of so-called transgressive variants (Givnish, 2010). These were defined, by Darlington & Mather (1949), as variants that 'fall outside the limits of variation defined by the parents and F_1 of the cross in respect of one or more characters'.

Rieseberg *et al.* (2007) have examined the role of hybridisation in the diversification of North American sunflowers (*Helianthus*). These studies have continued and extended the classic investigations begun in the 1940s by Charles Heiser and associates (Heiser, 1949b, 1979; Heiser *et al.*, 1969). Rieseberg *et al.* (2007) report that in the 'invasion of Texas by the common sunflower (*Helianthus annuus*) . . . introgression of chromosomal segments from locally adapted species may have facilitated range expansion'. And as a consequence the 'colonisation of sand dune, desert floor and salt marsh habitats by three hybrid sunflower species was made possible by selection on extreme or "transgressive" phenotypes generated by hybridization' (Rieseberg *et al.*, 2007). These experiments have continued and consider a crucially important question.

Can it be confirmed that adaptive traits are transferred from one species to another by introgression?

In an earlier review Rieseberg & Wendel (1993) make the point that evidence of the introgression of molecular marker genes between hybridising species may be readily demonstrated in experimental studies, but because such transfers might be selectively neutral, testing the adaptive significance of introgressed traits requires careful investigation and the evaluation of several lines of evidence.

In a very useful critique, Vekemans (2010) notes that robust validation requires 'evidence for genetic introgression from donor to recipient taxon; evidence that phenotypes of the recipient taxon at fitness-related traits have shifted towards those of the donor; and evidence of natural selection has favored recipient genotypes with phenotypes shifted towards the donor'.

As noted above, there is excellent evidence for the introgressive transfer of adaptive traits between *Iris fulva* and *I. hexagona* (see Arnold, Bouck & Cornman, 2004). A later paper by Martin, Bouck & Arnold (2006) charts the transfer of flood tolerance traits from *I. fulva* to *I. brevicaulis*.

However, it is helpful at this point to examine the case of the transfer of traits from *Helianthus debilis* to *H. annuus* in a little more detail. Overall the balance of evidence is very persuasive. First, marker experiments have confirmed the transfer of genetic material from *H. debilis* to *H. annuus* in the wild (Scascitelli, Cognet & Adams, 2010). To examine questions concerning adaptive variation, two common garden trials were set up with samples of parental taxa, and naturally occurring hybrids. Also included were synthesised hybrids – first-generation hybrids obtained by backcrossing – F_1 plants of the cross *H. debilis* × *H. annuus* with *H. annuus* as pollen parent (so-called BC1 hybrids). Ten ecophysiological, phenological and plant architecture traits were examined in the garden trial (Whitney, Randell & Rieseberg, 2010). The fitness of individual plants was estimated from the amount of viable seed produced. It was discovered that: '8 out of 10 traits showed patterns consistent with introgression from *H. debilis* into *H. annuus* and suggest that *H. debilis*-like traits allowing rapid growth and reproduction before summer heat and drought, have been favoured in the hybrid range'. Moreover, 'evidence suggests that natural selection currently favours BC1 hybrids with *H. debilis* branching patterns'.

'Altogether the results carefully demonstrate that adaptive introgression has resulted in the transfer of traits . . . from *H. debilis* to the genomic background of *H. annuus*. More importantly they show multiple traits related to adaptation to abiotic features of the environment can be transferred together' (Vekemans, 2010).

Speciation: where future advances might come

Earlier in the chapter we discussed the concept of 'reinforcement' involving the strengthening of reproductive barriers between incipient species in sympatric situations. Further studies, especially in the wild population, are necessary to investigate this aspect of speciation (Langerhans & Riesch, 2013). In addition, they stress, along with several other authors (Coyne & Orr, 2004; Hendry, 2009; Nosil, Funk & Ortiz-Barrientos, 2009), that there are many advantages in studying 'ongoing speciation' in 'young species'. 'Such systems represent the most direct way of evaluating the mechanism actually driving speciation, as "older" species may display a number of isolating mechanisms that may have evolved after, not during, speciation.' However, they recognise that there are some difficulties in recognising early-stage speciation and, as we have seen above, there is a lively debate about whether ecotypes are an earlier stage in evolution. Nonetheless, Langerhans & Riesch (2013) consider that one potentially revealing approach is to investigate, in detail, cases of 'parallel or non-parallel responses to shared environmental gradients'.

An excellent example is provided by the case of the parallel evolution of dwarf ecotypes (less than 4 m tall on exposed headlands) of the forest tree *Eucalyptus globulus* (typically 15–60 m tall at maturity in forests). Evidence from 12 microsatellite markers and chloroplast DNA reveals that the three dwarf populations were not genetically related to one another, but were related to adjacent tall trees (Foster *et al.*, 2007). It was also discovered that dwarf and tall ecotypes differed in flowering time. Overall, the study showed that 'small marginal populations of eucalypts are capable of developing reproductive isolation from nearby larger populations through differences in flowering times and/or minor spatial separation, making parapatric speciation possible'.

Other case studies of polyphyletically developed ecotypes are beginning to reveal some of the early processes in speciation (Langerhans & Riesch, 2013): for example, interior and coastal ecotypes of *Mimulus guttatus* that differ in size, flowering time, various morphological features and salt tolerance (Lowry, Rockwood and Willis, 2008); and contrasting ecotypes growing on heavy-metal-contaminated or uncontaminated soils in *Lasthenia californica* (Osevik *et al.*, 2012).

The role of hitchhiking in speciation

Presgraves (2013) characterises the commonly held view that speciation 'is an accident that happens as populations adapt to different environments and incidentally come to differ in ways that render them reproductively incompatible'. However,

> recent work on the molecular genetics of speciation has raised another possibility – the genes that cause hybrid sterility and lethality often come to differ between species not because of adaptation to the external ecological environment, but because of internal evolutionary arms races between selfish

> genetic elements and the genes of the plant genome . . . The notion that selfish genes are exotic curiosities is now giving way to a realization that selfish genes are common and diverse, each generation probing for transmission advantages at the expense of their bearers, fuelling evolutionary arms races and, not infrequently, contributing to the genetic divergence that drives speciation.

The detailed case for this view, and how it might be tested, is explored in Rieseberg & Blackman (2010), Presgraves (2010) and Johnson (2010).

Next generation sequencing technologies

Salmon & Ainouche (2010) evaluate the molecular tools available to study DNA methylation in the investigation of epigenetic changes that occur in hybrid and polyploidy plants. More recently, Twyford & Ennos (2012) consider the next generation sequencing technologies. These raise exciting possibilities in the study of epigenetics, speciation, hybridisation and introgression. It is forecast that it will be possible to examine, at reasonable cost, large sections of the genomes of non-model species. They point out that making the best use of these new sets of data presents a formidable challenge for computational analysis but also in interpretation. Plant genomes have many repetitive elements and are 'laden with transposable elements'. In addition, recurrent rounds of polyploidy have added to genomic complexity.

With regard to hybridisation and introgression, these new methodologies offer the prospect of investigating how frequent it is, and how introgressed elements fare in new genomic backgrounds. It has been postulated that very subtle patterns of introgression may occur and that such cryptic effects may be very important in evolution (Rieseberg & Wendel, 1993). Moreover, it might also be possible to test hypotheses concerning 'cryptic' introgression and ancient introgressive events (see below).

Zones of introgression: are they ephemeral or long-standing?

Buggs (2007) provides an important review of apparent hybrid zones in animal and plant populations. He considers the factors that might govern the boundary changes of such zones, the importance of zone movement as a possible indicator of environmental change, and the strengths and weaknesses of the methods available to study such issues.

An excellent pioneering example provides evidence of the stability of a hybrid zone. Hybridisation and introgression between the two introduced species *Carduus nutans* ($2n = 16$) and *C. acanthoides* ($2n = 22$) in Grey County, Ontario, was investigated in 1951 and involved cytological and morphological studies of adult plants and progenies of field-collected seed (Moore & Mulligan, 1956, 1964). Warwick *et al.* (1989) revisited the original and other sites and studied population variation using molecular and chemical markers. The results suggest that introgression may well be bidirectional and that a stable zone of hybridisation had changed little in the past 30 years. This case study points to the importance of long-term studies of hybridising populations.

Introgression: a key concept in microevolution

In 1949, Anderson's highly influential book entitled *Introgressive Hybridisation* was published. Since that time the concept of introgression has become an important topic of research.
The database Web of Science is a key to papers in the scientific literature. Between 1949 and 2012 over 7000 papers have been published containing introgression as a keyword in studies of a wide variety of different organisms. About 2000 refer to researches on plants.

Taxonomic considerations

Models of allopatric speciation might suggest that evolution produces entities that share an evolutionary history, and which, in time, become perfectly reproductively isolated from each other. However, taxonomists have found it difficult to delimit species in some genera, and the root cause, in some cases, is rampant hybridisation. Thus, it is abundantly clear that, in many plant groups, the reproductive barriers between species are imperfect, with over 25% of plants known to hybridise with other species (Mallet, 2005). The outcome of such hybridisation varies greatly, but some taxa recognised as species are still 'permeable filters to gene flow'.

The recent *Hybrid Flora of the British Isles* (Stace, Preston & Pearman, 2015) sheds new light on the question of the frequency of species hybridisation by providing detailed accounts of reliably reported hybrids (909) and their distribution (388 maps). This remarkable book is the first of its kind and 'will surely set the standard for future hybrid floras' (Abbott, 2015, in the foreword to the flora).

Introgression: its role in evolution

Despite decades of research, the precise role that introgression might play in evolution has long been debated (Rieseberg, 1995). Twyford & Ennos (2012) provide a very important commentary on recent opinion.

a) Some see introgression as providing no more than evolutionary noise.
b) Others consider it a 'potentially creative evolutionary process allowing genetic novelties to accumulate faster than through mutation alone'.
c) However, one of the residual difficulties is establishing the cause(s) of particular patterns of variation. Are they the result of 1) 'recent introgression'; 2) 'ancient introgression'; or 3) 'incomplete lineage sorting' – meaning that ancestral variation has not been sorted into genetically separate lineages during speciation prior to interspecific hybridisation. Twyford & Ennos (2012) point out that 'the genetic signature of incomplete lineage sorting is the same as ancient introgression soon after speciation'. One approach to investigating this problem is to examine allelic diversity near hybrid swarms (Arnold, 1992). He argues that 'one signature of recent introgression is higher allelic diversity near hybrid swarms and a cline of introgressed alleles as one samples away from them'.
d) Another issue to face is the complexity involved in establishing the adaptive significance of introgressed genes. For ease of comparison, some experimenters devised groupings of hybrid plants for examination of potentially adaptive responses. Clearly, the fitness of individual plants must be estimated to get a fuller picture of the microevolutionary significance of introgression (Arnold, Ballerini & Brothers, 2012).
e) Given the active debate about introgression, one conclusion is inescapable. Some models of speciation visualise diverging dichotomous branching patterns of speciation. Hybridisation and introgression run counter to this simple modelling, pointing instead to the complex reticulate patterns of speciation in many plant groups.

A number of other aspects of introgression are discussed elsewhere in this book. The possible long-term effects are considered, especially whether it is possible to detect ancient introgression. Secondly, hybridisation and introgression in species of human-influenced habitats are considered, in particular crop–weed interactions, introgression between GM plants and their wild relatives, and the part played by hybridisation and introgression in the success of invasive species.

In this chapter we have considered the evidence for allopatric speciation. In the next chapter we consider evidence for abrupt speciation and its implications for pattern and process in plant evolution.

14 Abrupt speciation

The processes of allopatric speciation discussed in Chapter 13 'may be responsible for many, if not most, speciation events' (Soltis & Soltis, 2009). However, new species also arise abruptly in a sympatric context, often as a consequence of species hybridisation, the 'evolutionary signature' of which is 'rampant' in many groups of plants (Soltis & Soltis, 2009). In this chapter, our account will reflect the major change of outlook that has emerged in the study of abrupt speciation processes. For example, Stebbins (1971) considered that 'polyploidy has contributed little to progressive evolution'. This view has been replaced by a radically different 'consensus view that polyploidy is a prominent force in plant evolution' (Wendel & Doyle, 2005).

Our present understanding of abrupt speciation is explored in three sections:

A. The emergence and testing of key concepts in the study of polypoidy
B. Recent insights into polyploidy from molecular studies
C. Other modes of abrupt speciation

A. The emergence and testing of key concepts in the study of polypoidy

For close on 100 years, botanists have been investigating abrupt speciation by polyploidy. Using the techniques available (and fashionable) in different periods, hypotheses about polyploid speciation have been tested. We consider the successes and limitations of the various approaches from cytogenetical and karyological methods (first developed in the nineteenth century), to chemical investigations of the 1960s. In presenting a number of case histories we review the emergence of key concepts and consider how early findings have been tested, confirmed, refined and extended by the use of an increasing range of molecular methods.

The concept of polyploidy: early cytogenetic studies

As we saw in Chapter 12, Winge's (1917) hypothesis that fertile derivatives could be derived from sterile hybrids by polyploidy was tested in a number of experimental studies.

A famous case of allopolyploidy was provided by *Primula kewensis*, which was discovered amongst seedlings of *P. floribunda* at Kew in 1899.
The proposition that *P. kewensis* was a hybrid between *P. floribunda* and *P. verticillata* was put to the test by making the hybrid experimentally, using *P. floribunda* as female parent. *P. kewensis* was morphologically intermediate between the parental stocks and had the same chromosome number $2n = 18$ (Digby, 1912; Newton & Pellew, 1929). Although meiosis was regular in the hybrid, the plants were sterile, presumably because of genetic imbalance. This sterile hybrid was vegetatively propagated and widely distributed as an ornamental garden plant. On three occasions, however, hybrid plants were observed to set good seed (in 1905 at the nurseries of Messrs Veitch, in 1923 at Kew Gardens, and in 1926 at the John Innes Horticultural Institution). In each case the progeny proved to be tetraploid and fertile. Also, in one original hybrid plant the investigators discovered that vegetative cells from the fertile stem were tetraploid, the parent plant itself being largely diploid with sterile inflorescences, showing that a sterile hybrid had become fertile by somatic doubling.

The fertile *Primula kewensis* behaves as a new species, morphologically similar to the sterile hybrid stocks from which it was derived, but distinct from both parents. However, in the account of *Primula* for *The New Royal Horticultural Society Dictionary of Gardening* (Huxley, Griffiths & Levy, 1992) some doubt has been expressed about the ancestry of *Primula kewensis* proposed by Newton & Pellew and, as far as we have been able to determine, this case has yet to be re-examined, using molecular markers.

It can readily be seen that, if the primary hybrid and allopolyploid derivatives arise in cultivation, they may be detected and their ancestry investigated. Many other examples of allopolyploids arising in experiments and in cultivation from sterile hybrids are known (e.g. Darlington, 1937; Grant, 1971, 1981; Lewis, 1980a, b, c), and essentially similar means were used to deduce ancestry.

Resynthesis of wild polyploids

In the determination of the origin of allopolyploids arising in experiments or in cultivation, the parental stocks may be obvious or the number of candidates limited. However, as there are often many diploid taxa in a genus, unravelling the ancestry of wild polyploids is altogether a more formidable undertaking. It is very instructive to examine the famous experiments of Müntzing (1930a, b), who studied the origin of the weedy species *Galeopsis tetrahit* ($2n = 4x = 32$). First, he made a careful study of six diploid species ($2n = 2x = 16$); from these, *G. pubescens* and *G. speciosa* were selected as closest in morphology to *G. tetrahit*. F_1 hybrids ($2n = 2x = 16$) were produced between *G. pubescens* (female) and *G. speciosa* (male), which proved to be highly, but not absolutely, sterile. After self-fertilisation these F_1 plants yielded strongly variable F_2 progeny, amongst which was a triploid plant ($2n = 3x = 24$), presumably arising from the union of one reduced and one unreduced gamete. This highly sterile triploid was backcrossed to *G. pubescens* and yielded one seed which germinated and grew to give a plant with $2n = 4x = 32$ chromosomes (presumably derived from a cross between an unreduced gamete from the triploid plant and *G. pubescens* pollen). Morphologically this tetraploid derivative was very like *G. tetrahit*. In a study of the 'status' of the experimentally produced tetraploid, Müntzing (1930a, b) discovered that there was no difficulty in crossing it with wild stocks of *G. tetrahit*. Moreover, fertile offspring were produced. Thus, Müntzing concluded that it is highly likely that the ancestors of present-day *G. tetrahit* arose from stocks ancestral to present-day *G. speciosa* and *G. pubescens*.

This classic textbook example has been re-examined using various AFLP, and cpDNA molecular markers (Bendiksby *et al.*, 2011). Müntzing's conclusions about the allopolyploid origin of *G. tetrahit* have been confirmed and extended. The origin of some of his stocks is unclear. Bendiksby *et al.* (2011) discovered that a distinct variant of *G. speciosa* (recognised as ssp. *sulphurea*, but sometimes treated as a species in its own right as *G. sulphurea*) was most likely to be the organelle parent of the tetraploid *G. tetrahit*. Müntzing (1932) suggested that *G. tetrahit* originated in central or eastern Europe: in contrast, the investigations of Bendiksby *et al.* (2011) narrowed the source of origin to the 'south-western Alps, where *G. sulphurea* is endemic'.

Müntzing's resynthesis process was essentially a two-stage process involving a triploid bridge. While his approach is conceptually sound, the method suffers from the weakness that the investigation is dependent upon chance events. Thus, in the *Galeopsis* experiments, one triploid plant was produced in the F_2 progeny and this might easily have been overlooked. Moreover, only a *single* tetraploid seed was produced on crossing this plant with *G. pubescens*, and this might have failed to grow. What was needed was a method of generating polyploids 'at will' from diploid stocks.

As long ago as 1904, the Czech botanist Nemeč reported that chromosome doubling in root cells could be induced by treatment with chloral hydrate or other

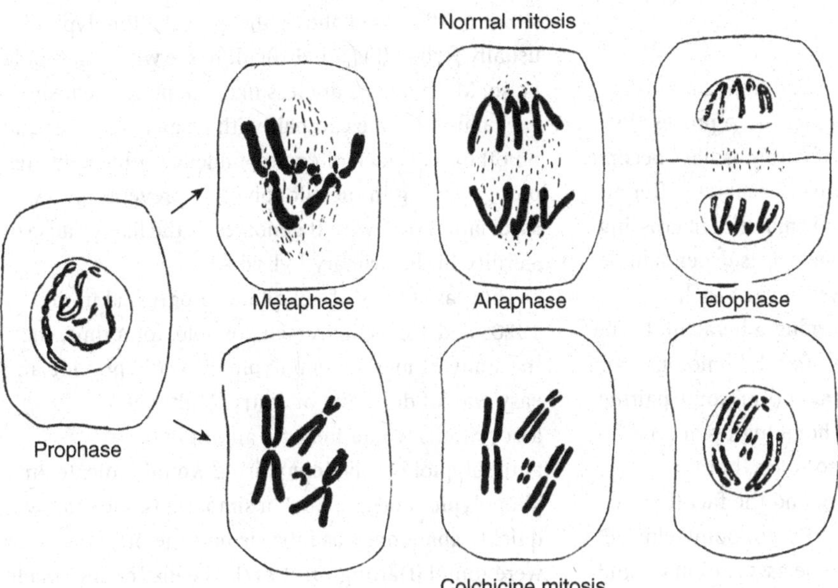

Fig. 14.1. The action of colchicine. (From Müntzing, 1961.) The normal mitosis gives rise to two cells each with four chromosomes. By the action of colchicine, in which the mechanism of movement of the chromosomes is anaesthetised, one cell with eight chromosomes is formed. Note the characteristically widely spaced chromatids in the metaphase with the colchicine treatment. If the cell with eight chromosomes is removed from the colchicine, a spindle may form at the next nuclear divisions producing daughter cells, both of which are polyploid. If cells remain in the colchicine, additional C-mitoses may take place with increments of polyploidy. For example, Onion (*Allium*) roots left in colchicine for four days have cells with more than 1000 chromosomes (Levan, 1938). On removal of roots from colchicine, competition between cells with different chromosome numbers occurs and a new root may grow out with diploid cells. For detailed information about colchicine, including its carcinogenic properties, see Eigsti & Dustin (1955).

narcotics. Blakeslee & Avery (1937), studying the effects of various substances and treatments on stem tissue, reported that the alkaloid colchicine had the property of inducing chromosome doubling. Different methods of application were devised. Seeds could be soaked in dilute colchicine solutions, or the alkaloid could be applied to plants as a lanolin mixture, in agar blocks, by application of drops of solution to bud tissue, or by atomised sprays. The cytological effect of colchicine is illustrated and discussed in Fig. 14.1.

The discovery that colchicine treatment could induce polyploidy was a major breakthrough in historic experimental taxonomic studies.
The technique has continued to be important, for example, in the investigation of the origin of Bread Wheat (see below), and in resynthesis experiments. Thus, colchicine-induced polyploidy has contributed to investigations aimed at assessing the structural, genetic and epigenetic changes that have occurred in contemporary polyploids, as it allows comparison of existing polyploids with newly synthesised counterparts artificially produced from parental stocks (Hegarty *et al.*, 2013).

The concepts of auto- and allopolyploidy

Autopolyploidy involves the multiplication of the same chromosome set. Thus a diploid, which has two like chromosome sets (genomes), could give rise to an autotetraploid with four such sets by chromosome doubling. Such a change could be represented symbolically as follows:

AA – – – – – – – – > AAAA.

Normal sexual reproduction in diploids involves the production of gametes by meiosis, a process in which the homologous pairs of chromosomes become associated together and eventually separate after an exchange of a portion of genetic material in crossing-over. This regular pairing at meiosis is dependent upon there being two, and two only, of each homologous chromosome, forming a bivalent. In the autotetraploid, four members of each homologue are present. Evidence suggests that chromosome pairing is only possible between two homologues at any particular point on the chromosomes, but the proximity of four homologues, and the fact that pairing may begin at several different points during the pairing process, results in the association of, and chiasma formation between, three or four chromosomes. Unpaired single chromosomes (univalents) may also remain. Groups of three or four chromosomes lead to chromosome structures known as multivalents, such associations being easily recognised in favourable cytological material.

Univalents may be segregated in different (unbalanced) numbers in the two products of the first division of meiosis, an event contributing to chromosome imbalance and infertility. Multivalent production sometimes results in a failure of normal separation of chromosomes: for example, bridges of chromosome material may be stretched across the division figures at anaphase I of meiosis, as multivalents 'attempt' disjunction. It is easy to see how such meiotic irregularity may lead to sterility in gametes. This may be detected on the male side, as a high proportion of irregular-sized, misshapen pollen grains are produced.

In contrast, the allopolyploid is the product of the addition of unlike chromosome sets, usually following hybridisation between two species. Thus, two diploid taxa AA and BB may yield an infertile hybrid AB, and the production of unreduced gametes from such a plant will give an allotetraploid of formula AABB. In contrast to the autopolyploid situation discussed above, the typical allopolyploid is usually fertile. It is not difficult to see why this should be so. Multivalents are less likely to be formed, since each chromosome can pair with its exact partner and no other. This lack of correspondence, which ensures proper pairing in the allopolyploid preventing the association of A with B genomes, is the likely cause of sterility in the primary hybrid AB.

Thus, as studies of polyploids progressed in the 1920s and 1930s, it seemed possible, for a time, that the study of meiosis in polyploids would provide an easy way of detecting ancestry. Multivalent associations would indicate an origin by autopolyploidy: bivalent pairing would indicate an allopolyploid origin. Such a simple classification was quickly abandoned as intermediate meiotic situations were found (Darlington, 1937). The distinction, made in Chapter 12, between autopolyploidy and allopolyploidy, though useful and clear enough in the extreme cases, now seems to be misleading when applied to the evolution of groups of polyploid taxa. The difficulty can be appreciated if we consider what we mean by a hybrid individual with A and B genome sets. We have seen in the previous chapters that ordinary diploid sexual species, with some degree of outbreeding, are genetically very variable.
The genomes of any two individuals of such a species are most unlikely to be identical. It is, therefore, a conventional oversimplification to represent such an individual as having identical genomes contributed by each parent, and it would be more realistic to write in such cases

$$AA' \xrightarrow{\text{doubling}} AAA'A'$$

to represent the origin of a polyploid derivative.
As soon as we do this, we see the nature of the difficulty. Is this situation to be described as autopolyploidy or as allopolyploidy? Clearly the answer hinges on our definition of these terms. If we restrict allopolyploidy to those cases where a *sterile species-hybrid* (represented by AB) gives rise to a fertile polyploid derivative (AABB), then all the other cases where the parents of the diploid belong to

the same species would be described as autopolyploid. This is an unsatisfactory situation. A better solution would be to use, as Stebbins (1947) suggested, a third term, 'segmental allopolyploidy', for all the intermediate cases where the parent diploid possesses some measure of chromosomal and genetic difference between its genome sets, but where its parents are sufficiently similar to be assigned to the same species.

It seems likely that many polyploids are of the segmental allopolyploid type and, because of their genomic constitution, multivalents and univalents are formed at meiosis; multivalent associations of chromosomes being 'chain-like' or 'ring-like'. What are the implications of multivalent production? In a very useful historic review of polyploidy Gibby (1981) writes:

> multivalents in themselves do not lead to infertility, for so long as segregation of the chromosomes during anaphase is balanced, then multivalent-forming polyploids are potentially fertile. The presence of trivalents and, to a lesser extent, chain quadrivalents may result in infertility following mis-orientation at metaphase I, but ring quadrivalents can give numerically equal segregation. It is the presence of univalents that leads to chromosome imbalance and infertility. In autotetraploid Rye, which shows the presence of trivalents and univalents as well as quadrivalents and bivalents, selection for improved fertility results in a decrease in the number of univalents and an increase in quadrivalent frequency (Hazarika & Rees, 1967).

Later in the chapter we will consider the important question of the genetic control of chromosome pairing, and the cytological behaviour and evolutionary potential of autopolyploids.

With regard to the behaviour of odd-numbered polyploids, Gibby writes that their fertility 'is usually reduced as a result of the presence of univalents or odd-numbered multivalents'. However, 'some triploids are fertile, but the gametes they produce are diploid, triploid etc. or aneuploid, i.e. with less or more than the normal number of chromosomes', giving rise to 'progeny with a variety of chromosome numbers'. This point is stressed by Tayalé & Parisod (2013) who note that triploids may produce a 'relatively large amount of viable, euploid gametes ($n = x$, $2x$, $3x$), regardless of their auto- or allopolyploid origin'. These gametes have the potential to form polyploids (Husband, 2004).

With regard to triploid seeds, Tayalé & Parisod (2013) note that they may abort, because of endosperm/embryo interactions. In the diploid plant 'the endosperm (i.e. the tissue made of one parental and two maternal genomes surrounding and providing nutrient to the embryo) seems sensitive to the dosage of parental genes' and may fail to develop properly 'in polyploids with misbalanced chromosome sets'. Detailed analysis of the impact of genome dosage on endosperm and seed development has been reviewed by Schatlowski & Köhler (2012).

The concept of genome analysis

Based on the models of chromosome pairing we have just outlined, the cytological study of polyploids and their hybrids has frequently yielded valuable evidence on ancestry. The basic idea of genome analysis may be appreciated by considering an example. If an allopolyploid ($2n = 4x = 28$) of genomic constitution AABB is crossed with a plant thought to be an ancestral diploid $2n = 2x = 14$, a triploid ($2n = 3x = 21$) is formed. Suppose at metaphase I of meiosis seven bivalents and seven univalents are seen in division figures. We could deduce that the diploid and tetraploid shared a genome in common, say genome 'A', by employing the following argument. In the triploid, the A genome from the tetraploid would form bivalents with the A genome originating from the diploid; the B genome would have no pairing partner and the chromosomes would remain as univalents. A search could be made for the donor of the B genome, by making triploids from crosses between the tetraploid and other diploids. In searching for such a plant, crosses might be made which would yield a triploid in which there was no pairing at meiosis, and the parental diploid could be supposed to have a genome different from either A or B, say C. This example is set out diagrammatically in Fig. 14.2.

Genome analysis: uncertainties about ancestry

Genome analysis through the study of pairing behaviour in hybrids etc. has been used to investigate relationships in a number of genera, e.g. *Polypodium* (in Europe: Manton, 1950; Shivas, 1961a, b); *Viola* (Moore, 1976: Fig. 14.3), and it has also proved to be of enormous value in studying the origins of important crop plants such as the Rape seed/Canola and its relatives (*Brassica* spp.), Cotton (*Gossypium*), Banana (*Musa*), Tobacco (*Nicotiana tabacum*), Potato (*Solanum tuberosum*) and Grape (*Vitis*) (see Simmonds, 1976; Zohary & Hopf, 1993; Smartt & Simmonds, 1995). However, experiments with Wheat and other species have revealed complications in the analysis of pairing behaviour in polyploids. In order to understand these, it is necessary first, to outline what has been discovered about the ancestry of Bread Wheat.

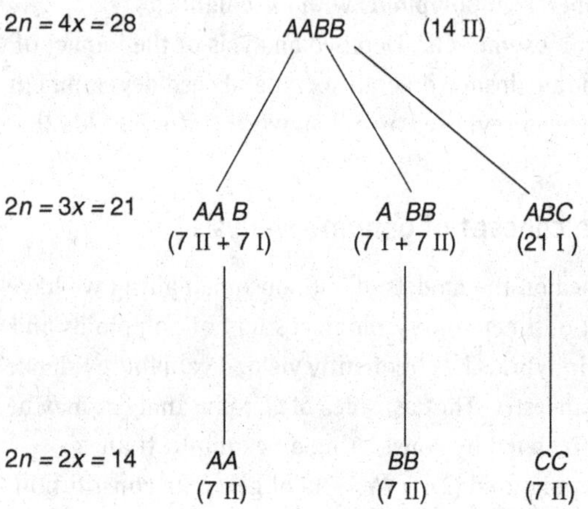

Fig. 14.2. Genome analysis of a presumed allopolyploid. Meiotic pairing noted in brackets (1 = a univalent; 11 = a bivalent).

After decades of experiments, observations and speculation, there seemed to be general agreement on the origin of Bread Wheat (*Triticum aestivum*; $2n = 42$; genomic constitution *AABBDD*) (Riley, 1965: Fig. 14.4).

McFadden & Sears (1946) had been successful in crossing tetraploid Wheat ($2n = 4x = 28$) with *Aegilops squarrosa* ($2n = 14$) and showed that when the triploid product was treated with colchicine, the synthetic hexaploid so formed resembled certain hexaploid Wheats. Sarkar & Stebbins (1956), after studying the patterns of variation in *Triticum* and *Aegilops*, considered that *A. speltoides* was the most likely donor of the 'B' genome. The logic of the argument used in their studies is of general interest. With knowledge of the morphology of a hybrid and one of its parents, and considering that most hybrids are intermediate between their parents in quantitative characteristics, then it should be possible to pick out the 'missing' parent from an array of taxa. This

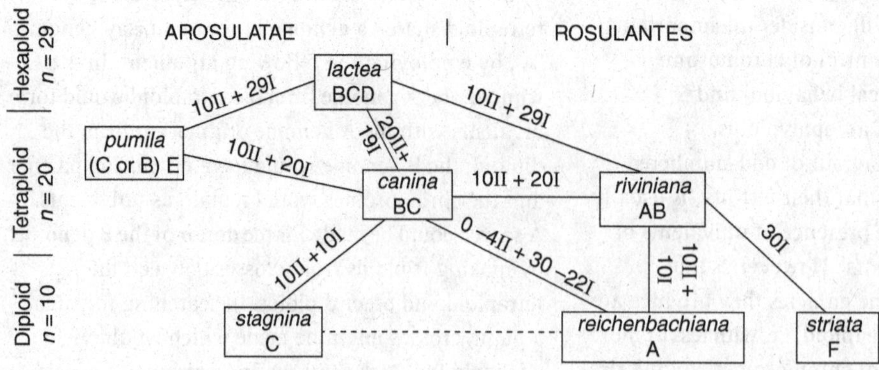

Fig. 14.3. Genomic constitutions of species of *Viola* subsection Rostratae, and chromosome pairing in hybrids. Unsuccessful crosses are indicated by broken lines. *Viola stagnina* is now called *V. persicifolia*. (Based partly on Moore & Harvey, 1961, from Moore, 1976.)

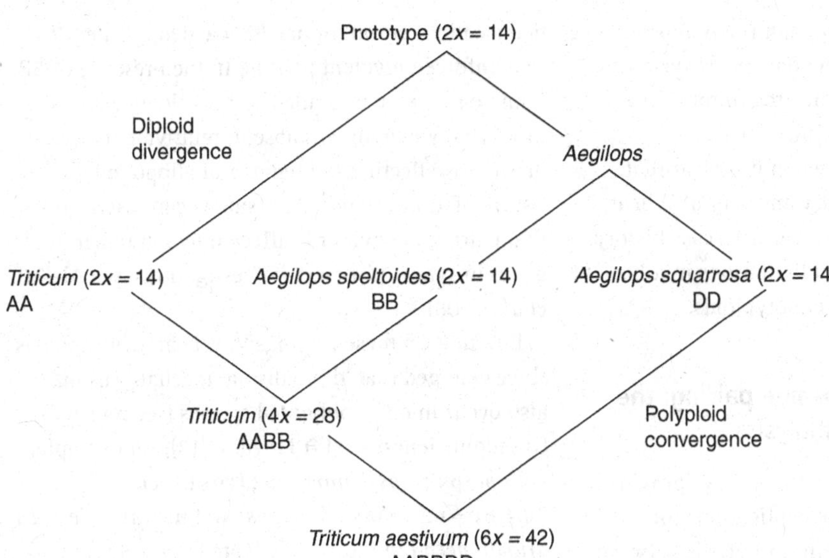

Fig. 14.4. Ancestry of Bread Wheats (*Triticum aestivum*): an early model. (From Riley, 1965.) (Note that in some of the literature *Aegilops squarrosa* is called *Triticum tauschii*.)

technique was called the method of 'extrapolated correlates' by its inventor Anderson (1949), and it has been used in many studies of introgression and other situations involving hybridisation.

In the case of Wheat, the variation in diploid and tetraploid *Triticum* was reasonably well known. That an *Aegilops* species was implicated in the ancestry seemed to be likely on cytological grounds. Early studies of karyotypes supported this idea (Riley, Unrau & Chapman, 1958). The Wheat 'story' displayed in Fig. 14.4 was widely quoted in the literature, often without giving the evidence. However, studies of meiosis (Kimber & Athwal, 1972), seed proteins (Johnson, 1972) and chromosome staining patterns (Gill & Kimber, 1974) cast doubt on *A. speltoides* as the source of the 'B' set.

Breiman & Graur (1995) reviewed all the evidence available in the 1990s. Molecular studies reveal that the tetraploid *Triticum dicoccum* (now called *T. turgidum*) shares the same chloroplast and mitochondrial DNA as *T. aestivum*, and therefore, given the maternal inheritance of these genomes, it may be concluded that *T. turgidum* was the female parent in the cross. However, studies of the restriction fragment profiles have failed to identify a diploid with the same mitochondrial DNA as *T. turgidum*. It seems very likely from these results that the donor of the 'B' nuclear genome also contributed the cytoplasmic genomes. Reviewing the question of the identity of the BB ancestor, Miller (1987) concluded that 'several possibilities exist for the origin of the B genome: the original donor may now be extinct, the donor may be a yet-undiscovered diploid species, the genome may be derived from more than one source, or a rearrangement of the DNA may have occurred since its incorporation into the tetraploid'. Furthermore, it is possible that very considerable hybridisation may have taken place at the tetraploid level, and that this may have modified the genomes to the point where the tracing of ancestry may be very difficult (Zohary & Feldman, 1962; Pazy & Zohary, 1965).

In the past few decades many researchers have continued to investigate the question of the origin of the B genome using a variety of techniques. Salse *et al.* (2008) examined the organisation of the seed storage locus in the A, B, D and S genomes. (Because of the uncertainty about the B genome, S is the designation given to the *Aegilops speltoides* genome.) Their intensive researches concluded that: '*A. speltoides* appears to be more evolutionary related to the B genome of *T. aestivum* than the A and D genomes'. However, 'the S genome of *A. speltoides* has diverged very early from the progenitor of the B genome which remains to be identified'.

Reflecting on decades of research it is clear that, while some elements of the early ancestry of Wheat are clear, others are as yet unknown. This case history highlights the complexities and limitations involved in the unravelling of ancestry in polyploids.

Genetic control of chromosome pairing: the implications for genome analysis

Studies of meiosis in hexaploid Wheat have provided important information that has implications for genome analysis. Pairing patterns in hybrids between diploid taxa suggested that some homologies between Wheat genomes might exist. Indeed, the scheme for the evolution of Wheat proposed that the taxa responsible for the formation of Wheat had a common ancestor and it might be supposed that, while some chromosome differentiation has taken place, some degree of residual homology remains between the genomes, i.e. the genomes could be said to be homoeologous. On the basis of these considerations *T. aestivum* is classified as a segmental allohexaploid. When meiosis in hexaploid Wheat is examined, however, pairing is seen to be strictly bivalent in character; pairing behaviour does not appear to reflect the presumed origin!

The classic experiments by Riley & Chapman (1958) were designed to investigate this phenomenon. In hexaploid Wheat (and many other polyploids) plants may survive and reproduce in the absence of a full chromosome complement, and plants with $2n = 40$, i.e. with a pair of chromosomes missing, can be produced experimentally. With $2n = 6x = 42$, 21 such types – known as nullisomics – are viable and were produced. Riley & Chapman (1958) discovered that, if a certain pair of chromosomes was absent (chromosome 5 of genome set B), then multivalent pairing occurred at meiosis. Studies of the other 20 nullisomics revealed strict bivalent pairing. They deduced that chromosome '5B' carried a gene (*Ph1*) that enforced bivalent pairing. In the presence of 5B, homologies were overruled and bivalents were produced: when 5B was absent, multivalents were produced reflecting the segmental allopolyploid nature of *T. aestivum*. It has since been discovered that pairing behaviour is affected by a number of genes in Wheat (for recent investigations, see Al-Kaff *et al.*, 2008).

Building on these findings in Wheat, evidence has since emerged that 'diploidising' mechanisms may also occur in other polyploid species (see review by Grandont, Jenczewskl & Lloyd, 2013), for example, *Lolium* spp.; *Gossypium* spp.; Oats (*Avena*; $2n = 6x = 42$); *Brassica napus*; and the grass *Festuca arundinacea* ($2n = 6x = 42$). More circumstantial evidence is available for *Aegilops, Hordeum, Nicotiana* and *Coffea* (Jenczewskl & Alix, 2004).

These findings have enormous implications for our understanding of speciation by polyploidy, and will be discussed at some length below. Here we are concerned with the strength and weaknesses of pairing-behaviour genome analysis.

In the past it was supposed that pairing behaviour was simply a rather mechanical affair, dictated by chromosome homologies alone. Such a view forms the rationale behind genome analysis. Now that the situation in Wheat and other plants is beginning to be revealed in its complexity, the basis of such genome analysis no longer seems so secure. How can we ever rule out the possibility that some degree of genetic control is being exercised in the pairing behaviour of polyploids? The experiments necessary to demonstrate the presence of a diploidising genetic system are time-consuming. They may not be technically feasible in plants with small rather undifferentiated chromosomes, since identification of different nullisomics is a necessary part of the analysis.

Another complexity must be raised at this point. It has been known for some time that certain polyploids producing some multivalent associations at meiosis are nonetheless fertile: e.g. *Agrostis canina,*

Arrhenatherum elatius, Dactylis glomerata and *Tradescantia virginiana*. After studying *Anthoxanthum odoratum* ($2n = 4x = 20$), Jones (1964) concluded that multivalent associations need not necessarily indicate a chaotic meiosis. It is possible to devise models of multivalent formation and separation that will yield balanced chromosome products, such systems being under genetic control. However, in other autopolyploids regular bivalent formation has been confirmed (Weiss & Maluszynska, 2000; Le Comber *et al*., 2010). This 'diploidization is still poorly understood . . . and both its genetic basis and adaptive value have to be firmly assessed' (Tayalé & Parisod, 2013).

Reviewing the available evidence, it is supposed, therefore, that, in polyploids, genetic control of meiotic behaviour may arise, taking the form either of a diploidising mechanism or of a system of coordinated multivalent formation and separation, both mechanisms yielding fertile offspring.

Given the possibility of genetic control of meiotic behaviour, the uncritical use of genome analysis must be avoided (see de Wet & Harlan, 1972).

Studies of karyotypes

Returning to the historic development of different techniques to test hypotheses concerning the origin of particular polyploids, sometimes, especially where chromosome sets are large and differentiated, an examination of karyotypic differences is often of great value in understanding the origin of particular polyploids (see, for example, the studies of *Holcus* by Jones, 1958).

An interesting example of the use of karyotypic information concerns the origin of the widespread polyploid weed *Poa annua* ($2n = 4x = 28$). Nannfeldt (1937) suggested that this plant was the allopolyploid derivative of the cross between diploids *P. supina* and *P. infirma*. In contrast, de Litardière (1939) considered that *P. annua* could be an autopolyploid derived from *P. infirma*. Crosses between *P. infirma* and *P. supina* were made by Tutin (1957) and yielded *P. annua*-like plants. Tutin suggested that during the Pleistocene period *P. supina*, a perennial mountain species, was probably driven down to lower altitudes and came into contact with the ephemeral grass *P. infirma* in the northern Mediterranean region. Crossing between the two taxa could have occurred, especially where mountains are close to the coast, giving rise to the tetraploid plant that is now found worldwide.

Nannfeldt (1937) showed that the karyotypes of *P. supina* and *P. infirma* were rather similar. However, Koshy (1968) made a detailed study of the karyotype of *Poa annua* and revealed that there are three particularly distinctive chromosomes, each present as a pair. Koshy draws the following conclusions: *Poa annua* does not have four identical sets of chromosomes, as would be required in an autotetraploid, nor does it have the sum of the karyotypes of *P. infirma* and *P. supina*. Either *P. annua* has undergone structural changes since its formation from *P. infirma* and *P. supina*, or it may be derived from either *P. infirma* or *P. supina* forming an allopolyploid derivative with another, as yet unknown, species of *Poa*. Clearly, karyological evidence on its own is not conclusive, as there is the possibility that the karyotype of a species may not be constant over time.

Two subsequent investigations have provided evidence concerning the origin of *P. annua*. In a glasshouse experiment, Darmency & Gasquez (1997) reported that a low level (0.2%) of spontaneous hybridisation occurred between *P. infirma* and *P. supina* when these species were grown together. Hybrids were diploid; resembled *P. annua* in morphology; and showed additivity in patterns of isozyme markers that were very similar to those in *P. annua*.

Studying nuclear and chloroplast markers, Moa & Huff (2012) have provided more critical evidence of the ancestry of *P. annua*. They concluded that 'the DNA sequences present within *P. annua* are inseparable from their respective orthologues within *P. supina* and *P. infirma*'. They also discovered that *P. infirma* was the maternal parent in the cross and at

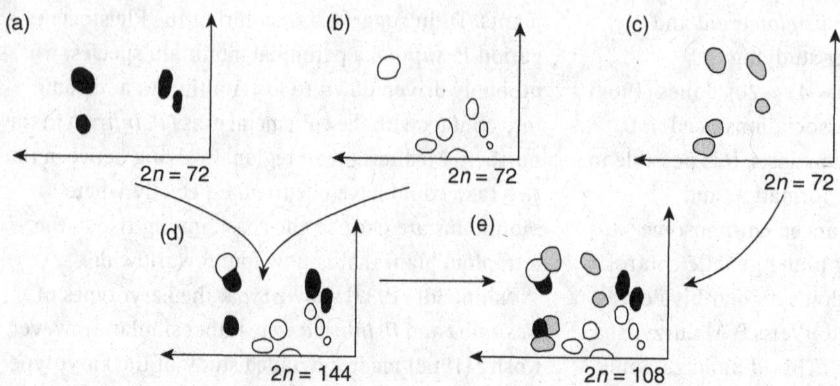

Fig. 14.5. Two-dimensional chromatograms of three species of *Asplenium* and their hybrids. The flavonoid pattern for *A. × kentuckiense* appears to combine the profiles of the three diploid species. This evidence adds weight to the hypothesis that *A. × kentuckiense* is the trigenomic allopolyploid derivative from the hybridisations indicated in the diagram. (a) *A. rhizophyllum*, (b) *A. montanum*, (c) *A. platyneuron*, (d) bigenomic allopolyploid *A. rhizophyllum × A. montanum*, (e) trigenomic allopolyploid *A. × kentuckiense* (*A. rhizophyllum × A. montanum × A. platyneuron*). (From Smith & Levin, 1963, in Heywood, 1976.) A more detailed study of the flavonoids in various *Asplenium* taxa has been undertaken that confirms the earlier findings (Harborne, Williams & Smith, 1973). Evidence from a study of isozymes corroborates the findings of the chemotaxonomists (Werth, Guttman & Eshbaugh, 1985a). Furthermore, detailed investigations of the allotetraploid *A. pinnatifidum* (= *A. rhizophyllum × A. montanum*) indicates that this widespread species has been formed more than once from its diploid progenitors (Werth, Guttman & Eshbaugh, 1985b). Since these investigations, polyploid speciation has been examined in a number of *Asplenium* species using molecular markers (see Perrie *et al.* 2010).

least two interspecific hybridisations gave rise to *P. annua.* They also consider it likely that 'chromosomal rearrangements' have occurred in the subsequent evolution of *P. annua.*

Chemical studies

Returning to our historical theme, in the early 1960s, chemical studies represented the frontier of research into the origin of polyploids. Evidence revealed that often species differ in chemical characters, and hybrids between them, and the polyploid derivatives from such hybrids, generally have 'additive'patterns of secondary chemical compounds. Thus, all the secondary chemical constituents found in species AA and species BB are likely to be present in the allopolyploid derivative AABB, providing a strong line of evidence with which to consider the ancestry of particular polyploids. A particularly elegant example of the use of chemical evidence is provided by studies of North American *Asplenium* species (Fig. 14.5).

In many experiments in the early 1960s, no attempt was made to identify the secondary chemical compounds separated by various techniques, and deductions were made on the basis of patterns alone. For example, Stebbins *et al.* (1963) found chromatographic evidence in support of the view that the tetraploid *Viola quercetorum* ($2n = 4x = 24$) is a polyploid hybrid derivative of the cross between the diploid ($2n = 2x = 12$) taxa *V. purpurea* ssp. *purpurea* and *V. aurea* ssp. *mohavensis*. In many of these classic studies a single individual served to represent a population or a taxon. Later investigations, however, suggested that it was important to consider variation within, as well as between, taxa.

While many chemotaxonomists studied 'patterns' of variation in unknown substances in extracts of plants, others made an attempt to identify the chemical compounds involved. For instance, the Red

Horse-Chestnut (*Aesculus* × *carnea*: $2n = 4x = 80$), which originated sometime before 1818 in Europe (Li, 1956), has long been considered to be the allopolyploid derivative from the European *A. hippocastanum* and the introduced North American species *A. pavia* (both taxa are diploid: $2n = 2x = 40$). Chromatographic studies of phenolic compounds (Hsiao & Li, 1973) offer support for this hypothesis. To the best of our knowledge, the proposed origin of *A.* × *carnea* has yet to be tested by molecular methods.

In situ hybridisation (ISH)

While classic cytogenetic approaches involving chromosome staining still have a role to play in botanical research, most informative results have now been obtained using a new cytogenetic technique, which exploits the fact that probes of DNA introduced appropriately to cytological preparations 'hybridise *in situ*' with chromosomal DNA, if both share homologous segments. Visualisation of the location of a probe/chromosome association is achieved by labelling the probe with radioactive or non-radioactive reporter molecules (Bennett, 1995).

Total genomic DNA may be used as a probe in chromosome preparations (genomic *in situ* hybridisation – GISH) or the probe may be labelled with fluoresent dye (fluorescence *in situ* hybridisation – FISH). In further developments of the technique, probes may be prepared of smaller DNA sequences, RNA or synthetic oligonucleotides. Using multiple probes labelled with different fluorochromes, it is possible to achieve multicoloured preparations. In addition, sometimes counterstaining with different coloured dyes is employed. These techniques are now widely used in biomedical research and screening, microbiology, etc. Chester *et al.* (2010a) provide details of the different methods, which are most effectively employed on taxa with large chromosomes.

In an influential early study, by Bennett, Kenton & Bennett (1992), the potential to test hypotheses concerning ancestry of polyploids was revealed. On the basis of the karyotypes, it has been proposed that *Milium vernale* ($2n = 8$; with 8 large 'L' chromosomes) is one of the parents of the allopolyploid grass *Milium montianum* ($2n = 22$ that has 8 'L' chromosomes visually matching those of *vernale* + 14 small chromosomes). To provide a molecular test of this hypothesis, a total DNA extract of *M. vernale* was made into an appropriately stained probe that was hybridised *in situ* to the root tip chromosomes of *M. montianum*. The probe hybridised to the eight 'L' chromosomes, confirming the presence of the *M. vernale* set of chromosomes in the allopolyploid set. Clearly, GISH provides a powerful tool for the analysis of polyploid ancestry. The results of the experiment also showed the potential of the technique for studying chromosome repatterning (see below). Here we note Soltis & Soltis's (1993) perceptive comment: 'significantly a few regions along the L chromosomes clearly lacked the hybridisation signal of the *M. vernale* probe, suggesting either (1) these sequences arose after the formation of the allopolyploid, or (2) a limited intergenomic transfer of DNA from chromosomes representing the second genome present in the allo(poly)ploid (unidentified) to the "L" chromosomes donated by *M. vernale*'.

A more recent investigation, using GISH and FISH techniques, has investigated the supposed allopolyploid origin of *Iris versicolor* ($2n = 108$) (Lim *et al.*, 2007). It was confirmed, as the celebrated geneticist Edgar Anderson had proposed many years ago, that this Iris inherited its genomes from *I. virginica* ($2n = 70$) and *I. setosa* ($2n = 38$).

FISH and GISH are now very important tools in the study of hybrids and polyploids.

Polytopic multiple origin of polyploids

In the past many botanists, perhaps subconsciously, assumed that a particular polyploid is formed only once, and therefore was genetically depauperate

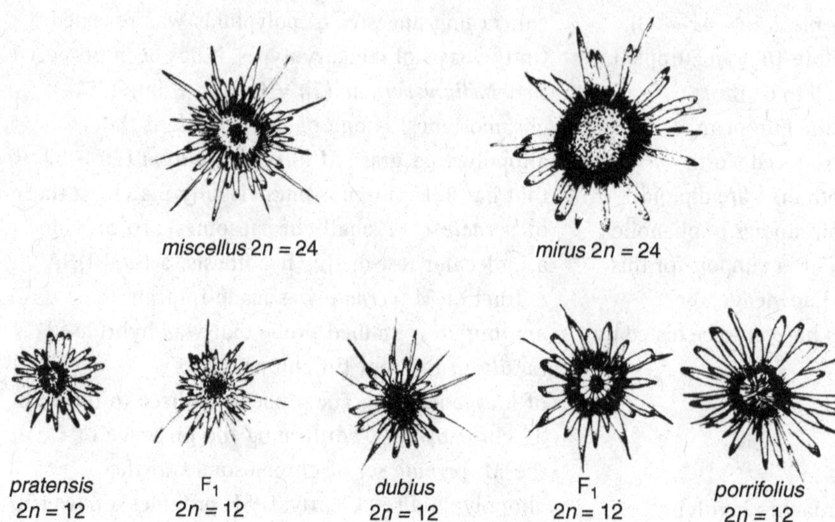

Fig. 14.6. *Tragopogon* species in the USA. (a) Bottom row: flowering heads of the diploid species of *Tragopogon*, *T. pratensis* (left), *T. dubius* (centre) and *T. porrifolius* (right). Upper row: flowering heads of the polyploid hybrid species, *T. miscellus* (left) and *T. mirus* (right). (From Ownbey, 1950.)

(Ehrendorfer, 1980; Favarger, 1984). But, if related diploid taxa regularly come into contact, there is now abundant evidence that allopolyploid derivatives may be produced repeatedly, and polytopically, i.e. at different sites (Soltis & Soltis, 1999; Tayalé & Parisod, 2013). This is particularly clear from the investigation of newly evolving polyploids (so called neo-polyploids) in *Tragopogon*, *Senecio* and *Spartina*.

***Tragopogon* species: multiple origins** Very clear evidence supporting the hypothesis of both recurrent and polytopic origin of allopolyploid species in nature comes from the study of introduced species of *Tragopogon* and their hybrids in North America. Three European species of this Composite genus occur in North America as weeds of roadsides and disturbed ground: *T. dubius*, *T. pratensis* and *T. porrifolius*. All these are diploid species with $2n = 12$ (Ownbey, 1950), and highly sterile F_1 hybrids between all the pairs are known in Europe, where the three species 'are largely allopatric and/or isolated ecologically' (Tate *et al.*, 2009). In the western USA where Ownbey studied them, he found it very easy to detect these hybrids by their failure to set good heads of seed. In four separate localities, however, he found small groups of fertile plants with the intermediate characters of the hybrids. These proved to be tetraploid with $2n = 24$. Morphologically, one of the polyploids appeared to have arisen from the cross *T. dubius* × *T. porrifolius* and the other from *T. dubius* × *T. pratensis*. Since these fertile allopolyploids are both morphologically distinct and genetically isolated from their parent species, Ownbey described them as new species: *T. mirus* and *T. miscellus* (Fig. 14.6).

Successive studies involving newly devised techniques supported Ownbey's ideas about the origin of *T. mirus* and *T. miscellus*. These investigations include the study of karyotypes (Ownbey & McCollum, 1953, 1954; Brehm & Ownbey, 1965); biochemical characteristics (Belzer & Ownbey, 1971); isozymes (Roose & Gottlieb, 1976) and DNA markers (literature reviewed by Soltis *et al.*, 2004, 2009, 2012). Of particular significance is the finding through molecular analyses that the two allopolyploid derivatives have karyotypes that are additive of their proposed diploid progenitors (Pires *et al.*, 2004a). Recently, in crossing experiments with the appropriate pair of parental species, *T. miscellus* and *T. mirus* have both been resynthesied by treating the F_1 hybrids with colchicine (Tate *et al.*, 2009). Interestingly, in these investigations allopolyploids between *T. pratensis* and *T. porrifolius* were also successfully produced. Such plants are unknown 'in the wild'.

The investigations of Ownbey and his associates suggested another interesting possibility, namely that *T. mirus* had arisen independently in three separate areas, and *T. miscellus* had been produced twice in separate localities. Recently, using molecular approaches, it has been possible to make critical tests of these suggestions. From the study of allozymes, Roose & Gottlieb (1976) obtained evidence that *T. mirus* had at least three independent origins, but their results were inconclusive concerning the possibility of multiple origins of *T. miscellus.* Subsequent studies of chloroplast DNA, which is maternally inherited, were very informative, as it has proved possible to determine which of the diploid species was the maternal parent in the production of a particular sample of the polyploid (Soltis & Soltis, 1989). From the DNA profiles produced by cleavage with restriction enzymes (RFLPs), it was discovered that populations of *T. miscellus* in the Pullman area in Washington State had *T. dubius* as the maternal parent. In contrast, all the other populations of *T. miscellus* had *T. pratensis* as the maternal parent. Evidence from RFLP analysis clearly indicated that all the populations of *T. mirus* had *T. porrifolius* as the female parent. However, an analysis of ribosomal DNA demonstrated that *T. mirus* had two independent origins (Soltis & Soltis, 1991).

A thorough search has revealed further populations of both *T. mirus* and *T. miscellus* and, combining the information from DNA and enzyme electrophoresis (a process that does not lead to a single definitive figure), it was concluded that there were likely to have been as many as 12 independent origins of *T. mirus* and a minimum of 2 and a maximum of 21 for *T. miscellus* (Cook *et al.*, 1998). Thus, the evidence for the multiple origins of *Tragopogon* allopolyploids on a local scale is very strong, and indeed recurrent formation of *T. mirus* would appear to have occurred in the Palouse region in Washington State and adjacent Idaho (Soltis *et al.*, 1995). Both polyploid species are appearing in other areas of the USA (Malinska *et al.*, 2011). Could these be the result of long-distance dispersal or are they the result of additional independent origins? In molecular studies of *T. mirus* these two alternative hypotheses have been investigated (Soltis *et al.*, 2012). The results offer strong support for the hypothesis of 'local formation of *T. mirus* from co-occurring populations of *T. dubius* and *T. porrifolius* in Arizona and Oregon'.

Using isozyme approaches Roose & Gottlieb (1976) made another important contribution, showing that 'the tetraploid *Tragopogon* species not only have additive profiles but also contain novel forms of enzymes not present in their diploid parents. This is because the subunits encoded by alleles from different diploid parents can form novel functional multimeric enzymes, perhaps with different activities and with the potential for different functions' (see Crawford *et al.*, 2014).

Novak, Soltis & Soltis (1991) have published an account of the history of *Tragopogon* in the western USA. Evidence suggests that *T. porrifolius* and *T. pratensis* were introduced about 1916, with *T. dubius* arriving a little later (c.1928). A field survey of populations revealed dramatic increases in the distribution of the new allopolyploids, since Ownbey's pioneering investigations. Moreover, it might be assumed that the appropriate pair of diploid species found in association with a 'new' polyploid are likely to be descendants of its parental species. However, in mixed populations, containing *T. mirus* and both parental diploids, the diploids present were not always of the correct genotype to have produced the allopolyploid. 'Only two of the tetraploid populations combined the RAPD marker profiles of the diploid progenitors occurring at the same site' (Cook *et al.*, 1998). Clearly, active dispersal of the plants is occurring. The survey also suggested that, as there were many primary diploid hybrids, there was the potential to produce further new polyploid derivatives.

The survey also provided information concerning population size of older populations. For instance, in 1949 two small populations of *T. miscellus* in

Moscow, Idaho, each had 30–35 individuals. In 1950, one of these had been reduced to 7 individuals, whereas the other had increased to 75 plants. A third population found at Moscow in 1954 was destroyed by the construction of a house and lawn. Clearly, each episode of polyploidy is likely to produce a single individual, and not all small populations are likely to survive. Finally, 'the survey' produced the revelation that several of the 25 non-native *Tragopogon* species obtained by Ownbey for his chromatographic work have formed hybrids in the garden where they were grown. Furthermore, a number of these species have escaped into the wild to join the three that have provided the raw material for one of the most fascinating situations of recent microevolution.

The question now arises: is the recurrent formation of allopolyploid derivatives in *Tragopogon* exceptional, or is the phenomenon found in other polyploid groups? Tate, Soltis & Soltis (2005) and Soltis & Soltis (2009) have reviewed the evidence, and report that nearly all polyploid species are likely to consist of populations that have arisen independently from their progenitor species. The recurrent formation of polyploids is such an important finding that we consider two European examples.

Senecio cambrensis: **multiple origins.** The allopolyploid self-compatible *Senecio cambrensis* ($2n = 60$) was first described from plants found on roadsides in Flintshire, North Wales, in 1953. It is similar to the sterile hybrid (*S.* × *baxteri*; $2n = 30$) between a species introduced into Britain (the self-incompatible *S. squalidus*; $2n = 20$) and the native self-compatible species, *S. vulgaris* ($2n = 40$). An artificial allopolyploid, which is closely similar to the wild plant, was synthesised by treating the synthetic hybrid *S. vulgaris* × *S. squalidus* with colchicine (see Rosser, 1953 for an account of experiments by Harland). Resynthesis has also been successfully accomplished by Weir & Ingram (1980). Extensive investigations suggest that the original crosses giving rise first to *S.* × *baxteri* and subsequently to *S. cambrensis* both involved *S. squalidus* as the pollen parent because the reciprocal cross is rarely successful (Hegarty *et al.*, 2011). While the route of resynthesis, via a triploid F_1, might indicate what happened in the wild, theoretically, there are other possibilities: e.g. the fusion of a normal reduced gamete of *S. vulgaris* with an unreduced gamete from *S. squalidus*. Alternatively, if, hypothetically, a tetraploid plant of *S. squalidus* occurred, this could have crossed with *S. vulgaris* to form the allohexaploid *S. cambrensis* (see Lowe, Harris & Ashton, 2009).

In 1982, *S. cambrensis* was discovered in another area: on demolition and redevelopment sites in Leith, Edinburgh (Abbott, Ingram & Noltie, 1983a, b; Ashton & Abbott, 1992; Hegarty, Abbott & Hiscock, 2012). Natural dispersal from North Wales is very unlikely. Examination of the chloroplast DNA of the *S. cambrensis* from the two sites revealed significantly different molecular profiles, indicating the multiple polytopic origin of this new allopolyploid species (Harris & Ingram, 1992). Support for this conclusion was also provided from studies of the variation in allozymes in *S. cambrensis* and its parental species (Ashton & Abbott, 1992).

Spartina townsendii **and** *S. anglica*: **multiple origins?** The second informative case history is provided by the studies of the famous polyploid '*Spartina townsendii*'. Huskins (1930) provides the classic account of this plant, but counting the small chromosomes of these plants proved a challenge. The situation was re-examined by Marchant (1963, 1967, 1968) who discovered the inaccuracy of Huskins' chromosome counts. Taking all the evidence at present available, it is likely that the fertile allopolyploid was produced, about 150 years ago, on tidal mud on Southampton Water in southern Britain, when the North American species *S. alterniflora*, $2n = 62$ (introduced in the nineteenth century, most probably in ship's ballast) hybridised with the native *S. maritima*, $2n = 60$. However, further research revealed that populations of so-called *S. townsendii* consisted either of a sterile hybrid, or the fertile allopolyploid or mixtures of the two. Taxonomic investigation revealed that *S.* × *townsendii* is the correct name for the sterile hybrid ($2n = 62$) first collected in 1870. The fertile allopolyploid ($2n = 120$, 122, 124), first noted in 1892, is now called *S. anglica*. Considering the chromosome numbers found in the genus, Marchant proposed that the base number of the genus is $x = 10$. Tetraploid ($2n = 40$) species have been reported elsewhere. The *Spartina* species we are concerned with here are hexaploid ($2n = 60 - 62$) or dodecaploid ($2n = 122$, 124).

Repeated attempts to produce the artificial hybrid between the supposed parents have been unsuccessful, and the attempted experimental production of polyploid derivatives from *S. alterniflora*, *S. maritima* and *S.* × *townsendii* by treatment with colchicine also failed (Marchant, 1968).

In an attempt to confirm the parentage of *S. anglica*, and to determine whether *S. anglica* might have been formed more than once, isozymes and seed proteins were examined by electrophoresis (Raybould, *et al.*, 1991a, b). The isozyme phenotypes discovered in these investigations of British material were those expected if the species was an allopolyploid derivative of *S. alterniflora* and *S. maritima*. Fixed heterozygosity (see below) was detected and very little genetic variability in the parental taxa and the allopolyploid. Investigations with molecular markers confirmed that *S. townsendii* has an 'aggregate of the diagnostic DNA fragments from *S. maritima* and *S. alterniflora*, thus confirming its hybrid origin'. Furthermore, 'the *S. townsendii* genotype was identical to most of the *S. anglica* individuals analysed, establishing the concordance of these two taxa' (Ayres & Strong, 2001). Refecting on more recent molecular studies, Ainouche *et al.* (2009) confirm that '*S. anglica* inherited the identical genome to *S.* × *townsendii*' and shows a lack of genetic diversity. Concerning the direction of the cross, investigation of cpDNA has established that *S. alterniflora* is the maternal parent of both *S. townsendii* and *S. anglica* ((Ferris *et al.*, 1997; Baumel *et al.*, 2001).

Another interesting issue has emerged. In an area of south-west France, near Bayonne, where *S. maritima* is found and *S. alterniflora* had been accidentally introduced, a variant of *Spartina* (*S.* × *neyrautii*) was discovered in 1892. It has the same chromosome number and meiotic behaviour as *S. townsendii*, but there are morphological differences between the two plants. Raybould *et al.* (1990) confirmed that they have the same isozyme phenotypes. It was suggested that the two sterile hybrids discovered last century in south-west France and southern Britain represent reciprocal hybridisations. However, investigation of maternally inherited chloroplast DNA revealed that the two F_1 hybrids shared the same chloroplast sequences as *S. alterniflora*, indicating that this species was the seed parent in both England and France. This finding strongly suggests multiple origin of the F_1 hybrid. Concerning the ancestry of *S. anglica*, present evidence confirms the conclusions of Gray, Marshall & Raybould (1991) that, as the polyploid is almost totally lacking in genetic variation, this may indicate a 'single origin'. However, they note the possibility that the plant could have multiple origins 'from uniform parents'.

Multiple origins: native species. The examples we have considered – *Tragopogon*, *Senecio* and *Spartina* – all involve at least one introduced parental species. It is important to stress that multiple origins have also been established in wild native plants. For instance, in studying the complex genus *Draba* in Scandinavia, Brochmann, Soltis & Soltis (1992a, b; Soltis & Soltis, 1993) discovered evidence for recurrent origins, in the narrow endemic octoploid *D. cacuminum* (3); and in *D. norvegica* (6); *D. lactea* (16); and *D. corymbosa* (6). (The number of sites is given in brackets.) Recurrent origins have also been proposed for Arctic polyploids of the genus *Saxifraga* (Brochmann *et al.*, 1998; Steen *et al.*, 2000).

Single origins. In contrast to these examples of recurrent origin, Soltis, Soltis & Tate (2004) note several cases where the evidence indicates the likelihood of a single origin of particular polyploids, e.g. Wheat, Peanut, *Spartina anglica* and *Arabidopsis suecica*.

B. Recent insights into polyploidy from molecular studies

Key questions about polyploidy and its significance

In the next sections a number of key questions concerning polyploidy are examined that have long intrigued biologists. How many species are polyploidy and given the complex reticulate patterns of variation found in polyploid groups, what are the taxonomic challenges in naming and classifying species? What mechanisms operate in nature that give rise abruptly to polyploid individuals from parental stocks? Once formed, what, if any, immediate genetic changes occur in polyploids? What ecological and reproductive 'hurdles' do polyploids face in becoming established in plant communities? What are the characteristics of polyploids as a group? What changes in breeding behaviour are associated with polyploidy? Is there any evidence for introgressive hybridisation in groups containing polyploids? And, in considering where our understanding of polyploidy is still incomplete, what is the evolutionary significance of polyploidy?

Molecular approaches have contributed to a flood of important new insights into polyploidy that often challenge the views of the past (Soltis, Visger & Soltis, 2014).

How many species are polyploid?

Evidence bearing on this question comes from a number of sources: morphological investigations, chromosome counts, and genome studies. First, it is essential to consider how species are delimited in polyploid groups. Then, the historic information on chromosome numbers is examined. Finally, we consider recent findings from the studies of plant genomes that have cast new light on the frequency of polyploidy.

The delimitation of taxa within polyploid groups

Polyploids are formed by the addition of like or unlike genomes. In experimental polyploids, an increase in chromosome number often yields a plant with larger nuclei and cells with larger diameters and volumes. Thus, the members of a polyploid series may differ in mean pollen and stomatal cell size, and measurement has proved to be a reliable way of separating plants of different chromosome numbers. However, while the study of cell size has been helpful in distinguishing chromosome races in some polyploid groups, similar comparisons in other groups do not show the so-called '*gigas*' effect (Stebbins, 1950; Davis & Heywood, 1963; Lewis, 1980a; Hegarty *et al.*, 2013).

In considering the frequency of polyploid species, it is important to acknowledge the effect of differences in taxonomic practice. Generally, whether a particular polyploid is recognised as 'a species' will depend upon the degree to which it differs morphologically from the taxa contributing its genomes. However, taxonomists investigating different groups employ different means to examine their material. For instance, while those studying mosses, liverworts and other 'lower plants' etc. use a microscope in their work, flowering plant taxonomists generally use a hand lens, and cell sizes are not routinely measured.

The degree of morphological difference detected between presumed ancestral taxa and a given polyploid may be considerable, slight or insignificant. In practice, if there are convenient morphological characters that can be used to distinguish polyploids, taxonomists generally treat them as species distinct from their parent taxa. However, in some cases, a particular polyploid (identified as such by counting the chromosome numbers) may lack any convenient morphological distinguishing features, and the taxonomist, considering practicalities, may then decide to treat the polyploid as an intraspecific variant rather than a distinct species. In such cases several chromosome numbers may be reported for a given species.

It is interesting to consider some of the now classical experimental investigations of species in the light of these comments. For example, *Leucanthemum vulgare* (*Chrysanthemum leucanthemum*), which we used to illustrate the biometricians' interest in 'local races' in Chapter 3, is now known to be complicated by the widespread occurrence of plants with different chromosome numbers that are to some extent morphologically separable on a number of quantitative characters (Favarger & Villard, 1965; Marchi *et al.*, 1983). In a rather similar way we now know that the kind of difference that Burkill found between Cambridge and Yorkshire populations of *Ficaria verna* (*Ranunculus ficaria*) (Chapter 3, Table 3.8) is explained, in part, by the presence of diploid and tetraploid plants of this common and variable species. Turning to the work of Turesson, we find again that part at least of the variability that he detected in common European species (such as *Achillea millefolium, Caltha palustris* and *Dactylis glomerata sens. lat.*) is attributable to the occurrence within these Linnaean species of more than one chromosome number. This does not, of course, in any way cast doubt upon his demonstration of ecotypic differentiation; it merely emphasises that experiment reveals that some species are highly complex entities. This point is beautifully made in the case of *Erophila verna*, which Jordan studied in such detail (see Chapter 2). A number of cytologists have studied this common variable annual weed, discovering the following chromosome numbers: $2n$ = 14, 24, 28, 30, 32, 34, 36, 38, 39, 40, 52, 54, 58, 64, 94 (see Winge, 1940).

Chromosome counts provide insights into the incidence of polyploidy

Botanists have been reporting the chromosome numbers of plants for more than 100 years. Historically, to make the scattered information accessible to biologists, 'Chromosome Atlases' have been produced (e.g. Tischler, 1950; Darlington & Wylie, 1955; Löve & Löve, 1961). The most up-to-date atlas, which attempts a worldwide listing of chromosome numbers of flowering plants, was published by Bolkhovskikh *et al.* (1969).

In addition, since the 1950s, there has been an Index of Plant Chromosome Numbers (*IPCN* project) begun by Marion Cave, University of California, Berkeley. Goldblatt (2007) discusses the history of the *IPCN*, which is now based at Missouri Botanic Garden. Chromosome numbers reported in the period 1986–7 are listed in Goldblatt & Johnson (1990), and details are given for counts reported in earlier years (web access to listings via www.tropicos.org/Project/IPCN).

However, considering the future of the Index, Goldblatt (2007) reports that the project is faltering, and has limited financial support. He questions whether 'chromosome information is passé and no longer relevant in today's climate of increasing enthusiasm for molecular systematics?'

Chromosome numbers are still being reported in the literature, often as local or regional surveys. For instance, as part of the *Flora Europaea* project, a verified list of chromosome numbers of European plants has been published (Moore, 1982). In addition, there have been heroic attempts to determine the chromosome numbers found in certain families. For example, Warwick & Al-Shehbaz (2006) provide a database on CD-Rom of chromosome numbers of the Brassicaceae. And a first compilation of chromosome numbers in the Asteraceae (Compositae was produced by Raven *et al.*, 1960). Nearly 40 years later, the 18th paper in the series was published by Carr *et al.* (1999).

Historic estimates of the incidence of polyploidy

While polyploidy is a common phenomenon in plants, it is apparently rare in animals, only being found in a small number of groups (White, 1978). Using chromosome numbers, calculations of the frequency of polyploid species in flowering plants and the proportion of polyploids have been published. However, those carrying out the surveys offered a word of caution. As we have seen above, calculations must take account of differences of opinion between taxonomists as to what constitutes a species. Furthermore, only a small fraction of the world's flora, mostly in temperate regions, has been studied cytologically. Also some, perhaps many, chromosome numbers published in the literature are incorrect. For example, Merxmüller & Grau checked the chromosome numbers listed in Tischer (1950). They discovered that only 23.8% of the numbers listed for the genus *Myosotis* were correct. Also, in many cases, no voucher specimens were available to check the identity of the specimens (see Merxmüller, 1970).

When the information for different higher plant groups is examined, a number of patterns emerge. The chromosome numbers of species in many genera show well-developed series, having simple multiples of a minimum or 'basic number'. For instance, in *Rumex* subgenus *Rumex* there is a series with the base number $x = 10$, which runs from $2n = 2x = 20$ for *Rumex sanguineus*, through $2n = 4x = 40$, in, for example, *R. obtusifolius*, up to $2n = 20x = 200$ in *R. hydrolapathum*. Other excellent examples of polyploid series based on a single basic number are provided by the species of the genus *Chrysanthemum sens. lat.* ($x = 9$): $2n = 18, 36, 54, 72, 90, 198$ (Tahara, 1915) and *Solanum* ($x = 12$): $2n = 24, 36, 48, 60, 72, 96, 120, 144$. However, in some cases the basic number may not be so obvious (Gibby, 1981). For instance, the Swede (*Brassica napus*; $2n = 38$) is not based on $x = 19$, but evidence suggests it incorporates the genomes of two parental species producing a so-called 'dibasic' tetraploid (Stebbins, 1971), originating from the cross between the Turnip (*B. rapa*; $2n = 2x = 20$) and the Cabbage (*B. oleracea*; $2n = 2x = 18$).

In some genera every step in a polyploid series is 'occupied'. But in others, we find that all the extant species have high chromosome numbers, and it is generally supposed that these plants are of ancient polyploid origin, the lower multiples of the base number having been lost. An extreme case has been discovered by Khandewal (1990) who discovered a variant of the fern *Ophioglossum reticulatum* with $2n$ = around 1200 (believed to be 96-ploid). The highest chromosome number in the dicotyledons is that of *Sedum suaveolens* ($2n$ = c.640, c.80-ploid; Uhl, 1978), while in the monocotyledons Johnson *et al.* (1989) discovered that *Voanioala gerardii* has $2n$ = c.596, which is c.50-ploid.

Müntzing (1936) and Darlington (1937) came to the conclusion that half of all angiosperms were polyploid. In another study, Stebbins (1971) calculated that some 30–35% of flowering plants are 'straightforward' polyploids fitting into polyploid series. However, as we shall see later in this chapter, some polyploids do not fit into simple series of multiples of a basic number. Working on the premise that plants with chromosome numbers in excess of $x = 13$ are polyploids, Grant (1971) came to the conclusion that 47% of the angiosperms are polyploids, the figure for dicotyledons being 43% and that for monocotyledons 58%. Others have suggested an even higher incidence of polyploidy. Considering the evidence then available, Goldblatt (1980) was of the opinion that many species with $n = 11$, $n = 10$ and even $n = 9$ have polyploidy in their ancestry, giving a figure of at least 70% for the incidence of polyploidy in the monocotyledons, and Lewis (1980a) concludes similarly that perhaps 70–80% of dicotyledonous species may be of polyploid origin.

Evidence from fossil plants

Given that cell size is often larger in polyploid species, Masterson (1981) compared stomata size in a range of fossil and extant species, and speculated that one or more polyploid events had occurred in over 70% of angiosperms.

Genetic evidence

To provide a more objective basis for considering the frequency of polyploidy, Gottlieb (1981b) studied patterns of isozyme inheritance to determine which chromosome numbers indicate polyploidy through the presence of multiple representatives of gene loci. Species in the Asteraceae (Compositae) with chromosome numbers $n = 4$, 5 and 9 were selected for investigation. It is possible that $x = 9$ is the original base number of the group and that 4 and 5 are the result of reduction in chromosome number, involving a process that we will consider later in this chapter. Alternatively, plants with $n = 9$ could be allopolyploid derivatives of crosses between species with $n = 4$ and $n = 5$. Polyploids have multiple representation of genomes, and therefore multiple representation of isozyme markers should be found if $n = 9$ is an allopolyploid. Gottlieb (1981b) studied the electrophoretic patterns of isozyme markers in a selection of Composites with $n = 4$, 5 and 9, found no multiplicity of isozymes in $2n = 9$, and concluded that, on this evidence, they were not allopolyploids.

Polyploidy in other plant groups

Polyploidy is apparently rare in conifers and unreported in cycads (Soltis, Soltis & Tate, 2004). In contrast, Grant (1971) calculates that, on the basis of chromosome numbers in excess of $x = 13$, 95% of fern species are polyploids. Polyploidy is common in mosses, but uncommon in liverworts (Wendel & Doyle, 2005).

Incidence of polyploidy: major new insights from the study of plant genomes of flowering plants Since 2000, with the publishing of the complete genome of *Arabidopsis thaliana*, the genomes of other taxa have been examined, including *Oryza, Populus, Vitis, Carica.* In considering the question of the incidence of polyploidy, it is important to recognise that all these genomes reveal evidence of genome duplication (Soltis, Soltis & Tate, 2009). For other taxa, partial genomic information (from ESTs) has become available, and ancient polyploidy (so-called palaeopolyploidy) has been identified in a number of plants including Maize (*Zea*), Soyabean (*Glycine*) and Cotton (*Gossypium*). Further investigations (see McGrath & Lynch, 2012 and references therein) have now revealed that 'all angiosperms, regardless of current genome size and chromosome number, have been affected by whole genome duplications (WGD), most of them repeatedly' (Fig. 6.4) (and see Weiss-Schneeweiss *et al.*, 2013; Soltis, Visger & Soltis, 2014).

In addition, McGrath & Lynch (2012) also note that 'a dramatic increase in species richness in several clades of angiosperms after an ancient WGD has been interpreted as indicative for a higher diversification in polyploids (Soltis *et al.*, 2009)'. However, the precise relationship between the frequency of polyploidy, diversification and species richness is still subject to ongoing debate (see Weiss-Schneeweiss *et al.*, 2013).

The origin of new polyploids: the role of somatic events and unreduced gametes

Considering the second of our important questions, it was Harlan & de Wet (1975) who pointed out that, in many accounts of polyploidy, the mechanism generating new polyploids is often left as a shadowy area or it is said that polyploids arise by 'hybridisation followed by chromosome doubling'. This rather ambiguous assertion could imply either that somatic doubling of chromosomes occurs in the primary diploid hybrid, giving rise to the polyploid derivative, or, more likely, that unreduced gametes are involved. Harlan & de Wet (1975) and de Wet (1980) review the copious literature on the subject, and conclude that in very few cases does somatic doubling seem to be implicated (as in *Primula kewensis*) and that unreduced gametes are of supreme importance. This is also the view of Thompson & Lumaret (1992) in their review of polyploidy. In high polyploids, with their multiple representation of genomes, variation in chromosome number is sometimes found (e.g. *Spartina anglica*, which has three chromosome numbers, $2n = 120$, 122 and 124). It is much more likely that these polyploid derivatives had their origin in meiotic 'events', rather than in somatic doubling.

Newer techniques, in particular the development of flow cytometry techniques, have provided a more critical means of determining the frequency of unreduced pollen and eggs (Bretagnolle & Thompson, 1995). Essentially, the technique involves the quantification of stained nuclear DNA by passing cell suspensions through a laser beam. If appropriate standards are employed – samples of leaf cells of individuals of known chromosome number – then it is possible to estimate the 'ploidy' level of individual cells/plants of unknown chromosome number.

Also, Maceira *et al.* (1992) devised a procedure involving allozyme markers to determine the occurrence and frequency of $2n$ pollen in diploid *Dactylis glomerata.* Progenies of controlled crosses ($4x$ crossed with $2x$) were examined for $4x$ progeny, arising from $2n$ egg from the tetraploid + $2n$ unreduced pollen from the diploid. *Dactylis* is mostly but not exclusively outcrossing, and allozyme markers permitted selfed and crossed $4x$ progeny to be identified. It was estimated that the average production of $2n$ pollen was 0.98%. Six genotypes produced exceptionally high frequencies of $2n$ pollen, ranging from 8% to 14%. Using the same method, De Haan *et al.* (1992) examined the occurrence and frequency of $2n$ eggs in the same species. On average the $2n$ egg frequency in fertile controlled crosses ($2x$ seed parent × $4x$ pollen parent) was 0.5%. The frequencies in individual plants were all less than 3.5%, but one plant proved exceptional with 26%.

Given this level of production of $2n$ gametes it might be expected that tetraploids could be produced either by the fusion of two $2n$ gametes from diploid plants, or by a two-step process involving fusion of two $2n$ gametes, one from a triploid, the other from a diploid.

Unreduced gametes: major insights from the studies of molecular genetics

The underlying genes and molecular mechanisms involved in unreduced gamete formation are at last being examined in *Arabidopsis thaliana.* Brownfield & Köhler (2011) review 'the recent discovery of several genes in which mutations give rise to a high frequency of unreduced gametes'. These owe their origin to *defects* in several aspects of meiotic processes. The paper gives details of mutations influencing early meiotic events: others cause nuclear restitution resulting from abnormalities of spindle orientation or disturbance in cytokinesis (the process of division of the cytoplasm). For instance, the gene *AtPS1* is implicated in the production of a high frequency of unreduced (diploid) pollen through abnormal spindle orientation during meiosis (d'Erfurth *et al.*, 2008). This gene has no effect on the formation of the female gamete. (For information into the genetics of unreduced gamete formation from intensive studies of *Arabidopsis*, see Schatlowski & Köhler, 2012.)

These new findings are of potential economic importance as they could lead to the wider exploitation of unreduced gametes in crop breeding. Molecular geneticists have yet to extend these investigations beyond the study of model plants: clearly, the frequency of unreduced gametes has important implications for the production of polyploids in nature. Also, the role of environmental factors in the production of unreduced gametes has yet to be properly investigated (Marble, 2004). As long ago as 1920, Hagerup (1932) suggested that unreduced gametes could be produced in greater frequency in plants subject to heat/cold shocks during development. Reviewing the literature, Tayalé & Parisod (2013) discovered that unreduced gametes may be more frequent in plants stressed by 'frost, wounding, herbivory, water or nutrient shortage'.

Turning, briefly, to another related issue, historically, it has been suggested, for instance by Clausen, Keck & Hiesey, that hybridisation itself might promote polyploidy in some way (Buggs, Soltis & Soltis, 2009). In a full review of present evidence, Buggs, Soltis & Soltis (2011) concluded that there is no 'currently persuasive evidence that hybridization between divergent parents serves as a driver for polyploidisation'.

Relative frequency of auto- and allopolyploidy

Historically, it was generally accepted that allopolyploidy predominates in plant evolution. In his classic book, *Variation and Evolution in Plants* Stebbins (1950) describes a lone example of autopolyploidy – the case of *Galax aphylla* (now *Galax urceolata*).
The traditional view has been that autopolyploidy is rare and 'maladaptive' (Buggs, Soltis & Soltis, 2011). Now, it is being suggested that we might have underestimated the frequency of autopolyploidy (Soltis *et al.*, 2009; Soltis, Visger & Soltis, 2014).

Stift *et al.* (2008) consider how autopolyploids behave cytologically and genetically. 'In extreme autotetraploids each chromosome has four homologous versions (denoted $A_1A_2A_3A_4$). Each chromosome may then pair randomly with any of its homologues in bivalents or quadrivalents during meiosis. This leads to tetrasomic inheritance i.e. all possible allelic combinations are produced in equal frequencies'. By examining segregation patterns using allozymes and other genetic markers tetrasomic inheritance has been confirmed in some suspected autotetraploids. Now it seems likely that autopolyploidy could be frequent, and, while some are plants that are unsuccessful in the wild, others may succeed (Soltis & Soltis 1990, 1993, 1999; Ramsey & Schemske 1998, 2002). Reviewing the current situation, Buggs, Soltis & Soltis (2011) acknowledge that autopolyploids are often difficult to detect and, as yet, we do not know how frequently they arise in nature. However, more information is likely to emerge if large-scale surveys using flow cytometry and other molecular approaches are brought to bear on the problem.

Some evolutionists consider that autopolyploids should be recognised as distinct species (see Soltis *et al.*, 2007). However, because they may be indistinguishable morphologically from their parental stock, autopolyploids are generally not given specific status by taxonomists. Sometimes, in floras, chromosome numbers are listed with the taxonomic descriptions. As such numbers are not assigned to any category (auto/allo/segmental), the underlying origin of polyploidy, in any particular case, is unclear, even if it has been investigated.

Unreduced gametes: is polyhaploidy important in plant evolution?

Raven & Thompson (1964) drew attention to the possible significance of the occasional phenomenon of 'polyhaploidy' – the production of functional plants of reduced chromosome number (e.g. $2x$ plants from $4x$ individuals etc.) from polyploids by automatic development of the egg cell without fertilisation (i.e. parthenogenetic development).
In a classic example, de Wet (1968) and de Wet & Harlan (1969, 1970) have described what they call diploid–tetraploid–dihaploid cycles in *Bothriochloa–Dichanthium*. In the words of Asker & Jerling (1992): 'in natural populations, tetraploid facultative apomictic plants predominate, but a small part of the population consists of sexual diploids. By fusion of unreduced gametes the diploids give rise to sexual autotetraploids, allowing gene exchange with apomictic tetraploids. By haploid parthenogenesis, the tetraploids give rise to new diploids (dihaploids), which are often fertile and vigorous.' Polyhaploids may be important in other apomictic groups, for Nogler (1984) detected a number of vigorous and fertile dihaploids in the progeny of hybrids of aposporous *Ranunculus auricomus* plants. Ornduff (1970) and de Wet (1971) consider the possible evolutionary implications of polyhaploidy. While polyhaploids have been detected in experiment and sometimes in wild populations, it remains an under-investigated phenomenon, and at present it is very difficult to evaluate its wider evolutionary significance.

Polyploids: their potential for evolutionary change

Buggs, Soltis & Soltis (2011) trace the development of historic ideas about 'what makes a successful polyploid' by examining the influential historic publications of Winge (1917), Darlington (1937), Clausen, Keck & Hiesey (1945), Stebbins (1950, 1971, 1974) and Grant (1981). Here, we consider an influential viewpoint that emerges from these studies.

Clausen and associates placed emphasis on the importance of balance between the contributing genomes. They argued that 'parents of an amphidiploid should be close enough related to produce a vigorous F_1 hybrid', which has 'harmonious and vigorous development'. For an allopolyploid to succeed through subsequent generations, however, the 'original balance' within the two genomes 'must remain unchanged'. Recombination could not be permitted between the two genomes 'since the balances that determine success or failure are intricate and delicate, they may be upset by slight genetic interchanges'. To prevent recombination, the chromosomes had to be different enough not to pair at meiosis. This generally meant that the two parental species had to be distantly related congeners.

Stebbins (1950) appears to have agreed with Clausen and his colleagues on the question of the 'relatedness' of the parental taxa contributing to a polyploid. Also, polyploidy was seen as an efficient means of conserving variation. In diploid organisms of heterozygous genotype *Aa* (where *A* is dominant to *a*), the allele *a* is sheltered from the immediate effects of selection and may survive in the population even if *aa* is deleterious. However, at meiosis, gametes containing either *A* or *a* are produced and selection may act on *a* at this stage. While the diploid state offers some 'shelter' for recessive alleles, polyploids, with their several genomes giving multiple representation at the *A* locus, have a greater capacity to store variation. Also, as a hybrid is likely to be heterozygous for many loci, chromosome duplication in the allopolyploid will produce fixed heterozygosity at these loci as only homologous pairing takes place.

In addition, as we have already seen, Stebbins saw polyploidy 'as a rare event', resulting in 'polyploid species that were genetically uniform' (Buggs, Soltis & Soltis, 2011). However, with the passage of time, natural polyploids, once formed, could change, through the action of mutation and selection, and as a consequence of chance events. This view was supported by the observation that fertility in experimentally produced polyploids increased, as 'raw' polyploids were subjected to artificial selection (Stebbins, 1950; de Wet, 1980).

Many of these ideas about the nature and potential of polyploids could not be critically examined until molecular techniques became available (Buggs *et al.*, 2011). Now, many historic views have been overturned.

Turning to the genetic consequences of polyploidy, the historic notion that the 'balance of the contributing genomes' must be maintained has been displaced by a realisation that major changes in the genomes of polyploids may occur shortly after their formation involving chromosomal changes and rearrangements, gene silencing etc. Thus, polyploid speciation is now recognised as a dynamic evolutionary process, with plants responding to the 'genomic shock' of the polyploid event in different ways (Buggs *et al.*, 2011). As we see below there is no standard response. The major review by Tayalé & Parisod (2013) tabulates 68 case studies of genetic reorganisation in synthetic or established auto- and allopolyploids. Here, we examine the results of a number of major areas of investigation.

Meiosis in polyploids

First, there have been major attempts to understand the molecular biology of chromosome behaviour at meiosis. Grancourt, Jenczewski & Lloyd (2013) provide an important overview of advances in our understanding of the whole process of meiosis in general, and in polyploids in particular. Despite the insights gained on the genetic control of pairing, for example in Wheat (by *Ph1* and other genes), we still have an imperfect understanding of chromosome pairing and the crossing-over involving Holliday junctions (Chapter 6). However, there is evidence that recombination may be increased in polyploids relative to diploids. For example this was clearly shown in a study of *Arabidopsis* diploids, and synthesised auto- and allopolyploids (Pecinka *et al.*, 2011).

Evidence for structural changes in polyploids

Turning to another area where there have been major recent advances in molecular studies of polyploids, there is evidence from a number of sources that chromosome changes may occur. Soltis & Soltis (1993) consider historic investigations, while Chester *et al.* (2010b) review recent advances.

In an early example, evidence for chromosome repatterning was discovered in studies of allopolyploid Tobacco (Parokonny *et al.*, 1994): the GISH technique revealing numerous intergenomic translocations. Using the same technique in a study of hexaploid Oats ($2n = 6x = 42$: genomes AACCDD), Chen & Armstrong (1994) discovered evidence for intergenomic interchanges between the A and C genomes. Chromosomal rearrangements both rapid and extensive have been detected in recent molecular studies of polyploids in *Triticum, Nicotiana* and *Brassica* (Paun *et al.*, 2007). Using FISH and GISH techniques, He *et al.* (2009) discovered intergenomic translocations and genomic rearrangements in studies of *Agrostis* tetraploids.

In the study of changes associated with polyploidy in *Brassica napus* (an allopolyploid originating c.7.9 and 14.6 Mya from *B. oleracea* and *B. rapa*), Pires *et al.* (2004b) have made use of the possibility of resynthesising allopolyploids from their parental taxa. Using RFLP and Simple Sequence Repeat markers, chromosome exchanges were detected in these synthetic neopolyploids (see Gaeta *et al.*, 2007).

Madlung & Wendel (2013) provide an important review of structural changes in polyploids. There is increasing evidence for a number of species that 'polyploidization has been associated with sequence loss' (Han *et al.*, 2005; Tate *et al.*, 2009). And, evidence is accumulating that 'recombination between homeologous chromosomes' occurs in some polyploids (Salmon *et al.*, 2010; Szadkowski *et al.*, 2010).

However, no intergenomic changes were detected in studies of *Gossypium* and *Spartina anglica*. The same conclusion was drawn by Guggisberg *et al.* (2008), who tested and confirmed the hypothesis that tetraploid *Primula egaliksensis* ($2n = 40$) had originated from the cross between *P. mistassinica* ($2n = 18$) and *P. nutans* ($2n = 22$). They discovered that all the chromosomes of the paternal taxa were present in the tetraploid, but there was no evidence of major intergenomic rearrangements.

Detailed investigations of *Tragopogon miscellus*, using the various molecular techniques including GISH, have revealed other types of change. Even though some plants had the 'typical' chromosome number ($2n = 24$), others were proved to be cytologically distinctive in having chromosome losses and gains even in plants that had the apparently 'normal' $2n = 24$ chromosomes. In one case a chromosome was missing and another was present three times i.e. $2n = 24 + 1 - 1$. In another case, material with $2n = 24 + 2 - 2$ was detected. These situations of 'uneven parental contributions' could have arisen by homeologous pairing and recombination (Chester *et al.*, 2010b).

Using next generation sequencing methods, Buggs *et al.* (2010b, 2012) have further explored genomic changes in *T. miscellus*, detecting loss of alleles, reciprocal chromosome monosomy/trisomy and homeologous recombination. Overall, from detailed investigations of *Tragopogon*, evidence has emerged for 'chromosomal translocations and other structural irregularities as well as aneuploidy' (Madlung & Wendel, 2013).

From a number of other case studies, including investigations of Wheat (Shaked *et al.*, 2001), there is evidence for the 'rapid and reproducible genome-wide removal of some, but not all redundant DNA sequences' (Paun *et al.*, 2007). Such removal may be mainly non-random (as in *Oryza* and *Arabidopsis*;

Paterson *et al.*, 2006), or mostly stochastic (as in the case of *Tragopogon miscellus*; Tate *et al.*, 2006). Furthermore, evidence suggests that with time 'through the deletion of less-adaptive or maladaptive loci' allopolyploids can become more 'diploidized' (Wu *et al.*, 2006).

Gene silencing: epigenetic alterations in gene expression

The parental taxa contributing to a new polyploid, with its multiple genomes, will share some gene copies. In such circumstances, influential models of genome evolution predict that some silencing and gene losses will occur (see Ohno, 1970). It is now possible to test these models using a variety of molecular approaches, including comparison of natural and their newly synthesised counterparts (for details see Lui & Wendel, 2003; Buggs *et al.*, 2009, 2010a, b; Salmon & Ainouche, 2010).

Investigations of *Spartina* provide an informative example of epigenetic changes involved in hybridisation and allopolyploidy (Salmon *et al.*, 2005; Parisod *et al.*, 2009; Parisod, Holderegger & Brochmann, 2010 and references therein). They investigated the hybrids (*S.* × *townsendii* and *S.* × *neyrautii*) between the native European species *S. maritima* and the introduced American species *S. alternifolia* and the allopolyploid derivative *S. anglica* (described earlier). Information on the methylation at restriction sites was investigated by employing the AFLP technique to study the genetic variation of the plants, together with a special variant amplification technique – the Methylation-Sensitive Amplification Procedure (MSAP). They discovered that '30% of the parental methylation patterns are altered in the hybrids, and the allopolyploid. This high level of epigenetic regulation might explain the morphological plasticity of *S. anglica* and its larger ecological amplitude.' Because they studied parental, hybrid and allopolyploid relatives they were able to deduce that 'hybridisation rather than genome doubling seems to have triggered most of the methylation changes observed in *Spartina anglica*'.

In studies of triploid F_1 hybrids of *Senecio* and their allopolyploid derivatives,

> a small but significant proportion of loci display nonadditive methylation in the hybrids, largely resulting from interspecific hybridization ... [However,] genome duplication [also] results in a secondary effect on methylation, with reversion to additivity at some loci and novel methylation status at others. These changes to methylation state in both F_1 triploids and their allohexaploid derivatives largely mirror the overall patterns of nonadditive gene expression observed in our previous microarray analyses and may play a causative role in generating those expression changes. We also observe differences in methylation state between different allopolyploid generations, predominantly in cases of additive methylation with regard to which parental methylation state is dominant ... [Thus] global changes to DNA methylation resulting from hybridization and genome duplication may serve as a source of epigenetic variation in natural populations, facilitating adaptive evolution. (Hegarty *et al.*, 2011)

Rapid genomic changes have also been detected in polyploidy Wheat (Lui *et al.*, 2009).

Reviewing the available evidence, Paun *et al.* (2007) concluded that 'duplicated genes are only rarely found to be expressed at equal levels; instead there is up- or down-regulation of one of the duplicated genes resulting in unequal expression or even epigenetic silencing of one copy. Silencing arising immediately after polyploid formation, in the absence of gene deletion, is probably epigenetically-induced via cytosine methylation.'

Madlung & Wendel (2013) also consider the evidence for changes in gene expression in polyploids. Their paper provides a range of examples. Here, we note their conclusion that 'gene expression alteration and various forms of non-additivity in allopolyploids are ubiquitous' and it is clear that polyploidy 'leads to genome-wide transcriptional rewiring. To what degree changes in RNA levels lead to altered protein abundance is beginning to be elucidated, setting the stage for understanding how duplicated genes lead to altered protein quantities and functions as well as metabolic flux through pathways.'

Transposable elements

As we saw in Chapter 6, flowering plant genomes contain transposable elements. In a classic paper, McClintock (1984) suggested that the formation of both auto- and allopolyploids can induce 'a genomic shock' and 'widespread genomic changes can result from higher activity of transposable elements' (quoted from Bento *et al.*, 2013). To recap, these elements can propagate either by 'cut and paste' replication (transposons); or by 'copy and paste' mechanisms (retrotransposons).

In a review of the significance of these transposable elements in polyploid plants, Paun *et al.* (2007) make a number of key points.

(a) 'Polyploid-induced demethylation and activation of dormant mobile elements may occur in newly formed polyploids' (e.g. in some experimental allopolyploid lines of *Arabidopsis thaliana*; Madlung *et al.*, 2005).
(b) 'Active transposable elements have the potential for insertional mutagenesis and changes in phenotype while altering local gene expression, but they may also facilitate rapid genomic reorganization in new polyploids.'
(c) 'Because polyploids contain duplicate copies of all genes, transposon insertions into "single" copy genes are less deleterious: thus transposable elements can multiply and persist far longer in polyploids than in diploids (Hegarty & Hiscock, 2005).'

The review by Bento *et al.* (2013) provides a full account of our current understanding of the role of retrotransposons in the 'genetic shock' induced in the formation of polyploidy species.

Polyploids: the implications of their recurrent formation

As we have seen above, polyploidy is often a recurrent phenomenon, in some cases with either parental taxon acting as the maternal parent.

Paun *et al.* (2007) point out that 'recurrent formation can create an array of genetically, ecologically, morphologically, and physiologically different genotypes/populations, among which subsequent gene flow, independent assortment, and recombination may produce additional variation (Soltis & Soltis, 1999)'. Moreover, it has been suggested that in allopolyploid populations of recurrent origin, differential silencing of duplicated genes in different plants may lead to reproductive isolation and speciation (Soltis *et al.*, 2005). In addition, different epigenetic modifications of homeologous genes in polyploid populations, particularly amongst populations of separate origin, may contribute to the molecular basis of natural variation and therefore play an important evolutionary role (Soltis *et al.*, 2004; Comai, 2005). It is clear, therefore, that recurrent polyploid formation can lead to derivatives that differ genetically, even though they are all given the same Latin name. And these diverse lineages after a period in geographical isolation could, in the course of time, become sympatric, with the possibility of further episodes of hybridisation/polyploidy (Modliszewski & Willis, 2012; Tayalé & Parisod, 2013). These possibilities have yet to be fully investigated. One major question has been considered in some detail, namely how neopolyploids evolve cytologically.

Polyploids: cytogenetic changes in the longer term

Stift *et al.* (2008) consider this important question, and what follows is drawn from their account. As we have seen above 'in extreme autotetraploids', where there are four homologous chromosomes, chromosomes pair randomly in bivalents or quadivalents leading to patterns of tetrasomic inheritance. In contrast, 'in extreme allotetraploids' there are two homologous sets of chromosomes. Where pairing is exclusively between homologues, patterns of disomic inheritance are found. However, there is evidence that in the course of evolution

> inheritance may shift from disomic to tetrasomic (or vice versa). In (tetrasomic) autopolyploids the four initially homologous chromosomes can differentiate into two sets of preferentially pairing chromosomes resulting in (cyto)genetic diploidisation (Soltis & Soltis, 1993; Ramsey & Schemske, 2002). In (disomic) allo-tetraploids meiotic pairing may not always be strictly preferential (Sybenga, 1996), so that crossing over between homeologous chromosmes (e.g. Udall, Quijada & Osborn, 2005) can homogenise the genome (Sybenga, 1996). A shift in inheritance patterns may take several generations with intermediate inheritance.

Clearly, the type of inheritance will greatly influence the segregation patterns in polyploids. Reflecting on the published studies, Stift and colleagues consider that normally 'in segregation studies only the completely disomic and tetrasomic inheritance models have been considered'. They examine statistical approaches to this problem, taking account of the work of a number of investigators, in studying the perennial tetraploids *Rorippa amphibia* and *R. sylvestris* and their hybrid *R.* × *anceps*. They employ a likelihood approach to the question: which type of genetics – disomic, tetrasomic or intermediate – best fits particular patterns of segregation? Grancourt, Jenczewski & Lloyd (2013) also consider the evolution of polyploids, especially the role of diploidisation. They conclude that 'it is still unclear the extent to which diploidization in auto- and allopolyploids converge'.

Becoming established: what 'hurdles' do polyploids face?

Turning now to other important questions concerning the evolutionary potential of polyploids, we consider the ecological and reproductive challenges facing the newly formed polyploid.

Thompson & Lumaret (1992) make the important point that 'the rate of successful establishment of polyploids is an entirely different matter from their rate of spontaneous origin', for it seems highly likely that single polyploid individuals may frequently be produced in populations and a 'new polyploid derivative' is likely to encounter competition with its diploid progenitor(s). However, many polyploids are hybrid in origin, and often exhibit transgressive segregation, i.e. variants are produced by genetic segregation that are morphologically and physiologically outside the range of those produced by the parental taxa. Thus, they may have a different ecological niche from their parental species and may be able to establish themselves in the wild (Fowler & Levin, 1984; Theodoridis *et al.*, 2013).

The arrival of lone polyploid individuals in ecosystems, however, must make them vulnerable to immediate or delayed extinction through natural selection or by chance events. For example, *S. cambrensis* was lost from the Edinburgh area as its site was taken over for development (Abbott & Forbes, 2002). Even established populations of polyploids may not necessarily flourish in the longer term. Thus, natural populations of *S. cambrensis* are reported to be in decline, perhaps as a result of human activities, despite the fact that they have high genetic diversity (Abbott, Ireland & Rogers, 2007). In contrast, *Spartina anglica* has flourished in many areas around the world, as the plant has been widely introduced to stabilise and extend salt marsh habitats. The species has been so successful that at many sites it has become a threat to the native species by changing water circulation patterns. It is now cited amongst the World's Worst 100 Invaders (Lowe *et al.*, 2000). The success of *S. anglica* is associated, as one of a number of factors, in the decline of its parent species at those sites where *Spartina* hybrids and allopolyploids originated. Both *S. maritima* and *S. alterniflora* have very low levels of genetic variability (Yannic

Baumel & Ainouche, 2004). However, decline in some British populations of *S. anglica* has also been recorded, as a result of the so-called 'die-back'. The cause of this condition has not been definitively determined: it could be 'age-related or due to competition with other species, where the habitat has been elevated following stabilisation' (Gray *et al.* 2004).

In addition to these ecological factors, a 'new polyploid' will encounter problems of reproduction. However, there is evidence that polyploidy is more common in perennial herbs that are capable of vegetative reproduction. Thus, a 'new polyploid' individual that is perennial might survive for a time. However, such a plant is likely to suffer 'minority-type disadvantage' in sexual reproduction (Levin, 1975a). For instance, a new allopolyploid may receive pollen from its parental diploids, leading to the production of sterile triploid progeny.

Characteristics of polyploids as a group

Historically, a number of generalisations have been made about diploids and polyploids, on the assumption that these classes are definable and fixed. Now, given the insight that all flowering plants have episodes of polyploidy in their ancestry, there are grounds for caution in comparing the properties of diploids and polyploids.

Polyploids: ecological considerations

It is often suggested that polyploids have a greater biochemical and genetic diversity than their diploid relatives, giving them a wider ecological tolerance, and wider geographical distributions. Weiss-Schneeweiss *et al.* (2013) have examined these propositions, by considering the best of a limited number of data sets available, namely, the 'percentages of polyploids in widespread European genera' (Stebbins & Dawe, 1987), 'the range of ecological requirements of Pyrenean plants' (Petit & Thompson, 1999) and the 'extent and mean geographical and ecological ranges in North American plants' (Martin & Husband, 2009). On the basis of this information they found no overall support for the view that polyploids had wider ecological tolerances and more extensive geographic distributions than their diploid realtives. Indeed, they found many exceptions. For example, intensive studies of *Biscutella laevigata* reveal a wide distribution of tetraploids, with diploids found in smaller areas, related to glacial refugia (Tremetsberger *et al.*, 2002; Parisod & Bescard, 2007). Further studies are clearly needed.

Polyploidy is often associated with a change in the breeding system

A. Self-fertilisation. Given the strength of minority disadvantage faced by a lone polyploid amongst numerous diploid relatives, it is likely that self-fertile polyploid derivatives would be at a selective advantage (Thompson & Lumaret, 1992).

It has frequently been pointed out that polyploids may have a higher level of self-compatibility than their related diploids (Stebbins, 1950; Mable, 2004; Husband *et al.*, 2008). Barringer (2007) searched the published literature, finding data on polyploidy/self-fertilisation for 235 species. His analyses supported the hypothesis for association between polyploidy and self-fertilisation.

It could be supposed that the formation of the polyploid preceeds the change to self-fertility, but such a model requires the near-simultaneous occurrence of two rare events. Given that many groups of self-fertilising species exhibit greater levels of polyploidy than in outcrossing taxa (Stebbins, 1957; Levin, 1975b; Grant, 1981; Ramsey & Schemske, 1998), a more likely scenario, explored by Barringe (2007), is that self-fertile diploids evolve and such plants are subsequently involved in polyploidy events. Thus, the previously acquired ability to self-fertilise in diploid stocks may facilitate the establishment of polyploid individuals, as minority cytotype disadvantage in reproduction is not an issue.

As we saw in Chapter 7, self-fertilisation can evolve in species with different breeding systems. Here we note a number of cases investigating the association of self-fertilisation with polyploidy. *Turnera* diploid and tetraploid plants are distylous and self-incompatible, while hexaploids are homostylous and self-compatible (Shore & Barrett, 1985). Self-fertility is also higher in polyploid species of *Paspalum* (Quarin & Hanna, 1980) and *Solanum* (Marks, 1966). In *Primula* (section Aleuritia) diploid species ($2n = 18$) are distylous and self-incompatible, while the polyploids in the group ($4x$, $6x$, $8x$, $14x$) are all homostylous and self-compatible (Richards, 1993). Naiki (2012) has investigated some groups of the Rubiaceae and Primulaceae, and reports that 'individuals with a lower ploidy level tended to have heterostyly and individuals with a higher ploidy level tended to be monomorphic.' In considering the evolution in such groups, Naiki suggests that the first stage in the 'breakdown of heterostyly was the recombination of the supergene (S) in a diploid heterostylous plant' followed later by polyploid formation.

We have seen in Chapter 7 how repeated self-fertilisation may lead to homozygous derivatives and inbreeding depression. In comparison with the diploid, population genetic models suggest that, in a polyploidy, the march to homozygosity with inbreeding may not be quite so rapid and this could be of great significance in the evolution of polyploid species. Thus, in a diploid plant of genotype *Aa*, selfing produces progeny distributed in the familiar Mendelian ratio 1*AA*: 2*Aa*: 1*aa* and 50% of the progeny are homozygous. In a tetraploid plant of genotype *AAaa*, in which the alleles are located near the centromere on different chromosomes and where the four homologous chromosomes separate at random in pairs, the ratio on selfing is

$$1/36AAAA: 8/36AAAa: 18/36AAaa: 8/36Aaaa: 1/36aaaa.$$

Segregation follows Mendelian principles but with a different ratio, yielding only 2/36 homozygous derivatives; 94.5% are heterozygotes of various genotypes.

Models have been devised in which selfing proceeds for several generations and where the genotype frequencies are not influenced by selection. Thus, to reduce the percentage of heterozygotes to less than 1% from a population initially wholly heterozygous (*Aa* in the diploid and *AAaa* in the autotetraploid) will take seven generations for the diploid but it will take 27 generations for the tetraploid. About 46 generations would be needed to achieve less than 1% heterozygosity in the autohexaploid (Parsons, 1959). Clearly inbreeding effects could be less severe in polyploids.

There is another complicating factor. Fixed heterozygosity is a phenomenon that occurs in polyploids of hybrid origin that have homeologous genomes. Taking a simple example, an allotetraploid has arisen from related parental taxa of genomic constitution AA × BB → AABB. Genomes A and B are different, but, because they are related, they retain some genetic information in common, i.e. they are homeologous. This commonality is indicated by their capacity to produce hybrids. Fixed heterozygosity occurs when, for a particular locus, the first genome has the homozygous dominant alleles AA, while at the same locus the second genome has the homozygous recessive alleles aa.

Because of preferential meiotic pairing of A with A and B with B in the amphidiploid, no crossing-over occurs between these two genomes, all the gametes produced will be of genotype Aa and all the progenies produced from crossing genotype AAaa × AAaa will be AAaa. No genotypes of AAAA or aaaa will be recovered in the cross; the progeny are all fixed heterozygotes. Given that many polyploids have high chromosome numbers representing a range of parental genomes, they can be seen as reservoirs of genetic variability, and because of the pairing behaviour, they often exhibit fixed heterozygosity. However, Brochmann *et al.* (2004) note that 'it is probable that the genetic variation stored as fixed heterozygosity within individual plants sometimes can be released via occasional pairing of homeologous chromosomes and translocation events'.

Another word of caution is needed here. As we have seen above, there is evidence that profound genetic changes occur in polyploids after their first formation, through processes of diploidisation, gene silencing etc. Such changes are likely to influence segregation, inbreeding depression etc.

B. Apomixis. Given the commonly observed correlation between polyploidy and apomictic reproduction (Richards, 1997), it is clear that newly formed polyploids may be released from minority disadvantage, if they are capable of vegetative or seed apomixis.

Historically, many have pointed out that the 'Achilles' heel' of 'new' polyploids is defective meiosis and consequent sterility (Briggs & Walters, 1997). Thus, hybridisation, whether between diploid or polyploid species, could produce more or less sterile clones. It is easy to see which classes are likely to show defective meiosis: those which have odd numbers of genomes (e.g. $3x$, $5x$), those which lack genomic homology and show multivalent formation with univalents, and those with abnormal, defective segregants because of genomic incompatibility. In sexual reproduction such hybrids are clearly at a selective disadvantage. However, the fitness of hybrids would be increased if they were able to reproduce apomictically, either by vegetative apomixis or agamospermy (seed apomixis).

Many polyploids are known to reproduce apomictically, whilst the related diploid taxa are sexual. Thus, within the variable species *Ficaria verna* (*Ranunculus ficaria*), sterile variants with bulbils (vegetative apomixis) are triploid ($2n = 3x = 24$) or tetraploid ($2n = 4x = 32$), whilst diploid plants ($2n = 2x = 16$) set seed by normal sexual means (Taylor & Markham, 1978).

Increasingly, modern molecular methods are being employed to study apomictic plants and their sexual relatives. Some focus on the study variation in progenies and populations, while others seek to determine phylogenetic relationships.

- In *Alchemilla*, very few of the hundreds of micro-species distinguished in Europe show any trace of sexuality and most have high polyploid chromosome numbers; they would appear to be ancient polyploids derived from sexual ancestors that are now extinct (Walters, 1972, 1986a). Modern molecular methods have the potential to shed light on these issues. At present, molecular investigations of *Alchemilla* are in their infancy (see Sepp *et al.*, 2000).
- North American diploid sexual and polyploid apomicts in the genus *Crataegus* have been investigated (Lo *et al.*, 2009, 2010).
- The variability *Hieracium* subgenus *Pilosella* introduced into New Zealand has been examined as part of the attempts to control these invasive plants (Houliston & Chapman, 2004; Trewick, Morgan-Richards & Chapman, 2004).
- There is evidence that European plants of the *Potentilla argentea* group are diploid selfers and outcrossing taxa, while hexaploids are apomictic (Paule, Shorbel & Dobes, 2011).
- Studies of hybridity, polyploidy and genetic variability have been carried out on the apomictic taxa of the *Ranunculus auricomus* complex (Pellino *et al.*, 2013).
- In North America spontaneous hybrids have been detected (using microsatellite and chloroplast markers) between native sexual species of *Rubus* and introduced exotic naturalised pseudogamous apomictic plants (Clark & Jasieniuk, 2012). These new hybrid variants are not yet invasive, but may become so in the future.
- In a study of European *Rubus* taxa, diploids have been shown to reproduce sexually, while triploids are obligately apomicic, and tetraploids show varying degrees of sexuality (Sarhanova *et al.*, 2012).
- In the genus *Sorbus*, three widespread and variable species in Europe are diploid and sexual, whilst other more restricted taxa, some of which have leaf-shape and other characters intermediate between two of the diploid species, are triploid or

tetraploid and reproduce apomictically. These apomictic taxa can very plausibly be derived by hybridisation and polyploidy from the diploids (Liljefors, 1953, 1955). Support for this view comes from studies of secondary chemicals (Challice & Kovanda, 1978), isozymes (Proctor, Proctor & Groenhof, 1989) and molecular investigations (Nelson-Jones, Briggs & Smith, 2002; Chester *et al.*, 2007; Feulner *et al.*, 2013).

- The *Taraxacum officinale* group contains sexual diploids and apomictic polyploids in complex patterns of reticulated evolution. New clones originate from hybridisation between sexual and apomictic plants (Majesky *et al.*, 2012).

The origin and significance of apomictic behaviour in polyploids Polyploidy, by itself, could be considered an infrequent event. For apomictic traits to appear by mutation immediately after polyploidisation would seem improbable. Reflecting on possibilities, perhaps the behavioural elements for apomixis might be found separately in different diploids. These could then perhaps be brought together in hybrids from which polyploids might arise. This scenario would suggest that diploid apomictic plants should be found; however, investigations have revealed that seed apomixis is apparently rare in diploid plants.

Clearly, the ease with which polyploid variants derived from sexually reproducing species can become agamospermic will depend upon the genetical and developmental control of apomixis (see Chapter 7). Perhaps the ongoing research into embryological development of flowering plants will shed some light on both 'normal' and variant apomictic behaviour.

In the past, apomixis has been seen as 'an escape from sterility' (Darlington, 1939). However, the positive adaptive significance of apomixis is now stressed. First, apomictic plants that produce good pollen can act as pollen parents in crosses with sexually reproducing species. Secondly, we might look upon the polyploid and apomictic lines of evolution more in terms of positive adaptation to particular kinds of habitat, than in terms of a negative escape from sterility. This is particularly true now that many polyploids are considered to reproduce by facultative rather than obligate, apomixis (Asker & Jerling, 1992; Schön *et al.*, 2009).

C. Transitions from hermaphrodite to dioecy etc. Polyploidy is associated with other trends in breeding behaviour. For example, in a study of 22 genera, Ashman, Kwok & Husband (2013) have explored in detail the underlying mechanisms that might lead to the evolutionary transition from combined sexes to separate sexes. In the genera studied they discovered that 'gender monomorphism (hermaphoditism, monoecy) is more common among diploid than polyploid species, whereas gender dimorphism (dioecy, gynodioecy, subdioecy) is more frequent among polyploidy species'. This paper considers the possible origin of sexual system polyploidy associations, and highlights how molecular and other approaches might be employed in their further investigation. However, the authors stress that in considering changes in breeding behaviour it is rarely clear whether polyploidy is the 'direct genetic cause' or 'consequence' of breeding system transitions.

Evidence for hybridisation between diploid and polyploid plants in the wild

The simple picture of an allopolyploid species arising suddenly and achieving at one bound fertility and genetic isolation is unlikely to be the whole story in the complex polyploid evolution of plants.
In particular the genetic isolation of allopolyploids must be questioned. In general, it appears that the post-zygotic barriers between different polyploidy levels are 'all but impermeable' (Tayalé & Parisod, 2013). However, gene flow, leading, in some cases, to introgression, has been detected in some cases.

The common *Dactylis glomerata* is a variable tetraploid with $2n = 4x = 28$ (Jones & Borrill, 1961). Cytologically, the plant is autopolyploid (Lumaret,

1988). Crossing *D. glomerata* and a diploid, *D. glomerata* ssp. *aschersoniana*, resulted in a triploid hybrid that is male-sterile, but partially fertile as female parent. Backcrosses of the triploid with ssp. *aschersoniana* were less successful than with the tetraploid *D. glomerata*. Female gametes produced by the triploid ranged in chromosome number from $n = 7$ to $n = 23$; in general those with $n = 14$ could function well in the backcross to the tetraploid. Thus, hybrid tetraploids (or near-tetraploids) could arise relatively easily and, moreover, showed a high fertility equal to that of wild *D. glomerata*.

Jones & Borrill then ask the important question: how likely is gene flow in natural populations? At first there was only fragmentary evidence. For example, Zohary & Nur (1959) reported that a deliberate search in an area in Israel where both diploid and tetraploid populations occur resulted in seven triploid plants in a total of 4000 examined. The intercrossibility of these triploids with diploids and tetraploids was studied and Zohary & Nur discovered that, considering the fertility of the product, gene flow was likely only from diploid to tetraploid level.

Levin (1978b) discusses a number of other examples of crossing between plants of different ploidy level. Although this is usually a unilateral occurrence from diploids via triploid to tetraploids, he presents the evidence from natural and experimental situations, e.g. in *Solanum*, *Viola* and *Papaver*, in which gene flow occurred in the reverse direction from tetraploid via triploid to diploid.

Stace (1993) has emphasised the important role of 'rare events' in the evolution of polyploid groups, in which sterile plants produced some gametes with the full haploid complements of chromosomes (so-called euploid gametes). Thus, in his studies of the *Festuca* and *Vulpia* groups, there is evidence of rapid introgression from tetraploid to hexaploid levels via highly sterile pentaploid intergeneric hybrids, through the rare production of euploid (in this case triploid) gametes. Stace cites a number of other published cases (e.g. *Ranunculus*, *Centaureum*, *Euphrasia* and *Pilosella*), which suggest that introgression can occur between species of different level, through the production of euploid gametes by highly sterile hybrids.

In considering the possibilities of gene flow between diploid and polyploid plants, models can be constructed which do not assume the involvement of plants with odd numbers of genomes – $3x$, $5x$ etc. For instance, in studying experimentally the variation of *Betula*, Johnson (1945) suggested that gene flow might occur directly between diploids and tetraploids via unreduced gametes of *B. pendula* ($2n = 28$) which could fuse with the normal ($n = 28$) gametes of *B. pubescens* ($2n = 56$) to give hybrid plants of tetraploid chromosome number. Elkington (1968) has also considered this possibility for species of *Betula*.

Not only is there evidence in particular cases of gene flow between plants of different ploidy, but there is also the important possibility of new hybridisation at the polyploid level. A case of particular interest was described by Fagerlind as early as 1937 (Fagerlind, 1937) in the common European genus *Galium*. The white-flowered *Galium mollugo* and the yellow-flowered *G. verum* are both represented in south-east Europe by diploid plants ($2n = 2x = 22$). These diploids are completely intersterile. In central and northern Europe, however, the common representatives of both species are tetraploid ($2n = 4x = 44$) and the hybrid between them shows almost normal meiosis and some degree of fertility. In this case effective sterility evolved at the earlier diploid level has been broken down at the tetraploid level.

Reviewing the evidence from recent studies, Chapman & Abbott (2009) note that great progress has been made in the study of introgression between plants with the same chromosome number, and there is now formal evidence for introgression between plants of different polyploidy level. But few investigations have faced the major question: is there any evidence of adaptively significant gene transfer between plants of different polyploid level?

Some revealing case histories have now been published. For example, introgression of genes across a ploidy barrier has been confirmed in investigations of many crop plants and their wild relatives, for example, Soybeans (*Glycine*, Doyle *et al.*, 2003) and Coffee (*Coffea arabica*, Herrera, D'Hont & Lashermes, 2007).

However, the possibility of introgression of adaptively significant traits across a ploidy barrier in wild plants has been little explored. One very convincing example of such introgression has emerged from the investigation of hybridisation in Britain between the native tetraploid *Senecio vulgaris*, which commonly has nonradiate flower heads, and the introduced rayed diploid *S. squalidus* (Chapman & Abbott, 2009). This has resulted in the 'origin of a variant form of *S. vulgaris* that produces radiate rather than nonradiate flower heads'. There is clear evidence that this transferred trait is adaptively significant, as the rayed variant of *S. vulgaris* 'is more attractive to pollinators and has a higher outcrossing rate' than the commoner rayless plant. For further details of the molecular characterisation of the genes see Chapman & Abbott (2009).

Reticulate patterns of variation in some groups

Some very complex reticulate patterns of evolution have been detected in some groups. A fine example of this is our expanding knowledge of *Elymus glaucus*, an allohexaploid member of the Wheat tribe (Mason-Gamer, 2004). There is evidence of at least 'five distinct genetic lineages coexisting within the species, acquired through a possible combination of allohexaploidy and introgression from within and beyond the Triticaceae'. Cytological studies suggest that nuclear genome donors may include *Hordeum*, *Pseudoroegneria*, while studies of cpDNA suggest that there are three potential donors: *Pseudoroegneria*, *Thinopyrum* and *Dasypyrum*. Genetic studies also indicate unexpected contributions from *Taeniatherum* and two other unidentified sources, one that falls outside the clade containing the Triticeae.

Ancient introgression. In Chapter 13, we discussed contemporary pattern, process and consequences of introgressive hybridisation, including the capture of chloroplasts following hybridisation between two species and repeated backcrossing of hybrid derivatives to one or both parental stocks. Climate changes have provoked range shifts of species in the past, and phylogenetic studies now provide evidence not only for the formation of new polyploids but also for ancient introgression between related species brought into sympatry. For instance, Kikuchi *et al.* (2010), investigating *Veratrum* species in Japan, provide evidence for the molecular signatures of ancient introgression and the capture of chloroplasts by one species from another. In another example, the complex origins and biogeographic distribution patterns of $2x$, $4x$, $5x$, $6x$, $7x$ cytotypes of the European montane species *Senecio carniolicus* have been investigated (Suda *et al.*, 2007). Intracytotype variation in nuclear DNA was detected in diploids and it is suggested that this may be the result of ancient introgression of western diploids with another species, *S. incanus*.

A further possible example of ancient introgression is suggested by the phytogeographical genetic structure of Japanese Beech (*Fagus crenata*) (Okaura & Harada, 2002).

Are polyploids more frequent in particular geographical areas?

One phenomenon that has excited interest since the early days of cytogenetic studies on plants is the apparent increase in frequency of polyploidy with increasing latitude in the Northern Hemisphere (Fig. 14.7). It is instructive to examine the history of this idea and then consider its current status.

The earliest reference to this correlation seems to have been made by the Swedish cytologist Täckholm (1922), in his impressive study of the genus *Rosa*. Hagerup (1932), Manton (1950), Morton (1966),

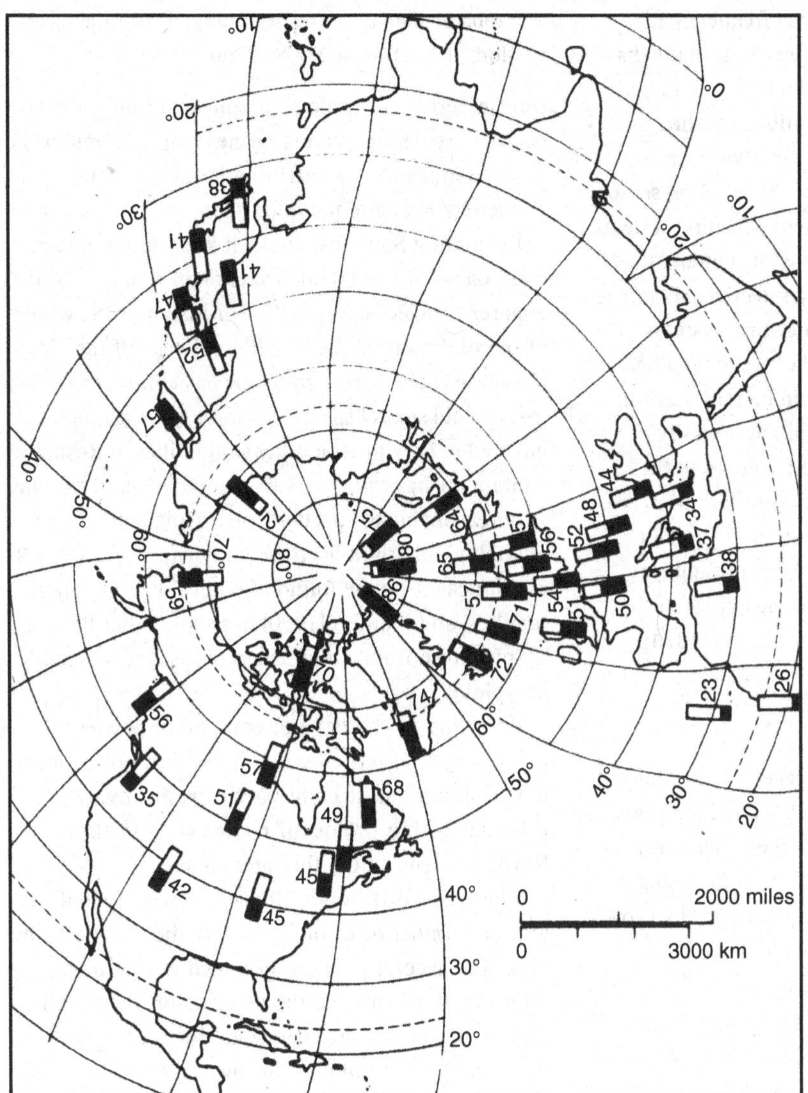

Fig. 14.7. Frequency of polyploids (given as percentage figures and indicated as black portion of columns) in floras of various territories in the Northern Hemisphere. (From Löve & Löve, 1974.)

Favarger (1967), and Stebbins (1984) have provided important reviews. Different hypotheses have been advanced to explain this phenomenon.

a. It has been proposed that polyploids have greater resistance to extreme climates (Hagerup, 1932), particularly cold at high latitude. Only few experimental tests of this proposition have been made; for example, in a study of diploid and tetraploid Rye (*Secale cereale*) the polyploid was less able to withstand warmer temperatures (Hall, 1972). Stebbins pointed out, however, that when the floras of different regions of the Pacific Coast of North America were compared, highest frequencies of polyploidy were found in the previously glaciated areas (52–54°N) than in a largely unglaciated area (63–71°N). As a consequence he proposed that there was a correlation between polyploidy and glaciation itself.

b. It has also been stressed that the frequency of polyploids may be related to their breeding behaviour (Crawford, 1985). Polyploidy is often associated either with a self-compatible breeding system or seed apomixis or vegetative apomixis/clonal growth. At high latitude the growing season is very short and plants with these breeding systems may, because of the higher level of reproductive assurance they provide, be at a selective advantage relative to plants with other breeding systems.
c. Stebbins (1950) argued that the greater ecological adaptability of polyploids could be important in the colonisation of post-glacial habitats
To account for the observed differences in the frequency of polyploids at different latitudes, Stebbins (1984) proposed a secondary contact hypothesis. This hypothesis relates present-day distribution of polyploids with the events of the recent geological past (see Pennington, 1974; Ritchie, 1987; Roberts, 1989).

Stebbins (1984) wrote:

> During the latter half of the Pliocene and beginning of the Pleiocene epochs, a period of 5–6 million years, many plant genera responded to the increasingly harsh conditions along the crests of newly rising mountain chains as well as at high latitude by evolving races and species, chiefly diploid, having increased tolerance of cold, and in the north, of long arctic nights. With the onset of the Pleistocene glaciations, alpine populations colonised lower altitudes and northern populations moved southwards. Many secondary contacts between different variants were repeatedly established and broken. Hybridisation between previously separated populations, accompanied or followed by either polyploidy or introgression at diploid levels, generated new races and species, some of which became adapted to the new conditions prevailing in regions vacated by the ice. These new races and species that now form the bulk of the arctic–alpine flora, originated during the entire period of a million years or more during which glaciers advanced and retreated, but some of them probably date from the beginning of the final recession, about 10–14000 years ago.

During periods of geological upheaval, such as the Quaternary period, related species that had previously evolved in isolation were brought together and formed hybrids and polyploids.

Expanding Stebbins' model it seems possible that secondary contact could also explain some cases of apparent ancient polyploidy. For instance, the whole of the subfamily Maloideae (Pomoideae) of the Rosaceae is characterised by the basic number $x = 17$, and diploid sexual species of *Sorbus*, for example, have $2n = 2x = 34$. The other subfamilies of Rosaceae contain the basic numbers $x = 8$ and $x = 9$, and it seems very reasonable to speculate on an ancient allopolyploid origin for the Pomoideae from $x = 8 + 9$. Similar cases can be found in other flowering plant families, and Manton (1950) has shown that the Tropical fern flora contains many cases of 'ancient' polyploidy.

Reverting to the general correlation between polyploidy and latitude, two important points must be stressed. First, it is by no means an invariable rule that within groups of related taxa in the Northern Hemisphere the diploids are the most southerly in distribution. Indeed, Lewis (1980a) gives a number of examples where the reverse is the case. The second point to be noted is that our interpretations involve the assumption that, within any group, polyploidy is essentially a one-way process, in which diploid or low-polyploid taxa are the parents of the higher allopolyploids. Whilst there is good reason to think of many of the patterns we see in this light, are we right in assuming that such an interpretation is valid for every case? Until recently, writers on evolution of higher plants have assumed that it was: ancient polyploid species such as make up the genus *Equisetum* certainly look like evolutionary relics that are eventually doomed, and, in the past, there was no obvious widespread mechanism for descending a polyploid series, as there is for ascending it. But, as we see below, mechanisms for the reduction in chromosome number have been discovered.

How have these ideas on the distribution of polyploids and significance and evolutionary potential of ancient polyploids stood the test of time?

Current views on the present-day distribution of polyploids

In a recent major study, Brochmann *et al.* (2004) have considered in detail the questions about the distribution of polyploidy in arctic plants. Here we note their conclusions.

They discovered that 'the frequency and level of polyploidy strongly increases northwards within the Arctic', but there is 'no clear-cut association between polyploidy and the degree of glaciation for the arctic flora as a whole'. There is, however, support for the hypothesis that 'polyploids are more successful than diploids in colonizing after deglaciation'. In addition, there is abundant evidence for the 'recurrent formation of arctic polyploids'. In some cases, 'low-level polyploids were formed after the last glaciation' with 'repeated and successively more high-level polyploidizations throughout the Quaternary'. They conclude that 'the evolutionary success of polyploids in the Arctic may be based on their fixed-heterozygous genomes, which buffer against inbreeding and genetic drift through periods of dramatic climate change'. However, Brochmann *et al.* (2004) note that rigid fixed heterozygosity may occasionally be broken down and new genetic variability released, through the occasional pairing of homeologous chromosomes and through mutational exchange of chromosome segments.

Apomictic taxa. *Ranunculus cassubicus* is a central European complex containing both diploids and polyploids. Paun, Stuessy & Hörandl (2006) have studied a wide range of samples using DNA fingerprint methods (AFLP and simple sequence repeats, SSR). They concluded that a hexaploid apomictic variant *R. carpaticola* has arisen by hybridisation between two sexual parents: an autotetraploid crossing with a diploid. They report that hybridisation may be correlated with the last glaciation. The sexual parents migrated south during the Last Glacial Maximum: then, as the climate became warmer in the post-glacial period, newly formed apomictic populations from a single hybridisation event may have expanded from the refuges more rapidly than sexual populations. Apomictic variants, with their fixed heterosis, could 'start a new population from a single colonist without being hampered by homozygosity and inbreeding depression, whereas sexual outcrossers would need at least two individuals and pollinators'. Tentative dating of the age of the apomictic, using data from the SSR sequence data, suggests that *R. carpaticola* was formed c.40 000 years ago, well within the 115 000–15 000 ^{14}C for the latest glacial period (Birks, 1986).

Studies of the North American Easter Daisy (*Townsendia hookeri*) have also yielded interesting results (Thompson & Whitton, 2006). '[The] broad range of sexual diploids was broken by the Wisconsin glaciation... with a recolonization of the post-glacial landscape by apomictic polyploids.' Molecular studies reveal that apomicts originated from sexual diploids at least four times. Moreover, the results indicated that there were two independent areas of refuges for the sexual species during the height of the glaciation, one to the north (Yukon/Beringia) and one to the south (Colorado/Wyoming).

In studies of other groups with asexual complexes, both the origin (from sexual parents sometimes by interspecific hybridisation) and subsequent distribution of asexual complexes has been linked to glacial events (see Asker & Jerling, 1992; Richards, 2003).

The implications of ancient polyploidy for studies of geographical distributions

While these recent findings broaden our current understanding of the post-glacial distribution of polyploids, new information has dramatically changed our outlook on the distribution of polyploids and the wider significance of polyploidy in plant evolution. As we discussed in Chapter 6, molecular studies have revealed that some contemporary diploid species have gene duplication in their genomes indicative of very widespread ancient polyploidy. Historically, there appeared to be a clear distinction between diploids and polyploids. Now, a more complex situation has emerged and a reappraisal of the distribution and evolutionary potential of polyploidy in plant evolution is being made. In considering geographical distribution of diploids/polyploids, researchers are now concentrating on patterns and processes in particular areas/genera, and are rather dismissive of the historic overarching generalisations made by cytotaxonomists in the past.

C. Other modes of abrupt speciation

So far we have examined evidence for abrupt speciation by polyploidy. This involves the evolution of plants with three, four, five, six or more complete chromosome sets, instead of the two as in diploids.

There is now evidence that new species may arise sympatrically by other processes than polyploidy, within populations of their parental species. Influenced by the results of experiments with animals, several models of abrupt speciation in plants have been proposed, e.g. involving changes in chromosome number, chromosome repatterning, changes in breeding behaviour and the stabilisation of hybrid derivatives. A number of important reviews provide a historical perspective on the development of theories of chromosomal evolution and its relation to abrupt sympatric speciation in plants (see Stebbins, 1950; White, 1978; Grant, 1981; Levin, 1993; Soltis & Soltis, 1995, 2009; Coyne & Orr, 2004; Lysak, 2014). We now consider a number of examples, where processes other than polyploidy have resulted in sympatric speciation.

Changes in chromosome number

In the past, it has generally been assumed that polyploidy presents an irreversible 'upward' pathway leading to higher chromosome numbers. The evidence that some living angiosperms have very low chromosome numbers, but yet have ancient WGD events in their ancestry, suggests that mechanisms of chromosome number reduction are very important in plant evolution. This is reflected in the observation that sometimes a genus has more than one basic number. Here, we will show, in outline, how one basic chromosome number may arise from another and how this process represents a mode of speciation quite different from polyploidy.

The model is based on the behaviour of centromeres. We have already discussed the role of these structures in nuclear division (Chapter 6). With the exception of certain diffuse centromere types (see below), cytological evidence reveals that chromosomes of plants have a defined centromere, which, together with the spindle fibres, ensures proper chromosome disjunction at meiosis and mitosis. The model makes the assumption that centromeres are formed from pre-existing centromeres. This is because plants require accurate mechanisms that allow faithful replication of the hereditary molecules in cell division. Another element of the model is the fact that plants with fewer or more chromosomes than the normal complement may be viable.

Many polyploids are tolerant to a degree of aneuploidy, a tolerance that may owe its origin to homologies between the 'different' genomes. Some duplication of genetic material is likely, which may cushion the plant against chromosome loss. Loss or gain of chromosomes may be much more damaging in diploids. Loss of a chromosome implies that the plant may have a portion of the normal DNA missing entirely from its genotype, and such plants are likely to be inviable. Given the sensitivity of diploids to chromosome loss (and perhaps gain also), changes in the base number requiring, as they do, losses or gains, seem difficult to explain.

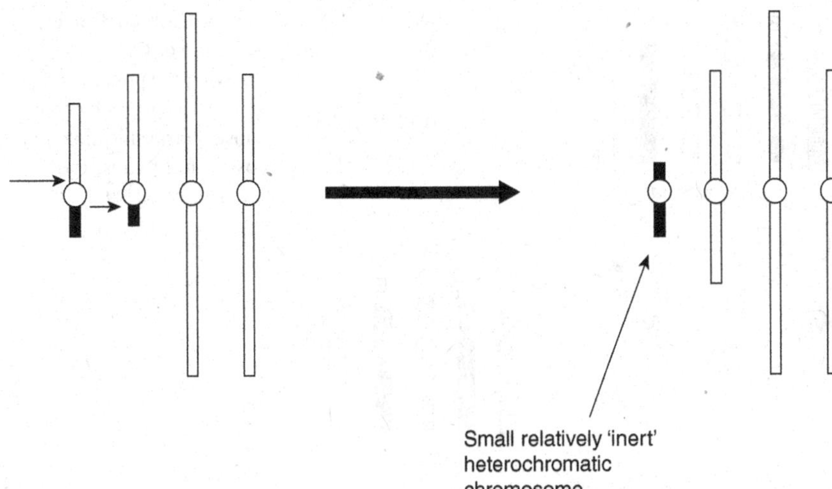

Fig. 14.8. Illustrating the possible mode of origin of a new basic chromosome number ($x = 3$) from $x = 4$, by translocation and loss of a small heterochromatic new chromosome. (From Moore, 1976.)

However, a simple model suggests a possible mechanism involving reciprocal translocations between two chromosomes producing a new larger chromosome, with two 'long' arms and a small mini-chromosome that is largely heterochromatic (Fig. 14.8).

In gamete formation the small, derived heterochromatic chromosome may be lost from the chromosome complement by misdivision of the chromosomes. As Darlington (1937) makes clear, 'a change in the number of chromosomes means a change in the number of centromeres'. An increase in chromosome number following reciprocal translocations seems less common, being reported, for example, in certain orchids (Cox *et al.*, 1998). Lysak (2014) notes that another outcome is possible: mini-chromosome may become part of the chromosome complement as B chromosomes, and, over time, these may accumulate sequences from the chromosomes and also from the plastid and mitochondrial genomes.

What is the outcome of reciprocal translocations and losses of mini-chromosomes? Crossing between parental and derivative taxa will result in hybrids that are likely to exhibit meiotic irregularities, such as multivalents, chromosome bridges etc. It is likely that an effective post-zygotic isolating mechanism has been created involving chromosome translocations and losses.

Clearly, there are enormous hurdles to surmount before gametes with reduced or increased chromosome numbers can give rise to a population with a different basic number. Many aneuploid individuals must fail to survive, but evidence suggests that changes in basic number have occurred in the evolution of plants and constitute an important class of speciation events by abrupt changes. In the current literature these alterations in chromosome numbers are sometimes referred to as dysploid changes (Yang *et al.*, 2014). Further changes are then possible, as species with changed chromosome numbers may then give rise to polyploid derivatives, based on the 'new' reduced or elevated number.

An excellent example of an aneuploid change is provided by the classic early studies of Kyhos (1965) on three species of the genus *Chaenactis* (Asteraceae). He studied the yellow-flowered *C. glabriuscula* ($2n = 12$), a western North American plant of mesic habitats, and two white-flowered Californian desert species, *C. stevioides* and *C. fremontii* (both $2n = 10$). By examining meiosis in the hybrids, both artificial and natural, between these three species, Kyhos was

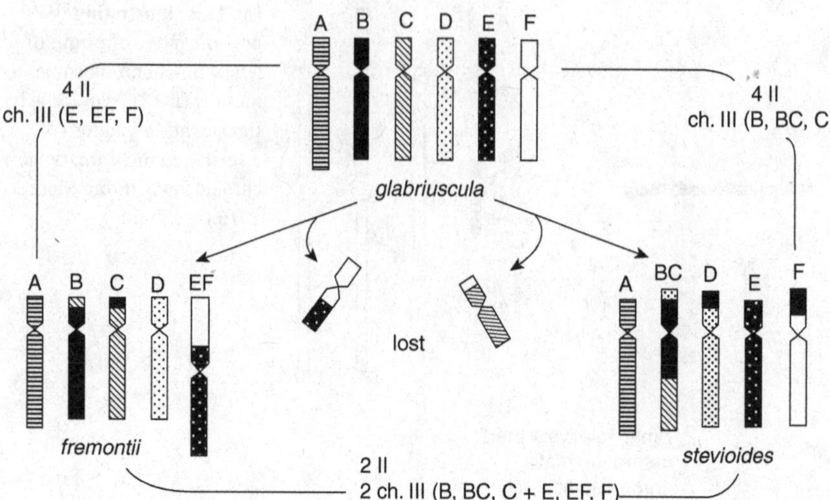

Fig. 14.9. Chromosome structure of *Chaenactis* showing the possible origin of related species by chromosome translocation and loss. (After Kyhos, 1965, from Moore, 1976.)

able to make a careful study of chromosome associations. He deduced that *C. stevioides* and *C. fremontii* had been independently produced from *C. glabriuscula* by processes involving chromosome translocations and loss (Fig. 14.9).

A number of other historic examples of changes in chromosome number have been reported in the literature, e.g. *Crepis fuliginosa* ($n = 3$) derived from *C. neglecta* ($n = 4$) or its near ancestor (Tobgy, 1943); *Crepis kotschyana* ($n = 4$) derived from an ancestor like *C. foetida* ($n = 5$) (Sherman, 1946); and *Haplopappus gracilis* ($n = 2$) from plants with $n = 4$ (Jackson, 1962, 1965). Further examples of both descending and ascending basic aneuploidy are given in Grant (1981).

Aneuploid reduction may also occur in polyploids, and several case histories are considered by Grant (1981). In some species both polyploids and aneuploids derived from them are found: for example, in the extreme case of *Claytonia virginica*, there are diploids with $2n = 12$, 14 and 16 and polyploids with the numbers $2n = 17$ to 37 inclusive, 40, 42, 44, 46, 48, 50, 72, 81, 85, 86, 87, 91, 93, 94, 96, 98, 102, 103, 104, 105, 110, 121, 173, 177 and 191 (Lewis, 1976). Some of the variation in chromosome number may be the result of aneuploidy, but it is likely that allopolyploidy has occurred repeatedly in this group; however, there is said to be relatively little taxonomic variation. Another taxonomic species with a spectacular array of chromosome numbers is *Cardamine pratensis*, with $2n = 16$, 24, 28, 30, 32, 33–37, 38, 40–46, 48, 52–55, 56, 57, 58, 59–63, 64, 67–71, 72, 73–96 (see Lövkvist, 1956).

Moving on to further empirical evidence for abrupt speciation in plants, the celebrated American geneticist Lewis and his associates studied the evolutionary relationships between a number of diploids in the genus *Clarkia* (Lewis, 1973).

The results of their work, which are based on analysis of meiosis in hybrids, are set out in Fig. 14.10. In some cases the evidence suggests an aneuploid origin of derived taxa with a reduction in chromosome number. However, as in the case of the origin of *C. lingulata* ($n = 9$) from *C. biloba* ($n = 8$), evidence suggests that chromosome rearrangement may be important, as well as an increase in chromosome number, for hybrids between these two species are sterile. In our discussion of gradual speciation (Chapter 13), we saw that allopatric differentiation often yielded derivatives showing significant genetic divergence, the mean genetic identity between such species being $I = 0.67$.

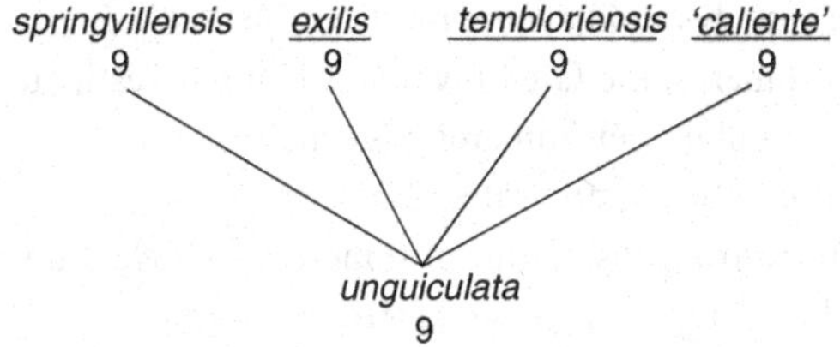

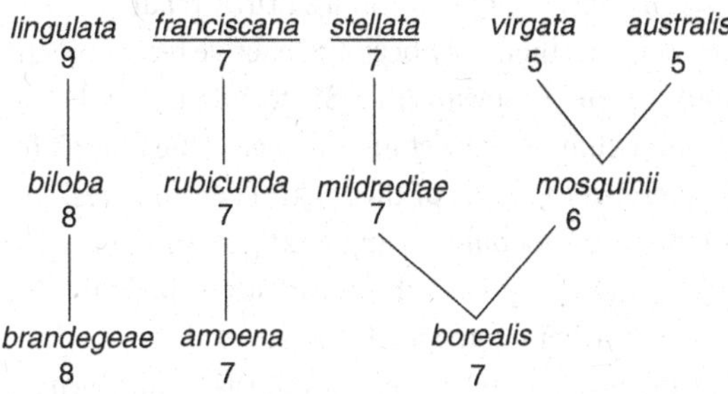

Fig. 14.10. Diagram showing the relationship of species of *Clarkia*. Haploid chromosome numbers are indicated. Predominantly self-pollinating species are underlined; all the others are normally outcrossed. (From Lewis, 1973.)

An electrophoretic study of *C. lingulata* and *C. biloba* revealed that progenitor-derivative species had a very high genetic identity ($I = 0.88$) indicating that 'a newly arisen species can be nearly identical genetically to its progenitor' (Gottlieb, 1986, in which there is a full review of speciation in *Clarkia*).

Speciation does not always involve a change in chromosome number. Evidence in support of this mode of speciation has been obtained by studying a number of pairs of species, one the supposed progenitor and the other the derived species. As judged by isozyme analysis, each of these pairs also had high genetic similarity. For instance, the self-compatible *Stephanomeria malheurensis* arose from the self-incompatible *S. exigua* subspecies *coronaria* (Gottlieb, 1973, 1979), and hybrids between the two species are sterile as a consequence of chromosome rearrangements. Other examples of progenitor-derivative species pairs, of high genetic similarity, have been investigated, e.g. *Gaura longiflora* and *G. demareei* (Gottlieb & Pilz, 1976); *Coreopsis nuecensoides* and *C. nucensis* (Crawford & Smith, 1982); and *Lasthenia minor* and *L. maritima* (Crawford, Ornduff & Vasey, 1985).

Turning to very recent investigations employing molecular approaches, an excellent example of the unravelling of important aspects of the chromosome phylogeny is provided by studies of the diploid species of Autumn Crocus *Prospero*, a genus found across the Mediterranean (see Jang *et al.*, 2013, from which this account is drawn). Despite the apparent taxonomic simplicity of the group, of only three taxonomic species, cytological studies, including fluorescence *in situ* hybridisation (FISH) employing rDNA probes, estimations of genome size, and studies of phylogenetic trees, have revealed a complex situation (Jang *et al.*, 2013). Here, we note the main findings.

Evidence has confirmed that the evolution of the group has involved a reduction of the putative ancestral basis number ($x = 7$ to $x = 6$) with another later reduction event ($x = 7$ to $x = 5$). The other analyses pinpoint details of chromosome losses, fusions and duplications in karyotypic change, with an increase in genomic size in some accessions. These studies of the diploids in *Prospera* are only part of the complexity found in the genus, for plants with basic chromosome numbers of $x = 4, 5, 6$ and 7 are ancestral to the evolution of an elaborate polyploidy complex of auto- and allopolyploids, with chromosome numbers ranging from $3x$ to about $20x$, with $4x$ and $6x$ being the most frequent.

Considering other examples, recent intensive studies of the evolution of the model species *Arabidiopsis thaliana*, have revealed that it has $n = 5$ chromosomes. Historically, it would not have been categorised as a polyploidy (Tate, Soltis & Soltis, 2005), yet there is evidence for two chromosome doublings and one tripling event in its remote ancestry. With regard to the cytological history of *A. thaliana* relative to other related species, Tang *et al.* (2008a, b) conclude that there is evidence for '9 to 10 chromosomal rearrangements in the past few million years since its divergence from *A. lyrata* (Rock Cress) and *Capsella rubella* (Pink Shepherd's purse), including condensation of 6 chromosomes into three, bringing the chromosome number from n = 8 to n = 5'. According to Alcázar *et al.* (2012), this change of chromosome number was achieved by a series of chromosome inversions, and reciprocal translocations.

Other experiments with *A. thaliana* and its relatives suggest an important evolutionary possibility (Matsushita *et al.*, 2012). Synthetic allopolyploids have been produced and these show a degree of karyological variability. A number of aneuploidy lines with different chromosome number variations have been produced and maintained through several generations. Could this be a means whereby significant evolutionary variation is introduced in wild allopolyploid populations?

Yang *et al.* (2014) have carried out very detailed investigations of the Cucumber ($2n = 2x = 14$) and its relatives, the majority of which have $2n = 2x = 24$ chromosomes, implying a reduction from $n = 12$ to $n = 7$. By employing next generation sequencing techniques, together with FISH mapping and genomic synteny-based modelling, fine-scale evidence of dysploid chromosome reduction was obtained. Remarkably, they detected 'at least 59 chromosome rearrangement events that led to the seven cucumber chromosomes, including five fusions, four translocations and 50 inversions'.

Nested chromosome insertion

Another route to chromosome number change has now been examined in some detail. 'The entire donor chromosome appears to be inserted into or near the centromere of a recipient chromosome' (Lysak, 2014). This so-called nested insertion has been discovered in very detailed investigations of grasses, and has contributed to a reduction in chromosome numbers, 'from the ancestral 12 chromosomes to 10 in *Sorghum*, seven in the Triticeae (e.g. *Aegilops*) or five chromosomes in *Brachypodium distachyon*'. The same process is likely to have occurred in the reduction in chromosome number $n = 20$ to $n = 10$ in Maize.

These dysploid events result in the 'loss' of centromeres, the fate of which is being investigated. Typical plant centromeres have an array of centromere-specific satellite repeats and retrotransposons, with centromere-specific histones and flanking repeat-rich heterochromatin. The models being tested suggest that if the extra centromeres are not lost through DNA repair mechanisms, then they become inactive by epigenetic modification. By means of FISH techniques and immuno-fluorescence studies it is possible to look for the presence/remains of these extra centromeres. Comparative genomics using next generation sequencing analysis of chromosomes is likely to provide further evidence of the fate of these centromeres, as well as changes in DNA colinearity that flow from translocations/insertions.

As well as centromere inactivation and loss, a key question is whether centromere repositioning is a possibility in the evolutionary changes of chromosomes. Inversion breaks in chromosome are one part of the story. Another model that is being tested envisages the maintenance of colinearity of the DNA, but gradual decay of old centromeres and epigenetic marks, and the emergence of a new centromere. Concerning the origin of such changes: is intrachromosomal recombination involved? Again next generation sequence methods should help to test this possibility.

It is worth stressing once again that the dysploid changes outlined here may also be associated with polyploidy and subsequent diploidisation: it is a matter of convenience that we have introduced these different processes separately.

Plants with diffuse centromeres

As we have seen, in most plant groups a definite centromere is present in each chromosome. In contrast, notably the Juncaceae (e.g. *Luzula*), centromeric activity is not localised. In the past, such plants were said to have 'diffuse' centromeres (Godward, 1985), now they are categorised as holocentric. In *Luzula*, Heckmann *et al.* (2013) note that chromosomes lack a primary constriction, and have instead 'holokinetic kinetophores' along almost the entire surface of the chromatids, 'to which the spindle fibres attach (Heckmann & Houben, 2012)'. In plants with this type of centromere, fragmentation and fusion of the chromosomes may occur; such plants may be viable, and a whole series of chromosome numbers may be generated thereby (a condition known as agmatoploidy).

In the large genus *Carex*, there is an exceptionally long series of chromosome numbers from $n = 6$ to $n = 66$ (Luceño & Castroviejo, 1991; Hipp, Rothrock & Roalson, 2009), with every number represented between 6 and 47. 'This chromosomal diversity is due in large part to the structure of the holocentric chromosomes: fragments that would not be heritable in organisms with monocentric centromeres have the potential to produce viable gametes' (Hipp, Rothrock & Roalson, 2009). In addition, more than 100 species in the genus are known to be made up of numerous cytotypes, with some taxa 'spanning a range of ten haploid (n) chromosomes from highest to lowest count' (see Cayouette & Morisset, 1986; Rothrock & Reznicek, 1998 and the many references cited therein). Only a few cases of autopolyploidy have been confirmed in the genus. Two cases of allopolyploidy have been proposed. Reviewing the present state of play, Hipp, Rothrock & Roalson (2009) consider that allopolyploidy is 'rare if present at all in *Carex*'. Thus, the cytological variabilty in the genus appears to be principally due to agmatoploid change.

In contrast, in the genus *Luzula*, there is not only evidence of orthodox polyploid series, but Nordenskiøld (1949, 1951, 1956, 1961) described in the *Luzula campestris* and *L. spicata* groups how diploid, tetraploid and octoploid races have chromosomes in descending order of size, a situation which is interpreted as being due to chromosome fragmentation, rather than multiplication of chromosome sets. More recently, Kuta *et al.* (2004) made chromosome counts and estimated DNA amounts in leaf tissue of *Luzula nivea*, *Luzula luzuloides* and *Luzula multiflora*. Chromosome numbers ranged from $2n = 8–24$ in *L. nivea* and *L. luzuloides* and $2n = 12–84$ in *L. multiflora*. Reviewing all the available evidence, in this genus, variability in chromosome numbers appeared to be the result of fission (agmatoploidy), aneusomaty (a situation where cells contain different numbers of chromosomes), chromosome fusion and polyploidy.

As in the case of dysploidy, next generation sequencing and advanced cytological techniques are likely to shed important new light on the evolution of species with holocentric chromosomes.

Speciation following hybridisation: homoploid speciation

It has been postulated that, in some circumstances, another mode of speciation occurs in plants. Hybrids between species are stabilised and become a new species, yielding a derivative with the same chromosome as its parents' that is reproductively isolated from them both (Grant, 1981; Rieseberg & Wendel, 1993; Soltis & Soltis, 2009; Abbott *et al.*, 2010). And it has been claimed that such derivatives could be the outcome of introgressive hybridisation (Rieseberg, 1997). There is an ongoing debate about the degree of parental genetic divergence (wide or narrow) found in cases of successful polyploidy and homoploidy (Buggs *et al.*, 2008).

In reviewing the evidence, Soltis & Soltis (2009) conclude that this so-called homoploid hybrid speciation 'appears to occur rarely' and is also 'difficult to detect, with perhaps only 20 good examples in the literature.' Here, briefly, we examine a number of case histories.

A. The hypothesis that *Stephanomeria diegensis* is derived from *S. exigua* and *S. virgata* has received support from studies of isozymes: markers from both parents are present in the derivative (Gallez & Gottlieb, 1982).
B. Molecular evidence supports the suggestion, based on morphological characters, that *Iris nelsonii* is

the stabilised hybrid derivative between three *Iris* species: *I. fulva, I. hexagona* and *I. brevicaulis.* With regard to its origin and characteristics Arnold & Bennett (1993) write: 'clonal reproduction by rhizomatous habit . . . may have facilitated the establishment of *I. nelsonii* ', which occupies a different niche from its parents, namely heavily shaded areas with deep water.

C. Several homoploid species have been detected and studied in detail in the genus *Helianthus.* The cultivated *Helianthus annuus* has hybridised with *Helianthus petiolaris* to give three neospecies; all are diploid derivatives in growing in more extreme and different habitats from the parental taxa. They are also reproductively isolated by chromosomal differences. Thus, *H. deserticola* occurs in very dry habitats in Nevada, Utah and Arizona; *H. anomalus* is found on sand dunes in Utah and Arizona; and *H. paradoxus* grows in saline wetlands in Texas and New Mexico. These new homoploid species, that are said to have originated between 60 000 and 200 000 years ago, have been investigated in detail by Rieseberg and associates (see Rieseberg, Van Fossen & Desrochers, 1995; Gross & Rieseberg, 2005; Rieseberg & Willis, 2007).

D. *Senecio squalidus* was brought from Mount Etna to Oxford Botanic Garden in the eighteenth century. Recent intensive field work in Italy, and investigations employing RAPD and ISSR techniques (James & Abbott, 2005), have revealed that '*S. squalidus* is genetically similar to hybrids between Sicilian natives *S. aethnensis* ($2n = 2x = 20$) and *S. chrysanthemifolius* ($2n = 2x = 20$) which form a large hybrid zone approximately midway up Mount Etna between the edge of their ranges: *S. chrysanthemifolius* occurs from sea level to *c.*700 m, while *S. aethnensis* occurs above 2500 m and the hybrid zone is found between *c.*1000–1800 m. Hybrid material on Mount Etna is interfertile with both *S. chrysanthemifolius* and *S. aethnensis,* but its removal to Britain over 300 years ago has allowed it to diverge sufficiently from its progenitors to give rise to a new homoploid hybrid species, *S. squalidus,* in allopatry'. It seems likely that the population genetics of *S. squalidus,* establishing in Oxford and subsequently spreading across Britain, has been highly influenced by founder effects, genetic drift and selection (Abbott *et al.*, 2010) and changes in gene expression (Hegarty *et al.*, 2009).

E. It has been proposed that homoploid hybrid species have evolved in other genera, for example, *Argyranthemum* (Brochmann *et al.*, 2000), *Pinus* (Song *et al.*, 2003) and *Sempervivum* (Klein & Kadereit, 2015).

Minority disadvantage

The importance of aneuploidy, dysploidy and homoploidy as modes of abrupt speciation has been clearly established. However, unless the new derivative establishes in an unoccupied specialised niche (as has been found in studies of *Iris nelsonii*), minority disadvantage must also apply, as it did in situations where new polyploids arise. Isolated individuals are at a reproductive disadvantage amongst the more numerous parental plants. How do these variants, produced in the first place as deviant individuals, come to form populations of a new species? Lewis (1973) has speculated on the likely steps involved in the evolution of species of *Clarkia,* calling the rapid process 'saltational speciation' due to 'catastrophic selection'.
The outline quoted below is from White (1978), who has condensed and slightly modified the detailed account of Lewis (1973).

> An exceptional drought reduces a normally outcrossing population to very few plants or eliminates it. The few survivors of founders that re-establish the population undergo self-pollination. Extensive chromosomal rearrangements occur; structural heterozygotes are partly sterile. Chance formation of a chromosomally monomorphic population [occurs]

from homozygous combinations of rearranged chromosomes. The 'neospecies' is genetically isolated from the parent species by hybrid sterility and because it is self-fertilizing.

Whether the process of saltational speciation, as envisaged in *Clarkia* by Lewis, is a general phenomenon in nature has yet to be determined.

Concluding remarks

The past decades have seen major advances in our understanding of abrupt speciation, by polyploidy, dysploidy and homoploidy following hybridisation.

While, for convenience, we discussed geographical divergence separately in Chapter 13, in reality, to understand the course of evolution in many lineages, the different modes of speciation must be considered together, as they are intimately connected, and may contribute to the complex patterns of variation in many groups (Tayalé & Parisod, 2013). Considering polyploidy, this involves 'mergers' of related 'more or less differentiated genomes' that earlier may have been subject to divergence under geographical isolation. These new polyploid lineages, with 'reshuffled genomes', may arise independently and repeatedly and may be partners in introgressive hybridisation with their diploid, and possibly other, relatives. If new polyploids succeed in establishing and spreading to produce isolated independent populations, another phase of divergence is likely to occur as they evolve in geographical isolation. Then, further mergers are possible in the future, if geographically isolated populations become sympatric. Moreover, changes in chromosome number may also occur. For convenience many elements of speciation have been studied in isolation. We should work towards a more holistic view of complexity of speciation.

In closing this chapter it is instructive to highlight once again the major advances in our recent knowledge of polyploidy. These are best appreciated considering the subject in its historical context. For more than 50 years ago, the writings of Professor George Ledyard Stebbins Jr, in his book *Variation and Evolution in Plants* and in other contributions, provided a basic highly influential framework of ideas about polyploidy. What follows is drawn from the major review of Soltis, Visger & Soltis (2014) that celebrates the many remarkable contributions Stebbins made to plant evolution.

Early views on polyploidy regarded it as a phenomenon of moderate frequency that was important 'over shallow evolutionary time with little long-term evolutionary impact . . . the evolutionary action was at the diploid level'. Polyploids were considered to have a single origin, resulting in plants of limited genetic variation and potential for adaptive evolution. With multiple genomes, there was considerable genetic buffering in polyploids, with low fixation of new mutations. Autopolyploidy was considered to be rare.

Now, there has been a paradigm shift in our understanding of the subject. Polyploids occur at high frequency, and ancient polyploidy is emerging as an important element in long-term evolution. Soltis, Visger & Soltis note that 'perhaps all eukaryotes possess genomes with considerable genetic redundancy, much of which is the results of (ancient) whole genome duplication (WGD)' and 'at least 50 independent ancient WGDs are distributed across flowering plant phylogenies'. WGDs are associated with diversification of the angiosperms, 'a burst of species richness appearing typically a few nodes after the WGD.' In contrast to the older views, it is now known that polyploids often have multiple origins, and there is increasing evidence of dynamic changes in polyploids, with evidence of chromosome changes, gene loss, alterations in gene expression, epigenetic and other changes. Moreover, autopolyploidy may not be rare, but common. While there is still much debate about the long-term evolutionary significance of polyploidy, it is clear that 'polyploidy can propel a population into a new adaptive sphere given the myriad changes that accompany chromosome doubling'.

The detailed evidence for such changes is presented by Soltis *et al.*, (2015) in a review honouring the scientific contributions of L. D. Gottlieb. As we have seen earlier, the researches of Gottlieb & Roose (1976) on *Tragopogon* revealed the possibility of 'novelty at the biochemical level' in newly formed polyploids. In a wide-ranging account of the evidence emerging from recent studies of this and other genera, there is increasing evidence for the 'dynamic nature of polyploid genomes – with alterations in gene content, gene number, gene arrangement, gene expression and transposon activity'. Such changes 'generate sufficient novelty that every individual in a polyploid population or species may be unique'. And 'while certain combinations of these features will undoubtedly be maladaptive, some unique combinations of newly generated variation may provide tremendous evolutionary potential and adaptive capabilities'. Also some key areas have been identified for future detailed investigation. For example, how does the alternative splicing in gene transcription function in polyploids with their multiple genomes?

Looking to the future, many important questions remain to be explored concerning the molecular genetic, biochemical, physiological and ecological changes associated with polyploidy, and the role of polyploidy in the evolution of breeding systems and isolating mechanisms. Further progress in the critical testing of the various models of speciation – geographic, polyploid, homoploid and dysploid – is certain, given the widening array of molecular tools being developed, in particular the recent development of next generation sequence methods. However, as the aim of studies in evolution is to understand species and speciation in nature, let us hope that the intensive laboratory-based approaches will be combined with detailed studies of wild plants. Are the processes of change and speciation, suggested by experiment, 'active' and important in the wild?

The importance of close examination of wild polyploids is emphasised by two recent insights. As we have seen, '*Tragopogon mirus* formed independently several times from *T. dubius* and *T. porrifolius*, while *T. miscellus* formed multiple times from *T. dubius* and *T. pratensis*' (Tate *et al.*, 2009). However, 'only *T. miscellus* has formed reciprocally in nature, and these reciprocally formed individuals can be distinguished morphologically'. Studies of wild populations have revealed that plants of *Tragopogon miscellus* have short- and long-ligule variants. Further research in which F_1 hybrids were made between the putative parents, followed by colchicine treatment, has confirmed the proposed origin of *T. miscellus* and *T. mirus* (Tate *et al.*, 2009) and revealed that 'the "short-liguled" form of *T. miscellus* has *T. pratensis* as the maternal progenitor, while the "long-liguled" form has *T. dubius* as the maternal parent'. The long-ligule variant is featured on the front cover of this book. In contrast, following colchicine experiments between the parental diploids, 'synthetic *T. mirus* was also formed reciprocally, but with no obvious morphological differences resulting from the direction of the cross'.

In another fascinating study (Lipman *et al.*, 2013), semi-fertile natural hybrids ($2n = 24$) have been discovered between *Tragopogon miscellus* and *T. mirus*. Five plants from Pullman have been investigated using advanced FISH and GISH techniques. 'None showed an additive F_1 chromosome complement, i.e. two sets of chromosomes from *T. dubius* and one set of chromosomes each from *T. porrifolius* and *T. pratensis*. No individuals shared an identical karyotype'. Lipman *et al.*'s detailed results suggest that the plants are moving away from a 'parentally additive F_1 karyotype to chromosome compositions that are mostly, or entirely, disomic'. There is therefore the possibility that eventually there may be 'elimination of chromosomes' and the production of 'a stabilized karyotype distinct from both allopolyploid parents'. These findings add another element of complexity to the patterns of reticulate evolution in the group, and show the importance of employing intensive laboratory investigations to study microevolutionary patterns and processes in wild populations.

15 The species concept

The 'species concept' in biology, as we saw in Chapter 2, has engaged the attention of most of the famous naturalists in earlier centuries, from Ray through Linnaeus to Darwin himself, and debate, argument and controversy continues to the present day (see Rieseberg, Wood & Baack, 2006; Wilkins, 2009; Mischler, 2010; Kunz, 2012).

Species as part of natural classifications

Kinds of organisms can be arranged in a hierarchy, each higher group containing one or more members of a lower group. A study of primitive biological classifications reveals that hierarchical classifications are a widespread feature of languages in general, and the particular hierarchy of genus and species, which modern biology uses, is really only a special case of a general linguistic phenomenon. The inescapable conclusion from such comparative studies is that hierarchical taxonomic classifications arise in human societies wherever they develop, and that the detail of the treatment found in folk taxonomies reflects the importance of the organisms concerned in the life of the particular tribe or culture. This view of taxonomy as the product of the human need to understand, describe and use the plants and animals applies to the Latin biological classification we use today.

Although we talk of associating species into genera and genera into families, that is not what happened in the early days of biological classification. It is indeed arguable that the idea of a kind of plant or animal envisaged by the ordinary person corresponds, in the history of classification, more closely to the modern genus than it does to species. It was these classical and medieval ideas of 'kinds' of plants that were available in the eighteenth century to Linnaeus, who stabilised the scientific names in the binomial form in which we still use them. So the Linnaean names (e.g. *Quercus* = Oak and *Fraxinus* = Ash) indicate the level of recognition of 'kinds' in the botany of medieval Europe. This is beautifully illustrated by the Carrot family (Apiaceae/Umbelliferae), many of which were familiar plants in classical times in Europe, mainly because they were cultivated for food or flavouring (e.g. *Daucus* = Carrot, *Pastinaca* = Parsnip) or because they were poisonous (e.g. *Conium* = Hemlock). All these familiar European plants were accurately described and given what later became their generic names, long before Linnaeus. For instance, Morison published a monograph on the Umbelliferae in Oxford in 1672. Then, as the exploration of the plants of the world proceeded, Linnaeus and his successors distinguished other 'kinds' of Oak, Ash, Carrot, etc., retaining the name of the genus for all the species so distinguished (Walters, 1961, 1962, 1986b).

Returning to the history of the species concept outlined in Chapter 2, and oversimplifying greatly, we could say that both Ray and Linnaeus thought that the genus and the species were 'God-given', a part of the natural order which could be named and described. Darwin, impressed by variation and change in evolution, saw both genera and species as mere convenient abstractions. Risking a generalisation, today the majority of biologists see the genus as an abstraction, but the species as 'real'. However, it is important to dissect the issues involved. In an important review, Stuessy (2009) recognises three different kinds of reality: mental, biological and evolutionary.

Mental reality of species

Many logicians and philosophers reflecting on the 'species' controversy tend to be impatient; they dismiss arguments about the reality of taxa as being conducted naively and in the wrong framework of conceptual thought, and point to the universality of the problem. The issue can be re-framed along the following lines. When we look at nature, are the 'units' we recognise and name already there to be recognised, or have we 'made' them in the process of looking? The 'naive realist' view is that they *are* all out there; but it does not need much training in the philosophy of science to see how unsatisfactory this view must be. Recognising even an individual tree or dog is a complex mental process, which cannot be independent of our previous experience. How much more so must the recognition of genera and species be 'subjective' in this sense. We *learn* to recognise both genera and species (and, for that matter, families as well): even if we wanted to, we could not begin without the accumulated wisdom and experience of our ancestors. Thus, biologists generally accept that species and other taxa have mental reality. However, for some, this is the *only* reality. For instance, the famous American botanist Bessey (1908) wrote: 'species have no actual existence in nature. They are mental concepts, and nothing more. They are conceived in order to save our selves the labor of thinking in terms of individuals.' Agnes Arber (1938), the celebrated plant morphologist, who examined the development of concepts employed by the writers of the sixteenth-century Herbals, wrote:

> The progression from the vague concepts of earlier writers to the sharp definition of genera and species to which we are now accustomed, has been in some ways a doubtful blessing. There is today . . . a tendency to treat these units as if they possessed concrete reality, whereas they are merely convenient abstractions, which make it easier for the human mind to cope with the endless multiplicity of living things.

Gilmour & Walters (1963) stress the same point when they forcefully argue that the appropriate question to confront is 'how many species is it convenient to recognise' in the taxonomy of groups of species.

Do species have biological reality? Considering whether species do indeed have biological reality, evidence from the study of 'folk taxonomies' is seen, by some, as important. These are the classifications made, quite independently, by aboriginal peoples isolated from the main cultural influences. If it can be shown that the biological units that are named and classified by a particular tribe in an obscure language are roughly equivalent to those recognised in modern biological taxonomy, the case for the 'biological reality' of taxa is strengthened. Gould (1979) surveys the results of a number of studies of this kind and is particularly impressed by the work of Berlin, Breedlove & Raven (1974, and earlier references cited), who investigated and analysed the plant classifications of the Tzeltal Indians of Mexico. These authors initially interpreted their data as showing relatively poor correspondence between folk names and Linnaean species but, after a more extensive study, which recognised unsatisfactory procedures in their earlier work, Berlin reversed his earlier view and agreed that 'there is a growing body of evidence that suggests that the fundamental taxa recognised in folk systematics correspond fairly closely with scientifically known species'. Correspondence between folk names and Linnaean species is especially close with higher animals, less close with angiosperms and relatively poor with 'lower' organisms, whether animals or plants. However, Stuessy (2009) considers that 'one might argue from these correspondences that people, whether culturally progressive or aboriginal, view the world in the same way'. However, this might 'tell us much about the cognitive powers of the human population but nothing about the biological reality of species'. For Stuessy, the critical observation is that 'species of organisms recognise each other in a consistent way and that the resulting life-forms are not completely continuous'. Therefore, he concludes that 'it seems obvious to me that species do have reality based on finding phenetic gaps in the living world that correspond largely to our formal species designations'.

Do species have evolutionary reality?

Given all the post-Darwinian evidence that has accumulated, Stuessy (2009) concludes that if species 'are admitted to be biologically real, then they must be judged to be evolutionarily real. They are clearly *products* of the evolutionary process'. Turning to another question, he queries whether 'species are necessary for evolutionary theory' and concludes that 'it is *convenient* to talk about the origin of new species when discussing the production of new diversity . . . but it is not essential for evolutionary theory'. The process *could be* formulated in terms of populations and the evolution of reproductive isolation alone, but clearly, as the study of evolution has developed, species and speciation are key elements in conceptional models.

Retreating from this complex, speculative field we now consider the current status of the Biological Species Concept. This concept, formulated many years ago, has proved highly influential in experimental studies of speciation (see Chapters 13 and 14). Here, we discuss the concept in relation to taxonomic investigations.

The Biological Species Concept

As we saw in Chapter 12, Mayr (1942) produced the most often quoted definition of the biological species. Biological species are 'groups of actually or potentially interbreeding natural populations which are reproductively isolated from other such groups'.

Many zoologists have followed Mayr in equating the taxonomic with the biological species in a variety of groups of animals. As we have seen, the Biological Species Concept has also been applied to plants. Clearly, the Biological Species Concept has been a most important model, drawing attention to the crucial role of isolating mechanisms in speciation. Indeed, some botanists have wished to replace the morphological species concept with that of Mayr.

How far has this endeavour been successful? Reviews of the concept are provided by Sokal & Crovello (1970), Raven (1976), Holsinger (1984), Jonsell (1984), Ruse (1987), King (1993), Stuessy (2009), Wilkin (2009) and Kunz (2012). Mayr (1982) has provided a further detailed account of the concept, but in our view he has not succeeded in silencing his critics, who, as we shall see, point to major problems in applying the concept in botanical taxonomy.

Jonsell (1984) regards the concept 'as a hypothesis open to testing'. He asks the very important question, 'has this hypothesis been carefully and unconditionally scrutinised? Or has this model been taken for granted to such an extent that there has been more of a trend to seek its confirmation, to make it fit the scheme, and to regard nonconforming observations as exceptions or anomalies?'

Detailed analysis of the Biological Species Concept in terms of its practical value in taxonomy has been provided in the classic paper of Sokal & Crovello (1970), who force attention on the impracticability of actually applying the concept in any concrete case. They consider each word or phrase in Mayr's definition.

'Groups' At the outset, setting the limits of the groups to be investigated, presents difficulties. How many taxa do we involve in the experimental study? In the case of *Elymus* investigated by Snyder (1951), as we saw in Chapter 13, hybrids between *E. glaucus* and species belonging to other genera such as *Agropyron, Hordeum* and *Sitanion* may well have influenced the variation pattern (Mason-Gamer, 2004; Tang, 2008). We have no way of telling, by looking at the taxonomic information before an experimental study is begun, where we should place the limits.

'Actual interbreeding' While it would be essential and prudent in any investigation to check any cases of presumed hybridisation in the wild, 'interbreeding' is often tested by setting up artificial crossing experiments. Many experiments have been carried out, the results of which have in some cases been set out as crossing polygons. It is clear from our previous comments that such crossing experiments provide information on interbreeding under some particular experimental conditions. How far can generalisations be made about groups from the reproductive behaviour of small samples? How far can experimental investigations reveal the likely behaviour in the field? Furthermore, they do not provide a proper test of whether pre-zygotic mechanisms might operate in nature.

'Potentially interbreeding natural populations' This phrase presents obvious difficulties. How does one proceed if two populations are allopatric? Holsinger (1984), summarising Mayr's views, states that 'To decide whether two allopatric populations that are somewhat different morphologically are part of the same species ... we assess the degree of difference between closely related non-interbreeding entities that occur sympatrically and determine whether the differences are less or greater than those between allopatric populations. If less, we regard them as the same species: if greater, as different species.' For many botanists this type of reasoning is not a proper basis for testing hypotheses, and 'potential interbreeding' is seen to be an unworkable criterion. In fact, in response to criticism, Mayr removed the words 'actually or potentially' in a later rewording of his definition (Mayr, 1969).

'Reproductively isolated from other such groups' Finally, how is 'reproductive isolation' to be assessed? Inferences have often been made. For example, as we saw in Chapter 12, breeding barriers have been detected between diploids and their related polyploids, and have been inferred in other cases where a polyploid series is found. While it is clear that there may be a minimum of two biological species – a parental diploid being in one biological species, while an allopolyploid derivative is in another – can we be confident that this is the full picture in the absence of any experiments? Clearly, it is possible that each may contain a number of biological species.

In the assessment of reproductive isolation, experiments have often been carried out to supplement the evidence from field and herbarium studies. But, in interpreting crossing experiments, what level of pollen stainability of F_1 and F_2 hybrids indicates an effective barrier to crossing: 50%: 5%? Clearly, subjective decisions enter into the assessment. It is not at all surprising that some botanists, who have produced elaborate 'crossing polygons' summarising their results, have not tried to recognise 'biological species' in their material.

The views of botanical taxonomists

While many zoologists have to a large extent accepted the Biological Species Concept for taxonomic studies, botanical authors have retained a healthy scepticism. Stace (1980) is perhaps typical. He writes, 'there have been many attempts to define a species, none totally successful', and proceeds to list four criteria that are used either singly or together by 'most taxonomists'. Only one of these is concerned with interbreeding, and even this has to be qualified at the outset with the phrase 'in sexual taxa'. Like most botanists, Stace recognises it as 'a fact of life' that our species will be 'equivalent only by designation and must therefore be regarded to a considerable degree as convenient categories to which a name can be attached'. He goes further: 'it is not realistic to consider species which are well differentiated on phenetic, genetic and distributional grounds as ideal or normal, and those whose taxonomic recognition poses great difficulty as non-ideal, abnormal or atypical'. This is a clear rejection of all attempts to apply a biological species concept in the taxonomic context. For a well-documented review of the controversy, which comes down uncompromisingly against any single species definition, we recommend the paper by Levin (1979). Heywood (1980) expresses a similar view.

Not all botanists would agree with Stace, Levin and Heywood. Löve over many years argued for a biological species concept to apply to all organisms, and frequently acted upon his principles in, for example, elevating to the rank of species variants that differ cytologically even if their morphological differences are negligible. However, with regard to apomictic taxa, even Löve, was obliged to recognise that they cannot be dealt with in this way. He says (1962), 'to classify them on the basis of reproductive isolation would lead to a confusion even greater than that created by the morphological–chorological method of study of these groups, since every individual is reproductively isolated from all its relatives'. When Löve wrote this opinion, it was assumed that apomictic plants were obligately so, and therefore the concept of biological species could not be applied. However, ideas about apomixis have now changed. As we saw in Chapter 7, it is now generally believed that obligate apomixis is rare and that many agamospermic groups are facultatively apomictic, and thus able to interbreed to some extent. Logically, some apomictic plants fall within the Biological Species Concept, and we can immediately appreciate the complexities involved in trying to define biological species in such groups.

With regard to species definitions in practical taxonomy, our own views are close to those of Stace (Briggs & Walters, 1997). Neither in theory nor in practice can we adopt as our definition of species any single criterion or even group of criteria.

The taxonomic process provides us with a hierarchical system of categories by means of which we can name, and therefore discuss, our material. It is not reasonable to assume that taxa must be equivalent: nothing in nature looks simple and there is no reason why we should expect it to be simple. Of course we tend to look for simplifying hypotheses, which enable us to understand what were previously independent phenomena, but we should not complain when the phenomena remain diverse.

Without doubt, those studying evolution will surely continue to employ the Biological Species Concept in their model-building. But, as the records in databases of current literature reveal, experimentalists are fascinated in testing hypotheses concerning the patterns and processes involved in speciation. They are not trying to 'perfect' taxonomy by defining biological species to replace the species of the taxonomist.

Different definitions of species

In closing this chapter, it is important to reflect once again on the different definitions of species devised by biologists (see Rosenberg, 1994; Ereshefsky, 2001). Some take the view that progress in biology may eventually make it possible to arrive at a definitive single correct species concept. But given the complexities of nature, the limitations of human cognitive abilities and the range of different endeavours in which species definitions have been devised – taxonomic, ecological, genetical, phylogenetical etc. – many biologists consider that the quest for a single species concept is misguided and futile. A powerful case can be made that we should accept the plurality of different entirely legitimate species concepts, as biologists investigate the amazing diversity of the microbial, fungal, animal and plant life.

16 Flowering plant evolution: advances, challenges and prospects

In a famous section of the *Origin* Darwin (1859) speculated on the evolution of the variety of organisms. He wrote:

> The affinities of all the beings of the same class have sometimes been represented by a great tree. I believe this simile largely speaks the truth. The green and budding twigs may represent existing species; and those produced during each former year may represent the long succession of extinct species . . . The limbs, divided into great branches, and these into lesser and lesser branches, were themselves once, when the tree was small, budding twigs; and this connexion of the former and present buds by ramifying branches may well represent the classification of all extinct and living species in groups subordinate to groups. (Darwin, 1859).

With regard to the 'testing' of this model, in a letter to T. H. Huxley, dated 26 September 1857 he expressed the view that: 'The time will come, I believe, though I shall not live to see it, when we shall have very fairly true genealogical trees of each great kingdom of Nature' (Darwin & Seward, 1903).

The devising of phylogenetic trees

Spectacular progress has recently been made in the study of phylogenetic trees from the DNA sequences of living organisms. In order to understand these powerful techniques it is important to consider the historical context in which they developed. But first, we examine the different sorts of classifications that biologists make, and consider the extent to which they might reveal evolutionary pathways.

Classifications

Gilmour (1937, 1940, 1951; Gilmour & Walters, 1963) has stressed that different classifications have been developed for different purposes. Two types of classification may be devised. First, there are the special-purpose classifications, sometimes called artificial classifications, which are based on one or a few characters. Thus, plants may be divided into trees, shrubs and herbs, the characters height and woodiness having been chosen *a priori*, i.e. before the assignment to class was made. As we saw in Chapter 2, Linnaeus produced a famous classification, his so-called Sexual System based on the number of parts of the flower (the number of stamens and the number of pistils). In this classification species of different families were placed in the same group. The only prediction we can make is that the members of each class are alike in the chosen characteristics.
An artificial classification, based on, for example, height and woodiness, is of course useful in some contexts, e.g. for gardeners, or ecologists interested in the distribution of the different life-forms.

By examining the same set of organisms, a natural classification may be devised, which groups together plants having many attributes in common, to provide a general-purpose classification that serves a much wider range of botanical purposes than the special-purpose one. Gilmour notes that the high information content of such a classification allows many predictions to be made about the properties of members of its constituent groupings. In the post-Linnaean period, natural classifications were increasingly developed using the hierarchical arrangement of Division; Class; Order; Family; Genus; Species; Subspecies; Variety; and Form. Additional intermediate categories were also introduced: Subfamily; Subgenus; Section; and Tribe (Stuessy, 2009). The basic concept underlying this type of classification is that of a nested system of classes within classes, which can also be visualised as an arrangement of group in group, box in box, parts in wholes (Stuessy, 2009; Stevens, 2012). A 'flagged system' is employed in which the 'rank terminations' used (categories ending in -ales, -aceae etc.) indicate the position of groups in the hierarchy (Steven, 2006). Such a scheme is 'useful in memory and communication. It improves memorisation by tapping in to the hierarchical structure of language (an extension of the noun–adjective structure of binomials).'

For reviews of the categories accepted by the *Code of Nomenclature*, see McNeill *et al.* (2006) and Brummitt (2006). In what follows we, somewhat reluctantly, refer to contemporary taxonomic practice as Linnaean. In Brummitt (1997) it is made clear that 'the structure of traditional biological classification is in fact based on Aristotelian logic (see Davis & Heywood, 1963). Apart from his emphasis on and re-evaluation of genera, Linnaeus, himself, contributed little to the conceptual development, *per se*, of biological classification.'

A good deal of the taxonomic framework we use today is pre-Darwinian (Cain, 1958). It has been argued that, in the post-Darwinian period, the craft of taxonomy was not affected; biologists continued to fit newly discovered organisms into an increasingly natural system. As before, close study of the material allowed taxonomists to use their expertise and intuition in recognising taxa and placing them within the classificatory system. However, increasingly the classification itself was reinterpreted and revised in order to reveal the presumed phylogenetic pathways (Davis & Heywood, 1963; Stuessy, 2009). Despite the poor fossil record of the flowering plants, the idea that there is only one true phylogeny is a lure that few biologists can resist (Sneath, 1988) and, in an attempt to find this 'true' phylogeny, many different arrangements of groups have been proposed at every level of the taxonomic hierarchy. For many biologists, the practice of taxonomy seemed to require its practitioners to produce phylogenetic schemes for every group. Indeed, some took the view that phylogenetic studies were a necessary route to classification. For example, Holttum (1967) wrote: 'if organisms have reached their present state by a process of evolution, it follows that they have a built-in classification, and man's problem is to find it. This is quite a different problem from that presented to the classifier of man-made objects.'

Here, we note two highly influential historical classificatory systems that were based on different views of the structure of the primitive angiosperm flower (see Davis & Heywood, 1963, for further details and references to the German literature).

(i) The Englerian approach proposed that the ancestral flower was unisexual, had no petals and was wind pollinated, as in the catkin-bearing *Betulaceae, Fagaceae* etc. This view was adopted by the celebrated German botanist Engler and set out in the *Syllabus* (1892) and the *Pflanzenfamilien* (1900 onwards).
(ii) In contrast, the Ranalian theory proposed that the primitive flower was hermaphrodite, and insect pollinated with numerous free petals and other parts.

In developing classifications from these two and other different starting points, taxonomists devised groups that were not avowedly phylogenetic, but arranged in a 'step-wise manner to form a generally progressional series' (Davis & Heywood, 1963). For example, the schema developed by Engler and Prantel proposed a series developing from naked catkin-like flowers → bract-like outer flower parts → development of petals and sepals → union of petals. Several evolutionary trends were also recognised by Bessey (1915), with others proposed by Hutchinson (1926, 1959). Twenty-nine of these judgements are listed in detail in Davis & Heywood (1963). The following are examples of the trends recognised: many floral parts preceded few; flowers with petals preceded those without petals: flowers with many parts are more primitive than those with few parts; having radially symmetrical flowers (actinomorphy) is a more primitive condition than bilateral symmetry (zygomorphy); having solitary flowers is more primitive than grouping flowers into inflorescences; and hermaphrodite flowers precede unisexual.

In the twentieth century, many different phylogenetic diagrams were developed by botanists (Bessey, 1915; Takhtajan, 1969, 1987; Stebbins, 1974; Dahlgren, 1987; Cronquist, 1988). Some were tree-like (Fig. 16.1), others networks.

Judd *et al.* (2007) make an important criticism of such portrayals. 'Extant groups were linked directly to other extant groups. (These diagrams are known as "minimum spanning-trees".) Such diagrams imply that groups that exist today are the ancestors of other groups that also currently exist, which doesn't make much sense in terms of evolutionary processes. A more evolutionary view would be to say that two extant groups are descended from a single extinct ancestor.' However, in defence of the earlier approaches, ancestor–descendant relations are problematic, where there is an imperfect fossil record.

Weighting characters

In arriving at different classifications taxonomists have always been selective in the characters they use, and such selectivity is known as weighting. Such weighting has always been practised in making classifications (Davis & Heywood, 1963; Stuessy, 2009), but in the post-Darwinian period it has often involved characters thought to be important indicators of evolutionary pathways. For example, recognising that floral parts are less variable than vegetative structures, taxonomists often weighted these characters *a priori* in constructing classifications and phylogenies. Sometimes, after a thorough study of the material, some characters – often those of presumed importance in evolution – were given special emphasis; this is so-called *a posteriori* weighting. Finally, some characters were not used at all, effectively weighting those that were used. In the post-Darwinian period many classifications, seeking to express evolutionary developments, have actually involved one or a few characters, and are therefore in Gilmourean terms artificial classifications. Put simply, two positions have emerged since the 1940s.

1. There are those who consider that, with increasing information, classifications may be successively modified from the earliest 'alpha-' to a more perfect 'omega-taxonomy', reflecting the way the plants evolved. Such a view was presented by Turrill (1940) in an influential volume of essays published in *The New Systematics* edited by Huxley. Stace (1989), for example, states that 'an omega-taxonomy' is almost by definition unattainable, but it 'is the distant goal at which taxonomists should aim'.
2. However, there has been serious criticism of this idea. As we have seen, generations of biologists have shaped a natural classification for organisms, and this broad map of variation is essential for biological science. 'Correct' and stable names are necessary as a means of communication and information retrieval, from books and databases, and frequent changes seriously impair the usefulness of the system. These pragmatic considerations were stressed by Gilmour and his colleagues (Gilmour & Gregor, 1939; Gilmour & Heslop-Harrison, 1954; Gilmour & Walters, 1963; Cullen & Walters, 2006).

Discussing a debate that has its roots in the last century might suggest that the issues involved are only of historic interest. But, as we shall see, we are dealing with what are still live controversial issues about the future development of taxonomy and the role of molecular phylogeny in classifications (Wheeler, 2008; Stuessy, 2009).

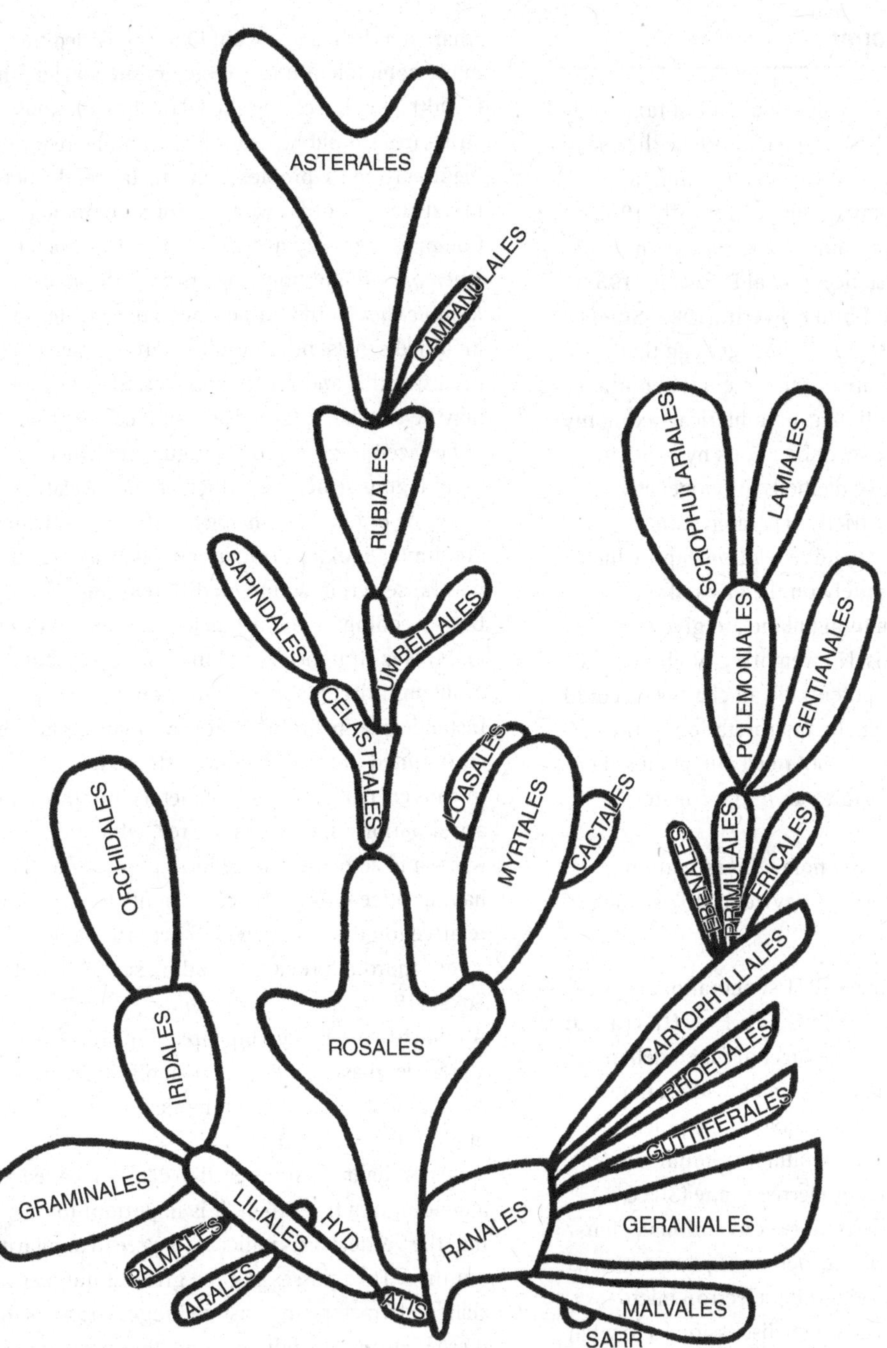

Fig. 16.1. Chart showing Bessey's views on the relationship of the orders of flowering plants. Abbreviations: ALIS = ALISMATALES; HYD = HYDRALES; SARR = SARRACENIALES. Relationship is indicated by position; the areas are approximately proportional to the number of species in the orders. (From Bessey, 1915.)

Numerical taxonomy

Returning to the historical development of taxonomy and phylogenetic studies, in the mid-1960s the use of computers by biologists increased and numerical approaches to taxonomy developed (Sneath, 1962, 1995). The following account draws on a number of major reviews of the subject (Sokal & Sneath, 1963; Sneath & Sokal, 1973; Dunn & Everitt, 1982; Stuessy, 1990, 2009; Pankhurst, 1991). Recognising that natural classifications are of fundamental importance to biologists, the practitioners of numerical taxonomy sought to produce such classifications by explicit, repeatable and objective methods. Phylogenetic speculation was to be strictly separated from procedures devised to examine relationships, which were to be evaluated purely on the basis of resemblances between living plants, to give so-called phenetic classifications. No weighting of characters was to be allowed and potentially all characters could be considered. Phylogenetic interpretation of the resulting relationships was not ruled out, but it did not play a part in the procedures used to estimate relationships.

A typical numerical taxonomic investigation involves the following steps (Heywood, 1976), each of which raises important issues.

1. A number of specimens (OTUs; operational taxonomic units) are chosen for study. How OTUs are to be chosen has been the subject of much debate.
2. The material is scored for a much wider range of characters than is usual in taxonomic practice. A total of 60 characters would be a minimum number, 80–100+ the preferred range (Stace, 1989). 'New', as well as more conventional, characters are considered (e.g. details of plant and seed surfaces revealed by scanning electron microscopy). There has been much discussion as to what constitutes a character. Numerical taxonomists attempted to use unit characters, i.e. characters that cannot be logically subdivided. Thus, the 'character' leaf shape would be subdivided into many component unit characters such as length, breadth, etc. In scoring the OTUs, homologous structures should be examined, and this requirement may cause problems. For instance, the petal-like structures of flowers are not all homologous. Comparative study has revealed that the 'petals' of *Anemone* are not homologous with those of *Ranunculus*. In the former they are regarded as modified sepals, but not in the latter. Stace (1989) discusses this and other examples. Stuessy (2009) may be consulted for a discussion of the problems to be faced in trying to distinguish between homologous structures, which are modifications of the same organ, and analogous structures, which are similar-looking structures evolved as a result of natural selection acting on different organs.
3. Binary coding of characters (+/–) is the next step and this is straightforward in many cases, but problems arise in some circumstances. For instance, plants of different flower colour may be represented in the OTUs under study. It is possible to convert red, white or blue petals into three binary situations: red versus not red, white versus not white and so on, but some have pointed out that having three binary choices weights flower colour relative to other characters (Stace, 1989). For a very thorough review of coding see Sneath & Sokal (1973).
4. A table of data is then prepared (Fig. 16.2) as a prelude to assessing overall similarity, using one of a number of different numerical means of estimation (Fig. 16.3). The results are displayed as a cluster diagram or a tree-like diagram called a dendrogram (Fig. 16.4). It is important to note that the vertical axis indicates degree of similarity, not time. Taxa possessing the greatest number of shared characters are clustered together: 'none of these [features] is individually either necessary or sufficient to define the group' (Stace, 1989).

The different ways of estimating similarity led to different dendrograms.

5. In making a classification from the resulting dendrograms, numerical taxonomists recommend that ranks appropriate to the different levels of similarity be chosen.

Characters	Taxa (OTUs)			
	A	B	C	D
1	+	+	−	NC
2	+	+	+	+
3	+	+	+	−
4	−	+	NC	NC
5	+	+	+	+
6	+	+	−	+
7	+	+	−	NC
8	NC	−	+	+
9	+	+	+	+
10	+	+	+	−
11	+	NC	−	NC
12	+	+	+	−

Fig. 16.2. Coded data table ($r \times n$ table). See text for details. (After Sneath, 1962, from Heywood, 1976.)

The influence of numerical taxonomy

In the development and refining of numerical taxonomic methods for estimating and displaying relationships, many research papers have been published. But, it is now clear that such approaches do not provide a practical alternative to traditional taxonomy (see Stuessy, 2009). While the direct impact of numerical taxonomy may be slight, nevertheless it caused biologists to reflect critically on the concepts and working methods used by taxonomists, and it also provoked much discussion about the relationship between classification and phylogenetics. However, there has been one highly influential consequence: some of the practitioners of numerical investigations were deflected from the strict confines of their original subject to begin, with others of different outlook, to devise, analyse and test models of phylogenetic pathways.

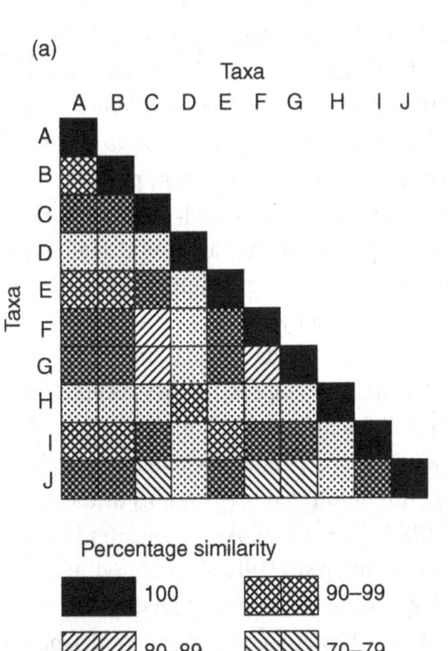

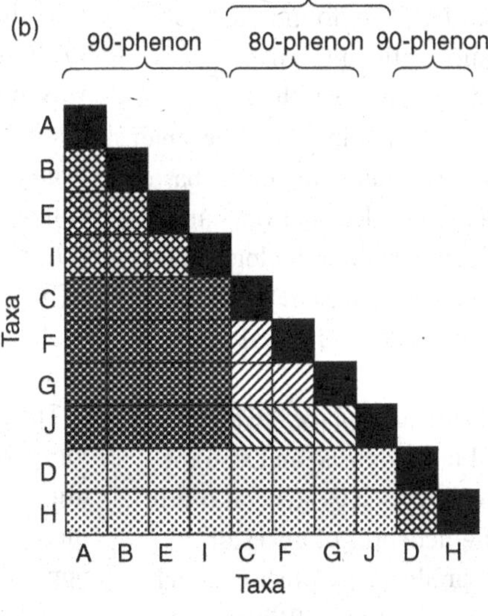

Fig. 16.3. (a) Schematic diagram showing a matrix of hypothetical similarity coefficients between pairs of groups (taxa); the magnitude of the coefficients is shown by the depth of shading. (b) The same coefficients arranged by placing similar taxa next to each other; this gives a triangle of high similarity values. Phenons are groups of desired rank (After Sneath, 1962, from Heywood, 1976.)

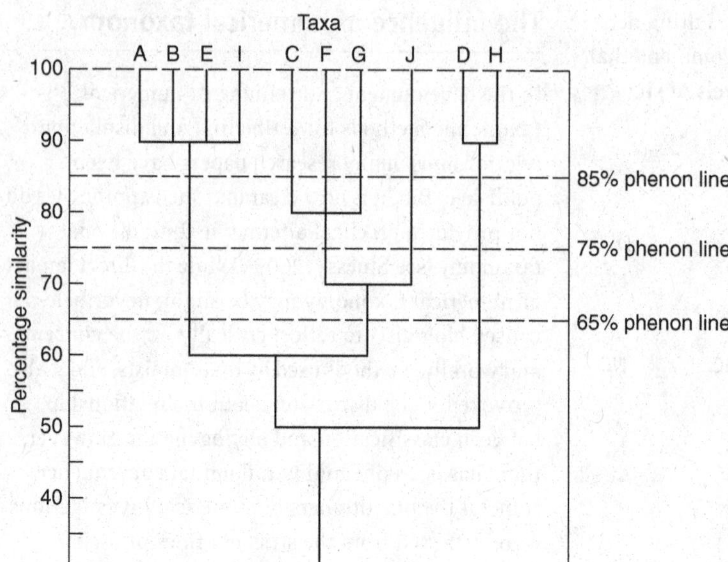

Fig. 16.4. A dendrogram representing the hypothetical hierarchy of groups (taxa), based on the data in Fig. 16.2. The ordinate indicates the magnitude of similarity coefficient at which stems join to form higher-ranking groups. Horizontal lines delimit groups of equal rank (per cent phenon lines). (After Sneath, 1962, from Heywood, 1976.)

Cladistics

In 1950, Hennig published a highly influential volume, *Grundzüge einer Theorie der phylogenetischen Systematik*. Many found the book very difficult and it was not until an English translation was produced in 1966, with the title *Phylogenetic Systematics*, that Hennig's views became more widely known. Hennig set out many of the basic principles of what Mayr (1969) called cladism, but it is important to recognise the role of other biologists; for example, the botanists Zimmermann (see Donoghue & Kadereit, 1992) and Wagner also contributed to the development of cladistics.

Cladistic investigations aim to replace the phenetic schemes of the numerical taxonomists, and also the intuitive phylogenetic trees devised by generations of biologists, with branching networks of ancestor/descendant relationships produced by precise models with clear assumptions and procedures (Wiley, 1981; Simpson, 1986; Stuessy, 1990, 2009; Forey, Humphries & Kitchen, 1993). At the heart of these procedures was a precise model of the speciation process: species A, by 'cleavage' giving rise to two new species B and C (Fig. 16.5).

Hennig's approach also introduced the concept of parsimony, which considered that the shortest hypothetical pathway of changes provided the most likely route of evolution. He stipulated that the aim was to identify natural clades (monophyletic groups). 'A clade is defined as a group of all the taxa that have been derived from a common ancestor, plus the common ancestor itself' (Graur & Li, 1999). However, it is important to stress that not all 'useful groupings of organisms are clades. For example, fishes and protozoa are not monophyletic but are seen as convenience groups or grades' (Stace, 1989).

The first cladistic investigations of plants employed morphological characters and led to a great deal of argument, sometimes acrimonious, about the aims, methods, interpretation and implications of these approaches (Patterson, 1987). As we see later in the chapter, it was from these cladistic investigations that modern molecular investigations of phylogeny developed. 'By the 1980s, statistical and model-based methods of inferring phylogenetic trees began to

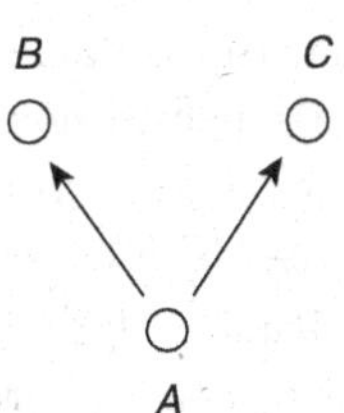

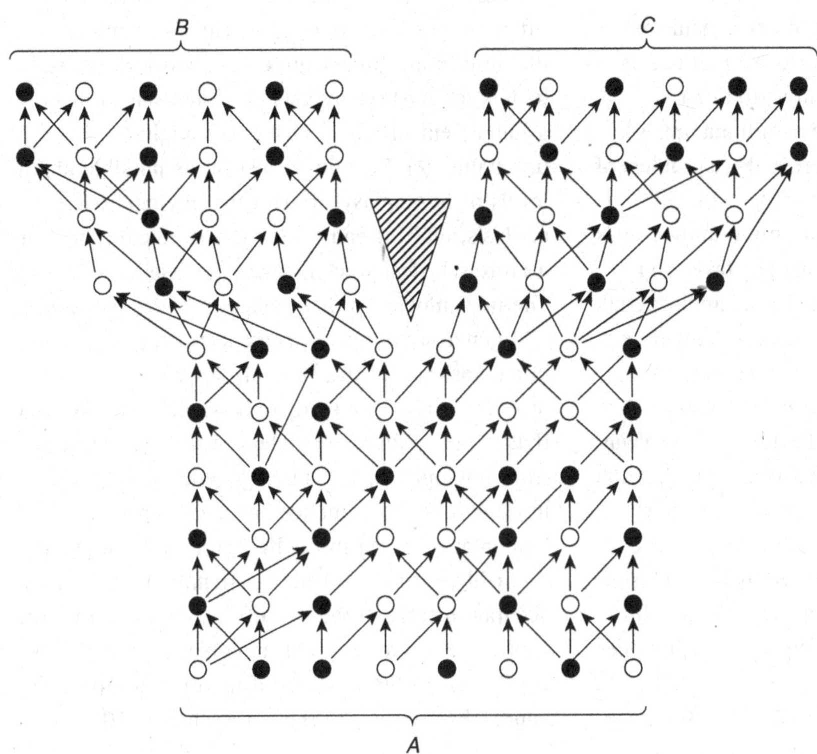

Fig. 16.5. Hennig's views on the process of species formation by 'cleavage' – species A giving rise to two new species B and C. (From Hennig, 1950.)

provide an alternative perspective to cladism' (Pagel, 2002).

Examining first the historic investigations of morphological characters, Stace (1989) emphasises that 'cladistics is basically a methodology that attempts to analyse phylogenetic data objectively (and hence produce an objective phylogenetic classification)'. In considering the phylogenetic history of a given group of organisms, characters are identified for which it is possible to determine primitive and advanced character-state. Then, for the group under study, the distinction is made between those characters that are *primitive and shared* in common between all the members of the group of taxa and which characters are *derived character states common to some but not all the taxa being investigated.* It is this latter group of derived character states that enables cladograms to be constructed, This approach, in its use of selected characters, is clearly different from the numerical taxonomic methods (discussed above), which avoid 'weighting' and employ as many characters as possible.

Considering this first step in the analysis, it is important to concede that the determination of advanced/derived and primitive states in organisms is not at all straightforward and may be controversial. How can the polarity of change be established: which character states are primitive and which derived/advanced?

First, it is important to stress that the terms primitive and derived are relative terms and the assessment of status depends upon the organisms under study. For example, the possession of hair is a derived state for mammals relative to other vertebrate groups, but primitive within mammals. In addition, as Stace stresses, there is also a problem of ensuring that homologous structures are being considered. For example, he points out that the leaves of different angiosperm groups may be taken to be homologous, but are those of a moss and angiosperm? He also emphasises the danger of false arguments, e.g. (a) common equals primitive; and (b) characters correlated with primitive states are themselves primitive. Two important lines of evidence offer a way out of this difficulty. First, in some cases, fossil plants of different ages may provide evidence of which characters are primitive. But usually, in cladistics, a second line of evidence is provided by the inclusion of one or more outgroups in cladistic investigations. These provide the means of rooting phylogenetic trees (see below).

Other challenges have to be faced in investigations of phylogenetic origins (Hall, 2011). Misleading evidence is to be expected.

> The basic premise of parsimony is that taxa sharing a common characteristic do so because they inherited that characteristic from a common ancestor. When conflicts with that assumption occur (and they often do), they are explained by reversal (a characteristic changed and then reverted back to its original state), convergence (unrelated taxa evolved the same characteristic independently), or parallelism (differing taxa may have similar embryological mechanisms that predispose a characteristic to develop in a certain way). These explanations are gathered together under the term homoplasy. (Hall, 2011)

Sanderson (2002) makes the very important point that 'a supposed homology' could 'actually be a homoplasy'. He considers four strategies to minimise the effects of homoplasy in phylogenetic investigations. (a) Very close study of the plants being investigated. 'Metaphorically speaking, increasing the "magnification on the microscope" often converts apparent similarities to evident dissimilarities.' (b) Examine the historical evidence (employed by taxonomists, systematists) to make a judgement of which characters might be unreliable. (c) 'Remain as neutral as possible about levels of homoplasy prior to the phylogenetic analysis, and instead let the evidence contained in a large set of characters speak for itself. The presumption is that evidence in the homologies will generally outweigh evidence in the homoplasies, and the homoplasies will be discovered after the fact.' (d) It is sometimes asserted that 'a newly discovered class of data is simply free from homoplasy', but 'phylogenetic analysis of multiple lines of evidence has always revealed homoplasy lurking in the background. Undoubtedly, this will continue to bedevil attempts to obtain an accurate picture of the tree of life.' It is important to emphasise, therefore, that homoplasy must also be considered in the analysis of taxa using molecular approaches (Wood, Burke & Rieseberg, 2005).

Molecular phylogenetic analysis

Properties of DNA allow new approaches to the investigation of phylogeny. Dawkins (2005) brilliantly characterises the possibilities as follows. While individual DNA molecules are not themselves preserved, 'the information in DNA can be preserved for ever but only by dint of frequent recopying. As long as the chain of reproducing life is not broken, its coded information is copied to a new molecule before the old molecule is destroyed.' Therefore, DNA sequences contain a

> kind of duplicated information which goes back an almost unimaginably large number of copying generations and which, with a little poetic licence, we can regard as the equivalent to a written text: a historical record that renews itself with astonishing accuracy for hundreds of millions of generations precisely because, like a writing system, it has a self-normalising alphabet. The DNA information in all living creatures has been handed down from remote ancestors with prodigious fidelity. The individual atoms of DNA are turning over continually, but the information they encode in the pattern of their arrangement is copied for millions, sometimes hundreds of millions, of years.

As a consequence, 'the DNA record is an almost unbelievably rich gift to the historian ... every single individual of every species carries, within its body, a long and detailed text'. While changes in DNA occur, they are 'seldom enough not to mess up the record yet often enough to furnish distinct labels'. Thus, DNA provides an 'archive', a 'message scripted in DNA that fell through the succession of sieves that is natural selection' providing a 'Genetic Book of the Dead; a descriptive record of ancestral worlds ... a palimpsest, many times overwritten'. As we shall see: 'large quantities of our ancestors' DNA information survives completely unchanged, some even from hundreds of millions of years ago', and this property of DNA has been brilliantly exploited in devising phylogenetic trees and other diagrams.

Molecular studies have revolutionised phylogenetic investigations, providing the means of testing a wide array of hypotheses generated by taxonomists in the post-Darwinian period (Judd et *al.*, 2007). Important evidence has been obtained: (a) concerning the placement of groups whose affinities were unknown or disputed; and (b) for defining the boundaries and evolutionary relationships of groups at different levels of the taxonomic hierarchy.

Considering the advantages of molecular sequence data, it is clear that DNA is the carrier of the genetic information throughout the living world. Unlike some morphological characters, the four different bases in DNA can be unambiguously identified and this provides many 'characters' in their sequences. Early proponents of molecular systematics claimed that the analysis of molecular data sets was more likely to reflect the 'true phylogeny' than morphological information. However, as we see below, much the same problems are encountered, including homoplasy (Hall, 2011): for example, a change in a base sequence, with later reversal back to the original sequence.

Looking first at historic investigations, in the 1960s, DNA–DNA hybridisation studies were informative in many animal experiments. In essence, the DNA molecule dissociates on heating. The DNA of a putatively related taxon is introduced and pairing assessed, on the assumption that the degree of pairing will reveal a relationship. However, this approach was much less successful in plant studies, because of 'looping' of DNA. Thus, low re-pairing of control samples occurred even where heat-treated DNA was mixed with DNA from the same taxon.

Some researchers have investigated nuclear and mitochondrial RFLPs, but chloroplast DNA has been selected for many investigations of different levels in the taxonomic hierarchy. Stuessy (2009) notes that 'literally hundreds of studies have been completed' providing a particularly important phase in the elucidation of plant relationships. Methods based first on RFLP and later on PCR techniques have been very widely used, but as sequencing DNA became cheaper and more accessible, sequence data sets are now commonly employed.

There have been many investigations of nuclear sequences, e.g. the DNA region that codes for ribosomal rRNA, which has several units 5S, 5.8S, 18S, 26S/28S, and also internal spacer ITS regions ITS-1 and ITS2. However, the most frequently studied sequences are those from the chloroplast, especially *rbcL*. This gene encodes for the large subunit of the photosynthetic enzyme Rubisco, which is almost universal in plants (excepting parasites and some saprophytes). This gene was studied, e.g. the groundbreaking phylogenetic studies of the major

Angiosperm groups by Chase *et al.* (1993). Now experts recommend that *rbcL* is best employed at the generic level and above. Other chloroplast markers have also been studied, e.g. *trnL*, *matK*, and these have proved useful for studying variation at the species and generic level. In the first studies of chloroplast genes, only one or two genes were examined in phylogenetic investigations. The current trend is to study 2–7 genes in combination. One issue potentially complicates the interpretation of chloroplast genes: parts of the genome have been transferred to the nucleus. Nuclear genes have also been used in molecular phylogenies of plants, especially ribosomal RNA genes that are present in tandem arrays of several hundred to thousands. These genes contain small internal transcribed spacers, and the ITS spacer region has been widely studied in examining the relationships between species. Mitochondrial DNA is less frequently studied, as there is a great deal of intramolecular recombination and a low level of base pair substitutions (Stuessy, 2009). Moreover, at replication homogenisation processes occur (concerted evolution). But in some cases mitochondria are not homogenised and there is variability within species or individuals.

In studies of an increasing number of plant groups, a combination of nuclear and organelle sequences has been investigated, e.g. the nuclear and chloroplast sequences, e.g. ITS and *rbcL*. In principle, information about DNA and amino acid sequences could be examined together, but, as the genetic code is degenerate, 'silent' changes in the DNA will not be detected.

Generation and analysis of molecular sequence data

There are many issues to consider in the design of investigations (see Wendel & Doyle, 1998). We note the critical importance of selecting appropriate taxa, genes and DNA sequences, and the satisfactory alignment of the sequences on which analysis is to be based. Judd *et al.* (2007) note that 'alignment is by far the most difficult part of using sequence data, and it is hard to automate . . . many computer programs will produce alignments, although in practice most systematists rely heavily on alignment "by eye"'. Also, it is particularly important to check the plant material, as misidentifications of plant material can occur. In the published account of the study, the location of voucher specimens in herbaria or other accessible collections should be provided (Goldblatt, Hoch & McCook, 1992).

Concerning the analysis of data sets, many different statistical approaches have been devised (see Table 16.1), some of which were initially developed for numerical taxonomic studies. While, historically, analyses of small data sets was often carried out by hand (see worked example in Stace, 1989), now, analyses are carried out using computer programs.

Cladograms

This account draws on the elegant account of the basis concepts by Vandamme (2003; Lemy, Salemi & Vandamme, 2009). In addition, Harrison & Langdale (2006) have written a useful step-by-step guide to phylogeny reconstruction.

Fig. 16.6a illustrates the structure of a rooted and an unrooted tree. Both trees have the same branching pattern – they are therefore said to have the same topology. The names of the elements – root, branch, node etc. – are suitably arboreal. External nodes A, B, C, D, E, F represent the living taxa of plants from which the samples were drawn for analysis. (In theoretical treatments of the subject but more rarely in reports of scientific findings, these taxa are called Operational Taxonomic Units (OTUs) or Evolutionary Units (EUs)). Comparable material is chosen for any particular analysis; thus, A–F may represent individuals, species, genera, families etc. involving the investigation of morphological traits or the DNA of related genes or genetic regions (Vandamme, 2003). In Fig. 16.6a A,

Table 16.1. **Methods of phylogenetic analysis. (From Vandamme, 2003)**

'The methods for constructing phylogenetic trees from molecular data can be grouped first according to whether the method uses discrete character states or a distance matrix of pairwise dissimilarities, and second according to whether the method clusters OTUs stepwise, resulting in only one best tree, or considers all theoretically possible trees.'

A. Exhaustive search investigating character states

Maximum parsimony (MP): 'aims to find the tree topology for a set of aligned sequences that can be explained with the smallest number of character changes (i.e. mutations). When all reasonable topologies have been evaluated, the tree that requires the minimum number of changes is chosen as the best tree.'

Maximum likelihood (ML): 'examines every reasonable tree topology and evaluates the support for each by examining every sequence position ... The most likely tree is chosen as the best tree ...The actual process is complex, especially because different tree topologies require different mathematical treatments'.

Distance Matrix: Fitch-Margoliash: 'is a distance-matrix method that evaluates all possible trees for the shortest overall branch length, using a specific algorithm that considers pairwise distances'.

B. Step-wise clustering of distance matrices

UPGMA **(unweighted pair group method with arithmetic means):** 'clustering is done by searching for the smallest value in the pairwise distance matrix. The newly formed cluster replaces the OTUs it represents in the distance matrix. The distances between the newly formed cluster and the OTUs and each remaining OTU are then calculated. This process is repeated until all OTUs are clustered.'

Neighbour-joining (NJ): this method 'minimises the length of internal branches and thus the length of the entire tree. So it can be regarded as parsimony applied to distance data'.

Comments on the different approaches.

Exhaustive search methods. Vandamme (2003) notes that: 'the number of possible trees, and thus the computing time, grows quickly as the number of taxa increases. With 6 OTUs the number of rooted trees is 954: with 10 OTUs this number rises to 34,459,425.' Clearly, 'only a subset of the possible trees can be examined. Confidence estimators are used to evaluate these trees. The most commonly used is the bootstrapping resampling method.'

Bootstrap analysis. Van de Peer (2003) provides the following details: 'the bootstrap analysis is a simple and effective technique to test the relative stability of groups within a phylogenetic tree'. It is a 'widely used sampling technique for estimating the statistical error in situations in which the underlying sampling distribution is either unknown or difficult to derive analytically... A new alignment is obtained from the original by randomly choosing columns from it. Each column in the alignment can be selected more than once or not at all until a new set of sequences, a bootstrap replicate, the same length as the original has been reconstructed ... For each reproduced (i.e. artificial) data set, a tree is constructed, and the proportion of each clade amongst all the bootstrap replicates is computed. This proportion is taken as the statistical confidence supporting the monophyly of the subset ... Between 200 and 2000 resamplings are usually recommended. Overall, under normal circumstances, considerable confidence can be given to branches or groups supported by more than 70 or 75%, conversely, branches supported by less than 70% should be treated with caution.'

An alternative resampling technique is called delete-half jackknifing. 'Jackknife randomly purges half the sites from the original sequences so that the new sequences will be half the original.' And this process is repeated

Table 16.1. **(cont.)**

many times to give many new samples. These new samplings of the original data set are 'subjected to the regular phylogenetic reconstruction. The frequency of subtrees are counted from reconstructed trees. If a subtree appears in all reconstructed trees, then the jackknife value is 100%: that is the strongest possible support for the subtree. As for bootstrapping, branches supported by a jackknifing value of less than 70% should be treated with caution.'
With the **step-wise clustering** the problem of creating 'too many trees' is avoided by examining local sub-trees first: then a single complete tree is produced. 'This method is fast and can accommodate large numbers of OTUs. Because they produce only one tree, confidence estimators . . . are not available'.
Super-trees
As the number and scope of molecular systematic investigations has developed, molecular systematists have devised methods of combining the results for individual trees to produce super-trees. Bininda-Emonds, Gittleman & Steel (2002) review the procedures, problems and future prospects in the devising of super-trees. They acknowledge that 'supertree construction has been criticized', as it analyses existing trees but does not produce any further primary data. Thus, it has been argued that 'supertrees present only a useful summary of source trees, rather than an accurate phylogenetic reconstruction'. While Bininda-Emonds, Gittleman & Steel accept that 'current supertree methods are not perfect', they conclude that they show 'good performance in simulation, suggesting that we can be reasonably confident in the phylogenetic estimates derived from them'. Considering the methodology, Bininda-Emonds, Gittleman & Steel express the hope that in the future super-trees will include not only molecular data sets but also morphological and fossil data. For up-to-date information on super-trees, Hodkinson & Parnell (2006), Sanderson (2007) and Wheeler (2012) may be consulted.

B and C form a cluster from the common ancestor H, and therefore 'are of monophyletic origin'. In contrast, C, D and E 'do not form a cluster' and on the evidence of this set of samples, 'they are not of monophyletic origin'. The tree has internal nodes G–K and these are referred to as 'hypothetical taxonomic units (HTUs)' in recognition of the fact that they are the 'hypothetical progenitors of the OTUs' under study. In some cases these HTUs may be represented in the fossil record. 'Branch lengths represent the amount of change between ancestor and its descendent' (Hall, 2011).

A tree may be rooted to indicate from left to right the direction of the evolutionary process by including in the analysis one or more OTUs, with samples of a related taxon as an outgroup. The root joins the outgroup to the rest of the OTUs – the ingroup. The choice of outgroup is not completely straightforward. If material is too closely related to members of the ingroup then it does not function as an appropriate outgroup. In avoiding this problem, it might be supposed that some distantly related taxon should be chosen. However, Vandamme points out that 'this may result in serious topological errors because sites may have become saturated with multiple mutations, by which information may have been erased'. In practice, some of these pitfalls are avoided if several outgroup taxa, carefully chosen, are included in an investigation. Vandamme (2003) stresses an important property of phylogenetic trees, namely 'that internal nodes can be rotated without altering the topology of a tree' (Fig. 16.6b). Helpfully, Stephens (2012) writes: 'phylogenetic trees are like mobiles, the only fixed points being the nodes'. Therefore, many legitimate representations of the tree are possible.

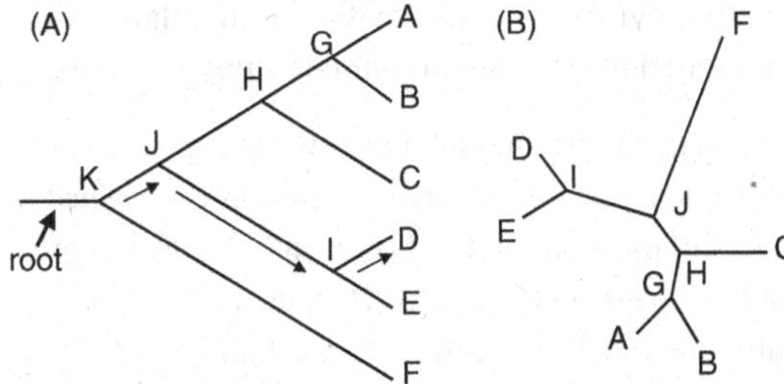

Fig. 16.6a. Structure of rooted (A) and unrooted (B) phylogenetic trees. Both trees have the same topology. The rooted tree is usually drawn with the root to the left. A, B, C, D, E and F are external nodes or OTUs. G, H, I, J and K are internal nodes or hypothetical taxonomic units (HTUs), with K as root node. The unrooted tree does not have a root node. The lines between the nodes are branches. The arrow indicates the direction of evolution in the rooted tree (e.g. from root K to external node D). The direction of evolution is not known in an unrooted tree. (From Vandamme, 2003.)

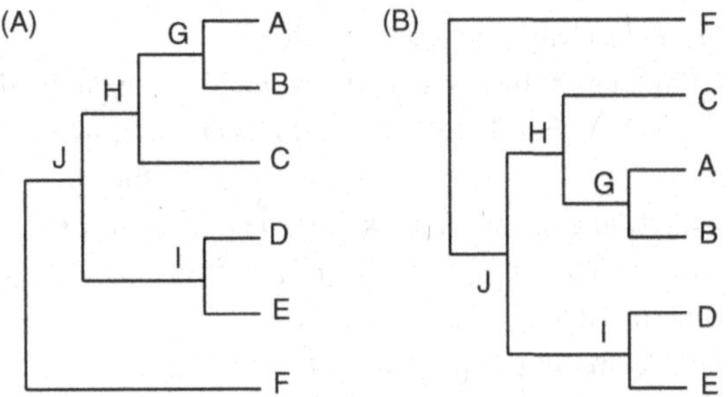

Fig. 16.6b. From a data set different trees may be produced as branches may be legitimately rotated at the nodes. This principle is illustrated by considering Figure (A) in 16.6a. This data set is represented here as trees (A) and (B). In these two diagrams, the internal nodes J and H have been rotated, but it is stressed that they have identical topologies. (From Vandamme, 2003.)

The findings of molecular systematics

This is not the place for a full critique of phylogenetic methods, but several important issues must be considered, not least the controversy surrounding some of the findings in the study of different groups. In addition, it is important to stress that despite the clear-cut nature of some analyses, others contain 'unresolved' elements and their interpretation is uncertain.

- Sometimes in phylogenetic studies the evolutionary relationships of the taxa under study do not resolve into a dichotomous system: i.e. there are more than two branches at a node. Sometimes this is because of insufficient information (so-called soft polytomies), but, in other circumstances, hard polytomies may be involved. For example, a non-dichotomous branching pattern may reflect the expansion of a species into a new range, with rapid independent speciation in different regions or areas of different ecology. Or independent colonisation of island habitats may occur, and several species may evolve more or less at the same time. Determining the cause(s) of polytomies, especially resolving hard polytomies, is extremely challenging, especially as the dating algorithms may be confounded by genetic drift, founder effects etc. In a search for solutions to this problem, a polytomy resolver, for dated phylogenies, has been devised by Kuhn, Mooers & Thomas (2011).
- The use of different techniques and the analyses on very tiny fractions of the total DNA of organisms can sometimes lead to different results. Stace (2010b) notes that 'in a recent analysis (Carlson, Mayer & Donoghue, 2009) of Dipsacaceae, for example, *Scabiosa sensu lato* was defined as a monotypic taxon using the evidence of cpDNA data, but polyphyletic using nuclear ITS evidence unless the genus *Sixalix* (including *S. atropurpurea*) is separated. *Coeloglossum* has been found embedded within *Dactylorhiza* following most analyses (e.g. Bateman *et al.*, 1997; Pridgeon *et al.*, 1997), but it was separated from it by the analysis of Devos *et al.* (2006)'.
- As we have seen in earlier chapters hybridisation and polyploidy are common in higher plants. Stace (2010) emphasises that 'hybrids, and polyploids derived from them, should in theory carry the DNA sequences of both parents, but this is often not true with regard to the two most commonly used regions

of DNA: chloroplast DNA and nuclear rDNA . . . chloroplasts are virtually always inherited from the female parent in angiosperms . . . and rDNA via a process known as concerted evolution most often represents only the female derived sequence (e.g. Franzke & Mummenhoff, 1999; Lihová *et al.*, 2004). Hence the phylogenies of hybridogeneous plants based on these regions are often in reality phylogenies of the female parents of the taxa being studied, rather than the taxa themselves, and moreover results from the two regions of DNA are not suitable as checks on each other as they exhibit the same parental directional bias.' Stace considers that 'ideally polyploids should be omitted from the initial analysis, to be added later when the relationships of the diploids have been clarified'. However, such a proposal may be difficult in practice given the complex polyploidy history of many apparently diploid taxa.

- Stace is also concerned about deductions made where no differences in DNA are detected in phylogenetic studies. 'For example, apparently no differences in DNA sequences have been detected between *Platanthera chlorantha* and *P. bifolia* (Bateman & Sexton, 2008) or between *Gentianella amarella* and *G. anglica* (Winfield & Parker, 2000). Experienced taxonomists, especially field botanists, are well aware that in each of these two examples two separate taxa are involved, and therefore that molecular differences must exist.'
- In some phylogenetic studies increasing the number of genes studied to include all three genomes (nucleus, plastid and mitochondrion) and enlarging the numbers of taxa studied has resolved previous difficulties. Thus, by this means Soltis *et al.* (2011) were able to clarify many important questions about deep-level relationships in the non-monocot angiosperms.
- We emphasise yet again that molecular systematic studies are not devised in isolation on random samples of plants, but often examine historic phylogenetic hypotheses derived from morphological and/or cytogenetic information accumulated over long periods of close taxonomic study of plants.

To keep pace with the flood of new phylogenetic trees, the Angiosperm Phylogeny Website was established in 2001 and information has been continuously updated. The latest version (13), dated 20 January 2014, is maintained by Hilary Davis. Very illuminating introductory material and examples of phylogenetic trees are provided by Professor P. F. Stevens (Angiosperm Phylogeny Website, 2014) of the University of Missouri, St Louis and Missouri Botanical Garden (www.mobot.org/MOBOT/research/APweb/).

He stresses that 'characterizations of all the orders and families of extant angiosperms and gymnosperms' have been produced, but 'our knowledge of the major clades of seed plants and the relationships within and between them are still somewhat in a state of flux, even if much of the broad outline is clear' (see A.P.G. III, 2009).

Pioneer studies In an early example that examined the relationships between *Clarkia* species proposed by Lewis & Lewis (1955), Sytsma & Gottlieb studied 29 restriction enzymes and examined restriction site variation in chloroplast DNA of the genus *Clarkia* (Onagraceae) (see Gottlieb, 1986). They did not include an outgroup, and employed several algorithms in the construction of trees. The shortest and most parsimonious tree is shown in Fig. 16.7. The figures on the internodes refer to the number of mutations shared by EUs distal to that point on the tree. Branch lengths do not indicate the degree of divergence of the groups or a timescale.

Layout of trees

Molecular systematic approaches have also been employed in the examination of the relationship of 'higher' groupings of plants. Baum & Smith (2012) provide a valuable account of the interpretation of trees.

We give an example that illustrates the layout of tree diagrams, showing the 'left to right'

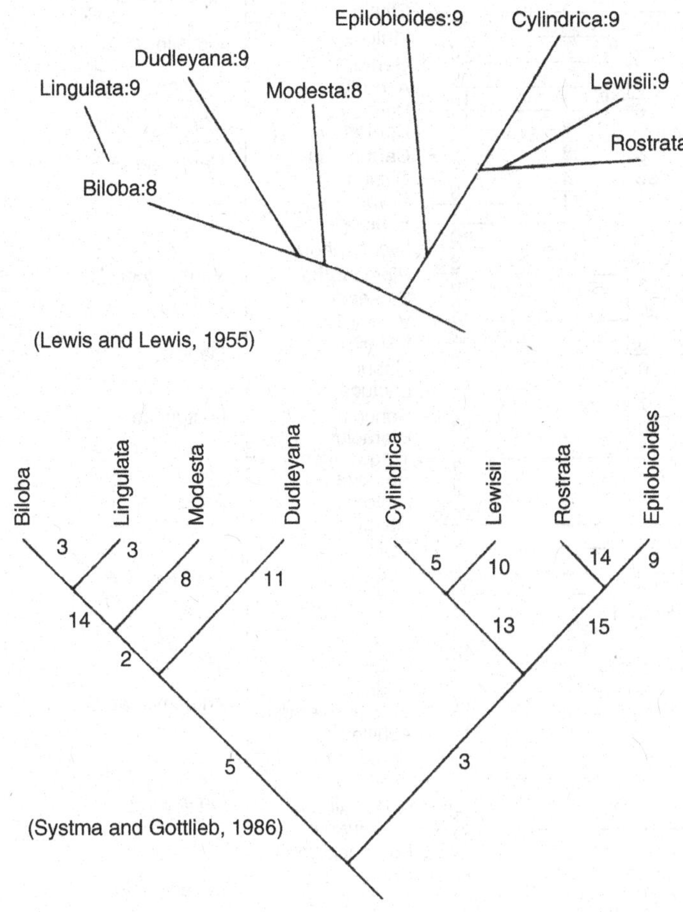

Fig. 16.7. The top diagram shows the phylogenetic relationships and gametic chromosome numbers of the eight species of *Clarkia* section *Peripetasma* proposed by Lewis & Lewis (1955), primarily on the basis of morphological evidence. The bottom diagram shows a phylogenetic tree of the same section discovered in classic investigations by Gottlieb and his colleagues. It is based on restriction endonuclease analysis of chloroplast DNA. This tree was generated from more than 100 restriction site mutations, and represents the shortest or most parsimonious tree. The numbers refer to mutations assigned to an internode in the tree. For example, *Clarkia epilobioides* and *C. rostrata* share 18 mutations of which 15 belong only to the branch leading to them. From their common ancestor, they differ by 9 and 14 mutations, respectively, and from each other by 23 mutations. (From Gottlieb, 1986.)

representation of the tree, the genetic changes detected, the information on bootstrap values (see Table 16.1), and their interpretation (Fig. 16.8).

Jansen *et al.* (1990, 1991) examined hypotheses concerning the tribes of the Compositae (Asteraceae). Representatives of 57 genera representing 15 tribes were studied. Chloroplast DNA from the different samples was treated with 11 restriction enzymes and a tree was constructed, using parsimony methods, on the basis of cleavage sites. Included in the sample, as an outgroup, were members of the related Barnadesiinae. There is evidence that this group has 'primitive' chloroplast DNA, as all other members of the Compositae have a more 'advanced' structure with a particular inverted segment (22 kilobases in length). Thus, the authors feel confident in using the Barnadesiinae as an outgroup. When trees were generated using Wagner parsimony algorithm, 20 equally parsimonious trees were produced. Using Dollo parsimony with 'bootstrap' analysis, a majority-rule consensus tree was produced (Fig. 16.8).

Angiosperm phylogeny

Investigations over the past decades have produced remarkable new evidence of the evolution and

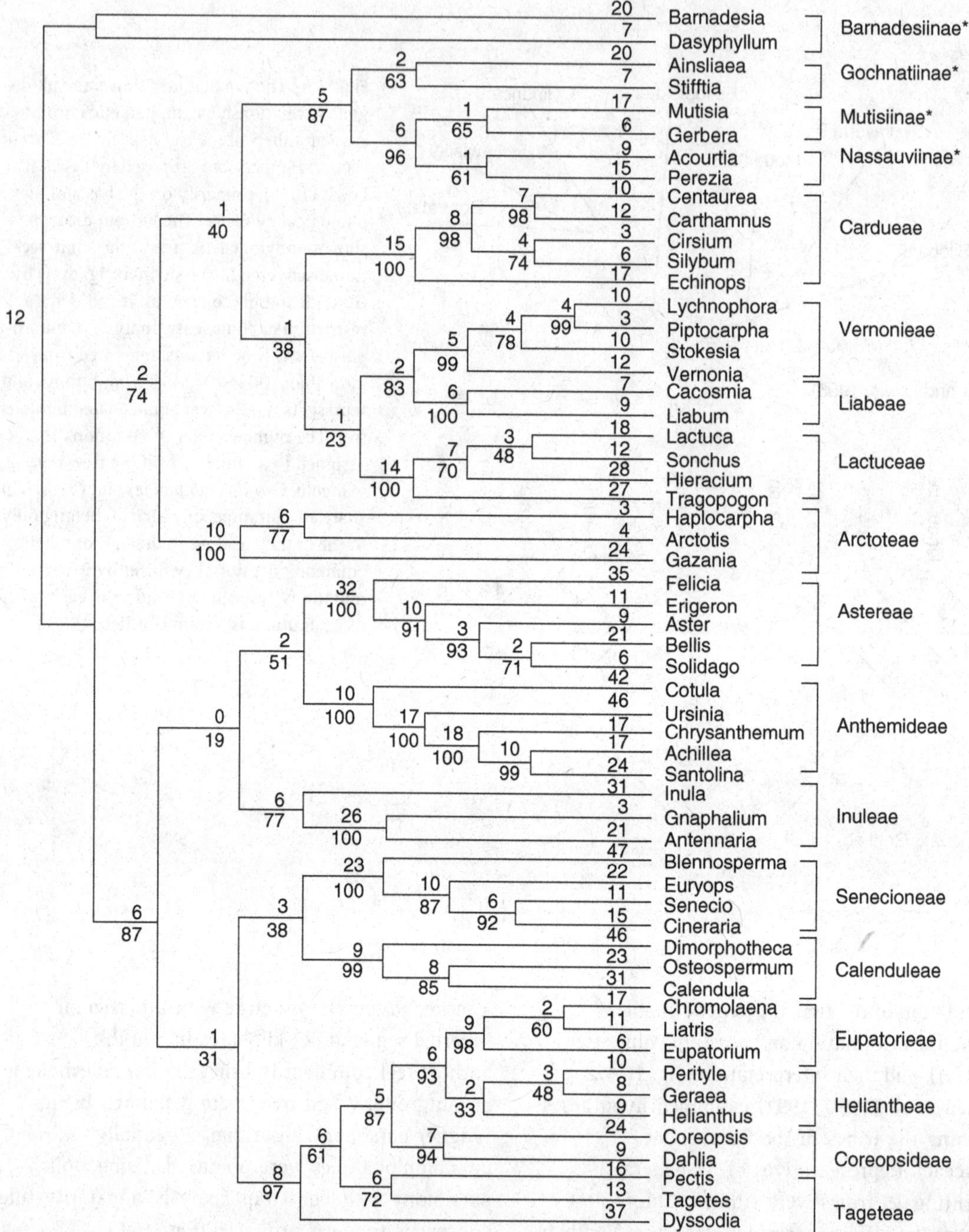

Fig. 16.8. Dollo tree summarising the phylogenetic relationships in the Asteraceae (Compositae) using 328 phylogenetically informative chloroplast DNA restriction site mutations. This is a majority-rule consensus tree based on a bootstrap analysis. The numbers above and below the nodes indicate the number of restriction site changes and the number of times that a monophyletic group occurred in 100 bootstrap replicates, respectively. In the bootstrap method, data points are sampled randomly, with replacement from the original data set until a new data set containing the original number of observations is obtained. Thus, some data points will not be included at all in a given bootstrap replication, others will be included more than once, and still others twice or more from the original data set (Swofford & Olsen, 1990). High bootstrap values indicate parts of the tree found in many sets of trees; lower values point to a higher degree of variability in the set for particular branches. Brackets show the current circumscription of 15 tribes, and four subtribes of the Mutisieae (*sensu* Cabrera, 1977) are indicated by asterisks. (From Jansen, Michaels & Palmer, 1991.)

relationships of Angiosperm groups (Fig. 6.4). As we discussed in Chapter 2, the two great groups of the angiosperms – the monocotyledons and the dicotyledons – were first recognised by John Ray. It might be supposed that this major division of the angiosperms occurred at the first evolution of the group. However, the first angiosperms would appear to have been technically dicots, from which the monocots and eudicots emerged (Soltis & Soltis, 2004). Molecular and morphological evidence supports the division of the dicotyledons into two groups:

1. A basal group, the ANITA grade (Amborellaceae, Nymphaeaceae, and Illiciaceae and their relatives) and the Magnoliid clade.
2. And the eudicots.

Of especial interest is the origin and position of the monocots in the angiosperm lineages (Fig. 6.4), which also shows our present understanding of the phylogenetic location of the various groups of the land plants (*Amborella* Genome Project, 2013).

Mapping additional information onto phylogenetic trees

In addition to bootstrap values, phylogenetic trees often include strategically placed information on a variety of other biological characteristics. This allows the investigator to explore different avenues of evolutionary change. For example the figures in Judd *et al.* (2007) include examples of different trees with overlaid information allowing the exploration of the evolutionary significance of differences in morphology, ecology, biochemistry, cytology (chromosome number) and genome designation. To illustrate this point we can consider a number of examples.

Systematists have long pondered the question: did some of the remarkable adaptations found in the angiosperms arise only once or were there multiple independent origins?

1. **Carnivory** Several different complex integrated systems for attracting, catching and digesting prey (from which nutrients are absorbed) have long attracted the interest of botanists. These take many forms: adhesive flypapers, pitchers, and traps of various types, including bladder and tank traps. Darwin (1875) considered that these different types of carnivory might have had independent origins. Early studies by Albert, Williams & Chase (1992), investigating the *rbcL* gene, concluded that different trap forms have arisen independently at least six times in different lineages of the angiosperms. More recent studies confirm the hypothesis of multiple origins of carnivory (see Soltis *et al.*, 2005; Chase *et al.*, 2009), but perhaps there have been fewer independent origins than initially proposed.
2. **Parasitism** About 4000 species of flowering plants in c.17 families are considered to be parasites (Nickrent *et al.*, 1998). Such plants obtain their nutrients either through symbiotic associations with mycorrhizae (mycotrophs), or by producing structures called haustoria that penetrate the tissues of host plants. Other species are hemiparasites that carry out some photosynthesis, but obtain nutrients and water from the host. Considering the haustorial parasites, it has proved difficult to resolve the phylogenetic relationships because molecular studies of angiosperms have relied, to a high degree, on chloroplast genes. Parasitic plants are non-photosynthetic and often have highly modified chloroplast genomes. Thus, with some exceptions, for example Orobanchaceae, it has not been possible to study the chloroplast genes that have been so revealing in the investigation of photosynthetic species. Progress has been made in studying mitochondrial genes (Barkman et al., 2007). The current view is that parasitism has evolved many times in the angiosperms, with a 'separate putative origin in 13 groups' including the Convolvulaceae, Orobanchaceae, Rafflesiaceae, Santalales etc. (Soltis *et al.*, 2007). Concerning particular parasitic plants, there has been a good deal of

speculation about their origin. Barkman *et al.* (2004) have considered the 'phylogenetic position' of *Rafflesia*, which produces the largest flowers in the world. Phylogenetic studies employing mitochondrial DNA place this genus in the order Malpighiales, which contains the 'poinsettias, violets and passionflowers'. (For the most up-to-date information on the placements of other parasitic plants, see Stevens, 2014.)

3. **Seed dispersal** Phylogenetic studies have also contributed greatly to our understanding of the evolution of seed dispersal. A very good example is provided by studies of the taxonomic and geographical distribution of seed dispersal by ants – so-called myrmecochory – which involves the provision of lipid-rich appendages to seeds (elaiosomes) to 'attract' ants and act as a 'reward' (Johnson, 1982). Lengyel *et al.* (2010) have examined the published phylogenetic reconstructions that include myrmecochorous species, and conclude that it is present in about 11 000 species (about 4.5% of all angiosperm species) in 334 genera, and 77 families, and published trees suggested that myrmecochory had at least 101, or possibly 147, independent origins. With regard to distribution and age they concluded that: 'Most myrmecochorous lineages were Australian, South African or northern temperate (Holoarctic).' Moreover, they discovered that seed dispersal involving ants has evolved in most of the major angiosperm lineages 'and is more frequent in younger families (crown group age <80 million years) than in older ones'. They suggest that the 'energetic costs of producing an elaiosome and the consistent selective benefits of myrmecochory explain the numerous evolutionary and developmental origins of myrmecochory in angiosperm plants' – the ants providing not only seed dispersal, but also protection from predators. In addition, seeds carried into nests are protected from fire and provided with safe nutrient-rich microsites in which to germinate. They conclude that myrmecochory represents 'one of the most dramatic examples of convergent evolution in biology'.
4. **Aquatic plants** Historically, it has proved difficult to understand exactly how aquatic species evolved from 'more conventional relatives', but considerable progress has been made using DNA phylogenies (Stephens, 2012). Stace (2010c) notes that in dealing with taxa with 'reduced morphology' taxonomists have been inclined to devise segregated groups. In molecular phylogenies such taxa are nested within (rather than being sister to) their 'unreduced' relatives (e.g. the Lemnaceae with the Araceae).
5. **Chemical pathways** In reviewing the evidence, Soltis *et al.* (2005) note that 'Some chemical pathways seem to be restricted to only one or a few lineages and must therefore have originated only one or a few times.' They discuss a number of examples.

a) Betalains – a particular class of red and yellow pigments – are found only in the core Caryophyllales clade.
b) The ability to produce glucosinates – which act as natural insecticides – has evolved only twice, once in the Brassicaceae and again in the Putranjivaceae (Malpighiales).
c) Symbiosis with nodule-forming nitrogen-fixing bacteria evolved only once in the Leguminosae.
d) Two photosynthetic pathways (C3 and C4) have been discovered in the angiosperms. C4 photosynthesis has 'evolved more than 60 times in at least 18 families of flowering plants. . . . This pathway provides an advantage where photorespiration levels are deleteriously high, especially in warm, dry, and/or saline habitats' (Sage, Christin & Edwards, 2011).
e) Crassulacean acid metabolism (CAM) is found in some groups of flowering plants. CAM is found in some groups of plants growing in extreme arid conditions. During the dark at low temperatures, when photosynthesis cannot function, carbon dioxide enters through open stomata and is 'fixed in organic acids' in the vacuoles. During the heat of the day the stomata are closed, minimising water

loss, yet carbon fixation through the light reactions of photosynthesis can occur, as 'stored' carbon dioxide is available. Crayn *et al.* (2004) have made a detailed study of the evolution of Neotropical Bromeliaceae using phylogenetic methods. They discovered evidence that the group evolved from a C3 ancestral plant, and that the epiphytic habit and CAM evolved 'a minimum of three times'.

Turning to the occurrence of CAM in the flowering plants as a group, there is evidence that it occurs in about 7% of plants, and that the condition has evolved many times with CAM having a range of different biochemical forms. While CAM is found in some plants in arid habitats and is typical in some succulent groups, interestingly, it is also found in a number of aquatic species, where carbon dioxide is limiting during the day.

Insights into evolutionary relationships provided by molecular phylogenetic studies

The excitement generated by molecular approaches to phylogeny has resulted in an enormous literature exploring the relationships between different groups (see Soltis *et al.*, 2005; Judd *et al.*, 2007). Here, we consider a number of new insights. Some of the examples are drawn from Stace (2010a, c), who has provided a commentary on the impact of molecular systematic studies on the revisions needed for the third edition of *New Flora of the British Isles* (Stace, 2010b).

Circumscription of taxa The results of molecular systematic studies have important implications for the circumscription of taxa, and monophyletic groupings have been produced for many orders, families and genera. Stephens (2012) makes the important point that 'taxa that are recognized formally should be monophyletic, that is, they should include all and only the descendants of a hypothesized common ancestor'. He continues: 'other things being equal, it is helpful if 1, taxa recognized formally are easily recognizable, 2, groups that are well-established in the literature are preserved, 3, the size of groups is taken into account . . . and 4, nomenclatural changes are minimized'.

Turning to some examples, the 'realignment' to create monophyletic groups is likely to produce many name changes. For instance, using the databases GrassBase and GrassWorld, Vorontsova & Simon (2012) estimate that '10–20% of Poaceae species will have to be moved to a different genus by the time the realignment is complete'. Regarding generic circumscriptions, in some cases molecular studies support the separation of genera, e.g. *Ficaria* from *Ranunculus; Rorippa* from *Nasturtium.* Other new molecular insights underlie the recommendation to amalgamate such groups as *Malva* and *Lavatera* (Ray, 1995), and *Artemisia* and *Seriphidium.* In many cases the traditionally recognised long-established families are well supported by molecular studies, e.g. Brassicaceae, Asteracerae etc., but in other cases molecular results suggest the need to reconsider the circumscription of familiar taxa, in some cases by dividing and in other cases by amalgamating groups, e.g. the semi-parasitic tribe Pedicularieae/ Rhinantheae transferred from Scrophulariaceae to the Orobanchaceae; the segregation of the Liliaceae into nine families dispersed into four orders; the joining of the Araliaceae with the Umbelliferae; the Apocyanaceae with the Asclepiadaceae; and changing the composition of the Labiatae/ Verbenaceae.

Some botanists have welcomed these changes to family circumscription. But as this has led to the merging and splitting of many families, such changes are often controversial and the cause of substantial disagreements (Stevens, 2012). Also, they may not take account of the views of the 'users of taxonomy'. For instance, Cullen & Walters (2006) recommend that taxonomists 'should take a very cautious approach to the new families (whether splits or mergers) . . . not accepting changes until or unless their usefulness has been clearly demonstrated'. They are concerned not to introduce 'unwelcome complications of upsetting a well-understood and effective reference and prediction system' provided by the present general-purpose classifications devised by the efforts of generations of taxonomists.

Locating the affinities of little-known groups The affinities of many rare, little-known and unusual taxa have been located using molecular approaches (Soltis *et al.*, 2005; Judd *et al.*, 2007). For example, taxonomists found it difficult to classify Aextoxicaceae, which contains a single species of dioecious tree endemic to Chile. Likewise there were questions concerning the placement of Berberidopsidaceae that has two genera of vines and shrubs found in South America and Australia. Molecular investigations reveal that these two families are related, although more investigation is needed to determine their position in the core Eudicots (Forest & Chase, 2009).

Surprising new insights into the relationship of families One of the most surprising discoveries is that in some cases there is a confirmed grouping of very dissimilar families. A good example is the unexpected membership of the Proteales (an early diverging order in the eudicots) (Forest & Chase, 2009). To the astonishment of many botanists, the Proteales contains the Nelumbonaceae (aquatic herbs including *Nelumbo*: the Water Lotus); the Proteaceae (a large Southern Hemisphere family of evergreen trees and shrubs); and the Platanaceae (a single genus of eight species of deciduous trees of the Northern Hemisphere).

Relocating wrongly-classified families A very good example is provided by the molecular studies of Saarela *et al.* (2007) who reveal that the

> Hydatellaceae, a small family of dwarf aquatics that were formerly interpreted as monocots, are instead a highly modified and previously unrecognised ancient lineage of angiosperms. Molecular phylogenetic analyses of multiple plastid genes and associated noncoding regions from the two genera of Hydatellaceae identify this overlooked family as the sister group of Nymphaeales. This surprising result is further corroborated by evidence from the nuclear gene phytochrome C (*PHYC*), and by numerous morphological characters. This indicates that water lilies are part of a larger lineage that evolved more extreme and diverse modifications for life in an aquatic habitat than previously recognised.

The ordering of taxa A linear sequence of taxa has been prepared by an International Botanical Consortium, the *Angiosperm Phylogeny Group* (APG), incorporating the results of recent studies (LAPG III, 2009). It might be supposed that such studies would provide a definitive linear order of groups for use in floras and herbaria etc. However, while the relationships of many of the higher groupings of flowering plants have now been very well explored, many relationships, particularly at the generic and species level, are not fully resolved. An additional factor must also be considered. Subjectivity is inherent in the ordering, because of the nature of the branching system of trees. At each bifurcation 'either portion may precede the other in sequence' (Stace, 2010b). Therefore, it is clear that different orderings may 'express the DNA data equally faithfully'. He considers the placement of the monocot group in the linear LAPG-III scheme. 'In the linear sequence, the pre-dicots [basal dicots] must start the angiosperms, followed by the eudicots and monocots. In LAPG-III the monocots are placed before the eudicots.' But in the preparation of his flora, Stace prefers to place 'the monocots at the end [of his flora], as this is a much more familiar arrangement' and because of the branching nature of the tree 'equally valid' (Stace, 2010b).

Employing phylogenetic trees to study coevolution Phylogenetic trees have also been employed in the investigation of coevolution involving plants/herbivores and also parasitoids.

There is evidence that herbivory is a potent selective force acting on plants, and that secondary metabolites and compounds are involved in 'plant defence': the evolution of plants and herbivores is therefore interrelated. The evolutionary relationship between plants/herbivores/parasitoids has been likened to an 'arms race', with adaptation by the plant followed by counter-adaptation of the herbivore. An interrelated arms race is simultaneously being played out between the host, herbivore and parasite. In considering models of arms races, valuable insights have come from the close comparative study of phylogenies of plants, herbivores and parasitoids with the aim of examining whether there is evidence of concordance or not. Figure 16.9 illustrates three possible hypothetical situations, where there is species-level matching, clade-for-clade correspondence and independent patterns (for further details, see Forister & Feldman, 2011). This approach provides a means of examining evolutionary interactions, and whether these take the form of species/species, species/guild level, or diffuse interactions. In addition, evidence may indicate that plant groups, through evolutionary innovation, might have escaped a particular arms race, allowing the opportunity for species radiation.

Futuyma & Agrawal (2009) have reviewed the research on plant/herbivore interactions, where the key insights of Ehrlich & Raven (1964) are widely acknowledged. They

> suggested that in response to herbivory a plant species may evolve a novel, highly effective chemical defense that enables escape from most or all of its associated herbivores. By an unspecified mechanism, this advantage enables the plant lineage to radiate into diverse species, which share the novel defense (hence related plants share similar chemistry). After some time, one or more insect species colonize this plant clade and adapt to it . . . these insects able to use 'empty niches' afforded by a diverse clade of chemically distinctive plants, themselves undergo adaptive radiation, as new species arise and adapt to different, but related, plants . . .

This leads to a 'great deal of biological diversity' through 'repetitions of such step-wise adaptive radiations through time', in both plant and herbivore (Futuyma & Agrawal, 2009).

Considering present evidence, Futuyma & Agrawal (2009) conclude that:

> phylogenetic reconstruction of ancestral states has revealed evidence for escalation in the potency or variety of plant lineages' chemical defences; however, escalation of defence has been moderated by trade-offs and alternative strategies (e.g. tolerance or defence by biotic agents, such as ants). There is still surprisingly scant evidence that novel defence traits reduce herbivory and that such evolutionary novelty spurs diversification. Consistent with the co-evolutionary hypothesis, there is some evidence that diversification of herbivores has lagged behind, but has nevertheless been temporally correlated with that of their host-plant clades, indicating colonisation and radiation in insects on diversifying plants. However there is still limited support for the role of host-plant shifts in insect diversification.

Another important aspect of the evolution of plant defences has been explored by Campbell & Kessler (2013). They make the distinction between constitutive strategies (defences expressed at all times) and inducible strategy, which 'may save the cost of trait expression in the absence of attackers by only inducing defences after initial attack (inducible strategy)'. Theory predicts that 'plants that grow in environments in which the probability of attack is variable' should exhibit inducible strategy and 'show greater phenotypically plastic response to herbivory'. Analysis of 56 species of Solanaceae provides evidence that 'the repeated, unidirectional transition from ancestral self-incompatibility (obligate outcrossing) to self-compatibility (increased inbreeding) leads to the evolution of an inducible (vs. constitutive) strategy of plant resistance to herbivores'. Thus, 'variation in plant sexual reproduction has broadly shaped the macroevolution of defence strategies across the Solanaceae'.

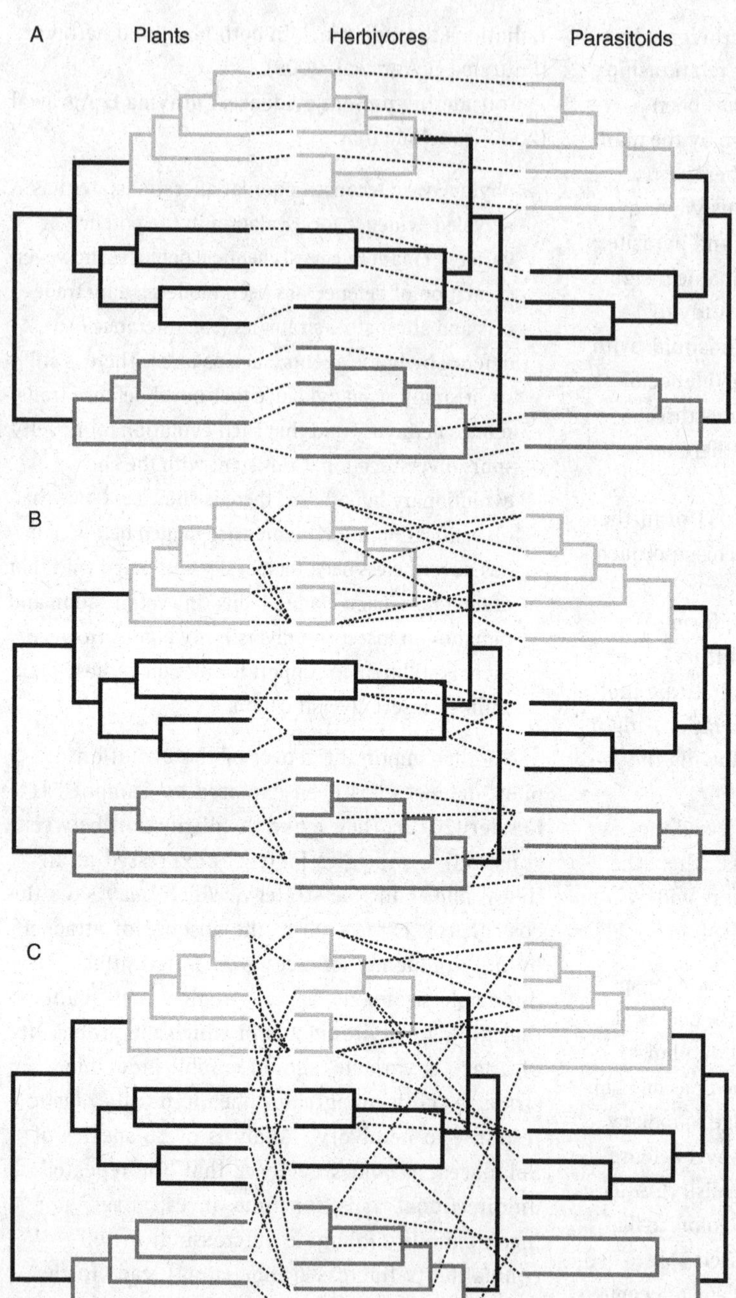

Fig. 16.9. Hypothetical phylogenies illustrating: (A) species-level matching of relationships across trophic levels, (B) 'clade-for clade' correspondence, and (C) independence across tropic levels. The first two models represent phylogenetic cascades. Different shading indicates the different clades within each phylogeny. (From Forister & Feldman, 2011.) Reprinted with permission.

Timescales and timetrees: the role of fossils and molecular clocks

Turning to another key issue, phylogenetic trees require an evolutionary timescale. This is provided by evidence from the fossil record and the use of molecular methods. Historically, Zuckerkandl & Pauling (1965) proposed that 'for any given protein, the rate of molecular evolution is approximately constant over time in all lineages, or in other words, that there exists a molecular clock'. If this is true it follows that such a clock can be used to 'reconstruct phylogenetic relationships among organisms' and estimate the timing of 'species divergence' (Graur & Li, 1999). The notion of a molecular clock is also predicted by the neutral theory of DNA evolution (Kimura, 1987). If mutations are neutral, by definition they will not affect fitness, and, over time, they will accumulate in a lineage in clock-like fashion at the rate of neutral mutations.

A number of statistical dating techniques have been devised that are called relaxed clock models (for details see Rutschman, 2006; Wheeler, 2012). The molecular clock in itself does not provide concrete dates: it is necessary to have independent evidence from fossils, or other dateable events to calibrate the clock. After calibration, phylogenetic trees become 'timetrees' (Kenrick, 2011).

Universal or local clocks?

Graur & Li (1999) conclude that while there is no evidence for universal clocks, 'there may be many local clocks that tick regularly for groups of closely related species'. Considering the evidence from plant studies, Stuessy (2009) confirms that 'no precise clock exists . . . some genes just mutate faster than others' (see Arbogast *et al.*, 2002; Donoghue & Smith, 2004; Rodriguez Trelles, Tarrio & Ayala, 2004). Thus, the clock model can still be helpful in studies of related plant groups, especially if the trees devised from molecular data can be calibrated with geologically determined dates, for instance from fossil finds, or correlated with specific dated geological events such as island formations. At least one fossil calibration is needed to 'set the basic rate at which the clock ticks' (Kendrick, 2011), but 'increasingly more are deployed to act as constraints on rate changes across different limbs of the tree' (Benton, Donoghue & Asher, 2009).

Clarke, Warnock & Donoghue (2011) consider the far from straightforward question of calibration of clocks using fossils.

1) The dating of fossils, and other geological events, is technically challenging. A number of lines of evidence are considered, including radiometric methods, in which the age of rocks is estimated by studying the rate of decay of the radioactive elements they contain (see Gradstein & Ogg, 2009).
2) How fossil-based calibrations are determined in the interpretation of phylogenetic trees is also a challenge. Fig. 16.10a shows the terms employed, and a proposal as to how calibrations might be achieved in the assignment of fossils to clades (Clarke, Warnock & Donoghue, 2011). Issues relating to dating of events in a phylogenetic history must be faced. Kenrick (2009) points out that: 'The first appearance of a group in the fossil record almost always post-dates its actual point of origin by some unknown amount.' In some cases there may be a number of relevant fossils to consider. The assignment of a single date to a node in a tree would therefore be misleading, and palaeobotanists provide some estimate of maximum and minimum limits – known as soft maximum and minimum constraints. Taking into account the fossil evidence, and its careful interpretation, Clarke *et al.* consider the results of a study of seven plastid genes and six molecular clock analyses in identifying 'minimum and maximum age bounds for 17 calibration points reflecting key nodes on the plant tree of life' (Fig. 16.10b).

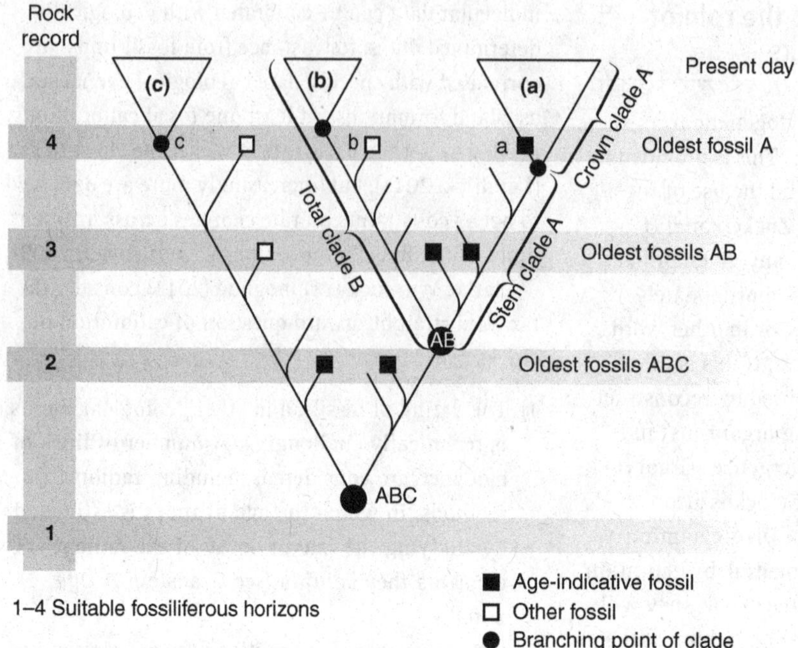

Fig. 16.10a. Definitions of terms that are used in assigning fossils to clades. The crown clade consists of all the living species and their most recent common ancestor, and this is preceded by a stem lineage of purely fossil forms that are closer to their crown clade than to another crown clade. The divergence or splitting point between a species in clade A and a species in clade B is the point AB. This is older than points of origin of crown clades A and B (indicated by points a and b). Fossils may belong to a crown clade or to a stem lineage, and cladistic evidence should indicate which. The crown clade and the stem clade for a particular lineage are together referred to as the total clade. Therefore, if calibrating the divergence between crown clades a and b, this inevitably means finding the oldest reliable fossil belonging to the total clade a and the total clade b: the oldest of which will provide the hard minimum constraint for the divergence between the two (point AB). Four fossiliferous horizons are indicated, the source of all relevant fossils. Fossiliferous horizon 1 that contains no fossils assignable to the clade ABC marks a maximum constraint (soft bound) on the age of the clade. Fossiliferous horizon 2 marks the maximum constraint on the age of the clade AB. Minimum constraints are indicated by the oldest fossils for ABC, AB and A. (From Clarke, Warnock & Donoghue, 2011.) Reprinted with permission.

Origin and age of the Angiosperms

Molecular phylogenetic studies have reignited an intense interest in the early angiosperms and their origins. There have also been advances in palaeobotanical techniques, involving very detailed studies of fossil pollen, flowers and seeds. For example, bulk sieving of Cretaceous deposits, and scanning electron and X-ray microscope studies, have led to the finding of hitherto unrecognised small fossil flowers in deposits. Friis, Crane & Pedersen (2011) provide an authoritative review of this new fossil evidence in historical context, considering the relationships between angiosperms and other plant groups, and between the major groups of the flowering plants. Here, we note a number of key issues.

A. There has been much speculation concerning the origin(s) of the angiosperms, with palaeobotanists exploring several hypotheses, namely, evolution of the flowering plants from a 'type of cycad', or

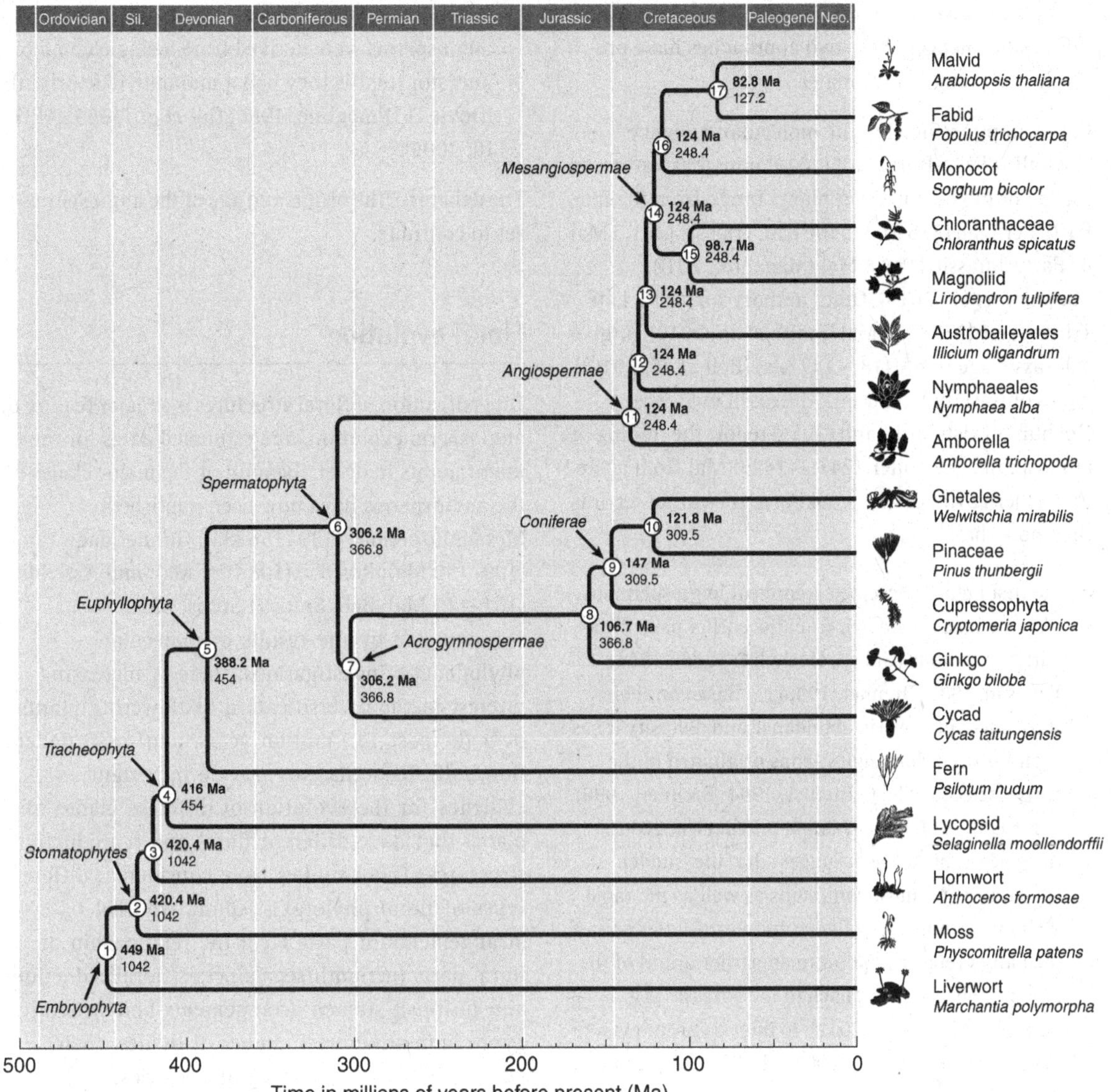

Fig. 16.10b. A representative tree of relationships between model representatives of the major land plant lineages whose plastid or nuclear genomes have been fully sequenced. The topology is based on a consensus of the most well-supported relationships as reviewed in recent literature. Calibrations are presented for 17 nodes, consisting of a hard minimum constraint (bold) and a soft maximum constraint (not bold) for each. The tree has been scaled to time on the basis of the minimum constraints (see text for further details). (From Clarke, Warnock & Donoghue, 2011.) Reprinted with permission.

'seed-bearing ferns', or a gymnosperm origin (Willis & McElwain, 2014). Friis, Crane & Pedersen (2011) consider all these possibilities in detail, and, essentially, their book reveals that the earliest phases of the evolution of flowering plants are still obscure.

B. With regard to the date of emergence of the flowering plants, two lines of evidence are available:

figures based on 'molecular clock estimates' and fossil evidence. These two approaches have produced a range of estimates.

For example, reviewing the molecular evidence Hochuli & Feist-Burkhardt (2013) note that 'estimates for the origin of flowering plants range from the late Early Permian (275 Ma) to the Late Triassic (221.5 Ma) or Early Jurassic (193.8 Ma) (Magallón, 2010; Magallón *et al.*, 2013). Other authors suggest a Late Triassic age (228–217 Ma) (Smith *et al.*, 2010), or give a Jurassic age range (183–147 Ma) (Bell *et al.*, 2010)'.

In contrast, in examining the fossil evidence, Hochuli & Feist-Burkhardt (2013) report the finding of angiosperm-like pollen (247.2–242.0 Ma) from a site in northern Switzerland. Looking at the earlier records they note that:

> the first broadly accepted records of angiospermous pollen grains are known from the earlier part of the Early Cretaceous (Gübeli, Hochuli & Wildi, 1984; Trevisan, 1988; Brenner, 1996) . . . Based on their subsequent increase in abundance and diversity it has been assumed that angiosperms originated in the Early Cretaceous (e.g., Hughes, 1994; Brenner, 1996; Soltis *et al.*, 2009; Friis, Crane & Pedersen, 2011). However, some authors suggest that the 'sudden appearance' on most continents as well as the rapid radiation of numerous clades during the latter part of the Early Cretaceous point to an earlier origin of the group, probably going back to the Jurassic (e.g., Stuessy, 2004; Doyle, 2012), to the Triassic or even to the Palaeozoic (Zavada, 2007; Clarke *et al.*, 2011) . . . But so far the search for unequivocal evidence (fossil flowers and carpels) from pre-Cretaceous sediments was unsuccessful.

C. Botanists have long pondered another fundamental question: are the flowering plants all evolved from a common ancestor? Willis & McElwain (2014) have considered the question and write: 'Despite early suggestions that angiosperms were of a polyphyletic origin (a number of different ancestors), almost all recent evidence (morphological and molecular) suggests that angiosperms were derived from a single common ancestor (that is they had a monophyletic origin) (Doyle & Donoghue, 1986; Qui *et al.*, 1999; APG III, 2009).

The debate on the origin and age of the angiosperms is set to continue.

Floral evolution

Diversification of floral structures is a major feature of angiosperm evolution, and estimated dates for the major points in diversification of the major clades of the angiosperms have now been published: Mesangiospermae (139–156 Ma); Gunneridae (109–139 Ma); Rosidae (108–121 Ma); and Asteridae (101–119 Ma) (Bell, Soltis & Soltis, 2010).

Stimulated by the results of molecular phylogenetic investigations, there is increasing interest in the diversification of flowering plants over the past 135 million years, and fossil pollen/flowers/fruits/seeds etc. provide important evidence for the evolution of different clades of plants that have different floral and reproductive structures. These studies have considered: different types of floral phylotaxis (spiral, whorled arrangements of parts etc.); the relationship of floral parts (perianth/sepals/petals/floral reductions and fusions); stamen arrangements and structures; gynoecium structure (number of ovules, their position etc.); the presence of nectaries; and a variety of types of fruits and seeds etc. (Friis, Pedersen & Crane, 2010; Enders, 2011).

Friis, Pedersen & Crane (2010) discuss the 'unexpected picture of diversity of early angiosperms in Early Cretaceous plant communities . . . some of the diversity included key lineages of early monocotyledons'. They note that the

> improved knowledge of fossil angiosperms from the Early Cretaceous has made some aspects of early angiosperm evolution much less mysterious, but

> a key remaining issue is the relationship of angiosperms to other seed plants. While this is still very uncertain, new fossils are constantly being added to the fossil record of angiosperms and potentially related seed plants . . . These discoveries may ultimately help to clarify the phylogenetic position of the angiosperms in the broader context of seed plant evolution. They may also help illuminate how key features of the angiosperm flower came together, and provide information needed to better understand the history of pollination and dispersal biology amongst angiosperms and their relatives.

However, Hu, Dilcher & Taylor (2012) emphasise that our 'understanding of the early phases of the evolution of angiosperm pollination is still limited and attempts to reconstruct the history of the interactions between angiosperms and pollinators are challenging'. They highlight a number of hypotheses in active consideration, including the following.

- Ancestral flowering plants were insect pollinated, or perhaps the early plants were both insect and wind pollinated (the ambophilous condition). In considering these possibilities, there is a debate about the properties of early pollen, involving its dryness or stickiness, and its capacity for 'clumping' that might favour insect pollination.
- From the mid-Cretaceous, more advanced 'pollination syndromes' appeared, involving the evolution of specialised floral and inflorescent characters associated with co-evolution with different pollinating agents.
- Wind pollination is a derived condition arising independently several times in angiosperm evolution (Friedman & Barrett, 2009).

Turning to another issue – the life-form(s) of the first angiosperms – some have suggested that they were forest shrubs, but others consider that they may have been aquatic herbs. This is yet another unresolved issue (see Soltis *et al.*, 2008).

Now flowering plants dominate most of the ecosystems on land. Looking at the mega-fossil record, in Europe, it is clear that flowering plant diversity increased during the Cretaceous, and many gymnosperms and ferns went extinct. Evidence suggests that flowering plants first came to dominance in fresh water lakes and associated wet lands (130–125 Mya); later (125–110 Mya) they are well represented in understorey flood plains, and, later still, they became dominant in natural levees, swamps and coastal swamps (100 Mya). These findings accord well with North American evidence (Soltis *et al.*, 2008).

Molecular genetics of floral evolution

Considering both the fossil evidence and the emerging results from molecular studies, Soltis *et al.* (2009) have provided a very valuable review of progress in the study of floral evolution, and what follows is drawn from this account.

Phylogenetic studies and fossil remains suggest that the first flowers were 'small to moderate size' and the 'perianth was probably undifferentiated'. Also, 'stamens were likely laminar' and 'the earliest carpels' were likely to have been 'closed by secretion'.

Advances in developmental genetics of the flower are being brought to bear on the evolution of the angiosperm flower and have the potential to contribute to an increased understanding of phylogenetic trees. It is emphasised that 'small changes in the timing or location of gene expression can lead to large changes in floral phenotype'; and 'developmental genetics is providing a host of candidate genes with which to test hypothesized effects and infer causation of floral diversity'.

An influential model has emerged from detailed investigations in *Antirrhinum*, *Arabidopsis*, etc. Around the time of the development of the flower the meristem is partitioned into several fields of gene activity. Soltis and colleagues note that the so-called ABC model 'describes the activities of transcription factors that regulate floral organ identity'.

This combinatorial model of gene activity posits that A function specifies sepals, the combined activity of A and B functions specifies petals, B and C functions together specify stamens, and C function alone specifies carpels. In *Arabidopsis, APETALA1 (AP1)* and *APETALA2 (AP2)* control the A function. *APETALA3 (AP3)* and *PISTILLATA (P1)* are B-function genes, and *AGAMOUS (AG)* is the C-function gene. Recognition of the role of *SEPALLATA (SEP)* genes in the specification of floral identities (E function; Pelaz *et al.*, 2000; Theissen, 2001; Ditta et al., 2004) has led to the revision of the 'ABC' model as the 'ABCE' model. The D function controls ovule identity and the D-function gene in *Arabidopsis* is *SEEDSTICK (STK)*.

With regard to the regulation of these genes, most of the key regulators of organ identity are MADS-box genes (Theissen & Melzer, 2007). Considering the affinities of these genes in lower groups of plants, it is important to note that there are 'clear orthologies between ABC genes from angiosperms and MADS-box genes from gymnosperms. However, the functions of these ABC orthologues in gymnosperms are uncertain' (Soltis *et al.*, 2009).

The ABC model is thought to have 'general applicability' even for distantly related flowering plants, as there is evidence that they have similar genes/expression patterns/flower mutants. Also, genes transferred by transgenic molecular techniques from one species to another (e.g. *Brassica* genes in tobacco) produce the predicted phenotype. Reflecting on the origin of the floral diversity of the angiosperms, Soltis and colleagues note that:

> despite numerous genetic-based models of the origin of the flower (Frohlich & Parker, 2000; Albert *et al.*, 2002; Theissen *et al.*, 2002), how that ancestral reproductive program was modified to yield a flower largely remains an abominable mystery itself. Much of floral diversity is represented by changes in number and arrangement of floral organs. Because the activities of the ABC regulators have little to do with expanding and declining numbers of floral organs (with the exception of organ absence, perhaps) other genes and genetic systems are needed to explain such differences.

Changes in floral symmetry are highly important in angiosperm evolution. It is of especial interest that studies of *Antirrhinum majus* have shown that floral symmetry is controlled by *CYCLOIDEA* and other genes. In addition, some genes in *Arabidopsis* have been shown to influence the numbers of parts (see Soltis *et al.*, 2009 for details). It is anticipated that detailed studies of the ABC and other models and indeed other non-model systems will provide further insights into floral evolution. Such studies are as yet 'in their infancy'.

These investigations on floral evolution explore the diversity of present-day genetic systems and developmental pathways. They increasingly reveal the importance of 'conserved developmental genetic processes' that underlie development in taxonomically and morphologically different taxa (Müller, 2008). Such investigations – so-called evo-devo approaches – have the potential to shed important light on genetic pathways and timing of significant evolutionary change in the plant kingdom. Given the wide floral variation in the flowering plants such studies are becoming increasingly revealing, with tree building playing a key role (Fig. 16.11).

As we have seen, some candidate genes have now been identified in studies of model plants aimed at understanding the origin of the flower (Soltis P *et al.*, 2009; Specht & Bartlett, 2009; Pires & Dolan, 2012). Also, as we have seen above, the part that ancient, and indeed modern, genome duplication has played in the evolution of the flowering plants is being investigated (De Bodt, Maere & van de Peer, 2005).

Many different investigative approaches are now brought to bear on the early evolution of the flowering plants, with major contributions from palaeobotany, phylogenetics, developmental biology

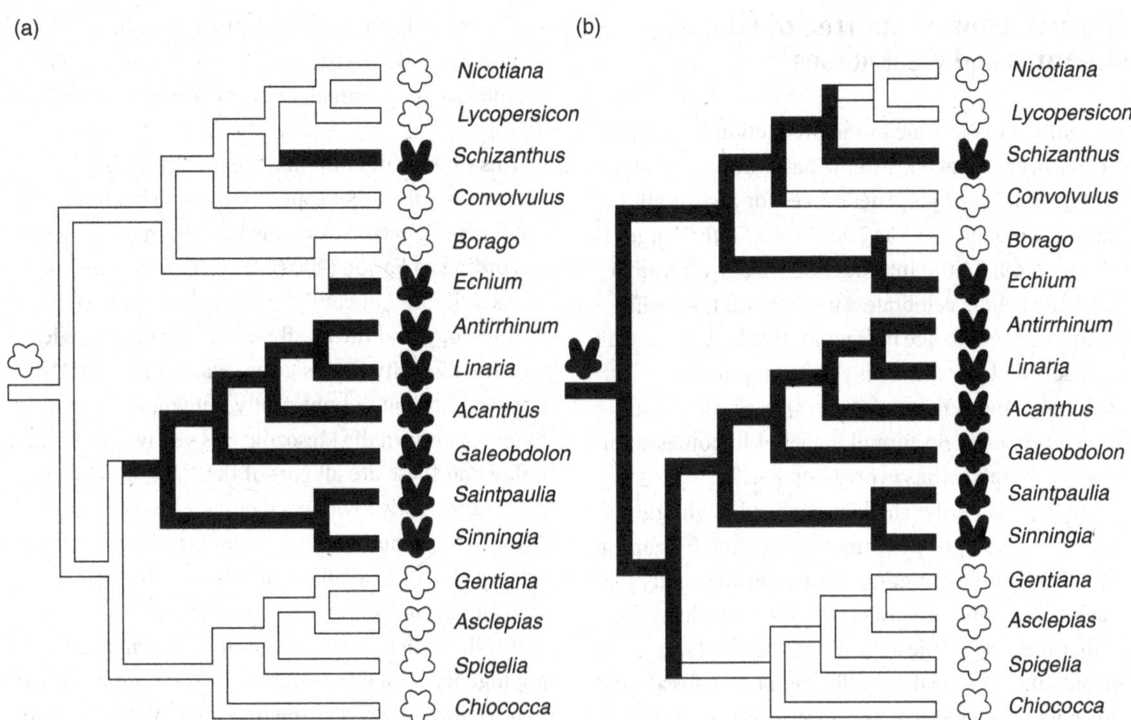

Fig. 16.11. Phylogenetic trees have an important role in the evaluation of evolutionary hypotheses. This is illustrated by investigations of the evolution of floral symmetry in the Asteridae, in which both actinomorphic (regular, radially symmetrical) and zygomorphic (irregular bilaterally symmetrical) flower types occur. Flower shapes in different genera are indicated in the diagram. Two scenarios have been evaluated by Coen & Nugent (1994). (a) A minimum of three changes to zygomorphy (black) are required when actinomorphy (white) is considered ancestral in Asteridae. When zygomorphy is ancestral in the group, phylogenetic studies indicated that there were three most-parsimonious reconstructions, each involving four changes. One of these – reconstruction (b) – is illustrated: it entails four independent changes from zygomorphy to actinomorphy.

With 'only-one step difference between these two scenarios', Donoghue, Ree & Baum (1998) raise the possibility that zygomorphy may have originated rather early in angiosperm evolution, and actinomorphy may have developed a number of times independently. They highlight three important considerations.

1. *Antirrhinum majus* has zygomorphic flowers. Peloric mutants of this have actinomorphic flowers and it has been shown that the gene CYCLOIDEA (CYC) controls the normal development to zygomorphy. Peloric mutations appear to be quite common in zygomorphic species. In contrast, mutations have not been described that yield zygomorphic flowers in otherwise actinomorphic plants. Has the transition from actinomorphy to zygomorphy occurred rarely?

2. In the evolution of the Asteridae there is evidence for reversals from zygomorphy to actinomorphy. But as these changes sometimes involve reduction in petal number and flower size, this may imply the action of mechanisms other than loss of CYCLOIDEA function.

3. An understanding of the developmental pathways involved in the different kinds of zygomorphic flowers will require detailed molecular genetic studies.

and developmental genetics (Soltis DE *et al.*, 2008). However, many would agree with Specht & Bartlett (2009) that the 'mode, mechanism, origin, and early diversification are still debated', and 'the sudden appearance of angiosperm diversity that Charles Darwin so famously hailed as an "abominable mystery" in 1879 to a large extent remains a mystery today'.

The metaphor of the Tree of Life: its strengths and weaknesses

A significant milestone in the production of a tree of life has been achieved with the publication of a major encyclopedia of phylogenetic trees for almost all groups of organisms: *The Time Tree of Life* (Hedges & Kumar, 2009). In the introduction (Hedges & Kumar, 2009) the editors celebrate a publication that 'will catalyse a Renaissance in comparative biology . . . will provide a biological timeline for comparison, prediction, and synchronisation with Earth history. In turn this will help formulate better hypotheses for how the biosphere has evolved on Earth'.

They reflect on the challenges ahead, highlighting the need to incorporate extinct species and fill gaps in the taxonomic coverage by more intensive study of protists, fungi, invertebrates etc. They also hope for further progress in the algorithms used to devise trees and advances in the date calibration using fossils and molecular clock approaches. They also acknowledge the difficult problems in unravelling the earliest evolutionary history of life on Earth.

How far do phylogenetic trees reveal the course of evolution?

While acknowledging the insights that molecular phylogenetic approaches have provided, a number of important issues must be considered.

Estimated trees and true trees The inference methods used in phylogenetic investigations assume that 'most parsimonious, minimum evolution, or maximum likelihood trees are the most preferable trees. However, there is no particular reason to believe that the trees that optimize our criteria are the true trees. Nature may not be parsimonious ... and nature is almost certainly more complicated than even our most complex models' (Pagel, 2002). It must be emphasised, therefore, that we do not know the true phylogenetic history of organisms. 'Inferring a phylogeny is an estimation procedure, in which a "best estimate" of the evolutionary history is made on the basis of incomplete information' (Grauer & Li, 1999).

This important point may be illustrated by considering the gymnosperms, for which a number of phylogenetic reconstructions have been published. According to Farjon (2007), these findings are 'at times strikingly in conflict with evidence from morphology and the fossil record'. In the geological past the diversity in this group was far greater. 'This diversity, although imperfectly known, of extinct gymnosperms in the Mesozoic was vastly greater than today and these are all part of the "true" phylogeny.' Currently we 'live with accidental leftovers from a long and complicated evolution', and these extant species alone are available to provide DNA for the devising of molecular phylogenetic trees.

It follows that however elaborate the molecular phylogenetic analyses might be, we cannot be certain that we have arrived at the true phylogenetic history. Judd *et al.* (2007) express the situation admirably. 'An evolutionary tree is simply a model or hypothesis, a best guess about the history of a group of plants. It follows that some guesses might be better, or at least more convincing than others'. Inevitably, there are often differences of opinion in the phylogenetic studies (Stephens, 2012). As the results of phylogenetic studies become available, some may survive the critical scrutiny by the scientific community; others may be superseded as new botanical information emerges. Current 'best estimates' therefore provide the stimulus for future research.

Considering the origin of the hypotheses that drive the current enthusiasm for molecular systematics, it is important to stress two points. (a) Molecular systematics does not operate in isolation, but builds on the foundations laid by generations of taxonomists. Thus, many of the hypotheses tested were first presented in the historic literature. (b) Interpretation of trees is only possible by using the names of categories of plants – orders, families etc. – that are the product of taxonomic insights. In addition to these 'formal' names a number of 'informal' names have also been employed, e.g. eudicots, etc. (c) Phylogenetic investigations require samples of named taxa for analysis, and these stand as representatives of various higher taxonomic groupings, families, orders etc. Questions about how truly representative such samples might be are difficult to answer and perhaps too rarely confronted in print. However, the issues involved are considered in Salisbury & Kim (2001). Refreshingly, in his introduction to the Angiosperm Phylogeny Website, Stephens (2012) makes specific mention of the problem. He emphasises that 'the sampling of nearly all molecular studies is incomplete', but goes on to write, 'but if sampling in molecular studies is less than we might wish – although fast improving – that of the morphological and chemical characters whose evolution we are interested in understanding is also often very poor'.

Evolutionary hypotheses and falsification Wiley & Lieberman (2011) discuss the nature of phylogenetic investigations. In brief, it is proposed that scientific experiments are designed with the concept of falsification in mind (Popper, 1965). Popper emphasises that a hypothesis cannot be *proved* by an experiment. In his view the design of an experiment is to test the null hypothesis: that there is no *difference* between treatment and control (see the worked example of the analysis of variance in Chapter 8).

Are phylogenetic investigations tests of hypotheses in a Popperian sense? Wiley & Lieberman (2011) consider that phylogenetic studies are singular 'one-off' investigations that are not amenable to falsification, but stress that the resulting trees are 'vulnerable to disconfirmation' by taking into account the full range of available botanical evidence. However, there is a difference of opinion about the status of phylogenetic trees. Pagel (2002) considers that 'we should treat all phylogenetic trees as hypotheses rather than real objects'. In contrast, Hall (2011) refutes the idea that a phylogenetic tree is a hypothesis. For an extended discussion of the philosophical issues involved in hypothesis testing in relation to phylogenetic studies see Wheeler (2012).

Species and gene trees These two tree-types may not be the same. A gene tree plots the divergence from an ancestral gene into two genes of different DNA sequence. In contrast, species trees contain an internal node that marks a speciation event, where an isolating mechanism has arisen. In recognition of this important point, there are different ways of analysing tree-like diagrams. 'Phylogenetic trees are measured in terms of substitutions or state changes, without an intrinsic time constraint' (Wakeley, 2008). However, for the analysis of gene trees, coalescent methods have also been developed (Lui *et al.*, 2009). Wakeley (2008) writes that: 'The emphasis in coalescent thinking is to view populations backwards in time, using the divergence observable in a population to estimate the time to a most recent common ancestor (MRCA); this ancestor is the point where gene genealogies come together, or "coalesce", in a single biological organism . . . coalescent trees are calculated in terms of time, set by a fixed mutation rate, and coalescent analyses therefore assume a molecular clock'. Wakeley (2008) presents a full discussion of the statistical and population biological theory underpinning coalescent analysis, together with reviews of practical applications.

Turning to species trees, Hall (2011) stresses that: 'in a sense, a phylogenetic tree is a naïve attempt at a simple explanation (mutation and genetic isolation) for a complex process (speciation). A phylogenetic tree assumes complete genetic isolation as species evolve by a process of mutation.' Furthermore, it is assumed that 'evolutionary speciation is a binary process resulting in the formation of two species from a single ancestral species' (see Hennig's highly influential diagram of the bifurcation process: Fig. 16.5). Also, at first sight, we might expect any phylogeny based on any gene to reflect the history of the organisms bearing that gene, but in fact this is not always true (Judd *et al.*, 2007).

Why might the gene trees be different from species trees? As we have seen in earlier chapters, various modes of speciation have been identified in plant evolution: Hennig's diagram may accurately reflect some of the speciation processes in plants, but not all. In some cases, evidence suggests that new species are budded off from existing species, and that both the ancestral and descendant species continue to exist (a process called quantum speciation by Grant, 1981). Lineage sorting is also a phenomenon that must be considered (Chapter 13). Also, processes occur that lead to reticulate rather than bifurcating evolution. Thus, genetic information may be transferred from one species to another through introgressive hybridisation or horizontal gene transfer (see below). Turning to other considerations, inconsistencies between phylogenetic trees using nuclear and chloroplast marker sequences may occur (Syvanen, 2012), that owe their origin to chloroplast capture following hybridisation and subsequent backcrossing of the hybrid to one or other of the parental species. In addition, there is abundant evidence that processes of allopolyploidy, which involve interspecific hybridisation, generate highly reticulate patterns of evolution in various groups of flowering plants and ferns (see, for example, in the case of the arctic–alpine *Cerastium alpinum*, Brysting *et al.*, 2011). An added complication is the recurrent origin of polyploids from the same taxa, but involving different genotypes. The implications of the reticulations introduced by ancient polyploidy must also be considered, as some of these events are likely to have involved allopolyploidy (Fig. 6.4).

One approach in the building of trees is to omit hybrids before the tree is devised, and then to add the necessary 'reticulations' later. Clearly such a procedure presupposes that we have already clarified the evolutionary history in groups where polyploidy is widespread. This is often not the case.

Horizontal transfer of genetic information One area of possible complexity in the interpretation of evolutionary trees has emerged over the past few decades. Horizontal transfer of genetic information (HGT) has been detected between many groups of organisms (Richardson & Palmer, 2007; Keeling & Palmer, 2008; Talianova & Janousek, 2011; Syvanen, 2012). HGT, sometimes referred to as lateral gene transfer, is defined as the 'movement of genetic information across normal mating barriers, between more distantly related organisms'. Therefore, HGT is distinct from the usual vertical transmission of genes from parent to offspring: genetic information being passed between organisms other than by descent. The very important additional finding is that the commonalities of biochemical process found in nature mean that 'transferred genes' can function in distantly related organisms, a fact exploited by the design and production of transgenic plants (Syvanen, 2012).

The significance of HGT and its likely frequency has been 'one of the most debated topics in evolutionary biology during the last two decades because it challenged our view of the evolutionary history of genomes' (Talianova & Janousek, 2012).

Considering bacterial groups, Gayon (2009) points out that '90% of the present terrestrial biomass is microbial', and 'the first three billion years of organic evolution were exclusively microbial'. Woese, considering the evidence emerging of the early history of life on Earth, writes: 'evolution at this early stage was probably a symphony of lateral gene flow'. There is an active debate about whether HGT was so dominant 'in early evolution that it is not possible to ascertain a tree of life for the deepest clades' (see Syvanen, 2012). Accumulating evidence suggests that 'for some parts of the history of life on Earth' the appropriate metaphor would be a reticulated 'ring' rather than a bifurcating tree of life (McInerney & Wilkinson, 2005).

Turning to the origin of the eukaryotes, there is now increasing evidence that in the course of evolution, HGT played a pivotal role through endosymbiosis (see Chapter 6). This involved the 'engulfment of bacteria [by the "proto-eukaryote"] that subsequently gave rise to plastids and mitochondria, an integration of their genetic material within the host genome, along with subsequent and still on-going migration of organelle DNA to the nucleus' (Talianova & Janousek, 2012). Thus, it is possible to see eukaryotes as 'chimeras built upon chimeras' (see Syvanen, 2011, for discussion of the evidence including the possibility of further secondary and tertiary endosymbiosis in some groups of organisms).

Sapp (2003) makes an important point about microbial evolution and the evolution of plants and animals. 'Classical evolutionary biology has been concerned with the last 560 million years of evolution: it is essentially about plants and animals. Accordingly its historians do not consider 85% of the Earth's evolutionary time – the evolution of bacteria ... Microbial evolutionists insist that symbiosis, mergers, and the transfer of genes between different kinds of microbes are cardinal mechanisms of evolutionary change.' The origin of the eukaryotes, by endosymbiosis, provides evidence that in some circumstances acquired characters are inherited (Sapp, 2003), and it seems likely that HGT provided an impetus for rapid evolution, more dramatic than the gradual change envisaged by many orthodox neo-Darwinians in the past. The use of the term 'Lamarckian' for such phenomena may be seen as misleading, for Lamarck himself had a different notion of the origin and significance of the inheritance of acquired characters.

A number of mechanisms of HGT are beginning to be revealed in higher plants. According to Stegemann *et al.* (2012), 'most well documented examples of plant-to-plant HGT concern the exchange of mitochondrial genes between species' (Syvanen, 2012, cites six cases). Accumulating evidence suggests that the mitochondrial HGT is particularly prevalent between plants where organisms are intimately associated or at least establish occasional cell-to-cell contact. The high propensity of mitochondrial genomes to engage in HGT could be related to two peculiar biological features of plant mitochondria: (i) they possess an active homologous recombination system; and (ii) they readily undergo organelle fusion, with mitochondria being physically connected and forming net-like structures.

Considering HGT from all sources, it would appear that cell-to-cell contact provides the best opportunity. For instance, the roots of many plants have mycorrhizal associations and these fungi are themselves surrounded by complex microbial communities (Frey-Klett, Garbaye & Tarkka, 2007). Such intimate association might provide opportunities for HGT. Parasitic associations are another arena in which HGT might occur with horizontal transfers 'from host to parasite lineage' (Barkman *et al.*, 2007). The research of Yoshida *et al.* (2010) provides strong evidence of the putative transfer of a gene (of unknown function) to the parasitic plant *Striga hermonthica* from its host *Sorghum* or a related grass species. 'Horizontal transfer of expressed genes' from host to parasite have been detected in studies of *Rafflesia cantleyi* and its obligate host *Tetrastigma rafflesiae* (Xi *et al.*, 2012).

Recent research suggests another possible route for HGT in plants. Natural grafts are sometimes found between unrelated plant species. Grafting experiments with *Nicotiana* species provide plausible evidence of chloroplast HGT, by showing 'that complete chloroplast genomes can travel across a graft junction from one species to another' (Stegemann *et al.*, 2012).

Rice *et al.* (2013) report another astonishing case of horizontal transfer, discovered in *Amborella trichopoda*, 'the sister species to all other extant angiosperms'. They report 'the complete mitochondrial genome sequence', and conclude that it 'contains six genome equivalents of foreign mitochondrial DNA, acquired from green algae, mosses and other angiosperms ... whole mitochondria [were captured] from diverse eukaryotes, followed by mitochondrial fusion'. They consider that 'Amborella's epiphytic load, propensity to produce suckers from wounds, and low rate of mitochondrial DNA loss probably all contribute to the high level of foreign DNA in its mitochondrial genome'.

Historic records reveal that grafting between distantly related species in the wild sometimes occurs (Seidel, 1879; Küster, 1899; Beddie, 1942). As we have seen, there is accumulating evidence of chloroplast transfer across grafted plants (Stegemann & Bock, 2009). A further remarkable case of HGT reveals that 'entire nuclear genomes can be transferred' (Fuentes *et al.*, 2014). This phenomenon was discovered in investigations of artificially prepared grafts between *Nicotiana glauca* and *N. tabacum.* In the experiments (some *in vitro* and others in a greenhouse) both species were transgenic, carrying antibiotic genetic markers. It was discovered that lines from the graft had the genetic material from both species, and genome size determinations by flow cytometry confirmed that fertile allopolyploid plants had been produced. Studies of inheritance confirmed that the plants were indeed polyploid derivatives – named *N. tabauca.* The new allopolyploid combined many traits of the two species.

An immediate question concerns the underlying transfer mechanism. Further investigations revealed that some of the derivatives had the plastid genome of *N. glauca* while others had the genome of *N. tabacum,* which argues against simple cell fusion. These studies suggest that grafting represents a potential asexual means of speciation, which bypasses the normal route of sexual hybridisation and polyploid formation. Moreover, such an asexual event might be possible between distantly related species. Whether grafting could be a route to polyploidy in wild plants has yet to be determined, but this investigation suggests that such a route is not implausible.

Many recent reviews agree that HGT is a significant but perhaps infrequent source of reticulation in eukaryotes evolution (Keeling & Palmer, 2008; Talianova & Janousek, 2011; Syvanen, 2012), but further studies are needed to clarify the situation, including consideration whether genetic transfers by HGT are adaptive and functional.

Plant evolution: limitations of the Tree of Life metaphor

Drawing together the various line of evidence, we have shown above that there are several well-established processes in plant evolution that produce reticulate patterns of variation, including hybridisation, introgression (both ongoing and ancient: Willyard, Conn & Liston, 2009), allopolyploidy, endosymbiosis and lateral gene transfer.

Furthermore, there are intriguing consequences of the evolution of complex interactions between organisms. As we saw in Chapter 6, typically, plant cells include other genomes – of virus, bacteria, fungi etc. These exist in a variety of relationships, including parasitism, infectiveness and symbiosis (Dupré, 2010). In discussing the concept of 'the polygenomic organism' he writes that: living systems . . . are extremely diverse and opportunistic compilations of elements from distinct sources'. These include associations 'between organisms of quite different species, or lineages . . . all of these cases contradict the common if seldom articulated assumption one genome, one organism'. Phylogenetic trees usually examine the evolution of plant taxa alone – a complex enough undertaking. Clearly, the evolution of 'polygenic' organisms in nature is much more complicated and is provoking new interest in coevolution (see for example Arnold *et al.* (2010), who stress that 'plant and fungal trees of life' are 'more akin to interwoven branches than to parallel bifurcations'. Given our increasing understanding of the reticulate nature of the evolutionary process and the complexity of species as they operate in nature, the metaphor of a bifurcating phylogenetic tree, while providing a valuable set of tools, can be seen as simplistic.

Considering the problems of reticulation in more detail, Hughes, Eastwood & Bailey (2006) reflect on some of the limitations of phylogenetic studies of plants in comparison with some animal groups. 'When we examine progress in constructing accurate phylogenies amongst closely related species the situation, at least for plants, is less encouraging. Partially-resolved gene and species trees are of limited use for studies of hybridization and polyploidy, character evolution and species diversification, speciation and domestication.' They are not optimistic that the problems of extreme reticulation can be resolved with current approaches. Thus, 'fully resolved divergent species trees should not necessarily be expected for many plant genera' (e.g. Linder & Rieseberg, 2004). For instance, as we saw in Chapter 14, Mason-Gamer (2004) showed that at least five different gene lineages exist within the single allohexaploid species *Elytrigia repens* (*Elymus repens*) (Poaceae) – some from distant genera even in other tribes of the family. Put bluntly, 'many interesting questions in plant evolutionary biology and biogeography are currently frustrated by lack of resolution towards the tips of the tree'. Considering a way forward, an important focus of future studies must be species-level phylogenies. Devising networks from molecular data sets is one important avenue that is being actively explored (Fig. 16.12).

At this point in the discussion of phylogenetics an issue of considerable interest should be raised, namely the 'weighting' given to data from studies of DNA. Some molecular biologists consider that there is 'sufficient' content in DNA alone for the elucidation of phylogenies, and are inclined to see information from this source as more important than any other. This has led to a situation familiar to students of the history of science. Donoghue & Sanderson (1992) clearly identify the problem and offer a strong recommendation for the future:

> Enthusiasm over a new source of evidence is understandable, as are exaggerated claims on its behalf. But too often in the history of systematics the rising popularity of one source of data takes place at the expense of another, which remains insufficiently explored . . . it is felt that the best way to promote new data is to find fault with the old, and what could be better than to claim that the old data are worthless? But rhetoric of this sort, and the fads that it encourages, are unhealthy from the standpoint of our common goal, namely reconstructing the phylogeny of plants . . . our efforts to construct phylogeny will be judged by their success in integrating *all* of our observations, which means that more attention should be devoted to combining molecular and phylogenetic evidence.

It should not be anticipated that this will be an easy task, for where morphological and molecular data have been considered, the two data sets may be to an extent irreconcilable (see Kadereit, 1994). Some of these difficulties arise from procedural matters, but Kadereit considers the possibility 'that incongruence may result from mutation of major morphogenetic genes leading to dramatic morphological divergence unaccompanied by equivalent change of the phylogenetic marker molecule(s) used'. There is evidence that sometimes mutations may have large phenotypic effect, but Kadereit notes that 'it remains unclear how often large changes have contributed to evolutionary change'. Clearly, the role of macromutations in evolution is still a live issue for molecular taxonomists. It is important to see such debates in their historical perspective, for as we have seen there were fierce debates on this very issue between the Mendelians and Darwinians in the early years of the twentieth century.

Angiosperm evolution: what role for saltational change?

There are some intriguing questions under active consideration concerning the 'major limbs' of phylogenetic trees of the angiosperms. In earlier chapters, we have identified a number of important processes – random mutation, recombination, chance events, selection etc. – that are involved in microevolutionary changes. Are these processes sufficient to explain macroevolution, namely evolution of the higher taxa that have their origin in the distant past?

In a major review, Kellogg (2002) notes that 'ancient developmental changes mark the origin of major clades'. Likewise, there are morphological

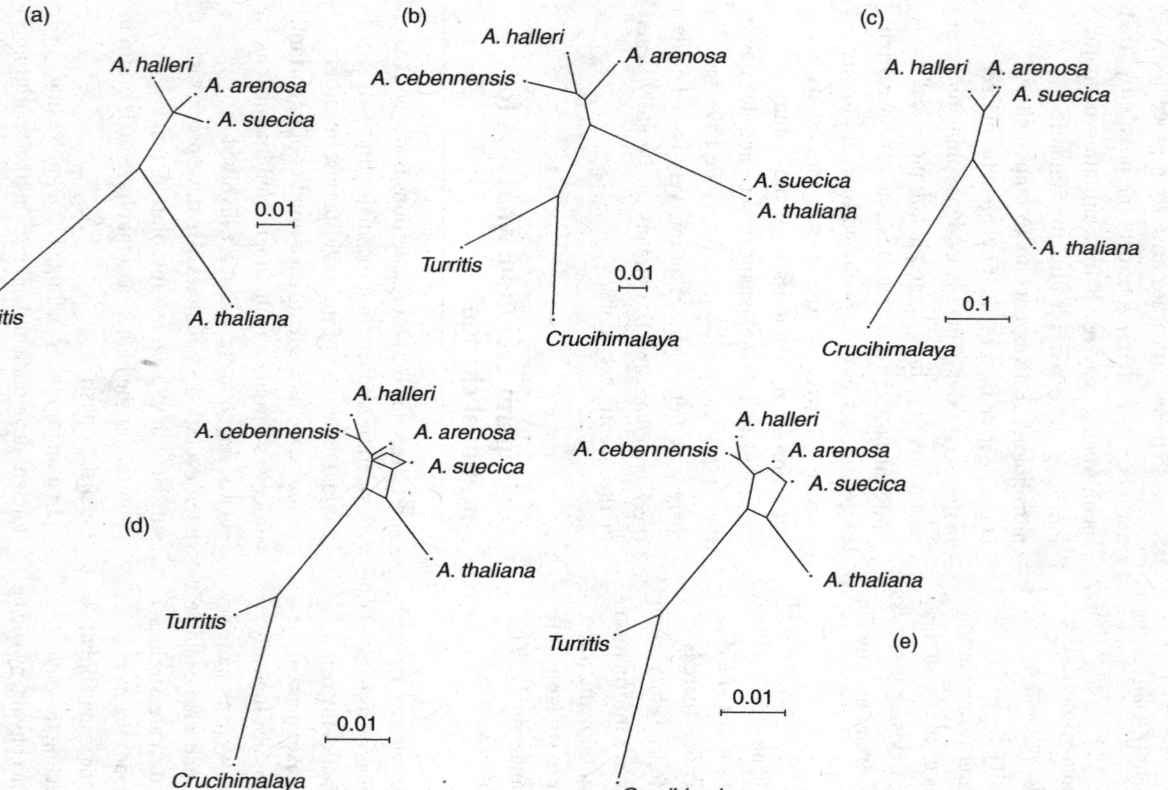

Fig. 16.12. Networks devised to investigate the reticulate evolution of *Arabidopsis* species and their hybrids. Gene trees for (a) alcohol dehydrogenase, (b) rDNA non-transcribed intergenic spacer region and (c) FRIGIDA. Supernetwork (d) for these three gene trees and their hybridisation network (e). (For further information on the construction of these trees see McBreen & Lockhart, 2006.)

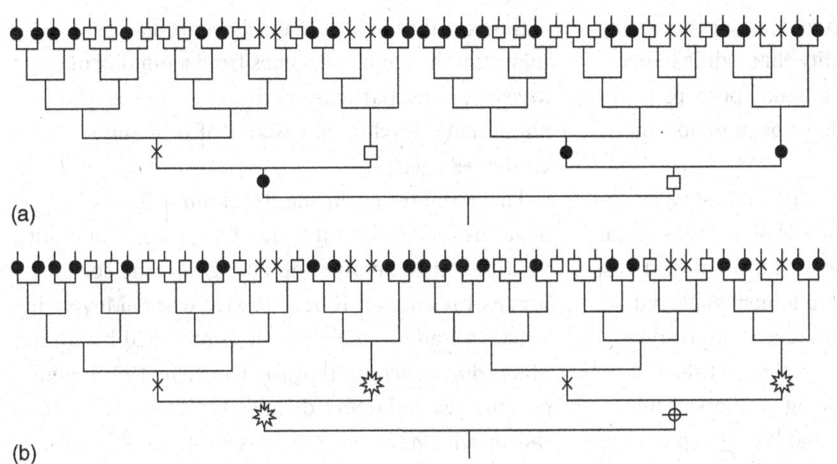

Fig. 16.13. Concepts of microevolution and macroevolution: hypothetical phylogenies. Symbols on branches indicate different types of mutations.

(a) Following the ideas of the Modern Synthesis, the types of mutations that mark recent branches should be the same kind and occur at the same frequency as those that mark deeper branches.

(b) Alternatively, deeper branches could be qualitatively different from those on recent branches. (From Kellogg, 2002.)

changes associated with more recent events. But, given the magnitude of some of the changes in the distant past, are there qualitative differences between these changes and those of more recent times? Kellogg (2002) visualises these two situations (Fig. 16.13).

Referring mainly to evidence from the evolution of the grasses, she considers the magnitude of changes in ancient nodes relative to those that are more recent. At first sight, the evidence appears to support the notion that there were qualitatively larger differences in the events in the distant past. But she points out that there are grounds for caution in reaching this conclusion. For example, Chase *et al.* (2000) suggest that 'the difference between deep and shallow nodes reflects selective superiority: all species that lacked the winning combination of traits have gone extinct'. Also of importance are issues relating to the possibility of reversal of characters, and 'gene-level changes could be completely uniform through evolutionary time, but perceived variation in phenotype may be far more episodic' (Kellogg, 2002).

In examining the question of macroevolution further, she writes: 'the key challenge will be to combine genetic approaches with the appropriate tools of molecular biology to determine the genetic basis of morphological characters'.

Thus, evo-devo approaches are being actively developed (Cronk, 2002, 2009), and explore the 'close relationship between developmental and evolutionary processes' (Theissen *et al.*, 2002). They point out that plants develop under genetic control from a single-celled zygote, and therefore evolutionary changes 'in developmental control genes may be a major starting point for evolutionary changes in morphology'. Understanding the phylogeny of developmental control genes should thus contribute significantly to understanding the evolution of plant and animal form' (Theissen *et al.*, 2002).

Evo-devo approaches are now well established, and are reinvigorating morphological studies and shedding light on the processes at work in macroevolution. Both artificial and natural mutants offer the possibility of insights into development. Amongst the natural 'mutants', meristic, and especially aberrant floral and vegetative variants (so-called terata) are frequently included (Bateman & Dimichele, 2002).

In considering macroevolution, a major focus, for some, is to explore the possibility that, while many evolutionary changes may be gradual, perhaps major saltational changes may also have been important in macroevolution.

Some invoke the concept of 'hopeful monsters' proposed by Goldschmidt (1940). Theissen (2006), for example, takes the view that several lines of evidence support the notion that: 'hopeful monsters played an important role during the origin of key innovations and novel body plans by saltational rather than gradual evolution', monsters being 'organisms with a profound mutant phenotype that have the potential to establish a new evolutionary lineage'. Theissen points out that historically in the 1930s–40s, when the synthetic theory 'the Modern Synthesis' was developed, it was considered 'that evolution occurs always gradually, and complex and unique structures evolve through an almost infinite number of generations subject to natural selection fine tuning these traits'. However, the importance of polypoidy and hybridisation was acknowledged in plant evolution especially by Darlington.

As we have seen in previous chapters, accumulating evidence points to the importance of abrupt changes in plant evolution. While many evolutionists are reluctant to use the term 'hopeful monster', there is increasing evidence, as Theissen argues, that 'novel morphological forms in evolution may result in just a few genes of large effect'.

He cites as evidence 'the most intensely studied case in plants', namely 'the domestication of Maize (*Zea mays* ssp. *mays*) from Teosinte (*Zea mays* ssp. *parviglumis*). During this process the female inflorescence (ear) of corn originated as an unprecedented novelty during changes at just about five gene loci (Doebley *et al.*, 1997; Wang *et al.*, 2005)'. However, it could be argued that changes in maize were due to the domestication of the crop and may not apply to wild species. It is of special interest therefore that Moritz & Kadereit (2001) point out that QTL analysis of wild species supports the view that major structural change can be the result of one or a few genetic loci.

Support for the idea that saltations may be important in evolution comes from a number of other investigations, particularly from the studies of abnormally developing variants of wild and cultivated species.

Those interested in the developing field of evo-devo are especially intrigued by homeotic mutants, i.e. those that involve the change of one set of organs for another (Coen, 1991; Coen & Meyerwitz, 1991). An admirable example is provided by studies of *Clarkia concinna* (Ford & Gottlieb, 1992), who provide the following details. Typical flowers of this species have parts in sets of 4: i.e. 4 sepals, 4 petals, 4 stamens and a 4-part ovary. A notable variant of this species is *bicalyx*, making up 20–30% of the population at Point Reyes, north of San Francisco, California, in which other individuals have normal flowers. Bicalyx has 'eight sepals, no petals and wild-type stamens and ovary with no reduction in fertility'. Crossing experiments revealed that the mutant bicalyx phenotype is governed by a single recessive gene. This example is particularly significant as it is involves a wild population, and reveals that homeotic change underlying a large morphological change is governed by a simple genetic situation.

Theissen (2006) reports another case history of homeotic variation in wild populations of *Capsella bursa-pastoris*, where 'all the petals are transformed into stamens'.

Turning from contemporary wild populations, another possible example of homeotic change underlies the evolution of a major plant group, the Rosaceae. Detailed morphological and phylogenetic investigations suggest that petals in this group were derived from stamens (Baum & Donoghue, 2002), leading to major and remarkable 'transference of function'.

In addition, intriguing evidence comes from the study of orchids, sometimes including teratological variants (see Rudall & Bateman, 2004). Here, and elsewhere in discussions of saltation, some key concepts are critical for the interpretation of major changes in basic morphology. Thus, in some cases, relative to the ancestral pattern, the position of organs may change (heterotropy). In others, changes in shape and size are interpreted as being the result of changes in the timing of events in the developmental pathways (heterochrony). In some evo-devo investigations, interpretation evokes neotony/juvenilism: variants being formed that exhibit arrested, slowed or delayed development.

Evo-devo approaches have the potential to illuminate our understanding of the role of saltation in plant evolution, but it is clear that the interpretation of evolutionary events long ago may be suggestive, but difficult to establish unequivocally. Box *et al.* (2008) make this point very clearly. In studying the repeated speciation in the tribe Orchidinae (Orchidaeae) 'earlier onset of growth . . . credibly explained the reduction in spur size and lateral lobing in *Gymnadenia odoratissima*'. But this is 'consistent with, but not proof of, saltational macroevolution operating via functional changes in one or more key developmental genes'.

Constraints in evolution

In considering the diversification of the flowering plants it is important to consider questions about the 'power' of selection. As Bradshaw (1989) has pointed out, it is 'tempting to follow Darwin's lead and to 'see no limit to its power'. However, as we have seen, it has become increasingly clear that selection cannot act unless the appropriate variation is available. Another constraint has also been proposed that may have very considerable influence on the course of flowering plant diversification. Gould & Lewontin (1979) make a very important observation that at any particular point in evolutionary history the *Bauplan* – the way the organism is constructed – represents a constraint on the evolution of that lineage. They also point out that evolutionists often 'atomise' the organism into a number of traits, each of which is explained in terms of a structure optimally designed by natural selection for its particular function; any sub-optimal functioning is then assumed to be due to a balance between competing demands. Each investigation ends with the telling of a plausible story invoking the force of selection. Plausibility is often the only test of validity. Gould & Lewontin take the view 'that organisms must be analysed as integrated wholes, with Baupläne so constrained by phylogenetic heritage pathways of development and general architecture that the constraints themselves become more interesting and more important in delimiting pathways of change than the selective forces that may mediate change when it occurs'.

Genome sequencing: prospects for further insights into phylogeny

So far relatively few genomes of plant species have been sequenced – mostly species of agricultural importance – and these have yielded very interesting information on the conservation or otherwise of gene sequences. As we saw in Chapter 6, there is evidence of gene conservation – synteny – between the genomes of grass species. But comparison between other species, which evidence suggests are more distantly related, reveals many changes to ancestral gene order. Recently the *Amborella* Genome Project (2013) reported 'the remarkable conservation of gene order (synteny) among the genomes of *Amborella* and other angiosperms . . . an ancestral gene set was inferred to contain at least 14,000 protein-coding genes' and 'relative to non-angiosperm seed plants, 1179 gene lineages first appeared in association with the origin of the angiosperms. These include genes important in flowering, wood formation, and responses to environmental stress.'

With the advent of next generation sequencing techniques, it is predicted that the genomes of many more species will become available and further insights into genomic evolution and phylogenetic relationships will be possible, by examining, in detail, sequence conservation and change.

Classification and the Tree of Life

This chapter began with a consideration of plant classifications. We return to this topic to discuss recent controversial developments in classification emerging from molecular phylogenetic studies. While there are some difficulties with the concept of the Tree of Life, it is still the guiding metaphor for many botanical studies, and it has provoked a fundamental reappraisal in plant classification.

Many molecular systematists have expressed dissatisfaction with the present classification systems, and have proposed radical new ideas that have provoked fierce heated debate. A new code of practice – the PhyloCode – has been promoted in influential publications (de Queiroz & Gauthier, 1994), and detailed proposals are being developed on the Internet (www.ohio.edu/phylocode/). Stevens (2006) provides an elegant review of the proposed system, in which its strengths and weaknesses are carefully explored. De Queiroz & Donoghue (2013) review important aspects of clade names and the nomenclature of living and fossil organisms etc.

To understand the dissatisfaction of some molecular systematists with the present system, it is important to reiterate a number of concepts that we have set out in previous chapters. Building on the work of Linnaeus (1758), generations of botanical taxonomists have recognised three key principles in the classification of plants.

First, species are given a unique binomial composed of a generic name with a species epithet. These binomials 'represent the noun–adjective combinations of ordinary language in a Latinised form' (Stevens, 2006). Secondly, these species names are those first validly published after an agreed starting date. And thirdly, these species and genera are part of a hierarchy of categories, in which the species are nested within genera, genera within families, families within orders etc. These principles are enshrined in an internationally agreed code of nomenclature for plants. In situations where a new species is found or reclassification takes place, agreement has been achieved on how to manage any changes of name that ensue.

Considering the purpose of a classification, it is important to stress that it serves a practical function (Stuessy, 2009). This crucial insight is admirably captured by Gilmour (1936/89), who wrote: 'it [the taxonomic process] is a tool by the aid of which the human mind can deal effectively with the almost infinite variety of the universe. It is not something inherent in the universe, but is, as it were, a conceptual order imposed on it by man for his own purposes.' This element of convenience is evident in all classifications. Thinking about the number and size of groups, Stevens (2012) stresses that 'major systems such as those of Linnaeus and Bentham & Hooker were constructed explicitly so as to ease the burden of memory, the latter ensuring that all groups in their classification were relatively small'. He speculates that perhaps fewer than 500 families 'is a reasonable goal at which to aim'.

Biological classifications have also served other purposes. In the Linnaean period, classifications were considered to reveal the patterns of life established by Special Creation. Proponents of the PhyloCode conclude therefore that the 'Linnaean system of nomenclature is non-evolutionary' (see Benton, 2000). But clearly this is not the case for, as Benton argues convincingly, in the post-Darwinian period, previously devised classifications have been successively modified, through changes in the circumscription and ordering of groups, to reflect advances in our knowledge of plant evolution. Moreover, the International Botanical Code is not 'fixed in stone', but, by agreement, allows classifications to 'evolve', as new botanical information has become available. Such changes in classification lead to name changes.

What does the PhyloCode aim to achieve? 'The proposals of phylogenetic nomenclature are to translate cladistic phylogenies directly into classifications' (Benton, 2000). According to Harris

(2005), 'the core proposition of the PhyloCode is to abandon Linnaean hierarchical ranks and recognize only species and clades ... the scheme does not dispense with hierarchical organization, as clades will be nested within one another according to phylogeny'. Given that under the existing Botanical code, name changes can arise during the course of taxonomic practice, it is claimed that the 'key advantage [of the PhyloCode] is that changes made in one part of the classification do not require altering other group names'. The naming of taxa under the PhyloCode system has still to be finalised, but binomials are likely to be abandoned. Benton notes that there are 13 distinct proposals for 'new names', many of which 'reduce the binomen to a uninomen'. Thus, for example, he notes that the name *Homo sapiens* could be abandoned in favour of 13 possibilities: '*Homo-sapiens, homo.sapiens, homosapiens, sapiens*1, or *sapiens*0127654'. With regard to flowering plants, Stevens (2006) points out that we will need 'some 325,000 or more uninomials' together with the names for clades. Moreover, 'if all we have are potentially millions of uninomials with no grammar as is explicit in a flagged hierarchy, additional aids to understanding will be needed. Rankless names can be understood *only* in the context of a tree – hardly convenient (Janovec *et al.*, 2003); other than monophyly, they convey no information'.

Questions about priority are important, as in some cases different research investigations have produced different trees. Rules for stabilising botanical names have not been finalised, but Benton (2000) understands that 'taxon terms can be made explicit by the first author who enters the fray ... an implication of the new system is that, once names have been defined by a taxonomist in explicit terms, that definition must remain. However, there is no agreement that such priority exists'.

In reviewing these radical proposals, Harris (2005) believes that the concept of the PhyloCode is perhaps receiving increasing support, but many botanists are blunt in their criticism. In recognition of the merits of the existing system, Harris emphasises that 'the Linnaean system effectively organises and conveys information about taxonomic categories at all levels of biological organization and that replacing this system does not justify redefining millions of species and higher taxonomic levels. If the PhyloCode is adopted, the change could mean reworking the names of 1.75 million species and counting.' The PhyloCode proposals would have enormous consequences for those attempting to conserve and document the Earth's biodiversity, through such projects as the Catalogue of Life (www.catalogueoflife.org). Likewise, there are profound implications for those writing floras, field guides, catalogues of pests, invasive organisms and diseases; and managing herbaria and Botanic Gardens.

If the PhyloCode project is implemented, Benton (2000) foresees 'endless vistas of fruitless time-wasting and bickering'. He acknowledges that 'the tenets of phylogenetic nomenclature have gained strong support among some vocal theoreticians', but notes the 'rigid principles for the legislative control of clade names and definitions'. Overall, he is disturbed that 'the consequences of this semantic maelstrom have not been worked out'. He concludes that 'in practice, phylogenetic nomenclature will be disastrous, promoting confusion and instability, and should be abandoned. It is based on a fundamental misunderstanding of the difference between a phylogeny (which is real) and a classification (which is utilitarian).'

Stuessy (2009) discusses the PhyloCode at some length. He concludes that 'most systematists are comfortable with the existing codes ... or believe that they can be adapted for naming clades ... Some people are testing the waters, so to speak, by publishing uninomials for species names (Pleijel, 1999; Fisher, 2006), but it is very doubtful that phylogenetic nomenclature will ever replace the existing modern, bacterial, botanical, and zoological codes.' In Stuessy's opinion, having parallel systems 'is not a good idea, because this can lead only to great confusion' (Blackwell, 2002; Dubois, 2007). Further, Stevens (2006) emphasises that, 'at this moment in history when the top priority of the systematics community is to inventory the organisms of the planet, advocating unnecessary name changes would be entirely counterproductive'. And having examined the proposals for the PhyloCode in great detail and in historical context, he wonders if there is a resolution to the problem but, has to conclude that 'the PhyloCode is cognitively, gravely, irredeemably flawed'.

While we see the force of arguments against the PhyloCode system set out above, it will be interesting to see how its proponents react to justified criticism. Can a mature practical form of their nomenclatural scheme be devised (Bryant & Cantino, 2002)?

Traditional taxonomy, cladism and molecular systematics

In closing this chapter, it is interesting to reflect on the response of taxonomists first to the cladistic approaches inspired by Hennig's work begun in the 1950s, and secondly, to the current wave of enthusiasm for molecular systematics.

As we have seen in our historical survey, some biologists believe that the evolution of the flowering plants will one day be effectively elucidated, and Turrill's (1938, 1940) concept of a perfected taxonomy be realised. Other biologists, however, have taken the contrary view. Stuessy (1990) writes, 'We do not know the true phylogeny for any group of organisms, nor will we ever know it.'

Thus, it is interesting to consider very carefully the historical impact of the investigations of the cladists. Before the full flowering molecular systematics, Cronquist (1987), the celebrated American systematist, who spent his life studying phylogeny, wrote a powerful critique of cladism in relation to taxonomy. He presents a challenging view:

> If taxonomy is to serve its historical and continuing function as a general-purpose system, then it cannot be held in thrall to debates on arcane matters that bear little if any relationship to putting together the things that are most alike and separating them progressively from things less alike. If the participants find such discourse [on cladistics] interesting and mentally rewarding, well and good, but let them then admit they are working towards a special-purpose system that cannot replace the general-purpose taxonomic system.

Faced with this very strong response, Donoghue & Catino (1988) responded:

> Cronquist's critique of cladism illuminates a basic decision that systematists face. On the one hand, we might choose to maintain phenetically defined taxa, even where these are found to be at odds with our best estimate of phylogeny. On the other, we might continually update our system of classification so that it accurately reflects what we know of phylogenetic relationships, even if this means abandoning traditional groups. In our opinion, the first option represents subjectivity and stagnation, while the second offers objectivity and progress.

Other sources of controversy have emerged (Brummitt, 1996, 1997, 2002, 2003). First, what is the relationship between a cladogram and a classification? Are they one and the same (as suggested for instance by Humphries & Funk, 1984) or different? In the *Origin*, Darwin himself wrote 'genealogy by itself does not give classification'. Brummitt takes up this key point in stressing that 'neither a cladogram nor a phylogenetic tree is a classification', and in the taxonomic process 'subjective decisions must always be taken to impose limits to taxa and their rank'. A second question involves the difficult issue of paraphyly about which there has been a great deal of debate (van Wyk, 2007; Zander, 2007; Hörandl, 2010; Podani, 2010a, b). Paraphyletic groups are those which include an ancestor and *some, but not all*, of its descendants (Fig. 16.14).

Brummitt notes that: 'In proposing his theory of phylogenetic systematics Willi Hennig argued that

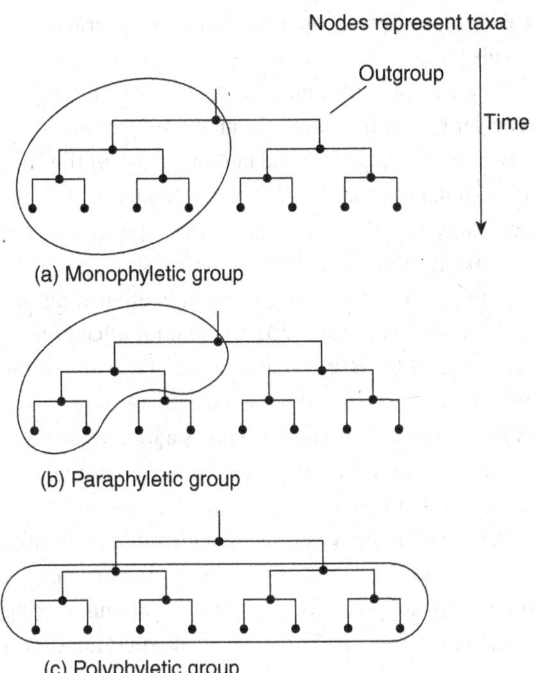

Fig. 16.14. Types of phyletic groups. (a) indicates a monophyletic group in which the descendants are grouped with a common ancestor. (b) A paraphyletic group is indicated in which all the descendants of a common ancestor are not included. (c) A polyphyletic group is shown which does not include an immediate ancestor. (From Pankhurst, 1991.)

no paraphyletic group should be accepted as a formal taxon in classification.' Cladistic research reveals that many groups recognised by taxonomists are paraphyletic. Clearly, accepting Hennig's views has practical consequences. Brummitt argues that, 'If paraphyletic taxa are not allowed, because every descendant taxon has to be of a lower rank than its ancestor and sunk into it, infinite regression occurs, and the whole classification telescopes back into its ancestral taxon.' Further, 'abandonment of Linnaean classification in favour of an attempt at a phylogenetic hierarchy must result in *loss* [author's italics] of ... information and some degree of predictivity, while causing undesirable taxonomic and nomenclatural upsets'. He holds the view that 'ideally taxonomists should be able to present both a classification, which may allow paraphyletic taxa, and a putative phylogeny ... A phylogeny and a classification are both desirable but have different functions, and should be allowed to exist side by side, interrelated but not interdependent, to give maximum predictivity.'

But argument has long continued. In discussing the relationship between phylogenetic studies, classification and practical taxonomy, we hope the debate will be set in its proper historical perspective. At first sight many contemporary biologists might consider that they are faced with a 'new set of issues'. However, as we have shown, arguments have raged many times in the past 'over essentially the same ground'. For instance, Winsor (1995) has provided an excellent account of the English debate about taxonomy and phylogeny in the period 1937–40. Then, many botanists 'asserted that taxonomy was a practical matter to be kept distinct from phylogenetic speculation': in contrast, many zoologists insisted that 'taxonomists must strive to represent evolution if they wished to be scientific'.

With the rise of molecular systematics many of these questions, so keenly debated in the past, have surfaced again with new urgency (Hörandl & Stuessy, 2010; Stuessy, 2010). How should the results of molecular investigations be reconciled with the traditional taxonomic approaches? These questions can be approached from different viewpoints: of evolutionists contemplating their molecular findings, or major users of the 'output' from taxonomy, such as botanists, horticulturalists, nurserymen, foresters, conservationists and landscape designers.

Brickell *et al.* (2008) are concerned that 'a fundamentalist approach to cladistic methodology, which requires that a classification should not include paraphyletic taxa', has resulted in the proposals 'to merge certain well-known and morphologically identifiable genera of horticultural interest. For instance,

- *Dionysia, Dodecatheon, Soldanella, Omphalogramma* and *Cortusa* into a broad genus *Primula*;
- *Hebe, Parahebe*, and *Chionohebe* into a broad *Veronica*;
- *Albuca, Drimia* and others into a broad *Ornithogalum*;
- *Fragaria* and *Alchemilla* into a broad *Potentilla*; . . .
- *Juniperus* nesting within *Cupressus*' and the possibility of
- 'sinking the Cactaceae into the Portulacaceae and the inclusion of *Cyclamen* in the Myrsinaceae'.

The 17 contributors to this letter to the editor of the journal *Taxon* emphasise that they are not 'against taxonomic change . . . but insist that horticulture needs a stable (though not static) classification and nomenclature'. According to Brummitt (quoted in this paper), the choice is stark: 'either one has a monophyletic system with an infinite number of nodes but no ranks, for which the *PhyloCode* is designed, or you have the Linnaean system with ranks at a few levels, and paraphyletic taxa.'

Thus, for those for whom the output of taxonomy serves practical purposes, the abandonment of the Linnaean approaches is unthinkable and unnecessary, especially as classifications can be adapted to accommodate, to a high degree, the flood of new phylogenetic insights. The way forward, for these biologists, therefore, is to acknowledge the importance of, and distinction between, the Linnaean classification that acts as a universal practical general-purpose classification and information retrieval system, and phylogenetic trees derived from molecular and other data, which may be viewed as special geometric constructions indicating the best guess at the evolutionary relationships of plants. For this group, there is no reason why these two different approaches to plant variation and evolution cannot coexist side by side. Moreover, if the formal nomenclature categories of the International Code of Nomenclature prove inadequate for the interpretation of trees, then informal categories can continue to be used, e.g. basal angiosperms, Eudicots, Asterids, Rosids etc.

Others have a different approach. For instance, Wheeler (2012) forcefully promotes a contrary viewpoint: 'groups must be monophyletic in the Hennigian sense' and 'classifications must match genealogy exactly . . . all systematists today, whether they like it or not, are Hennigian cladists'.

In the light of these sentiments it is interesting to read the views of Stace (2010c), who produced the *New Flora of the British Isles* third edition, one of the first floras to be based on the most up-to-date phylogenetic information. He takes a determined position on molecular phylogenetics. 'We can be confident that molecular classifications based on DNA base sequences will not be bettered in the future, but will endure for centuries to come and become universally adopted.' But in the practical undertaking of writing a flora, he stresses the following important guiding principles, especially 'pragmatic compromise'.

- Extremely distinctive taxa that markedly change the circumscription of the group to which they are closest should be considered candidates for separate recognition, leaving a more tightly defined, albeit paraphyletic, taxon. I see no merit in the dogma that all taxa must be monophyletic.
- Relatively weak molecular evidence should not be relied upon to change old classifications; changes should be made only once the data are unequivocal.
- Degrees of similarity/difference in DNA should not be used as an absolute criterion of relationships, only a relative one.
- There is always scope for argument and disagreement. Decisions should be reached by considering a great range of evidence in addition to the molecular data.

Drawing this important discussion to a close, Stevens (2012) makes the crucial point that reaching a

consensus is important, since what we know of angiosperm phylogeny allows a very large number of classifications to be based on it, and unfortunately, 'nature' does not dictate what the classification should be. All classifications are constructed by humans to communicate particular aspects of groups and relationships, they are a means to an end, not an end in themselves... our goals as systematists are surely to produce robust hypotheses of relationships, to understand the evolution of morphology, and the like – but not to argue *ad nauseam* whether something 'should' be a family or subfamily. That way surely, lies madness, and worse: the discredit of our discipline.

The debate continues.

17 Historical biogeography

One of the areas of plant evolutionary studies most enriched and stimulated by advances in molecular phylogeny is historical plant biogeography. 'Earth history has profoundly influenced the geographic ranges of species . . . evolutionary histories of areas and lineages are tightly coupled' (Ree & Smith, 2008). This long-standing botanical discipline has now been invigorated by advances in plate tectonics; the investigation of patterns and processes of dispersal (including the increasing use of analytical models); and the development of phylogeography (Avise, 2000). Phylogeography is an offshoot of the cladistic approaches devised by Hennig. One of the key advances was the realisation that phylogenetic diagrams, which have taxonomic entities at their branch tips, can be overlaid, or considered, together with geographical information, such as country of origin of each taxon, or geographical origins can be substituted for taxon names. Two classes of branching structures are then available for analysis, one with taxa and another, of the same data set, but with geographical details. Also, information on timeframes can be included. These are particularly revealing when estimates employing relaxed molecular clocks are calibrated against fossil evidence (Wikström, Savolainen & Chase, 2007; Magallón & Castillo, 2009). In addition, molecular evidence can be combined with information on the environments and ecosystems of the past.

These approaches, together with new advances in our understanding of Earth's history, have shed new light on many questions that have long intrigued botanists about the role of natural selection and chance events in dispersal, speciation, adaptive radiation, hybridisation and extinction. To appreciate the significant progress that has been made, a brief historical account of several puzzling biogeographical phenomena is helpful (see Lomolino *et al.*, 2005). Here we concentrate on a number of key issues, as they influence plant evolution.

The Deluge and Noah's Ark

In a fascinating book, Browne (1983) reveals that, in the seventeenth century many scholars, convinced of the literal truth of the story of Noah's Ark, raised a number of questions. How did all the animals fit into the Ark? How were they fed, housed and the dung removed? In search of answers, the German Jesuit priest Kircher produced a representation of the Ark as a long rectangular box-like structure of three decks, shaped more like a modern cruise liner than the traditional pyramidal structure commonly reproduced in children's toys and in works of art. Kircher's Ark had only a limited payload for he listed only '130 species of animals, 30 pairs of snakes and 150 different kinds of birds' (Browne, 1983).

As exploration of the remoter parts of the Earth progressed, the numbers of animal species rapidly increased. For example, in the eighteenth century, Linnaeus listed 5600 animal species and, in the years that followed, many more species were discovered, and it became more and more difficult to see how all the animals could be accommodated in the Ark.

The Deluge and plant distribution

While Linnaeus was sceptical about the Ark, he took the story of the Deluge seriously, and Browne (1983) sets out his view that:

> All living beings, including mankind, had their origins on a high mountain at the time primeval waters were beginning to subside... Taking an example from his botanical predecessor, J.P de Tournefort, he suggested that the paradisiacal mountain presented a wide range of ecological conditions, arranged in belts from tropical through temperate to polar zones. If they are sufficiently high, he argued, mountains of the present day show exactly this kind of climatic stratification. It is perhaps significant that Linnaeus took Mount Ararat as a suitable authoritative, modern example of successive zones of temperature. Although the foothills were rooted in deserts of the plain, Ararat undeniably carried permanent snow at its summit; it represented in summary the various climatic zones that circle the northern hemisphere ... From reindeer and arctic lichens at the summit to tropical palms and monkeys at the foot, every plant found its 'proper soil' and every animal its 'proper climate'... then as the primeval waters receded, Linnaeus envisaged organisms moving in turn to latitudes and environments where they were to remain for the rest of time.

Essentially, these notions contain the idea of a centre of origin.

Eighteenth- and nineteenth-century investigations of plant geography

Many explorers contributed to knowledge not only about the distribution of individual species of plants, but also to the recognition of distinctive vegetation realms, regions and zones, defined by climatic and physical barriers (see Lomolino *et al.*, 2005, for an analysis of the contributions of Willdenow, Humbolt, Wallace, Sclater, Agassiz etc.). Famous voyages also provided vital information on the biogeography of plants, e.g. Johann Reinhold Forster who sailed with Captain James Cook (exploration in the South Pacific: HMS *Endeavour* 1768–71); Charles Darwin (voyage around the world: HMS *Beagle* 1831–6); and Joseph Dalton Hooker (to the Southern Ocean and the Antarctic: HMS *Erebus* and HMS *Terror* 1839–43).

These new insights into the geography of plants were extended far beyond the local and regional, and raised some fundamental questions about the distribution of plants worldwide. In particular, remarkably disjunct distributions were discovered. Three examples illustrate the point.

1) *Adansonia*: has a species in Australia, another in Africa and six species in Madagascar.
2) *Nothofagus*: species in Chile, Argentina, Australia, New Zealand, New Guinea, New Caledonia. Fossils of this genus have been found in Antarctica.
3) *Magnolia*: species in east and south-east Asia and North, Central and South America, and the West Indies.

How could such strikingly disjunct distributions have arisen? Here we concentrate on the emergence of three key hypotheses to account for the patterns of extreme disjunction in the distribution of plant groups.

Long-range dispersal: early investigations

Darwin was very interested in the possibility of long-range dispersal, and in Chapter XII of the *Origin* discussed his experiments on the ability of seeds to survive submersion in seawater. He also studied published information on ocean currents and decided that seeds of some species might survive long enough in seawater to float to another country. He considered how driftwood might transport seeds and fruits in the compacted earth trapped behind stones in root systems. He found that viable seeds were found in the crops of dead pigeons floated in artificial seawater for 30 days. After experimentation and the collection of information about the food tract of birds and calculations of the transit time of their ingested food, he noted that 'the crops of birds do not secrete gastric juice, and do not, as I know by trial, injure in the least the germination of seeds'. He also considered the evidence for 'transport of seeds' on the feet and feathers of birds. He concluded that 'living birds can hardly fail to be highly effective agents in the transportation of seeds'. In addition, he examined reports of the transport of earth, stones, brushwood and debris by icebergs that might be important vehicles for the transport of seeds. Surveying all the evidence he had accumulated, he wrote: 'considering that these several means of transport, and that other means, which without doubt remain to be discovered, have been in action year after year for tens of thousands of years, it would, I think, be a marvelous fact if many plants have not become widely transported'.

However, Darwin pondered on the enormous distances between some continents, for example the New and the Old World. He thought that the means of long-range dispersal he had explored would not have been sufficient to account for transport, say, across the Atlantic. Taking into account the fact that 'our continents have long remained in nearly the same relative position, though subjected to great oscillations of [sea] level', he speculated that in the distant past, when the climate was warmer, there were land bridges between the continents. In his words, 'during these warmer periods the northern parts of the Old and New Worlds will have been almost continuously united by land, serving as a bridge, since rendered impassable by cold, for the intermigration of their inhabitants'.

Land bridges: historic ideas

Support for the land bridge hypothesis came from Edward Forbes, Joseph Hooker and Charles Lyell, all celebrated scientists working in the nineteenth century. They believed that, while long-distance dispersal might occur, it could not account for the extreme disjunctions in plant genera. Several hypothetical land bridges were proposed, including South America to Africa (Brazil to Guinea); Africa–India–Madagascar etc. Hooker stressed the affinities of plants from the Southern Hemisphere and considered that there must be undiscovered ancient submerged land bridges, and continents that had formerly been emergent.

The debate about land bridges continued in the early decades of the twentieth century. Lomolino *et al.* (2005) write: 'investigators proposed an incredible number of short-lived land bridges (or island archipelagos) now vanished; former continents, now sunken; or once-joined continents, now drifted apart. During this period tempers often flared as investigators debated their explanations for how groups arrived in such places as Australia, the Galápagos, and the Hawaiian Islands'. Also, in the writing of monographs, a common concern was to establish centres of origin, from which they spread to new areas.

Continental drift

Since the 1960s, a complete revolution has taken place in our views on the positions of continents. As we have seen, Darwin, in company with most geologists, believed that the continents had been fixed in their present positions, although their outlines were not always exactly the same because of sea level changes. However, the first notions of continental drift can be found as early as 1596, when 'the Flemish geographer and cartographer Abraham Octelius speculated in his *Thesaurus Geographicus* that the Americas were "torn away from Europe and Africa . . . by earthquakes and floods"'; and others contributed to the debate in the 19th century (Lomolino *et al.*, 2005).

It was Wegener (1912) who championed the theory that the continents were once joined into a single giant landmass and subsequently moved apart to their present positions. He placed great emphasis on the 'geometric fit' of certain landmasses, e.g. of South America and Africa, and also pointed out the similarity in their rocks at presumed points of previous contact. Wegener believed that the centrifugal forces of the Earth's rotation were the cause of continental drift, a theory that others found implausible. At first his ideas were ridiculed, but advances in the Earth sciences, especially plate tectonics, now offer overwhelming support for the theory of continental drift, but by other 'mechanisms' (the website of the US Geological Survey has excellent coloured maps of plate tectonics).

Research for military purposes revealed no sunken land bridges but giant underwater ridge systems with relatively young rocks (none were older than the Jurassic). Studies of these underwater structures revealed that lava flows out to produce 'stripes' on either side of the ridge. Thus, younger rocks are close to the ridge: older rocks further away. Over time, the magnetic polarity of these rocks has switched, giving stripes on either side of ocean ridges with normal or reversed polarity, and this polarity can be used to determine the latitude in which the rocks formed. The continents themselves sit on 7–8 plates (depending how they are defined) that move inexorably, but slowly, from the ocean ridges. The mid-Atlantic Ridge grows about 10–40 mm a^{-1} – a growth rate roughly the same as the growth of fingernails. At other places, such as the eastern Pacific Ocean, the rate of growth is faster, about the rate at which hair grows. Where the plates come together, their motion leads to earthquake and volcanic activity (including volcanic island formation) and mountain building. For example, the Himalayas were formed about 40–50 Mya, when 'India' collided with Southern Asia. The movement of the plates can be viewed as a conveyor belt system. With the area of the globe remaining more or less the same, and spreading of the ocean floor occurring, at other points of contact between plates, subduction is occurring, with material being carried down and lost in the molten mantle of the Earth.

A number of texts provide comprehensive analysis of the evidence for continental drift, which continues to be an extremely active area of research (for insight into the historical development of plate tectonics see Hallam, 1973; Briggs, 1987; Scotese, 2004; Kearey, Klepeis & Vine, 2009; Frisch, Meschede & Blakey, 2011; Ferrari & Guiseppi, 2011). What drives the movement of the plates is not yet clear. Here, we consider the implications of these important findings for plant biogeography and evolution of the flowering plants. In making analysis, significant information is provided by maps that have been prepared showing the relative positions of the different landmasses through all the different geological periods from the Precambrian (660 million years before the present) until the present day and forward into the future. There are also excellent animations of continental drift; see for example:

- www.tectonics.caltech.edu/outreach/animations drift.html
- www.ucmp.berkeley.edu/geology/tectonics.html

Pangaea and the geographical origin of the angiosperms

As we have seen in Chapter 16, there is much debate about the date of origin of the flowering plants. However, there is increasing evidence that the period, when the angiosperms diversified and became dominant members of ecosystems, coincided with the beginning of the break-up of the relatively continuous landmass of Pangaea c.180 Mya, with the development of the Tethys Sea separating Laurasia from Gondwana, and the subsequent separation of South America, Africa and peninsular India etc. (Fig. 17.1).

Historically, as we have seen, there has been much interest in attempting to discover 'centres of origin' of various groups of plants, especially cultivated plants (Schwanitz, 1966; Ladizinsky, 1998). Regarding the geographical origin of the angiosperms, there have been a number of opinions. For instance, taking into account the break-up of Pangaea, Doyle, writing in 1978, considered that 'it would be premature to localise the origin of angiosperms on one side of the Tethys Sea rather than the other'. In contrast, Cronquist (1988) considered that the flowering plants arose in Gondwana.

More evidence, both fossil and molecular, is now becoming available, but, as yet, it is not possible to determine the site of origin of the angiosperms. Indeed, a definitive determination may be impossible.

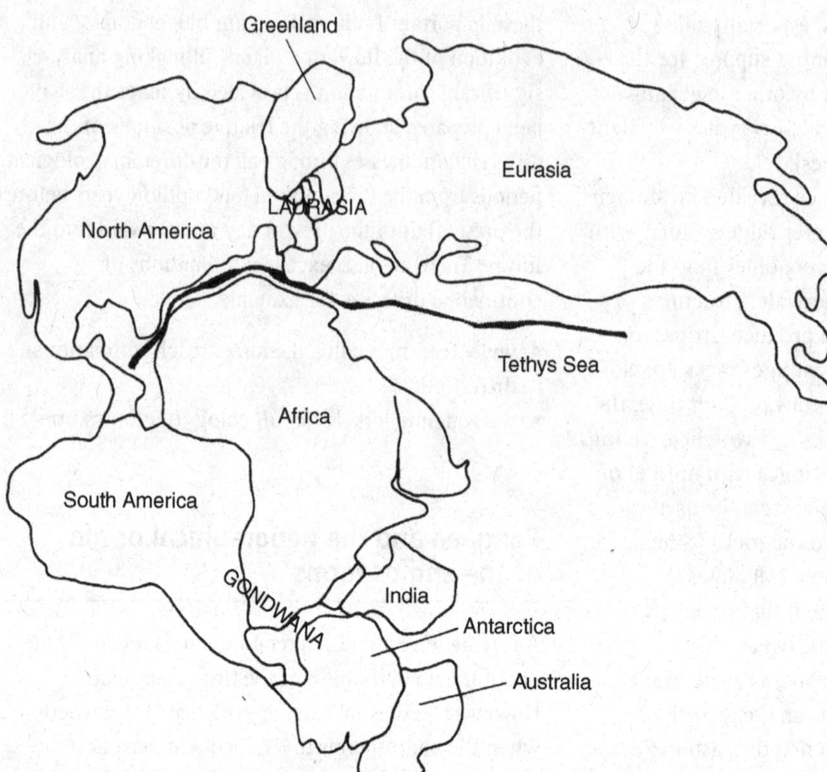

Fig. 17.1. The break-up of Pangaea during the Mesozoic Era. In the Late Triassic, rifting began to separate the northern region of Laurasia from southern Gondwana, and the Tethys Sea began to open between the two areas. During the Jurassic Period, Pangaea was torn in two by rifting, as North America separated from South America and began to break away from Africa. (From Avers, 1989.)

However, using the modern geographical names for territories, MacDonald (2003) lists three possible areas for the centre(s) of origin of the flowering plants.

A) A tropical origin with later migration towards the poles, from an 'African/South American' centre.
B) Origin in the South Pacific region (Malay area, eastern Australia, New Guinea, New Caledonia). This hypothesis is supported by the discovery in New Caledonia of the shrub *Amborella trichopoda*, which, phylogenetic analysis reveals, is a sister group to the rest of the flowering plants.
C) Another possible site of origin is China where significant ancient fossils of *Archaefructus liaoningensis* (c.125 million years old) have been discovered.

Ancient extinctions

As we have seen above, biologists seek to investigate three major influences on plant biogeography, namely dispersal, speciation and extinction. With the recognition that many contemporary rare and endemic species are under threat from human activities (Chapter 18), 'mass extinction' has become a topic of great interest.

Raup (1991) speculates that as many as 5 to 50 billion species may have existed on the planet since its origin, and perhaps as many as 50 million species exist today. However, there is increasing evidence that, in several geological periods, widespread extinction of species and higher groups has occurred.

1) Cretaceous–Tertiary 65.5 Mya (The K–T event – an abbreviation of the German for Cretaceous–Kreide: more recently called the Cretaceous-Palaeogene);
2) Triassic–Jurassic 201 Mya;
3) Permian–Triassic 252 Mya;
4) Late Devonian c.375–360 Mya;
5) Ordovician–Silurian c.450–440 Mya.

Major extinction events occurred in the Permian (in which perhaps as many as 75% of species died out), and at the Cretaceous–Tertiary (K–T) boundary.

Did these extinctions occur simply by natural selection, as we envisage in the microevolution of contemporary populations?

The K–T extinction

Alvarez *et al.* (1980) put forward a radical hypothesis: the mass extinction at the Cretaceous–Tertiary (K–T) boundary 65.5 Mya was caused by the impact on Earth of an extra-terrestrial body. Evidence for their view came from the finding that at many sites around the globe, a layer of clay, rich in the element iridium, has been found sealed in the rocks of the K–T boundary. Iridium is much more common in asteroids and meteorites than on the Earth's surface. Because this suggestion had come from physicists, who were working outside the geological community, this hypothesis was, at first, treated with some scepticism (Nichols & Johnson, 2008). However, at the northern edge of the Yucatán peninsula in Mexico, a crater of the appropriate size and age was discovered, pointing to impact with an asteroid of *c.* 10 km diameter.

Some geologists have continued to dispute the asteroid impact hypothesis as the sole cause of this mass extinction (Glen, 1994), and have proposed that other factors – sea level and climatic changes, or major episodes of volcanism – acting singly or in combination, might have contributed (Archibald & Fastovsky, 2004). There is also speculation as to whether there might have been multiple impacts by extra-terrestrial bodies. However, at a special meeting of 41 international experts, it was agreed that the asteroid was the most likely cause of the K–T extinction (Schulte *et al.*, 2010).

Many scientists have considered the possible effects of an asteroid hitting the Earth (see Nichols & Johnson, 2008). Evidence points to severe immediate effects worldwide. Crucially, impact is likely to have produced so much atmospheric dust that sunlight reaching the Earth's surface was reduced (by an estimated 10–20%). Given the rock type in the area of impact it is also likely that sulphuric acidic rain was produced. These aerosols would be expected to have severe effects on photosynthesis both on land and at sea. Food chains were catastrophically affected, with losses of herbivores and predators feeding on them. Some have speculated that firestorms may have also occurred, and, if so, these could have triggered a temporary greenhouse effect (see Chapter 18). While there is no evidence of widespread firestorms in the site studied in detail in western North America, they might have occurred elsewhere (Nichols & Johnson, 2008).

Considering the biological effects of the asteroid impact, certain groups – dinosaurs, some cephalopods, planktonic foraminifera etc. – became extinct. Mammals (rat-size or smaller) and dinosaurs coexisted for more than 100 million years but, with the extinction of the non-avian dinosaurs, mammals came to dominate the vertebrate world (Kemp, 2005). Clearly, with the loss of the non-avian dinosaurs, the further evolution of the mammals and other groups that survived the mass extinction event must have greatly influenced the course of plant evolution. At first there was no evidence for extinctions of angiosperm groups (Hughes, 1994). However, McElwain & Punyasena (2007) reviewing recent studies note that:

> five mass extinction events have punctuated the geological record of marine invertebrate life. They are characterised by faunal extinction rates and magnitudes that far exceed those observed elsewhere in the

geological record. Despite compelling evidence that these extinction events were probably driven by dramatic global environmental change, they were originally thought to have little macroecological or evolutionary consequence for terrestrial plants. Now new high-resolution regional palaeoecological studies are beginning to challenge this orthodoxy, providing evidence for extensive ecological upheaval, high species-level turnover and recovery intervals lasting millions of years.

There is also new information from a close study of sites in western North America. Here there is strong support (through the examination of fossil leaves and pollen) for mass extinction of plants and devastating disruptions to plant communities (Nichols & Johnson, 2008).

Considering the results of the detailed study of the K–T mass extinction, what are the implications for our general understanding of plant extinction? Four important points should be stressed.

A. It is clear that some species were in a sense unlucky, being in the wrong place at the wrong time (Raup, 1991).
B. Given that other mass extinctions have been recorded in the geological record, Raup (1991) has proposed that they were also caused by extraterrestrial impacts. This suggestion is hotly debated (see Glen, 1994; Archibald & Fastovsky, 2004).
C. Extraordinary advances have been made in our understanding of the factors influencing microevolution. However, in considering the evolution over geological timescales, we must face the certainty that truly extraordinary events have taken place that are beyond the realms of ordinary experience. At several points in this book we consider whether the present is a sufficient guide to the past: here is an instance where there is clear evidence that this is not the case.
D. A more general question to be faced is whether in evolution, continual step-wise change in populations and species is the norm. Van Valen (1976) proposed that such continual evolution is necessary if organisms are to survive and reproduce in continually changing biotic and climatic environments. This concept was inspired by the character of the Red Queen in Lewis Carroll's *Through the Looking Glass*, who said, 'it takes all the running you can do, to keep in the same place'.

In his classic hypothetical evolutionary tree published in the *Origin*, Darwin illustrated situations where some lineages continued to diverge over many generations (Fig. 17.2). However, others indicated in the diagram either survived without new speciation or went extinct.

It is of considerable interest that some modern plant genera appear in essentially the same form as ancient fossils (Fig. 17.3). In contrast to these ancient 'conservative stocks', new species, such as the grass *Spartina anglica*, have originated, as it were, in the 'last second' on a geological timescale.

However, the crucial unresolved question is the *balance* of gradual and abrupt events in particular cases (Williamson, 1981). Important debate on this issue has been initiated by the 'punctuated equilibrium' concept proposed by Gould & Eldredge (1977, 1993). This challenges the view, held by some, that continual change (so-called phyletic gradualism) is the norm, considering that, as reflected in the fossil record, most species changes very little or else fluctuated mildly in morphology with no apparent direction. While change may occur from generation to generation, this change is not cumulative. Thus, for most species it is proposed that periods of so-called 'stasis' are the rule, with significant change – punctuated change – happening more rarely and rapidly at periods of major upheavals such as mass extinctions. These two polar views are illustrated in Fig. 17.4.

Evidence in favour of the punctuated model has been found for a number of animal groups (Gould & Eldredge, 1993; Gould, 2007). However, critical appraisal of the concept of punctuated equilibrium has come from many quarters (Maynard Smith, 1983; Sterelny, 2007; Sepkoski, 2009; Sepkoski & Ruse,

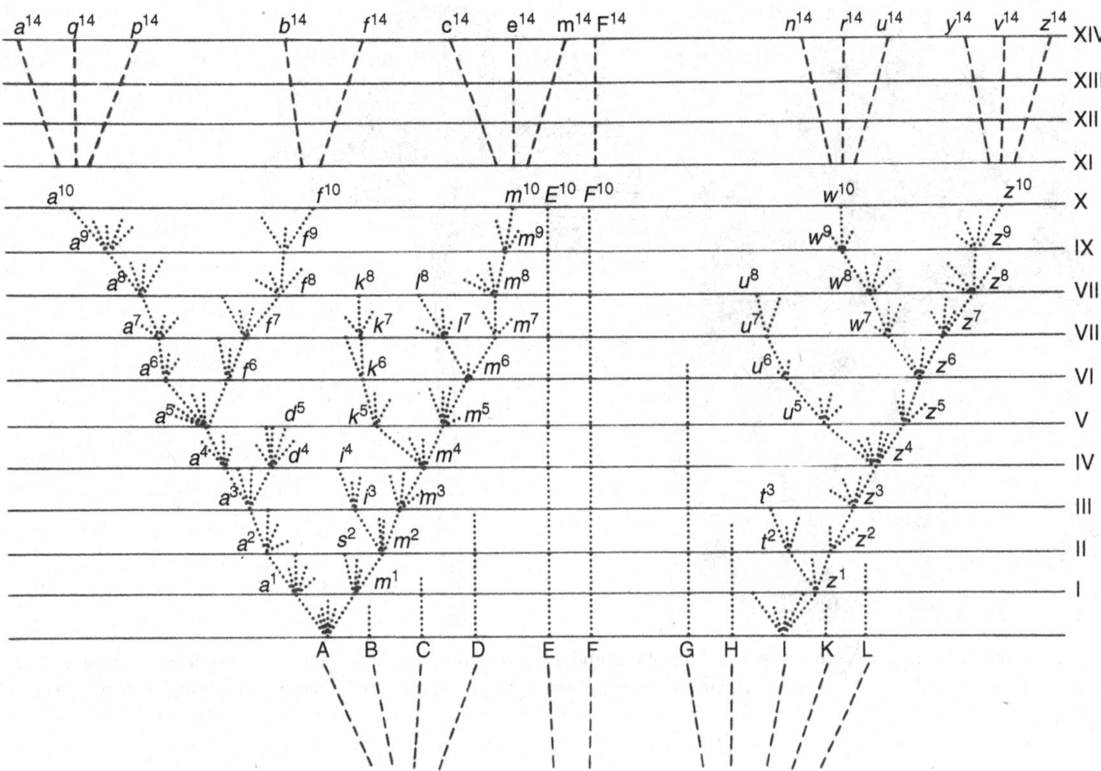

Fig. 17.2. A hypothetical evolutionary tree from Darwin's *On the Origin of Species*, illustrating descent with modification through natural selection. Eleven different species (A–I, K, L) of a genus (shown at the bottom of the illustration) diverge over time, shown by horizontal lines labelled with Roman numerals and each representing 1000 or more generations. Some of the original species, such as A and I, diverge more than others and produce new varieties, such as *a1*, *m1* and *z*1. These new varieties in turn diverge and perhaps give rise to new species, such as *p*14, *b*14 and *n*14, after thousands of generations. Darwin believed that some species diverge enough to produce new species. Others, such as E and F, remain relatively unchanged or, more likely, go extinct (B, C, D, G, H, K, L). (From Avers, 1989.)

2009). They have emphasised that, for most groups of organisms, the fossils record is too imperfect to resolve questions about gradual step-wise versus stasis with episodes of abrupt change.

However, the fossil record is not the only source of evidence. Genomic studies of a range of angiosperm groups reveal that abrupt change occurred in the distant past through ancient polyploidy. Using phylogenetic approaches in the dating gene duplications from a range of flowering plant groups, Fawcett *et al.* (2009) have made a very interesting proposal. 'Ancient polyploidy' occurred in several plant lineages (including the Fabaceae, Asteraceae, eudicots and monocots), facilitating the survival and diversification of 'these species-rich plant clades' in the Cretaceous–Tertiary K–T mass extinction event (Soltis & Burleigh, 2009). If this conclusion is correct, this points to a major evolutionary role for polyploidy, very different from the earlier view (see Soltis & Soltis, 1993) that polyploidy was an evolutionary dead end.

Commenting on the work of Fawcett and associates, Soltis & Burleigh (2009) note the difficulty of detecting and dating gene duplications indicative of polyploidy, but they feel that this 'exciting' speculative hypothesis enlarges the possible

Fig. 17.3. A fossil leaf of *Ginkgo* from Jurassic rocks (left) and a living leaf of *G. biloba* (right). Details of the taxonomy and former distribution of *Ginkgo* and related plants are given in Tralau, 1968. (Fossil: photograph from Natural History Museum, London.)

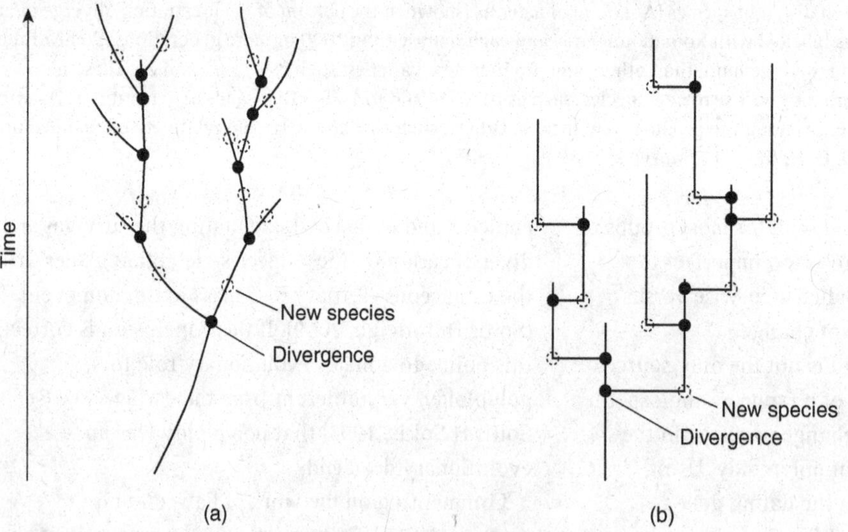

Fig. 17.4. Speciation by gradual divergence (a), as compared with the pattern of evolutionary stasis and burst of speciation postulated by the punctuated equilibrium model (b). (From Avers, 1989.)

evolutionary role of polyploidy. However, they observe that

> ultimately it may be impossible to determine whether polyploidy enabled plant lineages to survive the massive K–T extinctions, but the adaptive hypotheses born from such speculation may be relevant to understanding the current patterns of diversity and perhaps patterns in the future. For example, if polyploidy allowed certain plant lineages to survive and adapt during the tremendous global changes of the K–T period, does polyploidy confer a similar advantage in the current period of global warming?

The Pleistocene

Turning to other major geological events, there is evidence that, over its long history, the Earth has been subject to at least five major ice ages. The latest – the Pleistocene – extended from 2 588 000 to 11 700 years ago. Many lines of evidence have now become available of the effects of this ice age period, through studies of changing landforms; analysis of stratification of deposits (including peat, muds, marine deposits, ice etc.); investigations of biological remains (such as leaves, seeds, fruits, pollen, animal remains etc.) preserved in anaerobic deposits; climatological analysis from air trapped in ice, and ocean core samples; and phylogenetic investigations. These increasingly multidisciplinary studies are revealing the fascinating biogeographical changes in the distribution of plants.

First, it is important to note that in the Pleistocene period the continents, as we see them on the map today, have been more or less in their present position, and that this geological period has been marked by repeated continental-wide glaciations. In the Northern Hemisphere, individual glaciers fused into huge ice sheets, e.g. the Cordilleran in North America and the Fennoscandian ice sheet across northern Europe. In the Southern Hemisphere, where there is less land in the temperate and subarctic zones, glaciations were confined to high mountains, in New Zealand, Tasmania, Chile etc. In Africa, glaciations affected the highest mountains in east Africa and the Atlas Mountains (Lomolino *et al.*, 2005). A number of glaciations have been described with intervening interglacials. During the glaciations, ice covered large areas (to perhaps 30% of the Earth's surface), with zones of permafrost at their margins. At the height of glaciations, huge volumes of water were contained in glaciers, and sea level fell across the globe, with many former 'islands' becoming connected to continental landmasses, in some cases facilitating the migration of plants and animals. In the interglacial periods, sea level rose, creating island systems along continental margins. This process is complicated by the fact that in many places the 'land' itself often rose as the weight of the ice was removed. Ice had major effects on the geomorphology of the land, with scouring of valleys, deposits of glacial debris impounding huge lake systems producing changes in river systems etc. During the glaciations, climate was influenced worldwide, as rainfall was reduced because of lower evaporation from the land and sea (Clark, 2009).

There has been considerable speculation about the cause(s) of the glaciations in the Pleistocene (Ehlers, Gibbard & Hughes, 2011). There is general agreement that Milankovitch cycles – named after their discoverer the Serbian Milutin Milankovitch – are a major factor. Lomolino *et al.* (2005) give details.

> Earth's orbit is not perfectly circular, but varies elliptically or 'eccentrically' with a period of 100,000 years. Second, the tilt of the Earth on its axis – its obliquity – varies from 22.1 degrees to 24.5 with a 41,000-year period. Finally, Earth's orientation, or procession, wanders, with the axis of the North Pole shifting from one 'north star' (presently Polaris of Ursa Minor) to another (Vega of Lyra) with a periodicity of approximately 22,000 years. The combined effect of these cyclical changes in the Earth's orbit causes substantial fluctuations in the amount and seasonal variation in the solar energy striking the Earth, ultimately

combining in a complex fashion to create the glacial–interglacial cycles and climatic reversals of the Pleistocene.

Having discussed a number of major geological and climatological influences on plant distribution, we now consider the impact of these on biogeography.

Advances in plate tectonics on the interpretation of plant distributions

The following account draws on the admirable accounts by De Queiroz (2005a, b). Reviewing progress in biogeography, he notes that: 'during the 1960s and 1970s, two developments ignited a revolution in historical biogeography . . . The first was the validation of plate-tectonics theory, which provided vicariance explanations on a global scale . . . The second important development was the spread of cladistic thinking. Cladistics provided an objective method for reconstructing phylogenetic relationships'. Four key insights concerning plate tectonics are important.

A. Many taxa on land sharing a common ancestry have become separated onto different landmasses by plate movement. The distribution patterns seen today, for example, in *Adansonia, Nothofagus*, are explicable in terms of their ancestors once occupying a common landmass, Gondwanaland, that broke apart, 'carrying' them to their present highly disjunct distributions separated by vast stretches of ocean. This is the classic vicariance pattern of occurrence, in which a once potentially widespread species distribution has become divided and the separated taxa evolve in isolation. Not only have living plants been transported by the movement of plates: rocks, and the fossils they contain, have also been subject to separation through continental drift. In parenthesis, we may note that biogeographic studies of *Nothofagus* are in agreement with one of the most supported theories of the sequence of the break-up of Gondwana (Swenson, Hill & McLoughlin, 2001). But, for completeness, it should be recognised that alternative sequences of events have been proposed by some geologists and biogeographers, involving vicariance, dispersal and extinction (see Sanmartín *et al.*, 2007, for details).
B. Some separations are older than others. Those floras, long separated geographically from others, are often rich in endemic species. For example, the Australian fauna and flora are highly distinctive, reflecting the c.50 million years of evolution in isolation. However, they retain some affinities with South America. For example there are marsupials on both continents indicating that they were, in earlier times, part of the same landmass.
C. The movement of continents may result in the 'rejoining' of landmasses after a period of separation, but with major landform changes. For example, India, once an island with biota evolving in isolation, collided with what is now the Asian landmass, creating the Himalayas. This 'rejoining event' provided new evolutionary opportunities and challenges for plants. The precursors of North and South America were once joined in Pangaea, a contact that was later severed. A land bridge between the two landmasses later formed c.3 Mya offering opportunities for migration of flora and fauna – the so-called Great American Interchange (but see below).
D. Islands are of different types. Some were once part of continental landmasses, separation being associated with changes in sea level. Such separation has often been temporary in geological terms. In contrast, oceanic islands may never have been connected to a continental landmass, but have arisen, far from continental sources of biota, as a consequence of volcanic activity associated with undersea trenches and ridges. Some island systems, such as the Galápagos, are associated with 'hot spots' of volcanic activity. Separate islands of different ages have been formed, and, as a consequence of plate movement, islands 'drift away' from the hot spot before a new island is born.

Against this geological background, vicariance biogeography flourished. Looking at the strengths of the approaches, De Queiroz (2005b) points out that they 'provided unifying explanations for the disjunct distributions of many taxa . . . In addition, vicariance hypotheses were clearly falsifiable: the importance of specific events could be refuted if the branching history of lineages did not match the hypothesised vicariance history.'

What was the reaction to 'long-range dispersal' theories? 'Cladistic biogeographers claimed that hypotheses of dispersal were not falsifiable *because all relationships could be explained by some dispersal hypothesis*' (my italics). Vicariance biogeographers often sounded contemptuous of a dispersalist biogeography; for example, Nelson (1979) described dispersalism as 'a science of the improbable, the rare, the mysterious, and the miraculous . . . Dispersal was conceded for oceanic islands, but, for cases that could be explained by either vicariance or dispersal, most biogeographers assumed that vicariance was the more probable explanation.'

However, over the past 20 years there has been a dramatic change of outlook. De Queiroz (2005b) writes:

> vicariance hypotheses require that speciation and the corresponding fragmentation of areas must occur at the same time, and thus that information on absolute timing of speciation events is crucial in evaluating such hypotheses. The new support for oceanic dispersal has come from information about the timing of speciation, fueled by the development of improved methods of DNA sequencing and estimating lineage divergence dates based on molecular sequences. In a typical case, sister taxa occur on opposite sides of an ocean barrier and the competing explanations are of a relatively old vicariance event or a more recent oceanic dispersal. In many such cases, the molecular divergence (e.g. in nucleotide or amino acid sequences) apparently is too small to be explained by vicariance. For example, Baum, Randall & Wendel (1998) found that DNA sequences for the internal transcribed spacer region for baobab trees (*Adansonia* spp.) suggested a divergence age of 5–23 million years for taxa in Africa versus Australia. Even the oldest estimate is far too recent to be explained by the tectonic separation of these continents, which occurred c.120 million years ago.

Turning to dispersal theories it is important to note that they are thought to be 'unfalsifiable' and thus 'unscientific' (De Queiroz, 2005). 'However, this can be countered by noting that, if plausible vicariance hypotheses are falsified, then dispersal is supported by default. In addition, specific dispersal hypotheses make predictions about divergence dates and locations of sister lineages, and thus, are subject to refutation.'

Modern phylogeographical investigations of plant distributions

Investigations test models exploring a range of factors. 'Modelling is rapidly becoming more sophisticated' (see Ronquist & Sanmartín, 2011). Their review discusses a wide range of approaches including the following.

- 'Ecological niche modelling (ENM) uses the association between observed species occurrences and environmental variables (e.g. temperature, precipitation) to predict species ranges.'
- Also, there are now 'sophisticated online tools for paleogeography and paleoclimate reconstruction using a wide range of information'.
- 'Combined with paleoclimate data, ENM can project range distributions into the past' (Smith & Donoghue, 2010).
- Other modelling approaches examine different areas of historical biogeography, including predicting species distributions in the past by *hindcasting* using climate envelope models (CEMs). To make the study of hindcasting, present distributions and forecasting tractable, geographic areas

under study are divided into grid systems. Considering hindcasting, it is important to note that reconstructing past climates employs proxy methods by studying evidence preserved in rocks and in mountain and polar ice sheets (e.g. analysis of trapped air); studies of tree rings (tree growth responds to climatic variables); and sediments (including studies of chemical signatures, coral growth rings, pollen, shells and macrofossils) (see Bradley, 1985; Cronin, 2010 for details of the methods of palaeoclimate reconstruction). These surrogate and proxy measures of past climates have limitations, but a great deal of research has extended and refined earlier studies. An assessment of present-day environmental parameters defining the niche of a species, considered together with the best estimates of past climate, allows, through modelling, prediction of its past distribution. Likewise, projections of future climate change, in conjunction with information on the niches of species in the present day, allows the devising of models for *forecasting* the effects of future climate change on species distributions.

- Gravity models have been developed for backcasting and forecasting such important areas as biological invasions of inland lakes (MacIsaac *et al.*, 2004).

Factors considered in modelling

a) As we see below in discussion of various case histories, long-range dispersal is a major factor to consider (Levin *et al.*, 2003).

b) So too are other key factors, such as climate change and continental drift, that have broken up a once widespread distribution.

c) The availability of land bridges in certain geological periods allowing the possibility of geographical dispersal. A much more nuanced view of this subject is now taken by biogeographers, with close study of situations where temporary marine regressions have facilitated dispersal.

d) Whether, under adverse climatic conditions, some taxa escaped extinction by surviving in 'refugia', to which they are currently confined or from which they expanded.

e) Whether species migrate, from refuges or elsewhere, by diffusion, or by the establishment of an expanding front of outliers with later filling-in to give a more continuous geographical distribution. Ronquist & Sanmartín (2011) consider a wide range of models, including diffusions, dispersal to 'islands', vicariance and reticulate models. Often biogeographers have selected certain extreme situations for study and these can be very revealing. Thus, many botanists have studied establishment of migrants on remote 'islands' of various types, physical islands in oceans and lakes, isolated lakes, sky-islands on mountain tops, areas of special rock type such as serpentine etc.

f) These studies reveal that while some species, such as weeds, are more or less cosmopolitan on land, others are restricted in their geographical distributions: they are said to be endemic – being restricted, say, to a single island, lake, or mountain top; or to a particular country or continent etc.: they do not occur naturally elsewhere. Biogeographers have paid special attention to endemics, drawing a distinction between palaeo-endemics, plants that were once more widespread, but now have restricted 'relict' distributions, and neo-endemics, species that have just arisen, such as recently evolved polyploids, e.g. *Senecio cambrensis*, *Spartina anglica* etc. A detailed understanding of historical plant geography provides a firm base on which to consider the current threats, to endemics and plants species, often by human activities, and how, through conservation measures, declines might be reversed (Chapter 20).

g) Modelling also considers how adaptive radiation occurs, and how species might decline and become extinct.

h) Also, current microevolutionary patterns and processes are considered, such as hybridisation, polyploidy and apomixis, and how these contribute to different patterns of geographical distribution.

i) In the study of biogeography, dating events is of prime importance, involving estimation by radiometric methods, and geological and fossil calibrations. For example, phylogenetic trees with molecular clock timelines are calibrated by fossils, and provide a major source of evidence for many recent investigations. In interpreting the findings of these studies it is important to recall the discussion in Chapter 16, where we explored the limitations of molecular clock analyses, and the complexities, not only in dating fossils, but also their use in the calibration of phylogenetic trees. These issues are clearly crucial to a critical understanding of the advantages and limitations of phylogeographic approaches.

In addition, Graur & Martin (2004) highlight an important additional concern about the employment of molecular clock estimates for dating divergence events.

> Because the appearance of accuracy has an irresistible allure, non-specialists frequently treat these estimates as factual . . . the illusion of precision was achieved mainly through the conversion of statistical estimates (which by definition possess standard errors, ranges and confidence intervals) into errorless numbers . . . The time estimates of even the most ancient divergence events were made to look deceptively precise.

Taking an overview of the present situation, phylogeographic research is benefiting greatly from the widespread use of modelling using sophisticated programming; advances in our understanding of vegetation and climate history, as revealed by the geological record and climatological research; and the exciting insights offered by all aspects of molecular research, in particular phylogenetics. However, as we see below, the subject is not without its challenges.

There is an enormous literature on phylogeography. Here, we examine a number of case histories testing crucial hypotheses, to illustrate important areas of research relevant to understanding plant evolution. We look first at some key factors studied in isolation, and consider case studies that employ a range of investigative approaches.

Long-range dispersal: new insights

As we have seen above, it has long been recognised that human activities are important in the intentional and accidental long-distance dispersal of plants. For example, the investigations of Wichmann *et al.* (2009) reveal that seeds may be widely dispersed on human footwear.

Debate about the role of non-human dispersal has a very long history (see Ridley, 1930; Carlquist, 1965; Cain, Milligan & Strand, 2000).

Historically, the potential for such dispersal has often been judged by the morphology of fruits and seeds, the distribution of taxa, and observations of the behaviour of animals. For instance, Carlquist (1965) estimated that about 20% of the flora of Easter Island arrived on the island as seed in mud, on the feet of birds. Porter (1983) suggested that mud attached to animals was responsible for 42/378 original disposal events (11%) on the Galápagos archipelago.

Direct investigation of long-range dispersal is very difficult. However, some of the speculations of the theorists have been examined. Thus, recent investigation by ornithologists, who ringed more than 10 000 birds on the Azores, Canaries and Galápagos, led them to conclude that the presence of mud on feet and feathers is likely to be 'extremely rare' (Nogales *et al.*, 2012). Instead, they draw attention to the possible importance of other dispersal mechanisms mentioned in the early literature, namely seed transportation in the guts of (a) omnivorous birds of large body size that make regular movements between islands; (b) specialist seed predators that regularly migrate; and (c) secondary dispersal of seeds/fruits by predatory wide-ranging migratory birds that regularly eat granivorous species of birds and lizards.

Recent studies of the gut content of migratory ducks, geese and swans suggest the great potential for successful long-range dispersal of seeds by birds (see Brochet *et al.*, 2009).

Molecular genetic studies are increasing our knowledge of the origin, relationships and means of dispersal of many taxa (Nogales *et al.*, 2012). Inferences about the history of populations/species etc. can be drawn from branching structures in trees and patterns of genetic variability (see below). Advances in other fields of research also offer new information relevant to ocean and wind dispersal (Gillespie *et al.*, 2011). In the past, it was often believed that dispersal by air and water was essentially a random process. However, modern investigations using satellites and research vessels are revealing further insights into 'prevailing tracks' of ocean currents and air movements. Ocean currents and drift patterns have been closely studied and information relevant to the rafting of plant propagules has been discovered. They are also providing more details of the seasonality and wind speeds associated with storms and hurricanes. Historically, many insights into the speed and routes of bird migration have come from the fitting and recovery of metal tags. The successful attachment of GPS loggers to birds also has the potential to extend our understanding of long-distance dispersal (see for example Lenz *et al.*, 2011, for tracking experiments with Trumpeter Hornbills).

As we shall see, in many research papers on biogeography, there is evidence that long-range dispersal has occurred in particular groups in the past. However, case histories are often (and should always be) interpreted in terms of probabilities rather than certainties.

Single and recurrent long-range dispersal

Emerson (2002) discusses a number of models of colonisation between mainland and islands of various sorts, and between the islands of an archipelago. He points out that the most parsimonious explanation for long-distance dispersal of a taxon from a mainland to an island is that such an event occurred only once. Until the advent of molecular studies such a hypothesis could not be properly tested. Now a number of important case histories have been published.

- There is a striking disjunction in the distribution of the genus *Empetrum* with species in the Northern Hemisphere (north of 40° latitude) and in southern South America (south of 36° latitude). In a study by Popp, Mirré & Brochmann (2011) a wide range of samples were collected from across the distribution range, and sequence data from two low copy nuclear and two plastid DNA regions were employed in the devising of phylogenetic trees. To provide a timeline, the analyses included fossil-calibrated relaxed molecular clock approaches. It was concluded that 'a single dispersal by a bird from north-western North America to southern-most South America taking place in the Mid-Pleistocene, is sufficient to explain the disjunction in crowberries'. Although the present migration patterns of migratory birds might not exactly coincide with those of the past, there are a number of species that might have facilitated the long-distance dispersal. The Whimbrel and the American Golden Plover are both reported to eat the berries of *Empetrum* before they migrate from their breeding grounds in North America to overwinter in South America.
- Romeiras *et al.* (2011) have examined the endemic members of the genus *Echium* in the Cape Verde archipelago, using nuclear and plastid markers. Their results 'suggest a recent single colonization event' with later 'inter-island colonization'.

However, Emerson (2002) points out that it is difficult to be sure that 'single event' migrations took place. Indeed, in some cases, molecular evidence provides direct persuasive evidence for multiple long-distance events. For example, the fern *Asplenium hookerianum* occurs on the Chatham Islands 800 km west of New

Zealand, where it is common. In a study of cpDNA sequences (*trnL-trnF*), two haplotypes were detected that differed by four mutational events. One of these haplotypes was also detected in samples from the New Zealand mainland. These findings support the notion that airborne spores of *Asplenium* have on more than one occasion been carried from New Zealand to Chatham Island (Shepherd, de Lange & Perrie, 2009).

- Investigations of other groups of plants have provided evidence for multiple colonisations: for example, *Hedera* to Macronesia (Vargas *et al.*, 1999); and *Asteriscus* and *Sonchus* alliances to Macaronesia (Kim, Crawford & Jansen, 1996a; Kim *et al.*, 1996b; Francico-Ortega *et al.*, 1999).

Evidence for back colonisation

While colonisation is most likely to be a one-way process from mainland to island or between the islands of an archipelago (Trusty *et al.*, 2005), models have also been tested that consider the possibility that back colonisation from island to mainland can occur (Emerson, 2002). For instance, in a study of the genus *Aeonium*, back colonisation from the Canary Islands to North Africa is a distinct possibility (Mes, van Bredrode & Hart, 1996; Jorgensen & Frydenberg, 1999). In studies of other groups, for instance, Caujape-Castells (2004) studying *Androcymbium*, recolonisation from the Canaries to Africa is a likely explanation of the available evidence. Looking at a further example, Jaén-Molinia *et al.* (2009) have studied a wide range of samples of the genus *Matthiola* from Madeira, the Canaries and North Africa using molecular phylogenetic methods. They conclude that Madeiran and Canaries taxa had independent origins, from Mediterranean and north-west African sources, respectively. With regard to the Canaries taxa, they report 'high genetic differentiation between populations of *Matthiola* in Fuerteventura and Lanzarote' and suggest that this finding supports the hypothesis of 'independent founder events from the same mainland congener to either island'. In addition, the phylogenetic relationships of the Canaries and Moroccan taxa suggest 'further back colonisation' from Canaries to the continent.

From which source(s) did long-distance migrants originate?

Phylogenetic evidence provides new insights into the origin of 'migrants'. Islands provide the ideal place to test hypotheses. For instance, it might be postulated that all the taxa colonising an island came from the same source. Molecular phylogenetic studies have provided major insights on this question. For example, a molecular phylogenetic investigation of internal transcribed spacers of the nuclear ribosomal DNA has been carried out on the endemic genus of the Galápagos, *Darwiniothamnus* (Asteraceae). Sampling included many related taxa from other areas (Andrus *et al.*, 2009). It was concluded that *D. lancifolius* and *D. tenuifolius* were related to *Erigeron bellidiastroides*, an herbaceous species endemic to Cuba. In contrast, *D. alternifolius* proved to be sister to two Chilean species of *Erigeron*. This shows that flora of the Galápagos has affinities with both North America (*c.*6% Mexican or Caribbean) and South America (the majority of the flora). Given the extreme isolation of the Galápagos islands (*c.*1000 km from the nearest mainland), long-distance dispersal, from South America and Caribbean sources, is a likely explanation. However, perhaps dispersal may also have involved former land areas that are presently drowned seamounts (Andrus *et al.*, 2009).

Turning to another example, Svalbard is an extremely remote isolated group of islands in the Arctic Ocean, situated between East Greenland, northern Scandinavia and north-west Russia. Sea ice connects Svalbard to these areas during the winter. Evidence suggests that at the Last Glacial Maximum the archipelago was almost entirely glaciated, and it is debated whether the flora survived *in situ*.

To determine the likely source of the plants colonising

the islands, Alsos *et al.* (2007) examined 4439 samples of 9 different species using DNA fingerprinting (AFLP). All the closest possible land sources were included in the sampling. The samples of *Arabis alpina* did not reveal enough genetic variation to permit analysis. *Saxifraga rivularis* proved to be genetically variable with private markers in the Svalbard populations, suggesting that this species may have survived the Last Glacial Maximum on the archipelago. For the other 7 species, there is clear evidence, from the genetic fingerprints, for the likely source of the long-distance colonists. A predominant source was north-western Russia. Colonisation from Greenland was detected; but Scandinavian sources were rare. (Codes % allocation to G = Greenland; S = Scandinavia; R = north-west Russia.)

Empetrum nigrum G 100,
Betula nana S 9, R two regions 41+ 3,
Cassiope tetragona three regions G 4+1+45, S 6
Dryas octopetala G 1, S 2, R two regions 11+ 67,
Vaccinium uliginosum G 12, R 76
Rubus chamaemorus R 64
Salix herbacea S 3, R 88

The design of the experiment did not make it possible to locate source areas precisely or estimate the frequency of long-distance dispersal. Considering the means of dispersal, 'possible dispersal vectors are wind (which may have carried propagules through the air or over snow and sea ice), drift wood and drifting sea ice, birds and mammals ... Dispersal from Russia may have been facilitated by driftwood. Bank erosion along Russian rivers routinely results in logs and other debris finding their way onto drifting sea ice, which reaches Svalbard by means of surface currents'. Considering the genetic diversity of the 7 species on Svalbard, multiple colonisation events are indicated ('minimum 6–38 plants of each species must have successfully established and survived in Svalbard, implying that many more propagules actually reached the archipelago'). It is concluded that 'the recurrent glacial cycles have probably been selected for a highly mobile arctic flora' distributed by wind and sea ice.

Disjunctions: long-distance dispersal or vicariance?

As we have seen, historically, botanists considering plant geographical disjunctions explained them either by the existence of previous land connections or by long-distance dispersal. Given the current insights from molecular phylogenetics, plate tectonics etc. it is important to stress again that contemporary biogeographers do not see these factors as exclusive alternatives: much more nuanced pictures of the origins of disjunction are emerging involving both vicariance and long-distance dispersal.

We may take as an example studies of groups of tropical plants on either side of the Atlantic. In a wide-ranging review of phylogenetic and other evidence, Renner (2004) considers that trans-Atlantic disjunctions at the family level may date back to the break-up of South America and Africa *c.*95 Mya. However, 'at least 110 genera contain species on both sides of the tropical Atlantic ... Molecular phylogenies and age estimates are available for 11 distinct genera, tribes, and species. Inferred directions and modes of dispersal can be related parsimoniously to water currents between Africa and South America and to exceptional westerly winds blowing from northeastern Brazil to northwest Africa'. She continues: 'tangled plant parts ("floating islands") are constantly carried into the tropical Atlantic from the deltas of the Congo, Senegal and Amazon rivers, and some enter the conveyer belt-like currents that transport debris (including plastic trash; Barnes, 2002) in either direction across the Atlantic. Because of their speed, equatorial currents can transport large floating objects with wind-exposed surfaces across the Atlantic in less than two weeks.' For further discussion of the 'floating island' model of dispersal, as it relates to other areas of the world, see Zhou, Yang & Yang (2006). For a critical review of this and other mechanisms of long-range dispersal, see Nathan *et al.* (2008).

Further intercontinental tropical disjunctions are reviewed by Givnish & Renner (2004). They stress that, to understand any particular tropical taxon disjunction, many factors must be considered including Gondwana break-up, immigration across landmasses and transoceanic dispersal etc.

The same important point is emphasised by studies of major disjunctions occurring amongst the taxa found both in dry environments of western North America and west Eurasia. Kadereit & Baldwin (2012) have reviewed 25 published cases. Different individual explanations have been devised to account for these extreme patterns of disjunction. In 9 groups 'phylogenetic and other evidence (often circumstantial) was found for parallel evolutionary shifts of widespread Northern Hemisphere lineages into dry environments in western Eurasia and Western North America'. For 6 taxa, and possibly 4 others, 'long-range dispersal' was postulated. For a further group, parallel evolution or long-range dispersal might explain the disjunction. In 3 out of the 25 cases, overland migration is invoked, as there is evidence for former land connections between the two landmasses.

Milne (2006), writing about the Eurasiatic/North American disjunctions in relation to the history of Tertiary relict genera, i.e. those that now occur disjunctly in part of Eurasia and North America, makes the point that 'vicariance of populations previously connected across land bridges is almost certainly the principal means by which disjunct distributions exhibited by Tertiary relict floras have arisen'. However, 'hard data from fossils or intraspecific variation will be required to exclude long distance dispersal explanations for disjunction in any individual genus'.

Turning to another example, Wallis & Trewick (2009) have recently reviewed the phylogenetic and other evidence for the origins of the New Zealand flora. They conclude that 'the biota of New Zealand has links with other southern lands (particularly Australia) that are more recent than the breakup of Gondwana. A compilation of molecular phylogenetic analyses of *c.* 100 plant and animal groups reveals that only 10% of these are even plausibly of archaic origin dating back to the vicariant splitting of Zealandia from Gondwana.' They continue: 'clearly, more and better data are needed to resolve the evolutionary origins and the phylogeographic pattern in the NZ [New Zealand] flora and fauna. We can at least say that most of the NZ flora derives from long-distance dispersal and the same may be true for animals.'

The investigation of 'divergence times' for taxa

Modern phylogenetic studies provide crucial evidence on lineage divergence by estimating the timing of branching events, through the use of fossil-calibrated molecular clocks. While such approaches have possible limitations, it has proved possible to obtain new insights into hitherto controversial issues. For example, Baldwin & Wagner (2010) review the disagreements about the phytogeographical links between America and the Hawaiian Islands, and age of island endemics. Molecular studies support the conclusion that, in many cases, long-distance transoceanic dispersal from temperate and boreal North America was involved in the evolution of the Hawaiian flora (Mints, Sanicles, Violets, Spurges, Schiedeas, Silver Swords). These findings point to the importance of the existence of sky-islands of temperate or boreal climate within these tropical islands. Given the *c.*4000 km separation of Hawaii from continental landmasses and the geological history of the Hawaiian Islands, the origin of major disjunctions by vicariance is ruled out. Baldwin & Wagner consider the likely role played by birds in long-distance dispersal. Moreover, evidence from the phylogenetic studies, involving calibrated clock analysis, points to a recent rather than an ancient origin of endemic lineages.

Temperate sky islands are also important in the evolution of taxa in the East African highlands and

Andes. For both these areas, there is evidence for adaptive radiation of endemics from ancestral migrants of temperate origin (Carlquist, 1974; Hughes & Eastwood, 2006).

Mediterranean island endemics: dating and ancestral area reconstruction

By using phylogenetic methods, Mansion *et al.* (2009) investigated the origin of the endemics of the Boraginaceae family found in the islands of the Canaries and Corsica/Sardinia. They carried out phylogenetic analyses using six cpDNA regions, with molecular dating calibrated with four fossil 'constraints'. Biogeographical reconstructions revealed that the *Echium* endemics of the Canaries (oldest island *c.*20 Mya) diverged during the Miocene (15.3 ± 5.4 Mya). The 'Corso-Sardinian endemic *Borago* species derived from their primarily North African sister clade' about 6.9 ± 3.6 Mya – a long time after the fragmentation of the continental landmass that gave rise to the islands *c.*30 Mya. The *Anchus* endemics of the same islands were found to have affinities with South African *Anchusa capensis*. This study points to an Anatolian – eastern Mediterranean – origin of the three genera.

The evolution of island endemics has long fascinated biologists. The evolution of breeding systems of island species has been discussed in Chapter 7. Three other phenomena are of particular interest. Relative to their continental relatives elsewhere, in some, but not all, cases, the fruits and seeds of island endemic species have lost or have diminished means of dispersal, and some island endemics exhibit extreme woodiness and/or gigantism. Furthermore, groups of islands and archipelagos are often the theatre of adaptive radiation, where related species have entered a range of different habitats. Sherwin Carlquist, who wrote about these topics in his classic books *Island Life* (1965) and *Island Biology* (1974), has a website, www.sherwincarlquist.com/island-biology.html, which has very informative text and coloured plates.

Phylogenetic studies of the opening of a land bridge

Formerly, geographically isolated, North and South American landmasses became connected by the Central American isthmus, around 3 Mya. Before that time the 'nascent isthmus consisted of a series of volcanic islands' or a 'peninsula of southern Central America' separated from South America by 'a deep marine channel' (Cody *et al.*, 2010). It was commonly believed that until the completion of this land connection the biotas of South America evolved 'in splendid isolation' – the 'Great American Biotic Exchange' of biota taking place when plants and animals could migrate overland to the north and to the south (Simpson, 1980; Stehli & Webb, 1985). However, recent phylogenetic investigations of plants have prompted a reassessment. Phylogenetic evidence (with clock estimates calibrated to fossil evidence) now suggests that some migration of the plant biota occurred, by long-distance dispersal, before the isthmus was fully formed (Cody *et al.*, 2010).

Migration: implications of specialisation

It has also been possible to investigate another area of great interest. Emerson (2002) has drawn attention to the 'once commonly held viewpoint that extreme ecological specialization is irreversible' and 'limits the potential to adapt to new conditions', including colonisation of new territories. In refutation of this dictum, he cites the following example. Some continental African and Asian species of the genus *Dalechampia* (Euphorbiaceae) have flowers offering a non-nutritive pollinator reward in the form of a resin used by specific bee species for nest construction. Phylogenetic investigations of the group, including taxa from the island of Madagascar, where the resin-collecting bees do not occur, have established that reversal from a specialised to generalist (beetle and bee) pollination system has occurred on the island (Armbruster & Baldwin, 1998).

Quaternary Ice Ages: plant survival, migration and extinction

While we have increasing insights into changes in the distant past, events in more recent times are more accessible and have been intensively studied. Thus, there have been many pioneer investigations of the effects of the oscillations of the Quaternary Ice Ages (Hewitt, 1996). Dramatic climatic changes had profound effects on the evolution of plant species. In the words of Hewitt (2000), plant and animal species had 'to move, adapt or go extinct'.

Comes & Kadereit (1998) provide an important review of early advances that have come from the use of molecular markers in conjunction with palaeobotanical studies. They acknowledge that major insights have come from the use of radiocarbon-dated pollen spectra data and other fossil material to produce distribution maps, showing how species 'spread from glacial refugia into previously uninhabitable regions that became available through increasing warmth and glacial retreat'. Evidence indicates that the response of species was individualistic, rather than ecosystems invading en masse, and, as a result, present-day floras result from chance associations. While the study of sub-fossil pollen, leaves, fruits etc. provides a broad framework for interpreting vegetation change in the Quaternary period, there is a major limitation: it is presently not possible to identify all material to species level.

Phylogenetic studies have provided a very valuable new avenue of research on refugia and migration routes. Comes & Kadereit (1998) note that chloroplast DNA (cpDNA), which is maternally inherited in most angiosperms and therefore transmitted through seeds, but not pollen, has many valuable attributes for this kind of research. cpDNA is subject

> to less swamping through interpopulational gene flow of the initial genetic structure, established during refugial survival and/or at colonisation. Hence, together with its low mutation rate, cpDNA is likely to retain historic structures (past migration routes, colonization dynamics) over longer periods than biparentally inherited genetic markers . . . For polymorphic cpDNA markers, therefore, opportunities for genetic reassortment in glacial refugia through drift should be greater, leading to more genetic differentiation among refugial source populations or their colonising descendants than nuclear markers.

Studies of cpDNA have provided many insights into the refugia and migration routes of forest trees from southern Europe into northern areas. In detecting probable migration routes from southern refugia, it is important to take into account the orientation of mountain blocks that may obstruct the free movement of species. Thus, in North America the Rockies run north–south; in comparison in Europe the Alps and the Pyrenees, which run west–east, present possible barriers to migration. Such barriers to dispersal may lead to species subdivision, and independent evolution in populations beyond the capacity for active gene flow.

Considering particular cases, pollen maps suggested that the Alps might have been a barrier to colonisation in the post-glacial migration of the European deciduous Oaks (*Quercus*) from refuges in the Iberian Peninsula, Italy and the Balkans. Studies with cpDNA not only confirmed that these regions provided refugia for oaks, but also indicated that the oaks from refugia in Italy did migrate to sites north of the Alps.

Turning to another example, early palaeoecological research on the Beech (*Fagus sylvatica*) suggested that refugia for this species were located in southern Italy and the Carpathians from which they migrated north. However, detailed studies of chloroplast and nuclear markers in relation to palaeobotanical evidence have

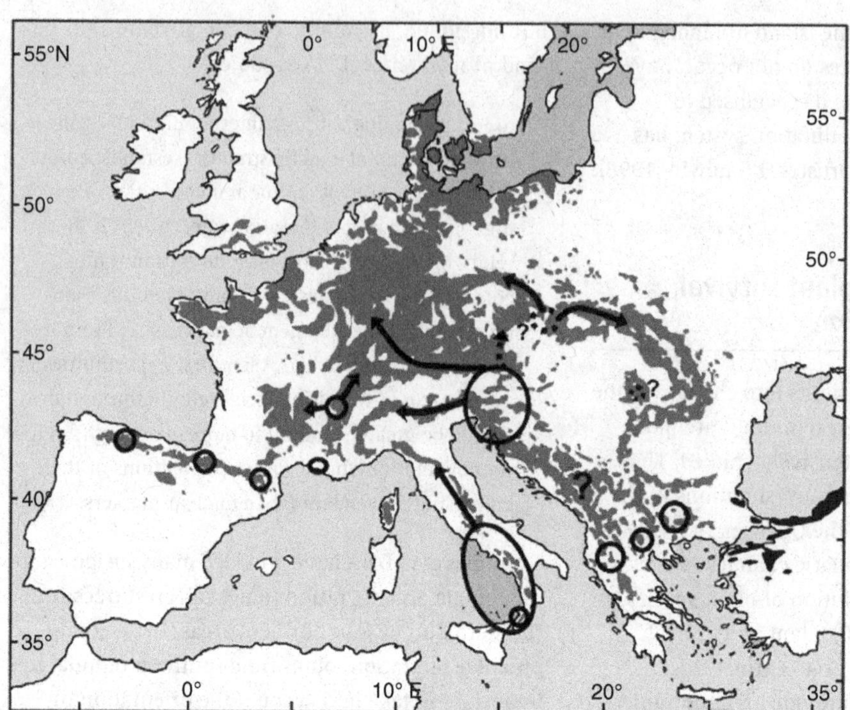

Fig. 17.5. Emerging complexities in the study of refugial areas and migration routes. Tentative map of refugial areas for *Fagus sylvatica* during the Last Glacial Maximum and main colonisation routes during the post-glacial period as indicated by a major investigation using molecular and other evidence. (From Magri *et al.*, 2006.) Present evidence points to several important conclusions. 1. From three separate areas in the Iberian Peninsula, beech extended only locally and did not contribute to the colonisation of the rest of Europe. 2. Populations from a small area of the southern French coastal zone expanded regionally. 3. Southern Italian beech populations spread northward but never reached Italy north of the Po valley. 4. Southern France was colonised by populations from refugia in the western Alps of France. 5. The Italian Alps were colonised by populations from the eastern Alps–Slovenia–Istria, and migration expanded westward to the Vosges, northern France, and England. 6. There is emerging evidence for refugia and migration pathways in eastern Europe. 7. There were refugia in Romania and the Balkans: beech from these areas did not spread northwards to colonise the rest of Europe.

yielded a much more complex picture (see Magri *et al.*, 2006, from which Fig. 17.5 is taken).

Through the presence of unique alleles or differences in cpDNA haplotype, molecular studies have also pinpointed previously unrecognised refugia for certain species, e.g. *Abies alba* (south-east France and southern Italy); *Fagus sylvatica* (southern Spain and Turkey); *Alnus glutinosa* (southern Spain and Turkey).

Concerning the genetic consequences of migration following the retreat of the ice, leading-edge expansion of species was probably not by diffusion – a step-by-step migration. Newly available territory is likely to have been colonised by successive, but rare, long-distance dispersal events, with later backfilling. As these migratory events may have involved small numbers of plants, repeated founding effects could have occurred leading to loss of alleles and homozygosity, with homogeneous populations persisting in some situations. This model of long-range dispersal from a refuge predicts a reduced genomic variability as the species extends its range. Confirmation of this general trend has been provided in studies of molecular markers in a range of species (Comes & Kadereit, 1998), for example:

allozymes: *Aslpenium* spp.; *Pinus contorta*
cpDNA: *Alnus glutinosa, Fagus sylvatica, Plantago media, Quercus* spp., *Tolmiea menziesii/Tellima grandiflora*
mtDNA: *Fagus crenata.*

While these examples provide support for the hypothesis that genetic variation may be lost as a species migrates from a refuge, it is important to note that the variation studied in investigations of cpDNA may be non-adaptive. It is unclear whether adaptive variation is also lost, as this variability is likely to be the key to successful migration from refuge to new territories.

Recent research on refuges and migration routes has examined regional areas in considerable detail. For instance, on the basis of molecular evidence, palaeobotanical and geological information, Schönswetter *et al.* (2005) concluded that glacial refugia were located to the south-west, southern, eastern and northern borders of the European Alps, together with refugia in the central areas, where evidence suggests that 'high-elevation plants survived the last glaciation on ice-free mountain tops'. In a study of the western Alps of Switzerland, Parisod (2008) also examined the evidence for post-glacial decolonisation. Using floristic and molecular information he concluded that refugia were located in ice-free areas (nunataks) within the Alps or in peripheral areas. For some species, three main migration pathways were identified including the Rhone and Rhine valleys. The genetic lineages of *Biscutella laevigata* were examined in detail and it was concluded that migration was important both from the Rhone valley and also across a southern transalpine pathway. Considering generalisations about the genetic loss of variability as the advancing front of the migration of a species from refuge, it is of particular interest that, in the case of *Biscutella*, current genetic variation was particularly high in the area where the two migration pathways came together.

The genetic consequences of range shifts in the Pleistocene have been examined in detail in *Arabis alpina* (Ehrich *et al.*, 2007). Using AFLP analyses, they examined the genetic structure of plants from a very large number of samples from the Arctic, Alps and East Africa. These studies indicated that there was

> virtually no diversity in the vast North Atlantic region, which was probably recolonised from a single refugial population, possibly located between the Alps and the northern ice sheets. In the European mountains genetic diversity was high and distinct genetic groups had a patchy and sometimes disjunct distribution. In the African Mountains genetic diversity was high [and] clearly structured . . . the fragmented structure in the European and African mountains indicated that *A. alpina* disperses little among established populations.

However, there was evidence for 'occasional long-distance dispersal events' in all regions. 'The lack of genetic diversity in the north may be explained by leading edge colonization by this pioneer plant in glacial forelands, closely following the retreating glaciers'.

Other researchers are attempting to locate and characterise refuges in particular regions. For instance, Médail & Diadema (2009) have identified 52 refugia in the Mediterranean region. They emphasise the role played by Mediterranean peninsulas, islands, and territories in Turkey etc. They stress that some of these sites might have been important as 'refugia', not just in the post-glacial but also for a period stretching back to the Tertiary.

In molecular investigations of Scottish alpine species (using AFLP techniques) Westergaard *et al.* (2008) concluded that 'Scottish populations of all species analysed, except *Saxifraga cernua*, belong to larger European/Eurasian genetic groups, and show little genetic distinctiveness'. Therefore, 'the ancestors of Scottish populations probably shared refugia with other northern European/Eurasian populations during the last glacial period'. However, in *Saxifraga cernua* and one population of *S. stellaris*, 'a unique genetic group restricted to Scotland' was detected.

Studies of refugia in the Caucasus region have also proved interesting. For example, molecular analysis of 167 globally distributed populations of *Arabidopsis thaliana* suggests the tentative possibility that this is the ancestral region for this species (Beck, Schmuths & Schaal, 2007). Also, it is possible that the expansion of the species 'coincides with the Eemian Interglacial (approximately 120,000 years ago)', when the species emerged from Pleistocene refugia in the region. Further investigations of *A. thaliana* reveal a shared history between North Africa and southern Europe (Brennan, 2014).

A great deal of progress has been made in the study of refugia in North America (see Calsbeek, Thompson & Richardson, 2003; Shafer *et al.*, 2010). Considering specific examples, Brunsfeld & Sullivan (2005) have studied cpDNA variability in *Cardamine constancei* and report in detail on a multi-compartmented glacial refugium in the Pacific Northwest.

Reflecting on the history and distribution of the Arctic flora in the light of the pioneering work of Eric Hultén, Abbott & Brochmann (2003) examined recent fossil, molecular and phytogeographical evidence. They conclude that there is strong support for Hultén's view that Beringia (north-east Russia/Alaska and the Yukon of north-west America) 'was a major northern refugium for arctic plants throughout the Quaternary'. However, 'there remain large gaps in our knowledge'.

Refugia in different parts of the world

The concept of refugia has been widely employed in studies of the biogeography of northern Europe, and North America. There is also increasing interest in refugia elsewhere. For instance, multiple refugia have been also been invoked in the interpretation of the post-glacial distributional history of Japanese Beech (*Fagus crenata*) by Okaura & Harada (2002); and the South American conifer *Fitzroya cupressoides* (Allnutt *et al.*, 1998).

Other investigations have considered the post-glacial history of whole floras of different regions. For instance, the sub-Antarctic region has been investigated, where each remote island group has a distinctive localised flora (Van der Putten *et al.*, 2010). Several lines of evidence suggest the survival of the native flowering plants 'in local refugia' during the Last Glacial Maximum (LGM) 'rather than a post-LGM colonization from more temperate distant landmasses' of South Africa, South America, Australia/New Zealand and India. Turning to another example, palaeobotanical investigations of the Icelandic flora suggest that many species survived in refugia on the island in the last glacial period (Rundgren & Ingólfsson, 1999).

Implications of refuges

In an important review, Stewart *et al.* (2010) draw together information on many important issues. They consider the present state of knowledge about the location and size of refuges.

The degree to which refuges were genetically isolated has been examined in studies of 100 populations of the wind-pollinated conifer Silver Fir (*Abies alba*) across sites in Europe (Liepeit, Bialozt & Ziegenhagen, 2002). Studying DNA markers, with contrasting modes of inheritance, different refuges were identified, but there was also evidence that pollen-mediated gene flow occurred between refugia. The Silver Fir was also found to contain the 'paternal lineages' of several *Abies* species (Ziegenhagen, Bialozt & Liepeit, 2004). For *Abies*, this finding overturns the simplistic view that geographically isolated refuges were necessarily genetically isolated from each other during the several glacial periods. With this result in mind, it is important to investigate the possibility of gene flow between refuges in other species.

Stewart *et al.* (2010) also review the evidence mounting for the existence not only of long-recognised refugial sites, but also the presence of so-called small, cryptic or micro-refugia (see, for example, Bordács *et al.*, 2002; Brus, 2010). First, there is some debate about whether these two types are essentially the same or different (Rull, 2010). Secondly, Birks & Willis (2008) consider the implications of these new insights. They point out that:

> the classical view of glacial stages is that alpine and arctic plants were widespread in the lowlands of central Europe and around the margins of continental and alpine ice-sheets, whereas trees were restricted to localised refugial areas in southern Europe and the Mediterranean basin. New palaeobotanical evidence in Europe suggests, however, that this classical view is incomplete and that tree distributional ranges during the glacial stages were more extensive and included many local areas of small populations in central and eastern Europe growing in so-called cryptic refugia. We argue that this concept of cryptic refugia is also applicable to arctic and alpine plants during temperate interglacial stages where small localised populations grow in naturally open habitats that are not beyond or above the forest limit.

As evidence accumulates that such cryptic refuges existed, Stewart *et al.* (2010) emphasise that the early estimates of the rate and pattern of migration of a particular species during the post-glacial period may have to be revised. They also raise pertinent questions about presently unresolved questions relating to what happened to population outside recognised refugia: how much extinction occurred? Crawford (2014) also considers the interesting question of extinctions, when he poses the question: where disjunctions in the current distribution of European arctic alpine plants exist, should we see these as 'merely accidents in dispersal' or are present-day populations 'the remnants of former widespread and continuous distributions' where plants have 'suffered either from physiological failure or other aspects of evolutionary maladaptation to a changing climate'?

Migration from refuges

Stewart *et al.* (2010) also confront the difficult issues in considering the migration of species: which genotypes are likely to have arrived at a refuge and how long species might have remained there. These factors have a bearing on the genetic diversity of species and their likely evolution in isolated refuges.

Also, we know little about the migration process from refuges. Species have exacting habitat requirements, and the abiotic and biotic characteristics of possible migration routes may, in actuality, allow passage to some species, but not to others. Thus, it is important to consider the processes of selection and adaptation involved in migration, as plants adapt to new neighbours and environments, and experience competition from other species that might lead to their competitive exclusion. In addition, stochastic events might be crucial in long-range dispersal. Exceptional 'sweepstakes routes' may be important, as propagules of some species (but not others) are, by chance, carried great distances across 'normal' barriers, during extreme weather events, such as hurricanes, and floods.

Investigations have tested models of other important factors in migration. For example, early migrants, establishing by long-distance dispersal, may still dominate today the areas they colonised some time ago. These ideas have been examined in studies of the long-lived species of Oak in the Loire region in western France. The distribution of six cpDNA variants was mapped in a 200 km × 300 km area. It was discovered that very large patches exist that are virtually fixed for a single haplotype, supporting the model of individual long-distance dispersal events, most probably by a widely foraging acorn-collecting bird, the Jay (*Garrulus glandarius*). Establishment was followed by population increase yielding patches of trees whose genotype reflected that of the original colonist(s) (Petit *et al.*, 1997).

Huck *et al.* (2009) have investigated another fascinating aspect of migration from refuges. They analysed population variability in *Meum arhamanticum*, using AFLP fingerprint techniques from samples drawn from across its entire native range – the high alpine regions in the Pyrenees, the

Alps, and Bulgaria. The hypothesis under investigation was whether central and northern highland populations are the result of migration from southern refugia. They discovered evidence that the species 'survived the last ice age in multiple refugia throughout its contemporary range and did not expand into areas north of the Alps from southern refugia. They detected 'regional-scale migration in northern, formerly periglacial, parts of the species range has resulted in the intermingling of populations. In contrast, southern populations are characterised by long-term isolation.'

Studies of the post-glacial migration of oaks into Britain have revealed further insights (Ferris *et al.*, 1995). Fossil pollen data suggest that oaks could arrive in England both by a western and eastern route from refugia in Iberia, southern Italy and the Balkans. In investigating whether the oaks of East Anglia could have been the result of 'multiple routes of colonization', Ferris and associates discovered that the chloroplast DNA of many East Anglian oaks was characterised by a 13-bp (base pair) duplication. The results suggested that 'the mutation occurred either in southern England, or during migration from the mainland, and became fixed in a source population from which East Anglia was colonized'. They also recognised the influence of human activities on oak populations, in concluding that the 'planting of non-native trees for roadside boundaries and in the grounds of old houses and estates, explains the absence of the marker from some East Anglian oaks'.

Considering another aspect of post-glacial migration, we are ignorant of evolutionary changes occurring at the rear 'trailing' edge of migrating populations adjusting to the impact of climate change (Thuiller *et al.*, 2008; Stewart *et al.*, 2010). It is postulated that 'shrinkage, dissection and extinction' may all occur together with severe bottleneck effects (Hewitt, 2000). In addition, Levin (2011) examines the possibility that mating system shifts may occur on the trailing edge: for example self-compatibility may be at a selective advantage if populations are greatly reduced in size. While we know little of the population effects of migration under climate change in the Quaternary, we have some information on how species are responding to the present climate changes (see Chapter 18).

We have concentrated on the post-glacial migration of species from refuges into territories released from the grip of ice. It is important to stress that there were successive advances and retreats of the ice, which forced successive responses from species. The outcome was that some species became extinct in territories where they had formerly grown successfully. For example, a number of tree taxa, *Carya, Cathaya, Tsuga, Pterocarya*, present in the early Pleistocene in Italy etc. are nowadays extinct in Europe (Corrado & Magri, 2011). Others successfully retreated repeatedly to refuges at the height of cold periods, followed by successive migrations to new territories, as the climate temporarily 'improved' before the next cold phase.

Prompted by the obvious realisation that the post-glacial period was only one phase of the history of lineages, some investigators have attempted to study, through the phylogenetic analysis of trees, the legacies of even earlier events. Commenting on a study of the Cork Oak (*Quercus suber*) by Magri *et al.* (2007), Hampe & Petit (2007) stress that phylogenetic trees retain the genetic imprint of ancient events – in the case of the Cork Oak, the Tertiary plate tectonics of the Mediterranean Basin and its islands.

A further revealing case study is provided by *Vaccinium uliginosum* (Eidesen *et al.*, 2007). This, and the investigations of an increasing number of species, establishes the firm link between evolutionary patterns and processes that have often been investigated in isolation in the past, namely a synoptic study of historical biogeography, polyploidy, refugia, migration routes and gene flow. A very complex history of this circumpolar species has been revealed by the use of cpDNA and nuclear markers. The results are too extensive to report in full.

Here, we note that there are diploids and tetraploids in the group. Molecular evidence is consistent with the proposal that there were glacial refuges in Beringia, western Siberia, southern European mountains, and in areas south/east of the Scandinavian and Laurentian ice sheets, and expansion from these refuges produced a number of contact zones.

The patterns of present-day variability in the group are not clear-cut, and it is suggested that 'a likely major driver causing inconsistencies is recent nuclear gene flow via unreduced pollen from diploids to tetraploids'.

We now turn to a number of important issues in plant biogeography that are illuminated by the use of advanced molecular and biogeographical methods.

The refugial hypothesis of Amazonian speciation

The concept of refuges has played an important role in the explanation of the great plant diversity in the forests of Amazonia. Thus, studying the distribution of bird species in the forests, Haffer (1969) discovered that some areas hold more endemic species than others. He proposed that during the ices ages, under more arid climatic conditions, forest areas contracted to isolated refuges surrounded by savanna. In his view, these isolated refuges were the theatres of allopatric speciation.

This refugial model was extended to cover not only birds, but also butterflies, frogs, lizards and some families of plants (Vanzolini & Williams, 1970; Haffer & Prance, 2001). However, for a number of years, important assumptions behind this model remained untested. Now, palaeobotanical research at many sites in Amazonia refutes the claim that savannas predominated in the period, and has discredited the view that ice age climates can be categorised as cold/wet/dry. As the refugial hypothesis of Amazonian speciation is not supported by the available evidence, many botanists now consider that the concept should be abandoned altogether (see Bush & de Oliveira, 2006; Bennett, Bhagwat & Willis, 2012).

Palaeoecology: insights from the study of ancient DNA

Willerslev & Cooper (2005) have reviewed this important active field of research. They note the '*post-mortem* instability of nucleic acids' involves 'spontaneous hydrolysis and oxidation'. At the molecular level there are 'strand breaks, baseless sites, miscoding lesions and crosslinks'. Sometimes ancient specimens do not contain any 'amplifiable endogenous DNA, while those that do possess only fragments in the 100–500 bp size range'. But, for some ancient remains, there are authenticated survivals of ancient DNA (aDNA). For example, from permafrost sites, 50 kyr old mammoth DNA has been retrieved, in addition to bison mtDNA greater than 65 kyr old. Also from permafrost, cpDNA has been recovered from plant material (*c.*300–400 kyr).

While prospects for the study of aDNA are greatly improving, Willerslev & Cooper (2005) write of a 'tarnished history' of study of aDNA. Many early spectacular claims have been 'either disproved or effectively discarded'; including claims that DNA sequences of plants have survived for millions of years (Goldenberg *et al.*, 1990; Soltis *et al.*, 1992). Often some of the supposed aDNA was microbial or fungal (Willerslev *et al.*, 2004). Also, specimens handled over many years often proved to be contaminated, especially fossil hominids. In addition, while PCR techniques have the advantage of allowing the amplification of small surviving specimens of DNA, the use of these methods has the potential to contaminate ancient materials, for in molecular laboratories PCR products from various investigations are likely to be unwittingly scattered around. There are also major difficulties in studying whether ancient material contains 'fossil' microorganisms; for example, our knowledge of extant bacteria is highly incomplete.

To avoid problems of contamination, Willerslev & Cooper (2005) stress that studies of aDNA require that 'laboratories must be completely isolated both physically and logistically, preferably in buildings free from all molecular biology research' with personnel moving only 'from ancient to modern laboratories'. These arrangements are as effective as high-tech positive pressure and UV irradiation systems. For critical determination, the safest approach is to cross-check the aDNA from different parts of the same specimen (e.g. teeth and bones), and above all have independent verification in findings in separate independent laboratories.

Two well-authenticated case studies point to the high potential in the study of aDNA in plants.

- aDNA has been recovered from ancient Maize cobs and offers important insights into plant domestication (see Jaenicke-Després *et al.*, 2003, and references therein). In molecular genetic studies, three genes that affect the morphology, seed protein and starch production have been identified. Investigations of aDNA show that 'these genes were largely homogenized as early as 4.4kyr ago by Mexican farmers, although one of the loci was not fixed until c.2 kyr ago. Encouragingly this study indicates that aDNA studies of selection will become increasingly possible as phenotypically important loci are identified by genome projects.'
- The study of sedimentary aDNA (so-called dirty DNA) has the potential to increase our understanding of palaeoecology. It is now possible 'to recover DNA from small samples (ca. 2 g) of sediments up to 300–400 kyr old', and examine the DNA of plants and animals, leading to information on 'sequences of herbs, shrubs trees, mosses and megafauna' (Willerslev *et al.*, 2004; Willerslev, Hansen & Poinar, 2004). Studies of pollen and other plant remains in dated deposits of various kinds have provided many insights into palaeoecology. Investigations of aDNA have the potential to provide new and important information on ice-age environments and ecosystems (Jørgensen *et al.*, 2011).

Multidisciplinary approaches in biogeography: two case histories

Post-glacial history in an alpine species *Primula hirsuta* is a widespread species in rocky crevices or stony pasture at 1400 (rarely 500) to 2800 m on silicaceaus rocks (Schorr *et al.*, 2012). There are two hypotheses to explain the survival of the species during glacial periods. The *tabula rasa* concept proposed that the species survived outside the alpine area and recolonised from peripheral areas after the glacial retreat. Alternatively, it is possible that the species survived on nunataks (ice-free areas above the glaciers). Fossil records could be used to discriminate between these two hypotheses, but such fossils are rare in the Alps. Furthermore, phylogenetic studies alone cannot discriminate between peripheral survival and nunataks, and refugial areas outside the present range can rarely be identified.

Potentially, species distribution models (SDMs) can provide new insights. These models are based on grid systems, and Schorr *et al.* (2012) and associates note that:

> SDMs quantify the relationships between species occurrence and geographic variation in the environment to describe the realized environmental niche of a species . . . The quantified niche can then be projected across a geographic area to map suitable environmental conditions for a species and predict potential distribution. Assuming that the species niche has remained unchanged over the time period of interest, the paleodistribution of species can be reconstructed by projecting SDMs to earlier time periods using paleo-climatic data.

The climatic data available, in this case, include monthly temperatures and precipitation variables describing annual trends, seasonal variability and extremes and potentially limiting factors. As it is recommended that several SDM techniques be employed in this type of investigation, Schorr *et al.* (2012) employed four different SDM techniques (see paper for details). In parallel studies, phylogenetic data from samples taken across the distribution of *Primula hirsuta* were also examined using AFLP techniques.

The combined approach of molecular and modelling approaches supports the hypothesis of widespread nunatak survival in the Central Alps, and on peripheral nunataks at the southern alpine margin, and the population genetic structure and diversity patterns were congruent with what is expected from nunatak survival.

Concerning the potential of combined approaches, it is clear that some of the potential refugial areas indicated by models might not be detected by phylogenetic methods. From the study of other species, the southern peripheral area between Lago Maggiore and Lago di Como has been identified as an important refugial area for several alpine species. Therefore, it is suspected that the modelling did not identify the possibility of this area as a refuge for *Primula hirsuta*. This may reflect issues with the modelling and/or sampling. But it is also possible that there was a niche shift in *Primula hirsuta* in the post-glacial period. Overall, Schorr *et al.* (2012) argue forcefully and convincingly that the combined methods of molecular analysis and advanced modelling provide a very promising approach to the unravelling of the post-glacial histories of species.

Post-glacial history of an arctic-alpine species In a second case study involving combined approaches, Alsos *et al.* (2009) carried out a detailed investigation of the post-glacial history of Dwarf Willow (*Salix herbacea*) studying samples from Europe, Greenland and North America. To investigate genetic structure, diversity and likely distribution routes, AFLPs were investigated. In addition, fossil records were examined, and modelling to define the present species niche, and predict the geographical distribution of the species at different times. Maxent version 3 was employed, using past and future climate data (for details see Alsos *et al.*, 2009). This allowed both hindcasting of past distributions and forecasting the likely effects of future climate change.

The results of the study are very extensive. Here, we note that fossils indicate that the species was present from before the Last Glacial Maximum in both North America and central Europe. Studies of fossils, and distribution modelling, indicate that Dwarf Willow had a 'continuous range in central Europe during the last glaciation', the area acting as a large refugium and dispersal corridor for later northwards colonisation. Such colonisation was likely to have been rapid, as high levels of genetic diversity were maintained. The AFLP results support the notion of a common post-glacial origin of populations in the east Atlantic and Alpine regions. Likewise, expansion was also on a broad front in North America, from refugia south of the Wisconsinan ice sheet. Evidence indicates that Dwarf Willow spread west from Europe to Iceland and East Greenland. West Greenland populations of the species have affinities with those of neighbouring eastern North America, indicating that the species arrived from two different sources.

With regard to the future of *Salix herbacea* populations, current genetic diversity is high in Scandinavia and southern Alpine populations. But considering the niche requirements of the species, climatic forecasting indicated that southern populations are at risk from future climate change.

Concluding remarks

The past few decades have seen remarkable progress in historical biogeography of plants. However, caution must be exercised in interpreting the past. Plausible scenarios suggested by various lines of evidence emerge rather than certainties.

Reflecting on modelling plant distributions, Nogués-Bravo (2009) writes: 'predicting past distributions of species climatic niches, hindcasting using climate envelope models is emerging as an exciting research area'. But there are significant challenges ahead in the prediction of the past distribution of species. These include acknowledging the complexities of the concept of the species niche: is it definable in terms of climatic factors? Moreover, it is important to question to what extent species occupy that niche in equilibrium with climate. Furthermore, are species stable in their niche requirements over the timeframes being considered in research investigations, often of the order of thousands of years? Moreover, there are uncertainties associated with the climatic data and modelling techniques.

Given that several modelling approaches are emerging, Nogués-Bravo (2009) advocates the use of more than one model in studying niches and palaeoclimates. He also makes a number of technical recommendations, including for better sampling, and independent validation of models.

Turning to other issues, new insights have provided an important stimulus for recent investigations of polyploidy, and further insights are likely, as attempts are made to investigate the pace of plant evolution, to see how the incidence and geographical distribution of polyploidy and evolutionary divergence are related to dated geological upheavals, such as those triggered by the impact on Earth of an extra-terrestrial body (KT boundary), and the effect of the Milankovitch cycles and especially ice ages (Bennett, 1997).

Another important area is also emerging. Instead of seeking very broad generalisations concerning the geographical distribution of polyploids (see Chapter 14), attention is now being focused on the eco-geographical setting for speciation in individual groups. As we have seen, evidence suggests that opportunities for speciation events occur when secondary contact occurs between previously isolated related taxa, principally as a result of climate change. Considering homoploid speciation and recognising the likely importance of transgressive properties of hybrid derivatives, it is clear that new ecological/geographical potentialities may arise, leading to ecological divergence from the parental taxa. Therefore, homoploid speciation may result from sympatry of related taxa. However, Kadereit (2015) points out that the 'speciation' event may not be complete when hybrid lineages are first formed: full speciation may involve the evolution of additional isolating mechanisms, perhaps over many generations. He suggests that in a period of rapid climate changes (such as occurred in the Quaternary period) parental lineages may have withdrawn from the contact zone as they tracked their ecological niches. Thus, further genetic changes could have occurred in isolated hybrid lineages, as the niche requirements of the parental taxa carried them outside the range of gene flow. If this was the case, then speciation in homoploids may have involved a secondary allopatric phase.

While the present distribution of many species is decisively influenced by human activities, it may be that critical study of others, living in more natural settings, could make it possible to test these ideas (Kadereit, 2015). More detailed eco-geographical information about species groups is needed, including insights from reciprocal transplant experiments. Also, hind-casting models could be employed. In addition, modern evo-devo approaches may make it possible to identify key genes involved on the ecological divergence of polyploid/homoploids from their parental taxa.

Concerning questions of dispersal, evidence now suggests that long-range events might be more frequent than hitherto recognised. Also, in contrast to previous views, the plant lineages found on particular landmasses could be comparatively recent in origin. These insights 'chime' with evolutionary insights from other fields (De Queiroz, 2004). He writes:

> in the past few decades, biologists have increasingly recognised the rapidity of biotic change, as evidenced by studies of speciation, mass extinction, evolution within populations, and distributional shifts within landmasses. The new support for widespread oceanic dispersal adds to this dynamic view of life on Earth. Disjunct distributions must, to some extent, be explained by the slow drift of tectonic plates. However, increasingly it appears that this pattern is overlaid and obscured by something resembling an airline map tracing the rafting, swimming, floating, and flying routes of countless transoceanic voyagers.

18 The evolutionary impact of human activities

This chapter examines the evolutionary influences of human activities, and considers the proposition that many species are threatened with extinction.

Humans: as animals practising extreme niche construction

Anatomically modern humans arose in Africa over 100 000 years ago, and by 40 000 years ago they were perhaps fully human in anatomy, behaviour and language (Diamond, 1992). Increasingly, as humans have migrated to almost all parts of the globe, they have influenced, modified, managed or transformed all the Earth's natural ecosystems by practising what has been termed 'niche construction'.

Leland (2002) notes that, traditionally, 'adaptation typically is regarded as a process by which natural selection molds organisms to fit a pre-established environmental template . . . yet in varying degrees, organisms choose their own habitats, choose and consume resources, generate detritus, construct important components of their own environments . . . destroy other components and construct environments for their offspring. Thus, organisms not only adapt to their environments but in part also construct them.' In the case of humans, niche construction may be seen, therefore, as an evolutionary strategy that aims to maximise survival and reproductive success.

Human environmental activities through the development of cultural practices have many intentional effects. As Western (2001) observes, 'the most universal and ancient features of our "humanscapes" arise from a conscious strategy to improve food supplies, provisions, safety, and comfort – or perhaps to create landscapes we prefer, given our savanna ancestry'. He points out, however, that there are many unintended side effects of human activities that lead to a multiplicity of consequences at every spatial scale, and his list includes: habitat and species loss; loss of keystone species; major changes to ecosystems resulting in reduced ecotones; truncated ecological gradients and changes to soils and accelerated erosion; introductions of invasive non-native species and diseases; side effects of fertilisers, pesticides, herbicides; over-harvesting of natural resources; nutrient leaching and eutrophication; pollution of soils, water and air; and 'global changes to the lithosphere, hydrosphere, atmosphere and climate'.

Human impact on the environment

The formal geological term for the last 11 700 BP years is the 'Holocene', but human niche construction has become so universal and dominating that the present era has been called the 'Anthropocene' (Ruddiman, 2013). There has been a great deal of debate about when this era might have begun. Some authorities point to the eighteenth-century Industrial Revolution in Britain, but others argue for an earlier date – the rise of agriculture (around 12 000 BC). However, other experts have considered the possibility that human activities in the pre-agricultural period are likely to have significantly altered some of the Earth's ecosystems, for example, through the use of fire, and the effects of hunting (leading to the loss especially of certain large mammals and bird species).

The evolutionary effects of human activities

Human activities impose selective influences on the plants in ecosystems. The evidence for this proposition is explored in detail in *Plant Microevolution and Conservation in Human-Influenced Ecosystems* (Briggs, 2009). Here, a brief examination of the concept is presented.
In humanscapes, some species/variants have the potential to thrive, while others might fail and become extinct. There are likely to be both winners and losers. This conclusion flows from the theory of natural selection proposed by Darwin (1859) in the *Origin*.

He wrote, 'The theory of natural selection is grounded on the belief that each new variety, and ultimately each new species, is produced and maintained by having some advantage over those with which it comes into competition; and the consequent extinction of the less favoured forms almost inevitably follows.' In reflecting on the domestication of plants and animals, Darwin (1868) made another important observation: 'man, therefore, may be said to have been trying an experiment on a gigantic scale'. Here, acknowledging the enormous and all-pervasive influences of humans on the Earth's ecosystems, we think there is a strong case for considering that humans are conducting an even larger experiment on the whole Earth and its biodiversity. It is certain that there will be a considerable evolutionary legacy from human influences on its plant life. The evolutionary impact of human activities has also been explored in a number of studies.

- In concluding that humans are the world's greatest evolutionary force, Palumbi (2001) draws his evidence from researches into changes in, for instance, 'antibiotic and human immunodeficiency virus (HIV)'; evolution of 'resistance to pesticides' in plants and insects; rapid evolutionary changes in invasive species, 'life-history change in commercial fisheries and pest adaptation to biological engineered [transgenic] products'.
- Molecular studies of the evolutionary impact of selective harvesting on wildlife have been examined, with dramatic changes in terms of changes in age structure, sex ratio, body size, horn length or antler size etc. being detected (Coltman, 2008).
- Examining studies of fish, ungulates, invertebrates and plants, Darimont *et al.* (2009) discovered that 'average phenotypic changes in 40 human-harvested systems are much more rapid than changes reported in studies examining not only natural ($n = 20$ systems) but also other human-driven ($n = 25$ systems) perturbations in the wild, outpacing them by >300% and 50% respectively'.
- International trade is responsible for about 30% of the threatened animal extinctions in developing nations: globalised international trade accelerates 'habitat degradation far removed from the place of consumption' (Lenzen *et al.*, 2012). Thus, many studies have linked export-intensive industries with biodiversity threats, for example, coffee growing in Mexico and Latin America, soya and beef production in Brazil, forestry and fishing in Papua New Guinea, palm oil plantations in Indonesia and Malaysia, and ornamental fish catching in Vietnam.

Assessing human impacts on ecosystems: sources of evidence

When humans were living as hunter-gatherers their impact on the Earth's ecosystems was at first perhaps insignificant. But, later, the influence and impact of human populations has increased dramatically (Woodwell, 1990; Diamond, 1992; McNeill, 2000; Dodds, 2008; Briggs, 2009; Goudie, 2009).
Archaeological evidence is very important in judging and understanding human migration, and the antiquity and scale of human influence. So too is the evidence obtained by the Quaternary specialist, who examines vegetation change as it is revealed in plant and animal remains preserved in anaerobic deposits of peat, and in

lake sediments, ice cores etc. Where studies have been carried out on dateable deposits, studies of the pollen/seeds they contain have provided information on forest clearance, and the history and extent of agricultural and urban-industrial development across the world. Modern techniques have also played a key role in the refinement of earlier techniques by providing new means of dating biological and other material. Also, completely new insights into biological and human history have emerged through the study of ancient DNA (Humel, 2003; Shapiro & Hofreiter, 2012).

Concerning more recent periods of history, historic documents, maps, herbarium specimens, and published evidence of all kinds are crucial to test hypotheses concerning significant environmental change, e.g. the impact of the accidental and deliberate introduction of animals and plants into new territories. Settlements attract place names and these, interpreted carefully, often indicate something of past history, e.g. in Britain those names containing the Old English element lēah, as in Hayley Wood, Cambridgeshire, suggest clearings in woodland (Rackham, 1975, 1980; Mills, 1993).

The study of present-day landmarks and the form of plants may provide important information on habitat change. For instance, the age of stands of trees and the history of forest fires can be determined by examining tree rings and fire scars.

Remote-sensing techniques have provided very important lines of evidence for critical study of hitherto intractable environmental problems. For example, they have proved invaluable in judging the extent of deforestation in different parts of the world, such as Brazil and Indonesia.

On a smaller scale, aerial photographs have been particularly valuable, especially where it has been possible to compare historic and recent images. They have been employed in the detailed analysis of archaeological sites, settlement history, changes in field systems, analysis of land reclamation, estimation of numbers of plants and animals, and the incidence of tree and crop diseases (Paine & Kiser, 2012).

The development of sophisticated instrumentation has provided accurate chemical and physical measurements of important environmental pollutants. Where investigations have been made repeatedly and over long periods of time they have provided important baseline information with which to compare contemporary measurements. Thus, they provide a timeline for judging the effectiveness or otherwise of environmental legislation through management. Also, a wide range of experimental approaches, based on statistical analysis, is now available.

The development of computer modelling allows the critical examination of past, present and future trends that illuminate many crucially important environmental issues. Thus, modelling provides a means of predicting, for example, the future growth of human populations and the effects of climate change. Instead of a single prediction, modelling allows a range of predictions to be made on the basis of different assumptions. It is important to stress that these approaches do not provide 'certainty' about the future, but they do provide valuable information with which to grapple with complex issues.

By using modelling techniques it has been suggested that an increase of human populations is likely from between '9.5 billion people in 2050 (range of projections 8.1–10.6 billion), and slightly over 10 billion in 2100 (range 6–16 billion)' (Ehrlich, Kareiva & Daily, 2012, citing current UN projections and other sources). This will lead to increased demand for land, water, and resources of all kinds.

From natural ecosystems to cultural landscapes

When the various lines of evidence are examined, what do they tell us about the human impact on the world's ecosystems in different historical periods?

Humans are 'social animals' and different regions of the world now have their own characteristic 'cultural landscapes', each of which

> reflects the different civilization, customs and artistic achievements of the people who produced it and live

within it. Today, there is an astonishing array of cultural landscapes ranging from the dwindling lands of the last remaining hunter-gatherers, through a multitude of different agricultural landscapes to the cityscapes and urban-industrial landscapes of modern civilizations. (Briggs, 2009)

Cultural landscapes are influenced by a number of important factors. Different crop plants are grown, and animals reared depending on local climate and soils. Cultural landscapes are not only defined by current influences, but also by events in the past. Sometimes cultural landscapes appear to have evolved by gradual change, but often abrupt events have been decisive, for instance, warfare, crop failure, exhausted soils, pestilence, climate change, the outbreak of human disease (for example the fourteenth-century outbreaks of the Black Death when perhaps one-third of the population of Europe perished).

Investigation of landscapes from a high vantage point such as an aircraft, or satellite images confirm that cultural landscapes predominate across huge areas of the world, and that the 'development of these areas has been at the expense of natural communities' (Briggs, 2009).

Given the extent of discernible cultural landscapes and the evidence of all-pervasive human influences across the globe (see below), it is questionable whether any wholly natural vegetation remains. Is there any wilderness left?

The context in which the 'wilderness' is employed, as for instance by many conservationists, is ill-defined, as are such concepts as the 'pristine environment', 'virgin forests', 'undamaged habitat', 'wild nature' etc. But seeking a definition, Oelschlaeger (1991) considers that the notion of wilderness 'emerged with the beginnings of agriculture, when the distinction was made between the "domestic" and the "wild", and these terms were applied to animals, plants and territories' (Briggs, 2009). Adams (2004) provides another helpful definition: 'the wild was un-tame, lying beyond the tended fields and managed woods'.

When overseas territories were first colonised by Europeans, in North and South America, Australia and elsewhere, they were often viewed as 'wilderness', notwithstanding the fact that such landscapes were inhabited. Explorers, and, later, colonists, failed to detect, or more likely to acknowledge, the cultural landscapes that lay before them, which bore the marks of a long history of human presence.

Considering the 'wilderness' status currently attributed to particular regions, it is important to examine archaeological and other evidence. For example, Willis, Gillson & Brncic (2004) concluded that 'large areas of what are often taken to be undisturbed rainforest in the Amazon, Congo and Indo-Malay regions have all been influenced by extensive prehistoric human activities, as part of former cultural landscapes' (see Briggs, 2009, for a review of the evidence).

The extent of human-modified ecosystems

In trying to estimate the extent of cultural landscapes, Nicholson (1970) made the important point that water covers two-thirds of the planet; about 10% of the land surface is ice, 10%, deserts, and a further 10% is inhospitable mountain and rock areas. Concerning the impact of humans on the rest of the land surface, there has been much debate about the definition of categories of land use, in particular the concepts of degraded, eroded and desert lands (Grove & Rackham, 2001; Imeson, 2012).

Vitousek *et al.* (1997) have assessed the current evidence and conclude that '50% of the Earth's land surface has been transformed or degraded and humans utilise c.39–50% of the biological productivity of the planet' (Briggs, 2009). The impact of domesticated animals is very considerable. In 1997, there were more than 4 billion sheep, goats, horses and pigs, together with 11 billion poultry (Pond & Pond, 2002). Concerning forests, Noble & Dirzo (1997) estimated that 'Earth's forested estate has shrunk by about a third (2 billion ha) since the rise of agriculture-based civilizations and continues to be eroded at a dramatic rate.' Currently 'about 10 million hectares of new land is cleared largely from forests'. They conclude that now about a third of the world's land surface (3.54 billion ha) is covered with forest, of which 150 million ha are plantations and another 500 million ha are actively managed. However, this is a 'considerable underestimate of the area of forest affected (and often dominated) by human activity as it excludes large areas affected by indigenous gardening, hunting and gathering, and indirect management such as changed fire regimes'.

In many areas of the world, ecosystems are strongly influenced by the deliberate use of fire as a tool to manage rangelands, grasslands, heaths etc. Fire suppression policies are also important: natural, accidental and deliberate, fires being quickly detected, and, in some cases, rapidly extinguished. In such circumstances huge fuel loads build up, leading to spectacular fires such as occurred in Yellowstone National Park in 1988.

Turning to the fate of other ecosystems, Briggs (2009) writes:

> Not only have forests been cleared, but also wetlands have been drained. Coastal areas have also been reclaimed for agriculture by the construction of sea walls and other defences. In addition, in many parts of the world, rivers have been diverted and dams built to provide water for domestic purpose, irrigation and power generation. Overall it is estimated that 'more than half the run-off water that is fresh and reasonably accessible' is put to human use (Vitousek *et al.*, 1997). And about two-thirds of the world's rivers are regulated in their flow, with about 36 000 dams worldwide (Humborg *et al.*, 1997). Only 2% of the rivers in the USA run naturally. Many rivers are used so extensively that very little water reaches the sea e.g. the Colorado, Ganges, Nile: Abramovitz, 1996) . . . In addition, huge quantities of water – both renewable and fossil – are drawn from underground (Gornitz, Rosenzweig & Hillel, 1997). Such management has a profound effect on freshwater aquatic and wetland systems.

Early in the history of cultural landscapes spectacular changes to land surfaces (field systems, terraces, irrigation, settlements, communication systems etc.) were produced by human toil. Now, through the development of machinery – at first simple, but later more and more complex and specialised – the impact of human activities is accelerating dramatically.

- Expansion of the land under cultivation is still taking place (e.g. deforestation in Indonesia for the planting of oil palm), and 'industrial' agriculture is being more widely adopted, with crop yields increased by the use of machinery, fertilisers, pesticides, herbicides, GM crops etc. Stock numbers are also increasing with the use of intensive husbandry.
- The development of specialist earth-moving and drilling equipment is bringing about profound and spectacular changes to the surface of the Earth. Mechanisation on a massive scale allows for the further exploitation of rivers, the expansion of cities and urban-industrial areas, and underpins the development of advanced power and transport systems. Such equipment also allows the building of enlarged harbour and dock facilities, sewage and irrigation systems, and provides a means for massive extraction of underground resources, such as gravel, clay, sand, minerals, water, oil and gas.
- Mountainous quantities of waste are produced in contemporary cultural landscapes. These range from domestic rubbish and sewage, to agricultural run-off, and the debris from mining. While some wastes are recycled others are not.
- Increasingly the surface of the Earth bears not only the discernible imprint of past human activities, but also the growing impact of current activities. Such is the scale of transformation that humans are now recognised as a geological force (Steffen *et al.*, 2011).
- Recent estimates indicate that some 43% of the Earth's land has been converted to agricultural and urban landscapes. 'This exceeds the physical

transformation that occurred at the last global-scale critical transition, when *c*.30% of Earth's surface went from being covered by glacial ice to being ice-free' (Barnosky *et al.*, 2012).

Plants: their different roles on the cultural landscape stage

Humans naturally classify plants, and have found it helpful to recognise different categories.

Native versus introduced species Native species are those that have colonised a particular area by their own natural means of dispersal. In contrast, introduced species have arrived in a new location either by accidental or deliberate human action. This is an apparently simple distinction, but in reality it is difficult to decide the status of many species. Fossil evidence, archaeological early records, herbarium specimens, independent dated evidence, and maps and surveys all provide useful clues. The detection of a species, as plant remains in post-glacial deposits, might provide definitive evidence, but it is not possible to identify all species from sub-fossil leaves, wood, pollen, seeds, fruits, etc. A knowledge of which plants are native provides a baseline with which to recognise narrow endemic species deserving of conservation, and helps to identify introduced species.

Wild versus cultivated plants Archaeological investigation into prehistory reveals that very many plant species have been 'exploited' for food, fibres, medicinal purposes etc. (Renfrew, 1973). Heywood & Stuart (1992) estimate that '5000 plant species are cultivated by humankind and twice as many harvested from the wild'. Another estimate, by Vietmeyer (1995), suggests that as many as *c*.20 000 plant species are used for food. In contrast, Solbrig (1994) emphasises the comparatively small number of cultivated plant species on which civilisation is based: 'twenty species of plant and five species of animal account for over 90% of human sustenance' and 'three cereal plants, wheat, rice and maize, account for 49% of human calorie intake'.

Motley, Zerega & Cross (2006) and Hancock (2012) provide important accounts of the evolution and origin of crop species, while Olsen & Wendel (2013) review recent genetic insights into crop domestication, through the study of evolutionary genomics and evo-devo. Olsen & Wendel emphasise that cultivated plants provide 'wonderful models for studying the evolutionary process' and show that the domestication process is associated with the evolution of a syndrome of characters distinguishing cultivated plants from their wild ancestors: greater productivity (more or bigger seed); ease of cropping (reduced seed dormancy, more erect plants with more dominant growth of central stem, reduced branching, leading to architecturally more compact plants; greater synchrony in flowering and/or fruiting; loss of natural seed dispersal leading to non-shattering of seeds/fruits); and greater palatability (loss of chemical and morphological defences leading to reduced toxicity relative to wild ancestors). Particular species/cultivars may exhibit some but not necessarily all these traits. The genetics of cultivated traits has been widely explored using Mendelian methods, QTL approaches, genome sequencing and screening. Many crop plants are polyploids and the analysis of their origin and subsequent change has provided many insights into plant evolution. Sometimes introgression events have perhaps played a key role, e.g. in the evolution of Rice (Sang & Ge, 2007; Zhao *et al.* 2010).

The genomic investigations of cultivated plants as 'model' species have allowed the development of techniques to study signatures of selection involved in domestication (see, for example, the control of Rice grain-filling and yield by a gene with a potential signature of domestication; Wang *et al.*, 2008). Olsen & Wendel characterise this approach as follows: 'Genetic drift acts strongly' in the domestication process, 'when a subset of individuals in the wild species become founders of the crop lineage; this is expected to result in a genome-wide reduction in genetic diversity'. Also

> selection is expected to differentially reduce diversity at the specific genes that control the traits subject to selection. As a favoured gene is driven to high frequency, much of the standing genetic variation around the target gene is removed from the population, creating a molecular signature of selection. The extent to which this deviation from neutrality is detectable depends upon many factors, including the mating system, strength of selection, population structure and recombination rates.

Compared to the rate of historical advances in domestication, the development of modern molecular and biotechnical approaches to crop improvement has the potential to accelerate the domestication process very greatly – so called super-domestication (Vaughan, Balazs & Heslop-Harrison, 2007).

Wild plants Tansley (1945) defined wildlife as 'all kinds of native animals and plants which maintain themselves without human assistance'.

He also included 'those introduced species which are securely established and reproduce themselves freely in the country'. The recognition of 'wildlife' as an important category of biodiversity, and the steps that should be taken to safeguard it, was part of Tansley's famous appeal for nature conservation in his book *Our Heritage of Wild Nature: A Plea for Organized Nature Conservation* (1945). Now that conservation of ecosystems and species is widely practised in many parts of the world, Tansley's definition should be revisited. Probably a great deal of what many would regard as important wildlife would be unable to maintain itself through establishment, growth and successful reproduction without human intervention to conserve and manage habitats (see below).

Wild versus crop plants Close examination reveals that it is not possible neatly to divide plants into wild and crop. For instance, many 'wild' species are used as famine food in times of need, and Raybould (1995) points out that it would be difficult to single out any species that has not been used at some time for some human purpose.

Ornamentals In many human cultural landscapes a wide diversity of ornamental species are grown in gardens and parks. While there is no comprehensive list of such species, Griffiths (1994) lists 60 000 different taxa in cultivation. The *European Garden Flora* provides information on over 17 000 species cultivated indoors and under glass in Europe (Tutin *et al.*, 1964–80). *The Plant Finder* lists c.55 000 taxa, including cultivars, commercially available in the UK. However, this does not reflect the totality of taxa in cultivation. For example, about 20 000 cultivars of *Rhododendron* are registered with the Royal Horticultural Society of the UK (personal communication from Dr J. C. Cullen).

Feral plants Cultivated crop plants and ornamentals sometimes 'escape', are introduced or become naturalised in the wild. By studying such cases using modern molecular approaches, Ellstrand *et al.* (2010) point out that the 'reconstruction' of feral histories is possible and offers an opportunity, at present under-exploited, for insights into evolutionary processes. It has been proposed that domestication might be relatively speedy and might involve a limited number of recessive traits (Gressel, 2005a, b). De-domestication to a feral state is possible with dominant traits being selected.

The Olive is very widely cultivated, but both wild and feral populations have been identified. Molecular approaches are providing insights into the origin and genetics of the cultivated plant together with the means of identifying and studying ancient wild and recent feral variants (Lumaret & Ouazzani, 2001; Ben-Ayed, 2014). The Olive is a very important feral species in Australia (Gressel, 2005a, b). In the late 1800s, Olive groves were extensively established in South Australia, but, with the collapse of the industry, cultivation was abandoned in many areas. Initially, olive seedlings were kept in check by sheep grazing, but the relocation of sheep to alternative pastures allowed the long-lived Olive, which has a very large seed bank, to thrive as a feral species invading many semi-natural habitats.

Rhododendron ponticum is another important feral, shrubby ornamental species in western Europe. It might be supposed that plants escaped from cultivation in parks and gardens, but the history of the species is more complex. While escapees are found in some localities, the plant was widely planted in woodland on country estates in the nineteenth century in the UK and Ireland, as cover for game birds. These plantings provided a bridgehead for the invasions of semi-natural woodland. The plant, which has toxic foliage and nectar, is capable of prolific vegetative spread from the rooting of lateral branches. It also produces abundant seed.

Weeds Weeds may be defined as plants growing in the wrong place. Holm *et al.* (1977) consider that fewer than 250 species have been recognised as troublesome, but clearly many more could be considered important weeds. No universal listing of weeds has been attempted, but many national lists have been assembled, for instance, a database of (a) noxious weeds of Australia www.weeds.org.au/noxious.htm; and (b) invasive and noxious weeds of the USA, produced by the US Department of Agriculture. Different suites of weed species are associated with different crop cultivations, e.g. arable, orchards, pasture, plantation crops etc.

Ruderals Human cultural landscapes are characterised by areas of highly disturbed wastelands, rubbish dumps etc. Ruderal species predominate here and the category includes many introduced annual weeds, which establish first, later being replaced by perennial and woody species.

Invasive species These are either native or more usually introduced species of plants and animals that invade natural or semi-natural ecosystems and thereby threaten native biological resources (Groves & di Castri, 1991; Cox, 2004; Sax, Stachowicz & Gaines, 2006; Richardson, 2011; Simberloff & Rejmánek, 2011). While there are no comprehensive worldwide lists of invasive species, IUCN SSC Invasive Species Specialist Group (ISSG) is a network of volunteers publishing lists of species in different areas and information on the control of invasive species. There is a central Cooperative Initiative on Invasive Alien Species on Islands (www.issg.org/cii). For particular countries comprehensive accounts have been produced (for example *Alien Plants of the British Isles* by Clements & Forster, 1994).

Species may have different status in different circumstances It is important to stress that a species can be classified into two or more of the categories we have just explored. For instance, some species may be classified as weeds in some situations and active steps are taken to control or eradicate them. But, in other circumstances, the same species may be promoted or tolerated in human cultural landscapes, as it provides food, for example, for birds or nectar for pollinating insects.

Interactions between plants

Domesticated–wild From an evolutionary viewpoint domestication is a recent phenomenon, and crop plants sometimes occur and hybridise with their wild relatives. For instance, Jarvis & Hodgkin (1999), Ellstrand, Prentice & Hancock (1999) and Ellstrand *et al.* (2013) summarise the evidence for natural hybridisation and introgression of crops and their wild relatives, including Wheat, Barley, Oats, Quinoa, Hops, Hemp, Potato, Common bean, Cowpea, Carrots, Squash, Tomato, Radish, Lettuce, Beet, Sunflower, Cabbage and Raspberries. The result of these interactions is not only the introgression of crop alleles into wild and weedy populations (including transgenic (GM) constructs), but also the movement of the alleles of wild plants into the crop (DeWoody *et al.*, 2010).

Crop-to-wild gene flow may also play a significant role in the evolution of weeds (Ellstand & Schierenbeck, 2000). A good example of crop–weedy–wild interactions is provided by Beet (*Beta vulgaris*) (van Dijk, 2004). In brief, cultivated Beet is generally biennial: it requires vernalisation to initiate flowering. The crop may be the leaves, or the swollen underground beets, harvested as a vegetable or for sugar production. In addition to the cultivated beets, wild ruderal beets occur in some inland sites: and these have an annual life cycle, especially in the Mediterranean Basin, where seed nurseries for beet cultivation elsewhere are located. Where these wild plants grow close to the cultivated seed nurseries, gene flow occurs and hybrid derivatives - annual weedy variants of beet - are later produced when the perennial Beet crop is grown commercially elsewhere. As crop and weed are of the same species it is very difficult to eradicate the annual weed beet, which has become a very serious weed. In addition, a conspecific wild beet is also found along coastal fringes of Northern Europe, and there is evidence, for example from France, where beet cultivation occurs near the coast, that these wild populations show the effects of gene flow, from cultivated to wild plants. For fuller details of the complex cytogenetic interactions of the different variants of *Beta*, see van Dijk (2004).

Hybridisation between taxa in the evolution of some invasive species Ellstand & Schierenbeck (2000, 2006) make a case that hybridisation may have acted as a stimulus for the evolution of invasiveness. Eighteen case histories of invasive taxa after intertaxon hybridisation are cited (including *Amelanchier erecta, Bromus hordeaceus, Helianthus annuus* spp. *texanus, Oenothera glazioviana, Senecio squalidus, Spartina anglica, Tragopogon mirus* and *T. miscellus*). In other cases invasive lineages have emerged from intertaxon hybridisation (*Avena barbata* ×*A. strigosa, Carpobrotus edulis* × *C. chilense, Lythrum salicaria* × *L. alatum, Raphanus raphanistrum* × *R. sativus, Spartina alterniflora* × *S. foliosa* etc.).

Interspecific hybridisation: a threat to some endangered species Gene flow from crop and ornamental plants to wild species has the potential to change the gene pools of wild plants and may contribute to their endangerment (Chapman & Burke, 2006).

For example, molecular studies have revealed that wild threatened populations of the Australian palm *Ptychosperma bleeseri* are being endangered by hybridisation with nearby congeneric taxa growing in urban gardens (Shapcott, 1998).

Do human activities present threats to biodiversity?

It is often asserted that humans are endangering species through: the loss, degradation and fragmentation of natural habitats; the over-exploitation of natural resources; the sometimes catastrophic effects of introduced plants, animals, fungi and other microorganisms; and the effects of environmental pollution, that results not only in localised loss of biodiversity, but also in the wider endangerment of species of plants and animals, through, for instance, acid rain and the effects of global climate change. We examine each of these categories of threat in turn and consider the evidence for the endangerment of plant species. Assessments of the scale of the threat to biodiversity have been widely discussed in the scientific literature, as well as in the press and on television. Some consider that estimates of loss or potential loss have often been exaggerated (Mann, 1991). For any particular situation, it is clearly important, therefore, to consider the strength of the available evidence.

Human influences: habitat loss and fragmentation

In the evolution of cultural landscapes, natural vegetation has been lost. Now, only fragments remain that have often been exploited or modified by human activities. To give an example, the history of deforestation and forest degradation in Australia illustrates some important general points.

Prior to European contact, aboriginal use of fire had a major influence on Australian ecosystems (Bliege Bird *et al.*, 2008). By examining many lines of evidence, it is estimated that some '30% of Australian land mass was covered by "forest" at the time of the first European colonization in the late 18th century' (Bradshaw, 2012). From this time onward, timber was widely harvested and the most fertile areas converted to agriculture. Deforestation occurred particularly near the coasts. 'In the 1950s, south western Australia was largely cleared for wheat production . . . since the 1970s, the greatest rate of forest clearance has been in south-eastern Queensland and northern New South Wales, although Victoria is the most cleared state. Today, degradation is occurring in the largely forested tropical north due to rapidly expanding invasive weed species and altered fire regimes' (Bradshaw, 2012).

In 1995, a detailed survey of the existing forests was carried out employing satellite imagery (Graetz, Olsson & Wilson, 1995). More than 38% of Australian forest has been completely destroyed since European settlement, with much of the remaining forest severely degraded. 'Forest cover is severely fragmented into small patches, especially in south-eastern Australia, with roads, urban development, agriculture and plantations, isolating existing fragments to the point that much of their biodiversity potential is severely compromised' (Gill & Williams, 1996). However, there has been much progress in conserving forests, and there have been no widespread clearances in recent decades. Strong protection is now in place for the tropical forests. 'However the legacy of deforestation and fragmentation that occurred prior to protection means that many extinctions are likely to occur over the coming decades' (Bradshaw, 2012).

The conversion of a large ecosystem into fragments has the potential to produce dramatic changes in ecosystems. Fragments are surrounded by other vegetation, and 'edge effects' occur; for example, the microclimate between forest fragment and matrix may be different with the edge of the forested fragment subject to different temperature, humidity, light, effects of storms etc. A multitude of different other fragment/matrix effects is possible. For instance, chemical spray drift may enter the fragment from nearby agricultural land; and invasive species may be more readily able to enter forested ecosystems (Myers, 1986).

In addition, the population biology and genetics of species are dramatically changed. As we saw in Chapter 9, gene flow between plants in different fragments may be interrupted or prevented, inbreeding may occur and the effects of genetic drift may increase etc. Over time, genetic erosion may occur in species in fragmented populations.

Human influences: introduced organisms

Mack & Lonsdale (2001) consider that introduction of plants involves three overlapping non-mutually exclusive phases.

The accidental phase Immigrants to new colonial territories often brought their own crop seed, bedding and fodder that contained weed and other contaminants. In addition, it is suspected that many species were accidentally introduced as a result of trade in timber, foodstuffs, wool, minerals, ores, and as seed in both wet or dry ship's ballast etc.

The utilitarian phase Once established, settlers deliberately introduced plants that were thought to be useful. Botanic Gardens have often played a key role, as in the introduction of Tea, Coffee, Palm oil and Rubber and their subsequent development as plantation crops (Brockway, 1979; Heywood, 1983). For instance, the Spanish and Portuguese colonisers introduced African grasses to South America (Parsons, 1970). Forage grasses from various countries were imported into Australia (Lonsdale, 1994). Large-scale agroforestry projects have involved the importation of a variety of trees to Chile, Brazil, Argentina and elsewhere (Richardson & Higgins, 1999). In some cases rooted plants were transferred with their ectomycorrhiza (Vellinga, Wolfe & Pringle, 2009).

The aesthetic phase Mack & Lonsdale observe that having reached a point where their survival was assured, immigrants began to import ornamental species for parks and gardens. In the nineteenth century, seed merchants in Britain, Germany and Japan were often the source of seeds (Mack, 1991). Heywood (2011) notes that Botanic Gardens have also played a key role in the introduction of invasive species. Thus, 'it is estimated that 80% of the current invasive alien plants of Europe were introduced as ornamentals or agricultural plants: seriously invasive plants deliberately introduced as ornamentals include Japanese Knotweed (*Fallopia japonica*), Summer Lilac (*Buddleja davidii*), Common Rhododendron (*Rhododendron ponticum*), and Giant Hogweed (*Heracleum mantegazzianum*)' (Heywood, 2011). Many ornamental plants introduced into North America, South Africa, Australia and elsewhere have also become invasive. For example, False Acacia (*Robinia pseudoacacia*), native of USA, has now become a serious invasive in parts of the USA and Europe. The Black Wattle (*Acacia mearnsii*), which originates from South Australia and Tasmania, is now a seriously invasive species, infesting 2.5 million ha of scrublands, gardens and savannas of South Africa.

Other introduced organisms The introduction of large herbivorous, omnivorous and carnivorous animals to new territories has had a major impact on their ecosystems (Cox, 2004; Sax, Stachowicz & Gains, 2006; Rotherham & Lambert, 2011). For example, grazing and (in some cases) trampling by introduced goats, horses, sheep, deer, donkeys, cattle, dogs, cats and rats has had a profound effect on the fauna and flora of oceanic islands (Caujapé-Castells *et al.*, 2010).

Introduced fungal pathogens are having an increasing impact on ecosystems (Rackham, 2008). Examining the evidence he concludes that 'the spread of pathogens is probably the most serious threat' to forests, wood pastures and savannas in Europe and elsewhere. Evidence of the potentially devastating effect of fungal pathogens is provided by classic cases of *Cryphonectria parasitica* that destroyed most of the American Chestnut (*Castanea dentata*) forests in New England (Dalgleish & Swihart, 2012); and the catastrophic effects of Dutch Elm disease in Europe (Rackham, 2008). The list of pathogenic diseases that pose a threat to plants, especially tree species, continues to grow. For example, Garbelotto & Pautasso (2012) review the serious impacts of pathogens on Mediterranean ecosystems:

- *Phytophthora* species, especially *P. ramorum*, affecting a range of species including Tanoaks (*Lithocarpus densiflorus*) in California; and *P. cinnamomi* that is lethal to many wild fruit tree crops and ornamental species in Australia.
- *Seiridium cardinale* – native to California – is now threatening a range of *Cupressus* species worldwide.
- *Fusarium circinatum* – originally thought to be from Mexico and/or southern Florida – is now affecting pines, especially the widely planted quick-growing *Pinus radiata*, a key plantation species worldwide.

There has been recent concern about emerging threats to tree species in northern Europe.

- *Ceratocystis platani* is a deadly fungal disease of plane trees in France, Germany and Greece. A native of eastern USA, it is thought that the fungus was introduced in munitions boxes brought to Europe in the Second World War.
- In recent years, Ash Die-back disease (first described as *Chalara fraxinea* but now known as *Hymenoscyphus pseudoalbidus*) has had a devastating impact on ash trees (*Fraxinus excelsior*) across Europe, and has now reached Britain.
- www.forestry.gov.uk has produced a number of tree disease alerts for the UK. These include many fungal diseases (including *Phytophthora austocedrae* on Juniper; and *Phytophthora ramorum* Sudden Oak Death), as well as newly discovered introduced insect pests.

It is clear that introduced fungal pathogens may have a devastating effect on many ecosystems through the loss or damage to keystone species. Also, many research papers have been published on the likely effects of climate change, including threats to plant health and diseases, but our present knowledge is imperfect and more research is urgently needed.

Introduced insect pests also have the potential to threaten individual species and thereby influence community ecology. There is a huge literature on the subject. Here, we note the potential for introductions through expanding global trade. Krushelnycky, Loope & Reimer (2005) report that 'inspection of cargo in one Hawaiian airport during only 20 weeks in 2000–2001 detected 125 alien insect species and 16 plant pathogens not known to be established in Hawaii' (quoted by Caujapé-Castells *et al.*, 2010, who also provide a review of the impacts of alien insects on island floras).

The ecological consequences of introduced species

Across the world large numbers of species have been introduced, and in areas where human activities predominate some native species have become invasive (Mooney & Cleland, 2001), and it has been estimated that invasive species have come to dominate 'c. 3% of the Earth's ice free surface' (Mack, 1985). Many other introductions are not yet invasive, but it is believed that some have the potential to become so.

Invasive species may have many effects on the ecosystems they invade. Case studies have revealed that in some situations invasive species have major impacts on native species through competition, predation, herbivory, disease and parasitism. Invasive species may have dramatic effects in ecosystems. For example,

- The aquatic fern *Salvinia molesta* – a native of South America – shades out and overgrows native species in aquatic ecosystems from the Philippines to Australia and New Zealand, and from southern Africa to south-east Asia (www.csiro.au/resources/salvinia-control).
- In Texas, Chinese Tallow (*Sapium sebiferum*) woodland has replaced prairie (Cox, 1999).
- Invasive species may increase the frequency and intensity of fires (Simberloff, 2001), as, for example, in grasslands of western USA, dominated by invasive populations of *Bromus tectorum*. In the Everglades of Florida huge areas are dominated by the introduced invasive Australian Paperbark (*Melaleuca quinquenervia*). There is increased fire risk where it dominates, as it produces highly flammable foliage and litter (Simberloff, 2001).
- The nutrient status of ecosystems may change as invading species become dominant, as for example in Hawaii where the nitrogen-fixing Firetree (*Myrica faya*) has become established. Sometimes invaders produce other changes in soils, e.g. where the invasive composite *Pilosella officinarum* (*Hieracium pilosella*) has become established in New Zealand, the soils become markedly more acidic (www.issg.org/database/species/ecology.asp).

Human influences: the effects of pollution

Human activities release a wide array of pollutants into air, soil, and fresh and marine environments. In many situations vegetation is damaged, with some species being more affected than others. Pollutants have increased, with the development of urban industrial areas. It is important to stress that the effects of pollution are not restricted to areas where the pollutants were first generated. Movement of air, dust and water across the globe carries the pollutants to distant territories, where their presence can be detected and their effects measured even in what appear to be pristine untouched ecosystems.

In the present context we emphasise the importance of testing hypotheses relating pollutants to specific plant damage. Thus, in field experiments, investigating damage, the alleged pollutant(s) must be identified, measured and confirmed in their effect. Also, while experimenters often investigate the effects of known quantities of single pollutants on plants, it must be appreciated that many pollutants react with each other, giving a range of primary and derivative toxic or damaging compounds (Jacobson, 2002; Fenger & Tjell, 2009).

The widespread and damaging effects of pollution have been acknowledged worldwide and in many countries legislation has been enacted to try to reduce the impact on ecosystems and human health, but still the complex effects of pollution, interacting with other factors, are widely evident (see the *World Atlas of Atmospheric Pollution*: Sokhi 2011). This is illustrated by studies of three major effects of pollution.

Eutrophication

Large ecosystem changes, some irreversible, are induced by an excess of nutrients that originate from the use of nitrogen compounds, phosphate and potassium (NPK) in intensive agricultural systems (Fener & Tjell, 2009). Nitrogen compounds are also introduced into the environment through the burning of fossil fuels. Human wastes of all kinds, and the effluents from domesticated animals, also lead to eutrophication of ecosystems.

Generally, 'excess fertility' leads to reduced biodiversity in ecosystems, and the dominance of a few highly productive species that can utilise the excessive amounts of nutrients. Thus, eutrophication has a major impact on nutrient-poor ecosystems, such as raised bogs and other types of mire, some dry grasslands, freshwater systems including some lakes, ground water and coastal sea regions. Eutrophication of aquatic environments has complex effects, but often leads to a loss of oxygen, due to high production of plankton, especially algal species. In highly eutrophic waters, blooms of toxic blue-green algae are often found.

Water-borne eutrophication has been long been recognised, but research has shown that gases and dust in atmospheric pollutions have significant effects, especially around point sources, such as intensive farms, highways and some industrial plants. The eutrophication caused by the excessive availability of nitrogen compounds was recognised many years ago. In contrast, considerable research was needed to confirm that phosphate pollution is a potent component of eutrophication, arising as a by-product of intensive agriculture, sewage and wastewater disposal, burning coal, straw etc.

Acid Rain

The phenomenon of acid rain provides an interesting example of the complexities revealed by research into pollution. It has been claimed that more than 50% of the conifers in Switzerland, for example, exhibit slight to severe defoliation, a phenomenon paralleled in forest trees in other parts of Europe and elsewhere (Briand *et al.*, 1989). It was suggested that forest decline was caused by sulphur dioxide pollution, produced by power stations, industrial plants and domestic installations. In addition, sulphur dioxide also reacts to produce weak sulphuric acid in rain, which not only causes immediate damage to plants, but also leads to other effects, such as promoting the release of toxic concentrations of aluminium ions in the soil. By the end of the twentieth century legislation to reduce sulphur dioxide levels proved successful in Europe and North America, and recent evidence suggests that some species have recovered following the introduction of legislation to reduce acid rain. For example, after the passage of the Clean Air Act in 1970, growth ring analysis of *Juniperus virginiana* in the Central Appalachian Mountains, USA, provides evidence of recovery (Thomas *et al.*, 2013). However, 'forest die-back' has continued to be a problem in some regions and further research has revealed many complexities.

Vehicle emissions also contribute another highly significant component of atmospheric pollution. They are a major source of damaging nitrogen oxides. In sunlight, nitrogen dioxide reacts to produce ozone, and experiments have shown that this gas damages the photosynthetic processes of plants. In addition, conifers exposed to high doses of ozone in summer may be more frost sensitive in the autumn. Furthermore, other pollutants and their interaction products also injure trees. More recently, other factors have emerged from research. There is plenty of evidence that pollution damages trees, but a review of all the factors contributing to 'forest decline' reveals a more complex picture, in which eutrophication, climatic stresses and tree diseases are also important (Strauss & Mainwaring, 1984; Mason, 1992). Fenger & Tjell (2009) point to eutrophication as a major factor in forest damage, leading to 'mineral deficiencies' in 'nitrogen and phosphate enriched' environments of Europe and North America. In other places – for instance, western USA and southern Europe – 'ozone may be the main cause of forest dieback'.

Human influences: global climate change

In 2010, the Geological Society of London placed a major policy document on the Internet (www.geolsoc.org.uk/climatechange). This showed that evidence from, for example, marine and lake sediments, ice sheets, fossil corals, stalagmites and fossil tree rings has provided evidence of the climates of the distant past. Thus, 'cores drilled through the ice sheets yield a record of polar temperatures and atmospheric composition ranging back to 120 000 years in Greenland and 800 000 years in Antarctica. Oceanic sediments preserve a record going back tens of millions of years and older sedimentary rocks extend this record to hundreds of millions of years. This vital baseline of knowledge about the past provides a context for estimating likely effects in the future'.

The Greenhouse Effect

To understand the present concern about climate change it is essential to understand the effect of certain gases that cause warming of the atmosphere. The Earth is warmed by incoming solar radiation and some of the energy is re-radiated. A portion of this energy is absorbed by gases in the atmosphere (for example by carbon dioxide, water vapour etc.), while the rest escapes into space. The carbon dioxide content of the air shows about a 40% increase since the industrial revolution (taken to be 1750) from 280 ppm in 1960 to 400 ppm in May 2013. This increase is associated with the burning of fossil fuels, deforestation, etc. Increased carbon dioxide results in the trapping of additional re-radiated energy, resulting in global warming – the so-called 'greenhouse effect'. In addition, a number of other pollutant gases are increasing due to human activity, and these are also contributing to the greenhouse effect. While their concentrations are much lower, their 'global warming potential' is greater than carbon dioxide, which has a global warming factor (GWF) of 1.

- Methane, an important source of which is rice paddy fields, has increased from pre-industrial levels of 715 ppb to more than 1732 ppb today. It has a GWF of 21 over a 100-year horizon. Other sources of methane are potentially very important. Methyl hydrates (solids in which methane is the centre of a crystalline molecule surrounded by water molecules) occur under Arctic permafrost, Antarctic ice and in some sedimentary deposits. Methane is released when the molecule is warmed. An active debate is in progress as to whether climate change will result in the release of enormous quantities of methane (see Ruppel, 2011) and thereby greatly increase the concentration of this highly significant greenhouse gas.
- Nitrous oxide, over a third of which comes from anthropogenic sources, has a GWF of 310 over a 100-year horizon.
- Chlorofluorocarbons (CFCs of different formulations have very high GWFs with values up to more than 10 000 over a 100-year horizon) and ozone are also potent greenhouse gases.

As greenhouse gases persist in the atmosphere (lifetimes in years: methane 12; nitrous oxide 120), they will continue to contribute to global warming for years to come.

Direct observation of climate change

From a comprehensive review of the available evidence, the IPCC Summary for Policy Makers (2013) gives the most up-to-date details and reaches a number of conclusions (with level of confidence attached to each statement). Here, quoting from the Summary for Policy Makers, we note that:

- 'Warming of the climate system is unequivocal, and since the 1950s, many of the observed changes are unprecedented over decades to millennia. The atmosphere and ocean have warmed. The amounts of snow and ice have diminished, sea level has risen, and the concentrations of greenhouse gases have increased.'
- 'Each of the last three decades has been successively warmer at the Earth's surface than any preceding decade since 1850. In the Northern Hemisphere, 1983–2012 was *likely* the warmest 30-year period of the last 1400 years (*medium confidence*).'
- 'Over the last two decades, the Greenland and Antarctic ice sheets have been losing mass, glaciers have continued to shrink almost worldwide, and Arctic sea ice and Northern Hemisphere spring snow cover have continued to decrease in extent (*high confidence*).'
- 'The rate of sea level rise since the mid-19th century has been larger than the mean rate during the previous two millennia (*high confidence*). Over the period 1901–2010, global mean sea level rose by 0.19 {0.17 to 0.21}m.'
- 'The atmospheric concentrations of carbon dioxide, and nitrous oxide have increased to levels unprecedented in at least the last 800 000 years. CO_2 concentrations have increased by 40% since pre-industrial times, primarily from fossil fuel emissions and secondarily from net landuse change emissions. The ocean has absorbed about 30% of the emitted anthropogenic carbon dioxide, causing ocean acidification.'

Predictions of future climate change

Different models of the climate system have been devised to predict the likely rise in temperatures etc. General circulation models have been refined and extended so that the effects of the ocean and vegetation change can be computed. Also, regional models have been produced.

The Coupled Model Intercomparison Project, Phase 5 (CMIP5) is an important international collaboration (IPCC, 2013). This project, and simulations with the Earth System Model (ESM), has investigated a number of scenarios studying the effects of increasing carbon dioxide in the atmosphere, specifically, carbon dioxide reaching 421 ppm (RCP2.6); 538 ppm (RCP4.5); 670 ppm (RCP6.0); and 936 pppm (RCP8.5) by the year 2100.

Important insights into future climate change, which could not be obtained by any other means, are offered by this modelling (Goudriaan *et al.*, 1999), but it is important to stress that the projections obtained 'are couched in the language of probabilities: they estimate likelihoods not certainties' (Briggs, 2009). Where possible, assessments given in the Reports of the IPCC are expressed 'probabilistically with a quantified level of likelihood' (from *exceptionally low* at one extreme to *virtually certain* at the other, with appropriate terms indicating intermediate status). These likelihoods are given in italics below.

In summary, the results indicate that: 'global surface temperatures change for the end of the 21st century is *likely* to exceed 1.5°C relative to 1850 to 1900 for all RCP scenarios except RCP2.6. It is *likely* to exceed 2°C for RCP6.0 and RCP8.5, and *more likely than not* to exceed 2°C for RCP4.5. Warming will continue beyond 2100 under all RCP scenarios except

RCP2.6.' Thus, 'continued emissions of greenhouse gases will cause further warming and long-lasting changes in all components of the climate system. Limiting climate change will require substantial and sustained reductions of greenhouse gas emissions.'

While in the past some scientists have considered that limiting global surface temperature to below 2°C (equivalent to *c.*450 ppm carbon dioxide) would avoid dangerous climate change, this view has been challenged by experts. While they still regard the 'two-degree boundary' a significant and potentially dangerous boundary to cross, they now consider that, if carbon dioxide reaches 450 ppm in the atmosphere, a rise in global surface temperature of 2°C is likely to be exceeded. A lower maximum level of 400 ppm has been set to be relatively certain that a rise of 2°C would be avoided.

Turning to other issues, 'changes in the global water cycle in response to warming over the 21st century will not be uniform. The contrast in precipitation between wet and dry regions and between wet and dry seasons will increase, although there will be regional exceptions.'

'Global mean sea level will continue to rise during the 21st century . . . the rate of sea level rise will *very likely* exceed that observed during 1971–2010 due to increased ocean warming and increased loss of mass from glaciers and ice sheets.'

Increasing climate change is resulting in more frequent extreme weather events, and as greenhouse gases are increasing, such events – floods, heatwaves, droughts etc. – are predicted to become more frequent (see Report of the Royal Society, *Resilience to Extreme Weather*, executive summary and other documents (https://royalsociety.org/~/media/policy/projects/resilience-climate-change/resilience-executive-summary.pdf).

Climate change: human influences

There has been a great deal of acrimonious debate and argument about the influence of human activities on climate change (see below). In the 2007 IPCC report it was concluded that 'most of the observed increase in globally average temperatures since the mid-20th century is *very likely* due to the observed increase in anthropogenic greenhouse gas concentrations'. The most recent IPCC report (2013) comes to a more decisive judgement about human influence on climate, in concluding that: 'It is *extremely likely* that human influence has been the dominant cause of the observed warming since the mid-20th century.'

Climate change sceptics and deniers

Climate change is one of the most important issues of our age. As we have seen above, the weight of scientific evidence offers the strongest support to the view that human activities are the major drivers of climate change, through the release of greenhouse gases, and, if dangerous climate instability is to be prevented, urgent steps must be taken to reduce such emissions. However, some deny that human activities are the major driver of climate change. Instead, they believe that natural climate cycles are responsible and that the Earth's climate will return to normality.

Oreskes (2004) reports that: 'Policy makers and the media, particularly in the United States, frequently assert that climate change science is highly uncertain. Some have used this as an argument against adopting strong measures to reduce greenhouse gas emissions.' Deniers have alleged that there is disagreement amongst climate scientists as to whether human activities are causing climate change. A number of studies strongly refute this assertion. For instance, '96–98% of the climate scientists most active in the field' support the tenets of ACC (Anthropogenic Climate Change) (Doran & Zimmerman, 2009). And amongst a survey of scientific papers on climate change, 97.1% 'endorsed the consensus position that humans are causing global warming' (Cook *et al.*, 2013).

Weart (2011) provides a revealing review of how climate change 'scepticism became denial' and Kemp,

Milne & Reay (2010), writing in the prestigious journal *Nature*, make the important distinction between deniers and sceptics. 'Denial of the science of climate change is eroding public understanding of the issue . . . Denialism is motivated by conviction rather than evidence. It has been applied to a wide range of issues, including evolution, and the link between HIV and Aids. Deniers use strategies that invoke conspiracies, quote fake experts, denigrate genuine experts, deploy evidence selectively and create impossible expectations of what research can deliver'. Kemp, Milne & Reay (2010) continue:

> by contrast, scepticism starts with an open mind, weighs evidence objectively and demands convincing evidence before accepting any claim. It contributes to the debate and forms the intellectual cornerstone of scientific enquiry. The public should understand the difference between deniers and sceptics, so that their trust in scientists is not threatened at a time when humanity needs us most . . . We do not need to speak with one voice about climate change, but we should stand together to defend proper scientific debate.

Recently the US National Academy of Sciences and the Royal Society have provided an important overview of *Climate Change: Evidence and Causes* written and reviewed by a UK–US team of leading climate scientists (Cicerone & Nurse, 2014). The preamble highlights the way scientists approach complex areas of research. They write:

> climate change is one of the defining issues of our time. It is now more certain than ever, based on many lines of evidence, that humans are changing Earth's climate . . . the evidence is clear. However, due to the nature of science, not every detail is ever totally settled or completely certain. Nor has every pertinent question yet been answered. Scientific evidence continues to be gathered around the world, and assumptions and findings about climate change continually analysed and tested . . . the publication makes clear what is well established, where consensus is growing, and where there is still uncertainty.

Important progress on climate change was made at the UN Conference in Paris in December 2015. 185 countries submitted non-legally binding voluntary pledges showing how they propose to limit climate change before 2020. However, these existing plans are likely to lead to at least 2.7°C above pre-industrial levels of warming this century, resulting in very severe climate change effects. To move forward from this unsustainable situation, it was agreed to limit global warming to well below 2°C (1.5°C above pre-industrial levels, if possible) by new improved pledges prepared for further regular meetings. Developed countries would provide financial support to developing countries. Whether this agreement heralds effective control and reduction in greenhouse gases remains to be seen.

Biological effects of climate change: species adapt, move or die

First, it is helpful to set out some theoretical responses of species to climate change. Then, we examine some of the empirical evidence of the effects attributed to climate change, both at the species and ecosystem levels.

Climate change presents new selection pressures

Climate change is likely to present very complex 'new' changes in the selection pressures impacting on ecosystems/species/populations. Some of these might be modelled as directional changes in temperature and rainfall etc. However, there is evidence that climate change also manifests itself in the form of extreme weather events, such as droughts, floods and storms. Species may respond initially to all climate changes through phenotypic plasticity (Anderson *et al.*, 2012). As we saw in earlier chapters, the extent of plasticity and flexibility is under genetic control. Our present knowledge of these phenomena suggests that there are clear limits to such responses, but they may be very important in providing a first adaptive response.

Our present understanding of climate change suggests that unless adequate measures are taken to limit and decrease greenhouse gases, more extreme directional selective forces are likely in the future. In such circumstances, species may reach the limits of their plastic or physiological responses, and only those populations/species in which genetic adaptation to the new conditions occurs will be able to survive (see Reyer *et al.*, 2013). Insights from many sources suggest that only those populations that contain the appropriate genetic variability will be able to make such responses (Bradshaw & McNeilly, 1991). If greenhouse gas emissions continue to increase, climate change is likely to accelerate and dramatic changes in climate are possible in short timeframes. Under such conditions those plants with very short generation times (and the appropriate genetic variability) are likely to respond more quickly than those that take decades to reach reproductive maturity after first establishment.

Another response to climate change, which does not necessarily exclude plastic and genetic components, is for species to colonise new areas by tracking suitable climatic regions. The extent to which a species is able to move to new territories depends upon the 'effectiveness' and distance of seed/fruit dispersal (Trakhtenbrot, Perry & Richardsson, 2005), and the degree to which a 'new site' has the appropriate climatic, abiotic and biotic conditions that allow the establishment, growth and reproductive success of the species over many generations. For many plants this will require the continuing interaction with specific pollinators/seed dispersers, which, it is predicted will themselves be influenced by climate change. Also, for many species, success in migration will depend upon the availability, outside the present geographic range of a species of highly specialised edaphic habitats such as serpentine soils.

The footprint of climate change

There is a huge literature on how species and ecosystems will respond to climate change. The Web of Science Database lists more than 128 500 research papers published in the period 1990–2013.

Here, in surveying anthropogenic effects on the plants, we examine some of the key evidence for species adaptation, migration or extinction in the face of climate change.

Phenological changes In response to climate change there has been alteration in the timing of biological events in those organisms that respond to spring air temperature. Walther (2010) reviewing the present evidence reveals that amongst 677 species a global meta-analysis revealed a 'significant mean advance of 2.3 days/decade' (Parmesan & Yohe, 2003), but not every species was equally responsive. Some species respond to another climatic cue, namely 'seasonal changes in photoperiod' (see Edwards & Richardson, 2004). The earlier arrival of spring is detectable in satellite images of vegetation zones by examining a 'greenness index' that is correlated with budburst. In addition, there has also been later arrival of autumn. Moreover, changes in rainfall patterns have influenced the phenology of plants in the tropics.

The effect on mutualistic associations between plants and animals has also been examined (Walther, 2010). For example, in some studies early flowering in plants was associated with earlier appearance dates of pollinators (Hegland *et al.*, 2009), but in others there was a 'phenological decoupling' of the date of flowering and insect arrival (Gordo & Sanz, 2005). Further study of this important issue is necessary before the mutualist responses associated with phenological change are clarified.

Altitudinal range changes There is evidence for altitudinal shifts in montane areas (Parmesan & Yohe, 2003). For instance, Vittoz *et al.* (2013) report on the response to climate change in Switzerland, where a 'noticeable increase in mean temperature has already been observed' with 'summer temperatures up to 4.8 K warmer expected by 2090'. Across all taxonomic groups, there is evidence of 'elevation shifts of distribution towards mountain summits, spread of thermophilous species, colonisation by new species from warmer areas and phenological shifts . . .

these changes will likely cause extinctions for alpine species (competition, loss of habitat) and lowland species (temperature, drought stress)'.

Peňuelas *et al.* (2007) have made a detailed investigation of the apparent altitudinal shifts in European beech forest in north-east Spain. Evidence came from historic photographs and demographic studies (age structure and recruitment patterns). They discovered that at lower altitudes Beech is being replaced by Holm Oak forests, and there is an 'upward shift' in the Beech forests, with recruitment above the former treeline. They conclude that this upward shift is a likely consequence of climate warming, but, as in many montane areas, changes in land use have also played a part. In the case of upward shifts of Beech forests, it was discovered that shepherds no longer burn the vegetation above the treeline, providing an opportunity for range expansion.

Range shifts have been reported in other regions. For example, in a study of the alpine Sikkim Himalaya, historical records combined with intensive present-day field work reveal that 'warming-driven geographical range shifts were recorded in 87% of 124 endemic plant species studies in the region . . . [with] a shift of 23–998 m in species' upper elevation limit' (Telwala *et al.*, 2013).

Using evidence of changes in forest plots and aerial photographs in the Green Mountains of Vermont, aerial photographs detected a 'rapid upward movement' of the northern-boreal hardwood forest ecotone in comparing data sets from 1962 to 2004. The upslope shift of this forests type shows that the forest typical of high elevation may 'be jeopardised by climate change sooner than anticipated' (Beckage *et al.*, 2008).

Latitudinal range changes Climate models predict that under climatic warming sessile species such as plants may migrate polewards with dramatic changes or extinctions in the trailing edge populations. A great deal of evidence supports this hypothesis (see Parmesan *et al.*, 1999; Warren *et al.* 2001; Parmesan & Yohe, 2003), but lags in response have been demonstrated in plant populations, and these have been attributed to limitations of dispersal and migration. Thus, butterflies are more responsive than plants in alpine ecosystems (Walther *et al.*, 2002).

Investigations of the Quivertree (*Aloe dichotoma*) in the Namib Desert have given important insights as to how this species is responding to climate change. This case history also provides a useful example of the research methods employed in testing hypotheses (Foden *et al.*, 2007). Field studies provided the initial impetus for the study. It was discovered that in some areas many plants had died, apparently as a result of drought. Then, Foden and colleagues made a detailed census of the plants throughout the entire range of the species and also extensive demographic data were collected. They report that

> individuals grow up to 10m tall and usually occur in dense populations of up to 10,000 trees. A long life span (at least 200 years and possibly up to 350 years), and a large geographical range (c. 200,000km^2) make this a useful subject for study of the impacts of long-term climate trends. In particular, because dead individuals decay relatively slowly *in situ*, often remaining standing for many years, it is possible to obtain a comparable measure of population mortality throughout the species' range.

They also examined photographs allowing the identification of individuals in images taken 41–98 years previously. The availability of extensive climate records for the region made it possible to model the impact of climate change on the future distribution of the species, using the bioclimatic niche modelling tool BIOMOD (McCarthy *et al.*, 2001; Thuiller, 2003). Foden and colleagues concluded that *A. dichotoma* 'is experiencing declines at its equatorial limits (i.e. its trailing edge; *sensu* Davis & Shaw, 2001) in response to anthropogenic climate change trends'. At the same time they 'report the species' failure to expand polewards in relation to its shifting climatic envelope . . . The geographic range of *Aloe dichotoma* is therefore apparently becoming progressively squeezed between an advancing zone of range contraction due to population declines, and a poleward zone of lagging range expansion' . . . This study and other evidence suggest that 'desert ecosystems are likely to become increasingly hostile to endemic biota'. They are likely to become 'more species-poor with intensifying global warming'.

Forecasting future changes in distribution

The study of *Aloe dichotoma* employed bioclimatic climate envelope approaches to explore the consequences of climate change (Araújo & Peterson, 2011). Such models have been discussed in an earlier chapter in relation to hindcasting in plant biogeography. Species Distribution Models (SDMs) have also been widely employed in the study of the likely effects of future climate change. It is important to look once again at these modelling approaches, and consider their strength and potential weaknesses.

In essence, SDMs examine the statistical correlation between the current distribution of species and such environmental variables as temperature, rainfall etc. Then, by examining projections of these variables under various climate change scenarios, future species ranges are predicted. The present and future distribution of a species is considered as presence/absence within the 'cells of a grid' for a specified geographical area.

Finer-scale 'niche' modelling has also been carried out. Taking what are considered to be the key physiological and abiotic factors defining the 'niche' of a species, the present distribution is examined using statistical correlation techniques, and a range of 'future distributions' are predicted under various scenarios of climate change.

In reviewing progress in modelling the effects of climate change, Araújo & Peterson (2011) consider the strengths/weaknesses, uses/misuses of the bioclimatic envelope and other SDM modelling approaches. Here, we note some key areas of debate.

A) In order to make it possible to carry out modelling, some simplifications and assumptions are necessary and these are not always clearly specified. For example, there is some evidence that for some species climate is a major determinant of distributions. In other cases, there is considerable debate about the relative importance of other factors. For example, the distribution of species may be highly influenced by such factors as barriers to dispersal and biotic interactions.

B) As we saw in the analysis of *Aloe dichotoma*, it is sometimes possible to consider the whole range of the species. However, in other cases the distributional data, on which the analysis is based, may be incomplete or defective.

C) Araújo & Peterson (2011) note that the assumption is often made that 'species distributions will frequently be at equilibrium with climate – that is, that species inhabit the entire spatial footprint of their habitable conditions'. This assumption may be unrealistic, 'since dispersal often constrains the potential of species from accessing habitable areas, and biotic interactions, including human impacts, may further prevent establishment in some areas'. Bioclimatic modelling has also been criticised for making 'simplified assumptions about dispersal and biotic interactions. Specifically, they have been criticised for assuming that species would either not disperse, or disperse without limitations, ignoring the biotic interactions shaping current and future distributions'.

D) Models assume that 'inherited physiological tolerances of species to environmental factors are conserved i.e. that the fundamental niche [of a species] does not change in the temporal frame involved in projections'. Given that some species may adapt (perhaps quickly) to climate change this assumption may not always be justified (see below).

E) The usefulness of the concept of the 'niche' has been widely debated amongst biologists (for a discussion of the origin(s) of the idea of 'niche' and critical appraisals of the concept, see Townsend Peterson, 2011). Here we note that modelling considers the realised niches of species.

F) Considering the 'future' ranges of species, it is important to recognise the extreme difficulty of predicting the suitability of habitat outside the present range of a species.

Despite these important caveats, Araújo & Peterson (2011) conclude that bioclimatic modelling can be a 'potentially powerful tool' that can 'place complex ideas in useful contexts', but stress that it is important to consider the most appropriate 'conceptual framework' and explore fully all underlying assumptions before the modelling process is carried out, and to set out these assumptions in research publications.

Adaptive responses to climate change

Climate change is predicted to cause temperature and other changes over decades. At first sight this timeframe makes the study of adaptation very difficult. However, dramatic events, such as droughts and changes in rainfall, are also predicted in many areas of the world. These climatic episodes will impose ecologically important selection pressures, and, as they act over shorter time, they provide an opportunity for investigation.

Pioneering investigations were carried out by Franks, Sim & Weis (2007) with the winter annual *Brassica rapa* in California. They were researching this species and collected seeds from specific sites in 1997, prior to a period of severe drought (in the years 2000–4). In 2004, seed samples were again collected from the same study sites, and ancestral and descendant lines (together with ancestral × descendant hybrids) were all grown under the same conditions, with steps being taken to minimise maternal effects (see Chapter 8). They discovered that 'the abbreviated growing seasons caused by the drought led to the evolution of earlier onset of flowering. Descendants bloomed earlier than ancestors, advancing first flowering by 1.9 days in one study population and 8.6 days in another. The intermediate flowering time of ancestor × descendant hybrids supports an additive genetic basis for divergence.' Of course it is not possible to be certain that any particular period of drought is connected to climate change, but droughts are predicted to be more frequent, if greenhouse gases continue to rise. Summarising their investigations, on *Brassica rapa*, Franks, Sim & Weis confirmed that 'summer drought selected for early flowering, that flowering time was heritable'. They concluded that 'natural selection for drought escape' appears to have occurred by 'adaptive evolution in a few generations'. In another paper, Franks & Weis (2008) report that adaptive change influenced other traits besides flowering time; descendants had thinner stems, fewer leaf nodes at time of flowering etc.

Given the serendipitous nature of their own investigation, Franks, Sim & Weis (2007) stress the value of sampling populations over long periods of time in studies of climate change. Storage of 'ancestral' seed stocks of various ages would provide an opportunity in some short-lived species to study the possibility of evolutionarily adaptive responses.

Restoration programmes have also provided an opportunity for studying natural selection. Fires burned in the Boise area of Idaho, USA, in August 2006, and, in restorations carried out in two sites, a seed mixture was sown, which included the cultivar 'Toe Jam Creek' of Squirreltail (*Elymus elymoides* spp.*californicus*). A common garden experiment was carried out over 2 years to compare plants of this cultivar (a) from original seed samples used in the restorations, with (b) samples of the plant that had grown to maturity the first summer after sowing in the restoration plots (Kulpa & Leger, 2012). They discovered evidence that directional selection had taken place 'favoring a correlated suite of traits in both field sites', namely 'small plant and seed size and earlier flowering phenology'.

In another approach to the study of adaptation the responses of plants of *Brassica napus* over four generations were examined under different experimental conditions provided by controlled growth chambers of a phytotron (in this case with two temperature regimes and two levels of atmospheric carbon dioxide – imposed individually and in combination). Rapid adaptive responses were detected, but overall a cautious interpretation of the results is made, as only small numbers of plants were grown.

Some other studies of adaptation of plants and animals to climate change have been published. It is very clear that designing experiments to test hypotheses concerning adaptive responses in species to climate change is technically and scientifically very demanding; see reviews by Alberto *et al.*, 2013a, b). For example, Hansen *et al.* (2012) review 44 published case histories and conclude that 'approximately half (43%) of these studies failed to rule out the alternative hypothesis of replacement by a different, better adapted population. Likewise, 34% of the studies based on phenotypic variation did not test for selection as opposed to drift.' They suggest a number of experimental designs and statistical tests to improve hypothesis testing, and point to the emergence of next generation sequencing techniques that will allow the monitoring of change at the genome level. Others have contributed valuable models and experimental investigations for testing various elements of adaptive change.

- Rafferty & Ives (2011) review plant–pollinator interactions.
- Schiffers *et al.* (2013) consider the interplay of gene flow and environmental heterogeneity.
- Muguia-Rosas *et al.* (2011) discuss the implications of the shift in flowering times for inter-specific relationships between plant species.
- Sexton, Strauss & Rice (2011) point out that, 'according to theory, gene flow to marginal populations' may have two outcomes: 'swamping peripheral populations with maladaptive gene flow' or 'aid[ing] adaptation to range limits' by 'enhancing genetic variability adaptation and reducing inbreeding depression'. After manipulating gene flow by crossing experiments in populations of the Californian species *Mimulus laciniatus*, Sexton and associates cultivated progenies in a common garden to examine fitness. They discovered that 'gene flow increases fitness at the warm edge of a species' range, and make the suggestion that appropriate controlled pollinations between populations might be an important option in the conservation of vulnerable species faced with climate change.

Ecosystem changes under climate change

Climate change has complex effects on ecosystems: for some species the climate will become hostile and they may be unable to survive, while for others, change of climate may be to their advantage. On the negative side, Ooi, Auld & Denham (2012) point out that changes in temperatures, rainfall and altered fire regimes have the potential to influence the persistence of seed banks in some species. Such resources are vital to maintaining viable seeds between episodes of recruitment in many species.

As a guide to other changes, Blois *et al.* (2013) have reviewed the associations and interrelationships of species during past episodes of climatic change. There is increasing evidence for 'disrupted ecosystems' and changing 'trophic interactions over time scales ranging from years to millennia', resulting in 'changing species relative abundances and geographic ranges, causing extinctions, and creating transient and novel communities dominated by generalist species and interactions'. They forecast that present-day climate disruptions are also likely to lead to complex biotic interactions.

Turning to another area of complex change, climate upheavals, through range shift, will lead to the arrival of new species, including pests, predators and diseases. Moreover, the distribution of weeds, pest and diseases is likely to change and impact on existing ecosystems (Helmann *et al.*, 2008). Here we consider some case studies that are beginning to reveal the complex effects of climate change on plants in different ecosystems.

A. **Climate-induced drought brings changes to ecosystems** One of the key effects of climate change is altered rainfall patterns and higher temperatures. In many parts of the world, drought is playing a key role in changes in plant communities. For instance, forest fires are increasing in western USA, especially in years with early snow-melt, which allow for large fires later in the year (Westerling *et al.*, 2006).

Other changes are also being reported. In south-western USA, drought induced die-back in Pinyon Pine (*Pinus edulis*) is occurring. The characteristic Pinyon–Juniper ecosystem is losing one of its major components in four states. Breshears *et al.* (2005) have examined the underlying factors, and conclude that lowered rainfall and higher temperatures are inducing climate-change-type droughts across four states, and that stressed trees are susceptible to insect attack and fungal diseases.

B. **Changes in insect herbivores** As yet our knowledge of the effects of climate change on insect/plant relationships is in its infancy, but one case study reveals some fascinating and far-reaching effects.

The Mountain Pine Beetle (*Dendroctonus ponderosae*) attacks mature boreal pine forest in western North America (Vreysen, Robinson & Hendrichs, 2007). It is a sub-cortical herbivore, with a mutualistic relationship with a pathogenic fungus that attacks the phloem tissue. In response, the pine secretes toxic resin. However, 'aggregation pheromones' are produced by the beetle, and the resultant mass attack often overwhelms the tree's defences. In the last century four major epidemics have been recorded, but between 1997 and 2007, 43 million hectares of forest have been affected in the USA, with millions more in Canada. Two factors have contributed to the severity of the outbreak. Fire suppression policies have not prevented fires and, when they have occurred, they have been very extensive because of the build-up of flammable debris etc. After such fires, the normal heterogeneous forest age structure is lost, as very large even-age stands dominated by Lodgepole Pine (*Pinus contorta*) develop. This species has cones sealed with resin: after fire, huge quantities of seed are released. When these extensive forest stands reach a certain maturity, mass attack by the beetle occurs.

Climate change has played a key role in the recent epidemic. If winter temperatures are cold enough the beetle larvae are killed, but a succession of mild winters has allowed greater survival of beetles not only generally, but also at higher altitudes (Régnière & Bentz, 2007). *Dendroctonus* is also increasing its range to the north, east and south-east and other pine species, such as Jack Pine (*Pinus banksiana*) and Loblolly Pine (*Pinus taeda*), may be vulnerable to attack. Thus, from the standpoint of the control and reduction of carbon dioxide emissions in the fight against climate change, forests are changing from 'carbon sinks' to a 'source' of carbon dioxide as dead and dying trees infected by the beetle decompose. Thus, infected forest increases are becoming a major carbon 'source' contributing significantly to increasing concentrations of an important greenhouse gas (Hoyle, 2008).

C. **The responses of invasive plant species to climate change** As we have seen, in response to climate change, plants, in general, will move – polewards and/or to higher altitude. There have been a number of investigations of invasives and weedy species that support this view. For instance, the Canadian authorities are concerned about troublesome weeds that have crossed or could cross the US border that runs along the 49°N latitude (Clements & Ditommoso, 2011). Nine species have been examined in some detail, and some of the adaptive traits have been identified that fit them for their further expansion into the cooler climates and shorter growing seasons in Canada. Amongst those investigated were the annual weed *Datura stramonium* (heavier seeds, earlier growth; Warwick, 1990a, b); *Echinochloa crus-galli* (more rapid life cycle; Potvin, 1986); and *Sorghum halepense* (southern populations perennial; northern populations annual; Warwick *et al.*, 1986).

In addition, another group of invasives has been recognised: those that have the potential for range expansion in North America (Clements & Ditommosa,

2011). This group includes *Abutilon theophrasti*, *Bromus tectorum*, *Buddleja davidii*, *Conyza canadensis*, *Impatiens glandulifera* and *Tamarix ramosissima*.

Considering range change in invasive plants in the Southern Hemisphere, a study in South Africa is informative (Dukes, 2011). Here, poleward migration is more limited than in North America because of location and shape of the landmass. A study suggests that an increase of 2°C would reduce the range in five invasive species to 63–92% of their current range.

Moving to more general considerations, changing climate is likely to favour those plant species with short generation times, broad environmental tolerances, the highest potential to establish in disturbed areas and capacity to migrate under changing conditions. Dukes (2011) points out that these characteristics are shared by many introduced species that have become invasive. He also confronts another very important point. Climate change is only one of a number of environmental factors that will impact on the future evolution of weeds and invasives: in many habitats complex responses are likely. For example, field trials suggest that under conditions of CO_2 enrichment the pestilential weed Canada Thistle (*Cirsium arvense*) could recover more quickly from glyphosate herbicide treatment (Ziska, Faulkner & Lydon, 2004).

Turning to another case study, investigations, over a 4-year period, were made on the effect of artificial warming on the highly invasive introduced grass *Bromus tectorum* at two field sites in Utah, USA (Zelikova *et al.*, 2013). This additional heating, which mimics the projected temperatures in the area under climate change, were achieved by infra-red radiant heaters to produce a +2°C and a +4°C warming above the plots, with dummy heaters in the control plots. They discovered that 'the timing of spring growth initiation, flowering, and summer senescence all advanced on warmed plots . . . the shift in phenology was progressively larger with greater warming'. But these large shifts in phenology were only evident following a wet winter and early spring. They also report that 'earlier green-up and development was associated with increases in *B. tectorum* biomass and reproductive output'. However, the experiment was extended, and in the following two dry years they saw 'no differences in phenology between warmed and ambient plots. In addition, warming had a generally negative effect on *B. tectorum* biomass and reproduction in dry years and this negative effect was significant in the plots that received the highest warming treatment.' Warming therefore did not produce a simple effect on *B. tectorum* and the results highlighted the 'importance of considering the interacting effects of temperature, precipitation, and site-specific characteristics such as soil texture'.

So far we have been considering how human activities have impacted on ecosystems. Now we turn to another issue: the number of endangered species.

How many species are threatened with extinction?

Historically, several attempts have been made to estimate the number of species threatened by human impacts on the environment. Some of these estimates were made in the period when the potential for climate change had not been appreciated or it was considered as a factor that might begin to operate a long time in the future. Given the projections for climate change in the coming decades, recent estimates have tried to take this factor into account.

Species–area relationships Estimates of the number of species threatened with extinction have been influenced by the highly influential study of species–area relationships examined in detail in *The Theory of Island Biogeography* (MacArthur & Wilson, 1967). Here, in brief, we note that studies of the biota of oceanic islands have revealed that large islands contain more species than small islands. Concerning natural vegetation on mainland areas, human activities, through the loss of natural habitats, have left 'islands of natural or more often semi-natural vegetation' of different size in a sea of developed land. What is the effect of such habitat loss on species loss?

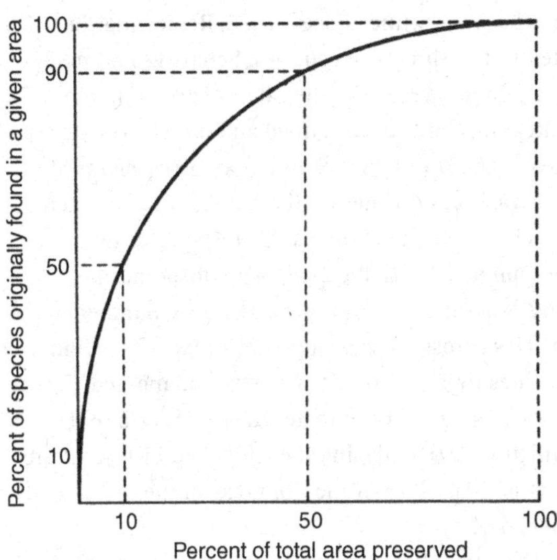

Fig. 18.1. The number of species present in an area increases asymptotically to a maximum value. As a result, if an area is reduced by 50%, the number of original endemic species going extinct may be 10%: if the habitat is reduced by 90%, the number of endemic species going extinct may be 50%. The shape of the curve is different for each region of the world and each group of species, but this diagram gives a general indication of the impact of habitat destruction on species extinction, and the persistence of species in the remaining habitat. (From Primack, 1993.)

The biogeographic models suggest that, if the area of original native habitat is reduced by 50%, then this will result in the extinction of a number of species restricted to that habitat. If the size of islands of natural habitat is further reduced, then additional extinctions will occur (Fig. 18.1 presents a hypothetical example).

On the basis of such arguments, Myers (1979) predicted that there would be an extinction spasm of 1 million species by the year 2000. Professor E. O. Wilson, the distinguished zoologist and renowned conservationist, addressing a meeting of the American Association for the Advancement of Science in 1995, calculated that, even with cautious assumptions, 'the number of species doomed each year was 27,000. Each day it was 74, and each hour three.'

Mann (1991) criticises these deductions from the model, posing the question: do mainland islands behave in the same way as oceanic islands? He considers that, as species in mainland territories may survive somewhere in the mosaic of derived habitats, a simple relationship between area lost and species made extinct must not be assumed. Furthermore, Mann points to the situation in eastern North America. In the colonisation and development of this area by colonial settlers, millions of acres of woodland were destroyed (see Williams, 1989), but no widespread species extinction was recorded. Also, Mann is concerned about the fact that a high proportion of the proposed losses in the estimates by Wilson and others are 'hypothetical species' not yet described by taxonomists. Thus, he ponders whether ecologists could be 'crying wolf', and could stand accused of overstating their case.

In a more recent assessment, Stork (2010) points out that 'for 30 years some have suggested that extinctions through tropical forest loss are occurring at a rate of up to 100 species a day and yet less than 1,200 extinctions have been recorded in the last 400 years'.

Assessment of extinction risk by experts using IUCN and other categories

Turning to another approach, the opinion of experts could be helpful, especially if there is information on species distributions, abundance and threats to habitats. Historical information may be available in well-studied areas in published papers, books, field notes and herbarium sheets. But, of course, the flora of many parts of the world is still imperfectly known.

In order to encourage the collation of information, to provide a stimulus for scientific study and to influence public and political opinion, the conservation status of species has been assessed by experts in many areas of the world. While some records are held nationally, global coverage is provided by the World Conservation Monitoring Centre (WCMC) in Cambridge, UK. The International Union for Conservation of Nature (www.iucn.org/) has devised a set of categories that are used to express the conservation status of individual species. Historically, the categories Extinct; Endangered; Vulnerable; Rare; and Indeterminate have been employed in the preparation of lists and 'Red Databooks', categorisation involving the assessment of status by experts, who judged the degree to which causal factors were endangering species. Now, these early categorisations have been modified and include the use of quantitative measures to determine status, by including in the assessment, where possible, of information on present and future estimates of (a) population size, (b) area occupied and (c) numbers of populations. Ideally, estimates of future risk are made within a specified timeframe. These redefined categories have been widely employed in assessing animal species (see www.eoearth.org July 2014 for a detailed account of the definitions and usage of the current IUCN categories). As we have seen, considering the status of plants presents some particular problems given our imperfect knowledge of many rare and potentially endangered species. Moreover, it is difficult sometimes, because of clonal growth, to define individual plants, and how to estimate future population numbers and population trends where species have seed banks.

Some countries have their own species categories and conservation organisation, e.g. the NatureServe Conservation Status Categories are used in the USA and Canada (Wilcove & Master, 2005). In addition to the WCMC databases, and IUCN's own online *Red List of Threatened Species* (www.iucnredlist.org/), 'Red Databooks' of rare and endangered species have been published for various regions. Generally, the IUCN categories are used, but, in addition, each species at risk may be given a threat number, and the higher the number the greater the threat. For example, the *British Red Data Book* (Perring & Farrell, 1983) ranks species in relation to past and present distribution, attractiveness, accessibility, how many extant sites there are in reserves etc.

Using the data sets held at WCMC, Smith *et al.* (1993) calculated that only *c.*0.3% of vascular plants had become extinct since *c.*1600. Reviewing the IUCN Red List of threatened species, Schatz (2009) notes that, for good reasons, the categories used in the IUCN categories have been refined and increased over the years, and this means that 'not all assessments are equivalent'. Considering the number of endangered species, the Red List notes '11,995 species of bryophytes, ferns, lycopods, gymnosperms and angiosperms out of an estimated total number of species across these groups of 379,881'. This figure equates to <3.2% of plants being endangered. Taking the wider category, threatened species, Walter & Gillett (1997) analysing the IUCN Red List, suggested that perhaps 13% of the flora was in the threatened category.

Threatened species: estimates using a proxy Pitman & Jørgensen (2002) have explored another means of estimating the size of the 'World's Threatened Flora'. They point to the lack of information on the conservation status of tropical plants in IUCN listings etc. They tested the hypothesis that 'the number of plant species endemic to a country is a reasonable proxy for the number of globally threatened plant species in that country'. They note that this relationship seems to be statistically significant in an analysis of the floras of European countries – endemic species in a country being a 'strong predictor' of the number of globally threatened species.

While we have imperfect knowledge of how to judge the risks to individual tropical species for IUCN listings etc., we have better (but clearly far from perfect) knowledge of national levels of endemism. Pitman & Jørgensen (2002) made a number of estimates of the numbers of threatened species, based on the numbers of endemic species (as a proxy for threatened) in 189 countries and hot spots. To arrive at global figures they used these 'tropical estimates' together with the actual number of currently threatened species from temperate countries. Estimates for the number of threatened species globally, calculated in this way, fall in the range 22–47%.

Extinction risk from climate change In the past few decades biologists have begun to appreciate the accelerating threat to endangered species from climate change. To try to estimate this threat, Thomas *et al.* (2004) determined climatic envelopes for a sample of plants and animals, and modelled future distributions for different climate scenarios. Then they estimated extinction threat, based on the 'species–area relationship'. This describes how 'the number of species relates to area ($S = cA^z$, where S is the number of species, A is area and c and z are constants). This relationship predicts adequately the number of species that become extinct or threatened when the area available to them is reduced by habitat destruction'. Taking all these results, which they stressed came from a study of a sample of regions covering c.20% of the Earth's land surface, Thomas and colleagues report that, on the mid-range climate warming scenarios for 2050, 15–37% of species in the sample of regions and taxa will be 'committed to extinction'. In presenting their findings they stress:

1. There are many unknowns in the estimation of extinction.
2. They report species 'committed to extinction', realising that 'time-lags between climate change and extinctions' are highly likely.
3. They stress that more studies of the likely rate of extinction are needed, if possible with the incorporation of land use in the analyses.
4. The analyses of climate envelopes should also take into account that perhaps some regions will have climatic conditions that do not have current analogues.

They conclude that anthropogenic climate change 'is likely to be the greatest threat in many if not most regions'. And 'returning to near pre-industrial global temperatures as quickly as possible could prevent much of the projected, but slower acting, climate related extinction from being realised'.

While many endangered species are likely to be at risk through climate change, changes in the distribution and frequency of species that are presently common are also predicted. Thus, Warren *et al.* (2013) report that: 'our global analysis of future climatic range change of common and widespread species shows that without mitigation, 57±6% of plants and 34±7% of animals are likely to lose ≥50% of their present climatic range by the 2080s'.They too point to the advantages of speedy reduction in greenhouse gases. They discovered that 'losses are reduced by 60% if emissions peak in 2016 or 40% if emissions peak in 2030. Thus, our analyses indicate that without mitigation, large range contractions can be expected even amongst common and widespread species, amounting to a substantial global reduction in biodiversity and ecosystem services by the end of this century.'

Endangered species: an overview

Clearly, the categories of risk in historic estimates of the numbers of endangered species were based on educated guesses, and we have to conclude that these lists are incomplete and imperfect. For instance, the threats to carnivorous plants have not been comprehensively assessed (Jennings & Rohr, 2011). Estimates using a proxy for threatened species have yielded interesting results, but Pitman & Jørgensen (2002) stress the importance of extending the IUCN Red Book listings to assemble a global database of endangered species, and calculate that this could be done for '< $100 per species per year' or '< $12.1 million' for all the hot spots of biodiversity. Stuart *et al.* (2010) also argue for a broader taxonomic base for the Red Lists to provide a more comprehensive coverage of biodiversity. Such a list would provide a solid basis for 'informing decisions globally, for example, in conservation planning, resource allocation, environmental impact assessments, monitoring biodiversity trends and enabling countries to develop national-level biodiversity indicators', that could act as a 'Barometer of Life'.

While estimates of threat by category listing by experts or indirect means have not provided definitive information on the numbers of threatened species globally, they have served a useful purpose in bringing the conservation message to politicians and the general public.

Turning to the most up-to-date studies of the problem, provided by Thomas and associates, it is clear that we are no nearer to being able to make accurate estimates of the number of endangered species, but the signs of crisis are becoming more apparent.

Concluding remarks

Archaeological investigations have revealed the antiquity of human influences in long-inhabited parts of the world. When humankind lived as hunter-gatherers, their impact on the world's ecosystems was probably small, but now, with burgeoning populations, almost everywhere across the world has been influenced to a greater or lesser extent by human activities. Now, perhaps, '2/3 of the world's terrestrial land has been modified for human use' (Mace & Purvis, 2008).

Despite the uncertainties about the number of species currently endangered, evidence supports the view that we are in the midst of a biodiversity crisis (see Wilson, 1985; Barnosky *et al.*, 2012), although there are often considerable difficulties in establishing whether any particular species has finally become extinct (see Collen, Purvis & Mace, 2010). In an unprecedented stocktaking of UK wildlife, compiled by 25 conservation groups, very serious declines in many species were revealed (State of Nature Report, 2013; see www.rspb.org.uk/Images/stateofnature). And, as we have seen, biodiversity loss is being driven by many interacting factors, some of which were recognised in Rachel Carson's exposé, in her famous book *Silent Spring* (1962), of the effects of indiscriminate use of pesticides.

However, as we have seen in this chapter, in addition to the many long-recognised threats to biodiversity, we now have to acknowledge the dramatic impact of accelerating climate change. Plant species whose ecosystems are subject to significant adverse climate change may not be able to adapt or disperse to ecologically appropriate new areas. Reaching the limits of phenotypic responses and biochemical adjustment, species may lack the genetic potential to adapt to changing conditions. Species may have very limited potential to migrate in tandem with the changing climatic conditions. The timeframe for possible response may not match the extended life cycle found in many species. Endangered species may have exacting habitat requirements in terms of soils, microclimatic conditions requirements and biotic interactions that cannot readily be matched in territories to which species could potentially migrate in the face of climate change. For example, species may require special soils such as serpentine, which only occur in isolated widely scattered localities. As vegetation zones move to higher altitudes in response to climate change, many alpine species with their exacting microclimatic requirements may become extinct, as their habitats are lost in a particular montane ecosystem. Furthermore, they may lack the dispersal capacity to migrate and establish in another area of alpine vegetation at higher elevation. Many species grow in low-lying coastal habitats and whole ecosystems may be threatened or lost as a consequence of sea level change in response to climate change. Species interrelations in ecosystems may change radically as some species respond adaptively to climate change, while others decline. Also, new species may arrive in an ecosystem, some of which might be invasive. At the same time, human influences on natural and semi-natural ecosystems are continuing to intensify as we increasingly manage (and mismanage) the natural environment to exploit its 'goods and services'.

In the next chapter we consider a key issue: will taxonomists be able to complete their task of describing and naming the world's plants, or will a very large number of species become extinct before they can be properly studied?

19 The taxonomic challenge ahead

We live in a species-rich world that is increasingly under human influence. As we have seen in earlier chapters, there is abundant evidence that many species are endangered, and this includes unknown organisms that have yet to be named and classified. If we are to appreciate the true extent of biodiversity, there is a formidable and urgent taxonomic task ahead (Mallet & Willmott, 2003; Bateman, 2011).

What are the prospects of the completion of a catalogue of life?

The challenges are discussed by Wheeler (2008), who comes to the conclusion that, at the very point in history where a major effort is required of taxonomists, there is a growing concern that taxonomy has suffered a 'decline in prestige and support'.

As we have seen in earlier chapters, taxonomy has a very long history. Wheeler considers that two twentieth-century developments, in particular, have challenged the status, funding and prospects of taxonomic studies. We have seen in earlier chapters that, in the 1920s and 1930s, biologists became increasingly interested in experimental and genetic investigations of evolutionary patterns and processes. Many botanists, initially trained as taxonomists, turned to these investigatory studies to study intraspecific variation, speciation etc. These approaches became known as Experimental Taxonomy or Biosystematics. Huxley (1940) in his classic book *New Systematics* hoped for the rise of an integrated subject, but Wheeler points out that some biologists, enthusiastic for the emerging specialisms, came to believe that taxonomy, being 'non-experimental', was not really a science, but was involved with 'subjectively naming and classifying species'. Thus, 'taxonomy involved arbitrary bookkeeping and pigeonholing practices and legalistic wrangling over scientific names, while systematics was experimental and intellectually exciting and expansive'. In parenthesis, we may note that this perception of taxonomy, as 'merely descriptive' and 'non-analytical', has been very strongly challenged. For example, Lipscombe, Platnick & Wheeler (2003) and Sluys (2013) argue very effectively that taxonomic study is 'hypothesis driven'.

Returning to Wheeler's analysis, he points to an even bigger revolution later in the century, when molecular and computational techniques were introduced to energise an already vigorous experimental tradition, providing the tools for major advances in all aspects of micro- and macro-evolution, and the emergence of molecular phylogenetics and modern biogeography. As there has always been a Darwinian struggle between competing scientists for funding, Wheeler argues that the success of these two twentieth-century approaches has been to 'dilute, detract from and eventually to decimate taxonomy'.

The current status of taxonomic studies has also been examined by Kim & Byrne (2006), who conclude that taxonomy

> has suffered a continuous decline in the latter half of the twentieth century, and has now reached a point of potential demise. At present, there are very few professional taxonomists, and trained local parataxonomists worldwide, while the need for, and demands on, taxonomic services by conservation and resource

> management communities are rapidly increasing. Systematic collections, the material basis of biodiversity information, have been neglected and abandoned, particularly in institutions of higher learning.

The situation in Britain has been investigated in a number of reports (see House of Lords Reports, 2010 for details). In the 2002 Report to the House of Lords, the contributors point out that 'the state of taxonomy and systematics in the UK is unsatisfactory – in some areas . . . to the point of crisis – and that more needs to be done to ensure the future health of the discipline'. Concerning the situation in 2010, the Third Inquiry into Systematic Biology by the House of Lords Science and Technology Committee (2010) reported that 'the number of taxonomists in UK universities is small. University-based taxonomists reaching retirement age have typically not been replaced by taxonomists, and this is the sector that has undergone the most marked decline since the mid-1990s'. Furthermore,

> three quarters of the 160 taxonomists interviewed considered their area of taxonomy in the UK to be in an 'unhealthy' state and the main stated cause for their concern was the apparent lack of succession planning in institutions that house taxonomic expertise. The definition of 'succession planning' apparent here is the like-for-like replacement of taxon-based expertise. Respondents were frustrated at this situation because mentoring a successor is a more efficient and more effective method of maintaining expertise levels than leaving a new taxonomist to acquire expertise in isolation.

The contributors continue: 'submissions to the House of Lords 2007–08 inquiry pointed to the near collapse of taxonomy in British universities. In their submission the Botanical Society of the British Isles commented that: "species identification skills and other field skills are hardly taught at undergraduate levels at universities".' However, they note the very considerable taxonomic expertise existing in certain amateur groups. The Report continues:

> The UK is not unusual in identifying a decline in the skilled taxonomist workforce. There are numerous reports that the number of trained and practising taxonomists is declining worldwide (e.g. Kim & Byrne, 2006) . . . The decline in numbers of taxonomists is most apparent and most consistent in developed countries, but the growing number of new species descriptions generated by taxonomists based in countries such as Brazil and China, perhaps indicates that the global picture is not one of uniform decline.

Turning to another issue concerning the funding of existing taxonomic institutions and their collections, Bebber *et al.* (2010) note that that the process of species discovery is not well publicised, in particular the crucial role of herbarium specimens. Thus, from a sample of new species described in the period 1970–2010, they discovered that 'only 16% were described within 5 years of being collected for the first time. The description of the remaining 84% involved much older specimens, with nearly a one-quarter of new species descriptions involving specimens >50 years old.' They emphasise that 'effort, funding and research focus should, therefore, be directed as much to examining extant herbarium material as collecting new material in the field'.

The renewal of taxonomy

Godfray (2002) takes the view that the Internet is the key to the future of taxonomy. He proposes a radical reinvention of the discipline. He emphasises that at present 'the taxonomy of a group of organisms does not reside in a single publication or single institution', but is scattered across different sources. The naming of new species is governed by what he calls 'venerable codes' of nomenclature. He suggests that the Internet is the ideal place to publish taxonomic treatments of groups, and that these should be written to serve the needs of the wide range of users. A first web revision would be produced for comments from the botanical community and any changes made, until a 'unitary taxonomy of the group' has been devised under the supervision of expert committees. This would become the consolidated baseline for future additions. New species and new taxonomic findings on synonymy and provisional material etc. could then be added to the accumulating web accounts. A major aim would be stability of names.

Taxonomists have responded, some positively (Bisby *et al.*, 2002), but others are more critical of Godfray's proposals (Thiele & Yeates, 2002). For example, Knapp *et al.* (2002) agree that 'web-provided taxonomy is clearly the way for the future', but they are not persuaded that 'we throw out the past mechanisms of doing systematics and begin anew in a revolutionary "brave new world" of unitary, web-based taxonomy, each group under the administration of an authoritarian body'. Working with the internationally agreed codes and conventions is more practical than 'beginning again'. In their view, 'some changes are clearly necessary', but what is needed for taxonomic practice is 'evolution, not revolution'. A very useful overview of this controversial area is provided by Godfray & Knapp (2004) whose joint introduction to a Royal Society Meeting explores both their common ground and their differences of opinion on key issues. In responding to the reactions to his proposals, Godfray (2002) acknowledges that 'taxonomy is a triumph of modern science – but its products could still be improved'.

Godfray and associates have taken the initial proposals to a new stage, in devising systems and methods to achieve consolidated taxonomies online through the CATE project (Godfray *et al.*, 2007; Clark *et al.*, 2009). They maintain that standards would need to be set, and Godfray envisages that groups of biologists will act as editorial boards to oversee the work. Some taxonomists, reacting to these suggestions, have concerns about the top-down oversight of the web-based project. They point out that books etc. are reviewed *after* production, while Godfray and associates (Clark *et al.*, 2009) propose open peer review *during* the process of preparing accounts. Critics of these proposals have expressed a major concern, namely that 'the complex process of revisionary taxonomy might become oversimplified', thereby 'giving the impression that there is one true taxonomy' (an opinion voiced by Thiele & Yeates, 2002). In defence of their own view, Godfray and colleagues point to the successful achievement of consensus taxonomies published by ink-on-paper, and make it clear that online taxonomies could present alternative viewpoints online. They emphasise that 'consensus therefore is neither intended to stifle dissent nor does it imply immutability'. Debate on these important issues continues.

From tentative beginnings, the Internet is now increasingly used for taxonomic and allied research, not only in botanical institutions, herbaria and museums, but also wherever there is internet access (Knapp, 2008). A wide variety of taxonomic information is now appearing on the web.

- Descriptions of new species, scans of herbarium specimens especially types, together with electronic versions of keys (Balakrishnan, 2005).
- Scans of historic taxonomic books, and papers.
- Apps of photographs and text etc. to enable field identification.
- Online catalogues and listings, e.g. World Checklist of Selected Plant Families (www.kew.org/wcsp/); Tree of Life (www.tolweb.org/tree/; Parr *et al*, 2011); the All Taxa Biodiversity Initiative (www.dlia.org/atbi/); web-base taxonomies (CATE project; www.cate-project.org/); the planetary Biodiversity Inventory projects (www.actin bioscience.org/biodiversity/page.html); and the International Plant Names index (IPNI, www.ipni.org) which 'began life as *Index Kewensis* with a legacy from Charles Darwin' (Knapp, Polaszek & Watson, 2007).
- There have been significant developments in botanical taxonomy. Knapp, Polaszek & Watson (2007) make a strong case for 'electronic publication of scientific names of plants'. Then, in a 'test case', Knapp (2010) published the names and details of four new *Solanum* species online in the

open-access journal *PLoS ONE*. 'Paper publication has been the gold standard' for botanical taxonomy since the time of Linnaeus (Cressey, 2010). To conform to the International Code of Botanical Nomenclature as it then stood, Knapp circulated identical printed copies of the paper to 10 major herbaria.

- Despite some expressed concerns about archiving and permanence of electronic files, the International Botanical Congress in Melbourne of 2011 agreed a major change, by removing the requirement to produce hard copy of papers describing new species. The meeting also questioned the role of Latin in botanical taxonomy (Flann, Turland & Monro, 2014). In the report of the meeting, Turland considers that 'permitting electronic-only publication is arguably the most important decision made in Melbourne, bringing taxonomy into the 21st century and the electronic age. As for Latin, it has become increasingly difficult to use and is often regarded as an irrelevant anachronism by modern scientists. The meeting clearly wanted an alternative.' Therefore, it was agreed that the short diagnosis that indicates how a new species differs from those already named, historically written in Latin, could now be either in Latin or English. It was hoped that these two changes will speed up the publication of new species and help those taxonomists working in countries where classical languages are rarely taught in schools. While many will welcome this use of English in taxonomy, there may be difficulties ahead. There could be a legitimate call to allow diagnoses to be written in other languages, such as Chinese or Spanish.

Barcoding: its history and potential in taxonomic investigations

Advances in molecular approaches have provoked other fierce debates. For instance, Tautz *et al.* (2003) consider that the results of molecular investigations might play a central role in taxonomy, and be the key to its reinvigoration. On the basis of pioneer work with several groups of plants and animals, it has been proposed that each species has a unique barcode of molecular information, and that these barcodes could be employed both in species identification and delimitation.

Historically, in the struggle to provide a satisfactory taxonomic treatment of bacteria, for which very few morphological characters are available, molecular techniques have been employed to discriminate between different isolates, but it is clear that barcodes have not provided the key to bacterial taxonomy. Put simply, this is because of the several ways DNA may be taken up by bacteria: (i) from the environment; (ii) through the transduction of foreign DNA into a given strain by infective bacteriophages; and (iii) conjugation between bacterial strains. Overall, lateral gene transfer is rife and the species concept used in the classification of higher plants cannot be generally applied to bacteria (see Nanney, 1982).

In an influential paper on a range of animal species, Hebert (2003) studied the sequence variability of a mitochondrial gene – cytochrome c oxidase subunit 1 (COI). The results supported the idea that specific DNA markers could help to resolve patterns of diversity, as different species of animals, including Lepidoptera, had unique sequences, which Hebert and associates called barcodes. This name was selected as the concept is familiar to the general public, as the retail trade has developed a barcoding identification system for different products employing '10 alternate numerals at 11 positions to generate 100 billion unique identifiers' (Hebert *et al.*, 2003).

Barcoding has now become a major topic of biological research attracting large amounts of funding. Over four hundred research papers have barcoding in their titles in the period 2003–10. Three major international websites report progress: Consortium for the Barcode of Life (CBOL; Smithsonian Institution Washington www.barcodeoflife.org); International Barcode of Life (iBOL; Biodiversity Institute of Ontario, Director Dr Hebert, www.ibol.org); and The Barcode of Life Datasystems (BOLD; www.boldsystems.org). It has also been predicted that an expensive Star Trek tricorder-style hand-held barcoding device will soon be available (Savolainen *et al.*, 2005).

Proposed advantages of barcoding Hebert *et al.* (2003) point out that taxonomic endeavour is 'no easy task' and that a 'community of 15000 taxonomists will be required, in perpetuity, to identify life, if our reliance on morphological diagnosis is to be sustained'. They note that conventional morphologically based taxonomies encounter a number of complex issues, e.g. phenotypic plasticity, the presence of cryptic species in some groups; and keys, based on morphological characteristics, do not make it possible to identify all life stages. Moreover, they take the view that the use of keys demands a high level of competence. Thus, the DNA barcoding system promises 'a better taxonomic resolution than that which could be achieved through morphological studies'. Also, it would provide 'a partial solution to the decline in traditional taxonomic knowledge', and the 'dwindling pool of taxonomists'. Not only is DNA barcoding proposed as a 'tool' to identify species, but also 'to define species boundaries and aid in species delimitation' (see Taylor & Harris, 2012).

Barcodes in practice DNA barcoding has proved practically helpful in a number of specific situations, making it possible, for instance, to identify not only juvenile life cycle stages in a number of organisms, but also fragmentary samples. Thus, it has proved crucial in investigating illegal trading in biological products, and the detection of fakes and contaminants, e.g. verifying the identity of fish for sale in supermarkets (Rasmussen *et al.*, 2009); the detection of non-label ingredients in tea (Stoeckle *et al.*, 2011); and looking for contamination and substitution in North American herbal products (Newmaster *et al.*, 2013).

In recent groundbreaking ecological studies, barcodes have also been employed in the taxonomic analysis of the gut contents of tropical insect herbivores in order to determine plant–herbivore networks (Garcia-Robledo *et al.*, 2013). They have proved to be a useful tool in assigning 'unidentified African rainforest trees to genus, but identification to a species was less reliable, especially in species-rich clades' (Parmentier *et al.*, 2013). Recent reviews point to other potential uses/pitfalls/shortcomings of barcoding in a number of situations: e.g. identifying seeds/fruits for seed banking (Nevill *et al.*, 2013); checking the identity of invasive plant species (Zhang *et al.*, 2013); and taxa of conservation interest (Krishnamurphy & Francis, 2012).

It is notoriously difficult to determine the composition and frequency of the microscopic taxa in microbial, fungal and marine communities, and molecular methods, akin to barcoding, are being increasingly and successfully applied to 'broad-brush' studies of these important aspects of ecosystems on fungal community analysis using high throughput sequencing of amplified markers (see, for example, the review by Lindahl *et al.*, 2013).

Questions about barcoding While barcoding has attracted many enthusiasts, and proved helpful in certain research projects, some biologists have been sceptical about the notion of a universal barcoding system for all the taxonomic groups (Taylor & Harris, 2012). First a critical question was confronted: would the mitochondrial gene – cytochrome c oxidase subunit 1 (COI) – employed in animal species prove to be suitable for all barcoding?

Would all the members of a species be expected to have the same barcode? If not, would intraspecific variability in Linnaean species complicate the interpretation of the results? With regard to sampling, barcoding only a single individual to represent a 'Linnaean' species could be seriously misleading. Seberg *et al.* (2003) stress this point when they write:

> Often one is forced to use one, or a few, carefully selected specimens as representatives of a taxon, but experience has repeatedly shown that this can be a major mistake. Deliberately using a single specimen,

as a representative of the taxon, will only create havoc in taxonomy, a fact long realised by taxonomists working with other types of data. Individual bases and DNA sequences are simply characters, tiny fragments of the lifecycle. It seems perverse to us to advocate using a DNA sequence as a mandatory identification tag for species, even as a first approximation.

In addition to these practical questions concerning sampling, other issues have been raised (see Taylor & Harris, 2012 for details). In analysing barcoding loci, tree-building methods have often been employed. Critics contend that, in the interpretation of tree structures, species delimitation is problematic. It has been suggested that a better approach is to examine the individual differences between the barcode sequences. Also, considering the raw barcode information, how should this be interpreted? Clearly, some level of judgement would be needed. Should such judgements be guided by statistical tests? Matz & Nielsen (2005) consider so, and have devised a likelihood test for species memberships based on DNA sequence data.

Requirements for a functioning barcode system For barcoding to function as a tool for identification, there must be a database of barcodes of existing recognised named species. Only when types and other significant specimens in herbaria and museums had been barcoded, could unknown specimens be identified. Systems would have to be devised to store voucher specimens of barcoded material, and a database of accumulating barcode records devised and frequently updated.

Could barcodes be used to recognise 'new' species? Some authorities assert that barcoding could provide a means of discovering new species and indeed revising the taxonomy of groups (see DeSalle, 2006). In response, others, such as Taylor & Harris (2012), protest 'that it would be naïve to describe a new species or infer a phylogeny without any corroborating evidence other than a single locus DNA sequence' and that 'barcoding should supplement morphological data for species description rather than replace it (Prendini, 2005)'. Some proponents of barcoding have conceded that, in the absence of other evidence, DNA evidence creates hypotheses regarding new species rather than 'discovering' them outright (Goldstein & DeSalle, 2011). However, the central question remains problematic: what level of barcode difference is to be regarded as taxonomically significant and where are significant boundaries to be drawn in the delimitation of taxa (Blaxter, 2004)? As we have seen in previous chapters, polyploidy and other cytological variability is common in some groups of plants. In the past it has been claimed that material differing in chromosome number should always be treated as different species (Löve, 1962). Likewise, it is now being debated whether specimens differing in barcode should be given specific rank. In response to these suggestions, taxonomists and the users of botanical taxonomic output (floras, check lists, keys etc.) point out that in the absence of morphological differentiating characters, it is impractical to treat variants differing in barcode alone as species.

What are the prospects for barcoding in plant groups? As we have seen, the first significant advances in barcode technology came in studies of bacterial and animal groups and provoked many of the debates we have just discussed.

Considering the first investigations of barcoding in plants, it was soon discovered that the COI mitochondral gene employed in early animal studies was inappropriate for barcoding flowering plants – there was insufficient sequence variability. Other marker genes, with the necessary variability, would need to be discovered. Two chloroplast genes *rbcL* and *matK* have figured widely in many studies, but other cpDNA gene loci have been investigated both individually and in combination –sometimes three or four markers have been employed together (Fazekas *et al.*, 2010). Hollingsworth, Graham & Little (2011) review progress during what is being seen as a 'trial period', as molecular biologists work towards the selection of appropriate genes.

In addition to studies of chloroplast genes, the 'internal transcribed spacers of nuclear ribosomal DNA' have been investigated, but some challenges have emerged as fungal contamination interferes with the determination of sequences in the plant material under examination, and departures have been detected from the usual concerted evolution of the marker. Instead of a single detectable sequence, divergent copies have been discovered in some plant groups. The search for appropriate nuclear genes continues.

Finding such nuclear genes is very important, as Chase *et al.* (2005) point out that the study of plastid genes alone will not allow for the barcode resolution in the case of taxa involved in hybridisation. They write: 'DNA barcodes based on uniparentally inherited markers can never reflect the complexity that exists in nature.' At present it remains an open question how barcoding will be resolved for species with diverse origins. The challenge is to devise an efficient barcode system that takes into account the complexities of allopolyploidy, introgression, homoploid speciation, apomictic behaviour etc.

In this context, it is noteworthy that in the recent investigation, using plastid markers, of the highly complex group of *Thymus* species, in only one species was a specific barcode resolved. For the other taxa, there were not sufficient 'barcoding gaps' to allow the resolution of species-specific barcodes (Federici *et al.*, 2013). Molecular issues are perhaps not the only reason for this difficulty; there are unresolved taxonomic difficulties in the group.

Barcoding herbarium material Continuing their critique, Chase *et al.* (2005) point out that it is impractical to collect live material of all the world's flora for barcoding. Named herbarium material is therefore of crucial importance. However, these historic specimens are often small and fragile, and there is concern about damaging them in taking material for analysis. Furthermore, such specimens usually contain only highly degraded DNA. But it has been discovered that some plastid and mitochondrial sequences may survive intact, and some limited progress had been made in barcoding herbarium specimens (see Sarkinen *et al.*, 2012).

In contrast, investigations with living material are advancing more quickly, leading to a bottleneck in the definitive identification of experimental material by expert taxonomists. In some cases, voucher specimens have not been retained for material that has been barcoded. In anticipation of the fact that it may be impossible to barcode some key type specimens, and to allow progress to be made in completing a barcode system, it has been suggested that material as close as possible to types be designated as epitypes. It remains to be seen how these issues are resolved, but it is clear that the taxonomic and nomenclatural status of epitypic material could be highly problematic.

Next generation sequencing and barcoding. The first genome sequences, determined for model plants, required a great deal of funding. Recently, advances in DNA sequencing, using next generation methods, have reduced the cost to an astonishing degree. While barcoding studies the sequences in single genes or small numbers of genes, next generation methods provide long sequences of DNA. McPherson (2009) suggests that these new advances may make barcoding obsolete.

Responding to this observation, Li *et al.* (2015) examine the possibility of using the sequences of entire plastid genomes as a super-barcode, in conjunction with the current barcodes, as a means of discriminating between closely related taxa.

Barcoding: a route to the reinvigoration of taxonomy?

We have examined some optimistic and also critical reactions to the techniques, implementation and goals of barcoding (DeSalle, Egan & Siddall, 2005). Exploring these proposals, Knapp (2008) notes that barcoding has some potential to encourage traditional taxonomy, and expresses the hope that the search for such 'universal identifiers' could provide 'more funds and impetus for taxonomy' (see Consortium for the Barcode of Life (COBOL: http://barcoding.si.edu/index_detail.htm)).

It will be very interesting to see if the current enthusiasm for barcoding provides a real stimulus and financial support for classical taxonomic work in institutions that are currently underfunded. At present, it is not clear whether a practical functioning system of barcoding named plants can be devised, through which unknown material can be accurately identified. Also, only time will tell whether the next generation sequencing methods will provide a practical alternative way forward in the molecular characterisation of plant species.

The status of taxonomy in an era dominated by molecular biology

Wagele *et al.* (2011) consider another important issue: they claim, with some justification, that taxonomists do not get sufficient credit for their work. Taxonomy underpins all biological research. The identity of species used in research must be unambiguously known. Accurate names are the key to the literature and communication of new findings. Yet, the authorities on which the names and classification are based are rarely mentioned in scientific papers. Journals are also selective in what they will publish. 'Most top-ranking evolutionary journals do not consider taxonomic revisions.' Moreover, 'it is considered unnecessary to cite original taxonomic descriptions and subsequent taxonomic revisions – the hypotheses behind the species names –even when those hypotheses crucially impact a given study and its design'. Indeed, 'more journals are willing to accept, and more authors choose to cite, papers that test the validity of taxonomic hypothesis with molecular data or papers promoting or using barcodes' (Agnarsson & Kuntner, 2007). Wagele *et al.* (2011) encourage editors, publishers and journal administrators to make sure that taxonomists receive the rightful acknowledgement for their work, which should be properly indexed in citation listings. Currently, more than 90% of all taxonomic journals are not indexed, so the overwhelming numbers of taxonomic citations are simply not counted. Their 'inclusion in the ISI^{sm} database will increase the awareness of the journal and guarantee a more accurate calculation of journal and author citation metrics'.

Agnarsson & Kuntner (2007) report on the aftermath of an imaginative initiative prompted in the USA to bring about a 'renaissance of taxonomy'. The NSF-PEET (National Science Foundation – Partnerships for Enhancing Expertise in Taxonomy) scheme, involving a range of taxonomic centres, trained a new generation of taxonomists. Investigating the consequences of this scheme, they discovered, crucially, that there appeared to be a 'lack of jobs and funding for taxonomists once trained'. Some of the young taxonomists reported that they may be 'unemployable because their dissertations did not include a molecular component'. The clear message from this and other studies is that, 'unless careers in taxonomy are available, the availability of training – no matter how good –will not prevent the loss of taxonomic expertise. Taxonomy needs more jobs.'

The prospects and status of taxonomy at a time of 'molecular dominance' are influenced by another serious issue. Taxonomists often produce their work in the form of monographs that examine in detail the taxa of a group across a wide geographical area. Such investigations review the literature and herbarium collections, establish the number of species, locate the type specimens, determine the correct names of plants (and which names are synonyms), identify endemic taxa, and for all taxa establish their geographical and historical status etc. In many ways this is the best format in which to publish taxonomy (Heywood, 2001). However, Agnarsson & Kuntner (2007) make it plain that, 'in the era of impact factors, a biologist who publishes a few large papers (inevitably in rather small journals) is at a disadvantage to one who publishes smaller articles more frequently'.

Will Earth's species all be named before they become extinct?

Costello, May & Stork (2013) have reviewed all the issues relating to the number of species and taxonomists, extinction rates, and how taxonomic productivity might be increased. Here, we report their contributions to the many ongoing debates.

- New estimates suggest that there are likely to be 'c.5 plus/minus 3 million species on Earth'. Previous estimates of 30–100 million based on potential deep-sea diversity and estimates of insect specificity now seem highly unlikely.
- While it is true that 'the numbers of taxonomists may be decreasing in some institutions of the countries that formerly led taxonomy', Costello, May & Stork write, 'new databases show that there are more taxonomists describing species than ever before', with an 'increase in the proportion of taxonomists based in the Southern Hemisphere and Asia-Pacific region'. They consider that this trend is 'appropriate because most species occur in these regions'.
- Turning to extinction rates, they emphasise that 'contemporary extinctions have not been as high as some have predicted'. In many cases conservation efforts have been successful, and some species, thought to be at risk of extinction are surviving in managed landscapes. But the case of some long-lived species is very revealing. Although severely challenged by changing conditions to the point where they may not be able to reproduce, they are still present in the wild. However, such plants, often perhaps living on borrowed time, are part of what is sometimes known as an 'extinction debt' of species, They appear to be 'committed to extinction' in the longer term, if conditions continue as at present (Gilbert & Levine, 2013). But some may perhaps be 'rescued, if conservation efforts successfully conserve their habitats by countering perceived threats'. Concerning extinction rates, it is very difficult to make predictions (Stork, 2010), as perhaps only <5% of species have been assessed for extinction risk. 'Most current models predict <5% extinction rate per decade, although the impact of climate change on extinctions is particularly uncertain because species may adapt and/or adjust their distributions ... If extinction rates are as high as 5% per decade, then regardless of how many species exist on Earth, more than half will be extinct within 150 years. If most species are unknown to science, then their extinction will also be unknown.' Costello, May & Stork calculate that given the current taxonomic effort: 'if there are 2 million species on Earth then most of them will have been described by 2040; if 5 million, by the year 2220'. However, if taxonomic effort could be greatly enhanced and the rate of extinction proves to be lower, then the rate of 'species description' could 'outpace extinction rate'.
- In making their own contribution to the debate about the revivification of taxonomy, Costello, May & Stork (2013) encourage the wider use of the Internet by taxonomists. They note a recent change in naming new species: 'codes of botanical and zoological nomenclature now accept descriptions of new species in publication in electronic-only journals' as well as in printed form. With regard to the peer review of such material, there is an active debate between those who see the need for some supervision of the taxonomic publishing process, and those who favour radical 'publish-as you-go' models, 'where a species' information would be online before its formal naming'.
- Overall, they stress that 'overestimates of how many species may exist on Earth and rates of extinction are self-defeating because they can make attempts to discover and conserve biodiversity appear hopeless.'
- Costello, May & Stork conclude that taxonomists are not in danger of extinction. 'They are increasing in number' and 'with modestly increased effort in taxonomy and conservation, most species could be discovered and protected from extinction'.

The views of Costello, May & Stork (2013) have been discussed in some detail. However, there have been challenges to their findings. For example, Mora, Rollo & Tittensor (2013), write:

> 'Costello *et al.* challenged the common view that many species are disappearing before they can be described. We suggest that their conclusion is overly optimistic because of a limited selection and interpretation of the available evidence that tends to overestimate rates of species description and underestimate the number of species on Earth and their current extinction rate . . . clearly, multiple sources of uncertainty remain in our knowledge of biodiversity and its rate of loss on Earth; however, the magnitude of the challenge ahead is considerable and should not be underestimated, because of the unique diversity of life on our planet and the services it provides for humankind are at stake'.

No doubt the debate on all the issues reviewed by Costello and colleagues will continue.

Having discussed the impact of human activities on the biosphere and examined the taxonomic and other challenges faced in trying to determine the number of species and the proportion that are threatened, in the next chapter we review efforts to conserve endangered species, and the ecosystems that contain them, in the face of increasing anthropogenic influence.

20 Conservation: from protection to restoration and beyond

It is clear that many thousands of species are at present vulnerable or endangered by human activities (Frankel & Soulé, 1981; Primack, 1993, 2010; Given, 1994; Meffe & Carroll, 1994; Frankel, Brown & Burdon, 1995). Broadly, two conservation options are available. Plants may either be conserved *ex situ*, in such places as Botanic Gardens, or *in situ* in their native habitats. Here we consider, briefly, the early history of conservation, and how theory and practice have changed – from protection, management, to reintroduction/restoration and creative conservation.

Ex situ conservation

Endangered species, including those at the very edge of extinction, are often conserved *ex situ* in Botanic Gardens and arboreta (Briggs, 2009; Hardwick *et al.*, 2011), of which there now more than 2700 worldwide (Ali & Trivedi, 2011). The scale of the holdings of Botanic Gardens is impressive: perhaps 25–30% of all vascular plants are represented in the collections (Wyse Jackson & Sutherland, 2000). Some gardens hold very large general collections (e.g. Kew with *c.* 10% of the world's plants), while others have specialist collections (e.g. The Arnold Arboretum, Boston, USA, grows several hundred species of temperate tree). Botanic Gardens Conservation International (BGCI), Richmond, England, organises and coordinates the conservation efforts of gardens (www.bgci.org/). Regrettably, most Botanic Gardens are located in temperate areas of the world, and it is costly to grow tropical plants in the glasshouses of Europe and North America. However, there are some notable Botanic Gardens in the tropics (Heywood & Wyse Jackson, 1991), and, if the number could be increased, it would be possible to conserve many tropical species of plants cheaply out of doors, and provide important centres for economic development and exploitation of plant biodiversity.

Other specialist gardens also play a key role in *ex situ* conservation. In England, the National Trust has now relocated its conservation activities from Knightshayes Court, Devon. It was necessary to move to another site because of an outbreak of Sudden Oak Death (caused by the fungus *Phytophthora ramorum*). The new *ex situ* facilities are designed to secure the future of the rare and endangered plant species that grow in the Trust's '200 gardens, 100 landscape parks' and 'in the many wild places it manages' (Morris, 2012).

With regard to other possibilities in conservation, Botanic Gardens sometimes propagate rare and unusual plants for sale, or make stock available to the commercial garden trade. If such species are widely grown in gardens, their future may be more secure. In addition, Botanic Gardens, by providing interpretive displays of living endangered material, are also well placed to educate the public on environmental issues and secure support for conservation projects (Maunder, Higgens & Culham, 2001).

Many plant species (with a capacity for vegetative propagation and self-fertility) are at first sight easier to conserve in gardens than are animals in zoos, where it often proves difficult to persuade them to breed. But, while gardens obviously have an important role to play in providing a last refuge for endangered species, problems of long-term conservation in gardens have to be faced.

A. While Botanic Gardens generally have the expertise to grow even the most exacting species, several major issues must be considered. The number of plants of any one species that can be grown in a Botanic Garden is strictly limited. This has consequences for the genetic variability of *ex situ* plants, relative to those in the wild. Genetic erosion is possible in small isolated populations, with short-lived species likely to be most strongly affected. The longer the period of their reproductive isolation in a garden, the greater the predicted magnitude of the genetic erosion (DeWoody *et al.*, 2010; Ensslin, Sandner & Matthies, 2011; Allendorf, Luikart & Aitken, 2013). Studies by Maunder (www.kew.org/conservation/cpdu/ssp.html) are revealing. *Echium pininana* is a threatened Canary Island endemic that is conserved in many Botanic Gardens. Molecular evidence suggests that the genetic diversity is higher in the wild than in the cultivated material. Also, there are feral populations associated with Botanic Gardens in the UK and France. The evidence suggests that these are derived from cultivated material, but their comparatively low variability may be due to bottleneck effects associated with the introduction of the species into gardens prior to their establishment outside cultivated areas.

Turning to another informative case study, the endemic species *Cochlearia polonica* was rescued from extinction in the 1970s, when 14 individuals were transplanted into a new *in situ* site in south Poland. To provide extra security, in the 1980s, 5 individuals were taken to establish an *ex situ* population in the Botanic Gardens, Warsaw (Rucińska & Puchalski, 2011). To investigate the effect of 18 years of cultivation in the garden, material from *ex situ* and *in situ* populations was examined using molecular markers. It was discovered that the *ex situ* population had only a portion of the genetic diversity of the new 'wild' population.

B. Turning to another crucial issue, many of the plants and existing collections are of unknown provenance. Labels may have been lost, and there are also problems of keeping good records. The move, by many Botanic Gardens, to grow plants of known wild origin is to be welcomed.

Some of the difficulties of labelling and good housekeeping are revealed in a recent study of the endangered species *Primula sieboldii* in Japan (Honjo *et al.*, 2008). They studied eight microsatellite loci and 'regional features' of cpDNA variation in devising 'assignment tests' of wild and cultivated stocks. They detected regional differences in wild samples. With regard to the *ex situ* plants, they concluded that 'the alleged origins' of 19 of the 29 garden stocks were confirmed. In contrast, evidence suggested that 5 stocks did not match those from the wild locality from which they were supposed to have been collected. It is concluded that plants may have been mixed up during cultivation, or mislabelled, as to origin, as they were transferred to other regions by commerce 'or whim'. Interestingly, in four *ex situ* collections, for which the alleged site of origin was confirmed, haplotypes were discovered that were not detected in the wild population from which they came in the past. Some wild populations of *Primula* are currently much reduced in size, with the potential for genetic erosion. The genetic variability in garden material might therefore be important in *in situ* restoration projects.

C. There is another issue of concern: if plants from different sources are grown side by side, hybrid seed may be produced, and this could potentially change the genetic variation of conservation collections. For instance, *Typha minima* is an endangered species that grows along alpine sections of the river Rhine (Galeuchet & Holderegger, 2005). In a study of isozyme variation of plants in Swiss Botanic Gardens, it was discovered that in two cases hybridisation had occurred between collections of different geographical origin.

Hybridisation between different species is also of concern in other Botanic Gardens. For example, Maunder *et al.* (2001) report that in a number of Botanic Gardens *Chionodoxa luciliae* has hybridised with *Scilla* species. Returning to Maunder's investigation of *Echium*, there is evidence that hybridisation between different species has occurred in some gardens.

D. Taking the longer view, if stocks are renewed repeatedly from seed, and small numbers of specimens used to provide seed in each of the successive generations, then the genetic variation represented in the original sample introduced into the garden may be successively reduced. Another key issue concerns the possibility of gardeners exercising unconscious selection on the plants they are cultivating in gardens or in specific *ex situ* conservation projects, leading to insidious domestication (Briggs, 2009). Until recently, this possibility has received little attention, although there is increasing evidence of genetic adaptation to captivity in animals (Frankham, 2008). Given the paucity of information about plants, two recent investigations are especially interesting.

Samples of *Cynoglossum officinale* from 12 Botanic Gardens and wild material from 5 natural populations in Germany were grown in a common garden and their genetic variability was examined using 8 nuclear microsatellite markers (Ensslin, Sandner & Matthies, 2011). They discovered that while the genetic variability of some of the garden material matched that of wild populations, plants from other gardens 'exhibited no genetic variation at all'. Furthermore, 'several lines of evidence indicated genetic changes in garden populations in response to cultivation. Seed dormancy was strongly reduced in garden populations, and, in response to nutrient addition, garden plants increased the size of their main inflorescence, while wild populations increased the number of inflorescences.' They considered that in the pursuit of showy plants 'gardeners may have collected seed from mainly tall plants with long main inflorescences. Such adaptations to cultivation could be maladaptive in nature.'

It seems certain, therefore, that plants taken from 'original wild habitats' and placed in new (and usually very different) human-made and human-managed environments of Botanic Gardens may be subject to unconscious selection (Zohary, 2004). The genetic and evolutionary consequences of such transfers deserve wider and more detailed investigation.

Lauterbach, Burkart & Gemeinholzer (2012) have carried out a very revealing study of the extent to which *ex situ* approaches conserve and possibly change genetic variation relative to that of the original source populations. They investigated the endangered *Silene otites* in German Botanic Gardens and wild sites. They were able to find three *ex situ* populations in the Botanic Gardens of Berlin, Marburg & Mainz, for which there were full records of their date of collection and site of origin in the wild. Using molecular markers (AFLP), they examined the genetic variability of the *ex situ* material with that of its wild source population after 20–36 years of isolation. They detected distinct differences, and concluded that the loss of diversity in the *ex situ* material was attributable to small population size and unconscious selection. Of the three Botanic Gardens, Maintz conserved the most genetic variability. Here, plants were grown in a 'near-natural dry grassland on sand with spontaneous rejuvenation, interspecific competition and overlapping generations of different ages'. In the other two gardens, plants were cultivated in 'a single-species bed without interspecific competition with biennial rejuvenation from the cultivated stock'.

Several important general principles emerge from these investigations. Lauterbach, Burkart & Gemeinholzer stress the importance of growing a minimum viable population (MVP) of 50 individuals in gardens, to discourage genetic decline (Krauss, Dixon & Dixon, 2002). Given that genetic and unconscious changes occur under cultivation in gardens, they also recommend

i) That conditions of cultivation in gardens should be as near natural as possible, as at Mainz.
ii) Care should be exercised to avoid selection especially at the germination and juvenile stage: gardeners should avoid choosing the quickest seedlings to appear or the tallest seedlings for transplantation.

iii) In the case of dioecious (separate male and female) plants steps should be taken to avoid biased sex ratios in conservation.

iv) If Botanic Garden material is to be used in restoration programmes, then the period in which stocks are cultivated in gardens should be kept to a minimum.

v) Where possible, and keeping material of different origin separate, successive transfers of wild material from the source population to the garden could be made to maintain or improve the genetic variability.

E. Finally, in considering the effectiveness of *ex situ* conservation, there is also the question of continuity of policy in gardens. The enthusiastic conservation efforts of one director might be difficult to fund by his/her successor in times of financial stringency.

Seed banks

Botanic Gardens may increase their contribution to conservation if they maintain modern seed storage facilities or have links with such centres. It has been discovered that the viability of seeds of many species may be extended if they're kept cold and dry and stored in moisture-proof conditions.
The recommended conditions are a maximum of 3–7% water content, and a sub-zero temperature (preferably −18°C) (Linington & Pritchard, 2001). However, not all species may be stored in this way, as about 25% of higher plants – many from tropical forests– have recalcitrant seeds that lack seed dormancy and/or cannot withstand desiccation. These represent 'the greatest conservation science challenge of the 21st century' (Li & Pritchard, 2009). There are also species with intermediate behaviour between orthodox and recalcitrant (Linington & Pritchard, 2001). Seed banks require assured and constant electricity supplies, with back-up facilities. In some countries seed banks have been lost, for example, on Fiji where ancient refrigerators broke down (Anon., 2002a).

For conservation purposes it is important to collect appropriate seed samples for storage, samples that adequately reflect the genetic variation of the original population. Falk & Holsinger (1991) present a theoretical framework for their recommendation to collect seeds from up to 5 populations per species, and from 10 to 50 individuals per population. Also, if all the seed cannot be safely collected in 1 year without affecting the reproductive success and size of the populations, they recommend that seed be collected over several seasons. Because stored seed, even at low-temperature storage, eventually loses its viability, it has been proposed that stocks be rejuvenated by growing samples in gardens to obtain new seed, steps having been taken to make sure that genetic variability is not lost in the rejuvenation process, with the plots so placed as to be free of any possibility of crossing with other stocks. However, rejuvenation is very expensive and it is cheaper, in some cases, to collect new samples in the wild, especially if several species are to be collected in the same area.

Seed from seed banks and propagated plant material are often made available by Botanic Gardens for reintroduction of a species back into the wild. Many Botanic Gardens, exploiting the capacity of plants to regenerate from small pieces in appropriate culture conditions, have set up micropropagation units. 'In essence, isolated surface-sterilised tissues are grown in liquid or agar culture in flasks or jars containing sterile media prepared from salts, sugars, vitamins and growth hormones (Sugii & Lamoureux, 2004). To an extent, the appropriate medium for each species is determined by trial and error' (Briggs, 2009). Many of the pioneering micromanipulation studies of wild plants were published in *Botanic Gardens Micropropagation News* from the Royal Botanic Gardens at Kew. Now, an increasing number of case studies have appeared in the scientific literature. For instance, the endangered endemic *Dianthus arenarius* ssp. *bohemicus* – of which there is only one population of about 30 clumps in the wild – has been successfully grown in nutrient solutions to provide rooted plants (Kovác, 1995). And following micromanipulation the orchid *Vanda spathulata* was successfully conserved at the Jawaharlal Nehru Tropical Botanic Garden and Research Institute, Kerala, India (Decruse *et al.*, 2003).

In the development of micropropagation technology, it has also been found possible to freeze pieces of plant tissue (called cryopreservation). Li & Pritchard (2009) report that 'plant vitrification solutions have now been used successfully to cryopreserve shoot tips and other plant tissues from >110 species across >80 genera'. They consider that 'cryo-preservation is the only large-scale, long-term option for *ex situ* conservation that are clonal or have recalcitrant seeds'.

The current state of micropropagation techniques is reviewed by Sarasan *et al.* (2006). They report major advances in sterilisation and other treatments to provide 'clean material', and see micropropagation becoming a routine approach for conservation of endangered plants. However, although successful restoration of populations of endangered species has been achieved using micromanipulated material, some important issues have emerged.

A. Some genotypes of a species may be more susceptible to micromanipulation than others. For instance, this phenomenon was detected in investigations of the critically endangered Australian species *Grevillea scapigera* (Krauss, Dixon & Dixon, 2002). Ye, Bunn & Dixon (2011) point out that 'biased selections of genotypes may inadvertently be allocated to a reintroduced population with the risk of genetic decline in subsequent generations'.
B. In some cases, micromanipulation of crop and wild material has produced sexually aberrant plants. For instance, Ye, Bunn & Dixon (2011) have reported male and female reproductive failure in such plants as the critically endangered Australian species *Rulingia* spp. 'Trigwell Bridge', of which only three plants remain in the wild. They recommend that, as part of restoration projects, micromanipulated stocks be screened for 'reproductive normality prior to release' by examining pollen viability, floral morphology and crosses between wild × manipulated material.
C. It has also been suggested that molecular markers could be used to confirm 'genetic fidelity' in micropropagation of conservation stocks (Gupta & Varshney, 1999).

Ex situ conservation: the future

Looking ahead many see a clear role for *ex situ* collections in the restoration and restocking of rare and endangered plants in the wild (Ali & Trivedi, 2011). Also, in the *ex situ* conservation of crops and their wild relatives, seed banks, DNA collections and micropropagation are crucial (Khoury, Laliberté & Guarino, 2010).

Thus, the largest seed banks have been constructed to conserve crops and other important plants (Li & Pritchard, 2009). Major organisations include The National Center for Genetic Resource Preservation, Fort Collins, USA; The National Centre for Crop Germplasm at the Chinese Academy of Agricultural Science in Beijing; and the Svalbard Global Seed Vault in the permafrost of Arctic Norway. Micropropagation is also seen as an important route for the conservation and commercial exploitation of tropical hardwood trees that have recalcitrant seeds (Pijul *et al.*, 2012). Seed banks are of crucial importance in the conservation of taxa from semi-arid biomes, for example in Latin America (León-Lobos *et al.*, 2012).

However, two major concerns are highlighted in a recent review (Godefroid *et al.*, 2011b). While 70% of the European flora is currently represented in storage, only 27% of European threatened species are represented. Pteridophytes and orchids are under-represented. Considering the number of accessions and seeds per accession as surrogates for genetic variability, Godefroid *et al.* (2011b) also make the important point that 'at least two thirds of the threatened species stored in European seed banks [are] likely to suffer from a too low genetic diversity in the collections'. A wider selection (an estimated 42%) of the threatened European flora is conserved *ex situ* in gardens, but there is concern about the quality of some collections (Sharrock & Jones, 2011).

Turning to another issue, some collections of plants, at present maintained out-of-doors of Botanic and other gardens, may be threatened, as global climate change could make their cultivation more difficult, or impossible. In the light of these concerns, Botanic Gardens are reviewing their scientific and other objectives: see for example the report on future directions at The Royal Botanic Gardens, Kew, where a 10-year programme is being developed to investigate 'plant based solutions to the challenges of climate change' (Ali & Trivedi, 2010).

Considering the long-term prospects for *ex situ* conservation of plants in the light of our brief review, three major conclusions must be confronted. These were identified many years ago by stalwarts of the early conservation movement, and there is a case to be made that these issues are still neither widely appreciated nor adequately investigated.

1. In their classic contribution to conservation principles and practice, Otto Frankel & Michael Soulé (1981) stress a crucial point: 'introduction into cultivation is likely to narrow the genetic diversity of a species and to change its variation pattern in response to drastically altered selection pressures'.
2. Considering the consequences of long-term cultivation, Professor Prance, a former Director of Kew Gardens, emphasises that *ex situ* conservation will 'halt or distort the natural process of evolution' (Prance, 2004).
3. And in a key insight regarding long-term conservation in gardens, Professor Peter Ashton, former Director of the Arnold Arboretum of Harvard University (Ashton, 1987), warns that 'for all practical intents *ex situ* species conservation leads irreversibly to domestication'.

As there are inherent limitations of *ex situ* approaches, we stress the overwhelming importance of *in situ* conservation. So far our discussion has focused largely on the conservation of species considered individually. With the aim of securing the future of all the interacting species of plants and animals in functioning ecosystems, legally protected reserve areas have been set up. How far has this aim been achieved?

The role of protected areas in countering the threat of extinction

Of the order of 161 000 protected areas of various sorts occupy about 10–15% of the world's land surface (Soutullo, 2010). Historically, it is important to realise that nature reserves and national parks were first set up to protect areas of outstanding natural beauty and landscape features (Briggs, 2009). For instance, Runte (1997) makes it clear that Yellowstone National Park, the first National Park in the USA, was 'not designed for wildlife'. As Sellars (1997) emphasises: 'scenery has provided the primary inspiration for national parks' and 'tourism has been their primary justification'.

Later, in the late nineteenth century and early twentieth century, many reserves were designated to protect particular groups of organisms; for example the National Trust has been involved with the Wicken Fen Nature Reserve, Cambridgeshire, England, since 1899, primarily because entomologists were interested in continuing their collecting of fenland Lepidoptera in the area.

Over the years, reserves have been established in many different areas. For instance, some are remnants of semi-natural communities – such as ancient coppiced woodland, species-rich grasslands and heathlands – often on low productivity land of marginal agricultural value (Meffe & Carroll, 1994). Such reserves may or may not be nationally designated. In some cases, local conservation groups have purchased areas or negotiated management agreements with landowners, to look after existing islands of distinctive biodiversity, found where traditional management – coppicing, haymaking, peat cutting, etc. – is still practised or, more usually, has declined or has been abandoned altogether.

At the other extreme, reserves have sometimes been set up in areas of what are regarded as pristine natural vegetation, such as tropical rainforests. However, it is clear that there is a fierce debate about whether 'pristine' natural vegetation still exists in the world.

Setting these arguments to one side, we face another arena of debate: how big should a reserve be and of what shape to effectively conserve the biodiversity it contains? The era of the creation of very large new parks and reserves has probably come to an end in many heavily populated parts of the world, but it is still possible to enlarge existing reserves, and in some cases designate new reserves in large territories such as in Australia (New, 2006).

Looking to the future, the United Nations Environment Programme (UNEP) reports that 'countries are on track to meet a 2020 goal established under the Convention on Biological Diversity to protect 17% of land areas, although reaching the 10% target for coastal and marine regions will require further efforts' (editorial in *Nature*, 10 December: Anon., 2014). However, the review confronts the reality that 'many protected areas are "paper parks", where hunting, fishing and habitat destruction continue apace because of lax enforcement. And most parks established so far do not protect the crucial areas – the ones full of threatened species and habitats, Nations are also investing much less on [environmental] protection than they were 15 years ago, after adjustments are made for inflation.'

Ideas about the design of nature reserves, with implications for their enlargement, modification and management, have been influenced by theoretical and experimental studies of island biogeography, which, as we have seen above, have considered the diversity found on areas of land of different size surrounded by water. In this account we indicate some of the main concepts as they affect nature reserve design; for a more comprehensive treatment of the influence of these ideas on conservation, see Schafer (1990). The general principle emerging from these approaches, which receives some support from theoretical and experimental studies, is that 'as area increases, so does the diversity of physical habitats and resources, which in turn support a larger number of species' (Meffe & Carroll, 1994).

Taking these ideas as a starting point, and noting that they have proved controversial and very difficult to test, it is generally agreed that, within a territory, large reserves are better for conserving biodiversity than small reserves (Fig. 20.1), and perhaps better able to accommodate the problems of introduced species than smaller areas. Modelling and experimental work have also attempted to resolve another question, whether it is better to have a single large reserve or several small reserves of equivalent total area (the so-called SLOSS debate). Another key issue is the extent to which reserves of different size, and also different shape, are affected by edge effects (a concept introduced earlier). For instance, where a reserve abuts on agricultural land, edge effects occur that are quite complex, and could include drift of herbicides and fertilisers, alterations of microclimatic factors such as light and temperature, as well as invasions of weedy species into the reserve. The degree to which edge effects affect the main communities contained in the reserve is influenced by the area–perimeter relationships. A single, large reserve suffers less edge effect than several smaller reserves of equal total area. Also, more or less circular reserves have much less edge than long, narrow reserves of the same area, which could be exposed to edge effects along their entire width.

However, we contend that the SLOSS debate should not be used as an argument against setting up reserves that do not conform to a supposed 'ideal' size and shape. Well-positioned small reserves may have a vital role in conserving rare species, e.g. butterflies. And even long, thin reserves have an important local role to play in any conservation in regions dominated by modern intensive agriculture. For example, Devil's

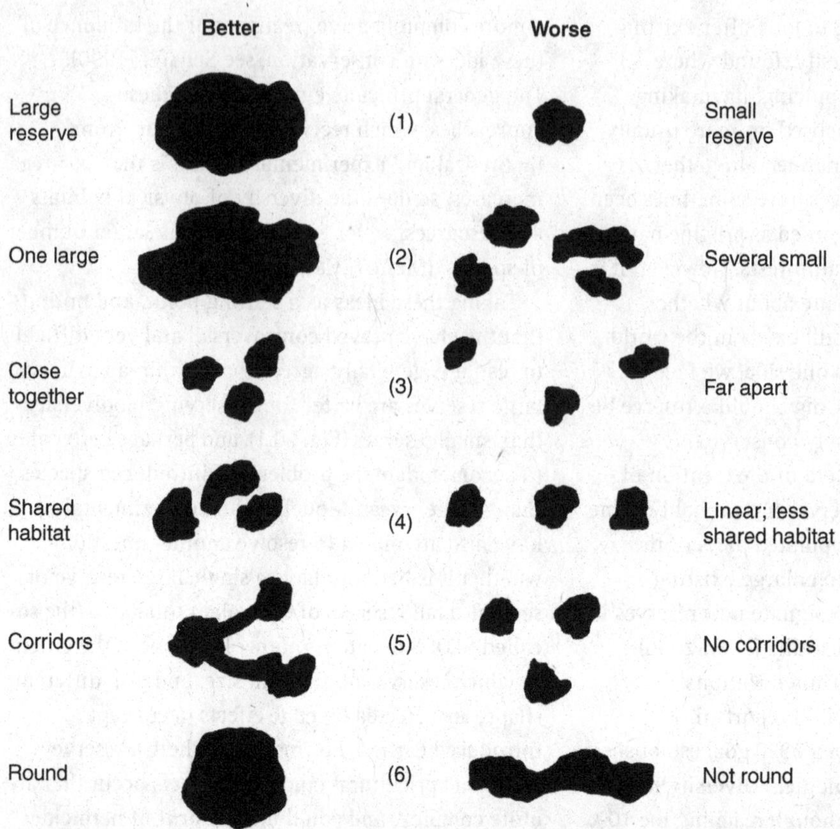

Fig. 20.1. Principles of reserve design that have been proposed based on theories of island biogeography. Imagine that the reserves are 'islands' of natural habitat surrounded by land that has been totally changed by human activity. The practical application of these principles is still under study. Principles 2 and 5 in particular have been the subject of much debate. (After Diamond, 1975, from Primack, 1993.)

Dyke, near Cambridge, England, is an important chalk grassland reserve (Walters, 1979). This archaeological site, dating from around the fourth to seventh centuries, consists of a 12 km (7.5 mile) long, linear, massive defensive earthwork and an associated ditch. Also, the degree to which edge effects should be minimised in reserves is a complex issue, as edges may contain their own suite of interesting rare species.

Another key issue emerging from island biogeography concerns the distance between reserves (Fig. 20.1). Looking first at a possible disadvantage of having reserves very close together, it is clear that pests, diseases and alien species might be able to migrate more easily between them. But there are also possible advantages, as genetic considerations suggest that reserves close enough to permit gene flow between the populations of particular species would be preferable to a collection of widely scattered reserves. In considering the implications of the distance between reserves in particular cases, it is becoming clear that it is unwise to estimate gene flow without proper research. As we saw in Chapter 10, molecular approaches are now offering new methods of investigation and considerable gene flow has been demonstrated in some species, but not in others.

Taking the question of reserve design one stage further, consideration should also be given to whether corridors of appropriate vegetation, between reserves, permit gene flow in particular species. These corridors might be envisaged as 'transit' corridors, along which, for example, seed dispersers might travel, or 'residence' corridors, in which plants may establish, grow and reproduce in the process of migrating from one reserve to another. A first point to note is that

narrow corridors are likely to suffer from extreme edge effects and, taking a specific case, one might ask: is there any evidence that strips of connecting woodland and hedgerows provide effective routes for gene flow between two ancient woodland 'island' nature reserves in eastern England? The characteristic plant species of ancient woodland are not very 'mobile', and observation suggests that some are unable to migrate between 'island woods' along connecting hedgerows (Peterken, 1981). Many conservationists have suggested that corridors should be set up between reserves to allow species to migrate, in response to climate change, to new areas appropriate to their climatic tolerances.
The effectiveness of such corridors, both now and in the future, has been called into question by Simberloff *et al.* (1992), for, while the movement of birds and mammals might be facilitated by corridors, it should not be assumed that they would automatically facilitate plant migration.

Reserves in particular areas are often very small, and it is increasingly being emphasised that what happens outside reserves is critical to the survival of much biodiversity. In considering the spatial relationship between reserves, conservationists often employ gap analysis, in an attempt to improve the effectiveness of existing protected areas (Fearnside & Ferraz, 1995; Rodrigues *et al.*, 2004; Scott & Schipper, 2006). By studying geographical, topographical and biological information, gap analysis allows the identification of territory, outside reserved areas, that is important for the success of particular conservation projects, where present land use management is inadequate to sustain particular species/ecosystems, or where the ecosystems present in the gaps do not support adequate migration/gene flow between reserves.

The aim of these analyses is to provide a strong case for expanding reserves in a strategic fashion to include irreplaceable ecosystems containing unique biodiversity. If it is not possible to take such areas into protective care, then discussions with landowners etc. might make it possible to secure the future of endangered species/threatened ecosystems by agreements on landscape management. Thus, where blocks of privately held land, and public forests exist adjacent to reserves, they could be managed to an agreed common plan. Such agreements provide increased habitats for wildlife, and in some cases effectively enlarge reserves or provide a buffer zone of sympathetically managed ground around or adjacent to them.

Systematic conservation planning, including gap analysis, is becoming a more and more complex area of research, as conservationists are being increasingly challenged to consider the socio-economic implications and cost-effectiveness of their proposals (Kukkala & Moilanen, 2013), and to incorporate in their analyses the likely effects of climate change.

Managing reserves to prevent extinction of species

Historically many reserves were established as 'preserves'. The idea was to isolate them from human activities to allow 'the balance of nature' to be restored (Pickett, Parker & Fiedler, 1992). This outlook is being displaced by the realisation that humans have decisively influenced the ecology of communities throughout most of the world. In these semi-natural communities active human intervention in the form of reserve management is necessary to conserve particular species.

This key concept emerged from the 1930s onwards. If endangered species in semi-natural vegetation are to be successfully conserved, traditional management of the areas should be continued or reimposed. 'Resort to precedent' was the key to management. For example, to conserve species-rich chalk grassland in Europe, which is the product of centuries of grazing (by sheep and rabbits in Britain), it is crucial to maintain the grass sward by controlling the invasion of the site by shrubs and trees. In such cases, active management is necessary, by continuing or re-establishing grazing regimes with appropriate animals, or by employing surrogate methods such as mowing, cutting or physical removal of plants.

The range of management techniques used in conservation is enormous (Sutherland & Hill, 1995; Perrow & Davy, 2002; Hobbs & Cramer, 2008). A few examples will indicate the range of options:

1. Site management and soils (e.g. control of water by impeding (through the building of bunds, dams) or increasing water flows (by dykes, ditches, watering etc.); alteration of the nutrient status (e.g. by cropping without fertiliser addition; by removal of top soils, or chemical treatment to change nutrient-rich cultivated land to nutrient-poor to match that of many natural ecosystems; by re-establishing fire regimes; and by decontaminating and restoring mining areas).
2. Management of vegetation (selective action involving sowing, planting, regular or occasional cutting, burning or weeding).
3. Management of animals (by selective culling; employing fences to prevent, restrict or intensify grazing; transfer of animals; veterinary interventions to protect endangered animals from diseases etc.).
4. Reintroducing plants and animals (both common and endangered). Animal conservationists often work on the assumption that if the vegetation is managed and restored the 'fauna will follow' (Hobbs & Cramer, 2008).
5. Controlling introduced species (by their physical removal, pesticide use, and introduction of other organisms to establish 'biological' means of control).
6. Management of human activities (by legal or agreed protection of the site and its biota, preventing or limiting access; controlling or denying the exploitation of natural resources, provision of educational and other facilities for visitors, etc.).

This sixth category presents some of the most difficult and politically complex issues. For instance, in tropical areas indigenous peoples live within or close to areas which have become reserves, and in their justified attempts to survive, often under conditions of abject poverty, the biodiversity of the area may be put at risk. Such situations provoke moral as well as scientific questions. How far should traditional land use be permitted? To what degree is the traditional use of resources compatible with conservation? Is it possible to devise a sustainable means of using resources, so that the long-term future of biodiversity is assured? In times past, local people were sometimes removed from the newly set-up National Parks (Runte, 1997; Sellars, 1997). Often they received no material benefits from having a reserve on their traditional lands, and in some cases became antagonistic to conservation efforts (Western, Russell & Cuthill, 2009). Many conservationists now consider that the people who live and work in or near nature reserves or National Parks – be they in India, Europe or elsewhere – must be more fully involved in decision-making and receive some income from conservation activities, perhaps from the sustainable development of resources or various kinds of ecotourism (Meffe & Carroll, 1994). In many countries conservation management involves payment to local farmers, who forgo the financial rewards of intensification of agriculture by agreeing to impose or continue with practices that encourage wildlife. The political, ecological and social conflicts involved in conservation and development have been studied worldwide (see below).

Restoration ecology

Humans' activities have caused great damage to plant communities, and another major change of conservation strategy began to emerge in the mid-twentieth century. Traditional management of reserves has continued, but, increasingly, restoration projects have been employed (Falk, Millar & Olwell, 1966) and then more 'creative conservation strategies' began to be more widely employed (Sheail, Treweek & Mountford, 1997).

For instance, major interventions have been carried out to restore and rehabilitate species-rich grasslands

(Crofts & Jefferson, 1994), woodland (Ferris-Kaan, 1995) and wetlands (Wheeler & Shaw, 1995). Sometimes these 'creative' approaches have enlarged existing reserves, but significantly they have also created *new* areas of managed ecosystems designed to encourage wildlife. Major schemes have also been undertaken to restore polluted rivers, damaged wetlands, degraded derelict land, docklands, and mining areas. And these have resulted in significantly increased habitats for wildlife.

Some very imaginative schemes have resulted in protected areas in urban settings as well as in the countryside. For instance, in Tower Hamlets, East End of London, a Victorian cemetery, where burials no longer take place, is now managed as a wooded local nature reserve (www.fothcp.org).

In another notable project, the London Wetland Centre has been produced from four disused Thames Water reservoirs, by the Wildfowl and Wetland Trust (www.wwt.org.uk). Opened in 2000, this reserve of wetlands and lagoons was created by breaking up the reservoirs and realigning 500 000 metric tons of debris and soil to create a 42 ha site on which 300 000 water plants and 27 000 trees were planted.

Creative conservation: community translocations

Sometimes, where essential civil engineering and excavation projects receive planning permission, ecosystems of high conservation value are put at risk of complete destruction, for instance in road building, quarrying, pipe laying, coal mining, peat extraction etc. Bullock (1998) has reviewed examples of 24 whole community translocations in Britain. Generally, intact communities and associated soil are lifted as a turf by hand or by machine, and these are re-laid on a specially selected prepared site. Sometimes the sites were given appropriate management, in the form of grazing, mowing and weed control. Examining these translocated ecosystems a few years later, Bullock concludes that there is a 'high risk that community translocation will not achieve the preservation, unchanged, of the complete community'. It remains to be seen what ecosystems develop from such transfers in the longer term.

Creative conservation: wildflower mixtures

Restorations often involve the establishment of common species, sown as commercial wildflower mixtures, sometimes certified as having originated from 'wild' collections brought into cultivation. Often, commercially grown herbaceous plants and tree saplings are used. Then, in some cases, rare species may be sown or transplanted into the 'restored' swards as small plants. However, commercially available stocks may turn out to have diverse and sometimes inappropriate origins. For instance, while the 'correct' species were sown in a project to restore chalk grassland in 1991 on the Gog Magog Hills, Cambridge, England, the wrong variants were unwittingly used in commercially grown seed stocks assembled to provide a wildflower mixture (Akeroyd, 1994). Critical taxonomic assessment revealed that the *Achillea millefolium* and *Centaurea cyanus* were most likely of horticultural origin, while the *Medicago lupulina, Trifolium pratense* and *Sanguisorba minor* ssp. *muricatus* were most probably agricultural strains. There was also evidence that 'seed impurities' were also included in the wildflower mixture: for instance *Ranunculus marginatus* var. *marginatus*, native of the Balkans and Crimea to Iran, appeared on the site.

Plants/seed of native provenance

Many restoration projects take their seeds, herbaceous plants and tree saplings from native sources. Given that species often exhibit ecotypic variation in relation to climate and edaphic factors, many ecologists consider that it is essential to plant material appropriate to the site – be it hay meadow, pasture, fenland etc. Thus, many practitioners recommend that restorations be carried out with local variants (McKay *et al.*, 2005). These may be taken directly from the 'wild' or from semi-natural managed habitats. For example, to restore hay meadows, seed is often collected by machine or by hand from species-rich hay meadows conserved as nature reserves (Perrow & Davy, 2002). This seed mixture, which contains seed of the 'hay ecotypes' of the many species found in ancient hay meadows, will, therefore, provide the seed for a more authentic restoration. In other cases, stocks are taken directly or indirectly from *ex situ* sources.

Match or mix? Some ecologists favour the use of native stocks to maximise the immediate chance of success of restoration projects by using stocks that are ecotypically adapted to the ecosystem under restoration. However, others take a different view. They suggest that stocks of mixed ancestry should be used. They emphasise that the aim of restoration is to establish populations that have an evolutionary future, by having the genetic variability to respond, through natural selection, to climatic and ecological changes as they occur. For those of this persuasion, restoration should be for the long term and must provide the means whereby populations can evolve *in the future* (Broadhurst *et al.*, 2008; Breed *et al.*, 2013). However, if stocks of mixed origin are brought together, in this way, maladapted progeny may result through outbreeding depression (Kramer & Havens, 2009; Forrest *et al.*, 2011; Frankham *et al.*, 2011). We examine the evidence for outbreeding depression below.

Manipulating and creating populations in an attempt to prevent extinction

As part of the 'creative' conservation enterprises, measures have been taken to reinforce or reintroduce declining populations of endangered species (Briggs, 2009). Zoologists have led the way in many practical conservation techniques, transfers being made to replenish existing populations, or to re-establish populations at sites where they once lived but have now become extinct (Bowles & Whelan, 1994; Meffe & Carroll, 1994; Bottin *et al.*, 2007; Maschinski & Haskins, 2012). Plant conservationists have also become interested in the possibility of restocking (augmenting), reintroductions and transfers in the management of both existing and lost populations, generally, within what is taken to be their native range (but see below). Restoration is now accepted by most conservationists as a valuable and necessary approach to conservation. But some naturalists, while accepting habitat management, still condemn the setting-up of 'new reintroduced' populations, as misusing valuable resources, devaluing natural populations, obscuring true and native distributions and likely to damage populations at source and maybe at point of transfer (Falk & Olwell, 1992). They draw attention to the fact that past transfers often were poorly documented and carried out without permission. Also, they are concerned about mitigation. If it becomes accepted by the general public that new habitats may be produced that are acceptable 'replicas' of present reserves, and that rare and endangered plants can be successfully translocated, then a developer, wishing to use and thereby damage an existing site and its ecosystem, could offer to produce a new identical ecosystem elsewhere in mitigation. Conservationists are rightly suspicious of such a 'bargain'.

Restocking (augmentation, reinforcement) of existing population(s)

In considering the possibility of augmentation, there are a number of research papers establishing baseline information for various species: for instance, investigations at Kirstenbosch Botanical Garden on the critically endangered cycad *Encephalartos latifrons* (Da Silva *et al.*, 2012); studies of the endangered Utah, endemic *Astragalus ampullaroides* (Breinholt *et al.*,

2009); and analysis of the extinction risk to the federally endangered *Ipomopsis sancti-spiritus* found in Holy Ghost Canyon in the Rocky Mountains (Maschinski, 2001). In some cases it has been decided that augmentation is not necessary, as in the case of *Castilleja levisecta* of the Pacific Northwest (Godt, Caplow & Hamrick, 2005; Lawrence & Kaye, 2011).

There are only a few well-documented cases of augmentation in the scientific literature (see www.kew.org/conservation/recplan.html and the reviews of Bowles & Whelan, 1994; Meffe & Carroll, 1994; Maschinski & Haskins, 2012). Two projects reveal the potential of this approach.

In 1975, the population of the rare mountain species *Saxifraga cespitosa* on Cwm Idwal, North Wales, had been reduced to 4 plants. From material collected from the population, plants were raised at Liverpool Botanic Garden, and, in 1978, the population was restocked with 130 mature plants, 195 small seedlings and 1300 seeds (Parker, 1982). In 1980, there were 48 mature plants in the population. In this case, the intention was to produce a population equal to the estimated population size in 1796. Perhaps it would have been more appropriate to aim for a minimum viable population.

Augmentation has been carried out in one of the fragmented populations of the endangered bird-pollinated long-lived Australian shrub *Banksia cuneata* (Coates, Williams & Madden, 2013). By increasing population size from 57 to 214 adult plants, there was a trend towards increased outbreeding.

Turning to another example, over a 10-year experiment, seeds of *Echinacea angustifolia* were collected from prairie remnants in Minnesota and 'overseeded' into nearby study plots of recently planted native grasses (Wagenius *et al.*, 2012).

Restorations using clonal plants from various sources

In some restorations clonal material has been employed. For instance, in reintroduction programmes along the River Isar, stem cuttings of the German False Tamarisk (*Myricaria germanica*) were employed, as the plant is difficult to propagate by seed (Koch & Kollmann, 2012).

Turning to another example, in the restoration of the dune systems along the shorelines of Lake Michigan and Lake Superior, American Beachgrass (*Ammophila breviligulata*) has been introduced. This species, which reproduces mostly by asexual means, was planted as clonally propagated material. By studying ISSR molecular markers, Fant *et.al.* (2007) discovered that native populations were genetically diverse, but restored areas, which had been planted with commercial clonal material, were monotypic. In restored areas, one genetic fingerprint indicative of commercially propagated stocks was widely detected. This genotype was not found in any native population. In addition, there was evidence that many of the other genotypes used in restoration were not of local origin. Fant *et al.* (2007) also criticised the planting arrangements in restored areas. To mimic the patterns of variability found in natural populations, restorations should be designed to plant out a range of local genotypes in mosaic patterns.

Re-establishment (reinstatement) of an extinct population

In many cases, extinct populations have been re-established in restoration projects. In the UK, the English Nature Recovery Programme has been set up with the aim of securing the future of very rare native species (Maunder & Ramsey, 1994). For example, the rare *Orchis laxiflora* has been successfully propagated in the laboratory (with its symbiotic fungus), and subsequently transferred to a new site in a nature reserve at Wakehurst Place, West Sussex, managed by the Royal Botanic Gardens, Kew (see Maunder & Ramsey, 1994).

Turning to another example, *Stephanomeria malheurensis* is known only from one locality in Oregon, USA, where it became extinct less than 20 years after its discovery, its original habitat having been damaged by fire followed by invasion by aggressive European introduced weed, *Bromus tectorum*. A restoration programme was initiated using *ex situ* material from the Berry Botanic Garden, Portland. Forty thousand seeds were produced by the new population in the first year, but population numbers fluctuated thereafter (Guerrant, 1992). Plants were present at the site in 2001, but in a report dated 2010 (see www.centerforplantconservation.org) it appears that numbers fluctuate annually within the fenced site and the plant may be declining in numbers.

Founding population(s) in new areas

Many plant species reintroductions (sometimes called translocations, or transplantations) have been initiated worldwide (see Milton *et al.*, 1999, and the valuable review by Godfroid *et al.*, 2011a). For example, at least 234 projects have been carried out in Europe. Here, we examine a very well-documented case that raises a number of key issues.

Demauro (1994) founded three new populations of the rare endemic self-incompatible *Hymenoxys acaulis* var. *glabra* in the Great Lakes area. As the species requires a mixture of S allele genotypes for the self-incompatibility system to function, the new populations were set up using plants of different mating types from different populations, together with F_1 and open-pollinated F_1 progenies. Planting arrangements were designed to maximise the functioning of the self-incompatibility system. The plan for the restoration was based on the MVP size of 1000, a figure large enough, it was thought, to conserve variability. In the event, 95% of the first transplants were lost through drought within two months, and a much larger initial population might have been more appropriate. In this case study, the use of composite stocks rather than material from a single source is easy to justify. As we have seen, many have recommended that restocking and reintroduction should only be attempted with stocks related to the site in question. In the case of *Hymenoxys* the problem with the breeding system could only be solved by using plants from different sources and also hybrid material.

Mix or match: inbreeding and outbreeding depression

We stress again that some geneticists have advocated mixed rather than matched plantings in restorations (Barrett & Kohn, 1991). In their view this will result in a widening of the range of genetic variation in the population and the action of selection may result in a population more fitted to changing circumstances. However, other conservationists, anticipating the possibility that wide crosses may sometimes be made – involving variants distinct at the ecotypic level – have pointed out the procedure may lead to serious outbreeding depression, a phenomenon discovered in animals, where many or all of the hybrid progeny may lack fitness.

Until recently, there was very little investigation of outbreeding depression in plants (see Lofflin & Kephart, 2005). An elegant case study has now been published by Forrest *et al.* (2011), who investigated populations of the endangered bird-pollinated Australian species *Grevillea mucronulata* Harwick. The results of pollen transfers were examined: (i) between neighbours in a series of 'home' populations; (ii) pollen brought to a 'home' population from an adjacent cluster of adults (30–50 m distant); (iii) pollen originating from a distant cluster (0.5 km distant); and (iv) open pollination. Plants were genotyped using molecular markers, and the performance of seedlings of different origin was compared. They discovered that '*Grevillea mucronulata* displayed evidence of inbreeding' (in crosses between near neighbours in 'home' populations). In comparison of seedlings derived from pollination from intermediate and distant pollen sources, 'intermediate' plants revealed

> consistently superior outcomes for most aspects of fitness including seed set, seed size, germination and

seedling growth . . . The superior outcome of intermediate pollen transfer and genetic differentiation of adjacent clusters suggests that in *G. mucronulata* selection disfavours matings among closely and distantly related neighbours. Moreover, the performance of open-pollinated seedlings was poor, implying that current mating patterns [within the home populations] are suboptimal.

While it would be unwise to generalise from the results of this investigation, it must be acknowledged that outbreeding depression could be a significant factor both in wild populations and also in restoration projects.

Founding new populations: seed v. plants

Given the high level of preparation necessary for transfers of plants in restoration projects, it is interesting to consider whether sowing seeds might be a cheaper, quicker and more reliable method. Such an approach was taken by Primack & Miao (1992), who set up experimental populations in Massachusetts, USA, by sowing seeds of four annual species, 40–600 m from their existing populations. At 3 places, new populations established and persisted for four generations, while at 7 sites small initial populations established but then died out. At 24 sites no establishment occurred. The authors consider that perhaps the chosen areas were unsuitable in some way, or perhaps not enough seed was sown. Given the very specific habitat requirements of many species, especially for germination and establishment, it should not be assumed that sowing seeds is a more reliable means of transferring species.

Restoration projects: issues and prospects

Given the imperilled state of many species, there is every reason to hope that more manipulations and transfers of endangered species will be attempted. Reviewing past practice what are the issues that must be confronted?

Species restoration projects: what counts as success?

A recent review by an international team of experts, from Australia, Belgium, Estonia, France, Germany, Italy, Norway, South Africa, Spain and the USA, has considered this question in detail by thoroughly searching the scientific literature and examining the responses to a questionnaire (Godefroid *et al.*, 2011a).

In a review of the success of the 249 reintroductions, they discovered that 'survival, flowering and fruiting rates of reintroduced plants are generally quite low (on average 52%, 19% and 16% respectively)'. Furthermore their analysis revealed 'a success rate decline in individual experiments with time'. They conclude that, in judging the success or otherwise of species restoration projects, too much attention has been paid to the immediate survival of plants and not enough to the reproductive success and subsequent recruitment of the next generation (Menges, 2008). Often what constitutes success, and how to measure it, is not considered. Godefroid and associates favour a more stringent test, such as that proposed by Primack & Dayton (1997) who consider that:

> A reintroduction can be considered truly successful only when a population is expanding in numbers and area, when individuals are flowering and fruiting, when a second and third generation of plants are appearing on their own, and the population gives every indication that it will persist into future decades. Further success would involve the population dispersing seeds into the surrounding countryside and producing satellite populations.

Recommendations for better restorations

Maschinski & Haskins (2012), Guerrant (2013) and Godefroid and associates make a number of recommendations for more effective restorations.

Whether these are all practical within the resources available is an important consideration.

- Guided by knowledge of the genetic variability and reproductive biology of the target species, the restoration should be planned as a scientific experiment and the results published (Sutherland *et al.*, 2004). Often details of introductions are referred to briefly only in reserve reports. Progress in this area would be more rapid if full details of introductions and their outcome were published in the scientific literature, so that success and failure can be properly analysed. For example, in the restoration of populations of Wild Grape (*Vitis vinifera* ssp. *silvestris*) along the Rhine Valley, initial failure was fully analysed, and provided the basis for undertaking further reintroductions (Arnold *et al.*, 2005).
- They recommend that restoration projects be carried out in protected sites. However, legislation protecting the wildlife may make experiments difficult. For instance, in Britain, the Wildlife and Countryside Act 1981 makes it illegal to uproot any plants without the owner's permission, and consent is also required to remove or introduce organisms at any designated Site of Special Scientific Interest (SSSI).
- One of the most difficult issues is the choice of 'suitable 'sites for potential transfers.
- If *ex situ* plants/seed are to be used in the restoration (Akeroyd & Wyse Jackson, 1995), then the time the plants are cultivated in the garden should be kept to a minimum to avoid the loss of genetic variability and unconscious selection. Hardwick *et al.* (2011) consider that Botanic Gardens should take a more active part in *in situ* conservation, and recommend that they 'produce hardy, restoration-ready plants rather than the tender nutrient-reliant individuals more commonly used for display or sale'. Gardens should move away from the traditional layout of lawns and flowerbeds towards creating more natural habitats (as for example in the Royal Botanic Garden, Jordan; and the Oman Botanic Garden, Muscat). In addition, active restoration efforts should be highlighted. An excellent example is the Washington Botanic Garden's Union Bay Natural Area, where visitors can see a former landfill site being restored to meadows, woods and wetlands.
- Given the high demographic costs of planting seeds in reintroduction projects (Maschinski, Baggs & Sacchi, 2004), conservationists should consider setting out young plants (as for instance in the conservation of *Dianthus morisianus* in Sardinia (Cogoni *et al.*, 2013)). Also, adequate initial management should be carried out to avoid immediate losses through desiccation and/or herbivory.
- They also recommend moving from tentative interventions to the reintroduction of larger numbers of individuals, guided by the recommendations of geneticists as to the size of minimum viable populations (MVPs) and population viability analysis (PVA).
- Using material from diverse populations – mix rather than match (Fant *et al.*, 2013).
- Making adequate site preparations.
- Checking the genetic fidelity of the restoration process to make sure that there is no genetic decline in the collection and propagation of seeds/plants and their translocated populations. This recommendation follows the findings in a number of genetic studies, e.g. the endangered Australian species, *Grevillea scapigera* (Krauss, Dixon & Dixon, 2002). This is one of the rarest plants in the world, with only 5 individuals known in the wild. Clonal propagation of 10 *ex situ* plants was followed by translocation to secure sites. Seedlings were also transplanted. In a molecular study (AFLP), Krauss and colleagues discovered that '8 clones, not 10, were present in the translocated population, 54% of all plants were of a single clone, and the F_1s were on average 22% more inbred and 20% less heterozygous than their parents, largely because 85% of all seed were the product of only 4 clones.'
- Carrying out long-term management and monitoring.

- As the long fate of many reintroductions often goes unrecorded, Drayton & Primack (2012) emphasise the importance of recording failures, to help with the design of better reintroductions. Their views were influenced by their attempts to introduce 8 perennial species into two reserves near Boston, MA, in 1994–5. Fifteen years later almost all of them had disappeared.

Having set out the case for well-designed experiments on habitat management and reintroductions, we should acknowledge the difficulties faced by those trying to carry out practical conservation. Very often, *ad hoc* decisions have to be made in situations of crisis and there is neither the time nor the resources to carry out long-term research. Action has to be taken quickly on the basis of the best available information. Sometimes, however, long-term funding for conservation does become available and then, we suggest, it is imperative to devise strategies to maximise the possibility of long-term success through well-thought-out experiments, the results of which are made available in the scientific literature (Soorae, 2011). Another issue also needs to be addressed. Restoration is sometimes seen as requiring a 'one-off intervention' (Hobbs & Cramer, 2008). While this approach may be useful in some cases, there is growing evidence that restoration projects need on-going monitoring, and in many (perhaps most) cases further interventions over the years.

Aims and objectives of conservation: looking back and considering the future

Given the challenges posed by climate change, population increase and sustainable development, many are now questioning the concepts that have dominated conservation management. Essentially, current approaches look to the past: the aim of management being to recreate historic ecosystems.

The first of the historic models is 'rewilding'. Championed by ecologists in both Europe (www.rewildingeurope.com) and North America, the underlying aim of conservation is to restore protected landscapes to their natural 'wilderness' state by minimising or withdrawing management (Navarro & Pereira, 2012). For instance, Aronson *et al.* (1993), citing the Society of Ecological Restoration, define restoration as 'intentional alterations of a site to establish a defined indigenous, historic ecosystem'. The essential message is to manage to 'let nature take its course'.

This very simple, beguiling prescription raises a number of fundamental issues (Choi, 2004). Considering any tract of land in the Northern Hemisphere in North America or Europe, we have abundant and increasing evidence, from many sources, that in the post-glacial period different ecosystems replaced each other as vegetation expanded northwards with the retreat of the ice. Therefore, we should ask the question: if we wish to 'restore' a particular ecosystem, which should we select as the model for the restoration at a particular site? (Jackson & Hobbs, 2009). We must also face the painful fact that the faithful recreation of wilderness is likely to be impossible as some of the species that played major roles in the past have become extinct in the area or even worldwide, e.g. top predators, passenger pigeon, moas, mastadons etc. (Jackson & Hobbs, 2009). Also, many invasive, pest and disease organisms are now widespread in protected lands, and, while their control is sometimes a possibility, their entire elimination is, in many cases, likely to be a practical impossibility (e.g. the total elimination of introduced fully naturalised species in New Zealand (30 mammals, 34 birds, 2000 invertebrates and 2200 plant introductions) is unachievable and sets limits to ecosystem restoration (Norton, 2009)).

As we have seen, the other influential concept underlying conservation is 'resort to precedent', which involves the reimposition of the appropriate traditional methods of land management (Hall, 2000; Briggs, 2009). Thus, the cultural practices of the past are reinstated, e.g. woodland management (coppicing, pollarding etc.) and grassland management (grazing regimes, fencing, mowing etc.). In the past, the aim of land management was to exploit the natural resources of the ecosystems to the fullest extent and achieve the highest yields in arable production, animal husbandry, forestry etc. Significant elements of wildlife were ruthlessly suppressed as weeds, vermin or pests. However, some 'wild' species survived these selection pressures, but often with diminished populations. Broadly, these are the 'remaining wildlife species' – birds, plants, insects – we seek to conserve today.

To conserve these species a change in attitude is needed. The success of management of, say, a grassland should not now be judged entirely by the number of sheep that it supports, but also how much wildlife it sustains, and in some countries subsidies are paid to those who manage their lands in agreed programmes of wildlife management. In this context, some propose that it is helpful to see wildlife as a specialised crop, conserved in concert with agricultural production (Mabey, 1980).

However, there are some serious issues to face. Atmospheric and environmental pollution has had major long-term effects on soils and water. Moreover, it is not always economic or practical for farmers and land managers to carry out some of this historic land management, and, where volunteers undertake the work, health and safety issues become important. In addition, in some countries, certain cultural practices of the past are less acceptable and sometimes outlawed, for instance, foxhunting with dogs in the UK. Also, land use is now very different from that of the nineteenth century: conservationists have to take account of mass-tourism, the behaviour of the general public when walking dogs, and the emergence of new sports such as skiing, off-road biking, surfing etc.

Turning to another issue, we might ask, is the site management for a particular reserve, presently based on the principles of 'resort to precedent', to be followed indefinitely? Is it envisaged that, centuries from now, coppicing of English woodlands will still be faithfully carried out, as in the historic past, and once reinstated, will such management be continued in perpetuity?

Assisted migration

Many conservationists are asking searching questions about the configuration of conservation management in the future, especially in the light of climate change, population increase and the increasing human use of natural resources (Schwartz *et al.*, 2012). In the future, both the geographical location of the 'climate envelope' and the ecosystem in which a species presently occurs are predicted to be radically altered by climate change. Thus, sites in which a species now thrives may not provide suitable habitat in the longer term.

This has profound implications for protected areas that over time may lose their characteristic fauna and flora. Will areas where all the charismatic biota have disappeared elsewhere or become extinct still be maintained, managed and cherished for whatever organisms they might come to contain (McDonald & Boucher, 2011)? Modelling the future distribution of the celebrated *Protea* species in South Africa has shown that '10% of species will be lost from reserves under 2050 climate change scenarios', but if they are able to migrate successfully, then some will 'arrive' at existing reserves from elsewhere (Hannah, 2011).

This example raises the important question: will species be able to migrate to the appropriate areas in the face of climate change? It has been estimated migration rates of 1000 m a^{-1} will be necessary to keep pace with human-induced climate warming over the coming century (Malcolm *et al.*, 2002).

The case of Brewer Spruce (*Picea breweriana*) demonstrates some general issues about the potential effect of climate change and the complexities now faced by conservationists who seek to conserve rare and endangered species (see Ledig, Rehfeldt & Jaquish, 2012, from which this account is drawn).

Brewer Spruce is endemic to the Klamath region of California and Oregon. Molecular studies reveal that it has moderate levels of genetic diversity, and, if present conditions were to persist, it is believed that with some management this outbreeding species could be expected to survive *in situ*.

However, projected climate change alters the situation entirely. The future distribution of the species – in 2030, 2060 and 2090 – has been predicted using three general circulation models, two projections for climate change and two sets of carbon emission scenarios (optimistic and pessimistic) for future climates. This modelling revealed that the species was likely to disappear from its present habitats by 2090: the 'climate niche' of *Picea breweriana* is 'projected to move north to British Columbia, the Yukon Territory and southeastern Alaska'. Ledig and colleagues consider that the prospect of Brewer Spruce adapting to new climatic conditions is 'not promising'. Many other forest tree species are in a similar position.

Could species adapt to the new conditions? Studies of *Pinus sylvestris* suggest such adaptation to climate change might take many years, perhaps up to 13 generations (see Rehfeld *et al.*, 2002). In the case of Brewer Spruce, a long-lived tree, such adaptation might be possible in a century or more, but significant climate change is predicted to occur in decades. Other iconic species are also threatened in western USA. The 'climatic niche' of Giant Sequoia (*Sequoiadendron giganteum*) is mostly moving upwards in the Sierra Nevada, while Port Orford Cedar (*Chamaecyparis lawsoniana*, more generally called the Lawson Cypress) is also 'shifting northwards along the coast'. Present evidence suggests that 'suitable habitat for both species is projected to largely disappear by the end of the century'. An additional uncertainty concerns the effects of 'extreme climatic events' predicted by studies of climate change.

In the case of Brewer Spruce, if no action is taken, extinction looks very likely. In the future, suitable habitat is predicted to be many kilometres north of its present distribution. It is unclear that the species will be able to disperse naturally to fill this new climatic niche, and assisted migration would seem to be the appropriate management. Given the probable speed of climate change, 'delay or inaction is not an option', if the future of the Brewer Spruce and many other species is to be made secure (Ledig, Rehfeld & Jaquish, 2012). The scale of the northward movement necessary is emphasised by the studies of McKenney *et al.* (2007, 2011), whose investigations were based on three climate models. They discovered that northwards movement for 130 tree species was of the order of 700 km by the end of the century.

While some authorities question the ability of tree species to migrate to new territories under climate change, for others this issue is still an open question. Kremer *et al.* (2012) consider that many tree species have 'evolved dispersal syndromes enabling the effective flow of genetic information across distant populations inhabiting contrasting environments'. However, they stress that we still do not have 'adequate tools to track long-distance pollen and seed dispersal'. Perhaps, given our present lack of knowledge about this crucial issue, assisted migration could be considered a precautionary measure.

Moving from the specific case of Brewer Spruce to more general issues, there are many concerns about practising assisted migration as part of species conservation.

The ethics of assisted migration has been debated (Sandler, 2009; Schwartz *et al.*, 2012). Some consider it a risky unpredictable intervention, one that introduces new players into existing ecosystems. New species assemblages will be produced to give communities with no previous analogues (Seddon, 2010). This raises some complex questions of how such ecosystems would be managed.

Ricciari & Simberloff (2008) recognise the good intentions of those who support assisted migration, but consider that it is not a viable strategy, risk assessments of such actions being fraught with difficulty and uncertainties. A major concern is that species assisted in migration could become invasive, detrimental, or disrupt ecosystems, especially if whole suites of plants and animals are involved (see Hoegh-Guldberg, 2008). All things considered,

Ricciari & Simberloff wonder whether species translocations are essentially 'ecological gambling' with 'risks that cannot be reliably estimated or anticipated'. Is the game essentially 'ecological roulette'? They cite the study of Mueller & Hellman (2008) who point to the serious negative effects of the introduction of some organisms for biological control. For many such introductions there have been positive effects, but for others unforeseen consequences are reported. For example, a seed-feeding weevil, introduced from Europe to control introduced *Carduus* species in the USA, caused an 80% reduction in seed production in a native thistle (*Carduus canescens*) in Nebraska.

At first sight, it looks paradoxical that a species threatened or endangered in its native range could become invasive when introduced elsewhere. However, some cases have been reported. For example, *Melaleuca quinquenervia* was introduced into the Everglades, Florida, where it has become a very serious invader. In Australia, in its native habitat, the species has been given the conservation status of 'threatened', and is receiving conservation management.

Another example is provided by the Monterey Pine (*Pinus radiata*), which is native to the coastal areas of California. Here, it is now occurs in only three wild stands and there is great concern for its survival. Elsewhere, the species is very widely grown, as a plantation crop, in many parts of the world (Australia, Chile, New Zealand, South Africa, Spain etc.); indeed, it has also become a very serious invasive species in Australia, South Africa and elsewhere.

Some concern has also been expressed about the 'certainty' of projections about climate change. As we saw in the previous chapter, climate modelling, on which predictions are made, provides projections, not certainties, about the future. They are best guesses at climate changes that cannot be obtained any other way. As more research and investigations are carried out the scientific consensus is that the footprints of climate change are already around us, and it would be foolish not to take the issue very seriously as one of the most important of our times.

Turning now to more specific questions, could species migrate naturally to new areas under climate change? While some ponder whether such assisted migration is necessary, others emphasise our general ignorance of natural dispersal and establishment, and wonder how many species will be able to migrate naturally in the face of climate change (Hampe, 2011).

Does evidence from fossil pollen in dated deposits provide information on the likely speed of natural migration in the past? 'The supposed rapid post-glacial migration of trees is often cited as reason for optimism' (Pearson, 2006), as rates of post-glacial migration of the order of 100–1000 m a^{-1} have been found. These calculations are based on the model of migration from a supposed northerly limit of the forest zone at some distance south of the ice (McLachlin, Hellman & Schwartz, 2007). However, this model of migration from a distant forest zone may be flawed (Pearson, 2006), as molecular studies of cpDNA suggest that low-density refugial populations of some tree species occurred within 500 m of the ice sheets, and later migration in the post-glacial period is likely to have come not only from distant refuges but also from those close by.

More importantly, the past may not be a good guide to the future. The potential for some plants to migrate in the present day is significantly different from that of the post-glacial period. Natural gene flow of many species may be insufficient to enable them to cross, unaided, formidable barriers posed by agricultural and urban/industrial land. In some cases, wild grazing animals might be the natural dispersers of seeds/fruits. This role may not be satisfactorily fulfilled by movements of domestic stock (Auffret, 2011).

Plants obviously differ in their capacity for effective migration in the modern landscapes. Short-lived species that colonise disturbed ground may be able to migrate, or be taken unwittingly, to 'new territory', as disturbance often accompanies human activities. However, migration by forest trees is less predictable, as the sites to which they might move are often already occupied by long-lived tree species, many in managed communities. Some opportunities may present themselves when disturbance events occur, but these may be part of human management, and the opportunity for colonisation may be restricted or denied.

Turning to more practical issues, as we have seen above, reintroductions within the historic ranges of plants have had an uncertain level of success (Godefroid *et al.*, 2011a). Assisted migrations may be even more difficult to achieve.

One area of complexity will be encountered: suitable sites for assisted migration of some species may be in an adjacent country. There are often constraints on the movement of plants across political boundaries. In addition, the management of plants transferred in assisted migration would have to be handed over to conservation authorities elsewhere. Returning to the case study of Brewer Spruce, trans-border cooperation with the Canadian authorities is being arranged. Discussions are in progress to set up two 'highly structured and managed trial plantations (less than 2 hectares in size)' under the control of the British Columbia Ministry of Forests, Lands and Natural Operations in Canada. Plantings of the Spruce will aim to capture the genetic diversity present in its existing populations. This model has been called quasi *in-situ* conservation (Volis & Blecher, 2010). It is a halfway house or hybrid between *ex situ* and *in situ* approaches. It has been widely employed in the conservation of animal populations (Redford, Jensen & Breheny, 2012). For example, managed wild populations of butterflies, turtles etc. have been supplemented with captive-raised young.

In this hybrid approach to conservation of plants, it is proposed that *in situ* sites are fenced, and protected, and some continuing management is provided, e.g. watered and weeded. Given the serious objections that some raise to the introduction of new species directly into supposed natural or semi-natural vegetation, there may be merit in locating assisted migrants in specially developed sites in or adjacent to National Parks or nature reserves.

But, currently, some National Parks and reserves do not permit the introduction of new taxa into their territories. If assisted migration becomes an accepted conservation practice, changes in legislation and management practice will need to be confronted (for example, the legal framework of legislation associated with the US Endangered Species Act; Shirley & Lamberti, 2010).

Turning to another aspect of climate change, projected sea level rise presents very complex issues. For example, the habitats of several endemic plant species on low-elevation islands in the Florida Keys are at risk of inundation (Maschinski *et al.*, 2011). Currently, there is suitable habitat on other islands in the Keys and on the Florida mainland, but rising sea levels also put these areas at risk. Indeed, sea level rise 'threatens to engulf the lower half of that State of Florida' (Overpeck & Weiss, 2009). Maschinski *et al.* (2011) consider that suitable areas for assisted migration must be at higher elevation, perhaps in the Caribbean. This would require agreement between countries.

Major dilemmas with past-orientated conservation models

We have discussed the emergence of restoration ecology and creative conservation, principally in relation to the future of endangered species. It is important to examine critically the models on which these endeavours are based. As we have seen, 'rewilding' and 'resort to precedent' are the two key concepts. It is important to stress that both look to the past. These concepts are being critically re-evaluated, and restoration towards the historic goals is being questioned (Jackson & Hobbs, 2009). For example, Choi (2007) writes, 'we need to admit our inability to restore an ecosystem to its very original state'. Nonetheless, he argues, 'ecological restoration is essential for mutual survival of both human and nature (Cairns 2002)'. It has a major part to play in our attempt to 'manage, conserve and repair (or enhance) the world's ecosystems'. He continues: 'although it is undoubtedly valid to learn from the past (Choi, 2004), our restoration efforts for the future need not be constrained by "historical-fidelity" (Throop, 2004)'. He argues that 'future-orientated restoration should focus on ecosystem functions rather than recomposition of species or the cosmetics of landscape surface ... our paradigm of ecological restoration needs to be redefined with functional rehabilitations for the future, not nostalgic recompositions of the past'.

Setting priorities in conservation

It is clear that many species are facing an uncertain future. The response of many conservationists is to demand that the future of all threatened species be secured. Others believe that, as resources for conservation are limited, some priorities will have to be established (Mace, Possingham & Leader-Williams, 2007). In the past, the establishment of priority has been based on a critical assessment of the degree of endangerment in particular cases. However, in making such judgements, it is clear that our knowledge of the distribution and population size for many plant species is seriously incomplete. In regions where the flora is well known, there has been an active debate about how to determine priority.

One salient point is that there are different kinds of rarity (Rabinowitz, Cairns & Dillon, 1986). In their study of the British Flora, three factors are considered: geographic range (broad or narrow); habitat specificity (wide or specialist); and local population (large or small). For example, considering the species selected from the British Flora by Rabinowitz and associates, Scottish Bird's Eye Primrose (*Primula scotica*), an endemic plant in the north of Scotland, has a narrow geographic range, a broad habitat specificity and somewhere in its range populations are large. In contrast, *Lloydia serotina*, known as the Snowdon Lily in Britain, has a narrow geographic range, restricted habitat specificity and everywhere its populations are small. Thus, the *Primula* is restricted in one dimension (geographic range), while *Lloydia* is restricted in all three dimensions.

The logic of Rabinowitz and associates in this paper is that we should pay more attention to the Snowdon Lily than the Primrose. We should note, however, that *Lloydia* has a wide circumpolar distribution, while *Primula scotica* is endemic to Britain.

Conservationists, almost without exception, consider that endemic taxa, which occur only in a particular localised area, should be given priority. Not only species, but also genera and even families, may be narrowly endemic. They are often represented by small endangered populations in the wild. On the other hand, they may be common in the limited areas where they occur. Major (1988) notes that high endemism is found on oceanic islands, roughly in proportion to their degree of isolation. In montane areas, the proportion of endemism increases with altitude. Tropical areas have the highest number of endemic species, genera and even families. Edaphic endemism also occurs where there are outcrops of specialised rocks, for example gabbro, serpentine, or heavy metal rich areas. A case has also been made for conserving not only the endemics developed as a result of long periods of speciation in geographic isolation but also the newly formed polyploid endemic species we discussed in Chapter 14, namely *Tragopogon mirus* and *T. miscellus* in western USA, and *Senecio cambrensis* in Britain. Holsinger & Gottlieb (1991) add to the list of priorities taxonomically distinctive groups, for example insectivorous plants. The relatives of agricultural and horticultural plants are also priority candidates for conservation. They cite as an example *Oryctes nevadensis*, small populations of which are found in a few localities in California and Nevada. This is the sole species of the genus (the so-called monotypic condition) and, as a member of the Solanaceae, the plant is a highly distinctive, distant relative of the Potato, Pepper, Tomato, Aubergine, Tobacco, etc. However, as *Orcytes* is a small, insignificant species, its conservation has been opposed. Holsinger & Gottlieb (1991) also suggest that phylogenetic diagrams might be helpful in deciding priorities. In their view, it would be more important to conserve species or genera on distant branches of phylogenetic trees than to select only closely related taxa on adjacent branches.

In assigning priority in the conservation of rare and endangered species, some botanists have assessed the present status and future prospects of populations, and have evoked the concept of triage. In wartime, those casualties most likely to survive have often been selected for medical treatment. Thus, in conservation terms it might be possible to classify endangered plants into three groups:

1) 'Stretcher cases' that are unlikely to survive in the wild despite intervention by conservationists. These species could be conserved in gardens.
2) Those whose prospects of survival in the wild are such that remedial action is not immediately required.
3) Those whose populations are likely to thrive if remedial action is taken.

There are, however, many conservationists who reject triage, considering that we should try to conserve all endangered species. The debate about triage will continue.

Creative conservation: economic and political considerations

In trying to make the strongest case for conservation, many have discussed utilitarian, ethical and aesthetic considerations. Out of these ideas has grown a major development, with the emphasis on the goods and services that ecosystems provide for humanity. McCauley (2006) writes:

> Probably the most important trend in conservation science at the moment is 'ecosystem services', typically seen as economic benefits provided by natural ecosystems ... The underlying assumption is that if scientists can identify ecosystem services, quantify their 'economic value', [then conservation will be brought] more in synchrony with market ideologies, [and] then the decision makers will recognize the folly of environmental destruction and work to safeguard nature.

A large number of ecosystem services have been identified. Constanza *et al.* (1997) group these into 17 major categories, including climate regulation, food supply, raw material, water supply and regulation, erosion control and sediment retention, soil formation, nutrient cycling, waste breakdown, pollination, biological control, refugia for species important for humanity such as fish etc. Ecosystems also provide recreational opportunities, as well as being cherished for their aesthetic, artistic and spiritual value. Constanza and colleagues have collected together estimates (and added new ones) of the value of these services and concluded that at a minimum 'for the entire biosphere, the value (most of which is outside the market) is estimated to be in the range US$ 16–54 trillion (10^{12}) per year, with an average of US$33 trillion per year'.

Faced with these sorts of figures, many ecologists have been reviewing their aims and objectives in conservation and restoration (Turner & Daily, 2008) and ask the question should the goals of conservation be reassessed? In the past, recommendations about priority in conservation of species and habitats have often been made with biodiversity as the crucial factor. Typically, conservationists, convinced of the merits of their case, have called for political and social support in search for appropriate funding to conserve endangered species and threatened ecosystems. As Naidoo *et al.* (2008) see it: 'in trying to stem biodiversity losses, ecologists and conservation biologists have focused on how conservation plans affect biological targets ... However, conservation plans cannot be implemented for free. By ignoring the cost side of conservation planning, ecologists and conservation biologists are missing great opportunities to achieve more efficiently conservation objectives in a world of limited conservation resources'.

A new field of 'conservation economics' is developing, in which costs are carefully identified and estimated (Balmford & Whitten, 2003; Palmer & Filoso, 2009). Often these approaches employ cost–benefit approaches.

What follows is drawn from Naidoo *et al.* (2008). They identify:

(a) 'Acquisition costs in securing 'property rights to a parcel of land' to be used for conservation';
(b) Management costs of staffing, establishing, and maintaining, through practical intervention, the ecosystems in a network of protected areas;
(c) Transactional costs associated with transfer of property rights;
(d) Damage costs associated with economic activities arising from the conservation programmes, such as damage to humans, crops and livestock by wild animals in areas adjacent to protected areas;
(e) Opportunity costs are also to be quantified, to secure 'a measure of what could have been gained via the next-best use of a resource had it not been put to its current [conservation] use'.

Looking at conservation from this perspective has profound implications. Naidoo and colleagues conclude: 'ecologists might be reluctant to let factors other than biology dictate their conservation priorities. Nevertheless, as soon as priority setting leaves the ivory tower, a host of real-world concerns, including the costs of conservation actions, must be considered.' Thus, increasingly, many see restoration and conservation primarily as a means of providing and enhancing vital goods and services for humanity. Anthropogenic concerns might trump biodiversity considerations (see May, 2011).

This move towards cost–benefit approaches is not without its critics.

In an impassioned plea, McCauley (2006) sees great dangers in what he views as the commodification of nature, and objects to 'approximating its monetary value'. He stresses that

> 'nature has an intrinsic value that makes it priceless, and this is reason enough to protect it . . . we view certain historical artefacts and pieces of art as priceless. Nature embodies the same kind of values we cherish in these man-made media . . . we must make it abundantly clear that our overall mission is to protect nature, not to make it turn a profit . . . nature conservation must be framed as a moral issue and argued as such by policy makers . . . we will make more progress in the long run by appealing to people's hearts rather than to their wallets.

However, some take a very positive position on the conservation of environmental services, seeing in this approach the key to safeguarding biodiversity. For instance, Mahli *et al.* (2008) examine ways of conserving the Amazon tropical rainforests. They note that 'the forest biome of Amazonia is one of the Earth's greatest biological treasures', containing perhaps 25% of all terrestrial species, and is facing the combined threats of deforestation and climate change through droughts and fires. In their view, perhaps the best way to protect Amazonia is to recognise the key role this vast area plays in mitigating climate change by sequestrating carbon, and, for the international community to provide the resources to protect this vital environmental service, and thereby secure the future of the biodiversity of the region.

Concluding remarks

In the face of huge challenges, conservationists can rightly claim credit for halting the extinction of some threatened and endangered species and attempting, and sometimes succeeding in their efforts, to protect significant ecosystems. In addition, we have seen that, through major ecological interventions of various kinds, restoration of very large areas of degraded, damaged ecosystems has been achieved. Also, entirely new ecosystems have been created to provide habitat for wildlife. While some see this as gardening on a large scale (Choi, 2004), others consider that 'creative conservation' has a key role to play in the encouragement of biodiversity. The successes of ecosystem management and restoration deserve to be widely acknowledged and celebrated, as, in many countries, degraded and damaged lands have been restored to produce cherished, valued and aesthetically pleasing landscapes once again, populated with significant wildlife.

While much progress has been made, much has still to be learned about species introductions, restoration and habitat creation. For example, Harris *et al.* (2009) highlight a major concern: we have as yet insufficient knowledge of the microbial elements of ecosystems (mycorrhizae, saprophytic, parasitic and disease organisms etc.) to be certain that accelerated restoration is feasible. Should we attempt to manipulate microbial communities so that they act to facilitate restoration, or should we take the view that microbial communities in the soil will naturally follow our manipulation of the plant communities by management and/or by selected new planting?

Furthermore, restoration ecology is a comparatively recent branch of the subject, and it remains to be seen how far ecosystems can be successfully restored in the long term, and whether complex communities, such as forests, can be quickly created on open land. This question is not merely of academic interest, for it has been proposed that in some circumstances ecosystems containing significant biodiversity should be surrendered to developers, with the concession that an equivalent area be created for wildlife, in so-called mitigation or off-setting agreements. Conservationists are rightly critical of such arrangements. They are not convinced that, for example, an ancient woodland in Cambridgeshire, with its long-documented history of management and rich diversity of characteristic species, could ever be recreated elsewhere. A visually agreeable planting scheme might be the best that could be achieved in off-setting schemes. Despite the concerns of conservationists, such schemes are increasingly being proposed and in some cases implemented.

Facing the many complex issues that land use change and ecosystem restoration provokes, Cairns, one of the early pioneers of restoration biology, perhaps reflected political and practical realities when he wrote: 'for both scientific and economic reasons, human society must rest satisfied with a *naturalistic assemblage* of plants and animals rather than a precise replication of the species that once inhabited the area' (my emphasis) (Cairns, 1998).

The future of restoration projects Progress in the species management and restoration of protected lands, and advances in *ex situ* conservation, must, however, be seen against the background of extreme uncertainty about the future. Climate change is an additional factor interacting with the major 'drivers of extinction' that were identified by pioneer conservationists (over-exploitation, loss and fragmentation of habitat, introduced species, climate change and pollution) (Mace & Purvis, 2008). In many cases these malign influences have not been reversed; indeed in many parts of the world, ecosystem change is accelerating (Millennium Ecosystem Assessment, 2005; www.unep.org/). Thus, narrowly distributed endemics are particularly at risk, especially those that live on oceanic islands. Hot spots of biodiversity are found in different parts of the world, e.g. tropical forests, but these are often located where present-day habitat change is greatest (Balmford & Long, 1994). Given the very widespread introduction of species across the globe by human activities, Lockwood & McKinney (2001) see a widespread homogenisation of floras, as suites of widespread species (some invasive alien taxa) come to dominate many ecosystems.

Looking at present evidence and future projections of climate change (IPCC, 2014), conservationists worldwide are coming to realise the looming threat to biodiversity and are wrestling with a key question – how to measure progress in reacting to such major changes (Mace, 2005). This is a challenge in uncharted waters, and adaptation is made more difficult as some still deny that anthropogenic climate change is occurring, despite the weight of scientific opinion.

Climate change raises crucial questions about the role of traditional approaches to nature conservation. Historically, the major response has been to set up reserves in areas where there is significant wildlife; these areas are then managed in an attempt to nullify or remove the threats that are endangering species. But now we must face an emerging reality. The notion that a system of geographically fixed reserves can secure the future of biodiversity is seriously undermined by the prospects of climate change, during which species must migrate, or adapt *in situ* or become extinct. Clearly, if effective international measures to reduce greenhouse gas emissions are taken in the coming decades, the existing protected lands might retain some of their charismatic animal and plant species. Regrettably, the journey towards international agreements and sustained action to reduce greenhouse gas emissions is uncertain, and we must face the reality of the potential limitation of the traditional reserves to sustain their charismatic species. However, they may have a continuing role to play, in conserving populations of once common species that are now in decline.

Adaptation of plant species in response to climate change Considering the evidence for *in situ* adaptation species in response to climate change, a number of case studies suggest this is possible and likely (see above), but there is doubt about the ability of declining and endangered species to respond adaptively to changing circumstances. With regard to their future survival, there is considerable debate as to whether many threatened species contain the necessary genetic variability to survive and adapt.

Of major concern is the fact that conservationists have not routinely incorporated evolutionary and genetic insights into planning for species conservation. For instance, Mace & Purvis (2008) note that conservation policies do not aim to maintain evolutionary processes: they focus instead on monitoring species numbers, their geographical distributions and the ecosystems that contain their niche. There is no regular investigation to measure and, if necessary, enlarge the genetic variability of populations with a view to ensuring their future adaptability and long-term capacity for evolution. The molecular tools now exist to carry out genetic monitoring, but the costs are likely to be significant, if routine monitoring is employed in management of all endangered species.

Migration: natural and assisted Reflecting on the subject of species migrations, studies of the post-glacial period show that many species moved out from refugia into lands released from the ice. Other species became extinct. Regarding the potential for migration in present times, it is important to emphasise, once again, that in the present day a maze of human-modified areas (from agricultural lands and forestry, to cities and industrial land) would have to be traversed by 'naturally' migrating plant species.

Considering whether such migration could be encouraged between protected areas – reserves and National Parks – it is not at all clear that corridors would prove effective migration routes allowing species to track their climatic envelopes. While appropriate corridors might be theoretically possible in some parts of the world, existing and projected land use does not permit the setting-up of corridors in the most populated areas. Indeed, conservationists have to appreciate that the land promoted by ecologists as a 'suitable wildlife corridor' is almost certain to be contested space: for others it is their traditional grazing grounds, or, in the developed world, a potential new housing development, golf course, plantation or theme park etc.

If climate change proceeds as models suggest, we must also face the reality that conservation concerns are likely to be subordinate to other pressing issues, as human societies grapple with a multitude of issues relating to human population growth, food supplies, water and energy, and sustainable resource use. Also, the effects of increasing climate change are predicted to lead to more frequent severe weather events, such as droughts, fires, storms, floods etc. Human adaptation to climate change is likely to be very complex and costly, especially if reductions in greenhouse gas emissions are not achieved in the next decade. Major reappraisal of human behaviour is needed. For instance, reviewing our present reliance on fossil fuels, Berners-Lee & Clark (2013) make the very strong case that, if we are to have any chance of reducing carbon emissions, we should not burn even half the present known reserves of oil, coal and gas.

Given the projected complexities of climate change, it is an open question whether conservation concerns, for example, assisted migration of species on a large and effective scale, would be given priority by the general public. Rather than being focused on the future of the Brewer Spruce, humanity will very likely be struggling with all the many diverse issues around human adaptation to climate disruption and other interacting environmental issues, such as concerns about food supplies, availability of water and energy, rising sea levels (threatening low-lying islands and major cities located near the coast), human migration from environmentally stressed areas, and the greater incidence of extreme weather conditions (droughts, flooding, fires) to name but a few (Gore, 2013).

For those who might contemplate the possibility of action to conserve ecosystems, would the price be too high? There are very few estimates of the likely cost, but some have been made. After predicting the loss of ranges of 74 plant species endemic to the forests of Madagascar under climate change (2000–80), Busch *et al.* (2012) concluded that there was likely to be considerable loss in species ranges. To fulfil the continuing objectives of conservation of these species, forest restoration comes at a high price. They calculated that:

> climate change increased the cost of achieving the conservation target by necessitating successively more costly management actions: additional management within existing protected areas (US$0–60/ha); avoidance of forest degradation (i.e., loss of biomass) in community-managed areas ($160–576/ha); avoidance of deforestation in unprotected areas ($252–1069/ha); and establishment of forest on non-forested land within protected areas ($802–2710/ha), in community-managed areas ($962–3226/ha), and in unprotected areas ($1054–3719/ha).

Busch and colleagues 'suggest that although forest restoration may be required for the conservation of some species as climate changes, it is more cost-effective to maintain existing forest wherever possible'.

Human influences and microevolutionary change
In earlier chapters we have presented convincing evidence for the action of natural selection in plant populations, in relation to climatic, edaphic and biotic influences. In addition, we have demonstrated how human activities impose selective forces on plants, e.g. evolution of different grassland ecotypes, heavy-metal- and pollution-tolerant plants and the emergence of pesticide-resistant genotypes. We have also shown how human activities have broken down geographical barriers between plants, leading to the evolution of invasive introduced taxa, hybridisation between native and related introduced species, and the evolution in some case of new polyploidy species. In addition, human activities have resulted in the breakdown of natural ecological barriers between species. In some groups of plants this has led to complex hybrid swarms and introgressive hybridisation.

Overall, niche construction in humans has led to 'unprecedented biotic mixing' (Mace & Purvis, 2008). Species that evolved in isolation have been brought together to produce new biotic associations, and this process is set to continue as global trade results in the transfer of plants, animal pests and diseases across the world. The adoption as a conservation strategy of assisted migration of endangered and other species could also contribute to the movement of species from place to place, resulting in the development of plant communities without historic counterparts. Furthermore, some ecologists, recognising the importance of the goods and services provided by the biosphere, have suggested that specially designed communities, with no current analogues, could be developed to sustain and protect these resources.

Evolution: human influences In drawing this chapter to a close, it is thought-provoking to consider the overall impact of human activities on the evolution of plants and animals (Briggs, 2009).

Some see humans as separate from the natural world, whose role is to employ the Earth's resources for their own use. Reflecting the insights of modern environmentalism, others see humans as having a stewardship role over the resources of the biosphere with the aim of sustainable use. In contrast, evolutionists see humans emerging, as an animal amongst animals, into a world shaped by Darwinian evolution. In the course of history, human civilisations have sought, and to some degree successfully, to modify and nullify the raw Darwinian processes that gave rise to us. Thus, birth control, medical and social interventions to protect the poor, the hungry, the sick and the elderly etc. are all accepted means of intervention that we rightly follow.

What has been the impact of human activities on plant life? As we have seen in the chapters of this book, there is growing evidence that 'new anthropogenic selection pressures' have been and are being imposed on the natural world, with some species emerging as winners or potential winners (biotic ecotypes, invasive plants, weeds, feral, cultivated and ornamental species), while others are potential losers –the declining, threatened and endangered species (Briggs, 2009). There is also evidence that the status of species as potential winners is not necessarily assured. The selection pressures imposed by human activities may change, sometimes gradually, often dramatically. Thus, many weed species, well adapted to former traditional agricultural practices, are now rendered rare or extinct through advances in seed-cleaning techniques and the widespread use of herbicides (Briggs, 2009).

In many ways conservation efforts seek, by management interventions, to subvert the natural and anthropogenic processes that lead to extinction. However, it has to be recognised that, in almost every ecosystem, including those selected and managed as reserves, humanity makes many decisions that impact on the evolutionary process. For instance, the fate of wild species is determined by decisions made on land and water resources, i.e. which areas are to be used in agriculture, transport facilities, city developments and industries. Generally, it is only areas of infertile soils that are assigned to reserves and National Parks etc. Therefore, it has to be recognised that the fate of species is dictated by how we choose to use the goods and services in ecosystems management.

There is another very important consideration that deserves more investigation: repeated and long-continued management, be it in agricultural lands or nature reserves, has microevolutionary consequences. Regular and sustained management exercised on some plant species has resulted in the evolution of our cultivated and garden plants thorugh processes of domestication. This process, which has deliberate and unconscious elements, has led to populations of plants that have been modified genetically from wild stocks, and to a high degree domesticates can survive and grow only under human management. Our conservation strategies involve the cultivation of many species *ex situ* in Botanic Gardens or the dedicated management of plants *in situ*.
If endangered species are to survive, these strategies, which are likely to involve more and more intensive management in the future, may see initial stocks of 'wild' plants evolve through many generations of care and unconscious selection along a track towards domestication.

Human activities: the long-term effects on the course of evolution Finally, we must face a major question that has occupied the minds of many evolutionists: what are the long-term consequences of human activities?

Threats to present-day species represent the current phase of a very long evolutionary drama, in which 'species extinction rates' may 'be 100–1000 times higher' than those indicated by the fossil record (Mace & Purvis, 2008). Thus, contemplating the 'Tree of Life' emerging from molecular phylogenetic studies, it is clear that, as a result of climate change, there is the potential for many plant species, evolved over a long geological history, to be lost. Indeed, as related species often have a similar ecology, it is possible that entire branches of the phylogenetic tree (whole clades) could become extinct. What we know of plant evolution suggests that 'new species' are unlikely to evolve in great numbers to take the place of those becoming extinct.

Considering the present crisis, Barnosky *et al.* (2011) confront the important question: Has the Earth's sixth mass extinction already arrived? In their analysis, they review the complexities of arriving at a simple answer. First they note that 'palaeontologists characterize mass extinctions as times when the Earth loses more than three-quarters of its species in a geologically short interval, as has happened only five times in the past 540 million years or so. Biologists now suggest that a sixth mass extinction may be under way.' Analysing the available evidence, Barnosky and colleagues conclude that:

> the recent loss of species is dramatic and serious but does not yet qualify as a mass extinction in the palaeontological sense of the Big Five. In historic times we have actually lost only a few per cent of assessed species (though we have no way of knowing how many species we have lost that had never been described). It is encouraging that there is still much of the World's biodiversity left to save, but daunting that doing so will require the reversal of many dire and escalating threats.

Reflecting on the contemporary threat to species, they consider that the loss of the

> critically endangered category would propel the world to a state of mass extinction that has previously been seen only five times in about 540 million years. Additional losses of species in the 'endangered' and 'vulnerable' categories could accomplish the sixth mass extinction in just a few centuries . . . this extinction trajectory would play out under conditions that resemble the 'perfect storm' that coincided with past mass extinctions: multiple, atypical high-intensity ecological stressors, including rapid, unusual climate change and highly elevated atmospheric CO_2. The huge difference between where we are now, and where we could easily be within a few generations, reveals the urgency of relieving the pressures that are pushing today's species towards extinction.

As a consequence of the multiple interacting factors we have just discussed, the fate of many species is uncertain and a catastrophic loss of species is a distinct possibility. Will those species lost be matched by a similar number of rapidly evolving new species? Present evidence does not support such a possibility. Speciation, in some cases, is likely to be gradual, and, while we have evidence for some recently evolved polyploid species such as *Spartina anglica*, *Tragopogon mirus* and *T. miscellus* and *Senecio cambrensis*, confirmed cases of new polyploids are very few in number. However, perhaps evolutionary novelties may appear, especially from the interactions involving introduced and invasive taxa.

Examining the geological record from another perspective, Myers & Knoll (2001) point out that extinctions in the past have been followed by diversification. Is re-diversification likely to occur quickly in contemporary ecosystems? Myers & Knoll think not: in their view, re-diversification will require a minimum of several million years, perhaps as many as 10 million, before evolution can re-establish the 'biological configurations and ecological circuitry' existing before the rise of human dominance. However, such a process would of course require the demise of humanity. Clearly, should this catastrophe be avoided, human activities will increasingly influence the course of evolution into the distant future. We are witnessing a fascinating phase in the anthropogenic impact on the world's biota, one that deserves to be closely studied, so that pattern and processes in the evolution of biodiversity can be better understood.

GLOSSARY

Our list is restricted to scientific expressions used repeatedly throughout the text. Many other technical terms, which are used only once or twice, are defined on first use; definitions of such terms may be sought via the index. Definitions, some of which have been simplified, have been taken from Heslop-Harrison (1953), Davis & Heywood (1963), Whitehouse (1973), Rieger, Michaelis & Green (1976), Heywood (1976), Stace (1980) and Schön, Martens & van Dijk (2009).

Adventitious embryony: The formation of somatic next to sexual embryos.

Agamospecies: Apomictic lineages with distinct morphological, ecological and geographical features, which have been classified formally as species.

Agamospermy: The production of seeds by asexual means.

Alleles (allelomorphs): Alternative forms of a gene which, on account of their corresponding positions on homologous chromosomes, are subject to Mendelian inheritance.

Allopatry: Of species or populations originating in or occurring in different geographical regions.

Allopolyploid: A polyploid originating through the addition of unlike chromosome sets, usually following hybridisation between two species.

Allozymes: Variant forms of an enzyme (i.e. differing in amino acid sequence) that are coded by different alleles at the same locus.

Aneuploid: Individuals having the different chromosomes of the set present in different numbers.

Apogamy: The phenomenon shown by some higher plants in which a gametophyte cell gives rise directly to a sporophyte, without the production of a zygote derived by fusion of gametes.

Apomixis: Reproduction that does not involve sexual processes. For some, but not all, biologists this includes vegetative propagation.

Apomixis: (in plants): asexual reproduction through seeds.

Apospory: The phenomenon shown by some higher plants in which a diploid embryo-sac is formed directly from a somatic cell of the nucellus or chalaza; an embryo is then formed without fertilisation. In addition to the normal reduced megagametophyte (n), a second but unreduced ($2n$) megagametophyte is formed from a non-spore cell (aposporous initial).

Artificial classification: A system of ordering based upon one or a few characters that gives a convenient arrangement of plants for some specific purpose. In such systems closely related taxa may be placed in different groupings.

Autogamy: Self-fertilisation: also called selfing: the fusion of egg cells and pollen grains produced by the same individual; persistent autogamy may result in an increase in homozygosity and division of a population into a number of pure lines.

Autopolyploid: A polyploid originating through the multiplication of the same chromosome set.

B chromosomes: Small chromosomes (frequently, but not always, heterochromatic) that are additional to the normal complement of A chromosomes.

Biological species: Groups of actually or potentially interbreeding natural populations, which are reproductively isolated from other such groups.

Bivalent: The associated pair of homologous chromosomes observed at prophase I in meiosis.

Chiasma (pl. chiasmata): An interchange occurring only at meiosis between chromatids derived from

homologous chromosomes. Chiasmata are the visible evidence of genetic crossing-over.
Classification: The ordering of plants into a hierarchy of classes to produce an arrangement that serves both to express the interrelationships of plants and acts as an information retrieval system.
Cleistogamy: Self-fertilisation within closed flowers.
Cline: A variational trend in space found in a population or series of populations of a species. Genetically, a gradual change in allele or genotypic frequencies at a given locus/loci, across the distributional range of a species population, and correlated with an environmental/geographic transition.
Clone: A group of independent individuals derived vegetatively from a single plant, and therefore of the same genotype.
Crossing-over: The occurrence of new combinations of linked characters following the process of exchange between homologous chromosomes at meiosis.
-deme terminology: Devised to bring clarity to a confusion of terms for units below, at and about the species level. (For a discussion of terms, including Turesson's experimental categories, see Stace, 1980). In cases where precise usage is necessary, the term species, and its prefixed derivatives, is to be reserved for taxonomic categories. Experimentalists could use the '-deme' terminology based on the neutral suffix-deme, which denotes a group of individuals of a specified taxon: The term 'deme' should not stand by itself. Terms are made by adding the appropriate prefixes.

The most commonly used terms are:
Groups of individuals of a specified taxon:
in a particular area (topodeme);
in a particular habitat (ecodeme);
with a particular chromosome condition (cytodeme);
within which free exchange of genes is possible [in a local area] (gamodeme: equivalent to the term Mendelian population);
which are believed to interbreed with a high level of freedom under a specified set of conditions (holo-gamodeme: approximately equivalent to the term biological species).

Despite the clear advantages of the '-deme' terminology it has been little used by biologists (Briggs & Block, 1981); indeed the most frequent use of the terminology is the incorrect employment of 'deme', without prefix, in the sense of gamodeme, by many zoologists.

Diploid: With two sets of chromosomes: the condition arising at fertilisation.
Diplospory: The phenomenon shown by some higher plants in which a diploid embryo-sac is formed directly from a megaspore mother-cell; an embryo is then formed without fertilisation. A normal reductional meiosis is replaced by a non-reductional division. Two unreduced megaspores ($2n$) are produced, of which one degenerates and the other develops into an unreduced gametophyte with an unreduced egg cell.
Directional selection: Selection occurring when the environment is changing in a systematic fashion, leading to a regular change, in a particular direction, of the adaptive characteristics of a population.
Disruptive selection: Selection that breaks up a homogeneous gamodeme into a number of differently adapted populations.
Dysploidy: Changes in chromosome base number resulting from losses or gains of individual chromosomes by fission and fusion events that often involve chromosome rearrangements.
Ecocline: A variational trend correlated with an ecological gradient.
Euchromatin: Parts of chromosomes showing the normal cycle of condensation and staining at nuclear divisions.
Euploid: A polyploid possessing a chromosome number that is an exact multiple of the basic number of the series.
Gametophyte: The haploid gamete-producing phase of the life cycle of plants.
Gametophytic apomixis: processes of diplospory and apospory that are strongly correlated with polyploidy.
Geitonogamy: Transfer of pollen from one flower of a plant to the stigma of another flower on the same plant, leading to self-pollination and the possibility of self-fertilisation.

Genecology: The study of intraspecific variation in plants in relation to environment.
Gene flow: The dispersal of genes in both gametes and zygotes, within and between breeding populations.
Genome: A single complete set of chromosomes. One such set is present in the gametes of diploid species; two genomes are found in the somatic cells. Polyploid cells contain more than two genomes.
Genotype: The totality of the genetic constitution of an individual.
Haploid: With a single set of chromosomes (one genome), such as occurs at gamete formation.
Heteroblastic change: The transition from a juvenile to an adult form accompanied by a more or less abrupt change in morphology.
Heterochromatin: Parts of chromosomes, or whole chromosomes that exhibit an abnormal degree of staining and/or contraction at nuclear divisions.
Heterozygote: A zygote or individual carrying two different alleles of a gene (e.g. Aa)
Holokinetic chromosomes: Have diffuse centromeres instead of a recognisable localised single centromere found in most plants.
Homoploid hybrids: Hybrids that have originated from diploid parental species with the same ploidy level as the parents.
Homozygote: A zygote or individual formed from the fusion of gametes carrying the same allele of a gene (e.g. AA or aa).
Introgressive hybridisation (introgression): Genetic modification of one species by another through the intermediacy of hybrids.
Isolating mechanisms: Factors that restrict the extent of gene flow between populations by preventing or reducing interbreeding.
Isozymes: Term used where more than one locus is involved in the genetics of enzyme variants. Also employed where the genetic basis of enzyme polymorphism is unknown.
Karyotype: The appearance and characteristics (shape, size etc.) of the somatic chromosomes at mitotic metaphase.
Linkage disequilibrium: reflects the difference between the expected frequencies of alleles under an assumption of their independence and those frequencies actually observed in experiment or in the field. [a complex concept: see Slatkin, M. (2008) *Nature Reviews Genetics*, 9, 477–85.]
MADS-box transcription factors: important family of regulators of cellular processes in plants. One of the best studied is involved in determining floral identity (as expressed in the ABC model).
Meiosis: A special nuclear division in which the chromosome number is halved.
Meristic variation: Variation in numbers of parts or of organs.
Mitosis: The nuclear division typical of somatic plant tissues in which a nucleus divides to produce two identical complements of chromosomes (and hence genes).
Multivalent: Association of more than two homologous chromosomes at meiosis, e.g. 3 = trivalent; 4 = quadrivalent.
Natural classification: A classification based upon overall resemblance, and serving a variety of purposes.
Phenetic classification: A classification based upon present-day resemblances and differences between plants.
Phenotype: The totality of characteristics of an individual; its appearance as a result of the interaction between genotype and environment.
Phenotypic plasticity: The ability of certain genotypes to respond to different environments by producing different phenotypes.
Phylogenetic classification: A classification showing the supposed evolution of groups.
Pleiotropism: The phenomenon shown by a gene that simultaneously influences more than one characteristic of the phenotype.
Polygenes: Genes of small individual effect which act jointly to produce quantitative genetic variation.
Polyhaploid: An organism with the gametic chromosome number arising by parthenogenesis in a polyploid, e.g. a diploid ($2x$) plant arising from the parthenogenetic development of an embryo of a tetraploid ($4x$).

Polymorphism: The occurrence of two or more distinct genetic variants of a species in a single habitat.

Polyploid: Having three or more sets of homologous chromosomes.

Pseudogamy: The phenomenon found in some apomictic plants, whereby pollination is necessary for seed development, even though no fertilisation of the egg-cell takes place.

Pure line: A lineage of individuals originating from a single homozygous ancestor.

Ramet: An individual belonging to a clone.

Ratio-cline: Clinal variation occurring in polymorphic species, in which successive populations show progressive change in the proportion of the variants.

Sporophyte: The diploid spore-producing phase of the life cycle of plants arising from the fertilisation of haploid gametes.

Sporophytic apomixis: Somatic embryos are formed within the sporophytic tissue that surrounds the gametophyte. These cells do not enter a gametophytic phase but remain sporophytic and produce an embryo directly (somatic embryo).

Stabilising selection: Selection favouring the average individuals of a population and eliminating extreme variants.

Sympatry: Of species or populations, originating in or occupying the same geographical area.

Taxon (pl. taxa): A classificatory unit of any rank: e.g. Daisy: *Bellis perennis* (species); *Bellis* (genus); Asteraceae(family).

Topocline: A geographical variational trend that is not necessarily correlated with an ecological gradient.

Trichomes: Outgrowths of hairs, glandular hairs, scales, papillae etc. that are of diverse structure and function.

Univalent: An unpaired chromosome at meiosis.

Variant: Any definable individual or group of individuals: a valuable neutral term.

Vivipary: The production of small plants or bulbils in place of flowers and seed.

REFERENCES

Ab-Shukor, N. A., Kay, Q. O. N., Stevens, D. P. & Skibinski, D. O. F. (1988). Salt tolerance in natural populations of *Trifolium repens*. *New Phytologist*, **109**, 483–9.

Abbott, R. J. & Brochmann, C. (2003). History and evolution of the arctic flora: in the footsteps of Eric Hultén. *Molecular Ecology*, 12, 299–313.

Abbott, R. J. & Forbes, D. G. (2002). Extinction of the Edinburgh lineage of the allopolyploid neospecies, *Senecio cambrensis* Rosser (Asteraceae). *Heredity*, **88**, 267–9.

Abbott, R. J., Ingram, R. & Noltie, H. J. (1983a). Discovery of *Senecio cambrensis* Rosser in Edinburgh. *Watsonia*, 14, 407–8.

Abbott, R. J., Noltie, H. J. & Ingram, R. (1983b). The origin and distribution of *Senecio cambrensis* Rosser in Edinburgh. *Transactions of the Botanical Society of Edinburgh*, **44**, 103–6.

Abbott, R. J., Ireland, H. E. & Rogers, H. J. (2007). Population decline despite high genetic diversity in the new allopolyploid species *Senecio cambrensis* (Asteraceae). *Molecular Ecology*, 16, 1023–33.

Abbott, R. J. *et al.* (2010). Homoploid hybrid speciation in action. *Taxon*, **59**, 1375–86.

Abbott, R. J. *et al.* (2012). Hybridization and speciation. *Journal of Evolutionary Biology*, **26**, 229–46.

Abramovitz, J. N. (1996). *Imperiled waters, impoverished future: the decline of freshwater ecosystems*. Washington DC: Worldwatch.

Adams, J. U. (2008). Transcriptome: connecting the genome to gene function. *Nature Education*, 1, 195.

Adams, W. M. (2004). *Against extinction: the story of conservation*. London & Sterling, VA: Earthspan.

Agnarsson, I. & Kuntner, M. (2007). Taxonomy in a changing world: seeking solutions for a science in crisis. *Systematic Biology*, **56**, 531–9.

Agrawal, A. A., Laforsch, C. & Tollrian, R. (1999). Transgenerational induction of defences in animals and plants. *Nature*, 401, 60–3.

Ågren, J. & Schemske, D. W. (2012). Reciprocal transplants demonstrate strong adaptive differentiation of the model organism *Arabidopsis thaliana* in its native range. *New Phytologist*, **194**, 1112–22.

Aguilar, R. *et al.* (2006). Plant reproductive susceptibility to habitat fragmentation: review and synthesis through a meta-analysis. *Ecological Letters*, **9**, 968–80.

Ahmed, S. *et al.* (2009). Wind-borne insects mediate directional pollen transfer between desert fig trees 160 kilometers apart. *Proceedings of the National Academy of Sciences, USA*, **106**, 20342–7.

Ainouche, M. L. *et al.* (2009). Hybridization, polyploidy and invasion: lessons from *Spartina* (Poaceae). *Biological Invasions*, **11**, 1159–73.

Akçakaya, H. R. (2002). RAMAS Metapop: Viability Analysis for Stage-structured Metapopulations (version 4.0). Setauket, NY: Applied Biomathematics.

Akeroyd, J. R. (1994). Some problems with introduced plants in the wild. In *The common ground of wild and cultivated plants*, ed. A. R. Perry & R. G. Ellis, pp. 31–40. Cardiff: National Museum of Wales.

Akeroyd, J. R. & Wyse Jackson, P. (1995). *A handbook for botanic gardens on the reintroduction of plants to the wild*. London: Botanic Gardens Conservation International.

Al-Hiyaly, S. E. K., McNeilly, T. & Bradshaw, A. D. (1988). The effect of zinc contamination from electricity pylons – evolution in a replicated situation. *New Phytologist*, **110**, 571–80.

Al-Hiyaly, S. E. K., McNeilly, T. & Bradshaw, A. D. (1990). The effect of zinc contamination from electricity pylons – contrasting patterns of evolution in five grass species. *New Phytologist*, **114**, 183–90.

Al-Hiyaly, S. E. K., McNeilly, T., Bradshaw, A. D. & Mortimer, A. M. (1993). The effect of zinc contamination from electricity pylons. Genetic constraints on selection for zinc tolerance. *Heredity*, **70**, 22–32.

Al-Kaff, N. *et al.* (2008). Detailed dissection of the chromosomal region containing the *Ph1* Locus in Wheat *Triticum aestivum*: with deletion mutants and expression profiling. *Annals of Botany*, 101, 863–72.

Albert, V. A., Williams, S. E. & Chase, M. W. (1992). Carnivorous plants: phylogeny and structural evolution. *Science*, 257, 1491–5.

Albert, V. A., Oppenheimer, D. & Lindqvist, C. (2002). Pleiotropy, redundancy and the evolution of flowers. *Trends in Plant Science*, 7, 297–301.

Alberto, F. J. *et al.* (2013a). Imprints of natural selection along environmental gradients in phenology-related genes of *Quercus petraea*. *Genetics*, 195, 495–512.

Alberto, F. J. *et al.* (2013b). Potential for evolutionary responses to climate change – evidence from tree populations. *Global Climate Change*, 19, 1645–61.

Alcázar, R. *et al.* (2012). Signals of speciation within *Arabidopsis thaliana* in comparison with its relatives. *Current Opinion in Plant Biology*, 15, 205–11.

Aldridge, S. (1996). *The thread of life*. Cambridge: Cambridge University Press.

Alexander, D. (2011). *The language of genetics*. London: Longman.

Ali, N. S. & Trivedi, C. (2010). *Plant diversity and climate change – a review of Kew's science activities relevant to climate change*. Richmond, UK: Royal Botanic Gardens, Kew.

Ali, N. S. & Trivedi, C. (2011). Botanic Gardens and climate change – an audit of scientific activities at the Royal Botanic Gardens, Kew. *Biodiversity Conservation*, 20, 295–307.

Allan, E. & Pannell, J. R. (2009). Rapid divergence in physiological and life-history traits between northern and southern populations of the British introduced neo-species, *Senecio squalidus*. *Oikos*, 118, 1053–61.

Allen, A. M. *et al.* (2011). Pollen–pistil interactions and self-incompatibility in the Asteraceae: new insights from studies of *Senecio squalidus* (Oxford Ragwort). *Annals of Botany*, 108, 687–98.

Allendorf, F. W, Luikart, G. & Aitken, S. N. (2013) *Conservation and the genetics of populations*, 2nd edn. Oxford: Blackwell Publishing.

Allison, L. A. (2007). *Fundamental molecular biology*. Malden, MA, Oxford & Carlton, Australia: Blackwell.

Allnutt, T. R. *et al.* (1998). Genetic variation in *Fitzroya cupressoides* (Alerce), a threatened South American conifer. *Molecular Ecology*, 8, 975–87.

Ally, D., Ritland, K. & Otto, S. P (2008). Can clone size serve as a proxy for clone age? An exploration using microsatellite divergence in *Populus tremuloides*. *Molecular Ecology*, 17, 4897–911.

Alonso-Blanco, C., Mendez-Vigo, B. & Koornneef, M. (2005). From phenotypic to molecular polymorphisms involved in naturally occurring variation of plant development. *International Journal of Developmental Biology*, 49, 717–32.

Alsos, I. G. *et al.* (2007). Frequent long-distance plant colonization in the changing Arctic. *Science*, 316, 1606–9.

Alsos, I. G. *et al.* (2009). Past and future range shifts and loss of diversity in Dwarf Willow (*Salix herbacea* L.) inferred from genetics, fossils, and modelling. *Global Ecology and Biogeography*, 18, 223–39.

Alston, R. E. & Turner, B. L. (1963a). *Biochemical systematics*. London & New York: Prentice Hall.

Alston, R. E. & Turner, B. L. (1963b). Natural hybridization among four species of *Baptisia* (Leguminosae). *American Journal of Botany*, 50, 159–73.

Altizer, S. M., Thrall, P. H. & Antonovics, J. (1998), Vector behaviour and transmission of anther-smut infection in *Silene alba*. *American Midland Naturalist*, 139, 147–63.

Alvarez, L. W., Alvarez, F. A., Asaro, F. & Michel, H. V. (1980). Extraterrestrial cause for the Cretaceous-Tertiary extinction. *Science*, 208, 1095–108.

Alvarez-Valin, F. (2002). Neutral theory. In *Encyclopedia of evolution*, vol. 2, 815–21, Oxford: Oxford University Press.

Amborella Genome Project (2013). The *Amborella* genome and the evolution of flowering plants. *Science*, 342, DOI:10.1126/science.1241089

Amsellen, L., Noyer, J. L, Le Bourgeois, T. & Hossaert-McKey, M. (2000). Comparison of genetic diversity of the invasive weed *Rubus alceifolius* Poir. (Rosaceae) in its native range and areas of introduction, using amplified fragment length polymorphism (AFLP) markers. *Molecular Ecology*, 9, 443–55.

Amsellem, L., Chevalier, M. H. & Hossaert-McKey, M. (2001). Ploidy level of the invasive weed *Rubus alceifolius* (Rosaceae) in its native range and in areas of introduction. *Plant Systematics and Evolution*, 228, 171–9.

Amsellem, L., Noyer, J.-L. & Hossaert-McKey, M. (2001). Evidences of a switch in the reproductive biology of *Rubus alceifolius* (Rosaceae) towards apomixis, between its native range and its area of introduction. *American Journal of Botany*, **88**, 2243–51.

Anderson, D. L. & East, I. J. (2008).The latest buzz about colony collapse disorder. *Science*, **319**, 724.

Anderson, E. (1949). *Introgressive hybridisation*. London: Chapman & Hall; New York: Wiley.

Anderson, E. (1953). Introgressive hybridization. *Biological Reviews*, **28**, 280–307.

Anderson, E. & Abbe, L. B. (1933). A comparative anatomical study of a mutant *Aquilegia*. *American Naturalist*, **67**, 380–4.

Anderson, G. *et al.* (2006). Reproductive biology of the dioecious Canary Islands endemic *Withania aristata* (Solanaceae). *American Journal of Botany*, **93**, 1295–1305.

Anderson, J. T. *et al.* (2012). Phenotypic plasticity and adaptive evolution contribute to advancing flowering phenology in response to climate change. *Proceedings of the Royal Society, B*, **279**, 3843–52.

Andreasen, K., Manktelow, M. & Razafimandimbison, S. G. (2009). Successful DNA amplification of a more than 200-year-old herbarium specimen: recovering genetic material from the Linnaean era. *Taxon*, **58**, 959–62.

Andrews, C. A. (2010). Natural selection, genetic drift, and gene flow do not act in isolation in natural populations. *Nature Education Knowledge*, **3**, 5.

Andrus, N. *et al.* (2009). Phylogenetics of *Darwiniothamnus* (Asteraceae: Astereae) – molecular evidence for multiple origins in the endemic flora of the Galápagos Islands. *Journal of Biogeography*, **36**, 1055–69.

Angiosperm Phylogeny Group (2009). An update of the Angiosperm Phylogeny Group classification for the orders and families of flowering plants: APGIII. *Botanical Journal of the Linnean Society*, **161**, 105–21.

Anon. (1965). *Iconographia Mendeliana*. Brno: The Moravian Museum.

Anon. (1994). *IUCN Red List categories*. Gland, Switzerland: IUCN Council.

Anon. (2002). Seed banks receive vital cash boost. *New Scientist*, 30 August.

Anon. (2014). Protect and serve. Nations must keep expanding conservation efforts to avoid a biodiversity crisis. Editorial in *Nature*, **616**, 144.

Antonovics, J. (1968). Evolution in closely adjacent plant populations. V. Evolution of self-fertility. *Heredity*, **23**, 219–38.

Antonovics, J. & Bradshaw, A. D. (1970). Evolution in closely adjacent plant populations. VIII. Clinal patterns at a mine boundary. *Heredity*, **25**, 349–62.

Antonovics, J., Bradshaw, A. D. & Turner, R. G. (1971). Heavy metal tolerance in plants. *Advances in Ecological Research*, **7**, 1–85.

Arabidopsis Genome Initiative (2000). Analysis of the genome sequence of the flowering plant *Arabidopsis thaliana*. *Nature*, **408**, 796–815.

Araújo, M. B. & Peterson, T. (2011). Uses and misuses of bioclimatic envelope modeling. *Ecology*, **93**, 1527–39.

Arber, A. (1938). *Herbals: their origin and evolution. A chapter in the history of Botany 1470–1670*, 2nd edn. Cambridge: Cambridge University Press.

Arbogast, B. S. *et al.* (2002). Estimating divergence times from molecular data on phylogenetic and population genetic timescales. *Annual Review of Ecology & Systematics*, **33**, 707–40.

Archibald, J. D. & Fastovsky, D. E. (2004). Dinosaur extinction. In *The Dinosauria*, ed. D. B. Weishampel *et al.*, pp. 672–684. Berkeley: University of California Press.

Armbruster, W. S. & Baldwin, B. G. (1998). Switch from specialized to generalized pollination. *Nature*, **394**, 632.

Armstrong, H. E., Armstrong, F. & Horton, E. (1912). Herbage studies 1. *Lotus corniculatus*, a cyanophoric plant. *Proceedings of the Royal Society, B*, **84**, 471–84.

Arnaud-Haond, S. *et al.* (2012). Implications of extreme life span in clonal organisms: millenary clones in meadows of the threatened seagrass *Posidonia oceanica*. *PLOS ONE*, **7**, e30454, DOI:10.1371/journal.pone.0030454.

Arnold, A. C. *et al.* (2005). Is there a future for wild grapevine (*Vitis vinifera* subsp. *silvestris*) in the Rhine Valley? *Biodiversity and Conservation*, **14**, 1507–23.

Arnold, A. E. *et al.* (2010). Interwoven branches of the plant and fungal trees of life. *New Phytologist*, **185**, 874–8.

Arnold, M. L. (1992). Natural hybridization as an evolutionary process. *Annual Review of Ecology and Systematics*, **23**, 237–61.

Arnold, M. L. (1993). *Iris nelsonii* (Iridaceae): origin and genetic composition of a homoploid hybrid species. *American Journal of Botany*, **80**, 577–83.

Arnold, M. L. & Bennett, B. D. (1993). Natural hybridization in Louisiana Irises: genetic variation and ecological determinants. In *Hybrid zones and the evolutionary process*, ed. R. G. Harrison, pp. 115–39. New York: Oxford University Press.

Arnold, M. L., Bouck, A. C. & Cornman, R. S. (2004). Verne Grant and Louisiana Irises: is there anything new under the sun? *New Phytologist*, **161**, 143–9.

Arnold, M. L., Ballerini, E. S. & Brothers, A. N. (2012). Hybrid fitness, adaptation and evolutionary diversification: lessons learned from Louisiana Irises. *Heredity*, **108**, 159–66.

Aronne, G. & Wilcock, C. C. (1994). First evidence of myrmecochory in fleshy-fruited shrubs of the Mediterranean region. *New Phytologist*, **127**, 781–8.

Aronson, J. *et al.* (1993). *Restoring natural capital: science, business and practice*. Washington DC, & London: Island Press.

Ashley, M. V. (2010). Plant parentage, pollination, and dispersal: how DNA microsatellites have altered the landscape. *Critical Reviews in Plant Sciences*, **29**, 148–61.

Ashman, T. L. *et al.* (2004). Pollen limitation of plant reproduction: ecological and evolutionary causes and consequences. *Ecology*, **85**, 2408–21.

Ashman, T. L., Kwok, A. & Husband, B. C. (2013). Revisiting the dioecy–polyploidy association: alternate pathways and research opportunities. *Cytogenetic and Genome Research*, **140**, 241–55.

Ashton, P. A. & Abbott, R. J. (1992). Multiple origins and genetic diversity in the newly arisen allopolyploid species, *Senecio cambrensis* Rosser (Compositae). *Heredity*, **68**, 25–32.

Ashton, P. S. (1987). Biological considerations in *in situ* vs *ex situ* plant conservation. In *Botanic gardens and the world conservation strategy*, ed. D. Bramwell, O. Hamann, V. Heywood & H. Synge, pp. 117–30. London: Academic Press for IUCN.

Asker, S. E. & Jerling, L. (1992). *Apomixis in plants*. Boca Raton, FL: CRC Press.

Assouad, M. W. *et al.*, (1978). Reproductive capacities in the sexual forms of the gynodiecious species *Thymus vulgaris*. *Botanical Journal of the Linnean Society*, **77**, 29–39.

Assunçáo, A. G. L., Schat, H. & Aarts, M. G. M. (2003). *Thlaspi caerulescens*, an attractive model species to study heavy metal hyperaccumulation in plants. *New Phytologist*, **159**, 351–60.

Aston, J. L. & Bradshaw, A. D. (1966). Evolution in closely adjacent plant populations II. *Agrostis stolonifera* in maritime habitats. *Heredity*, **21**, 649–64.

Atwell, S. *et al.* (2010). Genome-wide association study of 107 phenotypes in *Arabidopsis thaliana* inbred lines. *Nature*, **465**, 627–31.

Atwood, S. S. & Sullivan, J. T. (1943). Inheritance of a cyanogenic glucoside and its hydrolysing enzyme in *Trifolium repens*. *Journal of Heredity*, **34**, 311–20.

Auffret, A. G. (2011). Can seed dispersal by human activity play a useful role for the conservation of European grasslands? *Applied Vegetation Science*, **14**, 291–303.

Avers, C. J. (1989). *Process and pattern in evolution*. New York: Oxford University Press.

Avise, J. C. (1994). *Molecular markers, natural history and evolution*. New York: Chapman & Hall.

Avise, J. C. (2000). *Phylogeography: the history and formation of species*. Cambridge, MA: Harvard University Press.

Ayala, F. J. (2012). *Evolution*. London: Quercus.

Ayazloo, M. & Bell, J. N. B. (1981). Studies on the tolerance to sulphur dioxide of grass populations in polluted areas. 1. Identification of tolerant populations. *New Phytologist*, **88**, 203–22.

Ayre, D. *et al.* (2010). The accumulation of genetic diversity within a canopy-stored seed bank. *Molecular Ecology*, **19**, 2640–50.

Ayres, D. R. & Strong, D. R. (2001). Origin and genetic diversity of *Spartina anglica* (Poaceae) using nuclear DNA markers. *American Journal of Botany*, **88**, 1863–7.

Bacles, C. F. E., Lowe, A. J. & Ennos, R. A. (2006). Effective seed dispersal across a fragmented landscape. *Science*, **311**, 628.

Bailey, J. P., Bímová, K. & Mandák, B. (2009). Asexual spread versus sexual reproduction and evolution in Japanese Knotweeds sets the stage for the 'Battle of the Clones'. *Biological Invasions*, **11**, 1189–1203.

Bailleul, D. *et al.* (2012). Seed spillage from grain trailers on road verges during oilseed rape harvest: an experimental survey. *PLOS ONE*, **7**, 1–7.

Baker, A. M., Barrett, S. C. H. & Thompson, J. D. (2000). Variation of pollen limitation in the early flowering Mediterranean geophyte *Narcissus assoanus* (Amaryllidaceae). *Oecologia*, **124**, 529–35.

Baker, H. (1966). The evolution, functioning and breakdown of heteromorphic incompatibility systems. I. The Plumbaginaceae. *Evolution*, **20**, 349–68.

Baker, H. G. (1947). Criteria of hybridity. *Nature*, **159**, 1–5.

Baker, H. G. (1951). Hybridization and natural gene-flow between higher plants. *Biological Reviews*, **26**, 302–37.

Baker, H. G. (1954). Report of meeting of British Ecological Society, April 1953. *Journal of Ecology*, **42**, 570–2.

Baker, H. G. (1955). Self-compatibility and establishment after 'long-distance' dispersal. *Evolution*, **9**, 347–9.

Baker, H. G. (1965). Characteristics and modes of origin of weeds. In *The genetics of colonizing species*, ed. H. G. Baker & G. L. Stebbins, pp. 147–72. New York: Academic Press.

Baker, H. G. (1967). Support for Baker's Law – as a rule. *Evolution*, **21**, 853–6.

Baker, H. G. (1974). The evolution of weeds. *Annual Review of Ecology & Systematics*, **5**, 1–24.

Baker, H. G. (1991). The continuing evolution of weeds. *Economic Botany*, **45**, 445–9.

Baker, H. G. & Cox, P. A. (1984). Further thoughts on dioecism and islands. *Annals of the Missouri Botanical Garden*, **71**, 244–53.

Bakker, E. G. *et al.* (2006). A genome-wide survey of R gene polymorphisms in *Arabidopsis*. *Plant and Cell*, **18**, 1803–18.

Balakrishnan, R. (2005). Species concepts, species boundaries and species identification: a view from the tropics. *Systematic Biology*, **54**, 689–93.

Balcombe, D. (2011). Invigorating plants. *Research Horizons*, **5**, 9 and http://phys.org/news/2011-07-invigorating.html

Baldwin, B. G. (2006). Contrasting patterns and processes of evolutionary change in the tarweed-silversword lineage: revisiting Clausen, Keck, and Heisey's findings. *Annals of the Missouri Botanical Garden*, **93**, 66–96.

Baldwin, B. G. & Wagner, W. L. (2010). Hawaiian angiosperm radiations of North American origin. *Annals of Botany*, **105**, 849–79.

Balmford, A. & Long, A. (1994). Avian endemism and forest loss. *Nature*, **372**, 623–4.

Balmford, A. & Whitten, T. (2003). Who should pay for tropical conservation, and how could the costs be met? *Oryx*, **37**, 238–50.

Bannister, M. H. (1965). Variation in the breeding system of *Pinus radiata*. In *The genetics of colonizing species*, ed. H. G. Baker & G. L. Stebbins, pp. 353–72. New York & London: Academic Press.

Bannister, P. (1976). *Introduction to physiological plant ecology*. Oxford: Blackwell.

Barcaccia, G. & Albertini, E. (2013). Apomixis in plant reproduction: a novel perspective on an old dilemma. *Plant Reproduction*, **26**, 159–79.

Barkley, T. M. (1966). A review of the origin and development of the florists' Cineraria, *Senecio cruentus*. *Economic Botany*, **20**, 386–95.

Barkman, T. J., Lim, S.-H., Mat Salleh, K. & Nais, J. (2004). Mitochondrial DNA sequences reveal the photosynthetic relatives of *Rafflesia*, the world's largest flower. *Proceedings of the National Academy of Sciences, USA*, **101**, 787–92.

Barkman, T. J. *et al.* (2007). Mitochondrial DNA suggests at least 11 origins of parasitism in angiosperms and reveals genomic chimerism in parasitic plants. *BMC Evolutionary Biology*, **7**, 248.

Barling, D. M. (1955). Some population studies in *Ranunculus bulbosus* L. *Journal of Ecology*, **43**, 207–18.

Barlow, B. A. & Wiens, D. (1977). Host parasite resemblance in Australian Mistletoes: the case for cryptic mimicry. *Evolution*, **31**, 69–84.

Barnes, D. K. A. (2002). Accumulation and fragmentation of plastic debris in global environments. *Philosophical Transactions of the Royal Society, B*, **364**, 1985–98.

Barnosky, A. D. *et al.* (2011). Has the Earth's sixth mass extinction already arrived? *Nature*, **471**, 51–7.

Barnosky, A. D. *et al.* (2012). Approaching a state shift in Earth's biosphere. *Nature*, **486**, 52–8.

Barrett, P. H. *et al.* (1987). *Charles Darwin notebooks 1836–1844*. Cambridge: Cambridge University Press and the British Museum (Natural History).

Barrett, S. C. H. (1980a) Sexual reproduction in *Eichhornia crassipes* (Water Hyacinth). I. Fertility of clones from diverse regions. *Journal of Applied Ecology*, **17**, 101–12.

Barrett, S. C. H (1980b). Sexual reproduction in *Eichhornia crassipes* (Water Hyacinth). II. Seed production in natural populations. *Journal of Applied Ecology*, **17**, 113–24.

Barrett, S. C. H. (1983). Crop mimicry in weeds. *Economic Botany*, **37**, 255–82.

Barrett, S. C. H. (1985). Floral trimorphism and monomorphism in continental and island populations of *Eichhornia paniculata* (Spreng.) Sols. (Pontederiaceae). *Biological Journal of the Linnean Society*, **25**, 41–60.

Barrett, S. C. H. (1987). Mimicry in plants. *Scientific American*, **255**, 76–83.

Barrett, S. C. H. (1988). Mating system evolution and speciation in heterostylous plants. In *Speciation and its consequences*, ed. D. Otte & J. A. Endler, pp. 257–83. Sunderland, MA: Sinauer.

Barrett, S. C. H. (1989a). The evolutionary breakdown of heterostyly. In *The evolutionary ecology of plants*, ed. J. H. Bock & Y. B. Linhart, pp.151–69. Boulder, CO: Westview Press.

Barrett, S. C. H. (1989b). Mating system evolution and speciation in heterostylous plants. In *Speciation and its consequences*, ed. D. Otte & J. A. Endler, pp. 257–83. Sunderland, MA: Sinauer

Barrett, S. C. H. (1992). Heterostylous genetic polymorphisms: model systems for evolutionary analysis. In *Evolution and function of heterostyly*, ed. S. C. H. Barrett, pp. 1–29. Heidelberg: Springer Verlag.

Barrett, S. C. H. (2010a). Darwin's legacy: the forms, function and sexual diversity of flowers. *Philosophical Transactions of the Royal Society of London, Ser. B.*, **365**, 351–68.

Barrett, S. C. H. (2010b). Understanding plant reproductive diversity. *Philosophical Transactions of the Royal Society of London, Ser. B.*, **365**, 99–109.

Barrett, S. C. H. & Charlesworth, D. (2007). David Graham Lloyd 20 June 1937–30 May 2006. *Biographical Memoirs of the Fellows of the Royal Society*, **53**, 203–21.

Barrett, S. C. H. & Harder, L. D. (1996). Ecology and evolution of plant mating. *Trends in Ecology & Evolution*, **11**, 73–9.

Barrett, S. C. H. & Kohn, J. R. (1991). Genetic and evolutionary consequences of small population size in plants: implications for conservation. In *Genetics and conservation of rare plants*, ed. D. A. Falk & K. E. Holsinger, pp. 3–30. New York: Oxford University Press.

Barrett, S. C. H. & Richardson, B. J. (1986). Genetic attributes of invading species. In *Ecology of biological invasions*, ed. R. H. Groves & J. J. Brown, pp. 21–33. Canberra: Australian Academy of Science.

Barrett, S. C. H. & Shore, J. S. (1990). Isozyme variation in colonising plants. In *Isozymes in plant biology*, ed. D. E. Soltis & P. S. Soltis, pp. 106–26. London: Chapman & Hall.

Barrett, S. C. H. & J. S. Shore. (2008). New insights on heterostyly: comparative biology, ecology and genetics. In *Self-incompatibility in flowering plants: evolution, diversity and mechanisms*, ed. V. Franklin-Tong, pp. 3–32. Berlin: Springer-Verlag.

Barrett, S. C. H., Dorken, M. E. and Case, A. L. (2001). A geographical context for the evolution of plant reproductive systems. In *Integrating ecological and evolutionary processes in a spatial context*, ed. J. Silvertown & J. Antonovics, pp. 341–64. Oxford: Blackwell.

Barrett, S. C. H., Colautti, R. I. & Eckert, C. G. (2008). Reproductive systems and evolution during biological invasion. *Molecular Ecology*, **17**: 373–83.

Barrett, S. C. H., Ness, R. W. & Vallejo-Marín M. (2009). Evolutionary pathways to self-fertilization in a tristylous plant species. *New Phytologist*, **183**, 546–56.

Barrett, S. C. H. *et al.* (2011). Plant reproductive systems and evolution during biological invasion. *Molecular Biology*, **17**, 373–83.

Barringer, B. C. (2007). Polyploidy and self-fertilization in flowering plants. *American Journal of Botany*, **94**, 1527–33.

Barros, M. D. C. & Dyer, T. A. (1988). Atrazine resistance in the grass *Poa annua* is due to a single base change in a chloroplast gene for the D1 protein of photosystem II. *Theoretical & Applied Genetics*, **75**, 610–16.

Barton, N. H. (2000). Genetic hitchhiking. *Philosophical Transactions of the Royal Society of London, B*, **355**, 1553–62.

Barton, N. H. & Charlesworth, B. (1984). Genetic revolutions, founder effects, and speciation. *Annual Review of Ecology and Systematics*, **15**, 133–64.

Bashford, A. & Levine, P. (2012). *The Oxford handbook of the history of eugenics*. New York & Oxford: Oxford University Press.

Bateman, R. M. (1999). Integrating molecular and morphological evidence of evolutionary radiations. In *Molecular systematics and plant evolution*, ed. P. M. Hollingsworth, R. M. Bateman & R. J. Gornall, pp. 432–71. London & New York: Taylor & Francis. Systematics Association Special Volume Series 57.

Bateman, R. M. (2011). The perils of addressing long-term challenges in a short-term world: making descriptive taxonomy predictive. In *Climate change, ecology and systematics*, ed. T. R. Hodkinson. *et al.*, pp.67–95. Cambridge: Cambridge University Press.

Bateman, R. M. & DiMichele, W. A. (2002). Generating and filtering major phenotypic novelties: neoGoldschmidtian saltation revisited. In *Developmental genetics and plant evolution*, ed. Q. C. B. Cronk, R. M. Bateman & J. A. Hawkin, pp. 109–59. London: Taylor & Francis.

Bateman, R. M. & Sexton, R. (2008). Is spur length of *Platanthera* species in the British Isles adaptively optimized or an evolutionary red herring? *Watsonia*, **27**, 1–21.

Bateman, R. M., Pridgeon, A. M. & Chase, M. W. (1997). Phylogenetics of subtribe Orchidineae (Orchidoideae, Orchidaceae) based on nuclear ITS sequences. 2. Infrageneric relationships and reclassification to achieve monophyly of *Orchis* sensu stricto. *Lindleyana*, **12**, 113–41.

Bateson, W. (1895a). The origin of the cultivated *Cineraria*. *Nature*, **51**, 605–7.

Bateson, W. (1895b). The origin of the cultivated *Cineraria*. *Nature*, **52**, 29, 103–4.

Bateson, W. (1897). Notes on hybrid Cinerarias produced by Mr Lynch and Miss Pertz. *Proceedings of the Cambridge Philosophical Society*, **9**, 308–9.

Bateson, W. (1909). *Mendel's principles of heredity*. London: Cambridge University Press; New York: Macmillan.

Bateson, W. (1913). *Problems of genetics*. London: Oxford University Press; New Haven, CT: Yale University Press.

Bateson, W. & Punnett, R. C. (1911). On gametic series involving reduplication of certain terms. *Journal of Genetics*, **1**, 293–302.

Bateson, W. & Saunders, E. R. (1902). Experimental studies in the physiology of heredity. *Report to the Evolution Committee of the Royal Society*, **1**, 1–160.

Bateson, W., Saunders, E. R. & Punnett, R. C. (1905). Experimental studies in the physiology of heredity. *Report to the Evolution Committee of the Royal Society*, **2**, 1–55, 80–99.

Battaglia, E. (1963). Apomixis. In *Recent advances in the embryology of Angiosperms*, ed. P. Maheshwari, ch. 8, pp. 221–64. Delhi: University of Delhi Press.

Baucom, R. S. (2016). The remarkable repeated evolution of herbicide resistance. *American Journal of Botany*, **103**, 181–3.

Baum, D. A. & Donoghue, M. J. (2002). Transference of function, heterotopy, and the evolution of plant development. In *Developmental genetics and plant evolution*, ed. Q. C. B. Cronk, R. M. Bateman & J. A. Hawkin, pp. 52–69. London: Taylor & Francis.

Baum, D. A. & Smith, S. D. (2012). *Tree-thinking: an introduction to phylogenetic biology*. Greenwood Village, CO: Roberts & Company.

Baum, D. A, Randall, L. S. & Wendel, J. F. (1998). Biogeography and floral evolution of Baobabs (*Adansonia*, Bombacaceae) as inferred from multiple data sets. *Systematic Biology*, **47**, 181–207.

Baumann, U. *et al.* (2000). Self-incompatibility in the grasses. *Annals of Botany*, **85**, 203–9.

Baumel, A., Ainouche, M. L. & Levasseur, J. E. (2001). Molecular investigations in populations of *Spartina anglica* C.E. Hubbard (Poaceae) invading coastal Brittany (France). *Molecular Ecology*, **10**, 1689–1701.

Bawa, K. S., Perry, D. R. & Beach, J. H. (1985). Reproductive biology of tropical lowland rain forest trees. I. Sexual systems and self-incompatibility mechanisms. *American Journal of Botany*, **72**, 331–45.

Bayer, R. J., Ritland, K. & Purdy, B. G. (1990). Evidence of partial apomixis in *Antennaria media* (Asteraceae: Inuleae) detected by segregation of genetic markers. *American Journal of Botany*, **77**, 1078–83.

Beardsell, D. V. *et al.* (1993). Reproductive biology of Australian Myrtaceae. *Australian Journal of Botany*, **41**, 511–26.

Beattie, A. (1978). Plant–animal interactions affecting gene flow in *Viola*. In *The pollination of flowers by insects*, ed. A. J. Richards, pp. 151–64, Linnean Society Symposium Series 6. London: Academic Press.

Bebber, D. P. *et al.* (2010). Herbaria are a major frontier for species discovery. *Proceedings of the National Academy of Sciences USA*, **107**, 22 169–71.

Beck, J. B., Schmuths, H. & Schaal, B. A. (2007). Native range genetic variation in *Arabidopsis thaliana* is strongly geographically structured and reflects Pleistocene glacial dynamics. *Molecular Ecology*, **17**, 902–15.

Beckage, B. *et al.* (2008). A rapid upward shift of a forest ecotone during 40 years of warming in the Green Mountains of Vermont. *Proceedings of the National Academy of Sciences, USA*, **104**, 4197–202.

Becker, C. *et al.* (2011). Spontaneous epigenetic variation in the *Arabidopsis thaliana* methylome. *Nature*, **480**, 245–9.

Beckman, N. G. & Rogers, H. S. (2013). Consequences of seed dispersal for plant recruitment in tropical forests: interactions within the seedscape. *Biotropica*, **45**, 666–81.

Becqemont, D. (2009). *Charles Darwin, 1837–1839: aux sources d'une découverte.* Paris: Editions Kimé.

Beddall, B. G. (1957). Historical notes on avian classification. *Systematic Zoology*, **6**, 129–36.

Beddall, B. G. (1988). Darwin and divergence: the Wallace connection. *Journal of the History of Biology*, **21**, 1–68.

Beddie, A. D. (1942). Natural root grafts in New Zealand trees. *Transactions and Proceedings of the Royal Society of New Zealand*, **71**, 199–203.

Behe, M. J. (1996). *Darwin's black box: the biochemical challenge to evolution.* New York: The Free Press.

Beilstein, M. A. *et al.* (2010). Dated molecular phylogenies indicate a Miocene origin for *Arabidopsis thaliana*. *Proceedings of the National Academy of Sciences, USA*, **107**, 18724–8.

Bell, C. D., Soltis, D. E. & Soltis, P. S. (2010). The age and diversification of the angiosperms re-revisited. *American Journal of Botany*, **97**, 1296–1303.

Bell, G. (1982). *The masterpiece of nature: the evolution and genetics of sexuality.* London: Croom Helm.

Belzer, N. F. & Ownbey, M. (1971). Chromatographic comparison of *Tragopogon* species and hybrids. *American Journal of Botany*, **58**, 791–802.

Ben-Ayed, R. I. *et al.* (2014). Genetic similarity among Tunisian olive cultivars and two unknown feral olive trees estimated through SSR markers. *Biochemical Genetics*, **52**, 258–68.

Bendiksby, M. *et al.* (2011). Allopolyploid origins of the *Galeopsis* tetraploids – revisiting Müntzing's classical textbook example using molecular tools. *New Phytologist*, **191**, 1150–67.

Bengtsson, B. O. (2009). Asex and evolution: a very large-scale overview. In *Lost sex: the evolutionary biology of parthenogenesis*, ed. I. Schön, K. Martens & P. van Dijk, pp. 1–19. Berlin: Springer Publications.

Bennett, J. H. (1983). *Natural selection, heredity and eugenics.* Oxford: Oxford University Press.

Bennett, K. D., Bhagwat, S. A. & Willis, K. J. (2012). Neotropical refugia. *The Holocene*, **22**, 1207–14.

Bennett, M. D. (1995). The development and use of genomic *in situ* hybridization (GISH) as a new tool in plant systematics. In *Kew Chromosome Conference IV*, ed. P. E. Brandham & M. D. Bennett, pp.167–83. Richmond, UK: Royal Botanic Gardens, Kew.

Bennett, S. T., Kenton, A. Y. & Bennett, M. D. (1992). Genomic *in situ* hybridization reveals the allopolyploid nature of *Milium montianum* (Graminae). *Chromosoma*, **101**, 420–4.

Benson, L. (1962). *Plant taxonomy.* New York: Ronald Press.

Bento, M. *et al.* (2013). Retrotransposons represent the most labile fraction for genomic rearrangements in polyploid plant species. *Cytogenetic and Genome Research*, DOI:10.1159/000353308

Benton, M. J. (2000). Stems, nodes, crown clades, and rank-free lists: is Linnaeus dead? *Biological Reviews*, **75**, 633–48.

Benton, M. J., Donoghue, P. C. J. & Asher, R. J. (2009). Calibrating and constraining molecular clocks. In *The timetree of life*, ed. S. B. Hedges and S. Kumar, pp. 35–86. Oxford: Oxford University Press.

Bergman, B. (1935). Zytologische Studien über sexuelles und asexuelles *Hieracium umbellatum*. *Hereditas*, **20**, 47–64.

Bergman, B. (1941). Studies on the embryo sac mother cell and its development in *Hieracium* subg. *Archieracium*. *Svensk botanisk Tidskrift*, **35**, 1–42.

Berlin, B., Breedlove, D. E. & Raven, P. H. (1974). *Principles of Tzeltal plant classification.* London & New York: Academic Press.

Berners-Lee, M. & Clark, D. (2013). *The burning question: we can't burn half the world's oil, coal and gas, so how do we quit?* London: Profile Books.

Berry, P. E., Tobe, H. & Gómez, J. A. (1991). Agamospermy and the loss of distyly in *Erythroxylum undulatum* (Erythroxylaceae) from Northern Venezuela. *American Journal of Botany*, **78**, 595–600.

Berry, R. J. (1977). *Inheritance and natural history.* London: Collins.

Bessey, C. E. (1908). The taxonomic aspect of the species questions. *American Naturalist*, **42**, 218–24.

Bessey, C. E. (1915). The phylogenetic taxonomy of flowering plants. *Annals of Missouri Botanic Garden*, **2**, 109–64.

Bhattachayya, M. *et al.* (1990). The wrinkled-seed character of pea described by Mendel is caused by a transposon-like insertion in a gene encoding starch-branching enzyme. *Cell*, **60**, 115–22.

Bhattachayya, M., Martin, C. & Smith, A. (1993). The importance of starch biosynthesis in the wrinkled seed shape character of peas studied by Mendel. *Plant Molecular Biology*, **22**, 525–31.

Bicknell, R. A. & Koltunow, A. M. (2004). Understanding apomixis: recent advances and remaining conundrums. *The Plant Cell*, **16**, S228–S245.

Bicknell, R. A., Lambie, S. C. & Butler, R. C. (2003). Quantification of progeny classes in two facultatively apomictic accessions of *Hieracium. Hereditas*, **138**, 11–20.

Billington, H. L. (1991). Effect of population size on genetic variability in a dioecious conifer. *Conservation Biology*, **5**, 115–19.

Bininda-Emonds, O. R. P., Gittleman, J. L. & Steel, M. A. (2002). The (super)tree of life. *Annual Review of Ecology and Systematics*, **33**, 265–89.

Birks, H. J. B. (1986). Late-Quaternary biotic changes in terrestrial and lacustrine environments, with particular reference to north-west Europe. In *Handbook of Holocene palaeoecology and palaeohydrology*, ed. E. Berglund, pp. 3–65. New York: Wiley.

Birks H. J. B. & Willis K. J. (2008) Alpine trees and refugia in Europe. *Plant Ecology and Diversity*, **1**, 147–60.

Bisby, F. A. *et al.* (2002). Taxonomy, at the click of a mouse. *Nature*, **418**, 367.

Bishop, J. A. & Cook, L. M. (1981) *Genetic consequences of man made change*. London: Academic Press.

Bishop, J. A. & Korn, M. E. (1969). Natural selection and cyanogenesis in White Clover, *Trifolium repens. Heredity*, **24**, 423–30.

Bishop, O. (1971). *Statistics for biology. A practical guide for the experimental biologist*, 2nd edn. London: Longmans.

Bittencourt Júnior, N. S., Gibbs, P. E. & Semir, J. (2003). Histological study of post-pollination events in *Spathodea campanulata* Beauv. (Bignoniaceae), a species with late-acting self-incompatibility. *Annals of Botany*, **91**, 827–34.

Bittrich, V. & Kadereit, J. (1988). Cytogenetical and geographical aspects of sterility in *Lysimachia nummularia. Nordic Journal of Botany*, **8**, 325–8.

Blackwell, W. H. (2002). One-hundred-year code déjà vu? *Taxon*, **51**, 151–4.

Blair, A. W. & Williamson, P. S. (2010). Pollen dispersal in Star Cactus (*Astrophytum asterias*). *Journal of Arid Environments*, **74**, 525–7.

Blakeslee, A. F. & Avery, A. G. (1937). Methods of inducing chromosome doubling in plants. *Journal of Heredity*, **28**, 393–411.

Blaxter, M. L. (2004). The promise of a DNA taxonomy. *Philosophical Transactions of the Royal Society of London, B*, **359**, 669–79.

Bleeker, W. & Hurka, H. (2001). Introgressive hybridization in *Rorippa* (Brassicaceae): gene flow and its consequences in natural and anthropogenic habitats. *Molecular Ecology*, **10**, 2013–22.

Bliege Bird, R. *et al.* (2008). The 'fire stick farming' hypothesis: Australian Aboriginal foraging strategies, biodiversity and anthropogenic fire mosaics. *Proceedings of the National Academy of Sciences, USA*, **105**, 14796–801.

Blois, J. L. *et al.* (2013). Climate change and the past, present, and future of biotic interactions. *Science*, **341**. 499–504.

Blunt, W. (2004). *Linnaeus: The Compleat Naturalist.* London: Frances Lincoln.

Bøcher, T. W. (1949). Racial divergences in *Prunella vulgaris* in relation to habitat and climate. *New Phytologist*, **48**, 285–314.

Bøcher, T. W. (1963). The study of ecotypical variation in relation to experimental morphology. *Regnum Vegetabile*, **27**, 10–16.

Bøcher, T. W. & Larsen, K. (1958). Geographical distribution of initiation of flowering, growth habit and other factors in *Holcus lanatus. Botaniska Notiser*, **3**, 289–300.

Bøcher, T. W. & Lewis, M. C. (1962). Experimental and cytological studies on plant species. 7, *Geranium sanguineum. Biologiske Skrifter*, **11**, 1–25.

Bohm, W. (1979). *Methods of studying root systems.* Berlin, Heidelberg & New York: Springer-Verlag.

Bolkhovskikh, Z., Grif, V., Matvejeva, T. & Zakharyeva, O. (1969). *Chromosome numbers of flowering plants.* Leningrad: Academy of Sciences of the USSR.

Bomblies, K. (2010). Doomed lovers: mechanisms of isolation and incompatibility in plant speciation. *Annual Review of Plant Biology*, **61**, 109–24.

Bomblies, K. & Weigel, D. (2010). *Arabidopsis* and relatives as models for the study of genetic and genomic incompatibilities. *Philosophical Transactions of the Royal Society of London B*, **365**, 1815–23.

Bomblies, K. *et al.* (2007). Autoimmune response as a mechanism for a Dobzhansky–Muller-type incompatibility syndrome in plants. *PLoS Biology*, **5**, 1962–72.

Bone, E. & Farres, A. (2001). Trends and rates of microevolution in plants. *Genetica*, **112–113**, 165–82.

Bonin, A., Ehrich, D. & Manel, S. (2007). Statistical analysis of amplified fragment length polymorphism data: a toolbox for molecular ecologists and evolutionists. *Molecular Ecology*, **16**, 3737–58.

Bonnier, G. (1890). Cultures expérimentales dans les Alpes et les Pyrénées. *Revue générale de Botanique*, **2**, 513–46.

Bonnier, G. (1895). Recherches expérimentales sur l'adaptation des plantes au climat Alpin. *Annales des Sciences naturelles (Botanique)*, **20**, 217–360.

Bonnier, G. (1920). Nouvelles observations sur les cultures expérimentales à diverses altitudes et cultures par semis. *Revue générale de Botanique*, **32**, 305–26.

Bordacs, S. *et al.* (2002). Chloroplast DNA variation of white oaks in northern Balkans and in the Carpathian Basin. *Forest Ecology and Management*, **156**, 197–209.

Borg, S. J. (1972). Variability of *Rhinanthus serotinus* (Schönh.) Oborny in relation to environment. Unpublished thesis, Rijksuniversiteit te Groningen.

Borges L. A., Sobrinho, M. S. & Lopes, A. V. (2009). Phenology, pollination, and breeding system of the threatened tree *Caesalpinia echinata* Lam. (Fabaceae), and a review of studies on the reproductive biology in the genus. *Flora*, **204**, 111–30.

Borgström, G. (1939). Formation of cleistogamic and chasmogamic flowers in Wild Violets as a photoperiodic response. *Nature*, **144**, 514–15.

Bosemark, N. O. (1954). On accessory chromosomes in *Festuca pratensis*. I. Cytological investigations. *Hereditas*, **40**, 346–76.

Bossdorf, O. *et al.* (2005). Phenotypic and genetic differentiation in native versus introduced plant populations. *Oecologia*, **144**, 1–11.

Boswell Syme, J. T. (ed.) (1866). *English botany; or Coloured figures of British plants*, 3rd edn., vol. **6**. London: Hardwicke.

Bothmer, R. von *et al.* (1971). Clonal variation in populations of *Anemone nemorosa* L. *Botaniska Notiser*, **124**, 505–19.

Bottin L., Le Cadre S., Quilichini A., Bardin P., Moret, J. *et al.* (2007) Re-establishment trials in endangered plants: a review and the example of *Arenaria grandiflora*, a species on the brink of extinction in the Parisian region (France). *Ecoscience*, **14**, 410–19, DOI:10.2980/1195-6860.

Bouck, A. C. *et al.* (2005). Genetic mapping of species boundaries in Louisiana Irises using IRRE retrotransposon display markers. *Genetics*, **171**, 1289–1303.

Bowlby, J. (1990). *Charles Darwin. A biography*. London: Hutchinson.

Bowler, P. (2008). Foreword. In *Natural selection and beyond: the intellectual legacy of Alfred Russel Wallace*, ed. C. H. Smith & G. Beccaloni. Oxford: Oxford University Press.

Bowler, P. J. (1989a). *The Mendelian revolution*. London: The Athlone Press.

Bowler, P. J. (1989b). *Evolution. The history of an idea*. Revised edition. Berkeley: University of California Press.

Bowler, P. J. (1990). *Charles Darwin; the man and his influence*. Oxford: Basil Blackwell.

Bowles, M. L. & Whelan, C. J. (1994). *Restoration of endangered species, conceptual issues, planning and implementation*. Cambridge: Cambridge University Press.

Box, J. F. (1978). *R. A. Fisher. The life of a scientist*. New York, Chichester, Brisbane & Toronto: Wiley.

Box, M.S. *et al.* (2008). Floral ontogenetic evidence of repeated speciation via paedomorphosis in subtribe Orchidinae (Orchidaceae). *Botanical Journal of the Linnean Society*, **157**, 429–54.

Brackman, A. C. (1980). *A delicate arrangement: the strange case of Charles Darwin and Alfred Russel Wallace*. New York: Times Books.

Bradley, R. S. (1985). *Quaternary paleoclimatology: methods of paleoclimatic reconstruction*. Boston: Allen & Unwin.

Bradshaw, A. D. (1959a). Population differentiation in *Agrostis tenuis* Sibth. I. Morphological differentiation. *New Phytologist*, **58**, 208–27.

Bradshaw, A. D. (1959b). Population differentiation in *Agrostis tenuis* Sibth. II. The incidence and significance of infection by *Epichloë typhina*. *New Phytologist*, **58**, 310–15.

Bradshaw, A. D. (1959c). Studies of variation in bent grass species. II. Variation within *Agrostis tenuis*. *Journal of the Sports Turf Research Institute*, **10**, 1–7.

Bradshaw, A. D. (1960). Population differentiation in *Agrostis tenuis* Sibth. III. Populations in varied environments. *New Phytologist*, **59**, 92–103.

Bradshaw, A. D. (1965). Evolutionary significance of phenotypic plasticity in plants. *Advances in Genetics*, **13**, 115–55.

Bradshaw, A. D. (1976). Pollution and evolution. In *Effects of air pollution on plants*, ed. T. A. Mansfield, pp. 135–59.

London, New York & Melbourne: Cambridge University Press.

Bradshaw, A. D. (1989). Is evolution fettered or free? *Transactions of the Botanical Society of Edinburgh*, **45**, 303–11.

Bradshaw, A. D. & McNeilly, T. (1981). *Evolution and pollution.* London: Arnold.

Bradshaw, A. D. & McNeilly, T. (1991). Evolutionary response to global climate change. *Annals of Botany*, **87** (Suppl.), 5–14.

Bradshaw, C. J. A. (2012). Little left to lose: deforestation and forest degradation in Australia since European colonization. *Journal of Plant Ecology*. **5**, 109–20.

Bradshaw, M. E. (1963a). Studies on *Alchemilla filicaulis* Bus., sensu lato and *A. minima* Walters. Introduction and I. Morphological variation in *A. filicaulis*, sensu lato. *Watsonia*, **5**, 304–20.

Bradshaw, M. E. (1963b). Studies on *Alchemilla filicaulis* Bus., sensu lato, and *A. minima* Walters. II. Cytology of *A. filicaulis*, sensu lato. *Watsonia*, **5**, 321–6.

Bradshaw, M. E. (1964). Studies on *Alchemilla filicaulis* Bus., sensu lato and *A. minima* Walters. III. *Alchemilla minima*. *Watsonia*, **6**, 76–81.

Brady, K. U., Kruckeberg, A. R. & Bradshaw, H. D., Jr (2005). Evolutionary ecology of plant adaptation to serpentine soils. *Annual Review of Ecology, Evolution, and Systematics*, **36**, 243–66.

Brand, C. J. & Waldron, L. R. (1910). Cold resistance of Alfalfa and some factors influencing it. *U.S. Department of Agriculture, Bureau of Plant Industry. Bulletin* 185, 1–80.

Brannigan, A. (1979). The reification of Mendel. *Social Studies of Science*, **9**, 423–54.

Brauner, S. & Gottlieb, L. D. (1987). A self-compatible plant of *Stephanomeria exigua* subsp. *coronaria* (Asteraceae) and its relevance to the origin of its self-pollinating derivative *S. malheurensis*. *Systematic Botany*, **12**, 299–304.

Breed, M. F. *et al.* (2013). Which provenance and where? Seed sourcing strategies for revegetation in a changing environment. *Conservation Genetics*, **14**, 1–10.

Brehm, B. G. & Ownbey, M. (1965). Variation in chromatographic patterns in the *Tragopogon dubius pratensis – porrifolius* complex (Compositae). *American Journal of Botany*, **52**, 81, 1–18.

Breiman, A. & Graur, D. (1995). Wheat evolution. *Israel Journal of Plant Sciences*, **43**, 85–98.

Breinholt, J. W. *et al.* (2009). Population structure of an endangered Utah endemic *Astragalus ampullaroides* (Fabaceae). *American Journal of Botany*, **96**, 661–7.

Brenchley, W. E. & Warington, K. (1969). *The Park Grass plots at Rothamsted, 1856–1949*. Harpenden: Rothamsted Experimental Station.

Brennan, A. C. (2014). The genetic structure of *Arabidopsis thaliana* in the south-western Mediterranean range reveals a shared history between North Africa and southern Europe. *BMC Plant Biology*, **14**:17, DOI:10.1186/1471-2229-14-17.

Brennan, A. C. *et al.* (2011). Sporophytic self-incompatibility in *Senecio squalidus* (Asteraceae): S allele dominance interactions and modifiers of cross-compatibility and selfing rates. *Heredity*, **106**, 113–23.

Brenner G. J. (1996). Evidence for the earliest stage of angiosperm pollen evolution: a paleoequatorial section from Israel. In *Flowering plant origin, evolution and phylogeny*, ed. D. W. Taylor & L. J. Hickey, pp.91–115. New York: Chapman and Hall.

Breshears, D. D. *et al.* (2005). Regional vegetation die-off in response to global-change-type drought. *Proceedings of the National Academy of Sciences, USA*, **102**, 15144–8.

Bretagnolle, F. & Thompson, J. D. (1995). Gametes with the somatic chromosome number: mechanisms of their formation and role in the evolution in autoployploid plants. *New Phytologist*, **129**, 1–22.

Briand, F. *et al.* (1989). *The Alps. A system under pressure.* Chambéry: International Union for Conservation of Nature.

Brickell, C. D. *et al.* (2008) Do the views of users of taxonomic output count for anything? *Taxon*, **57**, 1047–8.

Briggs, D. (2009). *Plant microevolution and conservation in human-influenced ecosystems.* Cambridge: Cambridge University Press.

Briggs, D. & Block, M. (1981). An investigation into the use of the '-deme' terminology. *New Phytologist*, **89**, 729–35.

Briggs, D. & Walters, S. M. (1997). *Plant variation and evolution*, 3rd edn. Cambridge: Cambridge University Press.

Briggs, J. C. (1987). *Biogeography and plate tectonics.* Amsterdam: Elsevier.

Broadhurst, L. M. *et al.* (2008). Seed supply for broadscale restoration: maximizing evolutionary potential. *Evolutionary Applications*, 1, 587–97.

Brochet, A. L. *et al.* (2009). The role of migratory ducks in the long-distance dispersal of native plants and the spread of exotic plants in Europe. *Ecography*, 32, 298–319.

Brochmann, C., Soltis, P. S. & Soltis, D. E. (1992a). Recurrent formation and polyphyly of Nordic polyploids in *Draba* (Brassicaceae). *American Journal of Botany*, 79, 673–88.

Brochmann, C., Soltis, P. S. & Soltis, D. E. (1992b). Multiple origins of the octoploid Scandinavian endemic *Draba cacuminum*: electrophoretic and morphological evidence. *Nordic Journal of Botany*, 12, 257–72.

Brochmann C. *et al.* (1998). Molecular evidence for polyploid origins in *Saxifraga* (Saxifragaceae): the narrow arctic endemic *S. svalbardensis* and its widespread allies. *American Journal of Botany*, 85, 135–43.

Brochmann, C. *et al.* (2000). Multiple diploid hybrid speciation of the Canary Island endemic *Argyranthemum sundingii* (Asteraceae). *Plant Systematics and Evolution*, 220, 77–92.

Brochmann, C. *et al.* (2004). Polyploidy in arctic plants. *Biological Journal of the Linnean Society*, 82, 521–36.

Brockway, L. H. (1979). *Science and colonial expansion: the role of the British Royal Botanic Gardens*. London: Academic Press.

Bronstein, J. L., Alarcón, R. & Geber, M. (2006). The evolution of plant–insect mutualisms. *New Phytologist*, 172, 412–28.

Brooks, J. L. (1983). *Just before the Origin: Alfred Russel Wallace's theory of evolution*. New York: Columbia University Press.

Brooks, R. R. *et al.* (1977). Detection of nickeliferous rocks by analysis of herbarium specimens of indicator plants. *Journal of Geochemistry Exploration*, 7, 49–57.

Brougham, R. W. & Harris, W. (1967). Rapidity and extent of changes in genotypic structure induced by grazing in a Ryegrass population. *New Zealand Journal of Agricultural Research*, 10, 56–65.

Brown, A. H. D. (1979). Enzyme polymorphism in plant populations. *Theoretical Population Biology*, 15, 1–42.

Brown, A. H. D. & Burdon, J. J. (1983). Multilocus diversity in an outbreeding weed, *Echium plantagineum* L. *Australian Journal of Biological Sciences*, 36, 503–9.

Brown, A. H. D. & Marshall, D. R. (1981). Evolutionary changes accompanying colonization in plants. In *Evolution today*, ed. G. C. E. Sudder & J. L. Reveal, pp. 351–63. Pittsburgh, PA: Hunt Institute for Botanical Documentation.

Brown, A. D. H. & Schoen, D. J. (1992). Plant population genetic structure and biological conservation. In *Conservation of biodiversity for sustainable development*, ed. O. T. Sandlund *et al.*, pp. 88–104. Oslo: Scandinavian University Press.

Brown, V. K. & Lawton, J. H. (1991). Herbivory and the evolution of leaf size and shape. *Philosophical Transactions of the Royal Society of London, B*, 333, 267–72.

Browne, J. (1983). *The Secular Ark; studies in the history of biogeography*. New Haven, CT, & London: Yale University Press.

Browne, J. (1995). *Charles Darwin: Voyaging. Volume 1 of a biography*. London: Pimlico. Random House.

Browne, J. (2002). *Charles Darwin: The power of place. Volume 2 of a biography*. London: Pimlico. Random House.

Brownfield, L. & Köhler, C. (2011).Unreduced gamete formation in plants: mechanisms and prospects. *Journal of Experimental Botany*, 62, 1659–68.

Brucker, R. M. & Bordenstein, S. R. (2013).Speciation by symbiosis. *Trends in Ecology & Evolution*, 27, 443–51.

Brummitt, R. K. (1996). In defence of paraphyletic taxa. In *The biodiversity of African plants*, ed. L. J. G. van der Maesen, X. M. van der Burgt & J. M. van Medenbach de Rooy, Proceedings XIVth AETFAT Congress, 22–27 August 1994, Wageningen, pp. 371–84. Dordrecht: Kluwer.

Brummitt, R. K. (1997). Taxonomy versus cladonomy, a fundamental controversy in biological systematics. *Taxon*, 46, 723–34.

Brummitt, R. K. (2002). How to chop up a tree. *Taxon*, 51, 31–41.

Brummitt, R. K. (2003). Further dogged defense of paraphyletic taxa. *Taxon*, 52, 803–4.

Brummitt, R. K. (2006). The democratic processes of plant nomenclature. In *Taxonomy and plant conservation: the cornerstone of the conservation and the sustainable use of plants*, ed. E. Leadlay & S, Jury, pp. 101–29. Cambridge: Cambridge University Press.

Brunsfeld, S. J. & Sullivan, J. (2005). A multi-compartmented glacial refugium in the northern Rocky Mountains: evidence from the phylogeography of *Cardamine constancei* (Brassicaceae). *Conservation Genetics*, **6**, 895–904.

Brus, R. (2010). Growing evidence for the existence of glacial refugia of European beech (*Fagus sylvatica* L.) in the south-eastern Alps and north-western Dinaric Alps. *Periodicum Biologorum*, **112**, 239–46.

Brush, S. G. (2009). Choosing selection. The revival of natural selection in Anglo-American evolutionary biology, 1930–1970. *Transactions of the American Philosophical Society*, **99** (3), 1–183. Philadelphia.

Brussard, P. F. (1997). A paradigm in conservation biology. *Science*, **277**, 527–8.

Bryant, H. N. & Cantino, P. D. (2002). A review of criticisms of phylogenetic nomenclature: is taxonomic freedom the fundamental issue? *Biological Review*, **77**, 39–55.

Brysting, A. K., Mathiesen, C. & Marcussen, T. (2011). Challenges in polyploid phylogenetic reconstruction: a case story from the arctic–alpine *Cerastium alpinum* complex. *Taxon*, **60**, 333–47.

Budiansky, S. (1995). *Nature's keepers. the new science of nature management.* London: Weidenfeld & Nicolson.

Buggs, R. J. A. (2007). Empirical studies of hybrid zone movement. *Heredity*, **99**, 301–12.

Buggs, R. J. A. *et al.* (2008). Does phylogenetic distance between parental genomes govern the success of polyploids? *Castanea*, **73**, 74–93.

Buggs, R. J. A., Soltis, P. S. & Soltis, D. E. (2009). Does hybridization between divergent progenitors drive whole-genome duplication? *Molecular Ecology*, **18**, 3334–9.

Buggs R. J. A. *et al.* (2009). Gene loss and silencing in *Tragopogon miscellus* (Asteraceae): comparison of natural and synthetic allotetraploids. *Heredity*, **103**, 73–81.

Buggs, R. J. A. *et al.* (2010a). Tissue-specific silencing of homoeologs in natural populations of the recent allopolyploid *Tragopogon mirus*. *New Phytologist*, **186**, 175–83.

Buggs, R. J. A. *et al.* (2010b). Characterization of duplicate gene evolution in the recent natural allopolyploid *Tragopogon miscellus* by next-generation sequencing and Sequenom iPLEX MassARRAY genotyping. *Molecular Ecology*, **19**, 132–46.

Buggs, R. J. A., Soltis, P. S. & Soltis, D. E. (2011). Biosystematic relationships and the formation of polyploids. *Taxon*, **60**, 324–32.

Buggs, R. J. A. *et al.* (2011). Transcriptomic shock generates evolutionary novelty in a newly formed, natural allopolyploid plant. *Current Biology*, **21**, 551–6.

Buggs, R. J. A. *et al.* (2012). Next-generation sequencing and genome evolution in allopolyploids. *American Journal of Botany*, **99**, 372–82.

Bullock, J. M. (1998). Community translocation in Britain: setting objectives and measuring consequences. *Biological Conservation*, **6**, 166–74.

Bulmer, M. (2003) *Francis Galton: pioneer of heredity and biometry.* Baltimore: John Hopkins University Press.

Bulmer, M. G. (1967). *Principles of statistics*, 2nd edn. London & Edinburgh: Oliver & Boyd.

Burchfield, J. D. (1990). *Lord Kelvin and the age of the Earth.* Chicago & London: University of Chicago Press.

Burd, M. (1994). Bateman's principle and plant reproduction: the role of pollen limitation in fruit and seed set. *Botanical Review*, **60**, 83–139.

Burdon, J. J. (1980). Intraspecific diversity in a natural population of *Trifolium repens*. *Journal of Ecology*, **68**, 717–35.

Burdon, J. J. (1987). *Disease and plant population biology.* Cambridge: Cambridge University Press.

Burdon, J. J., Marshall, D. R. & Groves, R. H. (1980). Isozyme variation in *Chondrilla juncea* in Australia. *Australian Journal of Botany*, **28**, 193–8.

Burgess, K. S. *et al.* (2005). Asymmetrical introgression between two *Morus* species (*M. alba, M. rubra*) that differ in abundance. *Molecular Ecology*, **14**, 3471–83.

Burkhardt, F. & Smith, S. (1985–). *The correspondence of Charles Darwin.* Cambridge: Cambridge University Press.

Burkhardt, F. & Smith, S. (1991). *The correspondence of Charles Darwin.* Volume 7 1858–9. Cambridge: Cambridge University Press.

Burkill, I. H. (1895). On the variations in number of stamens and carpels. *Journal of the Linnean Society (Botany)*, **31**, 216–45.

Busch, J. *et al.* (2012). Climate change and the cost of conserving species in Madagascar. *Conservation Biology*, **26**, 408–19.

Bush, E. J. & Barrett, S. C. H. (1993). Genetics of mine invasions by *Deschampsia cespitosa* (Poaceae). *Canadian Journal of Botany – Revue Canadienne de Botanique*, **71**, 1336–48.

Bush, M. B. & de Oliveira, P. E. (2006). The rise and fall of the Refugial Hypothesis of Amazonian Speciation: a paleo-ecological perspective. *Biota Neotropica*, **6**, 1–17.

Busi, R. *et al.* (2013). Herbicide-resistant weeds: from research and knowledge to future needs. *Evolutionary Applications*, **6**, 1218–21.

Cahn, M. A. & Harper, J. L. (1976a). The biology of the leaf mark polymorphism in *Trifolium repens.* 1. Distribution of phenotypes at a local scale. *Heredity*, **37**, 309–25.

Cahn, M. A. & Harper, J. L. (1976b). The biology of the leaf mark polymorphism in *Trifolium repens.* 2. Evidence for the selection of marks by rumen fistulated sheep. *Heredity*, **37**, 327–33.

Cain, A. J. (1958). Logic and memory in Linnaeus's system of taxonomy. *Proceedings of the Linnean Society of London*, **169**, 144–63.

Cain, A. J. (1996). John Ray on 'Accidents'. *Archives of Natural History*, **23**, 343–68.

Cain, A. J. (1999). John Ray on species. *Archives of Natural History*, **26**, 223–38.

Cain, A. J. (2008). The post-Linnaean development of taxonomy. *Proceedings of the Linnean Society of London*, **170**, 234–44.

Cain, J. & Ruse, M. (2009). Descended from Darwin. Insights into the history of evolutionary studies, 1900–1970. *Transactions of the American Philosophical Society*, **99** (1), 1–386.

Cain, M. L., Milligan, B. G. & Strand, A. E. (2000). Long-distance seed dispersal in plant populations. *American Journal of Botany*, **87**, 1217–27.

Cairns, J. (1998). Ecological restoration. In *Encyclopedia of ecology & environmental management*, ed. P. Calow, pp. 217–19. Oxford: Blackwell.

Cairns, J. (2002). Rationale for restoration. In *Handbook of ecological restoration*, ed. M. R. Perrow & A. J. Davy, pp.10–23. Cambridge: Cambridge University Press.

Caisse, M. & Antonovics, J. (1978). Evolution in closely adjacent plant populations. IX. Evolution of reproductive isolation in clinal populations. *Heredity*, **40**, 371–84.

Callender, L. A. (1988). Gregor Mendel: an opponent of descent with modification. *History of Science*, **26**, 41–75.

Calsbeek, B. *et al.* (2011). Comparing the genetic architecture and potential response to selection of invasive and native populations of Reed Canary Grass. *Evolutionary Applications*, **4**, 726–35.

Calsbeek, R., Thompson, J. R. & Richardson, J. E. (2003). A phylogenetic analysis of Rhamnaceae using rbcL and trnL-F plastid DNA sequences. *Molecular Ecology*, **12**, 1021–9.

Caltagirone, L. E. (1981). Landmark examples in classical biological control. *Annual Review of Entomology*, **26**, 213–32.

Camacho, J. P. M. (2005). B chromosomes. In *The evolution of the genome*, ed. T. R. Gregory, pp. 223–86, London: Elsevier Academic Press.

Camp, W. H. & Gilly, C. L. (1943). The structure and origin of species. *Brittonia*, **4**, 323–85.

Campbell, R. C. (1967). *Statistics for biologists.* London & New York: Cambridge University Press. [2nd edn: 1974].

Campbell, S. A & Kessler, A. (2013). Plant mating system transitions drive the macroevolution of defence strategies. *Proceedings of the National Academy of Sciences, USA*, **110**, 3973–8.

Carlo, T. A., Tewksbury, J. J. & del Rio, C. M. (2009). A new method to track seed dispersal and recruitment using N-15 isotope enrichment. *Ecology*, **90**, 3516–25.

Carlquist, S. (1965). *Island life. A natural history of the islands of the world.* New York: Natural History Press.

Carlquist, S. (1974). *Island biology.* New York & London: Columbia University Press.

Carlson, S., Mayer, V. & Michael, J. (2009). Phylogenetic relationships, taxonomy, and morphological evolution in Dipsacaceae (Dipsacales) inferred by DNA sequence data. *Taxon*, **58**, 1075–91.

Carney, S. E., Cruzan, M. B. & Arnold, M. L. (1994). Reproductive interactions between hybridizing Irises: analyses of pollen-tube growth and fertilization success. *American Journal of Botany*, **81**, 1169–75.

Caron, G. E. & Leblanc, R. (1992). Pollen contamination on a small black spruce seedling seed orchard for 3 consecutive years. *Forest Ecology & Management*, **53**, 587–92.

Carr, G. D. *et al.* (1999). Chromosome numbers in Compositae. XVIII. *American Journal of Botany*, **86**, 1003–13.

Carson, R. (1962). *Silent spring.* London: Hamish Hamilton.

Catcheside, D. G. (1939). A position effect in *Oenothera. Journal of Genetics*, **38**, 345–52.

Catcheside, D. G. (1947). The *P*-locus position effect in *Oenothera. Journal of Genetics*, 48, 31–42.

Caujapé-Castells, J. (2004). Boomerangs of biodiversity? The interchange of biodiversity between mainland north Africa and the Canary Islands as inferred from cpDNA RFLPs in genus *Androcymbium. Botánica Macaronésica*, 25, 53–69.

Caujapé-Castells, J. *et al.* (2010). Conservation of oceanic island floras: present and future global challenges. *Perspectives in Plant Ecology, Evolution and Systematics*, 12, 107–29.

Cayouette, J. E. & Morisset, P. (1986). Chromosome studies on the *Carex salina* complex (Cyperaceae section Cryptocarpae) in northeastern North America. *Cytologia*, 51, 817–56.

Chadwick, M. J. & Salt, J. K. (1969). Population differentiation within *Agrostis tenuis* L. in response to colliery spoil substrate factors. *Nature*, 224, 186.

Chaing, G. C. K. *et al.* (2009). Major flowering time gene, *FLOWERING LOCUS C*, regulates seed germination in *Arabidopsis thaliana. Proceedings of the National Academy of Sciences, USA*, 106, 11661–6.

Chakraborty, R., Meagher, T. R. & Smouse, P. E. (1988). Parentage analysis with genetic markers in natural populations. I. The expected proportion of offspring with unambiguous paternity. *Genetics*, 118, 527–36.

Challice, J. & Kovanda, M. (1978). Chemotaxonomic survey of the genus *Sorbus* in Europe. *Naturwissenschaften*, 65, 111–12.

Chandler, S. & Dunwell, J. M. (2008). Gene flow, risk assessment and the environmental release of transgenic plants. *Critical Reviews in Plant Sciences*, 27, 25–49.

Chanway, C. P. Holl, F. B. & Turkington, R. (1989). Effect of *Rhizobium leguminosarum* biovar. *trifolii* genotype on specificity between *Trifolium repens* and *Lolium perenne. Journal of Ecology*, 77, 1150–60.

Chapman, M. A. & Abbott, R. J. (2009). Introgression of fitness genes across a ploidy barrier. *New Phytologist*, 186, 63–71.

Chapman, M. A. & Burke, J. M. (2006). Letting the gene out of the bottle: the population genetics of genetically modified crops. *New Phytologist*, 170, 429–43.

Charlesworth, B. (2009). Effective population size and patterns of molecular evolution and variation. *Nature Reviews of Genetics*, 10, 195–205.

Charlesworth, D. (1985). Distribution of dioecy and self-incompatibility in angiosperms. In *Evolution – essays in honour of John Maynard Smith*, ed. P. J. Greenwood & M. Slatkin, pp. 237–68. Cambridge: Cambridge University Press.

Charlesworth, D. (2002). Plant sex determination and sex chromosomes. *Heredity*, 88, 94–101.

Charlesworth, D. (2006). Evolution of plant breeding systems. *Current Biology*, 16, R726–R735.

Charlesworth, D. (2010). Self-incompatibility. F1000 Report Biology, 2,68. Published online, DOI:10.3410/B2-68.

Charlesworth, D. & Charlesworth, B. (1979). The evolutionary genetics of sexual systems in flowering plants. *Proceedings of the Royal Society London, B*, 205, 513–30.

Charlesworth, D. & Charlesworth, B. (1987). Inbreeding depression and its evolutionary consequences. *Annual Review of Ecology and Systematics*, 18, 237–68.

Charlesworth, D. & Willis, J. H. (2009). The genetics of inbreeding depression. *Nature*, 10, 783–96.

Charlesworth, D. *et al.* (2005). Plant self-incompatibility systems: a molecular evolutionary perspective. *New Phytologist*, 168, 61–9.

Charnov, E. L. (1988). Foreword. In *Plant reproductive ecology*, ed. J. Lovett Doust & L. Lovett Doust, pp. ix–x. New York: Oxford University Press.

Chase, M. W. *et al.* (1993). Phylogenetics of seed plants: an analysis of nucleotide-sequences from the plastid gene *rbcL. Annals of the Missouri Botanical Garden*, 80, 528–80.

Chase, M. W., Fay, M. F. & Savolainen, V. (2000). Higher-level classification in the angiosperms: new insights from the perspective of DNA sequence data. *Taxon*, 49, 685–704.

Chase, M. W. *et al.* (2005). Land plants and DNA barcodes: short-term and long-term goals. *Philosophical Transactions of the Royal Society of London, B*, 359, 1889–95.

Chase, M. W. *et al.* (2009). Murderous plants: Victorian Gothic, Darwin and modern insight into vegetable carnivory. *Botanical Journal of the Linnean Society*, 161, 329–56.

Chen, J.-Q. *et al.* (2008). Over-expression of OsDREB genes leads to enhanced drought tolerance in rice. *Biotechnology Letters*, 30, 2191–8.

Chen, J. *et al.* (2012). Disentangling the roles of history and local selection in shaping clinal variation in allele frequencies and gene expression in Norway Spruce (*Picea abies*). *Genetics*, **191**, 865–81.

Chen, J. *et al.* (2014). Clinal variation at phenology-related genes in Spruce; parallel evolution in FTL2 and Gigantea? *Genetics*, **197**, 1025–38.

Chen, Q. & Armstrong, K. (1994). Genomic *in situ* hybridization in *Avena sativa*. *Genome*, **37**, 607–12.

Cheplick, G. P. & Quinn, J. A. (1982). *Amphicarpum purshii* and the 'pessimistic strategy' in amphicarpic annuals with subterranean fruit. *Oecologia*, **52**, 327–32.

Cheplick, G. P. & Quinn, J. A. (1983). The shift in aerial/subterranean fruit ratio in *Amphicarpum purshii*: causes and significance. *Oecologia*, **57**, 374–9.

Cheptou, P. O. (2012). Clarifying Baker's Law. *Annals of Botany*, **109**, 633–41.

Chester, M. *et al.* (2007). Parentage of endemic *Sorbus* L. (Rosaceae) species in the British Isles. *Botanical Journal of the Linnean Society*, **154**, 291–304.

Chester, M. *et al.* (2010a). Review of the application of modern cytological methods (FISH/GISH) to the study of reticulation (Polyploidy/Hybridisation). *Genes* (Basel), **1**, 166–92, DOI:10.3390/genes1020166.

Chester, M. *et al.* (2010b). Extensive chromosomal variation in a recently formed natural allopolyploid species, *Tragopogon miscellus* (Asteraceae). *Proceedings of the National Academy of Sciences, USA*, **109**, 1176–81.

Choi, Y. D. (2004). Theories for ecological restoration in changing environment: towards 'futuristic' restoration. *Ecological Research*, **19**, 75–81.

Choi, Y. D. (2007). Restoration ecology to the future: a call for a new paradigm. *Restoration Ecology*, **15**, 351–3.

Chomorro, S. *et al.* (2012). Pollination patterns and plant breeding systems in the Galápagos: a review. *Annals of Botany*, **110**, 1489–1501.

Chung, M. G., Hamrick, J. L., Jones, S. B. & Derda, G. S. (1991). Isozyme variation within and among populations of *Hosta* (Liliaceae) in Korea. *Systematic Botany*, **16**, 667–84.

Church, S. A. & Taylor, D. R. (2007). The evolution of reproductive isolation in spatially structured populations. *Evolution*, **56**, 1859–62.

Cicerone, R. J. & Nurse, P. (2014). *Climate change: evidence and causes*. US National Academy of Sciences and the Royal Society.

Claessen, D. *et al.* (2005a). Which traits promote persistence of feral GM crops? Part 1: implications of environmental stochasticity. *Oikos*, **110**, 20–29.

Claessen, D. *et al.* (2005b). Which traits promote persistence of feral GM crops? Part 2: implications of metapopulation structure. *Oikos*, **110**, 30–42.

Clapham, A. R., Tutin, T. G. & Warburg, E. F. (1981). *Excursion flora of the British Isles*. London: Cambridge University Press.

Clark, B. R. *et al.* (2009). Taxonomy as an escience. *Philosophical Transactions of the Royal Society of London, B*, **367**, 953–66.

Clark, D. (2005). *Molecular biology*. Burlington, MA, & London: Elsevier.

Clark, L. V. & Jasieniuk, M. (2012). Spontaneous hybrids between native and exotic *Rubus* in the Western United States produce offspring both by apomixis and by sexual recombination. *Heredity*, **109**, 320–8.

Clark, P. *et al.* (2009). The Last Glacial Maximum. *Science*, **325**, 710–14.

Clarke, G. M. (1980). *Statistics and experimental design*. London: Edward Arnold.

Clarke, J. T., Warnock, R. C. M. & Donoghue, P. C. J. (2011). Establishing a timescale for plant evolution. *New Phytologist*, **192**, 266–301.

Clausen, J. (1951). *Stages in the evolution of plant species*. London: Oxford University Press; Ithaca, NY: Cornell University Press.

Clausen, J. & Hiesey, W. M. (1958). *Experimental studies on the nature of species. IV. Genetic structure of ecological races*. Carnegie Institution of Washington Publication No. 615, Washington DC.

Clausen, J., Keck, D. D. & Hiesey, W. M. (1939). The concept of species based on experiment. *American Journal of Botany*, **26**, 103–6.

Clausen, J., Keck, D. D. & Hiesey, W. M. (1940). *Experimental studies on the nature of species. I. The effect of varied environments on Western North American plants*. Carnegie Institution of Washington Publication No. 520, pp. 1–452, Washington DC.

Clausen, J., Keck, D. D. & Hiesey, W. M. (1941). Experimental taxonomy. *Carnegie Institution of Washington Year Book*, **40**, 160–70.

Clausen, J., Keck, D. D. & Hiesey, W. M. (1945). Experimental studies on the nature of species. 11. Plant evolution through amphiploidy and autoploidy, with examples from the *Madiinae*. *Carnegie Institute of Washington Publication* 564. Washington DC.

Clausen, R. E. & Goodspeed, T. H. (1925). Interspecific hybridization in *Nicotiana*. II. A tetraploid *glutinosa-tabacum* hybrid, an experimental verification of Winge's hypothesis. *Genetics*, 10, 279–84.

Clements, D. R. & Ditommosa, A. (2011). Climate change and weed adaptation: can evolution of invasive plants lead to greater range expansion than forecasted? *Weed Research*, 51, 227–40.

Clements, E. J. & Foster, M. C. (1994). *Alien plants in the British Isles*. London: Botanical Society of the British Isles.

Clements, F. E., Martin, E. V. & Long, F. L. (1950). *Adaptation and origin in the plant world. The role of environment in evolution*. Waltham, MA: Chronica Britanica Co.

Coates, D. J. & Byrne, M. (2005). Genetic variation in plant populations: assessing cause and pattern. In *Plant diversity and evolution: genotypic and phenotypic variation in higher plants*, ed. R. J. Henry, pp.139–64. Cambridge, MA, & Wallingford, UK: CABI Publishing.

Coates, D. J, Williams, M. R. & Madden, S. (2013). Temporal and spatial mating-system variation in fragmented populations of *Banksia cuneata*, a rare bird-pollinated long-lived plant. *Australian Journal of Botany*, 61, 235–42.

Cochran, W. G. (1963). *Sampling techniques*, 2nd edn. New York: Wiley.

Cock, A. & Forsdyke, D. R. (2008). *Treasure your exceptions. The science and life of William Bateson*. New York: Springer-Verlag Inc.

Cockburn, A. (1991). *An introduction to evolutionary ecology*. Oxford: Blackwell.

Cody, S. *et al.* (2010). The great American biotic interchange revisited. *Ecography*, 33, 326–32.

Coen, E. S. (1991). The role of homeotic genes in flower development and evolution. *Annual Review of Plant Physiology and Plant Molecular Biology*, 42, 241–79.

Coen, E. S. & Meyerwitz, E. M. (1991). The war of the whorls: genetic interactions controlling flower development. *Nature*, 353, 31–7.

Coen, E. S. & Nugent, J. (1994). Evolution of flowers and inflorescences. *Development*, 107 (Suppl.), 107–16.

Cogoni, D. *et al.* (2013). The effectiveness of plant conservation measures: the *Dianthus morisianus* reintroduction. *Oryx*, 47, 203–6.

Colautti, R., Lee, C. R. & Mitchell-Olds, T. (2012). Origin, fate, and architecture of ecologically relevant genetic variation. *Current Opinion in Plant Biology*, 15, 199–204.

Colautti, R. I. & Barrett, S. C. H. (2011). Population divergence along lines of genetic variance and covariance in the invasive plant *Lythrum salicaria* in eastern North America. *Evolution*, 65, 2514–29.

Colautti, R. I., Maron, J. L. & Barrett, S. C. H. (2009). Common garden comparisons of native and introduced plant populations: latitudinal clines can obscure evolutionary inferences. *Evolutionary Applications*, 2, 187–99.

Colautti, R. I., White, N. A. & Barrett, S. C. H. (2010). Variation of self-incompatibility within invasive populations of purple loosestrife (*Lythrum salicaria* L.) from eastern North America. *International Journal of Plant Sciences*, 171, 158–66.

Colautti, R. I., Eckert, C. G & Barrett, S. C. H. (2010). Evolutionary constraints on adaptive evolution during range expansion in an invasive plant. *Proceedings of the Royal Society of London, B*, 277, 1799–1806.

Colbach, N. & Sache, I. (2001). Blackgrass (*Alopecurus myosuroides* Huds.) seed dispersal from a single plant and its consequences on weed infestation. *Ecological Modelling*, 139, 201–19.

Cole, C. T. (2003). Genetic variation in rare and common plants. *Annual Review of Ecology, Evolution and Systematics*, 34, 213–37.

Collen, B., Purvis, A. & Mace, G. M. (2010). When is a species really extinct? Testing extinction inference from a sighting record to inform conservation assessment. *Diversity and Distributions*, 16, 755–64.

Collins, J. L. (1927). A low temperature type of albinism in Barley. *Journal of Heredity*, 33, 82–6.

Coltman, D. W. (2008). Molecular ecological approaches to studying the evolutionary impact of selective harvesting in wildlife. *Molecular Ecology*, 17, 221–35.

Comai, L. (2005). The advantages and disadvantages of being polyploid. *Nature Reviews Genetics*, 6, 836–46.

Comes, H. P. & Kadereit, J. W. (1998). The effect of Quaternary climatic changes on plant distribution and evolution. *Trends in Plant Science*, 3, 432–8.

Constanza, R. *et al.* (1997). The value of the World's ecosystem services and natural capital. *Nature*, 387, 253–60.

Cook, C. D. K. (1968). Phenotypic plasticity with particular reference to three amphibious plant species. In *Modern methods in plant taxonomy*, ed. V. H. Heywood, pp.97–111. London: Academic Press.

Cook, J. *et al.* (2013). Quantifying the consensus on anthropogenic global warming in the scientific literature. *Environmental Research Letters*, **8**, 1–7.

Cook, L. M. *et al.* (1998). Multiple independent formations of *Tragopogon* tetraploids (Asteraceae): evidence from RAPD markers. *Molecular Ecology*, **7**, 1293–1302.

Cook, S. A. (1962). Genetic system, variation and adaptation in *Eschscholzia californica*. *Evolution*, **16**, 278–99.

Cook, S. A. & Johnson, M. P. (1968). Adaptation to heterogeneous environments. I. Variation in heterophylly in *Ranunculus flammula* L. *Evolution*, **22**, 496–516.

Cooke, T.J. (2006). Do Fibonacci numbers reveal the involvement of geometric imperatives or biological interactions in phyllotaxis? *Botanical Journal of the Linnean Society*, **150**, 3–24.

Corcos, A. F. & Monaghan, F. V. (1990). Mendel's work and its rediscovery: a new perspective. *Critical Reviews in Plant Sciences*, **9**, 197–212.

Corkhill, L. (1942). Cyanogenesis in white clover (*Trifolium repens* L.) V. The inheritance of cyanogenesis. *New Zealand Journal of Science & Technology, B*, **23**, 178–93.

Corrado, P. & Magri, D. (2011). A late Early Pleistocene pollen record from Fontana Ranuccio (central Italy). *Journal of Quaternary Science*, **26**, 335–44.

Correns, C. (1909). Vererbungsversuche mit blass (gelb) grunen und buntblättrigen Sippen bei *Mirabilis jalapa, Urtica pilulifera*, und *Lunularia annua*. *Zeitschrift für Vererbungslehre*, **1**, 291–329.

Correns, C. (1913). Selbststerilität und Individualstoffe. *Biologisches Zentralblatt*, **33**, 389–443.

Corsi, P. (1988). *The age of Lamarck: evolutionary theories in France 1790–1830*. Berkeley, Los Angeles & London: University of California Press.

Costello, M. J., May, R. M. & Stork, N. E. (2013). Can we name Earth's species before they go extinct? *Science*, **339**, 413–16.

Cott, H. B. (1940). *Adaptive coloration in animals*. London: Methuen.

Coughtrey, P. J. & Martin, M. H. (1978). Tolerance of *Holcus lanatus* to lead, zinc and cadmium in factorial combination. *New Phytologist*, **81**, 147–54.

Cousens, R. & Mortimer, M. (1995). *Dynamics of weed populations*. Cambridge: Cambridge University Press.

Cowan, R. S. (1972). Francis Galton's statistical ideas: the influence of eugenics. *Isis*, **63**, 509–28.

Cox, A. V. *et al.* (1998). Genome size and karyotype evolution in the Slipper Orchids (Cypripedioideae: Orchidaceae). *American Journal of Botany*, **85**, 681–7.

Cox, G. W. (1999). *Alien species in North America and Hawaii*. Washington DC: Island Press.

Cox, G. W. (2004). *Alien species and evolution: the evolutionary ecology of exotic plants, animals, microbes, and interacting native species*. Washington DC: Island Press.

Cox, P. A. (1988). Monomorphic and dimorphic sexual strategies: a modular approach. In *Plant reproductive ecology*, ed. J. Lovett Doust & L. Lovett Doust, pp. 80–97. New York: Oxford University Press.

Cox-Foster, D. L. *et al.* (2007). A metagenomic survey of microbes in honey bee colony collapse disorder. *Science*, **318**, 283–7.

Coyne, J. A. (1994). Ernst Mayr and the origin of species. *Evolution*, **48**, 19–30.

Coyne, J. A. (1994). Recognising species. *Nature*, **364**, 298.

Coyne, J. A. & Lande, R. (1985). The genetic basis of species differences in plants. *The American Naturalist*, **126**, 141–5.

Coyne, J. A. & Orr, H. A. (2004). *Speciation*. Sunderland, MA: Sinauer Associates.

Cracraft, J. (1983). Species concepts and speciation analysis. *Current Ornithology*, **1**, 159–87.

Craig, N. L. *et al.* (2010). *Molecular biology: principles of genome function*. Oxford & New York: Oxford University Press.

Craig, P. (2005). *Centennial history of the Carnegie Institution of Washington*. Vol. IV. *The Department of Plant Biology*. Cambridge: Cambridge University Press.

Crawford, D. J. (1989). Enzyme electrophoresis and plant systematics. In *Isozymes in plant biology*, ed. D. E. Soltis & P. S. Soltis, pp. 146–64. London: Chapman & Hall.

Crawford, D. J. (1990). *Plant molecular systematics*. New York: Wiley.

Crawford, D. J. & Smith, E. B. (1982). Allozyme variation in *Coreopsis nuecensoides* and *C. nucensis* (Compositae), a progenitor-derivative species pair. *Evolution*, **36**, 379–86.

Crawford, D. J., Ornduff, R. & Vasey, M. C. (1985). Allozyme variation within and between *Lasthenia minor* and its derivative species, *Lasthenia maritima* (Asteraceae). *American Journal of Botany*, **72**, 1177–84.

Crawford, D. J. *et al.* (2010). Mixed mating in the 'obligately outcrossing' *Tolpis* (Asteraceae) of the Canary Islands. *Plant Species Biology*, **25**, 114–19.

Crawford, D. J. *et al.* (2014). Contemporary and future studies in plant speciation, morphological/floral evolution and polyploidy: honouring the scientific contributions of Leslie D. Gottlieb to plant evolutionary biology. *Philosophical Transactions of the Royal Society, B*, **369**, http://rstb.royalsocietypublishing.org/content/369/1648/20130341.full.html#ref-list-1

Crawford, R. M. M. (1985). *Studies in plant survival*. Oxford: Blackwell.

Crawford, R. M. M. (2014). Gaps in maps: disjunctions in European plant distributions. *New Journal of Botany*, **4**, 64–75.

Crawford, T. J. (1984). What is a population? In *Evolutionary ecology*, ed. B. Sharrocks, pp. 135–73. Oxford: Blackwell.

Crawford, T. J. & Jones, D. A. (1986). Variation in the colour of the keel petals in *Lotus corniculatus* L., 2. Clines in Yorkshire and adjacent counties. *Watsonia*, **16**, 15–19.

Crawford-Sidebotham, T. J. (1971). Studies of aspects of slug behaviour and the relation between molluscs and cyanogenic plants. Unpublished PhD thesis, University of Birmingham.

Cressey, D. (2010). Linnaeus meets the Internet. *Nature* online 5 May 2010, DOI:10.1038/news.2010.221.

Crew, F. A. E. (1966). Mendelism comes to England. In *G. Mendel Memorial Symposium. 1865–1965*, ed. M. Sosna, pp. 15–30. Prague: Academia Publishing House of the Czechoslovak Academy of Sciences.

Crick, F. (1988). *What mad pursuit: a personal view of scientific discovery*. London: Weidenfeld & Nicolson.

Crofts, A. & Jefferson, R. G. (1994). *The lowland grassland management handbook*. English Nature/The Wildlife Trusts.

Cronin, T. N. (2010). *Paleoclimates: understanding climate change past and present*. New York: Columbia University Press.

Cronk, Q. C. B. (2009) *The molecular organography of plants*. Oxford: Oxford University Press.

Cronk, Q. C. B. & Ojeda, I. (2008). Bird-pollinated flowers in an evolutionary and molecular context. *Journal of Experimental Botany*, **59**, 715–27.

Cronk, Q. C. B., Bateman R. M. and Hawkins, J. A. (2002). *Developmental genetics and plant evolution*. London: Taylor & Francis.

Cronquist, A. (1987). A botanical critique of cladism. *Botanical Review*, **53**, 1–52.

Cronquist, A. (1988). *The evolution and classification of flowering plants*, 2nd edn. New York: New York Botanic Garden.

Crooks, J. A. & Soulé, M. E. (1999). Lag times in population explosions of invasive species: causes and implications. In *Invasive species and biodiversity management*, ed. O. T. Sundlund, P. J. Schei & Å. Viken, pp.103–25. Dordrecht, Boston & London: Kluwer.

Crosby, A. W. (1986). *Ecological imperialism: the biological expansion of Europe, 900–1900*. Cambridge: Cambridge University Press.

Crouzet, P. & Hohn, B. (2007). Transgenic plants. In *Handbook of plant science*, ed. K. Roberts, pp. 612–18. Chichester: Wiley.

Cruden, R. W. & Hermann-Parker, S. M. (1977). Temporal dioecism: an alternative to dioecism? *Evolution*, **31**, 863–6.

Cullen, J. & Walters, S. M. (2006). Flowering plant families: how many do we need? In *Taxonomy and plant conservation*, ed. E. Leadley & S. L. Jury, pp. 45–90. Cambridge: Cambridge University Press.

Culley, T. M. & Klooster, M. R. (2007). The cleistogamous breeding system: a review of its frequency, evolution and ecology in angiosperms. *The Botanical Review*, **73**, 1–30.

Curtis, O. F. & Clark D. G. (1950). *An introduction to plant physiology*. London, New York & Toronto: McGraw-Hill.

d'Erfurth, I. *et al.* (2008). Mutations in *AtPS1* (*Arabidopsis thaliana Parallel Spindle 1*) lead to the production of diploid pollen grains. *PLoS Genet*, 4: e1000274, DOI:10.1371/journal.pgen.1000274.

Da Silva, J. M. *et al.* (2012). Population genetics and conservation of critically small cycad populations: a case study of the Albany Cycad, *Encephalartos latifrons* (Lehmann). *Biological Journal of the Linnean Society*, **105**, 293–308.

Daday, H. (1954a). Gene frequencies in wild populations of *Trifolium repens*. I. Distribution by latitude. *Heredity*, **8**, 61–78.

Daday, H. (1954b). Gene frequencies in wild populations of *Trifolium repens*. II. Distribution by altitude. *Heredity*, **8**, 377–84.

Daday, H. (1958). Gene frequencies in wild populations of *Trifolium repens* L. III. World distribution. *Heredity*, **12**, 169–84.

Daday, H. (1965). Gene frequencies in wild populations of *Trifolium repens* L. IV. Mechanism of natural selection. *Heredity*, **20**, 355–65.

Dafni, A. (1992). *Pollination ecology: a practical approach*. Oxford: Oxford University Press.

Dahlgren, G. (1987). An updated angiosperm classification. *Botanical Journal of the Linnean Society*, **100**, 197–203.

Dahlgren, K. V. O. (1922). Selbststerilität interhalb Klonen von *Lysimachia nummularia*. *Hereditas*, **3**, 200–10.

Dalgleish, H. J. & Swihart, R. K. (2012). American chestnut past and future: implications of restoration for resource pulses and consumer populations in the eastern U.S. *Restoration Ecology*, **20**, 490–7.

Darimont, C. T. *et al.* (2009). Human predators outpace other agents of trait change in the wild. *Proceedings of the National Academy of Sciences, USA*, **106**, 952–4.

Darlington, C. D. (1937). *Recent advances in cytology*, 2nd edn. London: Churchill.

Darlington, C. D. (1939). *The evolution of genetic systems*. Cambridge: Cambridge University Press.

Darlington, C. D. (1956). *Chromosome botany*. London: Allen & Unwin.

Darlington, C. D. (1963). *Chromosome botany and the origins of cultivated plants*. London: Allen & Unwin.

Darlington, C. D. & Mather, K. (1949). *Elements of genetics*. London: Allen & Unwin.

Darlington, C. D. & Wylie, A. P. (1955). *Chromosome atlas of flowering plants*, 2nd edn. London: Allen & Unwin.

Darmency, H. & Gasquez, J. (1997). Spontaneous hybridization of the putative ancestors of the allotetraploid *Poa annua*. *New Phytologist*, **136**, 497–501.

Darwin, C. (1859). *On the origin of species by means of natural selection*, 1st edn. London: Murray. [6th edn: 1872]

Darwin, C. (1862). *On the various contrivances by which British and foreign orchids are fertilised by insects and on the good effects of crossing*. London: Murray.

Darwin, C. (1868). *The variation of plants and animals under domestication*. London: Murray.

Darwin, C. (1871a). Pangenesis. *Nature*, **3**, 502–3.

Darwin, C. (1871b). *The descent of man, and selection in relation to sex*. Part 2, ed. P. H. Barrett & R. B. Freeman (1989). London: William Pickering.

Darwin, C. (1875). *Insectivorous plants*. London: John Murray.

Darwin. C. (1876). *The effects of cross- and self-fertilisation in the vegetable kingdom*. London: Murray.

Darwin, C. (1877a). *The different forms of flowers of the same species*. London: Murray.

Darwin, C. (1877b). *The various contrivances by which orchids are fertilised by insects*, 2nd edn. London: Murray.

Darwin, C. & Wallace, A. (1858). On the tendency of species to form varieties; and on the perpetuation of varieties and species by natural means of selection. *Proceedings of the Linnean Society of London*, **3**, 45–62.

Darwin, F. (ed.) (1909a) *The foundations of the origin of species. A sketch written in 1842 by Charles Darwin*. Cambridge: Cambridge University Press.

Darwin, F. (ed.) (1909b) *The foundations of the origin of species. Two essays written in 1842 and 1844 by Charles Darwin*. London: Cambridge University Press.

Darwin, F. & Seward, A. C. (1903). *More letters of Charles Darwin*, 2 vols. London: Murray.

Davenport, C. B. (1904). *Statistical methods with special reference to biological variation*, 2nd edn. London: Chapman & Hall; New York: Wiley.

David, F. N. (1971). *A first course in statistics*, 2nd edn. London: Griffin.

Davidson, J. F. (1947). The polygonal graph for simultaneous portrayal of several variables in population analysis. *Madroño*, **9**, 105–10.

Davies, H. M. (2010). Review article: Commercialization of whole-plant systems for biomanufacturing of protein products: evolution and prospects. *Plant Biotechnology Journal*, **8**, 845–61.

Davies, M. S. (1975). Physiological differences among populations of *Anthoxanthum odoratum* collected from the Park Grass experiment. IV. Response to potassium and magnesium. *Journal of Applied Ecology*, **12**, 953–64.

Davies, M. S. (1993). Rapid evolution in plant populations. In *Evolutionary patterns and processes*, ed. D. R. Lees & D. Edwards. Linnean Society Symposium Series, 14, pp. 172–88. London: Published for the Linnean Society by Academic Press.

Davies, M. S. & Snaydon, R. W. (1973a). Physiological differences among populations of *Anthoxanthum odoratum* collected from the Park Grass experiment. I. Response to calcium. *Journal of Applied Ecology*, **10**, 33–45.

Davies, M. S. & Snaydon, R. W. (1973b). Physiological differences among populations of *Anthoxanthum odoratum* collected from the Park Grass experiment. II. Response to aluminium. *Journal of Applied Ecology*, **10**, 47–55.

Davies, M. S. & Snaydon, R. W. (1974). Physiological differences among populations of *Anthoxanthum odoratum* collected from the Park Grass experiment. III. Response to phosphate. *Journal of Applied Ecology*, **11**, 699–707.

Davies, M. S. & Snaydon, R. W. (1976). Rapid population differentiation in a mosaic environment. III. Measures of selection pressures. *Heredity*, **36**, 59–66.

Davies, R. (2008). *The Darwin conspiracy. origins of a scientific crime.* London: Golden Square Books.

Davies, T. M. & Snaydon, R. W. (1989). An assessment of the spaced-plant trial technique. *Heredity*, **63**, 37–45.

Davies, W. E. (1963). Leaf markings in *Trifolium repens.* In *Teaching genetics in school and university*, ed. C. D. Darlington & A. D. Bradshaw, pp. 94–8. Edinburgh: Oliver & Boyd.

Davis, H. *et al.* (2004) An Allee effect at the front of a plant invasion: *Spartina* in a Pacific estuary. *Journal of Ecology*, **92**, 321–7.

Davis, J. I. (1995). Species concepts and phylogenetic analysis – introduction. *Systematic Botany*, **20**, 555–9.

Davis, M. B. & Shaw, R. G. (2001). Range shifts and adaptive responses to Quaternary climate change. *Science*, **292**, 673–9.

Davis, P. H. & Heywood, V. H. (1963). *Principles of angiosperm taxonomy.* Edinburgh: Oliver & Boyd; New York: Van Nostrand.

Davison, A. W. & Reiling, K. (1995). A rapid change in ozone resistance of *Plantago major* after summers with high ozone concentrations. *New Phytologist*, **131**, 337–44.

Davy, A. J. & Jeffries, R. L. (1981). Approaches to the monitoring of rare plant populations. In *The biological aspects of rare plant conservation*, ed. H. Synge, pp. 219–32. London: Wiley.

Dawkins, R. (2003). *A devil's chaplain.* London: Weidenfeld & Nicolson.

Dawkins, R. (2005). *The ancestor's tale: a pilgrimage to the dawn of life.* London: Weidenfeld & Nicolson.

Dawkins, R. (2009). *The greatest show on Earth: the evidence for evolution.* London: Bantam Press.

Dawson, C. D. R. (1941). Tetrasomic inheritance in *Lotus corniculatus* L. *Journal of Genetics*, **42**, 49–72.

De Beer, G. (ed.) (1960–1). Darwin's notebooks on transmutation of species I–IV. *Bulletin of British Museum (Natural History)*, Historical Series 2, Nos. 2–6.

De Beer, G. (1963). *Charles Darwin.* London: Nelson.

De Beer, G. (1964). *Atlas of evolution.* London: Nelson.

De Bodt, S., Maere, S. & Van de Peer, Y. (2005). Genome duplication and the origin of angiosperms. *Trends in Ecology and Evolution*, **20**, 591–7.

De Haan, A. *et al.* (1992). Production of 2*n* gametes in diploid subspecies of *Dactylis glomerata* L. 2. Occurrence and frequency of 2*n* eggs. *Annals of Botany*, **69**, 345–50.

De Nettancourt, D. (1977). *Incompatibility in angiosperms.* Berlin, Heidelberg & New York: Springer Verlag.

De Pamphilis, C. W. & Palmer, J. D. (1990) Loss of photosynthetic and chlororespiratory genes from the plastid genome of a parasitic plant. *Nature*, **348**, 337–9.

De Queiroz, A. (2005a). The resurrection of oceanic dispersal in historical biogeography. *Annual Review of Ecology and Systematics*, **26**, 373–401.

De Queiroz, A. (2005b). The resurrection of oceanic dispersal in historical biogeography. *Trends in Ecology and Evolution*, **20**, 68–73.

de Queiroz, K. & Donoghue, M. J. (2013). Phylogenetic nomenclature, hierarchical information, and testability. *Systematic Biology*, **62**, 167–74.

de Queiroz, K. & Gauthier, J. (1994). Toward a phylogenetic system of biological nomenclature. *Trends in Research in Ecology and Evolution*, **9**, 27–31.

de Vilmorin, P. (1910). Recherches sur l'hérédité Mendélienne. *Compte Rendu Hebdomadaire des Séances de l Académie des Sciences*, Paris, **151**, 548–51.

de Vilmorin, P. (1911). Etude sur la caractère adhérence des grains entre eux chez 'le Pois, Chenille'. *4th International Conference on Genetics*, Paris, 368–72.

de Vilmorin, P. & Bateson, W. (1911). A case of gametic coupling in *Pisum*. *Proceedings of the Royal Society, B*, **84**, 9–11.

De Vries, H. (1894). Uber halbe Galton-Kurven als Zeichnen diskontinurlichen Variation. *Bericht der Deutschen Botanischen Gesellschaft*, **12**, 197–207.

De Vries, H. (1897). Monstruosités héréditaires offertes en échange aux jardins botaniques. *Botanisch Jaarboek*, **9**, 80–93.

De Vries, H. (1905). *Species and varieties: their origin by mutation*. Chicago: Open Court Publishing Co.

de Wet, J. M. J. (1968), Diploid–tetraploid–haploid cycles and the origin of variability in *Dichanthium* agamospecies. *Evolution*, **22**, 394–7.

de Wet, J. M. J. (1971). Reversible tetraploidy as an evolutionary mechanism. *Evolution*, **25**, 545–8.

de Wet, J. M. J. (1980). Origins of polyploids. In *Polyploidy*, ed. W. H. Lewis, pp. 3–15. New York & London: Plenum Press.

de Wet, J. M. J. & Harlan, J. R. (1970). Apomixis, polyploidy and speciation in *Dichanthium*. *Evolution*, **24**, 270–7.

de Wet, J. M. J. & Harlan, J. R. (1972). Chromosome pairing and phylogenetic affinities. *Taxon*, **21**, 67–70.

De Witte L. C. & Stöcklin, J. (2010). Longevity of clonal plants: why it matters and how to measure it. *Annals of Botany*, **106**, 849–57.

Decruse, S. W. *et al.* (2003). Micropropagation and ecorestoration of *Vanda spathulata*, an exquisite orchid. *Plant Cell, Tissue & Organ Culture*, **72**, 199–202.

Demauro, M. M. (1993). Relationship of breeding system to rarity in the Lakeside Daisy (*Hymenoxys acaulis* var. *glabra*). *Conservation Biology*, **7**, 542–50.

Demauro, M. M. (1994). Development and implementation of a recovery program for the federal threatened Lakeside Daisy (*Hymenoxys acaulis* var. *glabra*). In *Restoration of endangered species*, ed. M. L. Bowles & C. J. Whelan, pp. 298–312. Cambridge: Cambridge University Press.

Dembski, W. A. & Ruse, M. (eds.) (2007a). *Debating design. From Darwin to DNA*. Cambridge: Cambridge University Press.

Dembski, W. A. & Ruse, M. (2007b). General introduction. In *Debating design. From Darwin to DNA*, ed. W. A. Dembski & M. Ruse, pp. 3–12. Cambridge: Cambridge University Press.

Dennis, A. J., Green, R. J. & Schupp, E. W. (2007). *Seed dispersal: theory and its application in a changing world*. Egham, Surrey: CABI.

Depew, D. J. & Weber, B. H. (1996). *Darwinism evolving: systems dynamics and the genealogy of natural selection*. Cambridge, MA: MIT Press, A Bradford Book.

Des Marais, D. L. & Rauscher, M. D. (2010). Parallel evolution at multiple levels in the origin of hummingbird pollinated flowers of *Ipomoea*. *Evolution*, **64**, 2044–54.

DeSalle, R. (2006). Species discovery versus species identification: response to Rubinoff. *Conservation Biology*, **20**, 1545–7.

DeSalle, R., Egan, M. G. & Siddall, M. (2005). The unholy trinity: taxonomy, species delimiting and DNA barcoding. *Philosophical Transactions of the Royal Society of London, B*, **360**, 1905–16, I.

Desmond, A. & Moore, J. (1991). *Darwin*. London: Michael Joseph.

Desmond, A. & Moore, J. (2009). *Darwin's sacred cause. race, slavery and the quest for human origins*. London: Allen Lane, an imprint of Penguin Books.

Dettner, K. & Liepert, C. (1994). Chemical mimicry and camouflage. *Annual Review of Entomology*, **39**, 129–54.

Devos, N. *et al.* (2006). On the monophyly of *Dactylorhiza* Necker ex Nevski (Orchidaceae): is *Coeloglossum viride* (L.) Hartman a *Dactylorhiza*? *Botanical Journal of the Linnean Society*, **152**, 261–9.

DeWoody, J. A. *et al.* (2010). *Molecular approaches in natural resource conservation and management*. Cambridge: Cambridge University Press.

Di Cesnola, A. P. (1904). Preliminary note on the protective value of colour in *Mantis religiosa*. *Biometrika*, **3**, 58–9.

Diamond, J. (1992). *The rise and fall of the third chimpanzee: how our animal heritage affects the way we live*. London: Vantage.

Diaz, A. & Macnair, M. R. (1998.) The effect of plant size on the expression of cleistogamy in *Mimulus nasutus*. *Functional Ecology*, **12**, 92–8.

Dick, C. W. (2001). Genetic rescue of remnant tropical trees by an alien pollinator. *Proceedings of the Royal Society of London, B*, **268**, 2391–6.

Digby, L. (1912). The cytology of *Primula kewensis* and of other related *Primula* hybrids. *Annals of Botany*, **26**, 357–88.

Dirzo, R. & Harper, J. L. (1982a). Experimental studies of slug–plant interactions. III. Differences in acceptability of individual plants of *Trifolium repens* to slugs and snails. *Journal of Ecology*, **70**, 101–17.

Dirzo, R. & Harper, J. L. (1982b). Experimental studies of slug–plant interactions. IV. The performance of cyanogenic and acyanogenic morphs of *Trifolium repens* in the field. *Journal of Ecology*, **70**, 119–38.

Ditta, G. *et al.* (2004). The *SEP4* gene of *Arabidopsis thaliana* functions in floral organ and meristem identity. *Current Biology*, **14**, 1935–40.

Dlugosch, K. M. & Parker, I. M. (2007). Molecular and life history trait variation across the native range of the invasive species *Hypericum canariense*: evidence for ancient patterns of colonization via pre-adaptation? *Molecular Ecology*, **16**, 4269–83.

Dlugosch, K. M. & Parker I. M. (2008a). Founding events in species invasions: genetic variation, adaptive evolution, and the role of multiple introductions. *Molecular Ecology*, **17**, 431–49.

Dlugosch, K. M. & Parker, I. M. (2008b). Invading populations of an ornamental shrub show rapid life history evolution despite genetic bottlenecks. *Ecology Letters*, **11**, 701–9.

Dobzhansky, T. (1935). A critique of the species concept in biology. *Philosophy of Science*, **2**, 344–55.

Dobzhansky, T. G. (1937). *Genetics and the origin of species*. New York: Columbia University Press.

Dobzhansky, T. G. (1941). *Genetics and the origin of species*, 2nd edn. New York: Columbia University Press.

Dobzhansky, T. G. (1951). *Genetics and the origin of species*, 3rd edn. New York: Columbia University Press.

Dodd, M. *et al.* (1995). Community stability – a 60-year record of trends and outbreaks in the occurrence of species in the Park Grass Experiment. *Journal of Ecology*, **83**, 277–85.

Dodds, W. K. (2008). *Humanity's footprint: momentum, impact, and our global environment*. New York: Columbia University Press.

Doebley, J., Stec, A. & Hubbard, L. (1997). The evolution of apical dominance in maize. *Nature*, **386**, 485–8.

Dommée, B., Assouad, M. W. & Valdeyron, G. (1978). Natural selection and gynodioecy in *Thymus vulgaris* L. *Botanical Journal of the Linnean Society*, **77**, 17–28.

Donoghue, M. J. & Catino, P. D. (1988). Paraphyly, ancestors, and the goals of taxonomy: a botanical defense of cladism. *The Botanical Review*, **54**, 107–28.

Donoghue, M. J. & Kadereit, J. W. (1992). Walter Zimmermann and the growth of phylogenetic theory. *Systematic Biology*, **41**, 74–85.

Donoghue, M. J. & Sandeson, M. J. (1992). The suitability of molecular and morphological evidence in reconstructing plant phylogeny. In *Molecular systematics of plants*, ed. P. S. Soltis, D. E. Soltis & J. J. Doyle, pp.340–68. New York: Chapman & Hall.

Donoghue, M. J. & Smith, S. A. (2004). Patterns in the assembly of temperate forests around the Northern Hemisphere. *Philosophical Transactions of the Royal Society, B*, **359**, 1633–44.

Donoghue, M. J., Ree, R. H. & Baum, D. A. (1998). Phylogeny and the evolution of flower symmetry in the Asteridae. *Trends in Plant Science*, **3**, 311–17.

Donohue, K. (2002). Germination timing influences natural selection on life-history characters in *Arabidopsis thaliana*. *Ecology*, **83**, 1006–16.

Doorenbos, J. (1965). Juvenile and adult phases in woody plants. In *Encyclopedia of plant physiology, XV/1*, ed. W. Ruhland, pp. 1222–35. Berlin: Springer.

Doran, P. T. & Zimmerman, M. K. (2009). Examining the scientific consensus on climate change. *Eos, Transactions, American Geophysical Union*, **90**, 22–3.

Douhovnikoff, V. & Dodd, R. S. (2003). Intra-clonal variation and a similarity threshold for identification of clones: application to *Salix exigua* using AFLP molecular markers. *Theoretical and Applied Genetics*, **106**, 1307–15.

Doyle, J. A. (1978). Origin of angiosperms. *Annual Review of Ecology & Systematics*, **9**, 365–92.

Doyle, J. A. (2012). Molecular and fossil evidence on the origin of angiosperms. *Annual Review of Earth and Planetary Sciences*, **40**, 301–26.

Doyle, J. A. & Donoghue, M. J. (1986). Phylogeny and the origin of angiosperms: an experimental cladistic approach. *Botanical Review*, **52**, 321–431.

Doyle, J. J. *et al.* (2003). Diploid and polyploid reticulate evolution throughout the history of the perennial soybeans (*Glycine* subgenus *Glycine*). *New Phytologist*, **161**, 121–32.

Drayton, B. & Primack, R. B. (2012). Success rate for reintroductions of eight perennial species after 15 years. *Restoration Ecology*, **20**, 299–303.

Dubois, A. (2007). Naming taxa from cladograms: some confusions, misleading statements, and necessary clarifications. *Cladistics*, **23**, 390–402.

Dukes, J. S. (2011). Climate change. In *Encyclopedia of biological invasions*, ed. D. Simberloff and M. Rejmanke, pp. 113–17. Encyclopedias of the Natural

World. No 3. Berkeley, CA: University of California Press.

Dunn, G. & Everitt, B. S. (1982). *An introduction to mathematical taxonomy.* Cambridge: Cambridge University Press.

Dunstan, W. R. & Henry, T. A. (1901). The nature and origin of the poison of *Lotus arabicus. Proceedings of the Royal Society of London,* **68**, 374–8.

Dupré, J. (2010). The polygenic organism. In *Nature after the genome,* ed. S. Parry & J. Dupré, pp. 19–31. Oxford: Blackwell.

Durka, W. *et al.* (2005). Molecular evidence for multiple introductions of Garlic Mustard (*Alliaria petiolata,* Brassicaceae) to North America. *Molecular Ecology,* **14**, 1697–1706.

Dyer, G. A. *et al.* (2009). Dispersal of transgenes through maize seed systems in Mexico. *PLoS ONE,* 4, e5734, DOI:10.1371/journal.pone.0005734.

East, E. M. (1913). Inheritance of flower size in crosses between species of *Nicotiana. Botanical Gazette,* **55**, 177–88.

East, E. M. & Mangelsdorf, A. J. (1925). A new interpretation of the hereditary behaviour of self-sterile plants. *Proceedings of the National Academy of Sciences, Washington,* **11**, 166–83.

Ebling, S. K. *et al.* (2011). Multiple common garden experiments suggest lack of local adaptation in an invasive ornamental plant. *Journal of Plant Ecology,* **4**, 209–20.

Eckert, C. G. (2000). Contributions of autogamy and geitonogamy to self-fertilization in a mass-flowering, clonal plant. *Ecology,* **81**, 532–42.

Eckert, C. G. & Barrett, S. C. H. (1993). Clonal reproduction and patterns of genotypic diversity in *Decodon verticillatum* (Lythraceae). *American Journal of Botany,* **80**, 1175–82.

Eckert, C. G., Manicacci, D. & Barrett, S. C. H. (1996). Genetic drift and founder effect in native versus introduced populations of an invading plant, *Lythrum salicaria* (Lythraceae). *Evolution,* **50**, 1512–19.

Eddy, S. R. (2012). The C-value paradox, junk DNA and ENCODE. *Current Biology,* **22**, R896.

Edwards, A. W. F. (1986). Are Mendel's results really too close? *Biological Reviews,* **61**, 295–312.

Edwards, M. & Richardson, A. J. (2004). Impact of climate change on marine pelagic phenology and trophic mismatch. *Nature,* **430**, 881–4.

Ehlers, B. K., & Bataillon, T. (2007). Inconsistent males and the maintenance of labile sex expression in subdioecious plants. *New Phytologist,* **174**, 194–211.

Ehlers, B. K., Maurice, S. & Bataillon, T. (2005). Sex inheritance in gynodioecious species: a polygenic view. *Proceedings of the Royal Society, B,* **272**, 1795–1802.

Ehlers, J., Gibbard, P. L. & Hughes, P. D. (2011). *Quaternary glaciations – extent and chronology: a closer look.* Amsterdam: Elsevier.

Ehrendorfer, F. (1980). Polyploidy and distribution. In *Polyploidy: biological relevance,* ed. W. H. Lewis, pp. 45–60. New York: Plenum Press.

Ehrenreich, I. M. & Purugganan, M. D. (2006). The molecular genetic basis of plant adaptation. *American Journal of Botany,* **93**, 953–62.

Ehrich, D. *et al.* (2007). Genetic consequences of Pleistocene range shifts: contrast between the Arctic, the Alps and the East African mountains. *Molecular Ecology,* **16**, 2542–59.

Ehrlich, P. R. & Raven, P. H. (1964). Butterflies and plants: a study in coevolution. *Evolution,* **18**, 586–608.

Ehrlich, P. R. & Raven, P. H. (1969). Differentiation of populations. *Science,* **165**, 1228–32.

Ehrlich, P. R, Kareiva, P. M. & Daily, G. C. (2012). Securing natural capital and expanding equity to rescale civilization. *Nature,* **486**, 68–73.

Eidesen, P. B. (2007). Nuclear vs. plastid data: complex Pleistocene history of a circumpolar key species. *Molecular Ecology,* **16**, 3902–25.

Eigsti, C. J. & Dustin, P. (1955). *Colchicine in agriculture, medicine, biology and chemistry.* Ames, IA: Iowa State College Press.

Elam, D. R. *et al.* (2007). Population size and relatedness affect fitness of a self-incompatible invasive plant. *Proceedings of the National Academy of Sciences, USA,* **104**, 549–52.

Elkington, T. T. (1968). Introgressive hybridization between *Betula nana* L. and *B. pubescens* Ehrh. in North-west Iceland. *New Phytologist,* **67**, 109–18.

Elliot, E. (1914), see Lamarck.

Ellis, W. M., Keymer, R. J. & Jones, D. A. (1977a). The effect of temperature on the polymorphism of cyanogenesis in *Lotus corniculatus* L. *Heredity,* **38**, 339–47.

Ellis, W. M., Keymer, R. J. & Jones, D. A. (1977b). On the polymorphism of cyanogenesis in *Lotus corniculatus* L. VIII. Ecological studies in Anglesey. *Heredity,* **39**, 45–65.

Ellstrand N. C. & Elam, D. R. (1993). Population genetic consequences of small population size: implications for plant conservation. *Annual Review of Ecology & Systematics*, **24**, 217–42.

Ellstrand, N. C. & Marshall, D. L. (1985a). Interpopulation gene flow by pollen in wild radish, *Raphanus sativus*. *The American Naturalist*, **126**, 606–16.

Ellstrand, N. C. & Marshall, D. L. (1985b). Variation in extent of multiple paternity among plants and populations of wild radish. *American Journal of Botany*, **72**, 876.

Ellstrand, N. C. & Schierenbeck, K. (2000). Hybridization as a stimulus for the evolution of invasiveness in plants? *Proceedings of the National Academy of Sciences, USA*, **97**, 7043–50.

Ellstrand, N. C. & Schierenbeck, K. A. (2006). Hybridization as a stimulus for the evolution of invasiveness in plants? *Euphytica*, **48**, 35–46.

Ellstrand, N. C., Whitkus, R. & Rieseberg, L. H. (1996). Distribution of spontaneous plant hybrids. *Proceedings of the National Academy of Sciences, USA*, **93**, 5090–3.

Ellstrand, N. C., Prentice, H. C. & Hancock, J. F. (1999). Gene flow and introgression from domesticated plants into their wild relatives. *Annual Review of Ecology & Systematics*, **30**, 539–63.

Ellstrand, N. C. *et al.* (2010). Crops gone wild: evolution of weeds and invasives from domesticated ancestors. *Evolutionary Applications*, **3**, 494–504.

Ellstrand, N. C. *et al.* (2013). Introgression of crop alleles into wild or weedy populations. *Annual Review of Ecology, Evolution & Systematics*, **44**, 325–45.

Elmqvist, T. (2000). Pollinator extinction in the Pacific Islands. *Conservation Biology*, **14**, 1237–9.

Elwell, A. L. *et al.* (2011). Separating parental environment from seed size effects on next generation growth and development in *Arabidopsis*. *Plant, Cell and Environment*, **34**, 291–301.

Emerson, B. C. (2002). Evolution on oceanic islands: molecular phylogenetic approaches to understanding pattern and process. *Molecular Ecology*, **11**, 951–66.

Emms, S. K. & Arnold, M. L. (1997). The effect of habitat on parental and hybrid fitness: transplant experiments with Louisiana Irises. *Evolution*, **51**, 1112–19.

Endersby, J. ed. (2009). Introduction, appendix and explanatory notes. In *On the origin of species by Charles Darwin*. Cambridge: Cambridge University Press.

Engeldow, F. L. (1950). Rowland Harry Biffin 1874–1949. *Obituary Notices of Fellows of the Royal Society*, **7**, 9–25.

Ennos, R. A. (1982). Association of the cyanogenic loci in White clover. *Genetic Research*, **40**, 65–72.

Ensslin, A., Sandner, T. M. & Matthies, D. (2011). Consequences of *ex situ* cultivation of plants: genetic diversity, fitness and adaptation of the monocarpic *Cynoglossum officinale* L. in botanic gardens. *Biological Conservation*, **144**, 272–8.

Epperson, B. K. (2007). Plant dispersal, neighbourhood size and isolation by distance. *Molecular Ecology*, **16**, 3854–65.

Ereshefsky, M. (2001). *The poverty of the Linnaean hierarchy: a philosophical study of biological taxonomy*. Cambridge: Cambridge University Press.

Erfmeier, A. & Bruelheide, H. (2011). Maintenance of high genetic diversity during invasion of *Rhododendron ponticum* in the British Isles. *International Journal of Plant Sciences*, **172**, 795–806.

Erikkson, G. (1983). Linnaeus the botanist. In *Linnaeus. The man and his work*, ed. T. Frangsmyr, pp.63–109. Berkeley: University of California Press.

Ernst, A. (1955). Self-fertility in monomorphic primulas. *Genetica*, **27**, 91–148.

Ernst, W. H. O. (1990). Mine vegetation in Europe. In *Heavy metal tolerance in plants: evolutionary aspects*, ed. A. J. Shaw, pp. 21–37. Boca Raton, FL: CRC Press.

Ernst, W. H. O. (1998a). Evolution of plants on soils anthropogenically contaminated by heavy metals. In *Plant evolution in man-made habitats*, ed. L. W. D. van Raamsdonk & J. C. M. den Nijs, pp. 13–27. Amsterdam: Hugo de Vries Laboratory.

Ernst, W. H. O. (1998b). Invasion, dispersal and ecology of the South African neophyte *Senecio inaequidens* in The Netherlands: from wool alien to railway and road alien. *Acta Botanica Neerlandica*, **47**, 131–51.

Ernst, W. H. O. (2006). Evolution of metal tolerance in higher plants. *Forest Snow and Landscape Research*, **80**, 251–74.

Escaravage, N. *et al.* (1998). Clonal diversity in a *Rhododendron ferrugineum* L. (Ericaceae) population inferred from AFLP markers. *Molecular Ecology*, **7**, 975–82.

Evans R. C. & Turkington, R. (1988). Maintenance of morphological variation in a biotically patchy environment. *New Phytologist*, **109**, 369–76.

Evenari, M. (1989). The history of research on white-green variegated plants. *Botanical Review*, **55**, 106–39.

Ewel, J. J. (1986). Invasibility: lessons from southern California. In *Ecology of biological invasions of North America and Hawaii*, ed. H. A. Mooney & J. A. Drake, pp. 214–39. New York: Springer Verlag.

Fagerlind, F. (1937). Embryologische, zytologische und bestäubungs-experimentelle Studien in der Familie Rubiaceae nebst Bemerkungen über einige Polyploiditätsprobleme. *Acta Horti Bergiani*, **11**, 195–470.

Fairbanks, D. J. (2008). Mendelian controversies: an update. In *Ending the Mendel–Fisher controversy*, ed. A. Franklin *et al.*, pp. 302–11. Pittsburgh, PA: University of Pittsburgh Press.

Falahati-Anbaran, M. *et al.* (2011). Genetic consequences of seed banks in the perennial herb *Arabidopsis lyrata* subsp. *petraea* (Brassicaceae). *American Journal of Botany*, **98**, 1475–85.

Falconer, D. S. (1952). *Introduction to quantitative genetics*, 2nd edn. London: Longmans.

Falconer, D. S. (1981). *Introduction to quantitative genetics*. 4th edn. London: Longmans.

Falk, D. A. & Holsinger, K. E. (1991). *Genetics and conservation of rare plants*. Oxford: Oxford University Press.

Falk, D. A. & Olwell, P. (1992). Scientific and policy considerations in restoration and reintroduction of endangered species. *Rhodora*, **94**, 287–315.

Falk, D. A., Millar, C. I. & Olwell, M. (1966). *Restoring diversity: strategies for reintroduction of endangered species*. Washington DC: Island Press.

Fant, D. A. *et al.* (2007). Genetic structure of threatened native populations and propagules used for restoration in a clonal species, American Beachgrass (*Ammophila breviligulata* Fern.) *Restoration Ecology*, **16**, 594–603.

Fant, J. B. *et al.* (2013). Genetics of reintroduced populations of the narrowly endemic thistle, *Cirsium pitcheri* (Asteraceae). *Botany-Botanique*, **91**, 301–8.

Farjon, A. (2007). In defence of a conifer taxonomy which recognises evolution. *Taxon*, **56**, 639–41.

Favarger, C. (1967). Cytologie et distribution des plantes. *Botanical Review*, **42**, 163–206.

Favarger, C. (1984). Cytogeography and biosystematics. In *Plant biosystematics*, ed. W. F. Grant, pp. 453–75. Toronto: Academic Press.

Favarger, C. & Villard, M. (1965). Nouvelles récherches cytotaxinomiques sur *Chrysanthemum leucanthemum* L. sens. lat. *Bericht der Schweizerischen Botanischen Gesellschaft*, **75**, 57–79.

Fawcett, J. A., Maerea, S. & Van de Peera, Y. (2009). Plants with double genomes might have had a better chance to survive the Cretaceous–Tertiary extinction event. *Proceedings of the National Academy of Sciences, USA*, **106**, 5737–42.

Fay, M. & Cowan, R. S. (2001). Plant microsatellites in *Cypripedium caleolus* (Orchidaceae): genetic fingerprints from herbarium specimens. *Lindleyana*, **16**, 151–6.

Fazekas, A. J. *et al.* (2010). Stopping the stutter: improvements in sequence quality from regions with mononucleotide repeats can increase the usefulness of noncoding regions for DNA barcoding. *Taxon*, **59**, 694–7.

Fearnside, P. M. & Ferraz, J. (1995). A conservation gap analysis of Brazil's Amazonian vegetation. *Conservation Biology*, **9**, 1134–47.

Federici, S. *et al.* (2013). DNA barcoding to analyse taxonomically complex groups in plants: the case of *Thymus* (Lamiaceae). *Botanical Journal of the Linnean Society*, **171**, 687–99.

Fenger, J. & Tjell, J. C. (2009). *Air pollution – from a local to global perspective*. Lyngby: Polyteknisk.

Ferguson, A. (1980). *Biochemical systematics and evolution*. Glasgow & London: Blackie.

Ferrari, D. M. & Guiseppi, A. R. (2011). *Geomorphology and plate tectonics*. New York: Nova Publishers.

Ferris, C. *et al.* (1995). Chloroplast DNA recognizes three refugial sources of European oaks and suggests independent eastern and western immigrations to Finland. *Heredity*, **80**, 584–93.

Ferris, C., King, R. A. & Gray, A. J. (1997). Molecular evidence for maternal parentage in the hybrid origin of *Spartina anglica* C. E. Hubbard. *Molecular Ecology*, **6**, 185–7.

Ferris-Kaan, R. (ed.) (1995). *The ecology of woodland creation*. Chichester & New York: Wiley.

Feuillet, C. *et al.* (2011). Crop genome sequencing: lessons and rationales. *Trends in Plant Science*, **16**, 77–88.

Feulner, M. *et al.* (2013). Floral scent and its correlation with AFLP data in *Sorbus*. *Organisms Diversity and Evolution*, **14**, 339–48.

Field, D. L. *et al.* (2010). Patterns of hybridization and asymmetrical gene flow in hybrid zones of the rare *Eucalyptus aggregata* and common *E. rubida*. *Heredity*, **106**, 841–53.

Fincham, J. R. S. (1983). *Genetics*. Bristol: Wright.

Fisher, K. (2006). Rank-free monography: a practical example from the moss clade *Leucophanella* (Calymperaceae). *Systematic Botany*, 31, 13–30.

Fisher, R. A. (1929). *The genetical theory of natural selection*, 2nd edn, reprinted 1958. London: Constable; New York: Dover Books.

Fisher, R. A. (1935). *The design of experiments*. Edinburgh & London: Oliver & Boyd.

Fisher, R. A. (1936). Has Mendel's work been rediscovered? *Annals of Science*, 1, 115–37.

Fisher, R. A. & Yates, F. (1963). *Statistical tables for biological, agricultural and medical research*, 6th edn. Edinburgh: Longman (Oliver & Boyd).

Flake, R. H., von Rudloff, E., & Turner, B. L. (1969). Quantitative study of clinal variation in *Juniperus virginiana* using terpenoid data. *Proceedings of the National Academy of Sciences, USA*, 64, 487–94.

Flann C., Turland N. M. & Monro, A. (2014). Report on botanical nomenclature – *Melbourne 2011 XVIII International Botanical Congress*, Melbourne: Nomenclature Section, 18–22 July 2011. *PhytoKeys*, 41, 1–289, DOI:10.3897/phytokeys.41.8398

Flowers, J. *et al.* (2009). Population genomics of the *Arabidopsis thaliana* flowering time gene network. *Molecular Biology and Evolution*, 26, 2475–86.

Flowers, J. M. & Purugganan, M. D. (2008). The evolution of plant genomes – scaling up from a population perspective. *Current Opinion in Genetics & Development*, 18, 565–70.

Focke, W. O. (1881). *Die Pflanzen-Mishlinge ein Beitrag zur Biologie der Gewächse*. Berlin: Gebrüder Borntraeger.

Foden, W. *et al.* (2007). A changing climate is eroding the geographical range of the Namib Desert tree *Aloe* through population declines and dispersal lags. *Diversity and Distributions*, 13, 645–53.

Ford, V. S. & Gottlieb, L. D. (1992). Bicalyx is a natural homeotic floral variant. *Nature*, 358, 671–3.

Forest, F. & Chase, M. W. (2009). *Eudicots*. In *The timetree of life*, ed. S. B. Hedges and S. Kumar, pp.169–176. New York: Oxford University Press.

Forey, P. L., Humphries, C. J. & Kitching, I. J. (1993). *Cladistics: a practical approach in systematics*. Oxford: Oxford University Press.

Forister, M. L. & Feldman, C. R. (2011). Phylogenetic cascades and the origins of tropical diversity. *Biotropica*, 43, 270–8.

Forrest C. N. *et al.* (2011). Tests for inbreeding and outbreeding depression and estimation of population differentiation in the bird-pollinated shrub *Grevillea mucronulata*. *Annals of Botany*, 108, 185–95.

Foster, S. A. *et al.* (2007). Parallel evolution of dwarf ecotypes in the forest tree *Eucalyptus globulus*. *New Phytologist*, 175, 370–80.

Fournier-Level, A. *et al.* (2011). A map of local adaptation in *Arabidopsis thaliana*. *Science*, 334, 86–9.

Fowler, N. & Levin, D. A. (1984). Ecological constraints on the establishment of a novel polyploid in competition with its diploid progenitor. *American Naturalist*, 124, 703–11.

Foxe, J. P. *et al.* (2010). Reconstructing origins of loss of self-incompatibility and selfing in North American *Arabidopsis lyrata*: a population genetic context. *Evolution*, 64, 3495–510.

Francico-Ortega, J. *et al.* (1999). Internal transcribed spacer sequence phylogeny of *Crambe* L. (Brassicaceae): molecular data reveal two Old World disjunctions. *Molecular Phylogenetics and Evolution*, 11, 361–80.

Frankel, O. H. & Soulé, M. E. (1981). *Conservation and evolution*. London: Cambridge University Press.

Frankel, O. H., Brown, A. H. D. & Burdon, J. J. (1995). *The conservation of plant biodiversity*. Cambridge: Cambridge University Press.

Frankham, R. (1995) Effective population size/adult population size ratios in wildlife: a review. *Genetics Research Cambridge*, 66, 95–107.

Frankham, R. (2008). Genetic adaptation to captivity in species conservation programs. *Molecular Ecology*, 17, 325–33.

Frankham, R., Ballou, J. D. & Briscoe, D. A. (2002). *Introduction to conservation genetics*. Cambridge: Cambridge University Press.

Frankham, R. J. D. *et al.* (2011). Predicting the probability of outbreeding depression. *Conservation Biology*, 25, 465–75.

Franklin, A. (2008) The Mendel–Fisher controversy. An overview. In A. Franklin *et al.*, *Ending the Mendel–Fisher controversy*, pp. 1–77. Pittsburgh, PA: University of Pittsburgh Press.

Franklin, A. *et al.* (2008). *Ending the Mendel–Fisher controversy.* Pittsburgh, PA: University of Pittsburgh Press.

Franklin, I. A. (1980). Evolutionary change in small populations. In *Conservation biology: an evolutionary-ecological perspective*, ed. M. E. Soulé & B. A. Wilcox, pp. 135–50. Sunderland, MA: Sinauer.

Franklin, I. R. & Frankham, R. (1998). How large must populations be to retain evolutionary potential? *Animal Conservation*, **1**, 69–70.

Franklin-Tong, V. E. & Franklin, F. C. H. (2003). The different mechanisms of gametophytic self-incompatibility. *Philosophical Transactions of the Royal Society of London, B*, **358**, 1025–32.

Franks, S. J. (2010). Genetics, evolution and conservation of island plants. *Journal of Plant Biology*, **2**, 481–8.

Franks, S. J. and Weis, A. E. (2008). A change in climate causes rapid evolution of multiple life-history traits and their interactions in an annual plant. *Journal of Evolutionary Biology*, **21**, 1321–34.

Franks, S. J., Sim, S. & Weis, A. E. (2007). Rapid response of flowering time by an annual plant in response to a climate fluctuation. *Proceedings of the National Academy of Sciences, USA*, **104**, 1278–82.

Franzke, A. & Mummenhoff, K. (1999). Recent hybrid speciation in *Cardamine* (Brassicaceae). Conversion of nuclear ribosomal ITS sequences *in statu nascendi*. *Theoretical and Applied Genetics*, **98**, 831–4.

Freckleton, R. P. & Watkinson, A. R. (2003). Are all plant populations metapopulations? *Journal of Ecology*, **91**, 321–4.

Frey-Klett, P., Garbaye, J. & Tarkka, M. (2007). The mycorrhiza helper bacteria revisited. *New Phytologist*, **176**, 22–36.

Friedman, J. & Barrett, S. C. H. (2009). Wind of change: new insights on the ecology and evolution of pollination and mating in wind-pollinated plants. *Annals of Botany*, **103**, 1515–27.

Friedman, S. T. & Adams, W. T. (1985). Estimation of gene flow into two seed orchards of Loblolly-pine (*Pinus taeda* L.). *Theoretical & Applied Genetics*, **69**, 609–15.

Friis, E. M., Pedersen, K.R. & Crane, P. R. (2010) Diversity in obscurity: fossil flowers and the early history of angiosperms. *Philosophical Transactions of the Royal Society*, **365**, 369–82.

Friis, E. M., Crane P. R. & Pedersen K. R. (2011). *Early flowers and angiosperm evolution.* Cambridge: Cambridge University Press.

Frisch, W., Meschede, M. & Blakey R. C. (2011). *Plate tectonics – continental drift and mountain building.* Heidelberg: Springer.

Fritsch, P. & Rieseberg, L. H. (1992). High outcrossing rates maintain male and hermaphrodite individuals in populations of the flowering plant *Datisca glomerata*. *Nature*, **359**, 633–6.

Frohlich, M. W. & Parker, D. S. (2000). The mostly male theory of flower evolutionary origins: from genes to fossils. *Systematic Botany*, **25**, 155–70.

Frost, H. B. (1938). Nuclear and juvenile characters in clonal varieties of *Citrus*. *Journal of Heredity*, **29**, 423–32.

Fröst, S. & Ising, G. (1968). An investigation into the phenolic compounds in *Vaccinium myrtillus* L. (Bilberries), *Vaccinium vitis-idaea* L. (Cowberries), and the hybrid between them *V. intermedium* Ruthe employing thin layer chromatography. *Hereditas*, **60**, 72–6.

Fryer, J. D. & Chancellor, R. J. (1979). Evidence of changing weed populations in arable land. *Proceedings of the 14th British Weed Control Conference*, pp. 958–64.

Fuentes, I. *et al.* (2014). Horizontal genome transfer as an asexual path to the formation of new species. *Nature*, **511**, 232–5.

Futuyma, D. J. & Agrawal, A. A. (2009). Macroevolution and the biological diversity of plants and herbivores. *Proceedings of the National Academy of Sciences, USA*, **106**, 18054–61.

Gaeta, R. T. *et al.* (2007). Genomic changes in resynthesized *Brassica napus* and their effect on gene expression and phenotype. *The Plant Cell*, **19**, 3403–17.

Gajewski, W. (1957). A cytogenetic study on the genus *Geum*. *Monographiae Botanicae*, No. 4.

Galeuchet, D. J. & Holderegger, R. (2005). (Erhaltung und Wiederansiedlung des Kleinen Rohrkolbens (*Typha minima*) – Vegetationsaufnahmen, Monitoring und genetische Herkunftsanalysen. *Botanica Helvetica*, **115**, 15–32.

Gallez, G. P. & Gottlieb, L. D. (1982). Genetic evidence for the hybrid origin of the diploid plant *Stephanomeria diegensis*. *Evolution*, **36**, 1158–67.

Galton, F. (1871). Experiments in pangenesis, by breeding from rabbits of a pure variety, into whose circulation blood taken from other varieties had previously been largely transfused. *Proceedings of the Royal Society of London*, **19**, 393–410.

Galton, F. (1876). The history of twins, as a criterion of the relative powers of nature and nurture. *Journal of the Royal Anthropological Institute*, 5, 391–406.

Galton, F. (1889). *Natural inheritance.* London & New York: Macmillan.

Galton, F. (1904). Eugenics. Its definition, scope and aims. In *Essays in eugenics*, pp. 35–43. London: Eugenics Education Society.

Galton, F. (1908). *Memories of my life.* London: Methuen.

Ganders, F. R. (1979). The biology of heterostyly. *New Zealand Journal of Botany*, **17**, 607–35.

Ganders, F. R. (1990). Altitudinal clines for cyanogenesis in introduced populations of white clover near Vancouver, Canada. *Heredity*, **64**, 387–90.

Garbelotto, M. & Pautasso, M. (2012). Impacts of exotic forest pathogens on Mediterranean ecosystems: four case studies. *European Journal of Plant Pathology*, **133**, 101–16.

Garcia, P. *et al.* (1989). Allelic and genotypic composition of ancestral Spanish and colonial Californian gene pools of *Avena barbata:* evolutionary implications. *Genetics*, **122**, 687–94.

Garcia-Robledo, C. *et al.* (2013). Tropical plant–herbivore networks: reconstructing species interactions using DNA barcodes. *PLOS ONE*, **8**, 1–10.

Garnier, A. *et al.* (2008). Measuring and modelling anthropogenic secondary seed dispersal along road verges for feral oilseed rape. *Basic and Applied Ecology*, **9**, 533–41.

Gartside, D. W. & McNeilly, T. (1974). The potential for evolution of metal tolerance in plants. III. Copper tolerance in normal populations of different species. *Heredity*, **32**, 335–48.

Gates, R. R. (1909). The stature and chromosomes of *Oenothera gigas* de Vries. *Archiv für Zellforschung*, **3**, 525–52.

Gay, P. A. (1960). A new method for the comparison of populations that contain hybrids. *New Phytologist*, **59**, 219–26.

Gayon, J. (2009). From Darwin to today in evolutionary biology. In *The Cambridge companion to Darwin*, ed. J. Hodge & G. Radick, pp. 227–301. Cambridge: Cambridge University Press.

Ge, Y. Z., Cheng, X. F., Hopkins, A. & Wang, Z. Y. (2007). Generation of transgenic *Lolium temulentum* plants by *Agrobacterium tumefaciens*-mediated transformation. *Plant Cell Reports*, 26, 783–9.

Gealy, D. R. *et al.* (2007). Implications of gene flow in the scale-up and commercial use of biotechnology-derived crop. *CAST Issue Paper* 37. Ames, IA: Council for Agricultural Science and Technology.

Geiger, R. (1965). *The climate near the ground.* Cambridge, MA: Harvard University Press.

Genton, B. J, Shykoff, J. A. & Giraud, T. (2005). High genetic diversity in French invasive populations of common ragweed, *Ambrosia artemisiifolia*, as a result of multiple sources of introduction. *Molecular Ecology*, **10**, 4275–85.

Gerstel, D. U. (1950). Self-incompatibility studies in Guayule. II. Inheritance. *Genetics*, **35**, 482–506.

Ghazoul, J. (2005). Pollen and seed dispersal among dispersed plants. *Biological Reviews*, **80**, 413–43.

Giakountis, A. *et al.* (2010). Distinct patterns of genetic variation alter flowering responses of *Arabidopsis* accessions to different daylengths. *Plant Physiology*, **152**, 177–91.

Gibby, M. (1981). Polyploidy and its evolutionary significance. In *The evolving biosphere*, ed. P. L. Forey, pp. 87–96. Cambridge: British Museum (Natural History) & Cambridge University Press.

Gilbert, B. & Levine, J. M. (2013). Plant invasions and extinction debts. *Proceedings of the National Academy of Sciences, USA*, **110**, 1744–9.

Gilbert, N. (2013) Case studies: a hard look at GM crops. Superweeds? Suicides? Stealthy genes? The true, the false and the still unknown about transgenic crops. *Nature*, **497**, 24–6.

Giles, B. E. & Goudet, J. (1997). Genetic differentiation in *Silene dioica* metapopulations: estimation of spatiotemporal effects in a successional plant species. *American Naturalist*, **149**, 507–26.

Giles, B. E., Lundqvist, E. & Goudet, J. (1998). Restricted gene flow and subpopulation differentiation in *Silene dioica*. *Heredity*, **80**, 715–23.

Gill, A. M. & Williams, J. E. (1996). Fire regimes and biodiversity: the effects of fragmentation of southeastern Australian eucalypt forests by urbanisation, agriculture and pine plantations. *Forest Ecology and Management*, **85**, 261–78

Gill, B. S. & Kimber, G. (1974). Giemsa C-banding and the evolution of Wheat. *Proceedings of the National Academy of Sciences, USA*, **71**, 4086–90.

Gill, L. *et al.* (2004). Phylogeography: English elm is a 2,000-year-old Roman clone. *Nature*, **431**, 1053.

Gillespie, R. G. *et al.* (2011). Long-distance dispersal: a framework for hypothesis testing. *Trends in Ecology and Evolution*, **27**, 47–56.

Gillham, N. W. (2001). *A life of Sir Francis Galton: from African exploration to the birth of eugenics*. Oxford: Oxford University Press.

Gilmour, J. S. L. (1936). Two early papers on classification. [Reprint 1989: *Plant Systematics and Evolution*, **167**, 97–107.]

Gilmour, J. S. L. (1937). A taxonomic problem. *Nature*, **139**, 1040.

Gilmour, J. S. L. (1940). Taxonomy and philosophy. In *The new systematics*, ed. J. Huxley, pp. 461–74. Oxford: Clarendon Press.

Gilmour, J. S. L. (1951). The development of taxonomic theory since 1851. *Nature*, **168**, 400–2.

Gilmour, J. S. L. & Gregor, J. W. (1939). Demes: a suggested new terminology. *Nature*, **144**, 333–4.

Gilmour, J. S. L. & Heslop-Harrison, J. (1954). The deme terminology and the units of micro-evolutionary change. *Genetica*, **27**, 147–61.

Gilmour, J. S. L. & Walters, S. M. (1963). Philosophy and classification. *Vistas in Botany*, **4**, 1–22.

Gilpin, M. E. & Soulé, M. E. (1986). Minimum viable populations: processes of species extinction. In *Conservation biology: the science of scarcity and diversity*, pp. 19–34. Sunderland, MA: Sinauer.

Given, D. R. (1994). *Principles and practice of plant conservation*. London: Chapman & Hall.

Givnish, T. (1979). On the adaptive significance of leaf form. In *Topics in plant population biology*, ed. O. T. Solbrig, S. Jain, G. B. Johnson & P. H. Raven, pp. 375–407. London: Columbia University Press.

Givnish, T. J. (2010). Ecology of plant speciation. *Taxon*, **59**, 1326–66.

Givnish T. J. & Renner, S. S. (2004). Tropical intercontinental disjunctions: Gondwana breakup, immigration from the Boreotropics, and transoceanic dispersal. *International Journal of Plant Sciences*, **165**(4 Suppl.), S1–S6.

Givnish, T. J. & Vermeij, G. J. (1976). Sizes and shapes of Liane leaves. *The American Naturalist*, **110**, 743–76.

Glass, B. (1959). Heredity and variation in the eighteenth century concept of the species. In *Forerunners of Darwin 1745–1859*, ed. B. Glass, O. Temkin & W. I. Straus, pp. 144–72. London: Oxford University Press.

Glen, W. (ed.) (1994). *The mass extinction debates*. Stanford, CA: Stanford University Press.

Glenn, T. C. I. & Schable, N. A. (2005). Isolating microsatellite DNA loci. *Methods in Enzymology*, **395**, 202–22.

Gliddon, C. & Saleem, M. (1985). Gene flow in *Trifolium repens* – an expanding genetic neighbourhood. In *Genetic differentiation and dispersal in plants*, ed. P. Jacquard, G. Heim & J. Antonovics, pp. 293–309. Berlin: Springer-Verlag.

Godefroid, S. *et al.* (2011a). How successful are plant species reintroductions? *Biological Conservation*, **144**, 672–82.

Godefroid, S. *et al.* (2011b). To what extent are threatened European plant species conserved in seed banks? *Biological Conservation*, **144**, 1494–8.

Godfray, H. C. J. (2002). Challenges for taxonomy. *Nature*, **417**, 17–19.

Godfray, H. C. J. & Knapp, S. (2004). Introduction. In 'Taxonomy for the twenty-first century'. *Philosophical Transactions of the Royal Society of London, B*, **359**, 559–69.

Godfray, H. C. J. *et al.* (2007). The web and the structure of taxonomy. *Systematic Biology*, **56**. 943–55.

Godt, M. J. Caplow, W. F. & Hamrick, J. L. (2005). Allozyme diversity in the federally threatened golden paintbrush, *Castilleja levisecta* (Scrophulariaceae). *Conservation Genetics*, **6**, 87–99.

Godward, M. B. E. (1985). The kinetochore. *International Review of Cytology*, **94**, 77–105.

Goebel, K. (1897). *Uber Jugendformen von Pflanzen und deren künstliche Wiederhervorrufung*. Sitzungsbericht der Mathematisch-Physikalischen Class der Königlich-Bayerischen Akademie der Wissenschaften, München, 26.

Goerke, H. (1989). *Carl von Linné*. Stuttgart: Wissenschaftliche Verlagsgesellschaft.

Göhre, V. & Paszkowski, U. (2006). Contribution of arbuscular mycorrhizal symbiosis to heavy metal phytoremediation. *Planta*, **223**, 1115–22.

Goldblatt, P. (1980). Polyploidy in angiosperms: monocotyledons. In *Polyploidy*, ed. W. H. Lewis, pp. 219–39. New York & London: Plenum Press.

Goldblatt, P. (2007). Index to plant chromosome numbers. *Taxon*, **56**, 984–986.

Goldblatt, P. & Johnson, D. E. (1990) *Index to plant chromosome numbers 1986–1987*. St Louis, MO: Missouri Botanical Garden.

Goldblatt, P., Hoch, P. C. & McCook, L. M. (1992). Documenting scientific data: the need for voucher specimens. *Annals of the Missouri Botanical Garden*, **79**, 969–70.

Goldschmidt, R. B. (1940). *The material basis of evolution*. New Haven, CT: Yale University Press.

Goldstein, P. Z. & DeSalle, R. (2011). Integrating DNA barcode data and taxonomic practice: determination, discovery, and description. *BioEssays*, **33**, 135–47.

Gordo, O. & Sanz, J. J. (2005). Phenology and climate change: a long-term study in a Mediterranean locality. *Oecologia*, **146**, 484–95.

Gore, A. (2013). *Earth in the balance*. London: Random House.

Gornall, R. J. (1999). Population genetic structure in agamospermous plants. In *Molecular systematics and plant evolution*, ed. P. M. Hollingsworth, R. M. Bateman & R. J. Gornall, pp. 118–38. London: Taylor & Francis.

Gornitz, V., Rosenzweig, D. & Hillel, D. (1997). Effects of anthropogenic intervention in the land hydrologic cycle on global sea level rise. *Global and Planetary Change*, **14**, 147–61.

Gottlieb, L. C. & Pilz, G. (1976). Genetic similarity between *Gaura longifolia* and its obligately outcrossing derivative *G. demareei*. *Systematic Botany*, **1**, 181–7.

Gottlieb, L. D. (1972). Levels of confidence in the analysis of hybridization in plants. *Annals of Missouri Botanic Garden*, **59**, 435–46.

Gottlieb, L. D. (1973). Genetic differentiation, sympatric speciation, and the origin of a diploid species of *Stephanomeria*. *American Journal of Botany*, **60**, 545–53.

Gottlieb, L. D. (1979). The origin of phenotype in a recently evolved species. In *Topics in plant population biology*, ed. O. T. Solbrig, S. Jain, G. B. Johnson & P. H. Raven, pp. 264–86. New York: Columbia University Press.

Gottlieb, L. D. (1981a). Electrophoretic evidence and plant populations. *Progress in Phytochemistry*, **7**, 1–46.

Gottlieb, L. D. (1981b). Gene number in species of Asteraceae that have different chromosome numbers. *Proceedings of the National Academy of Sciences USA*, **78**, 3726–9.

Gottlieb, L. D. (1982). Conservation and duplication of isozymes in plants. *Science*, **216**, 373–80.

Gottlieb, L. D. (1984). Isozyme evidence and problem solving in plant systematics. In *Plant biosystematics*, ed. W. F. Grant, pp. 343–57. Toronto: Academic Press.

Gottlieb, L. D. (1986). Genetic differentiation, speciation and phylogeny in *Clarkia* (Onagraceae). In *Modern aspects of species*, ed. K. Iwatsuki, P. H. Raven & W. J. Bock, pp. 145–60. Tokyo: University of Tokyo Press.

Gottlieb, L. D. (2003). Rethinking classic examples of recent speciation in plants. *New Phytologist*, DOI: 10.1046/j.1469-8137.2003.00922.x

Gottlieb, L. D., Warwick, S. I. & Ford, V. S. (1985). Morphological and electrophoretic divergence between *Layia discoidea* and *L. glandulosa*. *Systematic Botany*, **10**, 484–95.

Goudie, A. (2009). *The human impact: man's role in environmental change*. Oxford: Blackwell.

Goudriaan, J. *et al.* (1999). Use of models in global change studies. In *The terrestrial biosphere and global change*, ed. B. Walker, W. Steffen, J. Canadell & J. Ingram, pp. 106–40. Cambridge: Cambridge University Press.

Gould, S. J. (1979). Species are not specious. *New Scientist*, **83**, 374–6.

Gould, S. J. (2007). *Punctuated equilibrium*. Cambridge, MA: Belknap Press.

Gould, S. J. & Eldredge, N. (1977). Punctuated equilibria: the tempo and mode of evolution reconsidered. *Paleobiology*, **3**, 115–51.

Gould, S. J. & Eldredge, N. (1993). Punctuated equilibrium comes of age. *Nature*, **366**, 223–7.

Gould, S. J. & Lewontin, R. G. (1979). The spandrels of San Marco and the panglossian paradigm: a critique of the adaptionist programme. *Proceedings of the Royal Society of London, B*, **205**, 581–8.

Govindaraju, D. R. (1988). Relationship between dispersal ability and levels of gene flow in plants. *Oikos*, **52**, 31–5.

Gradstein, F. M. & Ogg, J. G. (2009). The geologic time scale. In *Tree of life*, ed. S. B. Hedges & S. Kumar, pp. 26–34. Oxford: Oxford University Press.

Graetz, R. D., Olsson, L. & Wilson, M. A. (1995). Interpreting a 9 year time-series of satellite observations for the Australian continent. *Proceedings of the Satellite Colloquium of the 10th International Congress of Photosynthesis*, Montpellier: International Society for Photogrammetry and Remote Sensing, 395–406.

Graham, B. F, Jr & Bormann, F. H. (1966). Natural root grafts. *The Botanical Review*, **32**, 255–92.

Grandont, L., Jenczewski, E. & Lloyd, A. (2013). Meiosis and its deviations in polyploidy plants. *Cytogenetic and Genome Research*, **140**, 171–84.

Grant, V. (1950). The protection of the ovules in flowering plants. *Evolution*, **4**, 179–201.

Grant, V. (1966). The selective origin of incompatibility barriers in the plant genus *Gilia. The American Naturalist*, 100, 99–118.

Grant, V. (1971). *Plant speciation.* New York & London: Columbia University Press.

Grant, V. (1975). *Genetics of flowering plants.* New York & London: Columbia University Press.

Grant, V. (1981). *Plant speciation.* 2nd edn. New York: Columbia University Press.

Grant, V. & Grant, K. A. (1965). *Flower pollination in the* Phlox *family.* New York: Columbia University Press.

Grant-Downton, R. T. & Dickinson, H. G. (2006). Epigenetics and its implications for plant biology. 2. The 'epigenetic epiphany': epigenetics, evolution and beyond. *Annals of Botany*, **97**, 11–27.

Graur, D. & Li, W.-H. (1999). *Fundamentals of molecular evolution*, 2nd edn. Sunderland, MA: Sinauer Associates.

Graur, D. & Li, W.-H. (2000). *Fundamentals of molecular evolution*, 2nd edn. Sunderland, MA: Sinauer Associates.

Graur, D. & Martin, W. (2004). Reading the entrails of chickens: molecular timescales of evolution and the illusion of precision. *Trends in Genetics*, **20**, 80–6.

Graur, D. *et al.* (2013). On the immortality of television sets: 'function' in the human genome according to the evolution-free gospel of ENCODE. *Genome Biology & Evolution*, **5**, 578–90.

Gray, A. (2004). Will *Spartina anglica* invade northwards with changing climate? In *Third International Conference on Invasive* Spartina. *San Francisco, CA, 8–10 November 2004.* San Francisco: Invasive Spartina Project.

Gray, A. J., Marshall, D. F. & Raybould, A. F. (1991). A century of evolution in *Spartina anglica. Advances in Ecological Research*, 21, 1–62.

Green, M. R. & Sambrook, J. (2012). *Molecular cloning: a laboratory manual.* Cold Spring Harbor, NY: Cold Spring Harbor Laboratory Press.

Green, R. H. (1979). *Sampling design and statistical methods for environmental biologists.* New York, Chichester, Brisbane & Toronto: Wiley.

Greene, E. L. (1909). Linnaeus as an evolutionist. *Proceedings of the Washington Academy of Sciences*, 11, 17–26.

Greenwood, M. S. (1986). Gene exchange in loblolly pine: the relation between pollination mechanism, female receptivity and pollen availability. *American Journal of Botany*, **73**, 1443–51.

Gregor, J. W. (1930). Experiments on the genetics of wild populations, I. *Plantago maritima. Journal of Genetics*, **22**, 15–25.

Gregor, J. W. (1931). Experimental delimitation of species. *New Phytologist*, **30**, 204–17.

Gregor, J. W. (1938). Experimental taxonomy. 2. Initial population differentiation in *Plantago maritima* in Britain. *New Phytologist*, 37, 15–49.

Gregor, J. W. (1939). Experimental taxonomy. 4. Population differentiation in North American and European Sea Plantains allied to *Plantago maritima* L. *New Phytologist*, **38**, 293–322.

Gregor, J. W. (1944). The ecotype. *Biological Reviews*, **19**, 20–30.

Gregor, J. W. (1946). Ecotypic differentiation. *New Phytologist*, **45**, 254–70.

Gregor, J. W. & Lang, J. M. S. (1950). Intra-colonial variation in plant size and habit in Sea Plantains. *New Phytologist*, **49**, 135–41.

Gregor, J. W., Davey, V. McM. & Lang, J. M. S. (1936). Experimental taxonomy. 1. Experimental garden technique in relation to the recognition of small taxonomic units. *New Phytologist*, **35**, 323–50.

Greig-Smith, P. (1964). *Quantitative plant ecology*, 2nd edn. London: Butterworth.

Greiner, S. *et al.* (2011). The role of plastids in plant speciation. *Molecular Ecology*, **20**, 671–91.

Gressel, J. (2005a). Introduction: the challenges of ferality. In *Crop ferality and volunteerism*, ed. J. Gressel, pp. 1–7. Boca Raton, FL: Taylor & Francis.

Gressel, J. (ed.) (2005b).*Crop ferality and volunteerism.* Boca Raton, FL: Taylor & Francis.

Griffiths, M. (1994). *Index of garden plants.* London: Macmillan.

Groot, J. & Boschuizen, R. (1970). A preliminary investigation into the genecology of *Plantago major* L. *Journal of Experimental Botany*, **21**, 835–41.

Gross, B. L. & Rieseberg, L. H. (2005). The ecological genetics of homoploid hybrid speciation. *Journal of Heredity*, **96**, 241–52.

Grotewold, E., Chappell, J. & Kellogg, E. (2015). *Plant genes, genomes and genetics.* Chichester: Wiley-Blackwell.

Grove, A. T. & Rackham, O. (2001). *The nature of Mediterranean Europe: an ecological history.* New Haven, CT, & London: Yale University Press.

Groves, R. H. & di Castri, F. (1991). *Biogeography of Mediterranean invasions.* Cambridge: Cambridge University Press.

Gruber, S. & Claupein, W. (2007). Fecundity of volunteer oilseed rape and estimation of potential gene dispersal by a practice-related model. *Agriculture, Ecosystems & Environment,* **119**, 401–8.

Gübeli, A., Hochuli, P. A. & Wildi, W. (1984). Lower Cretaceous turbiditic sediment from the Rif chain (northern Morocco), palynology, stratigraphy setting. *Geologisches Rundschau,* **73**, 1081–114.

Guerrant, E. O. (1992). Genetic and demographic considerations in the sampling and reintroduction of rare plants. In *Conservation biology,* ed. P. L. Fiedler & S. K. Jain, pp. 321–44. New York: Chapman & Hall.

Guerrant, E. O. (2013). The value and propriety of reintroduction as a conservation tool for rare plants. *Botany-Botanique,* **91**, v–x.

Gugerli, F., Parducci, L. & Petit, R. J. (2005). Ancient plant DNA: review and prospects. *New Phytologist,* **166**, 409–18.

Guggisberg, A. *et al.* (2008). Genomic origin and organization of the allopolyploid *Primula egaliksensis* investigated by *in situ* hybridization. *Annals of Botany,* **101**, 919–27.

Guggisberg, A. *et al.* (2013). Transcriptome divergence between introduced and native populations of Canada thistle, *Cirsium arvense. New Phytologist,* **199**, 595–608.

Guignard, L. (1891). Nouvelles études sur la fécondation. *Annales des Sciences naturelles (Botanique),* **14**, 163–296.

Gunther, R. W. T. (1928). *Further correspondence of John Ray.* London & New York: Oxford University Press.

Guo, L.-L. *et al.* (2011). Evolution of the S-locus region in *Arabidopsis* relatives. *Plant Physiology,* **157**, 937–46.

Gupta, P. K. & Varshney, R. K. (1999) Molecular markers for genetic fidelity during micropropagation and conservation. *Current Science,* **76**, 1308–10.

Gustafsson, Å. (1946). Apomixis in higher plants. 1. The mechanism of apomixis. *Acta Universitatis Lundensis,* **42** (3), 1–67.

Gustafsson, Å. (1947a). Apomixis in higher plants. 2. The causal aspect of apomixis. *Acta Universitatis Lundensis,* **43**(3), 69–179.

Gustafsson, Å. (1947b). Apomixis in higher plants. 3. Biotype and species formation. *Acta Universitatis Lundensis,* **43**(12), 183–370.

Gustafsson, L. & Gustafsson, P. (1994). Low genetic variation in Swedish populations of the rare species *Vicia pisiformis* (Fabaceae) revealed with RFLP & RAPD. *Plant Systematics & Evolution,* **189**, 133–48.

Gwynn-Jones, D., Lee, J. A. & Callaghan, T. V. (1997) Effects of enhanced UV-B and elevated carbon dioxide concentrations on a sub-arctic forest ecosystem. *Plant Ecology,* **128**, 243–49.

Habel, J. C. & Zachos, F. E. (2012). Habitat fragmentation versus fragmented habitats. *Biodiversity and Conservation,* **21**, 2987–90.

Haeckel, E. H. P. A. (1876). *The history of creation.* (Translation revised by E. R. Lankester.) London: Routledge.

Haffer, J. (1969). Speciation in Amazonian forest birds. *Science,* **165**, 131–7.

Haffer, J. (2007). *Ornithology, evolution, and philosophy: the life and science of Ernst Mayr 1904–2005.* San Francisco: Ignatius Press.

Haffer, J. & Prance, G. T. (2001). Climatic forcing of evolution in Amazonia during the Cenozoic: on the refuge theory of biotic differentiation *Amazoniana,* **16**, 579–607.

Hagerup, O. (1932). Uber polyploidie in Beziehung zu Klima, Okologie, und Phyogenie – Chromosomenzahlen aus Timbuktu. *Hereditas,* **16**, 19–40.

Håkansson, S. (1983). Competition and production in short-lived crop-weed stands: density effects. *Uppsala: Department of Plant Husbandry. Report* No. 127. Swedish University of Agricultural Sciences.

Haldane, J. B. S. (1932). *The causes of evolution.* London: Longmans.

Hall, B. G. (2011). *Phylogeny for physiologists: a pragmatic guide to building trees: phylogenetic trees made easy. A how-to manual,* 2nd edn. Sunderland, MA: Sinauer.

Hall, M. (2000). Comparing damages: Italian and American concepts of restoration. In *Methods and approaches in forest history,* ed. M. Agnoletti & S. Anderson, pp. 165–72. Wallingford, UK: CAB International.

Hall, O. (1972). Oxygen requirements of root meristems in diploid and autotetraploid Rye. *Hereditas*, **70**, 69–74.

Hallam, A. (1973). *A revolution in the earth sciences: from continental drift to plate tectonics*. Oxford: Clarendon Press.

Hampe, A. (2011). Plants on the move: the role of seed dispersal and initial population establishment for climate-driven range expansions. *Acta Oecologica – International Journal of Ecology*, **37**, 666–73.

Hampe, A. & Petit, R. J. (2007). Ever deeper phylogeographies: trees retain the genetic imprint of Tertiary plate tectonics. *Molecular Ecology*, **16**, 5113–14.

Hamrick J. L. (1990). Isozymes and the analysis of genetic structure in plant populations. In *Isozymes in plant biology*, ed. D. E. Soltis & P. S. Soltis, pp. 87–105. London: Chapman & Hall.

Hamrick, J. L. & Godt, M. J. W. (1989). Allozyme diversity in plant species. In *Plant population genetics, breeding and genetic resources*, ed. A. D. H. Brown, M. T. Clegg, A. L. Kahler & B. S. Weir, pp. 43–63. Sunderland, MA: Sinauer.

Hamrick, J. L. & Loveless, M. D. (1989). Associations between the breeding system and the genetic structure of tropical tree populations. In *The evolutionary ecology of plants*, ed. J. Bock & Y. B. Linhart, pp. 129–46. Boulder, CO: Westview Press.

Hamrick, J. L., Linhart, Y. B. & Mitton, J. B. (1979). Relationships between life history characteristics and electrophoretically detectable genetic variation in plants. *Annual Review of Ecology & Systematics*, **10**, 173–200.

Han, F. *et al.* (2005). Rapid and repeatable elimination of a parental genome-specific DNA repeat (pGc1R–1a) in newly synthesized wheat allopolyploids. *Genetics*, **170**, 1239–45.

Hancock, A. M. *et al.* (2011). Adaptation to climate across the *Arabidopsis thaliana* genome. *Science*, **334**, 83–6.

Hancock, C. N. *et al.* (2003). The S-locus and unilateral incompatibility. *Transactions of the Royal Society, London*, **358**, 1133–40.

Hancock, J. F. (2012). *Plant evolution and the origin of crop species*, 3rd edn. Wallingford: CABI.

Hannah, L. (2011). Climate change, connectivity, and conservation success. *Conservation Biology*, **25**, 1139–42.

Hansen, M. M. *et al.* (2012). Monitoring adaptive genetic responses to environmental change. *Molecular Ecology*, **21**, 1311–29.

Hao, Y.-Q. *et al.* (2012). The role of late-acting self-incompatibility and early-acting inbreeding depression in governing female fertility in monkshood, *Aconitum kusnezoffii*. *PLOS ONE*, DOI:10.1371/journal.pone.0047034ey

Harberd, D. J. (1957). The within population variance in genecological trials. *New Phytologist*, **56**, 269–80.

Harberd, D. J. (1958). Progress and prospects in genecology. *Record of the Scottish Breeding Station*, **1958**, 52–60.

Harberd, D. J. (1961). Observations on population structure and longevity of *Festuca rubra*. L. *New Phytologist*, **60**, 184–206.

Harberd, D. J. (1962). Some observations on natural clones in *Festuca ovina*. *New Phytologist*, **61**, 85–100.

Harberd, D. J. (1963). Observations on natural clones of *Trifolium repens* L. *New Phytologist*, **62**, 198–204.

Harborne, J. B. (1993). *Introduction to ecological biochemistry*, 4th edn. San Diego, CA: Academic Press.

Harborne, J. B. & Turner, B. L. (1984). *Plant chemosystematics*. Orlando, FL: Academic Press.

Harborne, J. B., Williams, C. A. & Smith, D. M. (1973). Species-specific Kaempferol derivatives in ferns of the Appalachian *Asplenium* complex. *Biochemical Systematics*, 1 (5), 1–4.

Hardesty, B. D. *et al.* (2006) Genetic evidence of frequent long-distance recruitment in a vertebrate-dispersed tree. *Ecology Letters*, **9**, 516–25.

Hardin, G. (1966). *Biology: its principles and implications*, 2nd edn. London & San Francisco: Freeman.

Hardin, J. Bertoni, G. & Kleinsmith, L. J. (2009). *Becker's world of the cell*, 8th edn. Boston: Benjamin Cummings.

Hardwick, K. A. *et al.* (2011). The role of botanic gardens in science and practice of ecological restoration. *Conservation Biology*, **25**, 265–75.

Hargreaves, A. L., Harder, L. D. & Johnson, S. D. (2009). Consumptive emasculation: the ecological and evolutionary consequences of pollen theft. *Biological Reviews of the Cambridge Philosophical Society*, **84**, 259–76.

Harlan, H. V. & Martini, M. L. (1938). The effect of natural selection on a mixture of barley varieties. *Journal of Agricultural Research*, **57**, 189–99.

Harlan, J. R. & de Wet, J. M. J. (1975). On Ö Winge and a prayer: The origins of polyploidy. *The Botanical Review*, **41**, 361–90.

Harley, J. L. & Harley, E. L. (1987). A check-list of mycorrhiza in the British Flora. *New Phytologist (Supplement)*, **105**, 1-102.

Harnesk, H. (2007). *Linnaeus: genius of Uppsala*. Uppsala: Hallgren & Fallgren.

Harper, J. L. (1977). *Population biology of plants*. London & New York: Academic Press.

Harper, J. L. (1978). The demography of plants with clonal growth. In *Structure and functioning of plant populations*, ed. A. H. J. Freyson & J. W. Waldendorp, pp. 27–48. Amsterdam, Oxford & New York: North Holland Publishing Company.

Harper, J. L. (1983). A Darwinian plant ecology. In *Evolution from molecules to men*, ed. D. S. Bendall, pp. 323–45. Cambridge: Cambridge University Press.

Harris, J. *et al.* (2009). Soil microbial communities and restoration ecology: facilitators or followers? *Science* **325**, 573–4.

Harris, R. (2005). Attacks on taxonomy. *American Scientist*, **93**, 311.

Harris, S. A. & Ingram, R. (1991). Chloroplast DNA and biosystematics: the effects of intraspecific diversity and plastid transmission. *Taxon*, **40**, 393–401.

Harris, S. A. & Ingram, R. (1992). Molecular systematics of the genus *Senecio* L. I. Hybridization in a British polyploid complex. *Heredity*, **69**, 1–10.

Harrison, C. J. & Langdale, J. A. (2006). A step by step guide to phylogeny reconstruction. *The Plant Journal*, **45**, 561–72.

Harshberger, J. W. (1901). The limits of variation in plants. *Proceedings of the National Academy of Sciences, USA*, **53**, 303–19.

Hartl, D. L. & Clark, A. G. (2004). *Principles of population genetics*, 3rd edn. Sutherland, MA: Sinauer.

Harvey, W. H. (1860). Darwin on the origin of species. *The Gardeners' Chronicle and Agricultural Gazette*, 18 February, 145–6.

Hathaway, W. H. (1962). Weighted hybrid index. *Evolution*, **16**, 1–10.

Hauser, T. P. & Loeschcke, V. (1994). Inbreeding depression and mating-distance dependent offspring fitness in large and small populations of *Lychnis flos-cuculi*. *Journal of Evolutionary Biology*, **7**, 609–22.

Hayes, W. (1964). *The genetics of bacteria and their viruses*. Oxford & Edinburgh: Blackwell.

Hayward, I. M. & Druce, G. C. (1919). *Adventive flora of Tweedside*. Arbroath, UK: Buncle.

Hazarika, M. H. & Rees, H. (1967). Genotypic control of chromosome behaviour in Rye. X. Chromosome pairing and fertility in autotetraploids. *Heredity*, **22**, 317–32.

He, S. *et al.* (2009). Dynamics of the evolution of the genus *Agrostis* revealed by GISH/FISH. *Crop Science*, **49**, 2285–90.

He, X.-J., Chen, T. & Zhu, J.-K. (2011). Regulation and function of DNA methylation in plants and animals. *Cell Research*, **21**, 442–65.

Heap, I. (2008). International survey of herbicide resistant weeds. www.weedscience.org/in.asp.

Hebert, P. D. N. (2003). Biological identification through DNA barcodes. *Philosophical Transactions of the Royal Society of London, B*, **270**, 313–21.

Heckmann, S. & Houben, A. (2012). Holokinetic centromeres. In *Plant centromere biology*, ed. J. Jiang & J. Birchler, pp. 83–94. Ames, IA: Wiley-Blackwell.

Heckmann, S. *et al.* (2013). The holocentric species *Luzula elegans* shows interplay between centromere and large-scale genome organization. *Plant Journal*, 73, 555–65.

Hedges, S. B. & S. Kumar, S. (2009). *The time tree of life*. New York: Oxford University Press.

Hegarty, M. J. and Hiscock, S. J. (2005). Hybrid speciation in plants: new insights from molecular studies. *New Phytologist*, **165**, 411–23.

Hegarty, M. J. *et al.* (2009). Extreme changes to gene expression associated with homoploid hybrid speciation. *Molecular Ecology*, DOI:10.1111/j.1365-294X.2008.04054.x

Hegarty, M. J. *et al.* (2011). Nonadditive changes in cytosine methylation as a consequence of hybridization and genome duplication in *Senecio* (Asteraceae). *Molecular Ecology*, **20**, 105–13.

Hegarty, M. J., Abbott, R. J. & Hiscock, S. J. (2012). Allopolyploid speciation in action: the origins and evolution of *Senecio cambrensis*. In *Polyploidy and genome evolution*, ed. P. S. Soltis & D. E. Soltis, pp. 245–70. Heidelberg: Springer.

Hegarty, M. *et al.* (2013). Lessons from natural and artificial polyploids in higher plants. *Cytogenetic and Genome Research*, **140**, 204–25.

Hegland, S. J. *et al.* (2009). How does climate warming affect plant–pollinator interactions? *Ecological Letters*, **12**, 184–95.

Heiser, C. B., Jr (1949a). Natural hybridization with particular reference to introgression. *The Botanical Review*, **15**, 645–87.

Heiser, C. B., Jr (1949b). Studies in the evolution of sunflower species *Helianthus annuus* and *H. bolanderi*. *University of California Publication in Botany*, **23**, 157–96.

Heiser, C. B., Jr (1973). Introgression re-examined. *The Botanical Review*, **39**, 347–66.

Heiser, C. B., Jr (1979). Hybrid populations of *Helianthus divaricatus* and *H. microcephalus* after 22 years. *Taxon*, **28**, 71–5.

Heiser, C. B., Jr *et al.* (1969). The North American Sunflower (*Helianthus*). *Memoirs of the Torrey Botanical Club*, **22**, 1–218.

Helmann, J. J. *et al.* (2008). Five potential consequences of climate change for invasive species. *Conservation Biology*, **22**, 534–43.

Hemming, M. N. & Trevaskis, B. (2011). Make hay while the sun shines: the role of MADS-box genes in temperature dependant seasonal flowering responses. *Plant Science*, **180**, 447–53.

Hendry, A. P. (2009). Evolutionary biology: speciation. *Nature*, **458**, 162–4.

Hennig, W. (1950). *Grundzüge einer Theorie der phylogenetischen Systematik*. Berlin: Deutscher Zentralverlag.

Hermanutz, L. A., Innes, D. J. & Weis, I. M. (1989). Clonal structure of Arctic Dwarf Birch (*Betula glandulosa*) at its northern limit. *American Journal of Botany*, **76**, 755–61.

Herrera, J. C., D'Hont, A. & Lashermes, P. (2007). Use of fluorescence *in situ* hybridization as a tool for introgression analysis and chromosome identification in coffee (*Coffea arabica* L.) *Genome*, **50**, 619–26.

Heslop-Harrison, J. (1953). *New concepts in flowering-plant taxonomy*. London: Heinemann.

Heslop-Harrison, J. (1964). Forty years of genecology. *Advances in Ecological Research*, **2**, 159–247.

Heslop-Harrison, J. (1978). *Cellular recognition systems in plants*. Studies in Biology, No. 100. London: Arnold.

Heslop-Harrison, J. S. (Pat) & Schmidt, T. (2007). Plant nuclear genome composition. In *Handbook of plant science*, ed. K. Roberts, pp. 565–72. Chichester: Wiley.

Heslop-Harrison, J. S. (Pat) & Schwarzacher, T. (2011). Organisation of the plant genome in chromosomes. *The Plant Journal*, **66**, 18–33.

Hesselman, H. (1919). Iakttagelser øver skogsträdpollens spridningsførmagå. *Meddelanden från Statens Skogsførsoksanstalt*, **16**, 27.

Hewitt, G. (1996). Some genetic consequences of ice ages, and their role in divergence and speciation. *Biological Journal of the Linnean Society*, **58**, 247–76.

Hewitt, G. (2000). The genetic legacy of the Quaternary ice ages. *Nature*, **405**, 907–13.

Heywood, V. H. (1976). *Plant taxonomy*, 2nd edn. Studies in Biology, No. 5. London: Arnold.

Heywood, V. H. (1980). The impact of Linnaeus on botanical taxonomy – past, present and future. *Veröffentlichungen der Joachim Jungius-Gesellschaft der Wissenschaften Hamburg*, **43**, 97–115.

Heywood, V. H. (1983). Botanic gardens and taxonomy: their economic role. *Bulletin of the Botanical Survey of India*, **25**, 134–47.

Heywood, V. H. (2001). Floristics and monography – an uncertain future? *Taxon*, **50**, 361–80.

Heywood, V. H. (2011). The role of botanic gardens as resource and introduction centres in the face of global change. *Biodiversity & Conservation*, **20**, 221–39.

Heywood, V. H. & Stuart, S. N. (1992). Species extinctions in tropical forests. In *Tropical deforestation*, ed. T. C. Whitmore & J. Sayer, pp. 91–117. London: Chapman & Hall.

Heywood, V. H. & Wyse Jackson, P. S. (1991). *Tropical botanic gardens: their role in conservation and development*. London: Academic Press.

Hierro, J. L., Maron, J. L. & Callaway, R. M. (2005). A biogeographical approach to plant invasions: the importance of studying exotics in their introduced and native range. *Journal of Ecology*, **93**, 5–15.

Hiesey, W. M. & Milner, H. W. (1965). Physiology of ecological races and species. *Annual Review of Plant Physiology*, **16**, 203–16.

Hillis, D. M., Moritz, C. & Mable, B. K. (1996). *Molecular systematics*, 2nd edn, Sunderland, MA: Sinauer.

Hinton, W. F. (1976). The evolution of insect-mediated self-pollination from an outcrossing system in *Calyptridium* (Portulacaceae). *American Journal of Botany*, **63**, 979–86.

Hipp, A. L., Rothrock, P. E. & Roalson, E. H. (2009). The evolution of chromosome arrangements in *Carex* (Cyperaceae). *The Botanical Review*, 75, 96–109.

Ho, P., Zhao, J. H. & Dayuan, X. (2009). Rethinking agro-biotechnological innovations in emerging economies: the case of Bt Cotton in China. *Journal of Peasant Studies*, 36, 345–64.

Hoballah, M. E. *et al.* (2007). Single gene-mediated shift in pollinator attraction in *Petunia*. *The Plant Cell*, 19, 779–90.

Hobbs, R. J. & Cramer, V. A. (2008). Restoration ecology: interventionist approaches for restoring and maintaining ecosystem function in the face of rapid environmental change. *Annual Review of Environment and Resources*, 33, 39–61.

Hochuli, P. & Feist-Burkhardt, S. (2013). Angiosperm-like pollen and *Afropollis* from the Middle Triassic (Anisian) of the Germanic Basin (Northern Switzerland). *Frontiers in Plant Science*, 4: 344.

Hodge, J. (2009). The notebook programmes and projects of Darwin's London years. In *The Cambridge companion to Darwin*, 2nd edn, ed. J. Hodge and G. Radick, pp. 44–72. Cambridge: Cambridge University Press.

Hodge, J. (2013). The origins of the *Origin:* Darwin's first thoughts about the tree of life and natural selection, 1837–1839. In *The Cambridge encyclopedia of Darwin and evolutionary thought*, ed. M. Ruse, pp. 64–71. Cambridge: Cambridge University Press.

Hodgins, K. A. & Barrett, S. C. H. (2008). Natural selection on floral traits through male and female function in wild populations of the heterostylous daffodil *Narcissus triandrus*. *Evolution*, 62, 1751–63.

Hodkinson, T. R. & Parnell, J. A. N. (2006). *Reconstructing the Tree of Life: taxonomy and systematics of species rich taxa*. Boca Raton, FL: CRC Press.

Hoebee, S. E. *et al.* (2007). Mating patterns and contemporary gene flow by pollen in a large continuous and a small isolated population of the scattered forest tree *Sorbus torminalis*. *Heredity*, 99, 47–55.

Hoegh-Guldberg, O. *et al.* (2008). Assisted colonization and rapid climate change. *Science*, 321, 345–6.

Hoffmann, H. (1881). Rückblick auf meine Variations-Versuche von 1855–80. *Botanische Zeitung*, 1881, 345–51, 361–7, 377–83, 393–9, 409–15, 424–31.

Holderegger, R., Kamm, U. & Gugerli, F. (2006). Adaptive vs. neutral genetic diversity: implications for landscape genetics. *Landscape Ecology*, 21, 797–807.

Holliday, R. J. & Putwain, P. D. (1977). Evolution of resistance to simazine in *Senecio vulgaris* L. *Weed Research*, 17, 291–6.

Holliday, R. J. & Putwain, P. D. (1980). Evolution of herbicide resistance in *Senecio vulgaris*: variation in susceptibility to simazine between and within populations. *Journal of Applied Ecology*, 17, 779–91.

Hollingsworth, M. L. & Bailey, J. P. (2000). Evidence for massive clonal growth in the invasive weed *Fallopia japonica* (Japanese Knotweed). *Botanical Journal of the Linnean Society*, 133, 463–72.

Hollingsworth, P. M., Graham, S. W. & Little, D. P. (2011). Choosing and using a plant barcode. *PLoS ONE*, 6, 1–13.

Holm, L. G. *et al.* (1977). *The world's worst weeds: distribution and biology*. Honolulu, HI: University of Honolulu Press.

Holmes, D. S. & Bougourd, S. M. (1991). B chromosome selection in *Allium schoenoprasum*. II. Experimental populations. *Heredity*, 67, 117–22.

Holsinger, K. E. (1984). The nature of biological species. *Philosophy of Science*, 51, 293–307.

Holsinger, K. E. & Gottlieb, L. D. (1991). Conservation of rare and endangered plants. In *Genetics and conservation of rare plants*, ed. D. A. Falk & K. E. Holsinger, pp. 195–208. New York: Oxford University Press.

Holttum, R. E. (1967). Comparative morphology, taxonomy and evolution. *Phytomorphology*, 17, 36–41.

Honjo, M. *et al.* (2008). Tracing the origins of stocks of the endangered species *Primula sieboldii* using nuclear microsatellites and chloroplast DNA. *Conservation Genetics*, 9, 1139–47.

Honnay, O. *et al.* (2005). Forest fragmentation effects on patch occupancy and population viability of herbaceous plant species. *New Phytologist*, 166, 723–36.

Honnay, O. *et al.* (2008). Can a seed bank maintain the genetic variation in the aboveground plant populations? *Oikos*, 117, 1–5.

Hooglander, N., Lumaret, R. & Bos, M. (1993). Inter-intra-specific variation of chloroplast DNA of European *Plantago* species. *Heredity*, 70, 322–34.

Hooykaas, P. J. J. (2007). Plant transformation. In *Handbook of plant science*, ed. K. Roberts, pp.665–70. Chichester: Wiley.

Hoquet, T. (2013). The evolution of the *Origin* (1859–1872). In *The Cambridge encyclopedia of Darwin and evolutionary thought*, ed. M. Ruse, pp 158–64. Cambridge: Cambridge University Press.

Hörandl, E. (2010). Beyond cladistics: extending evolutionary classifications into deeper time levels. *Taxon*, **59**, 45–350.

Hörandl, E. & Paun, O. (2007). Patterns and sources of genetic diversity in apomictic plants: implications for evolutionary potentials. In *Apomixis: evolution, mechanisms and perspectives*, ed. E. Hörandl, U. Grossniklaus, P. J. Van Dijk & T. Sharbel, pp. 169–94. Ruggell, Liechtenstein: Gantner.

Hörandl, E. & Stuessy, T. F. (2010). Paraphyletic groups as natural units of biological classification. *Taxon*, **59**, 1641–53.

Hörandl, E. *et al.* (2007). *Apomixis: evolution, mechanisms and perspectives*. Ruggell, Liechtenstein: Gantner.

Horsman, D. C., Roberts, T. M. & Bradshaw, A. D. (1979). Studies on the effect of sulphur dioxide on perennial ryegrass (*Lolium perenne*, L.). II. Evolution of sulphur dioxide tolerance. *Journal of Experimental Botany*, **30**, 495–501.

Hort, A. (1938). *The Critica botanica of Linnaeus*. London: Ray Society, British Museum.

Houben, A. *et al.* (2001). The genomic complexity of micro B chromosomes of *Brachycome dichromosomatica*. *Chromosoma*, **110**, 451–9.

Houben, A., Nasuda, S. & Endo, T. R. (2011). Plant B chromosomes, *Methods in Molecular Biology*, **701**, 97–111.

Houliston, G. J. & Chapman, H. M. (2004). Reproductive strategy and population variability in the facultative apomict *Hieracium pilosella* (Asteraceae). *American Journal of Botany*, **91**, 37–44.

House of Lords (2010). *Report into systematics and taxonomy by the House of Lords Science and Technology Committee*. London: HMSO.

Howe, C. J. (2007). Chloroplast genome. In *Handbook of plant science*, ed. K. Roberts, pp. 585–587. Chichester: Wiley.

Hoyle, B. (2008). Plight of the pines. *Nature Reports Climate Change*, 24 April 2008, DOI:10.1038/climate.2008.35.

Hsiao, J.-Y. & Li, H.-L. (1973). Chromatographic studies on the Red Horsechestnut (*Aesculus* × *carnea*) and its putative parent species. *Brittonia*, **25**, 57–63.

Hu, S., Dilcher, D. L. & Taylor, D. W. (2012). Pollen evidence for the pollination biology of early flowering plants. In *Evolution of the plant–pollinator relationships*, ed. S. Patiny, pp.165–218. Cambridge: Cambridge University Press.

Hu, S.-J. *et al.* (2011). Hybridization and asymmetric introgression between *Cypripedium tibeticum* and *C. yunnanense* in Shangrila County, Yunnan Province, China. *Nordic Journal of Botany*, **29**, 625–31.

Hu, Y., Poh, H. M. & Chua, N. H. (2006). The *Arabidopsis ARGOS-like* gene regulates cell expansion during organ growth. *Plant Journal*, **47**, 1–9.

Huang, J. *et al.* (2010). Functional analysis of the *Arabidopsis PAL* gene family in plant growth, development, and response to environmental stress. *Plant Physiology*, **153**, 1526–38.

Huck, S. *et al.* (2009). Range-wide phylogeography of the European temperate-montane herbaceous plant *Meum athamanticum* Jacq.: evidence for periglacial persistence. *Journal of Biogeography*, **36**, 1588–99.

Hudson, R. R., Keitman, M. & Aguadé, M. *et al.* (1987). A test of neutral molecular evolution based on nucleotide data. *Genetics*, **116**, 153–9.

Hughes, A. (1959). *A history of cytology*. London & New York: Abelard-Schuman.

Hughes, C. & Eastwood, R. (2006). Island radiation on a continental scale: exceptional rates of plant diversification after uplift of the Andes. *Proceedings of the National Academy of Sciences, USA*, **103**, 10334–9.

Hughes, C. E., Eastwood, R. J. & Bailey, C. D. (2006). From famine to feast? Selecting nuclear DNA sequence loci for plant species-level phylogeny reconstruction. *Philosophical Transactions of the Royal Society, B*, **361**, 211–25.

Hughes, M. A. (1991). The cyanogenic polymorphism in *Trifolium repens* L. (white clover). *Heredity*, **66**, 105–15.

Hughes, M. B. & Babcock, E. B. (1950). Self-incompatibility in *Crepis foetida* L. subsp. *rhoeadifolia*. *Genetics*, **35**, 570–88.

Hughes, N. F. (1994). *The enigma of angiosperm origins.* Cambridge: Cambridge University Press.

Humborg, C. *et al.* (1997). Effect of Danube river dam on Black Sea biogeochemistry and ecosystem structure. *Nature,* **386**, 385.

Humeau, L., Pailler, T. & Thompson, J. D. (1999). Cryptic dioecy and leaky dioecy in endemic species of *Dombeya* (Sterculiaceae) on La Réunion. *American Journal of Botany,* **86**, 1437–47.

Hummel, S. (2003). *Ancient DNA: typing methods, strategies and applications.* Heidelberg: Springer.

Humphries, C. J. & Funk, V. A. (1984). Cladistic methodology. In *Current concepts in plant taxonomy,* ed. V. H. Heywood & D. M. Moore, pp. 323–62. London: Systematics Association Special Volume 25, Academic Press.

Husband, B. C. (2004). The role of triploid hybrids in the evolutionary dynamics of mixed-ploidy populations. *Biological Journal of the Linnean Society,* **82**, 537–46.

Husband, B. C. *et al.* (2008). Mating consequences of polyploid evolution in flowering plants: current trends and insights from synthetic polyploids. *International Journal of Plant Science,* **169**, 195–206.

Huskins, C. L. (1930). The origin of *Spartina townsendii. Genetica,* **12**, 531–8.

Hussey, M. A. *et al.* (1991). Incluence of photoperiod on the frequency of sexual embryo sacs in facultative apomictic Buffelgrass. *Euphytica,* **54**, 141–5.

Hutchings, M. J. (1989). Population biology and conservation of *Ophrys sphegodes.* In *Methods of orchid conservation,* ed. H. W. Pritchard, pp. 101–15. Cambridge: Cambridge University Press.

Hutchinson, A. H. (1936). The polygonal representation of polyphase phenomena. *Transactions of the Royal Society of Canada,* Ser. 3, Sect. V, **30**, 19–26.

Hutchinson, J. (1926). *The families of flowering plants,* vol. 1, *Dicotyledons.* London: Macmillan.

Hutchinson, J. (1959) *The families of flowering plants,* 2nd edn. Oxford: Oxford University Press.

Hutchinson, T. C. (1967). Ecotype differentiation in *Teucrium scorodonia* with respect to susceptibility to lime-induced chlorosis and to shade factors. *New Phytologist,* **66**, 439–53.

Huxley, A., Griffiths M. & Levy, M. (1992). *The new Royal Horticultural Society dictionary of gardening.* London: Macmillan Press.

Huxley, C. R. (1991). Ants and plants: a diversity of interactions. In *Ant–plant interactions,* ed. C. R. Huxley & D. F. Cutler, pp. 1–11. Oxford: Oxford University Press.

Huxley, C. R. & Cutler, D. F. (eds.) (1991). *Ant–plant interactions.* Oxford: Oxford University Press.

Huxley, J. S. (1938). Clines: an auxiliary taxonomic principle. *Nature,* **142**, 219–20.

Huxley, J. S. (ed.) (1940). *The new systematics.* Oxford: Clarendon Press.

Huxley, J. S. (1942). *Evolution: the modern synthesis.* London: Allen & Unwin.

Igic, B., Lande, R. & Kohn, J. R. (2008). Loss of self-incompatibility and its evolutionary consequences. *International Journal of Plant Sciences,* **169**, 93–104.

Iltis, H. (1932). *Life of Mendel.* London: Allen & Unwin. Reprinted 1966 New York: Hafner.

Imerson, A. (2012). *Desertification, land degradation and sustainability.* London: Wiley.

Ingrouille, M. J. & Stace, C. A. (1985). Pattern of variation of agamospermous *Limonium* (Plumbaginaceae) in the British Isles. *Nordic Journal of Botany,* **5**, 113–25.

Ingvarsson, P. K. & Giles, B. E. (1999). Kin-structured colonization and small-scale genetic differentiation in *Silene dioica. Evolution,* **53**, 605–11.

IPCC, Intergovernmental Panel on Climate Change (2013). Summary for policy makers. 5th Assessment Report: In *Climate change 2013: the physical science basis* (AR5). (IPCC website www.ipcc.ch and the IPCC WGI AR5 website www.climatechange2013.org.)

IPCC, Intergovernmental Panel on Climate Change (2014). Summary for policy makers. In *Climate change 2014: impacts, adaptation, and vulnerability. Part A: global and sectoral aspects.* 5th Assessment Report of the IPCC, ed. C. B. Field *et al.*, pp.1–32. Cambridge & New York: Cambridge University Press.

Irwin, R. E., Adler, L. S. & Brody. A. K. (2004). The dual role of floral traits: pollinator attraction and plant defense. *Ecology,* **85**, 1503–11.

Jablonka, E. (2013). Epigenetic plasticity: the responsive germline. *Progress in Biophysics and Molecular Biology,* **111**, 99–107.

Jablonka, E. & Lamb, M. J. (1999). *Epigenetic inheritance and evolution: the Lamarckian dimension.* Oxford: Oxford University Press.

Jablonka, E. & Lamb, M. J (2010). Transgenerational epigenetic inheritance. In *Evolution – the extended synthesis,* ed. M. Pigliucci & G. B. Müller, pp. 137–74. Cambridge, MA, & London: The MIT Press.

Jackson, R. C. (1962). Interspecific hybridization in *Haplopappus* and its bearing on chromosome evolution in the Blepharodon section. *American Journal of Botany*, **49**, 119–32.

Jackson, R. C. (1965). A cytogenetic study of a three-paired race of *Haplopappus gracilis. American Journal of Botany*, **52**, 946–53.

Jackson, S. T. & Hobbs, R. J. (2009). Ecological restoration in the light of ecological history. *Science*, **325**, 567–8.

Jacobson, M. Z. (2002). *Atmospheric pollution: history, science, and regulation.* New York: Cambridge University Press.

Jaén-Molinia, R. *et al.* (2009). The molecular phylogeny of *Matthiola* R. Br. (Brassicaceae) inferred from ITS sequences, with special emphasis on the Macaronesian endemics. *Molecular Phylogenetics and Evolution*, **53**, 972–81.

Jaenicke-Després, V. *et al.* (2003). Early allelic selection in maize as revealed by ancient DNA. *Science*, **302**, 1206–8.

Jain, S. K. & Martins, P. S. (1979). Ecological genetics of the colonizing ability of Rose Clover (*Trifolium hirtum* All.). *American Journal of Botany*, **66**, 361–6.

Jäkäläniemi, A., Tuomi, J., Siikamäki, P. & Kilpiä, A. (2005). Colonization-extinction and patch dynamic of the perennial riparian plant, *Silene tatarica. Journal of Ecology*, **93**, 670–80.

Jakubowski, A. R., Casier, M. D. & Jackson, R. D. (2011). Has selection for improved agronomic traits made reed canarygrass invasive? *PLoS ONE*, **6**, e25757.

James, J. K. & Abbott, R. J. (2005). Recent, allopatric, homoploid hybrid speciation: the origin of *Senecio squalidus* (Asteraceae) in the British Isles from a hybrid zone on Mount Etna, Sicily. *Evolution*, **59**, 2533–47.

Jameson, D. L. (ed.) (1977). *Evolutionary genetics.* Stroudsburg, PA.: Dowden, Hutchinson & Ross.

Jang, T.-S. *et al.* (2013). Chromosomal diversification and karyotype evolution of diploids in the cytologically diverse genus *Prospero* (Hyacinthaceae). *BMC Evolutionary Biology*, **13**, Article ARTN 136.

Janovec, J. P., Clark, L. G. & Mori, S. A (2003). Is the Neotropical Flora ready for the phylocode? *The Botanical Review*, **69**, 22–43.

Jansen, P. A., Bongers, F. & Hemerik, L. (2004). Seed mass and mast seeding enhance dispersal by a neotropical scatter-hoarding rodent. *Ecological Monographs*, **74**, 569–89.

Jansen, R. K., Holsinger, K. E., Michaels, H. J. & Palmer, J. D. (1990). Phylogenetic analysis of restriction site data at higher taxonomic levels: an example from the Asteraceae. *Evolution*, **44**, 2089–105.

Jansen, R. K., Michaels, H. J. & Palmer, J. D. (1991). Phylogeny and character evolution in the Asteraceae based on chloroplast DNA restriction site mapping. *Systematic Botany*, **16**, 98–115.

Janzen, D. H. (1966). Coevolution of mutualism between ants and acacias in Central America. *Evolution*, **3**, 249–75.

Janzen, D. J. (1977). What are dandelions and aphids? *American Naturalist*, **111**, 586–9.

Janzen, D. H. (2001). Latent extinction: the living dead. In *Encyclopedia of biodiversity*, vol. 3, ed. S. A. Levin, pp. 689–99. London: Academic Press.

Jarvis, D. I. & Hodgkin, T. (1999). Wild relatives and crop cultivars: detecting natural introgression and farmer selection of new genetic combinations in agroecosystems. *Molecular Ecology*, **8**, S159–S173.

Jenczewski, E. & Alix, K. (2004). From diploids to allopolyploids: the emergence of efficient pairing control genes in plants. *Critical Reviews in the Plant Sciences*, **23**, 21–45.

Jenkin, F. (1867). Unsigned review of Darwin's 'On the origin of species'. *The North British Review*, July, 277–318.

Jennersten, O. (1988). Pollination in *Dianthus deltoides* (Caryophyllaceae): effects of habitat fragmentation on visitation and seed set. *Conservation Biology*, **2**, 359–66.

Jennings, D. E. & Rohr, D. E. (2011). A review of conservation threats to carnivorous plants. *Biological Conservation*, **144**, 1356–63.

Jensen, I. & Bogh, H. (1941). On conditions influencing the danger of crossing in the case of wind-pollinated cultivated plants. *Tidsskrift for Planteavl*, **46**, 238–66.

Jentschke, G. & Godbold, D. (2000). Metal toxicity and ectomycorrhizas. *Physiologia Plantarum*, **109**, 107–16.

Jersáková, J. *et al.* (2006). Mechanisms and evolution of deceptive pollination in orchids. *Biological Reviews of the Cambridge Philosophical Society*, **81**, 219–35.

Johannsen, W. (1909). *Elemente der exakten Erblichkeitslehre.* Jena: Fischer.

Johannsen, W. (1911). The genotype concept in heredity. *American Naturalist*, **45**, 129–59.

Johanson, U. *et al.* (2000). Molecular analysis of FRIGIDA, a major determinant of natural variation in *Arabidopsis* flowering time. *Science*, **290**, 344–7.

Johnson, B. L. (1972). Protein electrophoretic profiles and the origin of the B genome of wheat. *Proceedings of the National Academy of Sciences, USA*, **69**, 1398–402.

Johnson, H. (1945). Interspecific hybridization within the genus *Betula*. *Hereditas*, **31**, 163–76.

Johnson, H. B. (1975). Plant pubescence: an ecological perspective. *Botanical Review*, **41**, 233–58.

Johnson, M. A. T., Kenton, A. Y., Bennett, M. D. & Bradham, P. E. (1989). *Voanioala gerardii* has the highest known chromosome number in the monocotyledons. *Genome*, **32**, 328–33.

Johnson, N. A. (2010). Hybrid incompatibility genes: remnants of a genomic battlefield? *Trends in Genetics*, **26**, 317–25.

Johnson, P. E. (1993). *Darwin on trial*. Washington DC: Regnery Gateway.

Johnson, S. D. (1992). Plant–animal relationships. In *The ecology of fynbos: nutrients, fire and diversity*, ed. R. Cowling, pp. 195–205. Cape Town: Oxford University Press.

Jones, A. G. *et al*. (2010) A practical guide to methods of parentage analysis. *Molecular Ecology Resources*, **10**, 6–30.

Jones, C. J. *et al*. (1997). Reproducibility testing of RAPD, AFLP and SSR markers in plants by a network of European laboratories. *Molecular Breeding*, **3**, 381–90.

Jones, C. J. *et al*. (1998). Reproducibility testing of RAPDs by a network of European laboratories. In *Molecular tools for screening biodiversity*, ed. A. Karp, P. G. Isaac & D. S. Ingram, pp. 176–80. London: Chapman & Hall.

Jones, D. A. (1962). Selective eating of the acyanogenic form of the plant *Lotus corniculatus* L. by various animals. *Nature*, **193**, 1109–10.

Jones, D. A. (1966). On the polymorphism of cyanogenesis in *Lotus corniculatus*. Selection by animals. *Canadian Journal of Genetics and Cytology*, **8**, 556–67.

Jones, D. A. (1972). Cyanogenic glycosides and their function. In *Phytochemical ecology*, ed. J. B. Harborne, pp. 103–24. London & New York: Academic Press.

Jones, D. A. (1973). Co-evolution and cyanogenesis. In *Taxonomy and ecology*, ed. V. H. Heywood, pp. 213–42. Systematics Association Special Volume No. 5. London & New York: Academic Press.

Jones, D. A., Keymer, R. J. & Ellis, W. M. (1978). Cyanogenesis in plants and animal feeding. In *Biochemical aspects of plant and animal coevolution*, ed. J. B. Harborne, pp. 21–34. London, New York & San Francisco: Academic Press.

Jones, D. F. (1924). The attainment of homozygosity in inbred strains of Maize. *Genetics*, **9**, 405–18.

Jones, K. (1958). Cytotaxonomic studies in *Holcus*. I. The chromosome complex in *Holcus mollis* L. *New Phytologist*, **57**, 191–210.

Jones, K. (1964). Chromosomes and the nature and origin of *Anthoxanthum odoratum* L. *Chromosoma*, **15**, 248–74.

Jones, K. & Borrill, M. (1961). Chromosomal status, gene exchange and evolution in *Dactylis*. 3. The role of the inter-ploid hybrids. *Genetica*, **32**, 296–322.

Jones, M. D. & Brooks, J. S. (1952). Effect of tree barriers on outcrossing in Corn. *Oklahoma Agricultural Experiment Station Bulletin*, No. T-45.

Jones, M. E. (1971a). The population genetics of *Arabidopsis thaliana*. I. The breeding system. *Heredity*, **27**, 39–50.

Jones, M. E. (1971b). The population genetics of *Arabidopsis thaliana*. II. Population structure, *Heredity*, **27**, 51–8.

Jones, M. E. (1971c). The population genetics of *Arabidopsis thaliana*. III. The effect of vernalisation. *Heredity*, **27**, 59–72.

Jones, R. *et al*. (2012). *The molecular life of plants*. Chichester: Wiley-Blackwell.

Jones, R. N. (1995). B chromosomes in plants. *New Phytologist*, **131**, 411–34.

Jones, R. N., Viegas, W. & Houben, A. (2008). A century of B chromosomes in plants: so what? *Annals of Botany*, **101**, 767–75.

Jones, S. (1999). *Almost like a whale*. London: Doubleday.

Jones, S. (2009). *Darwin's island: the Galápagos in the garden of England*. London: Little, Brown.

Jonsell, B. (1978). Linnaeus's views on plant classification and evolution. *Botaniska Notiser*, **131**, 523–30.

Jonsell, B. (1984). The biological species concept reexamined. In *Plant biosystematics*, ed. W. F. Grant, pp. 159–68. Toronto: Academic Press.

Jordan, A. (1864). *Diagnoses d'espèces nouvelles ou méconnues pour servir de matériaux à une flore réformée de la France et des Contrées voisines*. Paris: Savy.

Jordanova, L. J. (1984). *Lamarck*. Oxford: Oxford University Press.

Jorgensen, T. H. & Frydenberg, J. (1999). Diversification in insular plants: inferring the phylogenetic relationship in *Aeonium* (Crassulaceae) using ITS sequences of nuclear ribosomal DNA. *Nordic Journal of Botany*, **19**, 613–21.

Judd, W. S. *et al.* (2007). *Plant systematics: a phylogenetic approach*, 3rd edn. Sunderland, MA: Sinauer.

Kadereit, J. W. (1994). Molecules and morphology, phylogenetics and genetics. *Botanica Acta*, **107**, 369–73.

Kadereit, J. W. (2015). The geography of hybrid speciation in plants. *Taxon*, DOI:10.12705/644.1.

Kadereit, J. W. & Baldwin, B. G. (2012). Western Eurasian-western North American disjunct plant taxa: the dry-adapted ends of formerly widespread north temperate mesic lineages – and examples of long-distance dispersal. *Taxon*, **61**, 3–17.

Kadereit, J. W. & Briggs, D. (1985). Speed of development of radiate and non-radiate plants of *Senecio vulgaris L.* from habitats subject to different degrees of weeding pressure. *New Phytologist*, **99**, 155–69.

Kane, N. C. & Rieseberg, L. H. (2008). Genetics and evolution of weedy *Helianthus annuus* populations: adaptation of an agricultural weed. *Molecular Ecology*, **17**, 384–94.

Kaplan, J. M (2010). Phenotypic plasticity and reaction norms. In *A companion to the philosophy of biology*, ed. S. Sarker & A. Plutyski, pp. 205–22. Malden, MA: Blackwell.

Karpechenko, G. D. (1927). Polyploid hybrids of *Raphanus sativus* L. × *Brassica oleracea* L. *Bulletin of Applied Botany & Plant Breeding (Leningrad)*, **17**, 305–410.

Karpechenko, G. D. (1928). Polyploid hybrids of *Raphanus sativus* L. × *Brassica oleracea* L. *Zeitschrift für induktive Abstrammungs Vererbungsiehre*, **39**, 1–7.

Karrenberg, S. & Favre, A. (2008). Genetic and ecological differentiation in the hybridizing campions *Silene dioica* and *S. latifolia*. *Evolution*, **62**, 763–73.

Karron, J. D. *et al.* (2012) New perspectives on the evolution of plant mating systems. *Annals of Botany*, **109**, 493–503.

Katori, T. *et al.* (2010). Dissecting the genetic control of natural variation in salt tolerance of *Arabidopsis thaliana* accessions. *Journal of Experimental Botany*, **61**, 1125–38.

Kauffman, M. J. Brodie, J. F. & Jules, E. S. (2010). Are wolves saving Yellowstone's aspen? A landscape-level test of a behaviorally mediated trophic cascade. *Ecology*, **91**, 2742–55.

Kawata, M. I., Murakami, K. & Ishikawa, T. (2009). Dispersal and persistence of genetically modified oilseed rape around Japanese harbors. *Environmental Science & Pollution Research International*, **16**, 120–6.

Kay, Q. O. N. (1978). The role of preferential and assortative pollination in the maintenance of flower colour polymorphisms. In *The pollination of flowers by insects*, ed. A. J. Richards, pp. 175–90. Linnean Society Symposium Series 6. London: Academic Press.

Kearey, P., Klepeis, K. A. & Vine, F. J. (2009). *Global tectonics*, 3rd edn. Oxford: Wiley-Blackwell.

Keeling, P. J. & Palmer, J. D. (2008). Horizontal gene transfer in eukaryotic evolution. *Nature Reviews Genetics*, **9**, 605–18.

Keller, S. R. & Taylor, D. R. (2008). History, chance, and adaptation during biological invasion: separating stochastic phenotypic evolution from response to selection. *Ecology Letters*, **8**, 852–6.

Kellogg, E. A. (2002) Are macroevolution and microevolution quantitatively different? Evidence from the Poaceae. In *Developmental genetics and plant evolution*, ed. Q. C. B. Cronk, R. M. Bateman & J. A. Hawkin, pp. 70–84. London: Taylor & Francis.

Kelly, L. J. & Leitch, I. J. (2011). Exploring giant plant genomes with next-generation sequencing technology. *Chromosome Research*, **19**, 939–53.

Kelvin, Lord (Sir W. Thompson) (1871). On geological time. *Transactions of the Geological Society of Glasgow*, **3**, 1–28.

Kemp, J., Milne, R. & Reay, D. S. (2010). Sceptics and deniers of climate change not to be confused. *Nature*, **464**, 673.

Kemp, T. S. (2005). *The origin and evolution of mammals.* New York: Oxford University Press.

Kemperman, J. A. & Barnes, B. V. (1976). Clone size in American aspens. *Canadian Journal of Botany*, **54**, 2603–7.

Kendall, M. G. & Plackett, R. L. (1977). *Studies in the history of statistics and probability*, vol. II. London: Griffin.

Kenrick, P. (2011). Timescales and timetrees. *New Phytologist*, **192**, 3–6.

Kerner, A. (1895). *The natural history of plants, their forms, growth, reproduction and distribution.* Translated and edited by F. W. Oliver. London: Blackie.

Kernick, M. D. (1961). Seed production of specific crops. In *Agricultural and horticultural seeds*, pp. 181–547. FAO Agriculture Studies No. 55.

Kettle, C. J. *et al.* (2012). Importance of demography and dispersal for the resilience and restoration of a critically endangered tropical conifer *Araucaria nemorosa*. *Diversity and Distributions*, **18**, 248–59.

Keymer, R. & Ellis, W. M. (1978). Experimental studies on plants of *Lotus corniculatus* L. from Anglesey polymorphic for cyanogenesis. *Heredity*, **40**, 189–206.

Keynes, M. (1993). *Sir Francis Galton, FRS. The legacy of his ideas*. London: Macmillan.

Khandelwal, S. (1990). Chromosome evolution in the genus *Ophioglossum* L. *Botanical Journal of the Linnean Society*, **102**, 205–17.

Khoury, C., Laliberté, B. & Guarino, L. (2010). Trends in *ex situ* conservation of plant genetic resources: a review of global crop and regional conservation strategies. *Genetic Resources and Crop Evolution*, **57**, 625–39.

Kihara, H. & Ono, T. (1926). Chromosomenzahlen und systematische Gruppierung der *Rumex*-Arten. *Zeitschrift für Zellforschung und mikroskopische Anatomie, Berlin, Wien*, **4**, 475–81.

Kikuchi, R. *et al.* (2010). Disjunct distribution of chloroplast DNA haplotypes in the understory perennial *Veratrum album* ssp. *oxysepalum* (Melanthiaceae) in Japan as a result of ancient introgression. *New Phytologist*, **188**, 879–91.

Kilian, J. *et al.* (2007). The AtGenExpress global stress expression data set: protocols, evaluation and model data analysis of UV-B light, drought and cold stress responses. *Plant Journal*, **5**, 347–63.

Kim, K. C. & Byrne, L. B. (2006). Biodiversity loss and the taxonomic bottleneck: emerging biodiversity science. *Ecological Research*, **21**, 794–810.

Kim, S.-C., Crawford, D. J. & Jansen, R. K. (1996). Phylogenetic relationships among the genera of the subtribe Sonchinae (Asteraceae): evidence from ITS sequences. *Systematic Botany*, **21**, 417–32.

Kim, S.-C. *et al.* (1996). A common origin for woody *Sonchus* and five related genera in the Macaronesian islands: molecular evidence for extensive radiation. *Proceedings of the National Academy of Sciences, USA*, **93**, 7743–8.

Kimber, G. & Athwal, R. S. (1972) A reassessment of the course of evolution of Wheat. *Proceedings of the National Academy of Sciences, USA*, **69**, 912–15.

Kimura, M. (1983). *The neutral theory of molecular evolution*. Cambridge: Cambridge University Press.

Kimura, M. (1987). Molecular evolutionary clock and neutral theory. *Journal of Molecular Evolution*, **26**, 24–33.

King, M. (1993). *Species evolution, the role of chromosome change*. Cambridge: Cambridge University Press.

King, R. C., Stansfield, W. D. & Mulligan, P. K. (2006). *A dictionary of genetics*, 7th edn. Oxford: Oxford University Press.

Kirk, J. T. O. & Tilney-Bassett, R. A. E. (1978). *The plastids. Their chemistry, structure, growth and inheritance*, 2nd edn. Amsterdam: Elsevier.

Kirschbaum, M. U. F. (1995). The temperature dependence of soil organic matter decomposition, and the effect of global warming on soil organic C storage. *Soil Biology & Biochemistry*, **27**, 753–60.

Kitajima, K. *et al.* (2006). Cultivar selection prior to introduction may increase invasiveness: evidence from *Ardisia crenata*. *Biological Invasions*, **8**, 1471–82.

Klaas, M. *et al.* (2011) Progress towards elucidating the mechanisms of self-incompatibility in the grasses: further insights from studies in *Lolium*. *Annals of Botany*, **108**, 677–85.

Klein, J. T. & Kadereit, J. W. (2015). Phylogeny, biogeography and evolution of edaphic association in the European oreophytes *Sempervivum* and *Jovibarba* (Crassulaceae). *International Journal of Plant Science*, **176**, 44–71.

Kliman, R., Sheehy, B. & Schultz, J. (2008). Genetic drift and effective population size. *Nature Education*, **1**, 3.

Klug, A. (2004). The discovery of the DNA double helix. *Journal of Molecular Biology*, **335**, 3–26.

Knapp, S. (2008). Naming nature: the future of the Linnaean system. *The Linnean Special Issue*, **8**, 161–7.

Knapp, S. (2010). Four new vining species of *Solanum* (Dulcamaroid clade) from montane habitats in Tropical America. *PLoS ONE*, 5(5), e10502.

Knapp, S. *et al.* (2002) Taxonomy needs evolution, not revolution. *Nature*, **419**, 559.

Knapp, S., Polaszek, A. & Watson, M. (2007). Spreading the word. *Nature*, **446**, 261–2.

Knight, T. M. *et al.* (2005). Pollen limitation of plant reproduction. Pattern and process. *Annual Review of Ecology, Evolution and Systematics*, **36**, 467–97.

Knispel, A. L. & McLachlan, S. M. (2010). Landscape-scale distribution and persistence of genetically modified oilseed rape (*Brassica napus*) in Manitoba, Canada. *Environmental Science & Pollution Research International*, **17**, 13–25.

Knispel, A. L. *et al.* (2008). Gene flow and multiple herbicide resistance in escaped canola populations. *Weed Science*, **56**, 72–80.

Knobloch, I. W. (1971). Intergeneric hybridization in flowering plants. *Taxon*, **21**, 97–103.

Knox, R. B. & Heslop-Harrison, J. (1963). Experimental control of aposporous apomixis in a grass of the Andropogoneae. *Botaniska Notiser*, 116, 127–41.

Koch, C. & Kollmann, J. (2012). Clonal reintroduction of endangered plant species: the case of the German False Tamarisk in pre-Alpine rivers. *Environmental Management*, 50, 217–25.

Koerner, L. (1999). *Linnaeus: nature and nation*. Cambridge, MA, & London: Harvard University Press.

Kohn, D. (1981). On the origin of the principle of diversity. *Science*, 213, 1105–8.

Kohn, D. (1985a). *The Darwinian heritage*. Princeton, NJ: Princeton University Press.

Kohn, D. (1985b). Darwin's principle of divergence as internal dialogue. In *The Darwinian heritage*, ed. D. Kohn, pp. 245–57. Princeton, NJ: Princeton University Press.

Kollmann, J. & Bañuelos, M. J. (2004). Latitudinal trends in growth and phenology of the invasive alien plant *Impatiens glandulifera* (Balsaminaceae). *Diversity & Distributions*, 10, 377–85.

Koornneef, M. & Schere, B. (2007) *Arabidopsis thaliana* as an experimental organism. In *Handbook of plant science*, ed. K. Roberts, pp. 684–9. Chichester: Wiley.

Kooyers, N. J. & Olsen, K. M (2012). Rapid evolution of an adaptive cyanogenesis cline in introduced North American white clover (*Trifolium repens* L.). *Molecular Ecology*, 21, 2455–68.

Kooyers, N. J. *et al.* (2014). Aridity shapes cyanogenesis cline evolution in white clover (*Trifolium repens* L.). *Molecular Ecology*, 23, 1053–70.

Koshy, T. K. (1968). Evolutionary origin of *Poa annua* L. in the light of karyotypic studies. *Canadian Journal of Genetics and Cytology*, 10, 112–18.

Kovác, J. (1995). Micropropagation of *Dianthus arenarius* subsp. *bohemicus* – an endangered endemic from the Czech Republic. *Botanic Gardens Micropropagation News*, 1, 105–7.

Kramer, A. T. & Havens, K. (2009). Plant conservation genetics in a changing world. *Trends in Plant Science*, 14, 599–607.

Krämer, U. (2010). Metal hyperaccumulation in plants. *Annual Review of Plant Biology*, 61, 517–34.

Krauss, S. L., Dixon, B. & Dixon, K. W. (2002). Rapid genetic decline in a translocated population of the endangered plant *Grevillea scapigera*. *Conservation Biology*, 16, 986–94.

Kremer, A. *et al.* (2012). Long-distance gene flow and adaptation of forest trees to rapid climate change. *Ecological Letters*, 15, 378–92.

Krishnamurphy, P. K. & Francis, R. A. (2012). A critical review of the utility of DNA barcoding in biodiversity conservation. *Biodiversity and Conservation*, 21, 1901–19.

Kroymann, J. *et al.* (2003). Evolutionary dynamics of an insect resistance quantitative trait locus. *Proceedings of the National Academy of Sciences, USA*, 100, suppl. 2, 14587–92.

Kruckeberg, A. R. (1951). Intraspecific variability in the response of certain native plant species to serpentine soil. *American Journal of Botany*, 38, 408–19.

Kruckeberg, A. R. (1954). The ecology of serpentine soils. III. Plant species in relation to serpentine soils. *Ecology*, 35, 267–74.

Krushelnycky, P. D., Loope, L. L., Reimer, N. J. (2005). The ecology, policy, and management of ants in Hawaii. *Proceedings of the Hawaiian Entomological Society*, 37, 1–25.

Kühl, S. (2013). *For the betterment of the race*. Translated by L. Schofer. New York: Palgrave Macmillan.

Kuhn, T. S., Mooers, A. Ø. & Thomas, G. H. (2011). A simple polytomy resolver for dated phylogenies. *Methods in Ecology and Evolution*, 2, 427–36.

Kujala, S. T. & Savolainen, O. (2012). Sequence variation patterns along a latitudinal cline in Scots Pine (*Pinus sylvestris*): signs of clinal adaptation? *Tree Genetics & Genomes*, 8, 1451–67.

Kukkala, A. S. & Moilanen, A. (2013). Core concepts of spatial prioritization in systematic conservation planning. *Biological Reviews*, 88, 443–64.

Kulpa, S. M. & Leger, E. A. (2012) Strong natural selection during plant restoration favors an unexpected suite of plant traits. *Ecological Applications*, 6, 510–23.

Kunz, W. (2012). *Do species exist? Principles of taxonomic classification*. Weinheim: Wiley-Blackwell.

Küster, E. (1899). Über Stammverwachsungen. *Jahrbücher für Wissenschaftliche Botanik, Berlin*, 33, 487–512.

Kuta, E. *et al.* (2004). Chromosome and nuclear DNA study on *Luzula* – a genus with holokinetic chromosomes. *Genome*, 47, 246–56.

Kyhos, D. W. (1965). The independent aneuploid origin of two species of *Chaenactis* (Compositae) from a common ancestor. *Evolution*, 19, 26–43.

Ladizinsky, G. (1998). *Plant evolution under domestication.* Dordrecht: Kluwer Academic Publishers.

Lai, Z. *et al.* (2005). Identification and mapping of SNPs from ESTs in sunflower. *Theoretical and Applied Genetics,* **111**, 1532–44.

Lake, J. A. & Rivera, M. C. (1994). The prokaryotic origin of eukaryotes. In *Evolution of microbial life*, ed. A. Roberts *et al.* Cambridge: Cambridge University Press.

Laland, K. N. (2002). Niche construction. In *Encyclopedia of evolution*, ed. M. Patel, pp. 821–3. Oxford & New York: Oxford University Press.

Lamarck, J. B. (1809). *Philosophie zoologique.* English translation, *Zoological philosophy*, translated by H. Elliot, published 1914, London & New York: Macmillan.

Lamb, H. H. (1970). Our changing climate. In *Flora of a changing Britain*, ed. F. H. Perring, pp. 11–24. Hampton, Middlesex: Botanical Society of the British Isles.

Lambers, H., Chapin, F. S., III & Pons, T. L. (2008). *Plant physiological ecology.* New York: Springer.

Lamprecht, H. (1961). Die Genekarte von *Pisum* bei normaler Struktur der Chromosomen. *Agri Hortique Genetica,* **19**, 360–401.

Lande, R. (1988). Genetics and demography in biological conservation. *Science*, **241**, 1455–60.

Lane, C. (1962). Notes on the Common Blue (*Polyommatus icarus*) egg laying and feeding on the cyanogenic strains of the Bird's-foot Trefoil (*Lotus corniculatus*). *Entomologist's Gazette*, **13**, 112–16.

Langerhans, R. B. & Riesch, R. (2013). Speciation by selection: a framework for understanding ecology's role in speciation. *Current Zoology*, **59**, 31–52.

Langlet, O. (1934). Om variationen hos tallen *Pinus sylvestris* och dess samband med climatet. *Meddelanden från Statens Skogsførsøksanstalt*, **27**, 87–93.

Langlet, O. (1971). Two hundred years genecology. *Taxon*, **20**, 653–722.

Lankau, R. A. (2012). Coevolution between invasive and native plants driven by chemical competition and soil biota. *Proceedings of the National Academy of Sciences, USA*, **109**, 11240–5.

Lankau, R. A. *et al.* (2009). Evolutionary limits ameliorate the negative impact of an invasive plant. *Proceedings of the National Academy of Sciences, USA*, **106**, 15362–7.

Lankester, E. (1848). *The correspondence of John Ray.* London: Ray Society, British Museum.

LAPG III (2009). The Linear Angiosperm Phylogeny Group (LAPG) III: a linear sequence of the families in APG III. *Botanical Journal of the Linnean* Society, **161**, 128–31.

Larcher, W. (2003). *Physiological plant ecology*, 4th edn. Berlin: Springer.

Larsen, E. C. (1947). Photoperiodic responses of geographical strains of *Andropogon scoparius. Botanical Gazette*, **109**, 132–50.

Lasso, E. (2008). The importance of setting the right genetic distance threshold for identification of clones using amplified fragment length polymorphism: a case study with five species in the tropical plant genus *Piper. Molecular Ecology Resources*, 8, 74–82, DOI: 10.1111/j.1471-8286.2007.01910.x

Laurie, D. A. & Bennett, M. D. (1985). Nuclear DNA content in the genera *Zea* and *Sorghum*: intergeneric, interspecific and intraspecific variation. *Heredity*, **55**, 307–13.

Lauterbach, D., Burkart, M. & Gemeinholzer. B. (2012). Rapid genetic differentiation between *ex situ* and their *in situ* source populations: an example of the endangered *Silene otitis* (Caryophyllaceae). *Botanic Journal of the Linnean Society*, **168**, 64–75.

Lavergne, S. & Molofsky, J. (2007). Increased genetic variation and evolutionary potential drive the success of an invasive grass. *Proceedings of the National Academy of Sciences, USA*, **104**, 3883–8.

Lawrence, B. A. & Kaye, T. N. (2011). Reintroduction of *Castilleja levisecta:* effects of ecological similarity, source population genetics and habitat quality. *Restoration Ecology*, **19**, 166–76.

Lawrence, W. E. (1945). Some ecotypic relations of *Deschampsia caespitosa. American Journal of Botany*, **32**, 298–314.

Lawrence, W. J. C. (1950). *Science and the glasshouse.* Edinburgh & London: Oliver & Boyd.

Le Comber, S. C. *et al.* (2010). Making a functional diploid: from polysomic to disomic inheritance. *New Phytologist*, **186**, 113–22.

Le Page, M. (2012). A brief history of the genome. *New Scientist*, **2882**, 30–5.

Lebaron, H. M. & Gressel, J. (eds.) (1982). *Herbicide resistance in plants.* New York: Wiley.

Ledig, F. T., Rehfeldt, G. E. & Jaquish, B. (2012). Projections of suitable habitat under climate change scenarios: implications for trans-boundary assisted colonization. *American Journal of Botany*, **99**, 1217–30.

Lee, A. (1902). Dr Ludwig on variation and correlation in plants. *Biometrika*, 1, 316–19.

Lee, P. L. M. *et al.* (2004). Comparison of genetic diversities in native and alien populations of hoary mustard (*Hirschfeldia incana* [L.] Lagreze-Fossat). *International Journal of Plant Sciences*, **165**, 833–43.

Lefèbvre, C. (1973). Outbreeding and inbreeding in a zinc-lead mine population of *Armeria maritima*. *Nature*, **243**, 96–7.

Lemey, P., Salemi, M. & Vandamme, A.-M. (eds.) (2009). *The phylogenetic handbook: a practical approach to phylogenetic analysis and hypothesis testing.* Cambridge: Cambridge University Press.

Lengyel, S. *et al.* (2010). Convergent evolution of seed dispersal by ants, and phylogeny and biogeography in flowering plants: a global survey. *Perspectives in Plant Ecology, Evolution and Systematics*, **12**, 43–55.

Lenz, J. *et al.* (2011). Seed-dispersal distributions by trumpeter hornbills in fragmented landscapes. *Proceedings of the Royal Society, B*, **278**, 2257–64.

Lenzen, M. *et al.* (2012). International trade drives biodiversity threats in developing nations. *Nature*, **486**, 109–12.

Leon-Lobos, P. *et al.* (2012). The role of *ex situ* seed banks in the conservation of plant diversity and in ecological restoration in Latin America. *Plant Ecology & Diversity*, 5, 245–58.

Lev-Yadun, S. & Inbar, M. (2002). Defensive ant, aphid and caterpillar mimicry in plants? *Biological Journal of the Linnean Society*, **77**, 393–8.

Levan, A. (1938). The effect of colchicine on root mitosis in *Allium*. *Hereditas*, **24**, 471–86.

Levey, D. J. & Sargent, S. (2000). A simple method for tracking vertebrate-dispersed seeds. *Ecology*, **81**, 267–74.

Levin, D. A. (1973). The role of trichomes in plant defense. *Quarterly Review of Biology*, **48**, 3–15.

Levin, D. A. (1975a). Minority cytotype exclusion in local plant populations. *Taxon*, **24**, 35–43.

Levin, D. A. (1975b). Pest pressure and recombination systems in plants. *American Naturalist*, **190**, 437–51.

Levin, D. A. (1978a). Pollinator behaviour and the breeding structure of plant populations. In *The pollination of flowers by insects*, ed. A. J. Richards, pp. 133–50. Linnean Society Symposium Series 6. London: Academic Press.

Levin, D. A. (1978b). The origin of isolating mechanisms in flowering plants. *Evolutionary Biology*, 11, 185–317.

Levin, D. A. (1979). The nature of plant species. *Science*, **204**, 381–4.

Levin, D. A. (1984). Immigration in plants: an exercise in the subjunctive. In *Perspectives on plant population ecology*, ed R. Dirzo & J. Sarukhán, pp. 242–60. Sunderland, MA: Sinauer.

Levin, D. A. (1985). Reproductive character displacement in *Phlox*. *Evolution*, **39**, 1275–81.

Levin, D. A. (1988). Local differentiation and the breeding structure of plant populations. In *Plant evolutionary biology*, ed. L. D. Gottlieb & S. K. Jain, pp. 305–29. London: Chapman & Hall.

Levin, D. A. (1993). Local speciation in plants: the rule not the exception. *Systematic Botany*, **18**, 197–208.

Levin, D. A. (2001a) The recurrent origin of plant races and species. *Systematic Botany*, **26**, 197–204.

Levin, D. A. (2001b). 50 years of plant speciation. *Taxon*, **50**, 69–91.

Levin, D. A. (2003). The ecological transition in speciation. *New Phytologist*, **161**, 91–6.

Levin, D. A. (2011). Mating system shifts on the trailing edge. *Annals of Botany*, 2011, 1–8, DOI:10.1093/aob/mcr159.

Levin, D. A. & Kerster, H. W. (1967). An analysis of interspecific pollen exchange in *Phlox*. *American Naturalist*, **101**, 387–400.

Levin, D. A. & Kerster, H. W. (1974). Gene flow in seed plants. *Evolutionary Biology*, **7**, 139–220.

Levin, S. A. *et al.* (2003). The ecology and evolution of dispersal: a theoretical perspective. *Annual Review of Ecology, Evolution and Systematics*, **34**, 575–604.

Lewis, D. (1979). *Sexual incompatibility in plants.* Studies in Biology, No. 110. London: Arnold.

Lewis, D. & Crowe, L. K. (1956). The genetics and evolution of gynodioecy. *Evolution*, **10**, 115–25.

Lewis, H. (1973). The origin of diploid neospecies in *Clarkia*. *The American Naturalist*, **107**, 161–70.

Lewis, H & Lewis, M. E. (1955). *The genus* Clarkia. Berkeley & Los Angeles: University of California Press.

Lewis, W. H. (1976). Temporal adaptation correlated with ploidy in *Claytonia virginica*. *Systematic Botany*, 1, 340–7.

Lewis, W. H. (ed.) (1980a). *Polyploidy.* London & New York: Plenum Press.

Lewis, W. H. (1980b). Polyploidy in species populations. In *Polyploidy,* ed. W. H. Lewis, pp.103–44. New York & London: Plenum Press.

Lewis, W. H. (1980c). Polyploidy in angiosperms: dicotyledons. In *Polyploidy,* ed. W. H. Lewis, pp. 241–68. New York & London: Plenum Press.

Li, D.-Z. & Pritchard, H. W. (2009). The science and economics of *ex situ* plant conservation. *Trends in Plant Science,* 14, 614–21.

Li, H. J. & Zhang, Z. B. (2003). Effect of rodents on acorn dispersal and survival of the Liaodong oak (*Quercus liaotungensis* Koidz.). *Forest Ecology & Management,* **176**, 387–96.

Li, H. J. & Zhang, Z. B. (2007). Effects of mast seeding and rodent abundance on seed predation and dispersal by rodents in *Prunus armeniaca* (Rosaceae). *Forest Ecology & Management,* **242**, 511–17.

Li, H. L. (1956). The story of the cultivated Horse-Chestnuts. *Morris Arboretum Bulletin,* **7**, 35–9.

Li, X. *et al.* (2015). Plant DNA barcoding: from gene to genome. *Biological Reviews,* **90**, 157–66.

Lieberman, B. S. & Kaesler, R. (2010). *Prehistoric life.* Oxford: Wiley-Blackwell.

Liepelt, S., Bialozyt, R. & Ziegenhagen, B. (2002). Wind-dispersed pollen mediates postglacial gene flow in refugia. *Proceedings of the National Academy of Sciences, USA,* **99**, 14590–4.

Lihová J. *et al.* (2004). Origin of the disjunct tetraploid *Cardamine amporitana* (Brassicaceae) assessed with nuclear and chloroplast DNA sequence data. *American Journal of Botany,* **9**, 1231–42.

Liljefors, A. (1953). Studies on propagation, embryology and pollination in *Sorbus. Acta Horti Bergiani,* **16**, 227–329.

Liljefors, A. (1955). Cytological studies in *Sorbus. Acta Horti Bergiani,* **17**, 47–113.

Lim, K. Y. *et al.* (2007). Parental origin and genome evolution in the allopolyploid *Iris versicolor. Annals of Botany,* **100**, 219–24.

Lindahl, B. D. *et al.* (2013). Fungal community analysis by high-throughput sequencing of amplified markers – a user guide. *New Phytologist,* **199**, 288–99.

Linde, M., Diel, S. & Neuffer, B. (2001). Flowering ecotypes of *Capsella bursa-pastoris* (L.) Medik. (Brassicaceae) analysed by a co-segregation of phenotypic characters (QTL) and molecular markers. *Annals of Botany,* **87**, 91–9.

Linder, C. R. & Rieseberg, L. H. (2004). Reconstructing patterns of reticulate evolution in plants. *American Journal of Botany,* **91**, 1700–8.

Linhart, V. B. & Baker, I. (1973). Intra-population differentiation of physiological response to flooding in a population of *Veronica peregrina. Nature,* **242**, 275–6.

Linhart, Y. B. (1988). Intra-population differentiation in annual plants III. The contrasting effects of intra- and inter-specific competition. *Evolution,* **42**, 1047–64.

Linhart, Y. B., Mitton, J. B., Sturgeon, K. B. & Davis, M. L. (1981). Genetic variation in space and time in a population of Ponderosa Pine. *Heredity,* **46**, 407–26.

Linington, S. H. & Pritchard, H. W. (2001). Gene banks. In *Encyclopedia of biodiversity,* vol. 3, ed. S. A. Levin, pp.165–81. San Diego, CA, & London: Academic Press.

Linnaeus, C. (1737, but not distributed until 1738). *Hortus cliffortianus.* Amsterdam.

Linnaeus, C. (Carl von Linné) (1737). *Critica botanica.* (English translation by A. Hort, 1938, London: Ray Society, British Museum.)

Linnaeus, C. (1744). Peloria. In *Amoenitates academicae* (1749–90). (See Stearn (1957) for details of the many editions.)

Linnaeus, C. (1749–90). *Amoenitates academicae.* (For details of the many editions see Stearn (1957).)

Linnaeus, C. (1751). *Linnaeus' Philosophia botanica.* Translated by S. Freer (2003). Oxford: Oxford University Press.

Linnaeus, C. (1751). *Philosophia botanica.* Stockholm.

Linnaeus, C. (1753). *Species plantarum.* (Facsimile edn 1957, London: Ray Society, British Museum.)

Linnaeus, C. (1762–3). *Species plantarum,* 2nd edn. Stockholm.

Linroth, S. (1983).The two faces of Linnaeus. In *Linnaeus. The man and his work,* ed. T. Frangsmyr, pp. 1–62. Berkeley: University of California Press.

Lipman, M. J. *et al.* (2013). Natural hybrids between *Tragopogon mirus* and *T. miscellus* (Asteraceae): a new perspective on karyotype changes following hybridization at the polyploid level. *American Journal of Botany,* **100**, 2016–22.

Lipscomb, D. L., Platnick, N. and Wheeler, Q. (2003). The intellectual content of taxonomy: a comment on DNA taxonomy. *Trends in Ecology and Evolution,* **18**, 65–6.

Litardière, R. de (1939). Sur les caractères chromosomiques et la systématique des Poa du group du *P. annua* L. *Revue de Cytologie et de Cytophysiologie végétales,* **4**, 82–5.

Lloyd, D. G. (1965). Evolution of self-compatibility and racial differentiation in *Leavenworthia* (Cruciferae). *Contributions from the Gray Herbarium*, **195**, 1–134.

Lloyd, D. G. (1975). The maintenance of gynodioecy and androdioecy in angiosperms. *Genetica*, **45**, 325–39.

Lloyd, D. G. (1979a). Some reproductive factors affecting the selection of self-fertilisation in plants. *American Naturalist*, **113**, 67–79.

Lloyd, D. G. (1979b). Evolution towards dioecy in heterostylous populations. *Plant Systematics and Evolution*, **131**, 71–80.

Lloyd, D. G. (1992). Self- and cross-fertilization in plants. II. The selection of self-fertilization. *International Journal of Plant Science*, **153**, 370–80.

Lloyd, D. G. & Webb, C. J. (1992). The evolution of heterostyly. In *Evolution and function of heterostyly*, ed. S. C. H. Barrett, pp. 151–78. Heidelberg: Springer-Verlag.

Lo, E. Y. Y., Stefanović, S. & Dickinson, T. A. (2009). Population genetic structure of diploid sexual and polyploidy apomictic hawthorns (*Crateagus*; Rosaceae) in the Pacific Northwest. *Molecular Ecology*, **18**, 1145–60.

Lo, E. Y. Y., Stefanović, S. & Dickinson, T. A (2010). Reconstructing reticulation history in a phylogenetic framework and the potential of allopatric speciation driven by polyploidy in an agamic complex in *Crataegus* (Rosaceae). *Evolution*, **64**, 3593–608.

Lockwood, J. & McKinney, M. (eds.) (2001). *Biotic homogenization*. New York: Kluwer Academic/Plenum Publishers.

Lofflin, D. L. & Kephart, S. R. (2005). Outbreeding, seedling establishment, and maladaptation in natural and reintroduced populations of rare and common *Silene douglasii* (Caryophyllaceae). *American Journal of Botany*, **92**, 1691–700.

Lomolino, M. V. *et al.* (2005). *Biogeography*, 4th edn. Sutherland, MA: Sinauer; London: Hatchette.

Lonsdale, W. M. (1994). Inviting trouble: introduced pasture species in northern Australia. *Australian Journal of Ecology*, **19**, 345–54.

Lookerman, D. J. & Jansen, R. K. (1995). The use of herbarium material for DNA studies. In *Sampling the green world*, ed. T. F. Stuessy, pp. 205–20. New York: Columbia University Press.

Löve, A. (1960). Biosystematics and the classification of apomicts. *Feddes Repertorium*, **62**, 136–48.

Löve, A. (1962). The biosystematic species concept. *Preslia*, **34**, 127–39.

Löve, A. & Löve, D. (1961). Chromosome numbers of Central and Northwest European plant species. *Opera Botanica*, **5**, 1–581.

Löve, A. & Löve, D. (1974). Origin and evolution of the arctic and alpine floras. In *Arctic and Alpine environments*, ed. J. D. Ives & R. G. Barry, pp. 571–603. London: Methuen.

Lovejoy, A. O. (1966). *The Great Chain of Being*. Cambridge, MA: Harvard University Press.

Lovett Doust, J. & Lovett Doust, L. (1988). *Plant reproductive ecology, patterns and strategies*. New York: Oxford University Press.

Lövkvist, B. (1956). The *Cardamine pratensis* complex. *Symbolae Botanicae Upsalienses*, **14**(2), 1–131.

Lövkvist, B. (1962). Chromosome and differentiation studies in flowering plants of Skåne, South Sweden. 1. General aspects. Type species with coastal differentiation. *Botaniska Notiser*, **115**, 261–87.

Lowe, A., Harris, S. & Ashton, P. (2009). *Ecological genetics: design, analysis and application*. London: Wiley.

Lowe, S. *et al.* (2000). *100 of the world's worst invasive alien species: a selection from the Global Invasive Species Database*. Auckland: The Invasive Species Specialist Group (ISSG), a specialist group of the Species Survival Commission (SSC) of the World Conservation Union (IUCN).

Lowry, D. B. (2012). Ecotypes and the controversy over stages in the formation of new species. *Biological Journal of the Linnean Society*, **106**, 241–57.

Lowry, D. B. (2012). Local adaptation in the model plant. *New Phytologist*, **194**, 888–90.

Lowry, D. B., Rockwood, R. C. & Willis, J. H. (2008). Ecological reproductive isolation of coast and inland races of *Mimulus guttatus*. *Evolution*, **62**, 2196–214.

Luceño, M. & Castroviejo, S. (1991). Agamatoploidy in *Carex laevigata* (Cyperaceae). Fusion and fission of chromosomes as the mechanism of cytogenetic evolution in Iberian populations. *Plant Systematics & Evolution*, **177**, 149–59.

Luckow, M. (1995). Species concepts : assumptions, methods and applications. *Systematic Botany*, **20**, 589–605.

Ludwig, F. (1895). Uber Variationskurven und Variationsflächen der Pflanzen. *Botanisches Zentralblatt*, **64**, 1–8, 33–41, 65–72, 97–105.

Ludwig F. (1901). Variationsstatistische Probleme und Materialen. *Biometrika*, **1**, 11–29.

Lui, B. & Wendel, J. F. (2003). Epigenetic phenomena and the evolution of plant allopolyploids. *Molecular Phylogenetics and Evolution*, **29**, 365–79.

Lui, B. *et al.* (2009). Rapid genomic changes in polyploid wheat and related species: implications for genome evolution and genetic improvement. *Journal of Genetics & Genomics*, **36**, 519–28.

Lui, G. *et al.* (2006). Comparison of genetic variability on populations of wild rice, *Oryza rufipogon*, plants and their soil seed banks. *Conservation Genetics*, **7**, 909–17.

Lui, L. *et al.* (2009). Coalescent methods for estimating phylogenetic trees. *Molecular Phylogenetics and Evolution*, **53**, 320–8.

Lumaret, R. (1984). The role of polyploidy in the adaptive significance of polymorphism at the GOT I locus in the *Dactylis glomerata* complex. *Heredity*, **52**, 153–69.

Lumaret, R. (1988). Cytology, genetics and evolution in the genus *Dactylis*. *Critical Reviews in Plant Sciences*, **7**, 55–91.

Lumaret, R. & Ouazzani, N. (2001). Ancient wild olives in Mediterranean forests. *Nature*, **413**, 700.

Lynch, M. & Walsh, B. (1997). *Genetics and analysis of quantitative traits*. Sunderland, MA: Sinauer.

Lynch, R. L. (1900). Hybrid Cinerarias. *Journal of the Royal Horticultural Society*, **24**, 269–74.

Lyons, E. & Freeling, M. (2008). How to usefully compare homologous plant genes and chromosomes as DNA sequences. *The Plant Journal*, **53**, 661–73.

Lysak, M. A. (2014). Live and let die: centromere loss during evolution of plant chromosomes. *New Phytologist*, **203**, 1082–9.

Lysak, M. A. *et al.* (2009). The dynamic ups and downs of genome size evolution in Brassicaceae. *Molecular Biology & Evolution*, **26**, 85–98.

Mabey, R. (1980). *The common ground. a place for nature in Britain's future*. London: Hutchinson.

MacArthur, R. H. & Wilson, E. O. (1967). *The theory of island biogeography*. Princeton, NJ: Princeton University Press.

MacDonald, G. M. (2003). *Biogeography: space, time and life*. New York: Wiley.

Mace, G. M. (2005). Biodiversity: an index of intactness. *Nature*, **434**, 32–3.

Mace, G. M. & Purvis, A. (2008). Evolutionary biology and practical conservation: bridging a widening gap. *Molecular Ecology*, **17**, 9–19.

Mace, G. M., Possingham, H. P. & Leader-Williams, N. (2007). Prioritizing choices in conservation. In *Key topics in conservation biology*, ed. D. W. Macdonald & K. Service, pp. 17–34. Oxford: Blackwell.

Maceira, N. O., De Haan, A. A., Lumaret, R., Billon, M. & Delay, J. (1992). Production of 2*n* gametes in diploid subspecies of *Dactylis glomerata* L. I. Occurrence and frequency of 2*n* pollen. *Annals of Botany*, **69**, 335–43.

MacIsaac, H. J. *et al.* (2004). Backcasting and forecasting biological invasions of inland lakes. *Ecological Applications*, **14**, 773–83.

Mack, R. N. (1985). Invading plants: their potential contribution to population biology. In *Studies in plant demography*, ed. J. White, pp. 127–42. London: Academic Press.

Mack, R. N. (1991). The commercial seed trade: an early disperser of weeds in the United States. *Economic Botany*, **45**, 257–73.

Mack, R. N. & Lonsdale, W. M. (2001). Humans as global dispersers: getting more than we bargained for. *BioScience*, **51**, 95–102.

Mack, R. N. *et al.* (2000). Biotic invasions: causes, epidemiology, global consequences and control. *Ecological Applications*, **10**, 689–710.

Macnair, M. R. (1983). The genetic control of copper tolerance in the Yellow Monkey Flower *Mimulus guttatus*. *Heredity*, **50**, 283–93.

Macnair, M. R. (1987). Heavy metal tolerance in plants: a model evolutionary system. *Trends in Ecology & Evolution*, **2**, 354–9.

Macnair, M. R. (1993). The genetics of metal tolerance in vascular plants. *New Phytologist*, **124**, 541–59.

Macnair, M. R. & Christie, P. (1983). Reproductive isolation as a pleiotropic effect of copper tolerance in *Mimulus guttatus*. *Heredity*, **50**, 295–302.

Macnair, M. R., Macnair, V. E. & Martin, B. E. (1989). Adaptive speciation in *Mimulus*: an ecological comparison of *M. cupriphilus* with its presumed progenitor, *M. guttatus*. *New Phytologist*, **112**, 269–79.

Macnair, M. R., Cumbes, Q. J. & Meharg, A. A. (1992). The genetics of arsenate tolerance in Yorkshire fog, *Holcus lanatus* L. *Heredity*, **69**, 325–35.

Maddox, B. (2002). *Rosalind Franklin: the dark lady of DNA*. London: Harper Collins.

Maddox, D. G. (1989). Clone structure in *Solidago altissima* populations: rhizome connections within genotypes. *American Journal of Botany*, **76**, 318–21.

Madlung, A. & Wendel, J. F. (2013). Genetic and epigenetic aspects of polyploid evolution in plants. *Cytogenetic and Genome Research*, **140**, 270–85.

Madlung, A. *et al.* (2005). Genomic changes in synthetic *Arabidopsis* polyploids. *Plant Journal*, **41**, 221–30.

Magallón S. (2010). Using fossils to break long branches in molecular dating: a comparison of relaxed clocks applied to the origin of angiosperms. *Systematic Biology*, **59**, 384–99.

Magallón, S. & Castillo, A. (2009). Angiosperm diversification through time. *American Journal of Botany*, **96**, 349–65.

Magallón, S., Hilu, K. W. & Quandt, D. (2013). Land plant evolutionary timeline: gene effects are secondary to fossil constraints in relaxed clock estimation of age and substitution rates. *American Journal of Botany*, **100**, 556–73.

Magee, B. (1973). *Popper*. Glasgow: Collins.

Magri, D. *et al.* (2006). A new scenario for the Quaternary history of European beech populations: palaeobotanical evidence and genetic consequences. *New Phytologist*, **171**, 199–221.

Magri, D. *et al.* (2007), The distribution of *Quercus suber* chloroplast haplotypes matches the palaeogeographical history of the western Mediterranean. *Molecular Ecology*, **16**, 5259–66.

Majeský, L. *et al.* (2012). The pattern of genetic variability in apomictic clones of *Taraxacum officinale* indicates the alternation of asexual and sexual histories of apomicts. *PLOS ONE*, 1 August, DOI: 10.1371/journal.pone.0041868

Major, J. (1988). Endemism; a botanical perspective. In *Analytical biogeography*, ed. A. A. Myers & P. S. Giller, pp. 117–46. London: Chapman & Hall.

Malcolm, J. R. *et al.* (2002). Estimated migration rates under scenarios of global climate change. *Journal of Biogeography*, **29**, 835–49.

Malhi, Y. *et al.* (2008). Climate change, deforestation, and the fate of the Amazon. *Science*, **319**, 169–72.

Malinska, H. *et al.* (2011). Ribosomal RNA genes evolution in *Tragopogon*: a story of New and Old World allotetraploids and the synthetic lines. *Taxon*, **60**, 348–54.

Mallet, J. (2005). Hybridization as an invasion of the genome. *Trends in Ecology and Evolution*, **20**, 229–37.

Mallet, J. & Willmott, K. (2003). Taxonomy: renaissance or Tower of Babel? *Trends in Ecology & Evolution*, **18**, 57–9.

Mann, C. C. (1991). Extinction: are ecologists crying wolf? *Science*, **235**, 736–8.

Mansion, G. *et al.* (2009). Origin of Mediterranean endemics in the Boraginales: integrative evidence from molecular dating and ancestral area reconstruction. *Journal of Biogeography*, **36**, 1282–96.

Manton, I. (1950). *Problems of cytology and evolution in the Pteridophyta*. London & New York: Cambridge University Press.

Mao, Q. & Huff, D. R. (2012). The evolutionary origin of *Poa annua* L. *Crop Science*, **52**, 1910–22.

Marble, B. K. (2004). Polyploidy and self-compatibility: is there an association? *New Phytologist*, **162**, 803–11.

Marchant, C. J. (1963). Corrected chromosome numbers for *Spartina ×townsendii* and its parent species. *Nature*, **199**, 299.

Marchant, C. J. (1967). Evolution in *Spartina* (Gramineae). 1. The history and morphology of the genus in Britain. *Journal of the Linnean Society (Botany)*, **60**, 1–24.

Marchant, C. J. (1968). Evolution in *Spartina* (Gramineae). 2. Chromosomes, basic relationships and the problem of *S. ×townsendii agg. Journal of the Linnean Society (Botany)*, **60**, 381–409.

Marchi, P., Illuminati, O., Macioce, A., Capineri, R. & D'Amato, G. (1983). Genome evolution and polyploidy in *Leucathemum vulgare* Lam. aggr. (Compositae). Karyotype analysis and DNA microdensitometry. *Caryologia*, **36**, 1–18.

Margulis, L. (1971). Symbiosis and evolution. *Scientific American*, **225**, 48–57.

Marion, G. M. *et al.* (1997). Open-top designs for manipulating field temperature in high-latitude ecosystems. *Global Change Biology*, **3** (Suppl. 1), 20–32.

Marks, G. E. (1966). The origin and significance of intraspecific polyploidy: experimental evidence from *Solanum chaeoense*. *Evolution*, **20**, 552–7.

Marsden-Jones, E. (1930). The genetics of *Geum intermedium* Willd. haud Ehrh. and its back-crosses. *Journal of Genetics*, **23**, 377–95.

Marsden-Jones, E. M. & Turrill, W. B. (1945). Report of the transplant experiments of the British Ecological Society. *Journal of Ecology*, **33**, 59–81. [See also earlier reports in the *Journal of Ecology*: **18**, 352; **21**, 268; **23**, 443; **25**, 189; **26**, 359 & 380.]

Marshall, D. R. & Brown, A. H. D. (1981). The evolution of apomixis. *Heredity*, **47**, 1–15.

Marshall, D. R. & Weiss, P. W. (1982). Isozyme variation within and among Australian populations of *Emex spinosa* (L.). *Campd. Australian Journal of Biological Sciences*, **35**, 327–32.

Martin, N. H., Bouck, C. & Arnold, M. L. (2006). Detecting adaptive trait introgression between *Iris fulva* and *I. brevicaulis* in highly selective field conditions. *Genetics*, **172**, 2481–9.

Martin, S. L. & Husband, B. C. (2009). Influence of phylogeny and ploidy on species ranges of North American angiosperms. *Journal of Ecology*, **97**, 913–22.

Maschinski, J. (2001). Extinction risk of *Ipomopsis sancti-spiritus* in the Holy Ghost Canyon with and without management intervention. In *Southwestern rare and endangered plants*, ed. J. Maschinski and L. Holter, pp. 206–12. Fort Collins, CO: US Department of Agriculture, Forest Service, Rocky Mountain Research Station.

Maschinski, J. & Haskins, K. E. (eds.) (2012). *Plant reintroduction in a changing climate: promises and perils.* Washington DC: Island Press.

Maschinski, J., Baggs, J. E. & Sacchi, C. F. (2004). Seedling recruitment and survival of an endangered limestone endemic in its natural habitat and experimental reintroduction sites. *American Journal of Botany*, **91**, 689–98.

Maschinski, J. *et al.* (2011). Sinking ships: conservation options for endemic taxa threatened by sea level rise. *Climate Change*, **107**, 147–67.

Mason, B. J. (1992). *Acid rain: its causes and its effects on island waters.* Oxford: Clarendon Press.

Mason-Gamer, R. J. (2004). Reticulate evolution, introgression, and intertribal gene capture in an allohexaploid grass. *Systematic Biology*, **53**, 25–37.

Massart, J. (1902). L'accomodation individuelle chez le *Polygonum amphibium. Bulletin de Jardin Botanique de l'Etat à Bruxelles*, **1**, 73–95.

Masterson, J. (1981). Stomatal size in fossil plants: evidence for polyploidy in majority of angiosperms. *Science*, **264**, 421–4.

Mather, K. (1943). Polygenic inheritance and natural selection. *Biological Reviews*, **18**, 32–64.

Mather, K. (1966). Breeding systems and response to selection. In *Reproductive biology and taxonomy of vascular plants*, ed. J. G. Hawkes, pp. 13–19. Conference report of Botanical Society of the British Isles. Oxford: Pergamon Press.

Matsushita, S. C. *et al.* (2012). Allopolyploidization lays the foundation for evolution of distinct populations: evidence from analysis of synthetic *Arabidopsis* allohexaploids. *Genetics*, DOI:10.1534/genetics.112.139295.

Matthew, P. (1831). Ideas on evolution, published in an Appendix to *On naval timber and arboriculture.* London.

Matthews, R. E. F. (1991). *Plant virology*, 3rd edn. New York: Academic Press.

Matz, M. V. & Nielsen, R. (2005). A likelihood ratio test for species membership based on DNA sequence data. *Philosophical Transactions of the Royal Society of London, B*, **359**, 1969–74.

Maunder, M. & Ramsay, M. (1994). The reintroduction of plants into the wild: an integrated approach to the conservation of native plants. In *The common ground of wild and cultivated plants*, ed. A. R. Perry & R. G. Ellis, pp. 81–8. Cardiff: National Museum of Wales.

Maunder, M. *et al.* (2004). Hybridization in *ex situ* plant collections; conservation concerns, liabilities and opportunities. In *Ex situ conservation: supporting species survival in the wild*, ed. E. O. Guerrant, K. Havens & M. Maunder, pp. 325–64. Washington DC, Covelo, CA, & London: Island Press.

Maunder, M., Higgens, S. & Culham, A. (2001). The effectiveness of botanic garden collections in supporting plant conservation: a European case history. *Biodiversity and Conservation*, **10**, 383–401.

Maurice, S. *et al.* (1993). The evolution of gender in hermaphrodites of gynodioecious populations with nucleo-cytoplasmic male-sterility. *Proceedings of the Royal Society of London, B*. **251**, 253–61.

Maxwell, B. D. & Mortimer, A. M. (1994). Selection for herbicide resistance. In *Herbicide resistance in plants: biology and biochemistry*, ed. S. B. Powles & J. A. M. Holtum, pp. 1–26. Boca Raton, FL, & London: Lewis Publishers.

May, R. M. (2011). Why should we be concerned about loss of biodiversity? *Comptes Rendus Biologies*, **334**, 346–50.

Mayer, M. S., Soltis, P. S. & Soltis, D. E. (1994). The evolution of the *Streptanthus glandulosus* complex (Cruciferae) – genetic divergence and gene flow in serpentine endemics. *American Journal of Botany*, **81**, 1288–99.

Maynard Smith, J. (1966). Sympatric speciation. *American Naturalist*, **100**, 637–50.

Maynard Smith, J. (1978). *The evolution of sex*. Cambridge: Cambridge University Press.

Maynard Smith, J. (1983). The genetics of stasis and punctuation. *Annual Review of Genetics*, **17**, 11–25.

Maynard Smith, J. (1989). *Evolutionary genetics*. New York: Oxford University Press.

Maynard Smith, J. & Haigh, J. (1974). The hitch-hiking effect of a favourable gene. *Genetic Research*, **23**, 23–35.

Maynard-Smith, J. & Szathmáry, E. (1995). *The major transitions in evolution*. Oxford: Oxford University Press.

Mayr, E. (1942). *Systematics and the origin of species*. New York: Columbia University Press.

Mayr, E. (1963). *Animal species and evolution*. London: Oxford University Press.

Mayr, E. (1969). *Principles of systematic zoology*. New York: McGraw-Hill.

Mayr, E. (1982). *The growth of biological thought: diversity, evolution and inheritance*. Cambridge, MA: Harvard University Press.

Mayr, E. & Provine, W. B. (eds.) (1980). *The evolutionary synthesis: perspectives on the unification of biology*. Cambridge, MA: Harvard University Press.

McBreen, K. & Lockhart, P. J. (2006). Reconstructing reticulate evolutionary histories of plants. *Trends in Plant Science*, **11**, 398–404.

McCarthy, J. J. *et al.* (2001) *Climate change 2001: impacts, adaptation, and vulnerability*. Contribution of Working Group II to the Third Assessment Report of the Intergovernmental Panel on Climate Change. New York: Cambridge University Press.

McCauley, D. E. (1994). Contrasting the distribution of chloroplast DNA and allozyme polymorphism among local populations of *Silene alba*: implications for studies of gene flow in plants. *Proceedings of the National Academy of Sciences, USA*, **91**, 8127–31.

McCauley, D. E. & Bailey, M. F. (2009). Recent advances in the study of gynodioecy: the interface of theory and empiricism. *Annals of Botany*, **104**, 611–20.

McCauley, D. J. (2006). Selling out to nature. *Nature*, **443**, 27–8.

McClintock, B. (1984). The significance of responses of the genome to challenge. *Science*, **226**, 792–801.

McCubbin, A. G. (2008). Heteromorphic self-incompatibility in *Primula*: twenty-first century tools promise to unravel a classic nineteenth century model system. In *Self-incompatibility in flowering plants, evolution, diversity and mechanisms*, ed. V. Franklin-Tong, pp. 286–308. Berlin & Heidelberg: Springer.

McDonald, J. H. & Kreitman, M. (1991). Adaptive protein evolution at the Adh locus in *Drosophila*. *Nature*, **351**, 652–4.

McDonald, R. I. & Boucher, T. M. (2011). Global development and the future of the protected area strategy. *Biological Conservation*, **144**, 383–92.

McElwain, J. C. & Punyasena, S. W. (2007). Mass extinction events and the plant fossil record. *Trends in Ecology and Evolution*, **22**, 548–57.

McFadden, E. S. & Sears, E. R. (1946). The origin of *Triticum spelta* and its free-threshing hexaploid relatives. Hybrids of synthetic *T. spelta* with cultivated hexaploids. *Journal of Heredity*, **37**, 81–9, 107–16.

McFadden, G. I. & van Dooren, G. G. (2004). Evolution: red algal genome affirms a common origin of all plastids. *Current Biology*, **14**, R514–R516.

Macfarlane, A. (1916). *Lectures on ten British mathematicians of the nineteenth century*. New York: Wiley; London: Chapman & Hall.

McGrath, C. L. & Lynch, M. (2012). Evolutionary significance of whole-genome duplication. In *Polyploidy and genome evolution*, ed. P. S. Soltis & D. E. Soltis pp.1–20. Heidelberg, New York, Dordrecht & London: Springer.

McInerney, J. O. & Wilkinson, M. (2005). New methods ring changes for the tree of life. *Trends in Ecology & Evolution*, **20**, 105–7.

McKay, J. K. *et al.* (2005) 'How local is local?' – A review of practical and conceptual issues in the genetics of restoration. *Restoration Ecology*, **13**, 432–40.

McKenney, D. W. *et al.* (2007). Potential impacts of climate change on the distribution of North American trees. *Bioscience*, **57**, 939–48.

McKenney, D. W. *et al.* (2011). Revisiting projected shifts in the climate envelopes of North American trees using updated general circulation models. *Global Change Biology*, **17**, 2720–30.

McLachlan, J. S., Hellmann, J. J.& Schwartz, M. W. (2007). A framework for debate of assisted migration in an era of climate change. *Conservation Biology*, **21**, 297–302.

McLean, R. C. & Ivimey-Cook, W. R. (1956). *Textbook of theoretical botany*. London, New York & Toronto: Longmans.

McLeish, J. & Snoad, B. (1962). *Looking at chromosomes*. London & New York: Macmillan. [2nd edn, 1972.]

McMillan, C. (1970). Photoperiod in *Xanthium* populations from Texas and Mexico. *American Journal of Botany*, **57**, 881–8.

McMillan, C. (1971). Photoperiod evidence in the introduction of *Xanthium* (Cocklebur) to Australia. *Science*, **171**, 1029–31.

McMullen, C. K. (1987). Breeding systems of selected Galápagos Islands angiosperms. *American Journal of Botany*, **74**, 1694–1705.

McNaughton, I. H. & Harper, J. L. (1960). The comparative biology of closely related species living in the same area. *New Phytologist*, **59**, 27–41.

McNeill, C. I. & Jain, S. K. (1983). Genetic differentiation studies and phylogenetic inference in the plant genus *Limnanthes* (Section Inflexae). *Theoretical and Applied Genetics*, **66**, 257–69.

McNeill, J. (1976). The taxonomy and evolution of weeds. *Weed Research*, **16**, 399–413.

McNeill, J. *et al.* (eds.) (2006). International Code of Botanical Nomenclature (Vienna Code): adopted by the Seventeenth International Botanical Congress Vienna, Austria, July 2005. *Regnum Vegetabile*, **146**, 1–568.

McNeill, J. R. (2000). *Something new under the sun.* London & New York: Allen Lane, The Penguin Press.

McNeilly, T. (1968). Evolution in closely adjacent plant populations. III. *Agrostis tenuis* on a small copper mine. *Heredity*, **23**, 99–108.

McNeilly, T. & Antonovics, J. (1968). Evolution in closely adjacent plant populations. IV. Barriers to gene flow. *Heredity*, **23**, 205–18.

McPherson, J. D. (2009). Next-generation gap. *Nature Methods*, **6**, S2–S5.

Meacher, T. R. (1986). Analysis of paternity within a natural population of *Chamaelirium luteum*. I. Identification of most-likely parents. *American Naturalist*, **127**, 199–215.

Meacher, T. R. & Thompson, E. (1987). Analysis of parentage for naturally established seedlings of *Chamaelirium luteum* (Liliacae). *Ecology*, **68**, 803–12.

Mead, R. (1988). *The design of experiments.* Cambridge: Cambridge University Press.

Médail, F. & Diadema, K. (2009). Glacial refugia influence plant diversity patterns in the Mediterranean Basin. *Journal of Biogeography*, **36**, 1333–45.

Medawar, P. (1984). *Pluto's Republic.* Oxford, New York: Oxford University Press.

Medawar, P. (1991). *The threat and the glory.* Oxford: Oxford University Press.

Meffe, G. K. & Carroll, C. R. (1994). *Principles of conservation biology.* Sunderland, MA: Sinauer.

Meharg, A. A. (2003). The mechanistic basis of interactions between mycorrhizal associations and toxic metal cations. *Mycological Research*, **107**, 1253–65.

Meharg, P. A., Cumbes, Q. J. & Macnair, M. R. (1993). Pre-adaptation of Yorkshire Fog *Holcus lanatus* L. (Poaceae) to arsenate tolerance. *Evolution*, **47**, 313–16.

Meirmans, P. G., Den Nijs, H. (J.) C. M. & Van Tienderen, P. H. (2006). Male sterility in triploid dandelions: asexual females vs asexual hermaphrodites. *Heredity*, **96**, 45–62.

Memon, A. R. & Schröder, P. (2009). Implications of metal accumulation mechanisms to phytoremediation. *Environmental Science and Pollution Research International*, **16**, 162–75.

Mena-Ali, J. I. & Stephenson, A. G. (2007). Segregation analyses of partial self-incompatibility in self and cross progeny of *Solanum carolinense* reveal a leaky S-allele. *Genetics*, **177**, 501–10.

Menchari, Y. *et al.* (2006). Weed response to herbicides: regional-scale distribution of herbicide resistance alleles in the grass weed *Alopecurus myosuroides*. *New Phytologist*, **171**, 861–74.

Mendel, G. (1866). Versuche über Planzenhybriden. *Verhandlungen des Naturforschenden Vereins in Brünn*, **4**, 3 44. [English translation in Bateson, W. (1909). *Mendel's principles of heredity.* London: Cambridge University Press; also Bennett, J. H. (ed.) (1965). *Experiments in plant hybridisation.* Edinburgh & London: Oliver & Boyd.]

Mendel, G. (1869). Uber einige aus künstlicher Befruchtung gewonnenen *Hieracium*-Bastarde. *Verhandlungen des naturforschen den Vereines in Brünn*, **8**.

Méndez-Vigo, B. *et al.* (2011). The flowering repressor *SVP* underlies a novel *Arabidopsis thaliana* QTL interacting with the genetic background. *PLoS, Biology*, 31 January 2013, DOI:10.1371/journal.pgen.1003289.

Menges, E. S. (1990). Population viability analysis for a rare plant. *Conservation Biology*, **5**, 158–64.

Menges, E. S. (1991). The application of minimum viable population theory to plants. In *Genetics and conservation of rare plants*, ed. D. A. Falk & K. E. Holsinger, pp. 45–61. Oxford: Oxford University Press.

Menges, E. S. (2008). Restoration demography and genetics of plants: when is translocation successful? *Australian Journal of Botany*, 56, 187–96.

Mengoni, A. *et al.* (2001). Characterization of nickel-resistant bacteria isolated from serpentine soil. *Environmental Microbiology*, 3, 691–8.

Mengoni, A. *et al.* (2003). Chloroplast genetic diversity and biogeography in the serpentine endemic Ni-hyperaccumulator *Alyssum bertolonii. New Phytologist*, 157, 349–56.

Mercer, K. L. & Wainwright, J. D. (2008). Gene flow from transgenic maize to landraces in Mexico: an analysis. *Agriculture, Ecosystems and Environment*, 126, 109–15.

Mereschkowsky, C. (1905). Über Natur und Ursprung der Chromatophoren im Planzenreich. *Biologisches Zentralblatt*, 25, 593–604.

Mergen, F. (1963). Ecotypic variation in *Pinus strobus. Ecology*, 44, 716–27.

Merrell, D. J. (1962). *Evolution and genetics: the modern theory of evolution.* New York: Holt, Rinehart & Winston.

Merxmüller, H. (1970). Biosystematics: still alive? Provocation of biosystematics. *Taxon*, 19, 140–5.

Mes, T. H. M. (1998). Character compatibility of molecular markers to distinguish asexual and sexual reproduction. *Molecular Ecology*, 7, 1719–27.

Mes, T. H. M., van Brederode, J. & Hart, H. (1996). Origin of the Woody Macaronesian Sempervivoideae and the phylogenetic position of the East African species of *Aeonium. Botanica Acta*, 109, 477–91.

Meyer, P. (2007). Gene silencing. In *Handbook of plant science*, ed. K. Roberts, pp. 627–34. Chichester: Wiley.

Michaelis, P. (1954). Cytoplasmic inheritance in *Epilobium* and its theoretical significance. *Advances in Genetics*, 6, 288–402.

Millener, L. H. (1961). Day length as related to vegetative development in *Ulex europaeus*. 1. The experimental approach. *New Phytologist*, 60, 339–54.

Miller, T. E. (1987). Systematics and evolution. In *Wheat breeding; its scientific basis*, ed. F. G. H. Lupton, pp.1–30. London: Chapman & Hall.

Mills, A. D. (1993). *English place names*. Oxford: Oxford University Press.

Milne, R. I. (2006). Northern hemisphere plant disjunctions: a window on Tertiary land bridges and climate change? *Annals of Botany*, 98, 465–72.

Milton, S. J. *et al.* (1999) A protocol for plant conservation by translocation in threatened lowland fynbos. *Conservation Biology*, 13, 735–43.

Minder, A. M, Rothenbuehler, C. & Widmer, A. (2007). Genetic structure of hybrid zones between *Silene latifolia* and *Silene dioica* (Caryophyllaceae): evidence for introgressive hybridization. *Molecular Ecology*, 16, 2504–16.

Minguzzi, C. & Vergnano, O. (1948). 11 contenuto di nichel nelle ceneri di *Alyssum bertolonii. Atti della Società Toscana de Scienze Naturali di Pisa*, 55, 49–74.

Mishler, B. D. (2010). Species are not uniquely real biological entities. In *Contemporary debates in philosophy of biology*, ed. F. Ayala & R. Arp, pp. 110–22. Singapore: Wiley-Blackwell.

Mitchell, R. S. (1968). Variation in the *Polygonum amphibium* complex and its taxonomic significance. *University of California Publications in Botany*, 45, 1–54.

Mitchell-Olds, T. & Schmidt, J. (2006). Genetic mechanisms and evolutionary significance of natural variation in *Arabidopsis. Nature*, 41, 947–52.

Mivart, St. G. (1871). *The genesis of species*, 2nd edn. London: Macmillan.

Modliszewski, J. L. & Willis, J. H. (2012). Allotetraploid *Mimulus sookensis* are highly interfertile despite independent origins. *Molecular Ecology*, 21, 5280–98.

Mogie, M. (1992). *The evolution of asexual reproduction in plants*. London: Chapman & Hall.

Mølgaard, P. (1976). *Plantago major* ssp. *major* and ssp. *pleiosperma*. Morphology, biology and ecology in Denmark. *Botanik Tidsskrift*, 71, 31–56.

Montesinos-Navarro, A. *et al.* (2011). *Arabidopsis thaliana* populations show clinal variation in a climatic gradient associated with altitude. *New Phytologist*, 189, 282–94.

Mooney, H. A. & Billings, W. D. (1961). Comparative physiological ecology of Arctic and Alpine populations of *Oxyria digyna. Ecological Monographs*, 31, 1–29.

Mooney, H. A. & Cleland, E. E. (2001). The evolutionary impact of invasive species. *Proceedings of the National Academy of Sciences, USA*, 98, 5446–51.

Moore, D. M. (1959). Population studies on *Viola lactea* Sm. and its wild hybrids. *Evolution*, 13, 318–32.

Moore, D. M. (1976). *Plant cytogenetics*. London: Chapman & Hall; New York: Wiley & Sons.

Moore, D. M. (1982). *Flora Europaea check-list and chromosome index*. London: Cambridge University Press.

Moore, D. M. & Harvey, M. J. (1961). Cytogenetic relationships of *Viola lactea* Sm. and other West European arosulate violets. *New Phytologist*, 60, 85–95.

Moore, R. J. & Mulligan, G. A. (1956). Natural hybridization between *Carduus acanthoides* and *Carduus nutans* in Ontario. *Canadian Journal of Botany*, 34, 71–85.

Moore, R. J. & Mulligan, G. A. (1964). Further studies on natural selection among hybrids of *Carduus acanthoides* and *Carduus nutans*. *Canadian Journal of Botany*, 42, 1605–13.

Morisset, P. & Boutin, C. (1984) The biosystematic importance of phenotypic plasticity. In *Plant systematics*, ed. W. F. Grant, pp. 293–306. Toronto: Academic Press.

Moritz, D. M. L. & Kadereit, J. W. (2001).The genetics of evolutionary change in *Senecio vulgaris* L.: a QTL mapping approach. *Plant Biology*, 3, 544–52.

Moro, C., Rollo, A. & Tittensor, D. P. (2013). Comment on 'Can we name Earth's species before they go extinct?' *Science*, 341, 237.

Morris, M. G. & Perring, F. H. (1974). *The British oak: its history and natural history*. Faringdon: Classey.

Morris, S. (2012). National Trust secret £700,000 complex keeps rare plants safe. *The Guardian*, 21 June.

Morrone, J. J. (2009). *Evolutionary biogeography: an integrative approach with case studies*. New York: Columbia University Press.

Morton, J. K. (1966). The role of polyploidy in the evolution of a tropical flora. In *Chromosomes today*, vol. 1, ed. C. D. Darlington & K. R. Lewis, pp.73–6, Edinburgh: Oliver & Boyd.

Motley, T. J., Zerega, N. & Cross, H. (2006). *Darwin's harvest*. New York: Columbia University Press.

Mueller, J. M. & Hellman, J. J. (2008). An assessment of invasion risk from assisted migration. *Conservation Biology*, 22, 562–7.

Muguia-Rosas, M. A. *et al.* (2011). Meta-analysis of phenotypic selection on flowering phenology suggests that early flowering plants are favoured. *Ecology Letters*, 14, 511–21.

Muller, G. (1977). Cross-fertilization in a conifer stand inferred from enzyme gene-markers in seeds. *Silvae Genetica*, 26, 223–6.

Müller, G. B. (2008). Evo-devo as a discipline. In *Evolving pathways: key themes in evolutionary developmental biology*, ed. A. Minelli and G. Fusco, pp. 3–29. Cambridge: Cambridge University Press.

Müntzing, A. (1930a). Uber Chromosomenvermehrung in *Galeopsis*-kreuzungen und ihre phylogenetische Bedeutung. *Hereditas*, 14, 153–72.

Müntzing, A. (1930b). Outlines to a genetic monograph of the genus *Galeopsis* with special reference to the nature and inheritance of partial sterility. *Hereditas*, 13, 185–341.

Müntzing, A. (1932). Cyto-genetic investigations on synthetic *Galeopsis tetrahit*. *Hereditas*, 16, 105–54.

Müntzing, A. (1936). The evolutionary significance of autopolyploidy. *Hereditas*, 21, 363–78.

Müntzing, A. (1961). *Genetic research*. Stockholm: L. T. Førlag.

Müntzing, A., Tedin, O. & Turesson, G. (1931). Field studies and experimental methods in taxonomy. *Hereditas*, 15, 1–12.

Murphy, C. E. & Lemerle, D. (2006). Continuous cropping systems and weed selection. *Euphytica*, 148, 61–73.

Murphy, S. D. *et al.* (2006). Promotion of weed species diversity and reduction of weed seedbanks with conservation tillage and crop rotation. *Weed Science*, 54, 69–77.

Myers, K. (1986). Introduced vertebrates in Australia, with emphasis on the mammals. In *Ecology of biological invasions*, ed. R. H. Groves & J. J. Burdon, pp. 120–36. Cambridge: Cambridge University Press.

Myers, N. (1979). *The sinking ark: a new look at the problem of disappearing species*. Oxford: Pergamon Press.

Myers, N. & Knoll, A. H. (2001). The biotic crisis and the future of evolution. *Proceedings of the National Academy of Sciences, USA*, 98, 5389–91.

Nägeli, C. von (1865). Die Bastardbindung im Pflanzenreiche. *Sitzungsbericht der Königlich-Bayerischen Akademie der Wissenschaften zu München Botanische Mitteilungen*, 2, 159–87.

Naidoo, R. *et al.* (2008). Integrating economic costs into conservation planning. *Trends in Ecology and Evolution*, 21, 681–7.

Naiki, A. (2012). Heterostyly and the possibility of its breakdown by polyploidization. *Plant Species Biology*, 27, 3–29.

Nanney, D. L. (1982). Genes and phenes in *Tetrahymena*. *BioScience*, 32, 783–8.

Nannfeldt, J. A. (1937). The chromosome numbers of *Poa*, Sect. Ochlopoa A. and Gr. and their taxonomical significance. *Botaniska Notiser*, 1937, 238–57.

Nascimento, C. W. A. & Xing, B. (2006). Phytoextraction: a review on enhanced metal availability and plant accumulation. *Scientia Agriccola, (Piracicaba, Brazil)*, 63, 299–311.

Nasrallah, M. E. *et al.* (2004). Natural variation in expression of self-incompatibility in *Arabidopsis thaliana*: implications for the evolution of selfing. *Proceedings of the National Academy of Sciences, USA*, **101**, 16070–4.

Nathan, R. *et al.* (2008). An emerging movement ecology paradigm. *Proceedings of the National Academy of Sciences, USA*, **105**, 19050–1.

Naumova, T. N. (1993). *Apomixis in Angiosperms; nucellar and integumental embryology*. Boca Raton, FL: CRC Press.

Navarro, L. M. & Pereira, H. M. (2012). Rewilding abandoned landscape in Europe. *Ecosystems*, **15**, 900–12.

Navashin, M. (1926). Variabilität des Zellkerns bei Crepis-Arten in Bezug auf die Artbildung. *Zeitschrift für Zellforschung und mikroskopische Anatomie*, **4**, 171–215.

Nei, M. (1972). Genetic distance between populations. *American Naturalist*, **106**, 283–93.

Nelson, A. P. (1967). Racial diversity in Californian *Prunella vulgaris*. *New Phytologist*, **66**, 707–46.

Nelson, G. (1979) From Candolle to Croizat: comments on the history of biogeography. *Journal of the History of Biology*, **11**, 269–305.

Nelson-Jones, E. B., Briggs, D. & Smith, A. C. (2002). The origin of intermediate species of the genus *Sorbus*. *Theoretical and Applied Genetics*, **105**, 953–63.

Neuffer, B. & Hurka, H. (1999). Colonization history and introduction dynamics of *Capsella bursa-pastoris* (Brassicaceae) in North America: isozymes and quantitative traits. *Molecular Ecology*, **8**, 1667–81.

Neuffer, B. & Linde, M. (1999). *Capsella bursa-pastoris*: colonisation and adaptation; a globetrotter conquers the world. In *Plant evolution in man-made habitats*, ed. L. W. D. van Raamsdonk & J. C. M. den Nijs, pp. 49–72. Amsterdam: Hugo de Vries Laboratory.

Neuhaus, D., Kühl, H., Kohl, J. G., Dörfel, P. & Börner, T. (1993). Investigations on the genetic diversity of *Phragmites* stands using genomic fingerprinting. *Aquatic Botany*, **45**, 357–64.

Nevill, P. G. *et al.* (2013). DNA barcoding for conservation, seed banking and ecological restoration of *Acacia* in the Midwest of Western Australia. *Molecular Ecology*, **13**, 1033–42.

New, J. K. (1958). A population study of *Spergula arvensis* 1. *Annals of Botany*, New Series, **22**, 457–77.

New, J. K. (1959). A population study of *Spergula arvensis* 2. *Annals of Botany*, New Series, **23**, 23–33.

New, J. K. (1978). Change and stability of clines in *Spergula arvensis* L. (Corn Spurrey) after 20 years. *Watsonia*, **12** (2), 137–43.

New, J. K. & Herriott, J. C. (1981). Moisture for germination as a factor affecting the distribution of the seedcoat morphs of *Spergula arvensis* L. *Watsonia*, **13**(4), 323–4.

New, T. R (2006). *Conservation biology in Australia: an introduction*. Oxford: Oxford University Press.

Newmaster, S. G. *et al.* (2013). DNA barcoding detects contamination in North American herbal products. *BMC Medicine*, **11**, 1741–7015.

Newton, W. C. F. & Pellew, C. (1929). *Primula kewensis* and its derivatives. *Journal of Genetics*, **20**, 405–66.

Nichols, D. J. & Johnson, K. R. (2008). *Plants and the K-T boundary*. Cambridge: Cambridge University Press.

Nicholson, M. (1970). *The environmental revolution: a guide for the new masters of the world*. London: Hodder & Stoughton.

Nickrent, D. L. *et al.* (1998). Molecular phylogenetic and evolutionary studies of parasitic plants. In *Plant molecular systematics II*, ed. D. E. Soltis, P. S. Soltis & J. J. Doyle, pp.211–41. Boston: Kluwer.

Nicolia, A. *et al.* (2013). An overview of the last 10 years of genetically engineered crop safety research. *Critical Reviews in Biotechnology*, **34**, 77–88.

Niemela, P. & Tuomi, J. (1987). Does the leaf morphology of some plants mimic caterpillar damage? *Oikos*, **50**, 256–7.

Nilsson-Ehle, E. (1909). Kreuzungsuntersuchungen an Hafer und Weizen. *Acta Universitatis Lundensis*, Ser. 2, 5(2), 1–122.

Njoku, E. (1956). Studies on the morphogenesis of leaves. II. The effect of light intensity on leaf shape in *Ipomoea caerulea*. *New Phytologist*, **55**, 91–110.

Noble, I. R. & Dirzo, R. (1997). Forests as human-dominated ecosystems. *Science*, **277**, 522–5.

Nogales, M. *et al.* (2012). Evidence for overlooked mechanisms of long-distance seed dispersal to and between oceanic islands. *New Phytologist*, **194**, 313–17.

Nogler, G. A. (1984). Gametophytic apomixis. In *Embryology of angiosperms*, ed. B. M. Johri, pp.475–518. Berlin: Springer Verlag.

Nogués-Bravo, D. (2009). Predicting the past distribution of species climatic niches. *Global Ecology and Biogeography*, **18**, 521–31.

Noirot, M., Couvet. D. & Hamon, S. (1997). Main role of self-pollination rate on reproductive allocations in pseudogamous apomicts. *Theoretical and Applied Genetics*, **95**, 479–83.

Nordenskiøld, H. (1949). The somatic chromosomes of some *Luzula* species. *Botaniska Notiser*, **1949**, 81–92.

Nordenskiøld, H. (1951). Cyto-taxonomical studies in the genus *Luzula* 1. Somatic chromosomes and chromosome numbers. *Hereditas*, **37**, 325–55.

Nordenskiøld, H. (1956). Cyto-taxonomical studies in the genus *Luzula*. 2. Hybridization experiments in the *campestris-multiflora* complex. *Hereditas*, **42**, 7–73.

Nordenskiøld, H. (1961). Tetrad analysis and the course of meiosis in three hybrids of *Luzula campestris*. *Hereditas*, **47**, 203–38.

Noret, N. *et al.* (2005). Palatability of *Thlaspi caerulescens* for snails: influence of zinc on glucosinolates. *New Phytologist*, **165**, 763–72.

North, H. *et al.* (2009). *Arabidopsis* seed secrets unravelled after a decade of genetic and omics-driven research. *The Plant Journal*, **61**, 971–81.

Norton, B. J. (1983). Fisher's entrance into evolutionary science: the role of eugenics. In *Dimensions of Darwinism*, ed. M. Grene, pp. 19–30. Cambridge: Cambridge University Press.

Norton, D. A. (2009). Species invasions and the limits to restoration: learning from the New Zealand experience. *Nature*, **325**, 569–73.

Nosil, P. & Feder, J. L. (2012). Genomic divergence during speciation: causes and consequences. *Philosophical Transactions of the Royal Society, B*, **367**, 332–42.

Nosil, P., Funk, D. J. & Ortiz-Barrientos, D. (2009). Divergent selection and heterogeneous genomic divergence. *Molecular Ecology*, **18**, 375–402.

Novak, J. & Mack, R. N. (1993). Genetic variation in *Bromus tectorum* (Poaceae): comparison between native and introduced populations. *Heredity*, **71**, 167–76.

Novak, S. J. & Mack, R. N. (1995). Allozyme diversity in the apomictic vine *Bryonia alba* (Cucurbitaceae): potential consequences of multiple introductions. *American Journal of Botany*, **82**, 1153–62.

Novak, S. J. & Mack, R. N. (2001). Tracing plant introduction and spread: genetic evidence from *Bromus tectorum* (cheatgrass). *Bioscience*, **51**, 114–22.

Novak, S. J., Soltis, D. E. & Soltis, P. S. (1991). Ownbey's Tragopogons – 40 years later. *American Journal of Botany*, **78**, 1586–1600.

Novak, S. J., Mack, R. N. & Soltis, P. S. (1993). Genetic variation in *Bromus tectorum* (Poaceae): introduction dynamics in North America. *Canadian Journal of Botany*, **71**, 1441–8.

Novy, A. Flory, S. L. & Hartman, J. M. (2013). Evidence for rapid evolution of phenology in an invasive grass. *Journal of Evolutionary Biology*, **26**, 443–50.

Núñez-Farfán, J. & Schlichting, C. D. (2005). Natural selection in *Potentilla glandulosa* revisited. *Evolutionary Ecology Research*, **7**, 105–19.

Nyberg Berglund, A. B., Dalgren, S. & Westerbergh, A. (2003). Evidence for parallel evolution and site-specific selection of serpentine tolerance in *Cerastium alpinum* during the colonization of Scandinavia. *New Phytologist*, **161**, 199–209.

Oelschlaeger, M. (1991). *The idea of wilderness: from prehistory to the age of ecology*. New Haven, CT: Yale University Press.

Ohno, S. (1970). *Evolution by gene duplication*. London: Allen & Unwin.

Okaura, T. & Harada, K. (2002). Phylogeographical structure revealed by chloroplast DNA variation in Japanese beech (*Fagus crenata* Blume). *Heredity*, **88**, 322–9.

Olby, R. C. (1979). Mendel no mendelian? *History of Science*, **17**, 53–72.

Olby, R. C. (1985). *Origins of Mendelism*, 2nd edn. Chicago: University of Chicago Press.

Olby, R. C. (2009). *Francis Crick: hunter of life's secrets*. Cold Spring Harbor, NY: Cold Spring Harbor Laboratory Press.

Olby, R. C. & Gautrey, P. (1968). Eleven references to Mendel before 1900. *Annals of Science*, **24**, 7–20.

Olivieri, L. (2001) The evolution of dispersal and other traits in metapopulation. In *Integrating ecology and evolution in a spatial context*, ed. J. Antonovics & J. Silvertown, pp. 245–68. Oxford: Blackwell Science.

Olmstead, R. G. & Palmer, J. D. (1994). Chloroplast DNA systematics: a review of methods and data analysis. *American Journal of Botany*, **81**, 1205–24.

Olsen, K. M. & Ungerer, M. C. (2008). Freezing tolerance and cyanogenesis in White Clover (*Trifolium repens* L. Fabaceae). *International Journal of Plant Sciences*, **169**, 1141–7.

Olsen, K. M. & Wendel, J. F. (2013). A bountiful harvest: genomic insights into crop domestication phenotypes. *Annual Review of Plant Biology*, **64**, 47–70.

Olsen, K. M., Sutherland, B. L. & Small, L. L. (2007). Molecular evolution of the Li/li chemical defence polymorphism in white clover (*Trifolium repens* L.). *Molecular Ecology*, **16**, 4180–93.

Olsen, K. M., Hsu, S.-C. & Small, L. L. (2008). Evidence on the molecular basis of the Ac/ac adaptive cyanogenesis polymorphism in white clover (*Trifolium repens L.*). *Genetics*, **179**, 517–26.

Olsen, K. M., Kooyers, N. J. & Small, L. L. (2013). Recurrent gene deletions and the evolution of adaptive cyanogenesis polymorphisms in white clover (*Trifolium repens* L.). *Molecular Ecology*, **22**, 724–38.

Olsen, K. M., Kooyers, N. J. & Small, L. L. (2014). Adaptive gains through repeated gene loss: parallel evolution of cyanogenesis polymorphisms in the genus *Trifolium* (Fabaceae). *Philosophical Transactions of the Royal Society, B*, **369**, 20130347, DOI:10.1098/rstb.2013.0347.

Ooi, M. K. J., Auld, T. D. & Denham, A. J. (2012). Projected soil temperature increase and seed dormancy response along an altitudinal gradient: implications for seed bank persistence under climate change. *Plant and Soil*, **353**, 289–303.

Oostermeijer, J. G. B., Den Nijs, J. C. M., Raijmann, L. E. L. & Menken, S. B. J. (1992). Population biology and management of the Marsh Gentian (*Gentiana pneumonanthe* L.), a rare species in the Netherlands. *Botanical Journal of the Linnean Society*, **108**, 117–30.

Orel, V. (1984). *Mendel*. Past Masters Series. Oxford: Oxford University Press.

Orel, V. (1996). *Gregor Mendel: the first geneticist*. New York: Oxford University Press.

Orel, V. & Matalová, A. (1983). *Gregor Mendel and the foundation of genetics*. Brno: Mendelianum of the Moravian Museum.

Oreskes, N. (2004). The scientific consensus on climate change. *Science*, **306**, 1686.

Ornduff, R. (1966). The origin of dioecism from heterostyly in *Nymphoides* (Menyanthaceae). *Evolution*, **20**, 309–14.

Ornduff, R. (1969). Reproductive biology in relation to systematics. *Taxon*, **18**, 121–33.

Ornduff, R. (1970). Pathways and patterns of evolution – a discussion. *Taxon*, **19**, 202–4.

Ortiz-Garcia, S. *et al.* (2005). Absence of detectable transgenes in local landraces of maize in Oaxaca, Mexico (2003–2004). *Proceedings of the National Academy of Sciences, USA*, **102**, 12338–43.

Osborn, H. F. (1894). *From the Greeks to Darwin: an outline of the development of the evolution idea.* London & New York: Macmillan.

Osevik, K. L. *et al.* (2012). Parallel ecological speciation in plants? *Hindawi Publishing Corporation International Journal of Ecology*, 2012, Article ID 939862, DOI:10.1155/2012/939862.

Ossowski, S. *et al.* (2010). The rate and molecular spectrum of spontaneous mutations in *Arabidopsis thaliana*. *Science*, **327**, 92–4.

Oswald, P. H. & Preston, C. D. (eds.) (2011). *John Ray's Cambridge catalogue* (1660) translated by P. H. Oswald, & C. D. Preston. London: Ray Society.

Ouborg, N. J. & Eriksson, O. (2004). Towards a metapopulation concept for plants. In *Ecology, genetics, and evolution of metapopulations*, ed. I. Hanski & O. E. Gaggiotti, pp. 447–69. Amsterdam: Elsevier.

Ouborg, N. J. & van Treuren, R. (1995). Variation in fitness-related characters among small and large populations of *Salvia pratensis*. *Journal of Ecology*, **83**, 369–80.

Overpeck, J. T., & Weiss, J. L. (2009) Projections of future sea level becoming more dire. *Proceedings of the National Academy of Sciences, USA*, **106**, 21461–2.

Ownbey, M. (1950). Natural hybridisation and amphidiploidy in the genus *Tragopogon*. *American Journal of Botany*, **37**, 487–99.

Ownbey, M. & McCollum, G. D. (1953). Cytoplasmic inheritance and reciprocal amphiploidy in *Tragopogon*. *American Journal of Botany*, **40**, 788–96.

Ownbey, M. & McCollum, G. D. (1954). The chromosome of *Tragopogon*. *Rhodora*, **56**, 7–21.

Ozias-Atkins, P. (2006). Apomixis: developmental characteristics and genetics. *Critical Review of Plant Sciences*, **25**, 199–214.

Pagel, M. (2002). Phylogenetic inference: methods. In *Oxford encyclopaedia of evolution*, ed. M. Pagel, pp. 895–904. Oxford: Oxford University Press.

Paine, D. P. & Kiser, J. D. (2012). *Aerial photography and image interpretation*, 3rd edn. Hoboken: Wiley.

Palmer, J. D. (1988). Intraspecific variation and multicircularity in *Brassica* mitochondrial DNAs. *Genetics*, **118**, 341–51.

Palmer, M. A. & Filoso, S. (2009). Restoration of ecosystems services for environmental markets. *Science*, **325**, 575–6.

Palumbi, S. R. (2001). *The evolution explosion: how humans cause rapid evolutionary change*. New York & London: Norton.

Pankhurst, R. J. (1991). *Practical taxonomic computing*. Cambridge: Cambridge University Press.

Pannell, J. (2008). Widespread functional androdioecy in *Mercurialis annua* L. (Euphorbiaceae). *Biological Journal of the Linnean Society*, **61**, 95–116.

Parisod, C. (2008). Postglacial recolonisation of plants in the western Alps of Switzerland. *Botanica Helvetica*, **118**, 1–12.

Parisod, C. & Bescard, G. (2007). Glacial *in situ* survival in the Western Alps and polytopic autopolyploidy in *Biscutella laevigata* L. (Brassicaceae). *Molecular Ecology*, **16**, 2755–67.

Parisod, C., Trippi, C. & Galland, N. (2005). Genetic variability and founder effect in the Pitcher Plant *Sarracenia purpurea* (Sarraceniaceae) in populations introduced into Switzerland: from inbreeding to invasion. *Annals of Botany*, **95**, 277–86.

Parisod, C. *et al.* (2009). Rapid structural and epigenetic reorganization near transposable elements in hybrid and allopolyploid genomes in *Spartina*. *New Phytologist*, **184**, 1003–15.

Parisod, C., Holderegger, R. & Brochmann, C. (2010). Evolutionary consequences of autopolyploidy. *New Phytologist*, **186**, 5–17.

Parker, D. M. (1982). The conservation, by restocking, of *Saxifraga cespitosa* in North Wales. *Watsonia*, **14**, 104–5.

Parker, I. M., Rodriguez, J. & Loik, M. E. (2002). An evolutionary approach to understanding the biology of invasions: local adaptation and general-purpose genotypes in the weed *Verbascum thapsus*. *Conservation Biology*, **17**, 59–72.

Parker, R. E. (1973). *Introductory statistics for biology*. London: Arnold.

Parks, J. C. & Werth, C. R. (1993). A study of spatial features of clones in a population of bracken fern, *Pteridium aquilinum* (Dennstaeditiaceae). *American Journal of Botany*, **80**, 537–44.

Parmentier, I. *et al.* (2013). How effective are DNA barcodes in the identification of African rainforest trees? *PLOS ONE*, **8**, 1–10.

Parmesan, C. & Yohe, G. (2003). A globally coherent fingerprint of climate change impacts across natural systems. *Nature*, **421**, 37–42.

Parmesan, C. *et al.* (1999). Poleward shifts in geographical ranges of butterfly species associated with regional warming. *Nature*, **399**, 579–83.

Parokonny, A. S., Kenton, A., Gleba, Y. Y. & Bennett, M. D. (1994). The fate of recombinant chromosomes and genome interaction in *Nicotiana* hybrids and their sexual progeny. *Theoretical and Applied Genetics*, **89**, 488–97.

Parr, C. S *et al.* (2011). Evolutionary informatics: unifying knowledge about the diversity of Life. *Trends in Ecology and Evolution*, **27**, 94–103.

Parsons, J. J. (1970). The Africanization of the New World tropical grasslands. *Tübinger Geographische Studien*, **34**, 141–53.

Parsons, P. A. (1959). Some problems in inbreeding and random mating in tetrasomics. *Agronomy Journal*, **51**, 465–7.

Paterniani, E. (1969). Selection for reproductive isolation between two populations of Maize, *Zea mays* L. *Evolution*, **23**, 534–47.

Paterson, A. H. *et al.* (2006). Many gene and domain families have convergent fates following independent whole-genome duplication events in *Arabidopsis, Oryza, Saccharomyces* and *Tetraodon*. *Trends in Genetics*, **22**, 597–602.

Patterson, C. (ed.) (1987). *Molecules and morphology in evolution: conflict or compromise?* Cambridge: Cambridge University Press.

Paule, J., Shorbel, A. & Dobes, C. (2011). Implications of hybridisation and cytotypic differentiation in speciation assessed by AFLP and plastid haplotypes – a case study of *Potentilla alpicola* La Soie. *BMC Evolutionary Biology*, **12**, 132.

Paun, O., Stuessy, T. F. & Hörandl, E. (2006). The role of hybridization, polyploidization and glaciation in the origin and evolution of the apomictic *Ranunculus cassubicus* complex. *New Phytologist*, **171**, 223–36.

Paun, O. *et al.* (2007). Genetic and epigenetic alterations after hybridization and genome doubling. *Taxon*, **56**, 649–56.

Pauwels. M. *et al.* (2005). Multiple origin of metallicolous populations of the pseudometallophyte *Arabidopsis halleri* (Brassicaceae) in central Europe: the cpDNA testimony. *Molecular Ecology*, **14**, 4403–14.

Pauwels, M. *et al.* (2008). Merging methods in molecular and ecological genetics to study the adaptation of plants to anthropogenic metal-polluted sites: implications for phytoremediation. *Molecular Ecology*, **17**, 108–19.

Pazy, B. & Zohary, D. (1965). The process of introgression between *Aegilops* polyploids: natural hybridization between *A. variabilis, A. ovata* and *A. biuncialis. Evolution*, **19**, 385–94.

Pearson, E. S. & Kendall, M. G. (1970). *Studies in the history of statistics and probability*. London: Griffin.

Pearson, K. (1900). *The grammar of science*, 2nd edn. London: Black.

Pearson, K. (1924). *The life, letters and labours of Francis Galton*, vol. II. Cambridge: Cambridge University Press.

Pearson, K. *et al.* (1903). Cooperative investigation on plants. 2. Variation and correlation in Lesser Celandine from diverse localities. *Biometrika* **2**, 145–64.

Pearson, R. G. (2006). Climate change and the migration capacity of species. *Trends in Ecology and Evlution*, **21**, 111–13.

Pecinka, A. *et al.* (2011). Polyploidization increases meiotic recombination frequency in *Arabidopsis. BMC Biology*, **9**, 24.

Peckham, M. (1959). *The origin of species by Charles Darwin*. A variorum text. London: Oxford University Press; Philadelphia: University of Pennsylvania Press.

Pelaz, S. *et al.* (2000). B and C floral organ identity functions require *SEPALLATA* MADS-box genes. *Nature*, **405**, 200–3.

Pellew, C. (1913). Note on gametic reduplication in *Pisum. Journal of Genetics*, **3**, 105–6.

Pellino, M. *et al.* (2013). Asexual genome evolution in the apomictic *Ranunculus auricomus* complex: examining the effects of hybridization and mutation accumulation. *Molecular Ecology*, DOI:10.1111/mec.12533

Pennington, W. (1974). *The history of British vegetation*, 2nd edn. London: English Universities Press.

Peñuelas, J. *et al.* (2007) Migration, invasion and decline: changes in recruitment and forest structure in a warming-linked shift of European beech forest in Catalonia (NE Spain). *Ecography*, **30**, 830–8.

Percy, D. M. & Cronk, Q. C. B. (1997). Conservation in relation to mating system in *Nesohedyotis arborea* (Rubiaceae), a rare endemic tree from St. Helena. *Biological Conservation*, **80**, 135–46.

Perrie, L. R. *et al.* (2010). Parallel polyploid speciation: distinct sympatric gene-pools of recurrently derived allo-octoploid *Asplenium* ferns. *Molecular Ecology*, **19**, 2916–32.

Perring, F. H. & Farrell, L. (1983). *British red data books*. Vol. 1, Vascular plants, 2nd edn. Lincoln: The Society for the Promotion of Nature Conservation with the financial support of the World Wildlife Fund.

Perring, F. H. & Walters, S. M. (1976). *Atlas of the British flora*, 2nd edn. Wakefield: EP Publishing.

Perrow M. R. & Davy A. J. (eds.) (2002). *Handbook of ecological restoration*. Cambridge: Cambridge University Press.

Peterken, G. F. (1981). *Woodland conservation and management*. London: Chapman & Hall.

Petit, C. & Thompson, J. D. (1999). Species diversity and ecological range in relation to ploidy level in the flora of the Pyrenees. *Evolutionary Ecology*, **13**, 45–66.

Petit, R. J. *et al.* (1997). Chloroplast DNA footprints of postglacial recolonization by oaks. *Proceedings of the National Academy of Sciences, USA*, **94**, 9996–10001.

Petit, R. J. *et al.* (2003). Glacial refugia: hotspots but not melting pots of genetic diversity. *Science*, **300**, 1563–5.

Peuke, A. D & Rennenberg, H. (2005). Phytoremediation. *Embo Reports*, **6**, 497–501.

Pharis, R. P. & Ferrell, W. K. (1966). Differences in drought resistance between coastal and inland sources of Douglas Fir. *Canadian Journal of Botany*, **44**, 1651–9.

Phy-Olsen, A. (2010). *Evolution, creationism, and intelligent design*. Westport, CT: Greenwood Press.

Pickett, S. T. A., Parker, V. T. & Fiedler, P. L. (1992). The new paradigm in ecology: implications for conservation above the species level. In *Conservation biology: the theory and practice of nature conservation, preservation and management*, ed. P. L. Fiedler & S. K. Jain, pp. 65–88. New York: Chapman & Hall.

Pietsch, T. W. (2012). *Trees of life: a visual history of evolution.* Baltimore, MD: Johns Hopkins University Press.

Pigliucci, M. (2001). *Phenotypic plasticity: beyond nature and nurture.* Baltimore, MD: Johns Hopkins University Press.

Pigliucci, M. (2002). *Denying evolution: creationism, scientism and the nature of science.* Sunderland, MA: Sinauer.

Pigliucci, M. (2010). Phenotypic plasticity. In *Evolution – the extended synthesis,* ed. M. Pigliucci & G. B. Müller, pp. 355–78. Cambridge, MA: MIT Press.

Pijul, P. M. *et al.* (2012). *In vitro* propagation of tropical hardwood tree species – a review 2001–2011. *Propagation of Ornamental Plants,* 12, 25–51.

Piñeyro-Nelson, A. *et al.* (2009). Transgenes in Mexican maize: molecular evidence and methodological considerations for GMO detection in landrace populations. *Molecular Ecology,* 18, 750–61.

Pires, C. J. *et al.* (2004a). Molecular cytogenetic analysis of recently evolved *Tragopogon* (Asteraceae) allopolyploids reveal a karyotype that is additive of the diploid progenitors. *American Journal of Botany,* **91**, 1022–35.

Pires, J. C. *et al.* (2004b). Flowering time divergence and genomic rearrangements in resynthesized *Brassica* polyploids (*Brassicaceae*). *Biological Journal of the Linnean Society,* **82**, 675–88.

Pires, N. D. & Dolan, L. (2012). Morphological evolution in land plants: new designs with old genes. *Philosophical Transactions of the Royal Society, B,* **367**, 508–18.

Pitman, N. C. A. & Jørgensen, P. M. (2002). Estimating the size of the threatened world flora. *Science,* **298**, 989.

Pivard, S. *et al.* (2008). Where do the feral oilseed rape populations come from? A large-scale study of their possible origin in a farmland area. *Journal of Applied Ecology,* **45**, 476–85.

Pleijel, F. (1999). Phylogenetic taxonomy, a farewell to species, and a revision of *Heteropodarke* (*Annelida, Polychaeta, Hesionidae*). *Systematic Biology,* **48**, 755–89.

Podani, J. (2010a). Monophyly and paraphyly: a discourse without end? *Taxon,* **59**, 1011–15.

Podani, J. (2010b). Taxonomy in evolutionary perspective. An essay on the relationships between taxonomy and evolutionary theory. *Synbiologia Hungarica,* 6, 1–42.

Pollard, A. J. (1980). Diversity of metal tolerances in *Plantago lanceolata* L. from the southeastern United States. *New Phytologist,* **86**, 109–17.

Pollard, A. J. & Baker, A. J. M. (1997). Deterrence of herbivory by zinc hyperaccumulation in *Thlaspi caerulescens* (Brassicaceae). *New Phytologist,* **135**, 655–8.

Pond, W. G. & Pond, K. R. (2002). *Introduction to animal science.* New York & Chichester, UK: Wiley.

Pope, O. A., Simpson, D. M. & Duncan, E. N. (1944). Effect of Corn barriers on natural crossing in Cotton. *Journal of Agricultural Research,* **68**, 347–61.

Popp, M., Mirré, V. & Brochmann, C. (2011). A single mid-Pleistocene long-distance dispersal by a bird can explain the extreme bipolar disjunction in crowberries (*Empetrum*). *Proceedings of the National Academy of Sciences, USA,* **108**, 6520–5.

Popper, K. (1963). *Conjectures and refutations: the growth of scientific knowledge.* London: Routledge.

Porter, D. M. (1983). Vascular plants of the Galápagos: origins and dispersal. In *Patterns of evolution in Galápagos organisms,* ed. R. I. Bowman, M. Berson & A. E. Levitan, pp. 33–96. San Francisco: American Association for the Advancement of Science.

Porter, T. M. (1986). *The rise of statistical thinking, 1820–1900.* Princeton, NJ: Princeton University Press.

Porter, T. M. (2004). *Karl Pearson: the scientific life in a statistical age.* Princeton, NJ: Princeton University Press.

Potato Genome Sequencing Consortium (2011). Genome sequence and analysis of the tuber crop potato. *Nature,* **475**, 189–95.

Potts, S. G. *et al.* (2010). Global pollinator declines: trends, impacts and drivers. *Trends in Ecology & Evolution,* **25**, 345–53.

Potvin, C. (1986). Biomass allocation and phenological differences among southern and northern populations of the C4 grass *Echinochloa crus-galli. Journal of Ecology,* **74**, 915–23.

Powles, S. B. (2008). Evolved glyphosate-resistant weeds around the world: lessons to be learnt. *Pest Management Science,* **64**, 360–5.

Powles, S. B. & Yu, Q. (2010). Evolution in action: plants resistant to herbicides. *Annual Review of Plant Biology,* **61**, 317–47.

Prance, G. T. (2004). Introduction. In *Ex situ plant conservation,* ed. E. O. Guerrant, K. Havens & M. Maunder, pp. xxiii–xxix. Washington DC, Covelo, CA, & London: Island Press.

Prendini, L. (2005). Comment on 'Identifying spiders through DNA barcodes'. *Canadian Journal of Zoology*, 83, 498–504.

Prentice, H. C. (1986). Climate and clinal variation in seed morphology of the White Campion, *Silene latifolia* (Caryophyllaceae). *Biological Journal of the Linnean Society*, 27, 179–89.

Presgraves, D. C. (2010). The molecular evolutionary basis of species formation. *Nature Reviews Genetics*, 11, 175–80.

Presgraves, D. C. (2013). Primer: hitchhiking to speciation. *PLoS Biology*, 11: e1001498.

Pridgeon A. M. *et al.* (1997). Phylogenetics of subtribe Orchidineae (Orchidoideae, Orchidaceae) based on nuclear ITS sequences. 1. Intergeneric relationships and polyphyly of *Orchis* sensu lato. *Lindleyana*, 12, 89–109.

Primack, R. B. (1993). *Essentials of conservation biology.* Sunderland, MA: Sinauer.

Primack, R. B. (2010). *Essentials of conservation biology*, 5th edn. Sunderland, MA: Sinauer Associates.

Primack, R. & Dayton, B. (1997). The experimental ecology of reintroduction. *Plant Talk*, 97, 25–8.

Primack R. B. & Miao, S. L. (1992). Dispersal can limit local plant distribution. *Conservation Biology*, 6, 513–19.

Prime, C. T. (1960). *Lords and ladies*. London: Collins.

Prober, S. M. & Brown, A. H. D. (1994). Conservation of the Grassy White Box woodlands: population genetics and fragmentation *of Eucalyptus albens. Conservation Biology*, 8, 1003–13.

Proctor, J. (1971a). The plant ecology of serpentine. II. Plant response to serpentine soils. *Journal of Ecology*, 59, 397–410.

Proctor, J. (1971b). The plant ecology of serpentine. III. The influence of a high magnesium/calcium ratio and high nickel and chromium levels in some British and Swedish serpentine soils. *Journal of Ecology*, 59, 827–42.

Proctor, M. C. F. & Yeo, P. F. (1973). *The pollination of flowers*. London: Collins.

Proctor, M. C. F., Proctor, M. E. & Groenhof, A. C. (1989). Evidence from peroxidase polymorphism on the taxonomy and reproduction of some *Sorbus* populations in south-west England. *New Phytologist*, 112, 569–75.

Proctor, M. C. F., Yeo, P. F. & Lack, A. J. (1996). *The natural history of pollination*. London: Harper Collins.

Provine, W. B. (1971). *The origins of theoretical population genetics*. Chicago & London: University of Chicago Press.

Provine, W. B. (1986). *Sewall Wright and evolutionary biology*. Chicago: University of Chicago Press.

Provine, W. B. (1987). *The origins of theoretical population genetics*, 2nd edn. Chicago: University of Chicago Press.

Putnam, A. R. & Tang, C.-S. (1986). *The science of allelopathy*. New York: Wiley.

Qiu Y.-L. *et al.* (1999). The earliest angiosperms: evidence from mitochondrial, plastid and nuclear genomes. *Nature*, 402, 404–7.

Quarin, C. L. (1986). Seasonal changes in the incidence of apomixis of diploid, triploid, and tetraploid plants of *Paspalum cromyorrizon. Euphytica*, 35, 515–22.

Quarin, C. L. & Hanna, W. W. (1980). Effect of three ploidy levels on meiosis and mode of reproduction in *Paspalum hexastachyum. Crop Science*, 20, 69–75.

Quattrocchio, F. *et al.* (1999). Molecular analysis of the *anthocyanin2* gene of *Petunia* and its role in the evolution of flower color. *The Plant Cell*, 11, 1433–44.

Queller, D. C. (1987). Sexual selection in flowering plants. In *Sexual selection: testing the alternatives*, ed. J. W. Bradbury & M. B. Andersson, pp. 165–79. Chichester: Wiley.

Quetelet, M. A. (1846). *Lettres à S.A.R. Ie Duc Régnant de Saxe-Coburg et Gotha, sur la théorie des probabilités, appliquée aux sciences morales et politiques*. Brussels. Translation by O. G. Downes (1849): Letters addressed to H.R.H. the Grand Duke of Saxe-Coburg and Gotha on the theory of probabilities as applied to the moral and political sciences. London: Charles & Edwin Layton.

Quinn, J. A. (1978). Plant ecotypes: ecological or evolutionary units. *Bulletin of the Torrey Botanical Club*, 105, 58–64.

Quintana-Ascencio, P. F. & Menges, E. S. (1996). Inferring metapopulation dynamics from patch-level incidence of Florida scrub plants. *Conservation Biology*, 10, 1210–19.

Quist, D. & Chapela, I. H. (2001). Transgenic DNA introgressed into traditional maize landraces in Oaxaca, Mexico. *Nature*, 414, 541–3.

Rabinowitz, D., Cairns, S. & Dillon, T. (1986). Seven forms of rarity and their frequency in the flora of the British

Isles. In *Conservation biology*, ed. M. E. Soulé, pp.182–204. Sunderland, MA: Sinauer.

Rackham, O. (1975). *Hayley Wood. Its history and ecology.* Cambridge: Cambridgeshire and Isle of Ely Naturalists Trust.

Rackham, O. (1980). *Ancient woodland: its history, vegetation and uses in England.* London: Arnold.

Rackham, O. (2008). Ancient woodlands: modern threats. *New Phytologist*, **180**, 571–86.

Radford, A. E., Dickison, W. C., Massey, J. R. & Bell, R. (1974). *Vascular plant systems.* New York: Harper & Row.

Rafferty, N. E. & Ives, A. R. (2011). Effects of experimental shifts in flowering phenology on plant–pollinator interactions. *Ecology Letters*, **14**, 69–74.

Rafinski, J. N. (1979). Geographic variability of flower colour in *Crocus scepusiensis* (Iridaceae). *Plant Systematics and Evolution*, **131**, 107–25.

Raijmann, L. E. L., van Leeuwen, N. C., Kersten, R., Oostermeijer, J. G. B., Den Nijs, H. C. N. & Menken, S. B. J. (1994). Genetic variation and outcrossing rate in relation to population size in *Gentiana pneumonathe* L. *Conservation Biology*, **8**, 1014–26.

Ramsbottom, J. (1938). Linnaeus and the species concept. *Proceedings of the Linnean Society of London*, **150**, 192–219.

Ramsey, J. & Schemske, D. W. (1998). Pathways, mechanisms, and rates of polyploid formation in flowering plants. *Annual Review of Ecology and Systematics*, **29**, 467–501.

Ramsey, J. & Schemske, D. W. (2002). Neopolyploidy in flowering plants. *Annual Review of Ecology & Systematics*, **33**, 589–639.

Ramsey, M. W., Cairns, S. C. & Vaughton, G. V. (1994). Geographic variation in morphological and reproductive characters of coastal and tableland populations of *Blandfordia grandiflora*. *Plant Systematics & Evolution*, **192**, 215–30.

Randolph, L. F., Nelson, I. S. & Plaisted, R. L. (1967). Negative evidence of introgression affecting the stability of Louisiana Iris species. *Cornell University Agriculture Experimental Station Memoir*, No. 398.

Ranker, T. A., Floyd, S. K. & Trapp, P. G. (1994). Multiple colonizations of *Asplenium adiantum-nigrum* onto the Hawaiian archipelago. *Evolution*, **48**, 1364–7.

Rasmussen, R. S. *et al.* (2009). DNA barcoding of commercially important salmon and trout species (*Oncorhynchus* and *Salmo*) from North America. *Journal of Agricultural and Food Chemistry*, **57**, 8379–85.

Raubeson, L. A. & Jansen, R. K. (2005). Chloroplast genomes of plants. In *Plant diversity and evolution: genotypic and phenotypic variation in higher plants*, ed. R. J. Henry, pp.45–68. Wallingford, UK; CABI Publishing.

Raup, D. M. (1991). *Extinction.* Oxford: Oxford University Press.

Raven, C. E. (1950). *John Ray: naturalist*, 2nd edn; reissued 1986. Cambridge: Cambridge University Press.

Raven, P. H. (1976). Systematics and plant population biology. *Systematic Botany*, **1**, 284–316.

Raven, P. H. & Thompson, H. J. (1964). Haploidy and angiosperm evolution. *The American Naturalist*, **98**, 251–2.

Raven, P. H. *et al.* (1960). Chromosome numbers in Compositae. I. Astereae. *American Journal of Botany*, **47**, 124–32.

Ravi, M. & Chan, S. W. L. (2010). Haploid plants produced by centromere-mediated genome elimination. *Nature*, **464**, 615–18.

Ray, J. (1691). *The Wisdom of God Manifested in the Works of Creation*, Facsimile edition of the 1826 edition published in 2005. London: Scion Publishing for The Ray Society to mark the 300th anniversary of John Ray's death in 1705.

Ray, M. F. (1995). Systematics of *Lavatera* and *Malva* (Malvaceae, Malveae) – a new perspective. *Plant Systematics and Evolution*, **198**, 25–53.

Raybould, A. F. (1995). Wild crops. In *Encyclopedia of environmental biology*, vol. 3, ed. W. A. Nierenberg, pp.551–65. San Diego, CA, & New York: Academic Press.

Raybould, A. F., Gray, A. J., Lawrence, M. J. & Marshall, D. F. (1990). The origin and taxonomy of *Spartina* × *neyrautii* Foucaud. *Watsonia*, **18**, 207–9.

Raybould, A. F. *et al.* (1991a). The evolution of *Spartina anglica.* C. E. Hubbard (Gramineae): origin and genetic variation. *Biological Journal of the Linnean Society*, **43**, 111–26.

Raybould, A. F. *et al.* (1991b). The evolution of *Spartina anglica* C. E. Hubbard (Gramineae): genetic variation and status of the parental species in Britain. *Biological Journal of the Linnean Society*, **44**, 369–80.

Rayner, A. A. (1969). *A first course in biometry for agricultural students.* Pietermaritzburg: University of Natal Press.

Redford, K. H., Jensen, D. B. & Breheny, J. J. (2012). Integrating the captive and the wild. *Science*, **338**, 1157–8.

Ree, R. H. & Smith, S. A. (2008). Maximum likelihood inference of geographic range evolution by dispersal, local extinction and cladogenesis. *Systematic Biology*, **57**, 4–14.

Reed, D. H. & Frankham, R. (2001). How closely correlated are molecular and quantitative measures of genetic variation? A meta-analysis. *Evolution*, **55**, 1095–1103.

Rees, H. & Hutchinson, J. (1973). Nuclear DNA variation due to B chromosomes. *Cold Spring Harbor in Quantitative Biology*, **38**, 175–82.

Rees, H. & Jones, R. N. (1977). *Chromosome genetics.* London: Arnold.

Régnière, J. & Bentz, B. J. (2007). Modeling cold tolerance in the mountain pine beetle, *Dendroctonus ponderosae*. *Journal of Insect Physiology*, **53**, 559–72.

Rehfeld, G. E. *et al.* (2002). Intraspecific responses to climate in *Pinus sylvestris*. *Global Change Biology*, **8**, 912–29.

Reichman, J. R *et al.* (2006). Establishment of transgenic herbicide-resistant creeping bentgrass (*Agrostis stolonifera* L.) in nonagronomic habitats. *Molecular Ecology*, **15**, 4243–55.

Reigosa, M. J., Pedrol, N. & González, L. (eds.) (2010). *Allelopathy: a physiological process with ecological implications.* Dordrecht: Springer.

Reiling, K. & Davison, A. W. (1992). Spatial variation in ozone resistance of British populations of *Plantago major* L. *New Phytologist*, **122**, 699–708.

Reinartz, J. A. & Les, D. H. (1994). Bottleneck-induced dissolution of self-incompatibility and breeding system consequences in *Aster furcatus*. *American Journal of Botany*, **81**, 446–55.

Renfrew, J. M. (1973). *Paleoethnobotany: the prehistoric food plants of the Near East and Europe.* London: Methuen.

Renner, S. S. (2004). Tropical trans-Atlantic disjunctions, sea surface currents, and wind patterns. *International Journal of Plant Sciences*, **165**, S23–S33.

Renner, S. S. (2014). The relative and absolute frequencies of angiosperm sexual systems: dioecy, monoecy, gynodioecy and an updated online database. *American Journal of Botany*, **101**, 1588–96.

Renner, S. S. & Ricklefs, R. E. (1995). Dioecy and its correlates in the flowering plants. *American Journal of Botany*, **82**, 596–606.

Reuther W., Batchelor L. D. & Webber H. J. (1968). *The citrus industry: California. Vol. 1.* Berkeley: University of California, Division of Agricultural Sciences.

Reyer, C. *et al.* (2013). A plant's perspective of extremes: terrestrial plant responses to climatic variability. *Global Change Biology*, **19**, 79–89.

Rhoné, B. *et al.* (2010). Evolution of flowering time in experimental wheat populations: a comprehensive approach to detect genetic signatures of natural selection. *Evolution*, **64**, 2110–25.

Ricciari, A. & Simberloff, D. (2008). Assisted colonization is not a viable conservation strategy. *Trends in Ecology and Evolution*, **24**, 248–53.

Rice, D. W. *et al.* (2013). Horizontal transfer of entire genomes via mitochondrial fusion in the angiosperm *Amborella*. *Science*, **342**, 1468–73.

Rice, E. L. (1984). *Allelopathy*, 2nd edn. London: Academic Press.

Rice, W. R. & Hostert, E. E. (1993). Laboratory experiments on speciation. *Evolution* **47**, 1637–53.

Richards, A. J. (1979). Reproduction in flowering plants. *Nature*, **278**, 306.

Richards, A. J. (1986). *Plant breeding systems.* London: George Allen & Unwin [2nd edn 1997].

Richards, A. J. (1993). *Primula.* London: Batsford.

Richards, A. J. (2003). Apomixis in flowering plants: an overview. *Philosophical Transactions of the Royal Society of London, B Biological Science*, **358**, 1085–93.

Richards, A. J. & Ibrahim, H. (1978). Estimation of neighbourhood size in two populations of *Primula veris*. In *The pollination of flowers by insects*, ed. A. J. Richards, pp. 165–74. Linnean Society Symposium Series 6. London: Academic Press.

Richardson, A. O. & Palmer, J. D. (2007). Horizontal gene transfer in plants. *Journal of Experimental Botany*, **58**, 1–9.

Richardson, D. M. (ed.) (2011). *Fifty years of invasion ecology: the legacy of Charles Elton.* Chichester: Wiley-Blackwell.

Richardson, D. M. & Higgins, S. I. (1999). Pines as invaders in the southern hemisphere. In *Ecology and biogeography of* Pinus, ed. D. M. Richardson, pp. 450–73. Cambridge: Cambridge University Press.

Ridgman, W. J. (1975). *Experimentation in biology.* Glasgow: Blackie.

Ridley, H. N. (1930). *The dispersal of plants throughout the world.* Ashford: L. Reeve & Co., Ltd.

Ridley, M. (1996). *Evolution*, 2nd edn. Cambridge, MA, & Oxford: Blackwell Scientific.

Rieger, R., Michaelis, A. & Green, M. M. (1976). *Glossary of genetics and cytogenetics*, 4th edn. Berlin, Heidelberg & New York: Springer-Verlag.

Rieseberg, L. H. (1995). The role of hybridization in evolution: old wine in new skins. *American Journal of Botany*, **82**, 944–53.

Rieseberg, L. H. (1997). Hybrid origins of plant species. *Annual Review of Ecology and Systematics*, **28**, 359–89.

Rieseberg, L. H. & Blackman, B. K. (2010). Speciation genes in plants. *Annals of Botany*, **106**, 439–55.

Rieseberg, L. H. & Ellstrand, N. C. (1993). What can molecular and morphological markers tell us about plant hybridization? *Critical Reviews in Plant Sciences*, **12**, 213–41.

Rieseberg, L. H. & Soltis, D. E. (1991). Phylogenetic consequences of cytoplasmic gene flow in plants. *Evolutionary Trends in Plants*, **5**, 65–84.

Rieseberg, L. H. & Wendel, J. F. (1993). Introgression and its consequences in plants. In *Hybrid zones and the evolutionary process*, ed. R. G. Harrison, pp. 70–109. New York: Oxford University Press.

Rieseberg, L. H. & Wendel, J. F. (2004). Plant speciation – rise of the poor cousins. *New Phytologist*, **161**, 3–8.

Rieseberg, L. H. & Willis, J. H. (2007). Plant speciation. *Science*, **317**, 910–14.

Rieseberg, L. H., Van Fossen, C. & Desrochers, M. (1995). Hybrid speciation accompanied by genomic reorganization in wild sunflowers. *Nature*, **375**, 313–16.

Rieseberg, L. H., Church, S. A. & Morjan, C. L. (2003). Integration of populations and differentiation of species. *New Phytologist*, **161**, 59–69.

Rieseberg, L. H., Wood, T. E. & Baack, E. J. (2006). The nature of plant species. *Nature*, **440**, 524–7.

Rieseberg, L. H. *et al.* (2007). Hybridization and the colonization of novel habitats by annual sunflowers. *Genetica*, **129**, 149–65.

Riley, H. P. (1938). A character analysis of colonies of *Iris fulva* and *Iris hexagona* var. *giganticaerulea* and natural hybrids. *American Journal of Botany*, **25**, 727–38.

Riley, R. (1965). Cytogenetics and the evolution of Wheat. In *Essays on crop plant evoluion*, ed. J. Hutchinson, pp. 103–22. London: Cambridge University Press.

Riley, R. & Chapman, V. (1958). Genetic control of the cytologically diploid behaviour of hexaploid Wheat. *Nature*, **183**, 713–15.

Riley, R., Unrau, J. & Chapman, V. (1958). Evidence on the origin of the B genome of Wheat. *Journal of Heredity*, **49**, 91–8.

Ritchie, J. C. (1955a). A natural hybrid in *Vaccinium*. 1. The structure, performance and chorology of the cross *Vaccinium intermedium* Ruthe. *New Phytologist*, **54**, 49–67.

Ritchie, J. C. (1955b). A natural hybrid in *Vaccinium*. 2. Genetic studies in *Vaccinium intermedium* Ruthe. *New Phytologist*, **54**, 320–35.

Ritchie, J. C. (1987). *Postglacial vegetation of Canada.* Cambridge: Cambridge University Press.

Rizvi, S. J. H. & Rizvi, V. (1992). *Allelopathy: basic and applied aspects.* London: Chapman & Hall.

Roach, D. A. & Wulff, R. D. (1987). Material effects in plants. *Annual Review of Ecology & Systematics*, **18**, 209–35.

Roberts, H. F. (1929). *Plant hybridisation before Mendel.* Princeton, NJ: Princeton University Press; London: Oxford University Press.

Roberts, N. (1989). *The Holocene: an environmental history.* Oxford: Blackwell.

Robledo-Aruncio, J. J. & Gil, L. (2005). Patterns of pollen dispersal in a small population of *Pinus sylvestris* L. revealed by total-exclusion paternity analysis. *Heredity*, **94**, 13–22.

Robson, G. C. & Richards, O. W. (1936). *The variation of animals in nature.* London, New York & Toronto: Longmans, Green and Co.

Rodrigues, A. S. L. *et al.* (2004). Global gap analysis: priority regions for expanding the global protected-area network. *BioScience*, **54**, 1092–1100.

Rodriguez-Trelles, F., Tarrio, R. & Ayala, F. J. (2004). Molecular clocks: whence and whither? In *Telling the evolutionary time: molecular clocks and the fossil record*, ed. P. C. J. Donoghue and M. P. Smith, pp. 5–26. London: Systematic Association Special Volume; CRC Press.

Rogstad, S. H., Nybom, H. & Schaal, B. A. (1991). The tetrapod DNA fingerprinting M13 repeat probe reveals genetic diversity and clonal growth in Quaking Aspen (*Populus tremuloides*, Salicaceae). *Plant Systematics & Evolution*, **175**, 115–23.

Roles, S. J. (1960). *Illustrations (Part II) to Flora of the British Isles*, Clapham, A. R., Tutin, T. G. & Warburg, E. F. Cambridge: Cambridge University Press.

Romeiras, M. M. *et al.* (2011). Origin and diversification of the genus *Echium* (Boraginaceae) in the Cape Verde archipelago. *Taxon*, **60**, 1375–85.

Romme, W. H. *et al.* (2005). Establishment, persistence, and growth of Aspen (*Populus tremuloides*) seedlings in Yellowstone National Park. *Ecology*, **86**, 404–18.

Ronquist, F. & Sanmartin, I. (2011): Phylogenetic methods in biogeography. *Annual Review of Ecology, Evolution and Systematics*, **42**, 441–64.

Roose, M. L. & Gottlieb, L. D. (1976). Genetic and biochemical consequences of polyploidy in *Tragopogon. Evolution*, **30**, 818–30.

Rose, M. R. & Oakley, T. H. (2007). The new biology: beyond the Modern Synthesis. *Biology Direct*, DOI:10.1186/1745-6150-2-30

Rosen, F. (1889). Systematische und biologische Beobachtungen über *Erophila verna. Botanische Zeitung*, **47**, 565–80, 581–91, 597–608, 613–20.

Rosenberg, A. (1994). *Instrumental biology or the disunity of science*. Chicago: Chicago University Press.

Ross-Craig, S. (1948–73). *Drawings of British plants*. London: Bell & Sons Ltd.

Rosser, E. M. (1953). A new British species of *Senecio. Watsonia*, **3**, 228–32.

Rotherham, I. D. & Lambert, R. A. (2011). Balancing species history, human culture and scientific insight: introduction and overview. In *Invasive and introduced plants and animals*, ed. I. D. Rotherham & R. A. Lambert, pp. 3–18. London: Earthscan Publishing.

Rothrock, P. E. & Reznicek, A. A. (1998). Chromosome numbers in *Carex* section Ovales (Cyperaceae): additions, variations, and corrections. *Sida*, **18**, 587–92.

Roux, C. *et al.* (2012). Recent and ancient signature of balancing selection around the S-locus in *Arabidopsis halleri* and *A. lyrata. Molecular Biology and Evolution*, **30**, 435–47.

Rowell, T. A. (1984). Further discoveries of the Fen Violet (*Viola persicifolia* Schreber) at Wicken Fen, Cambridgeshire. *Watsonia*, **15**, 122–3.

Rowell, T. A., Walters, S. M. and Harvey, H. J. (1982). The rediscovery of the Fen Violet, *Viola persicifolia* Schreber, at Wicken Fen, Cambridgeshire. *Watsonia*, **14**, 183–4.

Rubledo-Arununcio, J. J. & Garcia, C. (2007). Estimation of the seed dispersal kernel from exact identification of source plants. *Molecular Ecology*, **16**, 5098–109.

Rucinska, A. & Puchalski, J. (2011). Comparative molecular studies on the genetic diversity of an *ex situ* garden collection and its source population of the critically endangered Polish endemic plant *Cochlearia polonica* E. Fröhlich. *Biodiversity & Conservation*, **20**, 401–13.

Ruckert, J. (1892). Zur Entwicklungs Geschichte des Ovarioleies bei Selachiern. *Anatomischer Anzeiger*, **7**, 107.

Rudall, P. J. & Bateman, R. M. (2004). Evolution of zygomorphy in monocot flowers: iterative patterns and developmental constraints. *New Phytologist*, **162**, 25–44.

Ruddiman, W. F. (2013) The Anthropocene. *Annual Review of Earth and Planetary Sciences*, **41**, 45–68.

Ruhsam, M. *et al.* (2010). Significant differences in outcrossing rate, self-incompatibility, and inbreeding depression between two widely hybridizing species of *Geum. Biological Journal of the Linnean Society*, **101**, 977–90.

Ruhsam, M., Hollingsworth, P. M. & Ennos, R. A. (2011) Early evolution in a hybrid swarm between outcrossing and selfing lineages in *Geum. Heredity*, **107**, 246–55, DOI:10.1038/hdy.2011.9.

Runte, A. (1997). *National Parks: the American experience*, 3rd edn. Lincoln, NE, & London: Nebraska Press.

Ruppel, C. D. (2011). Methane hydrates and contemporary climate change. *Nature Education Knowledge*, **3**, 29.

Ruse, M. (1987). Biological species: natural kinds, individuals or what? *British Journal of the Philosophy of Science*, **38**, 225–42.

Ruse, M. (2003). *Darwin and design: does evolution have a purpose?* Cambridge, MA, & London: Harvard University Press.

Ruse, M. (2008). *Charles Darwin*. Oxford: Blackwell.

Ruse, M. (2013). *The Cambridge encyclopedia of Darwin and evolutionary thought*. Cambridge: Cambridge University Press.

Rushton, B. S. (1978). *Quercus robur* L. and *Quercus petraea* (Matt.) Liebl: a multivariate approach to the hybrid problem. 1. Data acquisition, analysis and interpretation. *Watsonia*, **12**, 81–101.

Rushton, B. S. (1979). *Quercus robur* L. and *Quercus petraea* (Matt.) Liebl.: a multivariate approach to the hybrid problem. 2. The geographical distribution of population types. *Watsonia*, 12, 209–24.

Russell, B. (1931). *The scientific outlook*. London: Allen & Unwin.

Rustad, L. E. *et al.* (2001). A meta-analysis of the response of soil respiration, net nitrogen mineralization, and above ground growth to experimental ecosystem warming. *Oecologia*, **126**, 543–62.

Rutschman, F. (2006). Molecular dating of phylogenetic trees: a brief review of current methods that estimate divergence times. *Diversity and Distributions*, **12**, 35–48.

Saarela, J. M. *et al.* (2007). Hydatellaceae identified as a new branch near the base of the angiosperm phylogenetic tree. *Nature*, **446**, 312–15.

Sachs, J. (1865). *Handbuch der Experimental-Physiologie der Pflanzen*. Leipzig: Verlag von Wilhelm Engelmann.

Sage, R. F., Christin, P. A. & Edwards, E. J. (2011). The C4 plant lineages of planet Earth. *Journal of Experimental Botany*, **62**, 3155–69.

Salisbury, B. A. & Kim, J. (2001). Ancestral state estimation and taxon sampling density. *Systematic Biology*, **50**, 557–64.

Salmon, A. & Ainouche, M. L. (2010). Polyploidy and DNA methylation: new tools available. *Molecular Ecology*, **19**, 213–15.

Salmon, A., Ainouche, M. L. & Wendel, J. F. (2005). Genetic and epigenetic consequences of recent hybridization and polyploidy in *Spartina* (Poaceae). *Molecular Ecology*, **14**, 1163–75.

Salmon, A. *et al.* (2010). Homoeologous non-reciprocal recombination in polyploid cotton. *New Phytologist*, **186**, 123–34.

Salmon, S. C. & Hanson, A. A. (1964). *The principles and practice of agricultural research*. London: Leonard Hill.

Salse, J. *et al.* (2008). Identification and characterization of shared duplications between Rice and Wheat provide new insight into grass genome evolution. *Plant and Cell*, **20**, 11–24.

Samis, K. E. *et al.* (2012). Latitudinal trends in climate drive flowering time clines in North American *Arabidopsis thaliana*. *Ecology and Evolution*, **2**, 1162–80.

Sanderson, M. J. (2002). Estimating absolute rates of molecular evolution and divergence times: a penalized likelihood approach. *Molecular Ecology*, **19**, 101–9.

Sanderson, M. J. (2007). Construction and annotation of large phylogenetic trees. *Australian Systematic Botany*, **20**, 287–301.

Sandler, R. (2009). The value of species and the ethical foundations of assisted colonization. *Conservation Biology*, **24**, 424–31.

Sang, T. & Ge, S. (2007). The puzzle of Rice domestication. *Journal of Integrative Biology*, **49**, 760–8.

Sanmartín, I., Wanntorp, L. & Winkworth, R. C. (2007). West wind drift revisited: testing for directional dispersal in the southern hemisphere using event-based tree fitting. *Journal of Biogeography*, **34**, 398–416.

Sapp, J. (2003). *Genesis: the evolution of biology*. Oxford & New York: Oxford University Press.

Sarasan, V. *et al.* (2006). Conservation in vitro of threatened plants – progress in the past decade. *In Vitro Cellular & Developmental Biology -Plant*, **42**, 206–14.

Sarhanova, P. *et al.* (2012). New insights into the variability of reproduction modes in European populations of *Rubus* subgen. *Rubus:* (how sexual are polyploid brambles? *Sex Plant Reproduction*, **25**, 319–35.

Sarkar, P. & Stebbins, G. L. (1956). Morphological evidence concerning the origin of the B genome in Wheat. *American Journal of Botany*, **43**, 297–304.

Särkinen, T. *et al.* (2012). How to open the treasure chest? Optimising DNA extraction from herbarium specimens. *PLOS ONE*, **7**, 1–9.

Saucy, F. *et al.* (1999). Preferences for acyanogenic white clover (*Trifolium repens*) in the vole *Arvicola terrestris*. I. Preliminary results with two varieties (Ladino and Aran) and complementary tests with the slugs *Arion ater* and *A. subfuscus*. *Journal of Chemical Ecology*, **25**, 1441–54.

Sauer, J. D. (1988). *Plant migration*. Berkeley: University of California Press.

Saunders, E. R. (1897). On discontinuous variation occurring in *Biscutella laevigata*. *Proceedings of the Royal Society, B*, **62**, 11–26.

Savolainen, V. *et al.* (2005). Towards writing the Encyclopedia of Life: an introduction to DNA barcoding. *Philosophical Transactions of the Royal Society of London, B*, **359**, 1805–11.

Sax, D. F., Stachowicz, J. J. & Gaines, S. D. (2006). *Species invasions: insights into ecology, evolution and biogeography*. Sunderland, MA: Sinauer.

Sayre, A. (1975). *Rosalind Franklin and DNA.* New York & London: Norton & Co.

Scascitelli, M. *et al.* (2010). Genome scan of hybridizing sunflowers from Texas (*Helianthus annuus* and *H. debilis*) reveals asymmetric patterns of introgression and small islands of genomic differentiation. *Molecular Ecology*, **19**, 521–41.

Scascitelli, M., Cognet, M. & Adams, K. L. (2010). An interspecific plant hybrid shows novel changes in parental splice forms of genes for splicing factors. *Genetics*, **184**, 975–83.

Schaal, B. A. (1980). Measurement of gene flow in *Lupinus texensis*. *Nature*, **284**, 450–1.

Schaal, B. A. (1988). Somatic variation and genetic structure in plant populations. In *Plant population ecology*, ed. A. J. Davy, M. J. Hutchings & A. R. Watkinson, pp. 47–58. Oxford: Blackwell.

Schaal, B. A., Leverich, W. J. & Rogstad, S. H. (1991). A comparison of methods for assessing genetic variation in plant conservation biology. In *Genetics and conservation of rare plants*, ed. D. A. Falk & K. E. Holsinger, pp. 123–34. New York: Oxford University Press.

Schaal, B. A., O'Kane, S. L. & Rogstad, S. H. (1991). DNA in plant populations. *Trends in Ecology and Evolution*, **6**, 329–33.

Schat, H., Vooijs, R. & Kuiper, E. (1996). Identical major gene loci for heavy metal tolerances that have independently evolved in different local populations and subspecies of *Silene vulgaris*. *Evolution*, **50**, 1888–95.

Schatlowski, N. & Köhler, C. (2012). Tearing down barriers: understanding the molecular mechanisms of interploidy hybridizations. *Journal of Experimental Botany*, **63**, 6059–67.

Schatz, G. E. (2009). Plants on the IUCN Red List: setting priorities to inform conservation. *Trends in Plant Science*, **14**, 638–42.

Schiffers, K. *et al.* (2013). Limited evolutionary rescue of locally adapted populations facing climate change. *Philosophical Transactions of the Royal Society, B*, **368**, 20120083.

Schlichting, C. D. (1986). The evolution of phenotypic plasticity in plants. *Annual Review of Ecology and Systematics*, **17**, 667–93.

Schlichting, C. D. & Levin, D. A. (1984). Phenotypic plasticity of annual *Phlox*: tests of some hypotheses. *American Journal of Botany*, **71**, 252–60.

Schlising, R. A. & Turpin, R. A. (1971). Hummingbird dispersal of *Delphinium cardinale* pollen treated with radioactive iodine. *American Journal of Botany*, **58**, 401–6.

Schmalhausen, I. I. (1949) *Factors of evolution*, trans. I. Dordick, ed. T. Dobzhansky. Philadelphia: Blakiston.

Schmidt, J. (1899). Om ydre faktorers indflydelse paa løvbladets anatomiske bygning hos en af vore strandplanter. *Botanisk Tidsskrift*, **22**, 145–65.

Schmitz, R. J. *et al.* (2011). Transgenerational epigenetic instability is a source of novel methylation variants. *Science*, **334**, 369–73.

Schnable, P. S. & Springer, N. M. (2013). Progress toward understanding heterosis in crop plants. *Annual Review of Plant Biology*, **64**, 71–88.

Schoen, D. J. & Lloyd, D. G. (1984). The selection of cleistogamy and heteromorphic diaspores. *Biological Journal of the Linnean Society*, **23**, 303–22.

Schön, I., Martens, K. & van Dijk, P. (eds.) (2009). *Lost sex: the evolutionary biology of parthenogenesis.* Dordrecht, Heidelberg, London & New York: Springer.

Schönswetter, P. *et al.* (2005). Molecular evidence for glacial refugia of mountain plants in the European Alps. *Molecular Ecology*, **14**, 3547–55.

Schorr, G. *et al.* (2012). Integrating species distribution models (SDMs) and phylogeography for two species of Alpine *Primula*. *Ecology and Evolution*, **2**, 1260–77.

Schrödinger, E. (1944). *What is life?* London: Cambridge University Press; New York: Macmillan.

Schulte, P. *et al.* (2010). The Chicxulub Asteroid impact and mass extinction at the Cretaceous–Paleogene boundary. *Science*, **327**, 1214–18.

Schupp, E. W., Jordano, P. & Gomaz, J. M. (2010). Seed dispersal effectiveness revisited: a conceptual review. *New Phytologist*, **188**, 333–53.

Schwaegerle, K. E. & Schaal, B. A. (1979). Genetic variability and founder effect in the Pitcher Plant *Sarracenia purpurea* L. *Evolution*, **33**, 1210–18.

Schwanitz, F. (1966). *The origin of cultivated plants.* Cambridge, MA.: Harvard University Press.

Schwartz, M. W. *et al.* (2012a). Managed relocation: integrating the scientific, regulatory, and ethical challenges. *Bioscience*, **62**, 732–43.

Schwartz, M. W. *et al.* (2012b). Predicting extinctions as a result of climate change. *Ecology*, **87**, 1611–15.

Schweber, S. S. (1977). The origin of the origin revisited. *Journal of the History of Biology*, 10, 229–316.

Scotese, C. R. (2004). Cenozoic and Mesozoic paleogeography: changing terrestrial biogeographic pathways. In *Frontiers of biogeography: new directions in the geography of nature*, ed. M. V. Lomolino & L. R. Heany, pp. 9–26. Sunderland, MA: Sinauer Associates.

Scott, J. M. & Schipper, J. (2006). Gap analysis: a spatial tool for conservation planning. In *Principles of conservation biology*, ed. M. J. Groom, G. K. Meffe & C. R. Carroll, 3rd edn, pp. 518–19. Sunderland, MA: Sinauer.

Seavey, S. R. & Bawa, K. S. (1986). Late-acting self-incompatibility in Angiosperms. *Botanical Review*, 52, 195–214.

Seberg, O. *et al.* (2003). Short cuts in systematics? A commentary on DNA-based taxonomy. *Trends in Taxonomy and Evolution*, 18, 63–5.

Seddon, P. J. (2010). From reintroduction to assisted colonization: moving along the conservation translocation spectrum. *Restoration Ecology*, 18, 796–802.

Seidel, C. F. (1879). Ueber Verwachsungen von Stämmen und Zweigen von Holzgewächsen und ihren Einfluss auf das Dickenwachsthum der betreffenden Theile. *Naturwissenschaftliche Gesellschaft Isis, Dresden, Sitzber*, 161–8.

Sellars, R. W. (1997). *Preserving nature in the National Parks: a history*. New Haven, CT, & London: Yale University Press.

Sepkoski, D. (2009). The origin and early reception of punctuated equilibrium. In *The paleobiological revolution: essays on the history of recent paleontology*, ed. D. Sepkoski and M. Ruse, pp. 301–25. Chicago: University of Chicago Press.

Sepkoski, D. & Ruse, M. (eds.) (2009). *The paleobiological revolution: essays on the history of recent paleontology*. Chicago: University of Chicago Press.

Sepp, S. *et al.* (2000). Genetic polymorphism detected with RAPA analysis and morphological variability in some microspecies of apomictic *Alchemilla*. *Anales Botanici Fennici*, 37, 105–23.

Sexton, J. P., Strauss, S. Y. & Rice, K. G. (2011). Gene flow increases fitness at the warm edge of a species' range. *Proceedings of the National Academy of Sciences, USA*, 108, 11704–9.

Shafer, A. B. A. *et al.* (2010). Of glaciers and refugia: a decade of study sheds new light on the phylogeography of northwestern North America. *Molecular Ecology*, 19, 4589–4621.

Shafer, C. L. (1990). *Nature reserves*. Washington DC: Smithsonian Institution Press.

Shaffer, M. L. (1981). Minimum population sizes for species conservation. *Bioscience*, 31, 131–4.

Shaked, H. *et al.* (2001). Sequence elimination and cytosine methylation are rapid and reproducible responses of the genome to wide hybridization and allopolyploidy in Wheat. *Plant and Cell*, 13, 1749–59.

Shapcott, A. (1998). The genetics of *Ptychosperma bleeseri*, a rare palm from the Northern Territory, Australia. *Biological Conservation*, 85, 203–9.

Shapiro, B. & Hofreiter, M. (eds.) (2012). *Ancient DNA: methods and protocols*. New York: Humana Press.

Sharbel, T. F. *et al.* (2005). Biogeographic distribution of polyploidy and B chromosomes in the apomictic *Boechera holboellii* complex. *Cytogenetic and Genome Research*, 109, 283–92.

Sharrock, S. & M. Jones. (2011). Saving Europe's threatened flora: progress towards GSPC Target 8 in Europe. *Biological Conservation*, 20, 325–33.

Sheail, J., Treweek, J. R. & Mountford, J. O. (1997). The UK transition from nature preservation to 'creative conservation'. *Environmental Conservation*, 24, 224–35.

Sheffield, E., Wolf, P. G., Rumsey, F. J., Robson, D. J., Ranker, T. A. & Challiner, S. M. (1993). Spatial distribution and reproductive behaviour of a triploid bracken (*Pteridium aquilinum*) clone in Britain. *Annals of Botany*, 72, 231–7.

Shepherd, L. D., De Lange, P. J. & Perrie, L. R. (2009). Multiple colonizations of a remote oceanic archipelago by one species: how common is long-distance dispersal? *Journal of Biogeography*, 36, 1972–7.

Sherman, M. (1946). Karyotype evolution: a cytogenetic study of seven species and six interspecific hybrids of *Crepis*. *University of California Publications in Botany*, 18, 369–408.

Shindo, C., Bernasconi, G. & Hardtke, C. S. (2007). Natural genetic variation in *Arabidopsis*: tools, traits and prospects. *Annals of Botany*, 99, 1043–54.

Shirley, P. D. & Lamberti, G. A. (2010). Assisted colonization under the US Endangered Species Act. *Conservation Letters*, 3, 45–52.

Shivas, M. G. (1961a). Contributions to the cytology and taxonomy of species of *Polypodium* in Europe and

America. 1. Cytology. *Journal of the Linnean Society*, **58**, 13–25.

Shivas, M. G. (1961b). Contributions to the cytology and taxonomy of species of *Polypodium* in Europe and America. 2. Taxonomy. *Journal of the Linnean Society*, **58**, 27–38.

Shore, J. S. & Barrett, S. C. H. (1985). The genetics of distyly and homostyly in *Turnera ulmifolia* L. (Turneraceae). *Heredity*, **55**, 167–74.

Sicard, A. & Lenhard, M. (2011). The selfing syndrome: a model for studying the genetic and evolutionary basis of morphological adaptation in plants. *Annals of Botany*, **107**, 1433–43.

Silvertown, J. (1984). Phenotypic variety in seed germination behaviour: the ontogeny and evolution of somatic polymorphism in seeds. *American Naturalist*, **124**, 1–16.

Silvertown, J. (1992). An experimental test of frequency-dependent fitness in mixtures of the two seed morphs of *Spergula arvensis*. *Acta Oecologica*, **13**, 627–34.

Silvertown, J. (2001). Plants stand still, but their genes don't: non-trivial consequences of the obvious. In *Integrating ecology and evolution in a spatial context*, ed. J. Silvertown & J. Antonovics, pp. 3–20. Oxford: Blackwell.

Silvertown, J. & Charlesworth, D. (2001). *Introduction to plant population biology*. Malden, MA, & Oxford: Blackwell Publishing.

Silvertown, J. W. & Lovett Doust, J. (1993). *Introduction to plant population biology*. Oxford: Blackwell.

Silvertown, J. *et al.* (2005). Reinforcement of reproductive isolation between adjacent populations in the Park Grass Experiment. *Heredity*, **95**, 198–205.

Silvertown, J. *et al.* (2010). Environmental myopia: a diagnosis and a remedy. *Trends in Ecology & Evolution*, **25**, 556–61.

Simberloff, D. (2001). Introduced species, effects and distribution of. In *Encyclopedia of biodiversity*, vol. **3**, ed. S. A. Levin, pp. 517–29. San Diego, CA: Academic Press.

Simberloff, D. (2013). *Invasive species: what everyone needs to know*. Oxford: Oxford University Press.

Simberloff, D. & Rejmánek, M. (eds.) (2011). *Encyclopedia of biological invasions*, Encyclopedias of the Natural World, No. 3. Berkeley, CA: University of California Press.

Simberloff, D. S., Farr, J. A., Cox, J. & Mehlman, D. W. (1992). Movement corridors: conservation bargains or poor investments? *Conservation Biology*, **6**, 493–504.

Simmonds, N. W. (ed.) (1976). *Evolution of crop plants*. London: Longmans.

Simpson, D. M. (1954). Natural cross-pollination in Cotton. *U.S. Department of Agriculture Technical Bulletin*, No. 1094.

Simpson, G. G. (1944). *Tempo and mode in evolution*. New York: Columbia University Press.

Simpson, G. G. (1961). *Principles of animal taxonomy. The species and lower categories*. New York: Columbia University Press.

Simpson, G. G. (1980). *Splendid isolation: the curious history of South American mammals*. New Haven, CT: Yale University Press.

Simpson, M. G. (1986). Phylogeny and structural evolution of plants. In *Fundamentals of plant systematics*, ed. A. E. Radford, pp. 217–48. New York: Harper & Row.

Simunek, M., Hossfeld, U. & Breidbach, O. (2012). 'Further development' of Mendel's legacy? Erich von Tschermak-Seysenegg in the context of the Mendelian-biometry controversy, 1901–1906. *Theory in Biosciences*, **131**, 243–52.

Sindu, A. S. & Singh, S. (1961). Studies on the agents of cross pollination of Cotton. *Indian Cotton Growing Review*, **15**, 341–53.

Siol, M., Wright, S. I. & Barrett, S. C. H. (2010). The population genomics of plant adaptation. *New Phytologist*, **188**, 313–32.

Sletvold, N. *et al.* (2010). Cost of trichome production and resistance to a specialist insect herbivore in *Arabidopsis lyrata*. *Evolutionary Ecology*, **24**, 1307–19.

Sluys, R. (2013). The unappreciated, fundamentally analytic nature of taxonomy and implications for the inventory of biodiversity. *Biodiversity and Conservation*, **22**, 1095–105.

Smartt, J. O. & Simmonds, N. W. (1995). *Evolution of crop plants*, 2nd edn. Harlow: Longmans.

Smith, A. (1965). The assessment of patterns of variation in *Festuca rubra* L. in relation to environmental gradients. *Scottish Plant Breeding Station Record*, 1965, 163–95.

Smith, A. (1972). The pattern of distribution of *Agrostis* and *Festuca* plants of various genotypes in a sward. *New Phytologist*, **71**, 937–45.

Smith, A. C. (1957). Fifty years of botanical nomenclature. *Brittonia*, **9**, 2–8.

Smith, A. M. *et al.* (2010). *Plant biology*. New York & Abingdon, UK: Garland Science, Taylor & Francis.

Smith, C. H. & Beccaloni, G. (eds.) (2008). *Natural selection and beyond: the intellectual legacy of Alfred Russel Wallace.* Oxford: Oxford University Press.

Smith, D. C., Nielsen, E. L. & Ahlgren, H. L. (1946). Variation in ecotypes of *Poa pratensis. The Botanical Gazette,* **108**, 143–66.

Smith, D. M. & Levin, D. A. (1963). A chromatographic study of reticulate evolution in the Appalachian *Asplenium* complex. *American Journal of Botany,* **50**, 952–8.

Smith, F. D. M. *et al.* (1993). How much do we know about the current extinction rate? *Trends in Ecology and Evolution,* **8**, 375–8.

Smith, G. L. (1963a). Studies in *Potentilla* L. 1. Embryological investigations into the mechanism of agamospermy in British *P. tabernaemontani* Aschers. *New Phytologist,* **62**, 264–82.

Smith, G. L. (1963b). Studies in *Potentilla* L. 2. Cytological aspects of apomixis in *P. crantzii* (Cr.) Beck ex Fritsch. *New Phytologist,* **62**, 283–300.

Smith, G. L. (1971). Studies in Potentilla L. 3. Variation in British *P. tabernaemontani* Aschers. and *P. crantzii* (Cr.) Beck ex Fritsch. *New Phytologist,* **70**, 607–18.

Smith, J. (1841). Notice of a plant which produces perfect seeds without any apparent action of pollen. *Transactions of the Linnean Society of London,* **18**, 509–12.

Smith, P. M. (1976). *The chemotaxonomy of plants.* London: Arnold.

Smith, S. (1960). The origin of the Origin. *Advancement of Science,* **16**, 391–401.

Smith, S. A. & Donoghue, M. J. (2010). Combining historical biogeography with niche modeling in the Caprifolium Clade of *Lonicera* (Caprifoliaceae, Dipsacales). *Systematic Biology,* **590**, 322–41.

Smith S. A., Beaulieu J. M. & Donoghue M. J. (2010). An uncorrelated relaxed-clock analysis suggests an earlier origin for flowering plants. *Proceedings of the National Academy of Sciences, USA,* **107**, 5897–902.

Smocovitis, V. B. (1996). *Unifying biology: the evolutionary synthesis and evolutionary biology.* Princeton, NJ: Princeton University Press.

Smocovitis, V. B. (2006). Keeping up with Dobzhansky: G. Ledyard Stebbins, Jr., plant evolution and the evolutionary synthesis. *History & Philosophy of the Life Sciences,* **28**, 9–48.

Smyth, C. A. & Hamrick, J. L. (1987). Realised gene flow via pollen in artificial populations of Musk Thistle, *Carduus nutans. Evolution,* **39**, 53–65.

Snaydon, R. W. (1970). Rapid population differentiation in a mosaic environment. 1. The response of *Anthoxanthum odoratum* populations to soils. *Evolution,* **24**, 257–69.

Snaydon, R. W. (1976). Genetic change within species. In *The Park Grass experiment on the effect of fertilisers and liming on the botanical composition of permanent grassland and on the yield of hay,* Thurston, J. M., Dyke, G. V. & Williams, E. D., Appendix. Harpenden: Rothamsted Experimental Station.

Snaydon, R. W. (1978). Genetic changes in pasture populations. In *Plant relations in pastures,* ed. J. R. Wilson, pp. 253–69. Melbourne: CSIRO.

Snaydon, R. W. & Davies, M. S. (1972). Rapid population differentiation in a mosaic environment. II. Morphological variation in *Anthoxanthum odoratum. Evolution,* **26**, 390–405.

Sneath, P. H. A. (1962). The construction of taxonomic groups. In *Microbial classification,* ed. G. C. Ainsworth & P. H. A. Sneath. pp. 289–332. Cambridge: Cambridge University Press.

Sneath, P. H. A. (1988). The phenetic and cladistic approaches. In *Prospects in systematics.* The Systematic Association Special Volume No. 36, ed. D. L. Hawksworth, pp. 252–73. Oxford: Clarendon Press.

Sneath, P. H. A. (1995). Thirty years of numerical taxonomy. *Systematic Biology,* **44**, 281–98.

Sneath, P. H. A. & Sokal, R. R. (1973). *Numerical taxonomy.* San Francisco: Freeman.

Snedecor, G. W. & Cochran, W. G. (1980). *Statistical methods,* 7th edn. Ames, IA: Iowa State University Press.

Snow, A. (2012). Illegal gene flow from transgenic creeping bent grass: the saga continues. *Molecular Ecology,* **21**, 4663–4.

Snow, A. A. *et al.* (2003). A Bt transgene reduces herbivory and enhances fecundity in wild sunflowers. *Ecological Applications,* **13**, 279–86.

Snyder, L. A. (1950). Morphological variability and hybrid development in *Elymus glaucus. American Journal of Botany,* **37**, 628–35.

Snyder, L. A. (1951). Cytology of inter-strain hybrids and the probable origin of variability in *Elymus glaucus. American Journal of Botany,* **38**, 195–202.

Sokal, R. R. & Crovello, T. J. (1970). The biological species concept: a critical evaluation. *American Naturalist*, **104**, 127–53.

Sokal, R. R. & Rohlf, F. J. (1969). *Biometry. The principles and practice of statistics in biological research.* [2nd edn 1981.] San Francisco: Freeman.

Sokal, R. R. & Sneath, P. H. A. (1963). *Principles of numerical taxonomy.* London & San Francisco: Freeman.

Sokhi, R. S. (2011). *World atlas of atmospheric pollution.* London, New York & Delhi: Anthem Press.

Solbrig, O. T. (1994). Biodiversity: an introduction. In *Biodiversity and global change*, ed. O. T. Solbrig, H. M. van Emden & P. G. W. J. van Oordt, pp. 13–20. Wallingford, UK: CAB International.

Solbrig, O. T. & Solbrig, D. J. (1979). *Introduction to population biology and evolution.* London: Addison-Wesley Publishing Company.

Soltis, D. E. & Burleigh, J. G. (2009). Surviving the K-T mass extinction: new perspectives of polyploidization in angiosperms. *Proceedings of the National Academy of Sciences, USA*, **106**, 5455–6.

Soltis, D. E. & Soltis, P. S. (1989). Allopolyploid speciation in *Tragopogon*: insights from chloroplast DNA. *American Journal of Botany*, **76**, 1119–24.

Soltis, D. E. & Soltis, P. S. (1990). *Isozymes in plant biology.* London: Chapman & Hall.

Soltis, D. E. & Soltis, P. S. (1993). Molecular data and the dynamic nature of polyploidy. *Critical Reviews in Plant Science*, **12**, 243–73.

Soltis, D. E. & Soltis, P. S. (1995). The dynamic nature of polyploid genomes. *Proceedings of the National Academy of Sciences, USA*, **92**, 8089–91.

Soltis, D. E. & Soltis, P. S. (1999). Polyploidy: recurrent formation and genome evolution. *Trends in Ecology and Evolution*, **14**, 348–52.

Soltis, P. S. & Soltis, D. E. (2004). The origin and diversification of angiosperms. *American Journal of Botany*, **91**, 1614–26.

Soltis, D. E., Soltis, P. S. & Tate, J. A. (2004). Advances in the study of polyploidy since *Plant speciation. New Phytologist*, **161**, 173–91.

Soltis, D. S. *et al.* (2005). *Phylogeny and evolution of angiosperms.* Sunderland, MA: Sinauer.

Soltis, D. E. *et al.* (2005). *Phylogeny, evolution, and classification of flowering plants.* Sunderland, MA: Sinauer.

Soltis, D. E. *et al.* (2007). Autopolyploidy in angiosperms: have we grossly underestimated the number of species? *Taxon*, **56**, 13–30.

Soltis, D. E. *et al.* (2008). Origin and early evolution of Angiosperms. *Annals of the New York Academy of Sciences*, **1133**, 3–25.

Soltis, D. E. *et al.* (2009). Polyploidy and angiosperm diversification. *American Journal of Botany*, **96**, 336–48.

Soltis, D. E. *et al.* (2010). What we still don't know about polyploidy. *Taxon*, **60**, 324–32.

Soltis, D. E. *et al.* (2011). Angiosperm phylogeny: 17 genes, 640 taxa. *American Journal of Botany*, **98**, 704–30.

Soltis, D. E. *et al.* (2012). The early stages of polyploidy: rapid and repeated evolution in *Tragopogon*. In *Polyploidy and genome evolution*, ed. P. S. Soltis & D. E. Soltis, pp. 271–92. Heidelberg: Springer.

Soltis, D. E., Visger, C. J. & Soltis, P. S. (2014). The polyploidy revolution then . . . and now: Stebbins revisited. *American Journal of Botany*, **101**, 1057–78.

Soltis, P. S. & Soltis, D. E (2009). The role of hybridization in plant speciation. *Annual Review of Plant Biology*, **60**, 561–88.

Soltis, P. S., Soltis, D. E. & Doyle, J. J. (1992). *Molecular systematics of plants.* New York: Chapman & Hall.

Soltis, P. S., Plunkett, G. M., Novak, S. J. & Soltis, D. E. (1995). Genetic variation in *Tragopogon* species: additional origins of the allopolyploids *T. mirus* and *T. miscellus* (Compositae). *American Journal of Botany*, **82**, 1329–41.

Soltis, P. S. & Soltis, D. E. (1991). Multiple origins of the allotetraploid *Tragopogon mirus* (Compositae): rDNA evidence. *Systematic Botany*, **16**, 407–13.

Soltis, P. S. *et al.* (2009). Floral variation and floral genetics in basal angiosperms. *American Journal of Botany*, **96**, 110–28.

Soltis, P. S. *et al.*, (2014). Polyploidy and novelty: Gottlieb's legacy. *Philosophical Transactions of the Royal Society*, **369**; 20130351.http://dx.doi.org/10.1098/rstb.2012.0351.

Solymosi P. & Lehoczki, E. (1989a). Characterization of a triple (atrazine-pyrazon-pyridate) resistant biotype of Common Lambs Quarters (*Chenopodium album* L.) *Journal of Plant Physiology*, **134**, 685–90.

Solymosi, P. & Lehoczki, E. (1989b). Co-resistance of atrazine-resistant *Chenopodium* and *Amaranthus* biotypes to other photosystem II inhibiting herbicides. *Zeitschrift für Naturforschung*, **44C**, 119–27.

Song, B. H. *et al.* (2003). Cytoplasmic composition in *Pinus densata* and population establishment of a diploid hybrid pine. *Molecular Ecology*, 12, 2995–3001.

Song, K. M., Osborn, T. C. & Williams, P. H. (1988). *Brassica* taxonomy based on nuclear restriction length polymorphisms (RFLPs). 1. Genome evolution of diploid and amphidiploid species. *Theoretical and Applied Genetics*, 75, 784–94.

Soorae, P. S. (ed.) (2011). *Global re-introduction perspectives: 2011. More case studies from around the globe.* Gland, Switzerland: IUCN/SSC.

Sørenson, T. & Gudjónsson, G. (1946). Spontaneous chromosome-aberrants in apomictic Taraxaca. *Biologiske Skrifter, K. Danske Videnskabernes Selskab*, 4, No. 2.

Sork, V. L. (1984). Examination of seed dispersal and survival in red oak, *Quercus rubra* (Fagaceae), using metal-tagged acorns. *Ecology*, 65, 1020–2.

Soukup, J. & Holec, J. (2004). Crop–wild interaction within the *Beta vulgaris* complex: agronomic aspects of weed beet in the Czech Republic. In *Introgression from genetically modified plants into wild relatives*, ed. H. C. M. den Nijs, D. Bartsch & J. Sweet, pp. 203–218. Wallingford, UK: CAB International.

Soulé, M. E. (1980). Thresholds for survival: maintaining fitness and evolutionary potential. In *Conservation: an evolutionary–ecological perspective*, ed. M. E. Soulé & B. A. Wilcox, pp. 151–70. Sunderland, MA: Sinauer.

Soutullo, A. (2010). Extent of the global network of terrestrial protected areas. *Conservation Biology*, 24, 362–3.

Specht, C. D. & Bartlett, M. E. (2009). Flower evolution: the origin and subsequent diversification of the angiosperm flower. *Annual Review of Ecology, Evolution & Systematics*, 40, 217–43.

Spencer, K. C. (1988). *Chemical mediation of coevolution.* San Diego, CA: Academic Press.

Sprengel, C. K. (1793). *Das entdeckte Geheimniss der Natur im Bau und in der Befruchtung der Blumen.* Berlin.

Srb, A. M. & Owen, R. D. (1958). *General genetics.* San Francisco: Freeman.

Stace, C. A. (1975). *Hybridization and the flora of the British Isles.* London: Academic Press.

Stace, C. A. (1980). *Plant taxonomy and biosystematics.* London: Edward Arnold.

Stace, C. A. (1989). *Plant taxonomy and biosystematics*, 2nd edn. London: Edward Arnold.

Stace, C. A. (1993). The importance of rare events in polyploid evolution. In *Evolutionary patterns and processes*, ed. D. R. Lees & D. Edwards. Linnean Society Symposium Series, 14, 157–68. London: Published for the Linnean Society by Academic Press.

Stace, C. A. (2010a). *New flora of the British Isles*, 3rd edn. Cambridge: Cambridge University Press.

Stace, C. A. (2010b). The new molecular classification: relevance to the flora of the British Isles. *Botanical Society of the British Isles: BSBI News*, 113, 4–6.

Stace, C. A. (2010c). Classification by molecules: what's in it for field botanists? *Watsonia*, 28: 103–22

Stace, C. A., Preston, C. D. & Pearman, D. A. (2015). *Hybrid flora of the British Isles.* Bristol: Botanical Society of Britain and Ireland.

Stapledon, R. G. (1928). Cocksfoot grass (*Dactylis glomerata* L.): ecotypes in relation to the biotic factor. *Journal of Ecology*, 16, 72–104.

Stearn, W. T. (1957). *Introduction to facsimile edition of Linnaeus' Species plantarum.* London: Ray Society, British Museum.

Stearns, S. C. (1992). *The evolution of life histories.* Oxford: Oxford University Press.

Stebbins, G. L. (1947). Types of polyploids: their classification and significance. *Advances in Genetics*, 1, 403–29.

Stebbins, G. L. (1950). *Variation and evolution in plants.* London: Oxford University Press; New York: Columbia University Press.

Stebbins, G. L. (1957). Self fertilization and population variability in the higher plants. *The American Naturalist*, 91, 337–54.

Stebbins, G. L. (1966). *Processes of organic evolution.* Englewood Cliffs, NJ: Prentice-Hall. [2nd edn 1971.]

Stebbins, G. L. (1971). *Chromosomal evolution in higher plants.* London: Arnold.

Stebbins, G. L. (1974). *Flowering plants. evolution above the species level.* London: Arnold; Cambridge, MA: Belknap Press.

Stebbins, G. L. (1984). Polyploidy and the distribution of the arctic–alpine flora: new evidence and a new approach. *Botanica Helvetica*, 94, 1–13.

Stebbins, G. L. & Dawe, J. C. (1987). Polyploidy and distribution in the European flora: a reappraisal. *Botanische Jahrbücher fur Systematik, Pflanzengeschichte und Pflanzengeographie*, 108, 343–54.

Stebbins, G. L. *et al.* (1963). Identification of the ancestry of an amphiploid *Viola* with the aid of paper chromatography. *American Journal of Botany*, 50, 830–9.

Steen, S. W. *et al.* (2000). Same parental species, but different taxa: molecular evidence for hybrid origins of the rare endemics *Saxifraga opdalensis* and *S. svalbardensis* (Saxifragaceae). *Botanical Journal of the Linnean Society*, **132**, 153–64.

Steffen, W. *et al.* (2011). The Anthropocene: conceptual and historical perspectives. *Philosophical Transactions of the Royal Society, A*, **369**, 842–67.

Stegemann, S. *et al.* (2012). Horizontal transfer of chloroplast genomes between plant species. *Proceedings of the National Academy of Sciences, USA*, **109**, 2434–8.

Stehli, F. G. & Webb, S. D. (eds.) (1985). *The great American biotic interchange*. New York & London: Plenum Press.

Steinger, T., Körner, C. & Schmid, B. (1996). Long-term persistence in a changing climate: DNA analysis suggests very old ages of clones of alpine *Carex curvula*. *Oecologia*, **105**, 94–9.

Stelleman, P. (1978). The possible role of insect visits in pollination of reputedly anemophilous plants, exemplified by *Plantago lanceolata* and syrphid flies. In *The pollination of flowers by insects*, ed. A. J. Richards, pp. 41–6. London: Academic Press.

Sterck, L. *et al.* (2007). How many genes are there in plants (... and why are they there)? *Current Opinion in Plant Biology*, **10**, 199–203.

Sterelny, K. (2007). *Dawkins vs Gould: survival of the fittest*. Cambridge: Icon Books.

Stern, C. & Sherwood, E. R. (1966). *The origin of genetics. A Mendel source book*. London & San Francisco: Freeman.

Sternberg, L. (1976). Growth forms of *Larrea tridentata*. *Madroño*, **23**, 408–17.

Stevens, P. F. (2006). An end to all things? – plants and their names. *Australian Systematic Botany*, **19**, 115–33.

Stevens, P. F. (2012). Angiosperm Phylogeny Website – Missouri Botanical Garden www.mobot.org/MOBOT/research/APweb/

Stewart, J. R. *et al.* (2010). Refugia revisited: individualistic responses of species in space and time. *Proceedings of the Royal Society of London, B*, **277**, 661–71.

Stewart, R. N. (1947). The morphology of somatic chromosomes in *Lilium*. *American Journal of Botany*, **34**, 9–26.

Stift, M. *et al.* (2008). Segregation models for disomic, tetrasomic and intermediate inheritance in tetraploids: a general procedure applied to *Rorippa* (Yellow Cress) microsatellite data. *Genetics*, **179**, 2113–23.

Stigler, S. M. (1986). *The history of statistics: the measurement of uncertainty before 1900*. Cambridge, MA: Belknap Press of Harvard University Press.

Stiller, J. W., Reel, D. C. & Johnson, J. C. (2003). A single origin of plastids revisited: convergent evolution in organellar gene content. *Journal of Phycology*, **39**, 95–105.

Stoeckle, M. Y. *et al.* (2011). Commercial teas highlight plant DNA barcode identification successes and obstacles. *Scientific Reports*, **1**, 42, DOI:101038/srep00042.

Stone, J. L., Thomson, J. D. & Dent-Acosta, S. J. (1995). Assessment of pollen viability in hand-pollination experiments: a review. *American Journal of Botany*, **82**, 1186–97.

Stork, N. E. (2010). Re-assessing current extinction rates. *Biodiversity and Conservation*, **19**, 357–71.

Stott, R. (2012). *Darwin's ghosts: in search of the first evolutionists*. London: Bloomsbury.

Stowe, M. K. (1988). Chemical mimicry. In *Chemical mediation of coevolution*, ed. K. C. Spencer, pp. 513–80. San Diego, CA: Academic Press.

Strasburg, J. L. *et al.* (2012). What can patterns of differentiation across plant genomes tell us about adaptation and speciation? *Philosophical Transactions of the Royal Society of London*, **367**, 364–73.

Strasburger, E. (1910). Chromosomenzahl. *Flora*, **100**, 398–446.

Strauss, W. & Mainwaring, S. J. (1984). *Air pollution*. London: Edward Arnold.

Streisfeld, M. A, Young, W. N. & Sobel, J. M. (2013). Divergent selection drives genetic differentiation in an R2R3-MYB transcription factor that contributes to incipient speciation in *Mimulus aurantiacus*. *PLOS Genetics*, **10**, DOI:10.1371/journal.pgen.1003385.

Strid, A. (1970). Studies in the Aegean flora. XVI. Biosystematics of the *Nigella arvensis* complex. With special reference to the problem of non-adaptive radiation. *Opera Botanica*, No. 28. Lund: Gleerup.

Stuart, A. (1984). *The ideas of sampling*. High Wycombe: Charles Griffin.

Stuart, S. N. *et al.* (2010). The barometer of life. *Science*, **328**, 177.

Stuessy, T. F. (1990). *Plant taxonomy: the systematic evaluation of comparative data.* New York: Columbia University Press.

Stuessy, T. F. (2004). A transitional-combinational theory of the origin of the angiosperms. *Taxon*, **53**, 3–16.

Stuessy, T. F. (2009). *Plant taxonomy: the systematic evaluation of comparative data*, 2nd edn. New York: Columbia University Press.

Stuessy, T. F. (2010). Paraphyly and the origin and classification of angiosperms. *Taxon*, **59**, 689–93.

Sturtevant, A. H. (1965). *A history of genetics.* New York: Harper Row.

Suda, J. *et al.* (2007). Complex distribution patterns of di-, tetra-, and hexaploid cytotypes in the European high mountain plant *Senecio carniolicus* (Asteraceae). *American Journal of Botany*, **94**, 1391–1401.

Sugii, N. & Lamoureux, C. (2004). Tissue culture as a conservation method: an empirical view from Hawaii. In *Ex situ plant conservation*, ed. E. O. Guerrant, Jr, K. Havens & M. Maunder, pp. 189–205. Washington DC, Covelo, CA, & London: Island Press.

Sultan, S. E. (1987). Evolutionary implications of phenotypic plasticity in plants. *Evolutionary Biology*, **21**, 127–78.

Sutherland, W. J. & Hill, D. A. (eds.) (1995). *Managing habitats for conservation.* Cambridge: Cambridge University Press.

Sutherland, W. J. *et al.* (2004). The need for evidence-based conservation. *Trends in Ecology & Evolution*, **19**, 305–8.

Sutton, W. S. (1902). On the morphology of the chromosome group in *Brachystola magna. Biological Bulletin. Marine Biological Laboratory, Woods Hole, Mass.*, **4**, 24–39.

Sutton, W. S. (1903). The chromosomes in heredity. *Biological Bulletin, Marine Biological Laboratory, Woods Hole, Mass.*, **4**, 231–48.

Swensen, S. M. *et al.* (1995). Genetic analysis of the endangered island endemic *Malacothamnus fasciculatus* (Nutt.) Green var. *nesioticus* (Rob.) Kearn (Malvaceae). *Conservation Biology*, **9**, 404–15.

Swenson, U., Hill, R. S. & McLoughlin, S. (2001). Biogeography of *Nothofagus* supports the sequence of Gondwana break-up. *Taxon*, **50**, 1025–41.

Swofford, D. L. & Olsen, G. J. (1990). Phylogenetic reconstruction. In *Molecular systematics*, ed. D. K. Hillis & C. Moritz, pp. 411–501. Sunderland, MA: Sinauer.

Sybenga, J. (1996). Aneuploid and other cytological tester sets in Rye. *Euphytica*, **89**, 143–51.

Sytsma, K. J. & Schaal, B. A. (1985). Phylogenetics of the *Lisianthius skinneri* (Gentianaceae) species complex in Panama utilizing DNA restriction fragment analysis. *Evolution*, **39**, 594–608.

Syvanen, M. (2012). Evolutionary implications of horizontal gene transfer. *Annual Review of Genetics*, **46**, 341–58.

Szadkowski, E. *et al.* (2010). The first meiosis of resynthesized *Brassica napus*, a genome blender. *New Phytologist*, **186**, 102–12.

Täckholm, G. (1922). Zytologische studien über die Gattung *Rosa. Acta Horti Bergiani*, **7**, 97–381.

Taggart, J. B., McNally, S. F. & Sharp, P. M. (1990). Genetic variability and differentiation among founder populations of the Pitcher Plant (*Sarracenia purpurea* L.) in Ireland. *Heredity*, **64**, 177–83.

Tahara, M. (1915). Cytological studies on *Chrysanthemum. Botanical Magazine (Tokyo)*, **29**, 48–50.

Takayama, S. I. & Isogai, A. (2005). Self-incompatibility in plants. *Annual Review of Plant Biology*, **56**, 467–8.

Takhtajan, A. (1969). *Flowering plants – origin and dispersal.* Authorised translation from the Russian by C. Jeffrey. Edinburgh: Oliver & Boyd.

Takhtajan, A. L. (1987). Sistema Magnoliofitov (Systema Magnoliophytorum). *Acad. Sciences, USSR, Nauka, Leningrad.* [In Russian: see Taxon, 37, 422–4 (1988) for details.]

Talianova, M. & Janousek, B. (2011). What can we learn from tobacco and other Solanaceae about horizontal DNA transfer? *American Journal of Botany*, **98**, 1231–42.

Tang, C. *et al.* (2007). The evolution of selfing in *Arabidopsis thaliana. Science*, **317**, 1070–2.

Tang, H. *et al.* (2008a). Synteny and collinearity in plant genomes. *Science*, **320**, 486–8.

Tang, H. *et al.* (2008b). Unraveling ancient hexaploidy through multiply-aligned angiosperm gene maps. *Genome Research*, **18**, 1944–54.

Tansley, A. G. (1945). *Our heritage of wild nature: a plea for organized nature conservation.* Cambridge: Cambridge University Press.

Tardif, B. & Morisset, P. (1991). Chromosomal C-band variation in *Allium schoenoprasum* (Lilaceae) in Eastern-North America. *Plant Systematics & Evolution*, **174**, 125–37.

Tate, J. Soltis, D. E. & Soltis, P. S. (2005). Polyploidy in plants. In *The evolution of the genome*, ed. T. R. Gregory, pp. 371–426. London: Elsevier Academic Press.

Tate, J. A. *et al.* (2006). Evolution and expression of homeologous loci in *Tragopogon miscellus* (Asteraceae), a recent and reciprocally formed allopolyploid. *Genetics*, 173, 1599–1611.

Tate, J. A *et al.* (2009). On the road to diploidization? Homoelog loss in independently formed populations of the allopolyploid *Tragopogon miscellus* (Asteraceae). *BMC Plant Biology*, 9, 80, DOI:10.1186/1471-2229-9-80

Tautz, D. *et al.* (2003). A plea for DNA taxonomy. *Trends in Ecology and Evolution*, 18, 70–4.

Tayalé, A. & Parisod, C. (2013). Natural pathways to polyploidy in plants and consequences for genome reorganization. *Cytogenetic and Genome Research*, DOI:10.1159/000351318.

Taylor, G. E., Jr & Murdy, W. H. (1975). Population differentiation of an annual plant species, *Geranium carolinianum* in response to sulfur dioxide. *Botanical Gazette*, 136, 212–15.

Taylor, G. E., Pitelka L. F. & Clegg, M. T. (1991). *Ecological genetics and air pollution*. New York: Springer-Verlag.

Taylor, H. R. & Harris, W. E. (2012). An emergent science on the brink of irrelevance: a review of the past 8 years of DNA barcoding. *Molecular Ecology Resources*, 12, 377–88.

Taylor, J. S. & Raes, J. (2005). Small-scale gene duplications. In *The evolution of the genome*, ed. T. R. Gregory, pp. 289–327. London: Elsevier Academic Press.

Taylor, K. & Markham, B. (1978). *Ranunculus ficaria* L. Biological flora of the British Isles. *Journal of Ecology*, 66, 1011–31.

Telwala, Y. *et al.* (2013). Climate-induced elevation range shifts and increase in plant species richness in a Himalayan biodiversity epicenter. *PLOS ONE*, 8, e57103.

Theaker, A. J. & Briggs, D. (1993). Genecological studies of groundsel (*Senecio vulgaris* L.). IV. Rate of development in plants from different habitat types. *New Phytologist*, 123, 185–94.

Theissen, G. (2001). Development of floral organ identity: stories from the MADS house. *Current Opinion in Plant Biology*, 4, 75–85.

Theissen, G. (2006). The proper place of hopeful monsters in evolutionary biology. *Theory in Biosciences*, 124, 349–69.

Theissen, G. & Melzer, R. (2007). Molecular mechanisms underlying origin and diversification of the angiosperm flower. *Annals of Botany*, 100, 603–19.

Theissen, G. *et al.* (2002). How land plants learned their floral ABCs: the role of MADS-box genes in the evolutionary origin of flowers. In *Developmental genetics and plant evolution*, ed. Q. C. B. Cronk, R. M. Bateman & J. A. Hawkins, pp. 173–205. London: Taylor & Francis.

Theobald, D. L. (2010). A formal test of the theory of universal common ancestry. *Nature*, 465, 219–22.

Theodoridis, S. *et al.* (2013). Divergent and narrower climatic niches characterize polyploid species of European primroses in *Primula* sect. *Aleuritia*. *Journal of Biogeography*, 40, 1278–89.

Thiele, K. & Yeates, D. (2002). Tension arises from duality at the heart of taxonomy. *Nature*, 419, 337.

Thiselton-Dyer, W. T. (1895a). Variation and specific stability. *Nature*, 51, 459–61.

Thiselton-Dyer, W. T. (1895b). Origin of the cultivated *Cineraria*. *Nature*, 52, 3–4, 78–9, 128–9.

Thoday, J. M. (1972). Disruptive selection. *Proceedings of the Royal Society of London, B*, 182, 109–43.

Thomas, C. D. *et al.* (2004). Extinction risk from climate change. *Nature*, 427, 145–8.

Thomas, D. A. & Barber, H. N. (1974). Studies of leaf characteristics of a cline of *Eucalyptus urnigera* from Mount Wellington, Tasmania. II. Reflection, transmission and absorption of radiation. *Australian Journal of Botany*, 22, 701–7.

Thomas, R. B. *et al.* (2013). Evidence of recovery of *Juniperus virginiana* trees from sulfur pollution after the Clean Air Act. *Proceedings of the National Academy of Sciences, USA*, 110, 15319–24.

Thomas, S. G. & Franklin-Tong, V. E. (2004). Self-incompatibility triggers programmed cell death in *Papaver* pollen. *Nature*, 429, 305–9.

Thompson, D. Q. *et al.* (1987). *Spread, impact, and control of Purple Loosestrife (*Lythrum salicaria*) in North American wetlands*. US Fish and Wildlife Service. Jamestown, ND: Northern Prairie Wildlife Research Center Online.

Thompson, J. D. & Lumaret, R. (1992). The evolutionary dynamics of polyploid plants: origins, establishment and persistence. *Trends in Ecology and Evolution*, 7, 302–6.

Thompson, J. N. (1998). Rapid evolution as an ecological process. *Trends in Ecology and Evolution*, 13, 329–32.

Thompson, S. L. & Whitton, J. (2006). Patterns of recurrent evolution and geographic parthenogenesis within apomictic Easter daisies (*Townsendia hookeri*). *Molecular Ecology*, 15, 3389–400.

Thompson, S. L. *et al.* (2008). Cryptic sex within male-sterile polyploid populations of the Easter Daisy, *Townsendia hookeri*. *International Journal of Plant Science*, 169, 183–93.

Thomson, J. D., Herre, E. A., Hamrick, J. L. & Stone, J. L. (1991). Genetic mosaics in Strangler Fig trees: implications for tropical conservation. *Science*, 254, 1214–16.

Thrall, P. H. *et al.* (1998). Metapopulation collapse: the consequences of limited gene flow in spatially structured populations. In *Modeling spatiotemporal dynamics in ecology*, ed. J. Bascompte & R. V. Sole, pp. 83–104. Berlin: Springer-Verlag.

Throop, W. (2004). A response to the article (Hobbs, 2004) 'Restoration Ecology: the challenge of social values and expectation'. *Frontiers in Ecology and the Environment*, 2, 47–8.

Thuiller, W. (2003). BIOMOD – optimizing predictions of species distributions and projecting potential future shifts under global change. *Global Change Biology*, 9, 1353–62.

Thuiller, W. *et al.* (2008). Predicting global change impacts on plant species' distributions: future challenges. *Perspectives in Plant Ecology, Evolution and Systematics*, 9, 137–52.

Thurston, J. M., Dyke, G. V. & Williams, E. D. (1976). *The Park Grass experiment on the effect of fertilisers and liming on the botanical composition of permanent grassland and on the yield of hay.* Harpenden: Rothamsted Experimental Station.

Tian, D. *et al.* (2002). Signature of balancing selection in *Arabidopsis*. *Proceedings of the National Academy of Sciences, USA*, 99, 11525–30.

Till, I. (1987). Variability of expression of cyanogenesis in White Clover (*Trifolium repens* L.). *Heredity*, 59, 265–71.

Tindall, K. R. & Kunkel, T. A. (1988). Fidelity of DNA synthesis by the *Thermus aquaticus* DNA polymerase. *Biochemistry*, 27, 6008–13.

Tischler, G. (1950). *Die Chromosomenzahlen der Gefässpflanzen Mitteleuropas.* 'S-Gravenhage: Junk.

Tobgy, H. A. (1943). A cytological study of *Crepis fuliginosa*, *C. neglecta* and their F_1 hybrid, and its bearing on the mechanism of phylogenetic reduction in chromosome number. *Journal of Genetics*, 45, 67–111.

Todhunter, I. (1873). *The conflict of studies and other essays.* London: Macmillan.

Toomajian, C. *et al.* (2006). A nonparametric test reveals selection for rapid flowering in the *Arabidopsis* genome. *PLoS Biology*, 4, 732–8.

Tower, W. L. (1902). Variation in the ray-flowers of *Chrysanthemum leucanthemum* L. at Yellow Springs, Green County, O, with remarks upon the determination of the modes. *Biometrika*, 1, 309–15.

Townsend Peterson, A. (2011). Ecological niche conservatism: a time-structured review of evidence. *Journal of Biogeography*, 38, 817–27.

Traill, L. W., Bradshaw, C. J. A. & Brook, B. W. (2007). Minimum viable population size: a meta-analysis of 30 years of published estimates. *Biological Conservation*, 139, 159–66.

Traill, L. W. *et al.* (2010). Pragmatic population viability targets in a rapidly changing world. *Biological Conservation*, 143, 28–34.

Trakhtenbrot, A. *et al.* (2005). The importance of long-distance dispersal in biodiversity conservation. *Diversity & Distributions*, 11, 173–81.

Tralau, H. (1968). Evolutionary trends in the genus *Ginkgo*. *Lethaia*, 1, 63–101.

Tremetsberger, K. *et al.* (2002). Infraspecific genetic variation in *Biscutella laevigata* (Brassicaceae): new focus on Irene Manton's hypothesis. *Plant Systematics and Evolution*, 233, 163–81.

Trevisan, L. (1988). Angiospermous pollen (monosulcate-trichotomosulcate phase) from very early Lower Cretaceous of Southern Tuscany (Italy): some aspects. In *Proceedings of the 7th International Palynological Congress, Brisbane, Australia*, Abstr. 165. Amsterdam: Elsevier.

Trewick, S. A., Morgan-Richards, M. & Chapman, H. M. (2004). Chloroplast DNA diversity of *Hieracium pilosella* (Asteraceae) introduced to New Zealand: reticulation, hybridization and invasion. *American Journal of Botany*, 91, 73–85.

Trusty, J. L. *et al.* (2005). Molecular phylogenetics of the Macaronesian-endemic genus *Bystropogon* (Lamiaceae): palaeo-islands, ecological shifts and interisland colonizations. *Molecular Ecology*, 14, 1177–89.

Tsuchimatsu, T. *et al.* (2010). Evolution of self-compatibility in *Arabidopsis* by a mutation in the male specificity gene. *Nature*, **464**, 1342–6.

Tubeuf, K. F. von. (1923). *Monographie der Mistel*. Munich & Berlin: Oldenbourg.

Tucker, M. R. & Koltuno, A. M. G. (2009). Sexual and asexual (apomictic) seed development in flowering plants: molecular, morphological and evolutionary relationships. *Functional Plant Biology*, **36**, 490–504.

Turesson, G. (1922a). The species and variety as ecological units. *Hereditas*, **3**, 100–13.

Turesson, G. (1922b). The genotypical response of the plant species to the habitat. *Hereditas*, **3**, 211–350.

Turesson, G. (1925). The plant species in relation to habitat and climate. *Hereditas*, **6**, 147–236.

Turesson, G. (1927a). Erbliche Transpirationsdifferenzen zwischen Ökotypen derselben Pflanzen Art. *Hereditas*, **11**, 193–206.

Turesson, G. (1927b). Untersuchungen über Grenzplasmolyse und Saugkraftwerte in verschiedenen Ökotypen derselben Art. *Jahrbücher für wissenschaftliche Botanik*, **66**, 723–47.

Turesson, G. (1930). The selective effect of climate upon the plant species. *Hereditas*, **14**, 99–152.

Turesson, G. (1943). Variation in the apomictic microspecies of *Alchemilla vulgaris* L. *Botaniska Notiser*, **1943**, 413–27.

Turesson, G. (1961). Habitat modifications in some widespread plant species. *Botaniska Notiser*, **114**, 435–52.

Turkington, R. (1989). The growth, distribution and neighbour relationships of *Trifolium repens* in a permanent pasture. V. The co-evolution of competitors. *Journal of Ecology*, **77**, 717–33.

Turkington, R. A. & Harper, J. L. (1979a). The growth, distribution and neighbour relationships of *Trifolium repens* in a permanent pasture. I. Ordination pattern and contact. *Journal of Ecology*, **67**, 201–18.

Turkington, R. A. & Harper, J. L. (1979b). The growth, distribution and neighbour relationships of *Trifolium repens* in a permanent pasture. II. Fine scale biotic differentiations. *Journal of Ecology*, **67**, 245–54.

Turnbull, C. (2011). Long-distance regulation of flowering time. *Journal of Experimental Botany*, **62**, 4399–4413.

Turner, I. M., Tan, H. T. W., Wee, Y. C., Ibrahim, A. B., Chew, P. T. & Corlett, R. T. (1994). A study of plant species extinction in Singapore: lessons for the conservation of tropical biodiversity. *Conservation Biology*, **8**, 705–12.

Turner, R. K. & Daily, G. C. (2008). The ecosystems services framework and natural capital conservation. *Environmental and Resource Economics*, **39**, 25–35.

Turner, T. L. *et al.* (2008). Population resequencing reveals local adaptation of *Arabidopsis lyrata* to serpentine soils. *Nature Genetics*, **42**, 260–3.

Turrill, W. B. (1938). The expansion of taxonomy with special reference to Spermatophyta. *Biological Reviews*, **13**, 342–73.

Turrill, W. B. (1940). Experimental and synthetic plant taxonomy. In *The new systematics*, ed. J. S. Huxley, pp. 47–71. Oxford: Clarendon Press.

Tutin, T. G. (1957). A contribution to the experimental taxonomy of *Poa annua* L. *Watsonia*, **4**, 110.

Tutin, T. G., Heywood, V. H., Burges, N. A., Moore, D. M., Valentine, D. H., Walters, S. M. & Webb, D. A. (1964–80). *Flora Europaea*. Cambridge: Cambridge University Press.

Twyford, A. D. & Ennos, R. A. (2012). Next-generation hybridization and introgression. *Heredity*, **108**, 179–89.

Udall, J. A., Quijada, P. A. & Osborn, T. C. (2005). Detection of chromosomal rearrangements derived from homoeologous recombination in four mapping populations of *Brassica napus* L. *Genetics*, **169**, 967–79.

Uesugi, R. *et al.* (2007). Restoration of genetic diversity from soil seed banks in a threatened aquatic plant, *Nymphoides peltata*. *Conservation Genetics*, **8**, 111–21.

Uhl, C. H. (1978). Chromosomes of Mexican *Sedum*, section Pachysedum. *Rhodora*, **80**, 491–512.

Uphof, J. C. Th. (1938). Cleistogamic flowers. *The Botanical Review*, **4**, 21–49.

Valente, L. P., Silva, M. C. C. & Jansen, L. E. T. (2012). Temporal control of epigenetic centromere specification. *Chromosome Research*, **20**, 481–92.

Valentine, D. H. (1941). Variation in *Viola riviniana* Rchb. *New Phytologist*, **40**, 189–209.

Valentine, D. H. (1956). Studies in British Primulas. V. The inheritance of seed incompatibility. *New Phytologist*, **55**, 305–18.

Valentine, D. H. (1975). Primula. In *Hybridization and the flora of the British Isles*, ed. C. A. Stace, pp. 346–8. London & New York: Academic Press.

Vallejo-Marín, M., Dorken M. E. & Barrett, S. C. H. (2010). The ecological and evolutionary consequences of clonality for plant mating. *Annual Reviews of Ecology, Evolution and Systematics*, 41, 193–213.

Van de Peer, Y. (2003). Phylogeny inference based on distance methods. In *The phylogenetic handbook: a practical guide to DNA and protein phylogeny*. ed. M. Salemi & A.-M. Vandamme, pp. 101–19. Cambridge: Cambridge University Press

van der Pilj, L. & Dodson, C. H. (1966). *Orchid flowers: their pollination and evolution*. Coral Gables, FL: University of Miami Press.

Van der Putten, W. H., Macel, M. & Visser, M. E. (2010). Predicting species distribution and abundance responses to climate change: why it is essential to include biotic interactions across trophic levels. *Philosophical Transactions of the Royal Society, B*, 365, 2025–34.

van Dijk, H. (2004). Gene exchange between wild and crop in *Beta vulgaris*: how easy is hybridization and what will happen in later generations? In *Introgression from genetically modified plants into wild relatives*, ed. H. C. M. den Nijs, D. Bartsch & J. Sweet, pp.53–61. Wallingford, UK: CAB International.

Van Dijk, P. J. (2003). Ecological and evolutionary opportunities of apomixis: insights from *Taraxacum* and *Chondrilla*. *Philosophical Transactions of the Royal Society of London, B*, 358, 1113–21.

Van Dijk, P. (2009). Apomixis: basics for non-botanists. In *Lost sex: the evolutionary biology of parthenogenesis*, ed. I. Schön, K. Martens & P. J. van Dijk, pp. 47–62. Berlin: Springer Publications.

van Dijk, P. & Bijlsma, R. (1994). Simulation of flowering time displacement between two cytotypes that form inviable hybrids. *Heredity*, 72, 522–35.

Van Dijk, P. J. & Vijverberg, K. (2005). The significance of apomixis in the evolution of the angiosperms: a reappraisal. In *Plant species-level systematics. New perspectives on pattern and process*, ed. F. Y. T. Bakker, L. W. Chatrou, B. Gravendeel & P. B. Pelser, pp. 101–16. *Regnum Vegetabile*, 143. Ruggell, Liechtenstein: A. R. G. Gantner Verlag.

van Groenendael, J. M. (1986). Life history characteristics of two ecotypes of *Plantago lanceolata*. *Acta Botanica Neerlandica*, 35, 71–86.

van Teuren, R. *et al.* (1991). The significance of genetic erosion in the process of extinction. I. Genetic differentiation in *Salvia pratensis* and *Scabiosa columbaria* in relation to population size. *Heredity*, 66, 181–9.

van Teuren, R. *et al.* (1993). The significance of genetic erosion in the process of extinction. IV. Inbreeding depression and heterosis effects caused by selfing and outcrossing in *Scabiosa columbaria*. *Evolution*, 47, 1669–80.

van Tienderen, P. H. & van der Toorn, J. (1991a). Genetic differentiation between populations of *Plantago lanceolata*. I. Local adaptation in three contrasting habitats. *Journal of Ecology*, 79, 27–42.

Van Tienderen, P. H. & van der Toorn, J. (1991b). Genetic differentiation between populations of *Plantago lanceolata*. II. Phenotypic selection in a transplant experiment in three contrasting habitats. *Journal of Ecology*, 79, 43–59.

Van Valen, L. (1976). Ecological species, multispecies, and oaks. *Taxon*, 25, 223–39.

van Wyk, A. E. (2007). The end justifies the means. *Taxon*, 56, 645–8.

Vandamme, A.-M. (2003). Basic concepts of molecular evolution. In *The phylogenetic handbook: a practical guide to DNA and protein phylogeny*, ed. M. Salemi & A.-M. Vandamme, pp. 1–23. Cambridge: Cambridge University Press.

Vangronsveld, J. *et al.* (2009). Phytoremediation of contaminated soils and groundwater: lessons from the field. *Environmental Science and Pollution Research*, 16, 765–94.

Vanzolini, P. E. & Williams, E. E. (1970). South American anoles: the geographic differentiation and evolution of the *Anolis chrysolepis* species group (Sauria, Iguanidae). *Arquivos de Zoologia (São Paulo)*, 19, 1–298.

Vargas, P. *et al.* (1999). Polyploid speciation in *Hedera* (Araliaceae): phylogenetic and biogeographic insights based on chromosome counts and ITS sequences. *Plant Systematics and Evolution*, 219, 165–79.

Vasek, F. C. (1980). Creosote Bush: long-lived clones in the Mojave Desert. *American Journal of Botany*, 67, 246–55.

Vashisht. D. *et al.* (2011). Natural variation of submergence tolerance amongst *Arabidopsis thaliana* accessions. *New Phytologist*, 190, 299–310.

Vaughan, D. A., Balazs, E. & Heslop-Harrison, J. S. (2007). From crop domestication to super-domestication. *Annals of Botany*, 100, 893–901.

Vaughton, G., Ramsey, M. & Johnson, S. D. (2011). Pollination and late-acting self-incompatibility in

Cyrtanthus breviflorus (Amaryllidaceae): implications for seed production. *Annals of Botany*, **106**, 547–55.

Vekeman, X. & Lefèbvre, C. (1997). On the evolution of heavy metal tolerant populations in *Armeria maritima*: evidence from allozyme variation and reproductive barriers. *Journal of Evolutionary Biology*, **10**, 175–91.

Vekemans, X. (2010). What's good for you may be good for me: evidence for adaptive introgression of multiple traits in wild sunflower. *New Phytologist*, **187**, 6–9.

Vekemans, X. & Hardy, O. J. (2004). New insights from fine-scale spatial genetic structure analyses in plant populations. *Molecular Ecology*, **13**, 921–35.

Vellinga, E. C., Wolfe, B. E. & Pringle, A. (2009).Global patterns of ectomycorrhiza introductions. *New Phytologist*, **181**, 960–73.

Venable, D. L. & Levin, D. A. (1985). Ecology of achene dimorphism in *Heterotheca latifolia*. I. Achene structure, germination and dispersal. *Journal of Ecology*, **73**, 133–45.

Vergeer, P., Wagemaker, N. & Ouborg, N. J. (2012). Evidence for an epigenetic role in inbreeding depression. *Biology Letters*, **8**, 798–801.

Verhoeven, K. J. F. *et al.* (2010). Stress-induced DNA methylation changes and their heritability in asexual dandelions. *New Phytologist*, **185**, 1108–18.

Vernon, H. M. (1903). *Variation in animals and plants*. London: Kegan Paul.

Verschaffelt, E. (1899). Galton's regression to mediocrity bij ongeslachtelijke verplanting. In *Livre Jubilaire dédié à Charles wan Bambeke*, pp. 1–5. Brussels: Lamerton.

Vesteg, M., Vacula, R. & Krajčovič, J. (2009). On the origin of chloroplasts, import mechanisms and chloroplast-targeted proteins, and loss of photosynthetic ability – review. *Folia Microbiologica*, **54**, 303–21.

Vickery, R. K. (1964). Barriers to gene exchange between members of the *Mimulus guttatus* complex (Scrophulariaceae). *Evolution*, **18**, 52–69.

Vietmeyer, N. (1995). Applying biodiversity. *Journal of the Federation of American Scientists*, **48**, 1–8.

Viette, M., Tettamanti, C. & Saucy, F. (2000). Preference for acyanogenic white clover (*Trifolium repens*) in the vole *Arvicola terrestris*. II. Generalization and further investigations. *Journal of Chemical Ecology*, **26**, 101–22.

Viktora, M., Savidge, R. A. & Rajora, O. P. (2011). Clonal and nonclonal genetic structure of subarctic Black Spruce (*Picea mariana*) populations in Yukon Territory. *Botany-Botanique*, 89, 133–40.

Vila-Aiub, M. M., Neve, P. & Powles, S. B. (2009). Fitness costs associated with evolved herbicide resistance alleles in plants. *New Phytologist*, **184**, 751–67.

Viosca, P., Jr (1935). The irises of southeastern Louisiana: a taxonomic and ecological interpretation. *Bulletin of the American Iris Society*, **57**, 3–56.

Vitousek, P. M. *et al.* (1997). Human domination of the Earth's ecosystems. *Science*, **277**, 494–9.

Vittoz, P. *et al.* (2013). Climate change impacts on biodiversity in Switzerland. *Journal for Nature Conservation*, **21**, 154–62.

Volis, S. & Blecher, M. (2010). Quasi *in situ*: a bridge between *ex situ* and *in situ* conservation of plants. *Biodiversity and Conservation*, **19**, 2441–54.

von der Lippe, M.& Kowarik, I. (2007). Crop spillage along roads: a factor of uncertainty in the containment of GMO. *Ecography*, **30**, 483–90.

Vorontsova, M. S. & Simon, B. K. (2012). Updating classifications to reflect monophyly: 10 to 20 percent of species names change in Poaceae. *Taxon*, **61**, 735–46.

Vorzimmer, P. J. (1972). *Charles Darwin: the years of controversy*. London: University of London Press.

Vouillamoz, J., Maigre, D. & Meredith, C. P. (2003). Microsatellite analysis of ancient alpine grape cultivars: Pedigree reconstruction of *Vitis vinifera* L. 'Cornalin du Valais'. *Theoretical and Applied Genetics*, **107**, 448–54.

Vreysen, M. J. B., Robinson, A. S. & Hendrichs, J. (2007). The Mountain Pine Beetle *Dendroctonus ponderosae* in western North America: potential for area-wide integrated management. 2007. In *Area-wide control of insect pests from research to field implementation*, ed. M. J. B. Vreysen, A. S. Robinson and J. Hendrichs, pp. 297–307. New York: Springer.

Waddington, C. H. (1966). Mendel and evolution. In *G. Mendel Memorial Symposium, 1865–1965*, ed. M. Sosna, pp. 145–50. Prague: Academia Publishing House of the Czechoslovak Academy of Sciences.

Wägele *et al.* (2011). The taxonomist – an endangered race. A practical proposal for its survival. *Frontiers in Zoology*, **8**, 25–30.

Wagenius, S. et al. (2012). Seedling recruitment in the long-lived perennial, *Echinacea angustifolia*: a 10-year experiment. *Restoration Ecology*, **20**, 352–9.

Wagner, M. (1868). *Die Darwin'sche Theorie und das Migrationgesetz der Organismen.* Leipzig: Duncker & Humblot.

Wakeley, J. (2008). *Coalescent theory: an introduction.* Greenwood Village, CO: Roberts & Company Publishers.

Waldron, L. R. (1912). Hardiness in successive Alfalfa generations. *The American Naturalist*, **46**, 463–9.

Waller, D. M. (1979).The relative costs of self- and cross-fertilized seeds in *Impatiens capensis* (Balsaminaceae). *American Journal of Botany*, **66**, 313–20.

Waller, D. M. (1984). Differences in fitness between seedlings derived from cleistogamous and chasogamous flowers in *Impatiens capensis. Evolution*, **38**, 427–40.

Waller, D. M., O'Malley, D. M. & Gawler, S. C. (1987). Genetic variation in the extreme endemic *Pedicularis furbishiae* (Scrophulariaceae). *Conservation Biology*, **1**, 335–40.

Walley, K. A., Khan, M. S. I. & Bradshaw, A. D. (1974). The potential for evolution of heavy metal tolerance in plants. I. Copper and zinc tolerance in *Agrostis tenuis. Heredity*, **32**, 309–19.

Wallis, G. P. & Trewick, S. A. (2009). New Zealand phylogeography: evolution on a small continent. *Molecular Ecology*, **18**, 3548–580.

Walsh, B. (2008). Using molecular markers for detecting domestication, improvement, and adaptation genes. *Euphytica*, **16**, 1–17.

Walsh, N. E. *et al.* (1997). Experimental manipulations of snow-depth: effects on nutrient content of caribou forage. *Global Change Biology*, **3**, 158–164.

Walter, K. S. & Gillett, H. J. (1997) *IUCN red list of threatened plants*, compiled by the World Conservation Monitoring Centre. Gland, Switzerland, & Cambridge: IUCN – The World Conservation Union.

Walters, S. M. (1961). The shaping of angiosperm taxonomy. *New Phytologist*, **60**, 74–84.

Walters, S. M. (1962). Generic and specific concepts in the European flora. *Preslia*, **34**, 207–26.

Walters, S. M. (1970). Dwarf variants of *Alchemilla* L. *Fragmenta Floristica et Geobotanica*, **16**, 91–8.

Walters, S. M. (1972). Endemism in the genus *Alchemilla* in Europe. In *Taxonomy, phytogeography and evolution*, ed. D. H. Valentine, pp. 301–5. London & New York: Academic Press.

Walters, S. M. (1979). Progress in biological conservation in Cambridge. In *Landscape towards 2000: conservation or desolation*, ed. D. Smith, pp. 56–8. London: Landscape Institute.

Walters, S. M. (1986a). *Alchemilla*: a challenge to biosystematists. *Acta Universitatus Upsaliensis, Symbolae Botanicae Upsalienses*, **XXVII**, 193–8.

Walters, S. M. (1986b). The name of the rose: a review of ideas on the European bias in angiosperm classification. *New Phytologist*, **104**, 527–46.

Walters, S. M. (1989a). Obituary of John Scott Lennox Gilmour. *Plant Systematics & Evolution*, **167**, 93–5.

Walters, S. M. (1989b). Experimental and orthodox taxonomic categories and the deme terminology. *Plant Systematics & Evolution*, **167**, 1–2.

Walters, S. M. & Stow, E. A. (2001). *Darwin's mentor: John Stevens Henslow, 1796–1861.* Cambridge: Cambridge University Press.

Walther, G.-R. (2010). Community and ecosystem responses to recent climate change. *Philosophical Transactions of the Royal Society, B*, **365**, 2019–24.

Walther, G.-R. *et al.* (2002). Ecological responses to recent climate change. *Nature*, **416**, 389–95.

Walther-Hellwig, K. & Frankl, R. (2003). Foraging habitats and foraging distances of bumblebees, *Bombus* spp. (Hym., Apidae), in an agricultural landscape. *Journal of Applied Entomology*, **124**, 299–306.

Wang, E. *et al.* (2008). Control of rice grain-filling and yield by a gene with a potential signature of domestication. *Nature Genetics*, **40**, 1370–4.

Wang, H. *et al.* (2005). The origin of the naked grains of Maize. *Nature*, **436**, 714–19.

Wang, W.-X. *et al.* (2007). MicroRNAs (miRNAs) and plant development. In *Handbook of plant science*, ed. K. Roberts, pp. 640–9. Chichester: Wiley.

Ward, D. B. (1974). The 'ignorant man' technique of sampling plant populations. *Taxon*, **23**, 325–30.

Ward, J. *et al.* (2000). Is atmospheric CO_2 a selective agent on model C_3 annuals? *Oecologia*, **123**, 330–41.

Warren, J. (2009). Extra petals in the buttercup (*Ranunculus repens*) provide a quick method to estimate the age of meadows. *Annals of Botany*, **104**, 785–8.

Warren, M. S. *et al.* (2001). Rapid responses of British butterflies to opposing forces of climate and habitat change. *Nature*, **414**, 65–9.

Warren, R. *et al.* (2013) Quantifying the benefit of early climate change mitigation in avoiding biodiversity loss *Nature Climate Change*, **3**, 678–82.

Warwick, S. I. (1990a). Allozyme and life history variation in five northwardly colonizing North American weed species. *Plant Systematics & Evolution*, 169, 41–54.

Warwick, S. I. (1990a). Genetic variation in weeds – with particular reference to Canadian agricultural weeds. In *Biological approaches and evolutionary trends in plants*, ed. S. Kawano, pp. 3–18. New York: Academic Press.

Warwick, S. I. (1991). Herbicide resistance in weedy plants: physiology and population biology. *Annual Review of Ecology & Systematics*, 22, 95–114.

Warwick, S. I. & Al-Shehbaz, I. A. (2006). Brassicaceae: chromosome number index and database on CD-Rom. *Plant Systematics and Evolution*, 259, 237–48.

Warwick, S. I. & Briggs, D. (1978a). The genecology of lawn weeds. I. Population differentiation in *Poa annua* L. in a mosaic environment of bowling green lawns and flower beds. *New Phytologist*, 81, 711–23.

Warwick, S. I. & Briggs, D. (1978b). The genecology of lawn weeds. II. Evidence for disruptive selection in *Poa annua* L. in a mosaic environment of bowling green lawns and flower beds. *New Phytologist*, 81, 725–37.

Warwick, S. I. & Briggs, D. (1979). The genecology of lawn weeds. III. Cultivation experiments with *Achillea millefolium* L., *Bellis perennis* L., *Plantago lanceolata* L., *Plantago major* L. and *Prunella vulgaris* L. collected from lawns and contrasting grassland habitats. *New Phytologist*, 83, 509–36.

Warwick, S. I. & Briggs, D. (1980a). The genecology of lawn weeds. IV. Adaptive significance of variation in *Bellis perennis* L. as revealed in a transplant experiment. *New Phytologist*, 85, 275–88.

Warwick, S. I. & Briggs, D. (1980b). The genecology of lawn weeds. V. The adaptive significance of different growth habit in lawn and roadside populations of *Plantago major* L. *New Phytologist*, 85, 289–300.

Warwick, S. I. & Briggs, D. (1980c). The genecology of lawn weeds. VI. The adaptive significance of variation in *Achillea millefolium* L. as investigated by transplant experiments. *New Phytologist*, 85, 451–60.

Warwick, S. I. & Gottlieb, L. D. (1985). Genetic divergence and geographic speciation in *Layia* (Compositae). *Evolution*, 39, 1236–41.

Warwick, S. I., Phillips, D., Andrews, C. (1986). Rhizome depth: the critical factor in winter survival of *Sorghum halepense* (L.) Pers. (Johnson grass). *Weed Research*, 26, 381–7.

Warwick, S. I., Thompson, B. K. & Black, L. D. (1987). Genetic variation in Canadian and European populations of the colonising weed species *Apera spica-venti*. *New Phytologist*, 106, 301–17.

Warwick, S. I., Bain, J. F., Wheatcroft, R. & Thompson, B. K. (1989). Hybridization and introgression in *Carduus nutans* and *C. acanthoides* re-examined. *Systematic Botany*, 14, 476–94.

Warwick, S. I., Beckie, H. J. & Hall, L. M. (2009). Gene flow, invasiveness, and ecological impact of gentically modified crops. *Annals of the New York Academy of Sciences*, 1168, 72–99.

Watrud, L. S. *et al.* (2004) Evidence for landscape-level, pollen-mediated gene flow from genetically modified creeping bentgrass with CP4 EPSPS as a marker. *Proceedings of the National Academy of Sciences, USA*, 101, 14533–8.

Watson, J. D. (1968). *The double helix*. London: Weidenfeld & Nicolson.

Watson, J. D. & Crick, F. H. C. (1953). A structure of deoxyribose nucleic acid. *Nature*, 171, 737–8.

Watson, J. D. *et al.* (2014). *Molecular biology of the gene*, 7th edn. Boston: Pearson.

Watson, P. J. (1969). Evolution in closely adjacent plant populations. VI. An entomophilous species, *Potentilla erecta*, in two contrasting habitats. *Heredity*, 24, 407–22.

Weart, S. (2011). Global warming: how skepticism became denial. *Bulletin of the Atomic Scientists*, 67, 41–50.

Weber B. H. & Depew, D. J. (2007). Darwinism, design, and complex systems dynamics. In *Debating design. From Darwin to DNA*, ed. W.A Dembski & M. Ruse, pp. 173–90. Cambridge: Cambridge University Press.

Weber, E. & Schmidt, B. (1998). Latitudinal population differentiation in two species of *Solidago* (Asteraceae) introduced into Europe. *American Journal of Botany*, 85, 1110–21.

Webster, S. D. (1988). *Ranunculus penicillatus* (Dumort.) Bab. in Great Britain and Ireland. *Watsonia*, 17, 1–22.

Weeden, N. F. & Gottlieb, L. D. (1979). Isolation of cytoplasmic enzymes from pollen. *Plant Physiology*, 66, 400–3.

Weeden, N. F. & Wendel, J. F. (1990) Genetics of plant isozymes. In *Isoenzymes in plant biology*, ed. D. E Soltis & P. S. Soltis, pp. 46–72. London: Chapman & Hall.

Wegener, A. (1912). Die Herausbildung der Grossformen der Erdrinde (Kontinente und Ozeane), auf geophysikalischer Grundlage. *Petermanns Geographische Mitteilungen*, **63**: 185–95, 253–6, 305–9. Presented at the annual meeting of the German Geological Society, Frankfurt am Main (6 January 1912).

Weigel, D. (2012). Natural variation in *Arabidopsis*: from molecular genetics to ecological genomics. *Plant Physiology*, **158**, 2–22.

Weigel, D. & Colot, V. (2012). Epialleles in plant evolution. *Genome Biology*, **13**, 249–54.

Weimark, H. (1945). Experimental taxonomy in *Aethusa cynapium*. *Botaniska Notiser*, 1945, 351–80.

Weinstein, A. (1977). How unknown was Mendel's paper? *Journal of the History of Biology*, **10**, 341–64.

Weir, J. & Ingram, R. (1980). Ray morphology and cytological investigations of *Senecio cambrensis* Rosser. *New Phytologist*, **86**, 237–41.

Weising, K. *et al.* (2005). *DNA fingerprinting in plants. Principles, methods and applications*, 2nd edn. London: Taylor & Francis.

Weismann, A. (1883). *Uber die Vererbung. English translation, On heredity* (1889), translated by A. E. Shipley: Oxford: Clarendon Press.

Weiss, H. & Maluszynska, J. (2000). Chromosomal rearrangement in autotetraploid plants of *Arabidopsis thaliana*. *Hereditas*, **133**, 255–61.

Weiss, S. F. (2013). *The Nazi symbiosis: human genetics and politics in the Third Reich*. Chicago: University of Chicago Press.

Weiss-Schneeweiss, H. *et al.* (2013). Evolutionary consequences, constraints and potential of polyploidy in plants. *Cytogenetic and Genome Research*, **140**, 137–50.

Weldon, W. F. R. (1895a). The origin of the cultivated Cineraria. *Nature*, **52**, 54, 104, 129.

Weldon, W. F. R. (1895b). Remarks on variation in animals and plants. *Proceedings of the Royal Society of London*, **57**, 379–82.

Weldon, W. F. R. (1898). Presidential address. Section D. Zoology. *Nature*, **58**, 499–506.

Weldon, W. F. R. (1902a). On the ambiguity of Mendel's categories. *Biometrika*, **2**, 44–55.

Weldon, W. F. R. (1902b). Seasonal changes in the characters of *Aster prenanthoides* Muhl. *Biometrika*, **2**, 113–14.

Wells, C. L.& Pigliucci, M. (2000). Adaptive phenotypic plasticity: the case of heterophylly in aquatic plants. *Perspectives in Plant Ecology, Evolution and Systematics*, **3**, 1–18.

Wells, W. C. (1818). An account of a White female, part of whose skin resembles that of a Negro. [Paper given at the Royal Society, 1813.] In *Two essays upon dew and single vision*. London.

Wendel, J. F. & Doyle, J. J. (1998). Phylogenetic incongruence: window into genome history and molecular evolution. In *Molecular systematics of plants* II, ed. P. Soltis and J. Doyle, pp. 265–96. Boston: Kluwer.

Wendel, J. F. & Doyle, J. J. (2005). Polyploidy and evolution in plants. In *Plant diversity and evolution*, ed. R. J. Henry, pp. 97–117. Wallingford, UK: CABI Publishing.

Wenny, D. G. (2000). Seed dispersal, seed predation, and seedling recruitment of a neotropical montane tree. *Ecological Monographs*, **70**, 331–51.

Werth, C. R., Riopel, J. L. & Gillespie, N. W. (1984). Genetic uniformity in an introduced population of Witchweed (*Striga asiatica*) in the United States. *Weed Science*, **32**, 645–8.

Werth, C. R., Guttman, S. I. & Eshbaugh, W. H. (1985a). Electrophoretic evidence of reticulate evolution in the Appalachian *Asplenium* complex. *Systematic Botany*, **10**, 184–92.

Werth, C. R., Guttman, S. I. & Eshbaugh, W. H. (1985b). Recurring origins of allopolyploid species in *Asplenium*. *Science*, **228**, 731–3.

Westerbergh, A. (1975). Serpentine and non-serpentine *Silene dioica* plants do not differ in nickel tolerance. *Plant and Soil*, **167**, 297–303.

Westergaard, K. B. *et al.* (2008). Genetic diversity and distinctiveness in Scottish alpine plants. *Plant Ecology & Diversity*, **1**, 329–38.

Westerling, A. L. *et al.* (2006). Warming and earlier spring increase Western U.S. forest wildfire activity. *Science*, **313**, 940–3.

Western, D. (2001). Human modified ecosystems and future evolution. *Proceedings of the National Academy of Sciences, USA*, **98**, 5458–65.

Western, D., Russell, S. & Cuthill, I. (2009). The status of wildlife in protected areas compared to non-protected areas of Kenya. *PLoS ONE*, **4**, 1–6.

Wettstein, R. von (1895). Der Saison-Dimorphismus als Ausgangpunkt für die Bildung neuer Arten im Pflanzenreich. *Berichte der Deutschen botanischen Gesellschaft*, **13**, 303–13.

Wheeler, B. D. & Shaw, S. C. (1995). *Restoration of damaged peatlands*. London: HMSO.

Wheeler, Q. D. (ed.) (2008). *The new taxonomy*. Systematics Association Special Volume, Series No. 76. Boca Raton, FL, & London: CRC Press.

Wheeler, W. C. (2012). Clocks and rates. In *Systematics: a course of lectures*. Wiley: published online: 17 May 2012, DOI:10.1002/9781118301081

White, G. M., Boshier, D. H. & Powell, W. (2002). Increased pollen flow counteracts fragmentation in a tropical dry forest: an example from *Swietenia humilis* Zuccarini. *Proceedings of the National Academy of Sciences, USA*, **99**, 2038–42.

White, M. J. D. (1978). *Modes of speciation*. San Francisco: Freeman & Company.

White, O. E. (1917). Inheritance studies in *Pisum*. 2. The present state of knowledge of heredity and variation in peas. *Proceedings of the American Philosophical Society*, **56**, 487–588.

Whitehouse, H. L. K. (1950). Multiple-allelomorph incompatibility of pollen and style in the evolution of the angiosperms. *Annals of Botany*, **14**, 199–216.

Whitehouse, H. L. K. (1959). Cross- and self-fertilisation in plants. In *Darwin's biological work*, ed. P. R. Bell, pp. 207–61. London: Cambridge University Press.

Whitehouse, H. L. K. (1965). *Towards an understanding of the mechanism of heredity*. London: Arnold. [3rd edn, 1973.]

Whitney, G. G. (1994). *From coastal wilderness to fruited plain: a history of environmental change in temperate North America, 1500 to the present*. Cambridge: Cambridge University Press.

Whitney, K. D., Randell, R. A. & Rieseberg, L. H. (2010). Adaptive introgression of abiotic tolerance traits in the sunflower *Helianthus annuus*. *New Phytologist*, **187**, 230–9.

Whitton, J. *et al.* (2008). The dynamic nature of apomixis in the angiosperms. *International Journal of Plant Science*, **169**, 169–82.

Wichmann, M. C. *et al.* (2009). Human-mediated dispersal of seeds over long distances. *Proceedings of the Royal Society, B*, **276**, 523–32.

Wickler, W. (1968). *Mimicry in plants and animals*. New York: McGraw-Hill.

Widén, B. (1991). Phenotypic selection on flowering phenology in *Senecio integrifolius*, a perennial herb. *Oikos*, **61**, 205–15.

Widén B. (1993). Demographic and genetic effects on reproduction as related to population size in a rare, perennial herb, *Senecio integrifolius*. *Biological Journal of the Linnean Society*, **50**, 179–95.

Widén, M. (1992). Sexual reproduction in a clonal, gynodioecious herb *Glechoma hederacea*. *Oikos*, **63**, 430–8.

Wiens, D. (1978). Mimicry in plants. *Evolutionary Biology*, **11**, 365–403.

Wikström, N., Savolainen, V. & Chase, M. W. (2007). Evolution of the angiosperms: calibrating the family tree. *Proceedings of the Royal Society, B*, **268**, 2211–20.

Wilcove, D. S. & Master, L. L. (2005). How many endangered species are there in the United States? *Frontiers of Ecology and the Environment*, **3**, 414–420.

Wiley, E. O. (1981). *Phylogenetics: the theory and practice of phylogenetic systematics*. New York: Wiley.

Wiley, E. O. & Lieberman, B. S. (2011) *Phylogenetics: the theory and practice of phylogenetic systematics*, 2nd edn. Hoboken: Wiley-Blackwell.

Wilkins, D. A. (1959). Sampling for genecology. *Record of the Scottish Plant Breeding Station*, 1959, 92–6.

Wilkins, D. A. (1960). Recognising adaptive variants. *Proceedings of the Linnean Society of London*, **171**, 122–6.

Wilkins, J. S. (2009). *Species: a history of the idea*. Berkeley, Los Angeles & London: University of California Press.

Willerslev, E. & Cooper, A. (2005). Ancient DNA. *Proceedings of the Royal Society, B*, **272**, 3–16.

Willerslev, E. *et al.* (2004). Long-term persistence of bacterial DNA. *Current Biology*, **14**, R9–R10.

Willerslev E., Hansen A. J. & Poinar H. N. (2004). Isolation of nucleic acids and cultures from ice and permafrost. *Trends in Ecology and Evolution*, **19**, 141–7.

Williams, K. & Gilbert, W. L. (1981). Insects as selective agents on plant vegetative morphology: egg mimicry reduces egg laying by butterflies. *Science*, **212**, 467–9.

Williams, M. (1989). *Americans and their forests*. Cambridge: Cambridge University Press.

Williamson, P. G. (1981). Morphological stasis and developmental constraint: real problems for Neo-Darwinism. *Nature*, **294**, 214–15.

Willis, K. & McElwain, J. (2014). *The evolution of plants*, 2nd edn. Oxford: Oxford University Press.

Willis, K. J., Gillson, L. & Brncic, T. M. (2004). How 'virgin' is virgin rainforest? *Science*, **304**, 402–3.

Willson, M. F. (1979). Sexual selection in plants. *American Naturalist*, **113**, 777–90.

Willson, M. F. (1983). *Plant reproductive ecology*. New York: Wiley.

Willyard, A., Conn, R. & Liston, A. (2009). Reticulate evolution and incomplete lineage sorting among the ponderosa pines. *Evolution*, **52**, 498–511.

Wilmott, A. J. (1949). Intraspecific categories of variation. In *British flowering plants and modern systematic methods*, ed. A. J. Wilmott, pp. 28–45. London: Botanical Society of the British Isles.

Wilson, E. O. (1985). The biological diversity crisis. *BioScience*, **35**, 700–6.

Wilson, G. B. & Bell, J. N. B. (1985). Studies of the tolerance to sulphur dioxide of grass populations in polluted areas. IV. The spatial relationship between tolerance and a point source of pollution. *New Phytologist*, **102**, 563–74.

Wilson, M. A., Gaut, B. & Clegg, M. T. (1990). Chloroplast DNA evolves slowly in the Palm family (Arecaceae). *Molecular Biology & Evolution*, **7**, 303–14.

Winchester, A. M. (1966). *Genetics*. Boston: Houghton.

Winfield, M. & Parker, J. (2000). A molecular analysis of *Gentianella* in Britain. *English Nature Species Recovery Programme/Plantlife Report*, No. 155.

Winge, Ø. (1917). The chromosomes, their numbers and general importance. *Comptes Rendus des Travaux du Laboratoire Carlsberg*, **13**, 131–275.

Winge, Ø. (1940). Taxonomic and evolutionary studies in *Erophila* based on cytogenetic investigations. *Comptes Rendus des Travaux du Laboratoire Carlsberg (Ser. Physiol.)*, **23**, 41–74.

Winkler, H. (1908). Uber Parthenogenesis und Apogamie im Pflanzenreich. *Progressus rei Botanicae*, **2**, 293–454.

Winkler, H. (1916). Uber die experimentelle Erzeugung von Pflanzen mit abweichenden Chromosomenzahlen. *Zeitschrift für Botanik*, **8**, 417–531.

Winsor, M. P. (1995). The English debate on taxonomy and phylogeny, 1937–1940. *History and Philosophy of the Life Sciences*, **17**, 227–52.

Wolff, K., Rogstad, S. H. & Schaal, B. A. (1994). Population and species variation of minisatellite DNA in *Plantago*. *Theoretical and Applied Genetics*, **87**, 733–40.

Woltereck, R. (1909). Weitere experimentelle Untersuchungen über Artveränderung, speziel über das Wesen quantitativer Artunterschiede bei Daphniden. *Verhandlungen der deutschen zoologischen Gesellschaft*, **19**, 110–73.

Wood, T. E., Burke, J. M. & Rieseberg, L. H. (2005). Parallel genotypic adaptation: when evolution repeats itself. *Genetica*, **123**, 157–70.

Woodell, S. R. J. (1965). Natural hybridization between the Cowslip (*Primula veris* L.) and the Primrose (*P. vulgaris* Huds.) in Britain. *Watsonia*, **6**, 190–202.

Woodson, R. E., Jr (1964). The geography of flower color in Butterflyweed. *Evolution*, **18**, 143–63.

Woodwell, G. M. (1990). *The Earth in transition: patterns and processes of biotic impoverishment*. Cambridge: Cambridge University Press.

Wookey, P. A. *et al.* (1993). Comparative responses of phenology and reproductive development to simulated environmental change in sub-Arctic and high-Arctic plants. *Oikos*, **67**, 490–502.

Wright, J. W. (1953). Pollen dispersion studies: some practical applications. *Journal of Forestry*, **51**, 114–18.

Wright, K. M. (2013). Indirect evolution of hybrid lethality due to linkage with selected locus in *Mimulus guttatus*. *PLOS Biology*: published 26 February 2013. DOI:10.1371/journal.pbio.1001497.

Wright, S. (1931). Evolution in Mendelian populations. *Genetics*, **16**, 97–159.

Wright, S. (1938). Size of population and breeding structure in evolution. *Science*, **87**, 430–1.

Wright, S. (1943). Isolation by distance. *Genetics*, **28**, 114–28.

Wright, S. (1946). Isolation by distance under diverse systems of mating. *Genetics*, **31**, 39–59.

Wright, S. (1950). Genetical structure of populations. *Nature*, **166**, 247–9.

Wright, S. (1966). Mendel's ratios. In *The origin of genetics*, ed. C. Stern & E. R. Sherwood, pp. 173–5. London & San Francisco: Freeman.

Wright, S. (1977). *Evolution and the genetics of populations*, vol. 3: *Experimental results and evolutionary deductions*. Chicago & London: University of Chicago Press.

Wright, S. I., Kalisz, S. & Slotte, T. (2013). Evolutionary consequences of self-fertilization in plants. *Proceedings of the Royal Society of London, B*, **280**, http://dx.doi.org/10.1098/rspb.2013.0133

Wu, J., Hettenhausen, C. & Baldwin, I. (2006). Evolution of proteinase inhibitor defenses in North American allopolyploid species of *Nicotiana*. *Planta*, DOI:10.1007/s00425-006-0256-6.

Wu, L. (1990). Colonization and establishment of plants in contaminated sites. In *Heavy metal tolerance in plants: evolutionary aspects*, ed. A. J. Shaw, pp. 269–84. Boca Raton, FL: CRC Press.

Wu., L., Bradshaw, A. D. & Thurman, D. A. (1975). The potential for evolution of heavy metal tolerance in plants. III. The rapid evolution of copper tolerance in *Agrostis stolonifera*. *Heredity*, **34**, 165–87.

Wu, L., Till-Bottraud, I. & Torres, A. (1987). Genetic differentiation in temperature-enforced seed dormancy among golf course populations of *Poa annua* L. *New Phytologist*, **107**, 623–31.

Wyatt, R. (1988). Phylogenetic aspects of the evolution of self-pollination. In *Plant evolutionary biology*, ed. L. D. Gottleib & S. K. Jain, pp. 109–31. London: Chapman & Hall.

Wyatt, R. E., Evans, A. & Sorenson, J. C. (1992). The evolution of self pollination in granite outcrop species of *Arenaria* (Caryophyllaceae). VI. Electrophoretically detectable genetic variation. *Systematic Botany*, **17**, 201–9.

Wyse Jackson, P. S. & Sutherland, L. A. (2000). *International agenda for botanic gardens in conservation*. Kew: Botanic Gardens Conservation International.

Xi, Z. *et al.* (2012). Horizontal transfer of expressed genes in a parasitic flowering plant. *BMC Genomics*, **13**, 227.

Xiao, Z. S., Jansen, P. A. & Zhang, Z. B. (2006). Using seed-tagging methods for assessing post-dispersal seed fate in rodent-dispersed trees. *Forest Ecology & Management*, **223**, 18–23.

Yang, L. *et al.* (2014). Next-generation sequencing, FISH mapping and synteny-based modeling reveal mechanisms of decreasing dysploidy in *Cucumis*. *Plant Journal*, **77**, 16–30.

Yannic, G., Baumel, A. & Ainouche, M. (2004). Uniformity of the nuclear and chloroplast genomes of *Spartina maritima* (Poaceae), a salt-marsh species in decline along the Western European Coast. *Heredity*, **93**, 182–8.

Yates, F. (1960). *Sampling methods for censuses and surveys*, 3rd edn. London: Griffin.

Yates, F. (1981). *Sampling methods for censuses and surveys*, 4th edn. London: Charles Griffin.

Ye, Q., Bunn, E. & Dixon, K. W. (2011). Failure of sexual reproduction found in micropropagated critically endangered plants prior to reintroduction: a cautionary tale. *Botanical Journal of the Linnean Society*, **165**, 278–84.

Yeo, P. F. (1975). Some aspects of heterostyly. *New Phytologist*, **75**, 147–53.

Yoder, J. B. *et al.* (2014). Genomic signature of adaptation to climate in *Medicago truncatula*. *Genetics*, **196**, 1263–75.

Yoshida, S. *et al.* (2010). Horizontal gene transfer by the parasitic plant *Striga hermonthica*. *Science*, **328**, 1128.

Young, A. G. & Merriam, H. G. (1994) Effects of forest fragmentation on the spatial genetic structure of *Acer saccharum* Marsh. (sugar maple) populations. *Heredity*, **72**, 201–8.

Young, M. & Edis, T. (2004). *Why intelligent design fails: a scientific critique of the new Creationism*. New Brunswick, NJ, & London: Rutgers University Press.

Youngner, V. B. (1960). Environmental control of initiation of the inflorescence, reproductive structures and proliferations in *Poa bulbosa*. *American Journal of Botany*, **47**, 753–7.

Yu, Q. *et al.* (2009). Distinct non-target-site mechanisms endow resistance to glyphosate, ACCase and ALS-inhibiting herbicides in multiple herbicide-resistant *Lolium rigidum* populations. *Planta*, **230**, 713–23.

Yule, G. U. (1902). Mendel's laws and their probable relations to intra-racial heredity. *New Phytologist*, **1**, 193–207, 222–38.

Zalapa, J. E., Brunet, J. & Guries, R. P. (2009). Patterns of hybridization and introgression between invasive *Ulmus pumila* (Ulmaceae) and native *U. rubra*. *American Journal of Botany*, **96**, 1116–28.

Zander, B. & Wiegleb, G. (1987). Biosystematische Untersuchungen an Populationen von *Ranunculus* subgen. Batrachium in Nordwest-Deutschland. *Botanische Jahrbücher für Systematik, Planzengeschichte und Planzengeographie*, **109**, 81–130.

Zander, R. H. (2007). Paraphyly and the species concept, a reply to Ebach et al. *Taxon*, **56**, 642–4.

Zangerl, A. R. & Berenbaum, M. R. (2005). Increase in toxicity of an invasive weed after reassociation with its coevolved herbivore. *Proceedings of the National Academy of Sciences, USA*, **102**, 15529–32.

Zapiola, M. L. & Mallory-Smith, C. A. (2012) Crossing the divide: gene flow produces intergeneric hybrid in feral transgenic Creeping Bentgrass population. *Molecular Ecology*, 21, 4672–80.

Zapiola, M. L. *et al.* (2008). Escape and establishment of transgenic glyphosate-resistant creeping bentgrass *Agrostis stolonifera* in Oregon, USA: a 4-year study. *Journal of Applied Ecology*, 45, 486–94.

Zavada, M. S. (2007). The identification of fossil angiosperm pollen and its bearing on the time and place of the origin of angiosperms. *Plant Systematics and Evolution*, 263, 117–34.

Zelikova T. J. *et al.* (2013). Eco-evolutionary responses of *Bromus tectorum* to climate change: implications for biological invasions. *Ecology & Evolution*, 3, 1374–87.

Zevenhuizen, E. (2000). Keeping and scrapping: the story of a Mendelian Lecture Plate of Hugo de Vries. *Annals of Science*, 57, 329–52.

Zhang, W. *et al.* (2013). Species-specific identification from incomplete sampling: applying DNA barcodes to monitoring invasive *Solanum* plants. *PLoS ONE*, 8, 1–7.

Zhao, K. Y. *et al.* (2010). Genomic diversity and introgression in *O. sativa* reveal the impact of domestication and breeding on the Rice genome. *PLoS ONE*, 5, e10780

Zhou, Z., Yang, X. & Yang, Q. (2006). Land bridge and long-distance dispersal – old views, new evidence. *Chinese Science Bulletin*, 51, 1030–8.

Ziegenhagen, B., Bialozyt, R. & Liepelt, S. (2004). Contrasting molecular markers reveal gene flow via pollen is much more effective than gene flow via seeds. In *Biological resources and migration*, ed. D. Werner, pp.239–51. Heidelberg: Springer.

Zietkiewicz, E., Rafalski, A. & Labuda, D. (1994). Genome fingerprinting by simple sequence repeat (Ssr)-anchored polymerase chain-reaction amplification. *Genomics* 20, 176–83.

Zirkle, C. (1941). Natural selection before the 'Origin of species'. *Proceedings of the American Philosophical Society*, 84, 71–123.

Zirkle, C. (1966). Some anomalies in the history of Mendelism. In *G. Mendel Memorial Symposium 1865–1965*, ed. M. Sosna, pp. 31–7. Prague: Academia Publishing House of the Czechoslovak Academy of Sciences.

Ziska, L. H., Faulkner, S. & Lydon, J. (2004). Changes in biomass and root : shoot ratio of field grown Canada thistle (*Cirsium arvense*), a noxious invasive weed, with elevated CO_2: implications for control with glyphosate. *Weed Science*, 52, 584–8.

Zohary, D. (2004). Unconscious selection and the evolution of domesticated plants. *Economic Botany*, 58, 5–10.

Zohary, D. & Feldman, M. (1962). Hybridisation between amphidiploids and the evolution of polyploids in the Wheat (*Aegilops triticum*) group. *Evolution*, 16, 44–61.

Zohary, D. & Hopf, M. (1993). *Domestication of plants in the Old World: the origin and spread of cultivated plants in West Asia, Europe and the Nile valley*, 2nd edn. Oxford: Clarendon.

Zohary, D. & Nur, V. (1959). Natural triploids in the Orchard Grass *Dactylis glomerata* polyploid complex and their significance for gene flow from diploid to tetraploid levels. *Evolution*, 13, 311–17.

Zuckerkandl, E. & Pauling, L. (1965). Molecules as documents of evolutionary history. *Journal of Theoretical Biology*, 8, 357–66.

INDEX

The index includes the Latin names of plants and animals investigated in the case studies analysed in the text. In the space available, it has not been possible to include the names of all the plants mentioned in the book. In general, common names of organisms are not listed. Figure page numbers in bold.

Printed in the United States
by Baker & Taylor Publisher Services